T0224315

Advances in Volcanology

An Official Book Series of the International Association of Volcanology and Chemistry of the Earth's Interior – IAVCEI, Barcelona, Spain

Series editor

Karoly Nemeth, Palmerston North, New Zealand

More information about this series at http://www.springer.com/series/11157

Carina J. Fearnley · Deanne K. Bird
Katharine Haynes · William J. McGuire
Gill Jolly
Editors

Observing the Volcano World

Volcano Crisis Communication

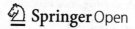 Springer Open

Editors
Carina J. Fearnley
Department of Science and Technology
University College London
London
UK

Deanne K. Bird
Institute of Life and Environmental
  Sciences
University of Iceland
Reykjavík
Iceland

Katharine Haynes
Department of Geography and Planning
Macquarie University
Sydney, NSW
Australia

William J. McGuire
UCL Hazard Centre, Department
  of Earth Sciences
University College London
London
UK

Gill Jolly
GNS Science
Lower Hutt
New Zealand

ISSN 2364-3277   ISSN 2364-3285   (electronic)
Advances in Volcanology
ISBN 978-3-030-09584-0   ISBN 978-3-319-44097-2   (eBook)
https://doi.org/10.1007/978-3-319-44097-2

Printed on acid-free paper

This Springer imprint is published by the registered company Springer International Publishing
AG part of Springer Nature
The registered company address is: Gewerbestrasse 11, 6330 Cham, Switzerland

Preface

During 2007–2008 I had the most wonderful privilege of interviewing over 93 people involved in the management of volcanic crisis in the USA as part of my Ph.D. research. From the United States Geological Survey, to the Federal Aviation Authority, the National Weather Service, and local media, I spent over 300 hours listening to years of experiences and stories about successes and failures, and lessons learnt from volcanic crises all around the world. I was overwhelmed with the experience and expertise I encountered. However, little of this 'experience or expertise' was published in either the academic or grey literature. I became acutely aware that in the future this knowledge could be lost, and that there was a need now to understand better how to manage a volcanic crisis. Yet, whilst actively publishing in their own research field, each scientist interviewed undervalued their tacit experiences and the contribution they could provide for future volcanic crises. Additionally, there was no clear place to publish these reflections. Publications on seismic studies, petrology, and new technological monitoring techniques are far more common, and perhaps historically a priority within the volcanic community. It is all too clear, however, that many volcanic disasters occur not as a result of uncertain and complex science, but frequently because of a breakdown in communication between the varying stakeholders, a weakness in management structures, and/or a lack of understanding of the risks involved. I felt a moral obligation to capture the knowledge and experiences, in their words, before they were lost for good.

Why is it so important? It is hoped this book will be the first of many that celebrates the challenging job of managing volcanic crises. Without scholarly work that reflects on volcanic crises around the world, how is it possible to identify trends, establish good practices, and help communities develop tools and systems to best mitigate volcanic hazards? This includes examining the warning process, communication between multiple stakeholders, and the difficulties involved in decision-making. As such there is a significant wealth of knowledge that is not yet documented that could be of significant value to a wide range of stakeholders. To date, most literature on volcanic crises lies in the grey literature – that is documents for the United Nations, international, and national meetings reviewing a crisis, and the odd memoir or report that lies buried in archives and libraries globally. It is this literature that provides insights into what actually happened during a crisis; not the analysis of data, but the story of what happened, by who, when, and what strategies worked and what did not. One of the most enlightening books I have read about volcanic crises is *Volcano Cowboys* by journalist Dick Thompson. This book tells some of the great stories about American volcanologists' personal experiences of working in various crises all over the world. Sadly out of print, this is one of the absolute treasures in the literature on volcanic crises. The events of Mount St Helens in 1980–86, and the eruption of Mt Pinatubo in 1991 produced two classic 'doorstop' books: *The 1980 eruptions of Mount St. Helens, Washington* and *Fire and Mud*, each with several chapters dedicated to the management of the crises, with reflective and analytical insights. Over the last three decades there have been numerous books published on various individual crises, such as *Surviving Galeras* by Stanley Williams, or *Fire from the Mountain* by Polly Pattullo, but none that try to provide some form of comparison between different crises.

It is clear there is a need to develop our knowledge about past events and document these, so that lessons identified and learnt can be shared, with the hope of developing robust understanding to aid future crises. It is unlikely there are best practices that can be shared globally, but certainly there is a need to share what works and doesn't work so each vulnerable area can make informed management decisions. There has been a steady and growing interest in Volcanic Crisis Communication, as exemplified by the World Organization of Volcano Observatories (WOVO) Volcano Observatories Best Practices Workshops, in recent years, and various academic research projects that have focused on risk communication and scientific advice globally. Another major platform to discuss volcanic crisis communication has been via the Cities on Volcanoes (CoV) conference series. Over my ten years of attending these conferences the focus has increasingly shifted from volcanological sciences towards an ever increasingly interdisciplinary perspective, engaging with those from any background interested in coming together to discuss how the varying cultural, political, economic, and legal contexts manage and respond to the unsolved epistemic uncertainties inherent in volcanology. It is clear progress is being made, from active scholars and practitioners, and undergraduates to leaders in their field; the interdisciplinarity that volcanoes force us to embrace is generating vital findings and a paradigm shift in the field. This book aims to capture, in a small way, this move.

Being part of the IAVCEI Book series *Advances in Volcanology*, our aim is make an advance on the topic of volcanic crisis communication. This book brings together authors from all over the globe who work with volcanoes, ranging from institutions (e.g. Volcanic Ash Advisory Centres, Civil Aviation, Weather services, Smithsonian Institute), to disaster practitioners (civil protection, emergency managers), observatory volcanologists/scientists, government & NGO officials and practitioners, the insurance sector, indigenous populations, and teachers/educators, and academics (from multiple disciplines). These authors have been asked to reflect on three key aspects of volcanic crises. First, the unique and wide-ranging nature of volcanic hazards that makes them a particularly challenging natural hazard to forecast and manage. Second, lessons learnt on how to best manage volcanic hazards based on a number of crises that have shaped our understanding. Third, the diverse and wide ranging aspects of communication involved in a crisis that bring together old practices and new technologies in an increasingly challenging and globalised world. Without this knowledge there is little scope to draw on established knowledge to move towards developing more robust volcanic crisis management, and to understand further how the volcano world is observed from a range of perspectives in different contexts around the world. The book is presented in these three parts, with a summary section for each written by the part editors: Part One was edited by William J. McGuire (UCL), Part Two by Gill Jolly (GNS) and myself, and Part Three by Deanne K. Bird (University of Iceland) and Katharine Haynes (Macquarie University). An introduction and summary to the book intend to provide valuable context, and a summary of the key findings from the chapters.

The editorial team was highly dedicated to raising the funds required to make the book open access so that everyone anywhere in the world would be able to read these stories, and hopefully in the future contribute new ones. We would like to thank very much all our generous sponsors, including:

- The Bournemouth University Disaster Management Centre, UK
- GNS Science, New Zealand
- Risk Frontiers, Australia
- The University of Auckland, New Zealand
- King's College London, UK
- Aon Benfield, Australia
- ICAO Meteorology Panel/Chief Meteorological Office, New Zealand
- Geophysical Institute, University of Alaska Fairbanks, USA

We owe our gratitude to Richard Gordon, John McAneney, Gill Jolly and Julia Becker, Jan Lindsay, Amy Donovan, Russell Blong, Peter Lechner, and Peter Webley for helping arrange this sponsorship.

We also recognise that many people would like to have contributed to the book but were unable to. We can only hope that this is just the beginning of a new dialogue, one that cuts across disciplines, stakeholders, and across different natural hazards to explore how volcanic crises can be better managed. This is just the start; there are many more stories to be told, some already known, and some yet to unfold.

This book took longer than any of us anticipated. Whilst this is common for academic publications, it was in fact a car crash that I was a victim in that led to a traumatic brain injury that added the most delay. Determined to complete the project I think my stubbornness certainly helped my miraculous recovery. During this time all five of our editorial team lost close family members, some got married, some had children, some changed jobs, and yet despite this, we were determined to publish the stories that needed to be told. I would like to take this opportunity to thank my editorial team for being so amazing, and keeping things going despite the many challenges we all faced. I would also like to thank our Springer editor, Johanna Schwartz who has been supportive from the very start of this adventure. Thanks must also be given to all the authors and reviewers for their patience, endless feedback and engagement.

I would also like to take this opportunity to thank the people who have provided endless support and love in my life; Gerlinde Fearnley, Geoffrey Fearnley (RIP), Nathan Farrell, the Fearnley and Thorosian families, Deborah Dixon, Phyllis Illari, Clive Prince, Lynn Picknett, Chris Kilburn, Annie Winson, Chiara Ambrosio, John Grattan, and the thriving community at UCL (particularly the Science and Technology Studies Department and the UCL Hazard Research Centre). As always there are plenty more folk to thank, but I think it is important to end by saying this book is dedicated to all those living with volcanic risk. It is hoped that this book will help reduce losses so that those who have suffered or died in previous volcanic crises did not do so in vain.

London, UK   Carina J. Fearnley

We Thank Our Sponsors

Deanne K. Bird has been supported by the Nordic Centre of Excellence for Resilience and Societal Security—NORDRESS, which is funded by the Nordic Societal Security Programme.

Contents

Part One
Adapting Warnings for Volcanic Hazards

William J. McGuire

Volcano Crisis Communication: Challenges and Solutions in the 21st Century

Carina Fearnley, Annie Elizabeth Grace Winson,
John Pallister and Robert Tilling

Abstract

This volume, *Observing the volcano world: volcanic crisis communication,* focuses at the point where the 'rubber hits the road', where the world of volcano-related sciences and all its uncertainties meet with the complex and ever-changing dynamics of our society, wherever and whenever this may be. Core to the issues addressed in this book is the idea of how volcanic crisis communication operates in practice and in theory. This chapter provides an overview of the evolution of thinking around the importance of volcanic crisis communication over the last century, bringing together studies on relevant case studies. Frequently, the mechanisms by which volcanic crisis communication occurs are via a number of key tools employed including: risk assessment, probabilistic analysis, early-warning systems, all of which assist in the decision-making procedures; that are compounded by ever-changing societal demands and needs. This chapter outlines some of the key challenges faced in managing responses to volcanic eruptions since the start of the 20th century,

C. Fearnley (✉)
Department of Science and Technology Studies,
University College London, Gower Street, London
WC1E 6BT, UK
e-mail: c.fearnley@ucl.ac.uk

A. E. G. Winson
Department of Geography and Earth Sciences,
Aberystwyth University, Aberystwyth, Ceredigion
SY23 3FL, UK

J. Pallister
U.S. Geological Survey, David A, Johnston
Cascades Volcano Observatory, 1300 SE Cardinal
Court, Building 10, Suite 100, Vancouver, WA
98683-9589, USA

R. Tilling
U.S. Geological Survey, Volcano Science Center,
345 Middlefield Rd, Menlo Park, CA 94025, USA

Advs in Volcanology (2018) 3–21
https://doi.org/10.1007/11157_2017_28
© The Author(s) 2017
Published Online: 06 December 2017

to explore what has been effective, what lessons have been learnt from key events, and what solutions we can discover. Adopting a holistic approach, this chapter aims to provide a contextual background for the following chapters in the volume that explore many of the elements discussed here in further detail. Finally, we consider the future, as many chapters in this book bring together a wealth of new knowledge that will enable further insights for investigation, experimentation, and development of future volcanic crisis communication.

1 Introduction: The Complexities of Volcanic Crisis Communication

With growing populations in volcanically vulnerable areas, it is likely that in the future more people will be affected by volcanic eruptions, most of whom will be busy with their daily concerns. The challenge today remains how to engage with a vulnerable population so that, when the time is right, appropriate actions are taken to mitigate loss of life and livelihood. If anything, the 21st century presents ever-increasing challenges to this goal. In part, this is demonstrated by the issues of mistrust and poor communication that emerged during the L'Aquila trial of five scientists and two emergency managers. These individuals were accused of making poor judgements on uncertainty that affected their communication to the public, and the risk-management actions the public took in response (Benessia and De Marchi 2017; Alexander 2014; Bretton et al. 2015). Whilst hugely complex, the L'Aquila case highlights the role that science plays within the broader field of crisis communication. As Sir Peter Gluckman, the Chief Science Adviser to the Prime Minister of New Zealand states (2014, p. 4):

> Science advice is not generally a matter of dealing with the easy issues that need technical solutions. Rather it is largely sought in dealing with sensitive matters of high public concern and inevitably associated with uncertainty and considerable scientific and political complexity.

Over the last 100 years scientists and various stakeholders have made significant progress in volcanic crisis communication. In this volume, *volcanic crisis communication* is the term used to encompass all forms of communication during a volcanic crisis: from the communication between monitoring equipment and scientists, to the interpretation and decision-making between scientists and, the communication between differing stakeholders on what actions to take and when, to name a few examples.

Volcano observation began in a structured way at the beginning of the 20th century. The earliest observatories were established in Asama, Japan and Hawaii, USA (Tilling 1989). As observations increased, the role of volcanologists in hazard management and mitigation grew. Progressive crises have imparted lessons to the volcanological community, helping to define different roles in these situations. While this led to great successes, in which volcanologists worked closely with civil-defence authorities, volcanic tragedies have also taken place, requiring reflection on how knowledge was communicated to stakeholders. For example, ineffective communication during the Nevado del Ruiz eruption in Colombia in 1985 resulted in the tragic catastrophe of over 23,000 deaths. This was not because of inadequate scientific knowledge or technology, but rather because local authorities and communities did not act on warnings (Hall 1990; Voight 1990). This was especially surprising as there had been a large effort to educate the population about the risks, and because an alert was issued in time. It is believed that a fundamental lack of understanding of terminology used in education campaigns led the community of Armero to perceive the risk less significantly. The tragedy highlighted the role of scientists in crisis response, and the need to effectively engage with stakeholders. It is important that all information is presented clearly and with recognition that the audience may not possess the same understanding of jargon that scientists are comfortable with.

Effective volcanic crisis communication is a fundamental component of the concepts of mitigation, disaster management, and disaster risk reduction. As part of this communication process, a number of tools have emerged that are regularly employed in volcanic crises that assist in the structure and formulation of communicative processes. This chapter focuses on four vital lessons learnt from key crisis events. First, advocating the need for resources to develop knowledge surrounding dangerous volcanoes and establishing potential threats via risk assessments. Second, the need to communicate the inherent scientific uncertainties in managing volcanic hazards, which has led to probabilistic analysis playing an ever-increasing role in crisis communication. Third, the value of providing warnings, typically through networks commonly known as early-warning systems. Finally, the intricate role of decision-making, increasingly assisted by various tools such as digital maps, automated messaging and alerting tools, as well as new policies and procedures to communicate data and knowledge. Together these lessons have generated a diverse range of volcanic-crisis communication around the world, shaped largely by the crises experienced to date and by the capabilities of the people and institutions engaged in volcano hazard analysis and warning.

2 Learning from the Past: Key Events that Shaped Crisis Communication

The Nevado del Ruiz disaster prompted a significant paradigm shift within the global volcanological community towards developing a keener understanding of local contexts when issuing volcanic warnings. This event, however, is not isolated. A number of other volcanic crises over the last 100 years have demonstrated the powerful influence of the social context on a crisis, and the need to continue to investigate crises where science and society come together in a pressured situation (see Table 1). Societal influence can be demonstrated by: the influence of political interference at Mt. Pelée, Martinique, 1902 (Scarth 2002); miscommunication between scientists and the media in Guadeloupe, 1976 (Fiske 1984); interactions between scientists and authorities in Montserrat, in 1995 (Druitt and Kokelaar 2002); differing levels of trust and understanding of the uncertainties and risks involved in volcanic crises (Haynes et al. 2008a, b); the importance of community leaders and past experience with volcanic crises (Andreastuti et al. 2017) and the ability for early warnings to successfully fulfil their purpose (Peterson and Tilling 1993), as key examples.

This volume addresses many key events that have shaped the paradigm of volcanic-crisis communication. Of the events listed in Table 1, three are especially noteworthy, well-studied case histories: La Soufrière (1976), El Chichón (1982), and Merapi (2010); for these, we have prepared detailed summaries from the many pertinent publications (see online supplementary materials). Acting as constant reminders, these events collectively have shaped practices around volcanic crisis communication.

One author of this chapter (Tilling) played a key role in many of these events in his capacity as head of the Volcano Programme at the United States Geological Survey (USGS). In 1989, Tilling identified five specific measures in volcano hazard mitigation to provide short- or long-term mitigation that collectively brings together the components required for effective volcanic management. He explored the relationships between these groups and their required actions in practice by identifying five key areas: (i) identification of high-risk volcanoes; (ii) hazard identification, assessment and zonation; (iii) volcano monitoring and eruption forecasting; (iv) engineering-oriented measures, and (v) volcanic emergency management (Fig. 1). It is important to note that the critical

Table 1 Overview of key events that have shaped volcanic crisis communication

Volcano, country	Date	Event	References
La Grande Soufrière, Guadeloupe	1976	Phreatic eruptions in Guadeloupe led to mass evacuations and very public disagreements between scientists	Fisk (1984), Komorowski et al. (2015)
Mount St. Helens, USA	1980	First successful implementation of volcano alert levels as a warning tool; first use of probabilistic event tree	Lipman and Mullineaux (1982), Newhall and Hoblitt (2002, Newhall and Pallister (2015)
Long Valley Caldera, USA	1980	First caldera unrest at Long Valley resulting in leaked news that eroded trust between the local communities and the scientists	Hill et al. (2017, Chap. "Volcanic Unrest and Hazard Communication in Long Valley Caldera, California")
El Chichón, Mexico	1982	Eruption kills 2000 attributable to lack of monitoring, background information and mixed messages from scientists to military emergency managers	Macías et al. (1997), Espíndola et al. (2002), Tilling (2009)
Nevado del Ruiz, Colombia	1985	Lahars kill 23,000 people. The realisation that science is not enough, it needs to be effectively communicated and understood	Hall (1990), Voight (1990)
Pinatubo, Philippines	1991	Eruption of Pinatubo, daily use of Volcano Early Warning Systems (VEWS) to alert public and trigger evacuations that saved tens of thousands of lives. Use of IAVCEI sponsored video to demonstrate types of hazard, based mainly on films by Maurice and Katia Krafft	Newhall and Punongbayan (1996), Punongbayan et al. (1996), Newhall and Solidum (2017, Chap. "Volcanic Hazard Communication at Pinatubo from 1991 to 2015")
Galeras, Colombia	1993	Retrospective analysis suggests need for a more robust appraisal of hazards and introduction of Bayesian Belief Networks to aid decision making	Aspinall et al. (2003), Garcia and Mendez-Fajury (2017, Chap. "If I Understand, I Am Understood: Experiences of Volcanic Risk Communication in Colombia")
Rabaul, Papua New Guinea	1994	Demonstrating the capacity for the public to self-evacuate and balancing the communication of uncertainty with safety	McKee et al. (2017, Chap. "Instrumental Volcano Surveillance and Community Awareness in the Lead-Up to the 1994 Eruptions at Rabaul, Papua New Guinea")
Soufriere Hills, Montserrat	1995-present	Communicating uncertainty in long running volcanic crises; use of Science Advisory Committee, Expert Elicitation and links between Volcanic Alert Level System (VALS) and mitigation actions; trust and its influence on risk communication	Aspinall et al. (2002), Haynes et al. (2008a)
Huila, Colombia	2006–2007	Effective communication and use of VEWS and VALS by INGEOMINAS (now SGC) that saves hundreds of lives from lahars	Santacoloma et al. (2011, Pulgarin et al. 2011, 2015), Garcia and Mendez-Fajury (2017, Chap. "If I Understand, I Am Understood: Experiences of Volcanic Risk Communication in Colombia")

(continued)

Table 1 (continued)

Volcano, country	Date	Event	References
Eyjafjallajökull, Iceland	2010	Demonstrating how volcanic ash can affect multiple countries and industries at once	Donovan and Oppenheimer (2012), Bird et al. (2017, Chap. "Crisis Coordination and Communication During the 2010 Eyjafjallajökull Eruption")
Merapi, Indonesia	2010	Preparedness and practice saves thousands despite rapid development of a major crisis	Surono et al. (2012), Mei et al. 2013)
Sinabung and Kelud, Indonesia	2010-present and 2014	Preparedness and differences in communication and culture at newly awakening vs. frequently erupting volcanoes; and the importance of local leaders and community engagement	Andreastuti et al. (2017, Chap. "Integrating Social and Physical Perspectives of Mitigation Policy and Practice in Indonesia")

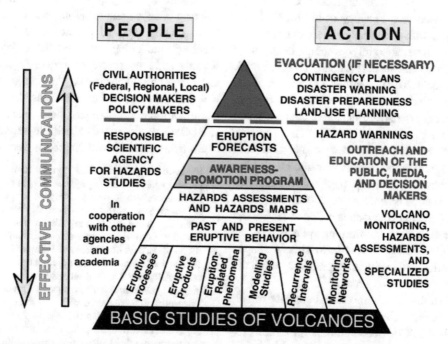

Fig. 1 Schematic diagram of an idealized program to reduce volcanic risk. The apex is separated from the rest of the triangle to emphasize that volcano scientists, while responsible for providing the best possible scientific information and advice, do not typically have knowledge of other key factors (e.g., socio-economic, cultural, political) and rarely have the authority to make final decisions regarding mitigation measures, including possible evacuation (modified from Tilling 1989, Fig. 1)

role of volcanic emergency management was identified as being undervalued, partly because of the complexities of society. By 1993, Peterson and Tilling demonstrated that volcano warnings were largely hindered by institutional weaknesses in emergency-response procedures and infrastructures, particularly the poor integration and sharing of critical information, as well as ineffective communications between scientists, decision-makers, and the affected populace. Communications clearly required more focus.

In the 1990s, there was significant focus on the communication of volcanic hazards for the aviation sector following two significant near-disasters and a major eruption that closed multiple airways: the first was the encounter of British Airways Flight 9 that encountered an ash cloud from Galunggung volcano in Java, the second was the near-loss of KLM flight 867 when it encountered the ash cloud of Redoubt volcano in 1989 (Guffanti and Miller 2013); and the eruption of Mount Pinatubo in 1991 that closed airports and airways across a wide region of the western Pacific (Tayag et al. 1996). In response to these crises, previous systems were refined and the schemes that are familiar today, such as the USGS Aviation colour code scheme, were devised (see Fearnley et al. 2012). However, even with variations of these systems now in place there are still questions as to their effectiveness and success globally (Winson et al. 2014; Papale 2017). Another key area of focus was the eruption of Soufriere Hills Volcano in Montserrat, extensively captured within Sparks and Young's memoir (2002). The ongoing eruption enabled significant exploration and experimentation, not just in volcanic monitoring and forecasting, but also in the governance of communication (Haynes et al. 2008a), the use of expert elicitation (Aspinall and Cooke 1998), and in generating maps and warning systems (Haynes et al. 2007). These events have closely shaped organisation and institutional practices within the volcanic and aviation sectors.

The publication 'Professional conduct of scientists during volcanic crises' emanating from the 1999 IAVCEI Subcommittee for Crisis Protocols provides guidance on what procedures and actions to take during a crisis (Newhall 1999). This simple yet powerful checklist is vital in minimizing communication pitfalls and is based on lessons identified by the committee's wealth of experiences during crises. The IAVCEI protocols seem to be standing the test of time, but society is dynamic and poses new challenges. This is particularly true in the context of: an ever-increasing population at risk, increasing pressures for global levels of warning, new forms

of technologies that aid scientific understanding, and communication—most recently via social media.

By the early 2000s, there was extensive focus around the interaction of volcanic events and cultures, wonderfully captured by Grattan and Torrance (2003, 2007). These two volumes explored a wealth of knowledge to better understand how culture is vital to the forms of communication that are fostered during volcanic crises, providing many a lesson learnt to shape future efforts. A surprising element of these books is the lesson that can be learnt from evaluating how early humans and civilisations 'sunk or swam' following volcanic crises.

Continuing the growth in key literature on volcanic crisis communication, in 2008, Barclay et al. (2008) explored the advances in understanding, modelling, and predicting volcanic hazards, and more recent techniques for reducing and mitigating volcanic risk. Providing valuable new insights, the article advocates the role of community-based disaster risk management (CBDRM) to aid effective risk communication. The article concludes with the following (p. 165):

> Evidence suggests that the current 'multidisciplinary' approach within physical science needs a broader scope to include sociological knowledge and techniques. Key areas where this approach might be applied are: (1) the understanding of the incentives that make governments and communities act to reduce volcanic risk; (2) improving the communication of volcanic uncertainties in volcanic emergency management and long-term planning and development. To be successful, volcanic risk reduction programmes will need to be placed within the context of other risk-related phenomena (e.g., other natural hazards, climate change) and aim to develop an all-risks reduction culture. We suggest that the greatest potential for achieving these two aims comes from deliberative inclusive processes and geographic information systems.

Areas highlighted specifically for further research included: (1) effectively conveying uncertain information, (2) methods for making decisions in uncertain situations, and (3) methods for dealing with dynamic and changing uncertainty without losing credibility and trust. Since 2008, a majority of research within volcanic crisis communication has focused on these three areas, and this work forms the foundations for this volume.

In an increasingly globalised world, recent volcanic disasters take on a new level of complexity. Whilst relatively minor loss of life may occur, particular events such as the 2010 Eyjafjallajökull eruption demonstrate the possibilities for significant global economic impacts. With ever increasing challenges to volcanic communities, ever more innovative solutions are needed to enhance crisis communication effectiveness for all sectors of society.

3　Key Solutions

Based on past events, we identify four key solutions to the challenges presented, alongside valuable lessons learnt: (i) assessing the threat; (ii) assessing and communicating uncertainty; (iii) establishing an early warning system; and (iv) developing and integrating decision-making tools.

3.1　Assessing the Threat

Attempts to assess the threat posed by volcanoes, relative to each other, began in the 1980s with three schemes created by: Bailey et al. (1983), Lowenstein and Talai (1984) and Yokoyama et al. (1984). The purpose of these was to identify the volcanoes most likely to generate destructive eruptions specific to the USA (Bailey et al. 1983) and Papua New Guinea (Lowenstein and Talai 1984) and, globally (UNESCO report, Yokoyama et al. (1984)). This would allow for preferential deployment of monitoring equipment for maximum threat mitigation. These three schemes were used as the basis for the U.S. National Volcano Early Warning System (NVEWS) (Ewert et al. 2005, 2007). NVEWS ranked 169 volcanoes in the USA in a combined assessment of 15 hazard and 9 exposure factors to generate a threat score. This scheme notably included a score for the potential exposure of aviation to an eruption of a specific volcano.

Threat scores allow volcanoes to be ranked against each other and thus enable recommendations for varying levels of monitoring (Ewert et al. 2005; Moran et al. 2008). Monitoring efforts are therefore focused on the volcanoes most likely to generate significant risk.

It is important to note that all authors of these types of ranking systems recognize that such comparisons are dependent on the existing quality and quantity of information. If little is known of a volcano, then it is difficult to accurately calculate its threat, except through global comparisons to analogue volcanoes and rapid investigations and monitoring installations during a crisis (i.e., playing "catch up"). For example, the global assessment prepared for UNESCO by Yokoyama et al. (1984) failed to recognize the potential for Pinatubo to produce a large, explosive eruption. This was not an oversight, but rather a reflection of what was known at the time. Less than a decade later, this volcano produced one of the largest recorded eruptions of the century, highlighting the need for vigilance and thorough assessment of any volcanoes near population centres, even if they have appeared dormant for hundreds of years. Although, there was a remarkably successful response that saved an estimated 20,000 lives (Newhall and Punongbayan 1996), it is now widely recognized that playing "catch up" is not the best solution, as it puts the response team in danger and it may not always result in a positive outcome. Consequently, the importance of developing a Volcano Early Warning System (VEWS) well in advance of a crisis, for all high-risk volcanoes, is now widely accepted as best practice. This was one of the major motivations for NVEWS that was first implemented in 2005, and is now used in a number of nations.

3.2　Assessing and Communicating Uncertainty

A principal challenge is that the degree of certainty in forecasting varies widely with time before (Newhall 2000). It is possible to forecast eruptions with relative certainty at an intermittently active volcano over time scales of centuries, highly uncertain at intermediate time spans of months to a few years, and with greatly improved certainty at short time spans of days to hours. Yet, in spite of these restrictions, forecasts of eruptions have become relatively common. Volcano observatories worldwide issue alert levels, many of which include qualitative statements about the probability of an eruption (e.g., it is "likely") within certain periods of time (e.g., "within days to weeks"). Repeated lava- dome eruptions were predicted successfully at Mount St. Helens (Swanson et al. 1983) and forecast at Montserrat (Voight et al. 1988) using changes in deformation and seismic rates. Similarly, based mainly on an escalation in seismicity and observations of physical changes at Pinatubo, the USGS-Philippine Institute of Volcanology and Seismology (PHIVOLCS) team estimated a 40% probability on 17 May 1991 for an eruption, 3 weeks before the eruption started. As levels and rates of unrest increased through early June, alert levels were used to warn that an eruption was likely to begin within 2 weeks and then within 24 hours (Punongbayan et al. 1996; Newhall and Pallister 2015). Based on seismic pattern-recognition during precursory activity and associated conceptual models of magma dynamics, successful forecasts have been made at many other volcanoes during the past several decades (McNutt 1996; Chouet 1996; White and McCausland 2016).

A problem with such forecasts, however, is that they typically use descriptive terms such as "likely" to convey the hazard (Doyle et al. 2014). This is a major shortcoming, because without a working understanding and effective communications of probability and uncertainty, emergency managers and the public may not be convinced of the potential hazard and urgency to take timely mitigation measures. In order to make forecasts more quantitative, probabilistic and statistical methods are now increasingly used. Probabilistic eruption forecasting typically utilizes Bayesian statistics, in which the probabilities of subsequent events depend on the outcomes of prior events; i.e., they are path

dependent and increase in magnitude as the path is realized and the volcano progresses toward an event. These methods typically assign initial (a priori) probabilities on the basis of historical statistics and then update them into a posteriori probabilities that are based on interpretation of monitoring data and on the physical and chemical processes that are thought to be controlling the system.

The Bayesian methods may be used in both long-term and short-term forecasting. The most common applications of statistics and uncertainty analyses in short-term eruption forecasting is the Bayesian Event Tree (Newhall and Hoblitt 2002), which considers probabilities and uncertainties of occurrence at each node in a tree-like time series leading to a potential eruption. Monitoring information is often combined with pre-determined patterns or thresholds and with conceptual models pertaining to the dynamics of magmatic systems to forecast outcomes of volcanic unrest. Current practitioners of Bayesian Event Tree (BET) analysis use either the Cooke-Aspinall method (Cooke 1991; Aspinall 2006) or the INGV (National Institute of Geophysics and Volcanology) method (Marzocchi et al. 2004, 2008), although there are other implementations (e.g., Sobradelo et al. 2014; Jolly et al. 2014; Newhall and Pallister 2015). In addition, Bayesian Belief Networks (BBN), another graphic method that does not require the same type of linear time progression as in BET systems, may be used effectively in some situations (e.g., Lindsay et al. 2010; Hincks et al. 2014; Aspinall and Woo 2014). All of these methods integrate some form of elicitation of opinions from a team of experts to assign probabilities and uncertainties based on monitoring data, past eruptive behaviour and conceptual models. They vary with respect to whether monitoring thresholds are defined in advance for the volcano in question and in how uncertainties are established. In comparison, the USGS/Volcano Disaster Assistance Programme (VDAP) team (Newhall and Pallister 2015) uses group discussion and consensus to assign nodal probabilities. In the INGV method, probability distributions are established for each node in the event tree

(Marzocchi et al. 2008). In this procedure, the parameters, weights, and thresholds are established through expert opinions, updated using data of past eruptions, and uncertainty is expressed as a probability density function for each node in the tree (Marzocchi and Bebbington 2012).

A daunting challenge for scientists who use any of these methods is to effectively communicate the results to emergency managers and the public; groups who are rarely well versed in statistics. A well-designed VEWS should utilize everyday terminology that is well-known to the population at risk, and be explicitly linked to any assigned numerical probabilities. For example, the USGS/VDAP team generally translates probabilities in terms of odds and rounds to the nearest 10%; e.g., "1 out of 3" or "9 out of 10" and terms such as "unlikely" are defined as <10%, "moderately likely" as 10–70% and "highly likely" as >70%.

3.3　Establishing an Early Warning System

Early-warning systems (EWS) are employed globally for a range of rapid onset hazards. The United Nations International Strategy for Disaster Reduction (UNISDR) recognises EWS as a core component of disaster risk reduction (DRR) measures both in the Hyogo Framework (2005) and the Sendai Framework for Disaster Risk Reduction (2015), stipulating the need to 'substantially increase the availability of and access to multi-hazard early warning systems and disaster risk information and assessments to the people by 2030' (UN ISDR 2015 p. 12). EWS can be defined as 'the set of capacities needed to generate and disseminate timely and meaningful warning information to enable individuals, communities and organizations threatened by a hazard to prepare and to act appropriately and in sufficient time to reduce the possibility of harm or loss' (UNISDR 2009, p. 12). This approach is comprised of four key sections: risk knowledge, monitoring and warning service, dissemination and communication, and response capacity (UNISDR PPEW 2006). This definition moves

away from a traditional approach to EWS, as merely technical warnings through a siren or other simple warning method.

According to Leonard et al. (2008), VEWS are composed of five key components (Fig. 2): the early warning system itself, planning, co-operation, education and participation, and exercises. It is widely accepted that VEWS are part of a broader framework of DRR measures including: scientific knowledge and limitation, education, technology capabilities, and policy. EWS are arguably the process by which many DRR measures are implemented, often within a broader mitigation strategy.

The process of developing a VEWS requires cooperation and communication not only across different cultures, but also different languages and political regimes. Garcia and Fearnley (2012) highlight that, whilst an EWS may have four key components as outlined by the UNISDR, it is often the links between these categories that are the focus of systemic failure. With multi-national volcanic events or hazards, these links are likely to be highly stressed. Whilst there are excellent studies on EWS (e.g., Mileti and Sorenson 1990; Kuppers and Zschau 2002; Basher 2006; Golnaraghi 2012), few look beyond the individual case study to focus on more international scale implications of a hazard event (Fig. 2).

It is possible to establish some of the complexities that VEWS have to deal with by applying the concept of classification of mitigation strategies to VEWS (Day and Fearnley 2015). This depends on how the VEWS has been designed. Responsive mitigation strategies prescribe actions after a hazard-source event has occurred, such as evacuations to avoid lahars, which require capacities to detect and quantify the hazard and to transmit warnings fast enough to enable at risk populations to decide and act effectively. Permanent mitigation strategies prescribe actions such as construction of SABO dams or land use restrictions: they are frequently both costly and "brittle" in that the actions work up to a design limit of hazard intensity or magnitude and then fail. Permanent warning systems exist on volcanoes, whereby a warning is

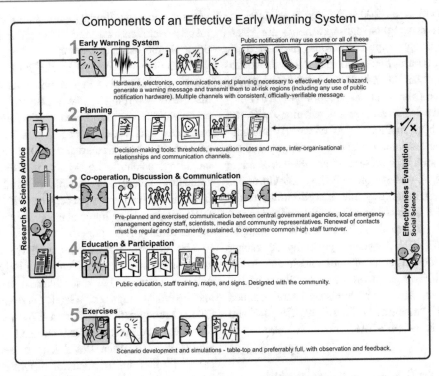

Fig. 2 Effective early-warning systems model with permission from Leonard et al. (2008, p. 204)

triggered, for example, by an automated lahar warning system. Anticipatory mitigation strategies, used in the mitigation of volcanic hazards more than for any other type of hazard, prescribe use of the interpretation of precursors to hazard source events as a basis for precautionary actions. However, challenges arise from uncertainties in hazard behaviour and in the interpretation of precursory signals. For example, evacuating vulnerable populations who live in areas susceptible to pyroclastic density currents prior to the onset of an eruption, pose hard questions about whether an early warning is based on forecasts, or on current activity and observations only, as well as our dependency on technology and statistical methods to make potentially life and death decisions.

Many countries operate so that their early warnings are based only and exclusively on scientific data and probabilistic forecasts. Other countries explicitly consider the social risks involved, alongside the scientific data and forecasts. There is potential for skewing of alert level

assignment, intentionally or unintentionally, when there is prior-knowledge of the risks involved, and when scientists rely upon non-probabilistic decision making (Fearnley 2013). Papale (2017) presents an argument that warnings may be flawed by implicit vested interests, and he recommends that observatories should rely on pre-established thresholds and communication of scientifically based probabilistic forecasts for hazard communication. Dependant on the context, differing approaches may be taken in either adopting a top-down (government led initiative) or bottom-up approach (driven by community based approaches).

What remains a challenge is to define whether a VEWS has been successful or not; this also depends on how success is measured. Paton et al. (1998) state that effectiveness of an integrated response can be constrained by communication and coordination across stakeholders, training experience, and organisational capabilities. It is imperative that all warning communication has one consistent message, with no contradiction to cause confusion. This is essential to establish trust between the public and other users that the information is correct (Mileti and Sorenson 1990). Further challenges can arise from the accumulation of multiple disasters, e.g., the impact of Typhoon Yunya in the Philippines during the 1991 Pinatubo eruption significantly exacerbated lahars, ashfall distribution and loading (Newhall and Punongbayan 1996). It is also challenging to determine the cost benefit of a VEWS prior to the impact of the event and as a result, many disregard the value of the system, particularly for events with a long-return frequency.

Science is a necessary evidence base for making decisions and has become a key component in EWS or Incident Command Systems (ICS). In some cases, EWS have become 'hazard-focused, linear, top-down, expert-driven systems, with little or no engagement of end-users or their representatives' (Basher 2006, p. 2712). However, there are many examples where major efforts are being made to engage with end users via community outreach and educational activities such as PHIVOLCS (Philippine Institute of Volcanology and Seismology), the USGS, and CVGHM (Center for Volcanology and Geological Hazard Mitigation). Typically, government institutions that manage potential disasters use simple prescriptive policy. Within this they recognise that decision-making is more complex and that local practitioners and vulnerable populations are increasingly managing disasters relevant to them using community-based warning and emergency response systems (UN ISDR PPEW 2006). Such community-based warning and response systems are based upon local capabilities and technologies where communities can have ownership, generating a bottom-up approach. Although initially considered a radical approach when introduced by Hewitt (1983), community-based early warning and response systems have gained momentum and have been proven effective and empowering during crises (Andreastuti et al. 2017). Subsequently it is suggested by the UN ISDR PPEW (2006) that these community-based approaches develop people-centric early warning and emergency response systems.

3.4 Decision-Making Tools

The way that people perceive information that has been communicated to them is vitally important, as it will shape how they frame problems and make decisions. There is significant progress in the role of various tools to assist in applying new knowledge making use of communicative products such as: map making, messages in preparedness products, infograms, and the simple verbal conveyance of crisis communication. Equally there are numerous new challenges and benefits to effective communication, For example there may be too little monitoring data, which increases the uncertainties in forecasts. In a few select cases where there are many different types of monitoring methods available, it may be difficult for scientists to synthesise all the information into a forecast in a timely manner. This suggests that there are optimal levels of monitoring and/or procedures for timely data processing and interpretation if the aim is to forecast future activity. Equally, the expansion of social media has opened lines of communication both to and from volcano observatories in new transparent and engaging ways, as seen via Twitter feeds, new citizen science apps, and community based monitoring (e.g., Stone et al. 2014), and in the sharing of knowledge. However, it also has placed pressure on the credibility of information, raising the risk of false data and interpretations that require careful management, and new levels of trust and engagement that must be built between the volcano observatories and the publics.

Maps are increasingly being used as a tool in conveying uncertainty, risk, and warnings. Volcano hazard maps are widely used to graphically portray the nature and extent of hazards and vulnerabilities and, in a few cases, the societal risk. Such maps may also be used to designate prohibited, restricted entry, or warning zones. They vary widely in style and content from nation to nation, and from volcano to volcano. In the most basic form, a volcano hazard map consists of hazard zones based on the underlying geology and history of past eruptions to define the extent of past flows and tephra falls. More sophisticated hazard maps utilize detailed geologic mapping and modelling of potential flow paths, often using Digital Elevation Models (DEMs) and statistical or numerical models that simulate flows of varying volume and duration. Some new approaches use automatic GIS-based systems that incorporate numerical model results and display the results in a GIS format (Felpeto et al. 2007) or that display the results of spatial probability for potential vent locations and flow inundation (Bevilacqua et al. 2015; Neri et al. 2015). These automated methods provide the capability to quickly modify the hazard map during a rapidly developing crisis. In addition, a new generation of numerical models have enabled near-real-time probabilistic forecast maps of ash cloud and ash fall hazards (Schwaiger et al. 2012 and references therein). Regardless of their degree of sophistication, hazard maps are a fundamental means to convey the spatial distribution of danger zones to emergency managers and the public. Although not everyone can effectively read a topographic map, shaded relief and 3D oblique projections using DEMs provide more effective means to communicate map information (Newhall 2000; Haynes et al. 2007).

To date there has been little evaluation of the influence of institutional organisation and the flow of information between different actors in a crisis when deciding what to do with the 'threat'. Fearnley (2013) investigated the role of decision-making in the USGS when assigning a volcano alert level, which established that informal communication is essential to enable key user groups to determine the extent of risk and likelihood of events. This was commonly achieved via face-to-face meetings, workshops and exercises, and telephone conversations, alongside web resources. Interactions are conducted in a multi-directional manner as various stakeholders may discuss relevant issues, moving away from typical one or two-way communication models. Evidence suggested that the ability to develop dialogue enabled key decision-makers

to gauge the volcano's behaviour and forecast in terms relevant to their own geographical, and temporal relations to the hazard. Today, observatories have developed a number of institutional communication tools, whether they are simply telephone calls or meetings that enable dialogue, or a one-way tool of information from the observatory via standardised messages targeted to specific users, such as the Volcano Activity Notice (VAN) or Volcano Observatory Notice for Aviation (VONA). Information can be communicated via daily, weekly, or monthly formal updates, status or information reports, or via Tweets, social networking, and the Smithsonian Weekly Updates. With so many options available it is up to the observatory and their stakeholders to establish what tools best serve their purpose.

In addition, during times of crisis, most observatories also participate in National Incident Command systems, or other similar civil protection procedures. For example, the USGS volcano observatories contribute scientific information to the National Incident Management System (NIMS), which was developed over many decades in response to inter-agency responses to wildfires, and is now used for all types of crises and disasters. The fundamental element of NIMS is the Incident Command System (ICS) system, which is used to structure and organize responses by federal, state and local agencies with responsibility for responding to natural as well as man-made crises and disasters. Figure 3 shows how the USGS contributes to the ICS system during volcanic crises. For example, in a disaster response, USGS scientists serve as technical advisors in the Planning Section to provide information about hazards (e.g., forecasts regarding eruptive activity, information about areas likely to be affected, extent and duration of impacts, etc.). They may also have a role in the Operations Section (e.g., in helping coordinate aviation operations). During an ICS response, a Joint Information Center (JIC) and a Joint Operations Center (JOC) are established. Through the JIC, press briefings and other media events are planned and conducted (Dreidger et al. 2004). The JIC and JOC are places where representatives of all involved agencies meet to coordinate information and crisis operations.

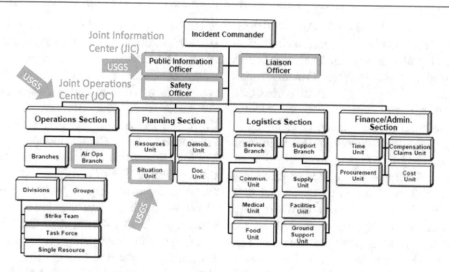

Fig. 3 USGS Volcano Observatories play a role as technical advisors in the U.S. Incident Command System (ICS). This is typically led by emergency-management agencies. Coordinates response and communication among multiple agencies and jurisdictions. (*Source* USGS)

4   Where Are We Now and What Are the New Challenges?

The above solutions are four of many that exist but are those of greatest focus currently within the field. This volume have been specifically crafted to build on prior research in the field and case studies from over the last 100 years (and sometimes beyond), to show how multidisciplinary approaches can be used to successfully manage a volcanic crisis, and that core to this are the communication processes. It is the intention to enable the next stage of understanding of volcanic crisis management in the 2020s to help navigate strong, easy, and effective communication. To do this, the book has three parts focusing on various lessons surrounding volcanic crisis communication.

First, it is well established that due to the longevity of hazards and the uncertainties in lead-time, and because of their numerous primary and secondary hazards, volcanoes pose a particular challenge. Some may argue volcanologists make too much of this distinction of volcanic crises being 'different' from other hazards and that in terms of the complexities of societal impacts and recovery, they are similar to earthquakes, hurricanes, tsunami and flooding,

etc. However, it is the very challenge of providing warnings with great uncertainties that makes volcanoes one of the most complex phenomena to manage and communicate. Volcanic hazards vary in location, scale and duration as explored independently in Part 1 of the volume. Hazards range from: volcanic bombs within close proximity of a vent (Fitzgerald et al. 2017, Chapter "The Communication and Risk Management of Volcanic Ballistic Hazards"), to pyroclastic flows whose impacts can also be proximal (Lavigne et al. 2017, Chapter "Mapping Hazard Zones, Rapid Warning Communication and Understanding Communities: Primary Ways to Mitigate Pyroclastic Flow Hazard"), lahars that can travel extensive distances often into non-volcanic terrain (Becker et al. 2017, Chapter "Organisational Response to the 2007 Ruapehu Crater Lake Dam-Break Lahar in New Zealand: Use of Communication in Creating an Effective Response"), volcanic gas hazards (Edmonds et al. 2017, Chapter "Volcanic Gases: Silent Killers") that can influence global climates (Donovan and Oppenheimer 2017, Chapter "Imagining the Unimaginable: Communicating Extreme Volcanic Risk"), and volcanic ash that can affect aviation (Lechner et al. 2017, Chapter

"Volcanic Ash and Aviation—The Challenges of Real-Time, Global Communication of a Natural Hazard") as well as local populations (Stewart et al. 2017, Chapter "Communication Demands of Volcanic Ashfall Events"). Volcanic hazards often evolve over time, becoming more or less intense, or changing in character e.g., from Plinian to Hawaiian style eruptions. It is this diverse nature that poses significant challenges to the idea of creating a single VEWS to communicate unrest and danger. In fact, there are numerous VEWS in place for volcanoes globally; many tailored for the specific hazard of a particular volcano, e.g., hydrothermal activity at Yellowstone (Erfurt-Cooper 2017, Chapter "Active Hydrothermal Features as Tourist Attractions"), or lahars on Mt Ruapehu (Becker et al. 2017, Chapter "Organisational Response to the 2007 Ruapehu Crater Lake Dam-Break Lahar in New Zealand: Use of Communication in Creating an Effective Response"). Part 1 explores the specific nuances each hazard presents to developing effective volcanic crisis communication, for specific or for a combination of hazards that may occur during a crisis.

In Part 2, the chapters discuss some of the key challenges involved in developing communication procedures and tools, and how these processes have evolved through the development of volcano observatories during the last 100 years. This is done by sharing and analysing some key lessons identified/learnt and best practices to improve the development and implementations of crisis communication following a chronological order (some are highlighted in Table 1). In essence, we are asking how can we move forward and develop more robust and effective early warning and, volcanic hazard, and risk communication. Using a range of international examples, Part 2 considers: small island states (Komorowski et al. 2017, Chapter "Challenges of Volcanic Crises on Small Islands States"), politically contested areas including Mt Cameroon (Marmol et al. 2017, Chapter "Investigating the management of geological hazards and risks in the Mt Cameroon area using Focus Group Discussions"; Miles et al. 2017, Chapter

"Blaming Active Volcanoes or Active Volcanic Blame? Volcanic Crisis Communication and Blame Management in the Cameroon"), challenges of institutional and culturally different approaches to communicating during crises in the Canaries and Italy (Solana et al. 2017, Chapter "Supporting the Development of Procedures for Communications During Volcanic Emergencies: Lessons Learnt from the Canary Islands (Spain) and Etna and Stromboli (Italy)"), extensive work to protect the millions of people that live under the shadow of Popocatépetl (De la Cruz-Reyna et al. 2017, Chapter "Challenges in Responding to a Sustained, Continuing Volcanic Crisis: The Case of Popocatépetl Volcano, Mexico, 1994-Present"), and, challenges of representing those living by a volcano as seen at Colima in Mexico (Cuevas-Muñiz and Gavilanes-Ruiz 2017, Chapter "Social Representation of Human Resettlement Associated with Risk from Volcán de Colima, Mexico") and at Eyjafjallajökull in Iceland (Bird et al. 2017, Chapter "Crisis Coordination and Communication During the 2010 Eyjafjallajökull Eruption"). There are old stories told with fresh eyes, old stories told for the first time, and some new stories that require humility to learn from.

Part 3 examines the numerous ways in which we communicate, not just across the science-society divide, but also across different disciplines including: religion (Chester et al. 2017, Chapter "Communicating Information on Eruptions and Their Impacts from the Earliest Times Until the Late Twentieth Century"), history and politics (Pyle 2017, Chapter "What Can We Learn from Records of Past Eruptions to Better Prepare for the Future?"). However we choose to communicate, whether via: oral histories (Procter et al. 2017, Chapter "Reflections from an Indigenous Community on Volcanic Event Management, Communications and Resilience"), social media (Sennert et al. 2017, Chapter "Role of Social Media and Networking in Volcanic Crises and Communication"), by drawing maps (Thompson et al. 2017, Chapter "More Than Meets the Eye: Volcanic Hazard Map Design and Visual Communication"), using satellite data (Webley and Watson 2017, Chapter "The role of geospatial technologies in communicating a more effective hazard assessment: application of GIS tools and remote sensing data"); is vital is to be effective. To achieve this several tools can be adopted: education from children to adults (Kitagawa 2017, Chapter "Living with an Active Volcano: Informal and Community Learning for Preparedness in South of Japan"; Sharpe 2017, Chapter "Learning to be practical: A guided learning approach to transform student community resilience when faced with natural hazard threats"), developing tools such as role-play to enhance the learning process (Dohoney et al. 2017, Chapter "Using Role-Play to Improve Students' Confidence and Perceptions of Communication in a Simulated Volcanic Crisis "), to practicing evacuation scenarios with emergency managers and communities (Hudson-Doyle 2017, Chapter "Decision-making: preventing miscommunication and creating sharing meaning between stakeholders"). Awareness of the challenges of communicating across cultures is also of great importance. Miscommunication is a frequent issue and there is a need to understand the psychological elements of decision-making and risk perception given these different cultural reference frames (Wilmshurst 2017, Chapter "There is no plastic in our volcano: A story about losing and finding a path to participatory volcanic risk management in Colombia"). Participatory methods have been highly successful to foster the participation of local communities (Cadag et al. 2017, Chapter "Participatory approaches to foster the participation of local communities in volcanic disaster risk reduction"), but there is often a need to find ways to bridge cultural differences between the scientists and end-users, often worlds apart (Newhall 2017, Chapter "Cultural Differences and the Importance of Trust Between Volcanologists and Partners in Volcanic Risk Mitigation"), or between the science and arts to explore differing understandings around volcanoes (Dixon and Beech 2017,

Chapter "Re-enchanting Volcanoes: The Rise, Fall, and Rise Again of Art and Aesthetics in the Making of Volcanic Knowledges"). Mistakes in managing these numerous complexities can restrict the maintenance of trust between the various stakeholders involved. We can negotiate these difficult interactions by developing tools to reduce the uncertainties and help decision making processes. Such tools include: statistical and probabilistic tools (Sobradelo and Marti 2017, Chapter "Using Statistics to Quantify and Communicate Uncertainty During Volcanic Crises"), establishing a robust EWS (Potter et al. 2017, Chapter "Challenges and Benefits of Standardising Early Warning Systems: A Case Study of New Zealand's Volcanic Alert Level System"), and using insurance to offset the risks (Blong et al. 2017, Chapter "Insurance and a Volcanic Crisis—A Tale of One (Big) Eruption, Two Insurers, and Innumerable Insureds"). Core to the communication process is the ability to make decisions about what to do, when to do it, and who is affected, what can be done, what resources need to be made available to support these decisions. An exemplary set up is the co-ordination between the USA, Japan, and Russia in managing airbourne ash hazards for aviation (Igarashi et al. 2017, Chapter "International Coordination in Managing Airborne Ash Hazards: Lessons from the Northern Pacific") where these three nations, at times politically and economically constrained, have managed to foster meaningful and successful coordination.

What all these chapters have in common, is that they demonstrate the value of communication and the open and timely sharing of knowledge, so finding a way to generate meaningful understanding; the need to keep both relationships and procedures strong and current; and the ability to cope with rapid changes in both society and volcanic activity. There have been many lessons learnt and many new tools are available to both volcanologists, emergency management practitioners, and the public; no doubt the future will present us with new challenges to overcome. The ability to adapt and evolve before, during and after a crisis is of utmost importance and this can only happen through open, honest and robust communication. This sounds simple and this volume provides evidence that simplicity and clarity are often key to successful outcomes in volcanic crisis.

References

Alexander DE (2014) Communicating earthquake risk to the public: the trial of the L'Aquila Seven. Nat Hazards 72(2):1159–1173. doi:10.1007/s11069-014-1062-2

Andreastuti S, Paripurno ET, Gunawan H, Budianto A, Syahbana D, Pallister J (2017) Character of community response to volcanic crises at Sinabung and Kelud volcanoes. J Volcanol Geoth Res (in press)

Aspinall WP (2006) Structured elicitation of expert judgement for probabilistic hazard and risk assessment in volcanic eruptions. In: Mader HM, Coles SG, Connor CB, Connor LJ (eds) Statistics in volcanology. IAVCEI Publications, ISBN 978-1-86239-208-3, pp 15–30

Aspinall W, Cooke RM (1998) Expert judgement and the Montserrat volcano eruption. In: Proceedings of the 4th international conference on probabilistic safety assessment and management PSAM4 (vol 3, pp 13–18)

Aspinall WP, Woo G (2014) Santorini unrest 2011–2012: an immediate Bayesian belief network analysis of eruption scenario probabilities for urgent decision support under uncertainty. J Appl Volcanol 3(1):12. doi:10.1186/s13617-014-0012-8

Aspinall WP, Loughlin SC, Michael FV, Miller AD, Norton GE, Rowley KC, Sparks RSJ, Young SR (2002) The montserrat volcano observatory: its evolution, organization, role and activities. Geol Soc, Lond, Mem 21:71–91

Aspinall WP, Woo G, Voight B, Baxter PJ (2003) Evidence-based volcanology: application to eruption crises. J Volcanol Geoth Res 128:273–285. doi:10.1016/S0377-0273(03)00260-9

Bailey RA, Beauchemin PR, Kapinos FP, Klick DW (1983) The volcano hazards program; objectives and long-range plans (No. 83–400). US Geological Survey

Barclay J, Haynes K, Mitchell T, Solana C, Teeuw R, Darnell A, Crosweller HS, Cole P Pyle D, Lowe C, and Fearnley C (2008) Framing volcanic risk communication within disaster risk reduction: finding ways for the social and physical sciences to work together, vol 305, Geological Society Special Publication, pp 163–177

Basher R (2006) Global early warning systems for natural hazards: systematic and people-centred. Philos Trans R Soc 364:2167–2182

Benessia A, De Marchi B (2017) When the earth shakes… and science with it. The management and communication of uncertainty in the L'Aquila earthquake. Futures

Bevilacqua A, Isaia R, Neri A, Vitale S, Aspinall WP, Bisson M, Flandoli F, Baxter P, Bertagnini A, Ongaro TE, Iannuzzi E, Pistolesi M, Rosi M (2015) Quantifying volcanic hazard at Campi Flegrei caldera (Italy) with uncertainty assessment: 1. Vent opening maps. J Geophys Res 120. doi:10.1002/2014JB011775

Bretton RJ, Gottsmann J, Aspinall WP, Christie R (2015) Implications of legal scrutiny processes (including the L'Aquila trial and other recent court cases) for future volcanic risk governance. J Appl Volcanol 4(1):18. doi:10.1186/s13617-015-0034-x

Chouet BA (1996) Long-period volcano seismicity: its source and use in eruption forecasting. Nature 380:309–316

Cooke RM (1991) Experts in uncertainty: opinion and subjective probability in science. Oxford University Press, New York

Day S, Fearnley C (2015) A classification of mitigation strategies for natural hazards: implications for the understanding of interactions between mitigation strategies. Nat Hazards 79(2):1219–1238. doi:10.1007/s11069-015-1899-z

Donovan A, and Oppenheimer C (2012) Governing the lithosphere: insights from Eyjafjallajökull concerning the role of scientists in supporting decision-making on active volcanoes. J Geophys Res: Solid Earth 117 (B3). doi:10.1029/2011JB009080

Doyle EEH, McClure J, Johnston DM, Paton D (2014) Communicating likelihoods and probabilities in forecasts of volcanic eruptions. J Volcanol Geoth Res 272:1–15. doi:10.1016/j.jvolgeores.2013.12.006

Driedger CL, Neal CA, Knappenberger TH, Needham DH, Harper RB, Steele WP (2004) Hazard information management during the autumn 2004 reawakening of Mount St. Helens Volcano, Washington. A Volcano Rekindled: the renewed eruption of Mount St. Helens, 2006, 505–519

Druitt TH and Kokelaar BP (eds) (2002) The eruption of Soufrière Hills volcano, Montserrat, from 1995 to 1999. Geological Society of London

Espíndola JM, Macías JL, Godínez L, Jiménez Z (2002) La erupción de 1982 del Volcán Chichónal, Chiapas, México. In: Lugo HJ, Inbar M (eds), Desastres naturales en América Latina: Mexico, DF, Fondo de Cultura Económica, pp 37–65

Ewert JW (2007) System for ranking relative threats of US volcanoes. Nat Hazards Rev 8(4):112–124

Ewert JW, Guffanti M, and Murray TL (2005) An assessment of volcanic threat and monitoring capabilities in the United States: framework for a National volcano early warning system (No. 2005–1164)

Fearnley CJ (2013) Assigning a volcano alert level: negotiating uncertainty, risk, and complexity in decision-making processes. Environ Plann a 45 (8):1891–1911. doi:10.1068/a4542

Fearnley CJ, McGuire WJ, Davies G, Twigg J (2012) Standardisation of the USGS Volcano Alert Level System (VALS): analysis and ramifications. Bull Volcanol 74(9):2023–2036. doi:10.1007/s00445-012-0645-6

Felpeto A, Marti J, Ortiz R (2007) Automatic GIS-based system for volcanic hazard assessment. J Volcanol Geoth Res 166:106–116. doi:10.1016/j.jvolgeores.2007.07.008

Fiske RS (eds) (1984) Volcanologists, journalists, and the concerned local public: a tale of two crises in the eastern Caribbean, National Research Council, Geophysics Study Committee, Explosive Volcanism, National Academy Press, Washington, DC, pp 110–121

Garcia C, Fearnley CJ (2012) Evaluating critical links in early warning systems for natural hazards. Environ Hazards 11(2):123–137. doi:10.1080/17477891.2011.609877

Gluckman P (2014) Evidence based policy: A quixotic challenge? Address given at the invitation of the science policy research unit. Brighton, UK: University of Sussex [January 21st, 2014; See http://www.pmcsa.org.nz/wp-content/uploads/Sussex_Jan-21_2014_Evidence-in-Policy_SPRU.pdf(Accessed 01 May 2016)]

Golnaraghi M (ed) (2012) Institutional partnerships in multi-hazard early warning systems: a compilation of seven national good practices and guiding principles. Springer Science and Business Media

Grattan J, Torrence R (2003) Natural disasters and cultural change. Routledge

Grattan J, Torrence R (eds) (2007) Living under the shadow: cultural impacts of volcanic eruptions. Routledge

Guffanti M, Miller TP (2013) A volcanic activity alert-level system for aviation: review of its development and application in Alaska. Nat Hazards 69 (3):1519–1533. doi:10.1007/s11069-013-0761-4

Hall ML (1990) Chronology of the principal scientific and governmental actions leading up to the November 13, 1985 eruption of Nevado-del-Ruiz, Colombia. J Volcanol Geoth Res 42:101–115. doi:10.1016/0377-0273(90)90072-N

Haynes K, Barclay J, Pidgeon N (2007) Volcanic hazard communication using maps: an evaluation of their effectiveness. Bull Volcanol 70:123–138. doi:10.1007/s00445-007-0124-7

Haynes K, Barclay J, Pidgeon N (2008a) The issue of trust and its influence on risk communication during a volcanic crisis. Bull Volcanol 70:605–621. doi:10.1007/s00445-007-0156-z

Haynes K, Barclay J, Pidgeon N (2008b) Whose reality counts? Factors affecting the perception of volcanic risk. J Volcanol Geoth Res 172(3):259–272. doi:10.1016/j.jvolgeores.2007.12.012

Hewitt K (1983) Interpretations of calamity from the viewpoint of human ecology. Allen and Unwin, Boston, p 304

Hincks TK, Komorowski JC, Sparks SR, Aspinall WP (2014) Retrospective analysis of uncertain eruption precursors at La Soufrière volcano, Guadeloupe, 1975–77: volcanic hazard assessment using a bayesian belief network approach. J Appl Volcanol 3(1):3. doi:10.1186/2191-5040-3-3

Jolly GE, Keys HJR, Procter JN, Deligne NI (2014) Overview of the co-ordinated risk-based approach to science and management response and recovery for the 2012 eruptions of Tongariro volcano, New Zealand. J Volcanol Geoth Res 286:184–207. doi:10.1016/j.jvolgeores.2014.08.028

Komorowski JC, Hincks T, Sparks RSJ, Aspinall W, The CASAVA ANR Project Consortium (2015) Improving crisis decision-making at times of uncertain volcanic unrest (Guadeloupe, 1976). In: Loughlin SC, Sparks RSJ, Brown SK, Jenkins SF, Vye-Brown C (eds) Global volcanic hazards and risk. Cambridge University Press, pp 255–261

Kuppers AN, Zschau J (2002) Early warning systems for natural disaster reduction. Springer, Berlin

Leonard GS, Johnston DM, Paton D, Christianson A, Becker J, Keys H (2008) Developing effective warning systems: ongoing research at Ruapehu volcano, New Zealand. J Volcanol Geoth Res 172:199–215. doi:10.1016/j.jvolgeores.2007.12.008

Lindsay J, Marzocchi W, Jolly G, Constantinescu R, Selva J, Sandri L (2010) Towards real-time eruption forecasting in the Auckland Volcanic Field: application of BET_EF during the New Zealand National Disaster Exercise 'Ruaumoko'. Bull Volcanol 72 (2):185–204. doi:10.1007/s00445-009-0311-9

Lipman PW, Mullineaux DR (1982) The 1980 eruptions of Mount St. Helens, Washington. Geological survey professional paper 1250. US GPO, Washington

Lowenstein PL, Talai B (1984) Volcanoes and volcanic hazards in Papua New Guinea. Rep 263:315–331

Macías JL, Espíndola JM, Taran Y, Sheridan MF, García PA (1997) Explosive volcanic activity during the last 3,500 years at El Chichón Volcano, México. In: International Association of Volcanology and Chemistry of the Earth's Interior, Puerto Vallarta, México, General Assembly, Field Guide, Excursion No. 6, Guadalajara, Jalisco, Gobierno del Estado del Jalisco, Secretaria General, Unidad Editorial, 53 pp

Marzocchi W, Bebbington MS (2012) Probabilistic eruption forecasting at short and long time scales. Bull Volcanol 74:1777–1805. doi:10.1007/s00445-012-0633-x

Marzocchi W, Sandri L, Gasparini P, Newhall C, Boschi E (2004) Quantifying probabilities of volcanic events: the example of volcanic hazard at Mount Vesuvius. J Geophys Res 109:B11201. doi:10.1029/2004JB003155

Marzocchi W, Sandri L, Selva J (2008) BET_EF: a probabilistic tool for long- and short-term eruption forecasting. Bull Volcanol 70:623–632. doi:10.1007/s00445-007-0157-y

McNutt SR (1996) Seismic monitoring and eruption forecasting of volcanoes: a review of the state-of-the-art and case histories. In: Scarpa L.A, Tilling RI (eds) Monitoring and mitigation of volcano hazards. Springer, Berlin, pp 99–196

Mei ETW, Lavigne F, Picquout A, de Bélizal E, Brunstein D, Grancher D, Sartohadi J, Cholik N, Vidal C (2013) Lessons learnt from the 2010 evacuations at Merapi volcano. J Volcanol Geoth Res 261:348–365. doi:10.1016/j.jvolgeores.2013.03.010

Mileti DS, and Sorenson JH (1990) Communication of emergency public warnings: a social science perspective and state-of-the-art assessment. Oak Ridge National Laboratory

Moran SC, Freymueller JT, LaHusen RG, McGee KA, Poland MP, Power JA, Schmidt DA, Schneider DJ, Stephens G, Werner CA, White RA (2008) Instrumentation recommendations for volcano monitoring at US volcanoes under the national volcano early warning system (no. 2008–5114), U.S. Geological Survey, Scientific Investigations Report, no. 2008-5114, pp 47

Neri A, Bevilacqua A, Ongaro TE, Isaia R, Aspinall WP, Bisson M, Flandoli F, Baxter P, Betagnini A, Iannuzzi E, Orsucci S, Pistolesi M, Rosi M, Vitale S (2015) Quantifying volcanic hazard at Campi Flegrei caldera (Italy) with uncertainty assessment: 2. Pyroclastic density current invasion maps. J Geophys Res 120. doi:10.1002/2014JB011776

Newhall C (1999) Professional conduct of scientists during volcanic crises. Bull Volcanol 60(5):323–334. doi:10.1007/PL00008908

Newhall C (2000) Volcano warnings. In: Sigurdsson H, Houghton B, McNutt S, Rymer H, Stix J (eds) Encyclopedia of Volcanoes. Academic Press, New York, pp 1185–1197

Newhall CG, Hoblitt RP (2002) Constructing event trees for volcanic crises. Bull Volcanol 64:3–20. doi:10.1007/s004450100173

Newhall CG, Pallister JS (2015) Using multiple data sets to populate probabilistic volcanic event trees. Chapter 8 in Papale P., volcanic hazards, risks and disasters. Elsevier, Berlin, pp 203–227

Newhall CG, Punongbayan R (eds) (1996) Fire and mud: eruptions and lahars of Mount Pinatubo, Philippines (p 1126). Philippine Institute of Volcanology and Seismology, Quezon City

Papale P (2017) Rational volcanic hazard forecasts and the use of volcanic alert levels. J Appl Volcanol 6:13

Paton D, Johnston D, Houghton B (1998) Organisational responses to a volcanic eruption. Disaster Prev Manage 7:5–13

Peterson DW, Tilling RI (1993) Interactions between scientists, civil authorities and the public at hazardous volcanoes. In: Kilburn CRJ, Luongo G (eds) Active Lavas. UCL Press, London, pp 339–365

Pulgarín B, Agudelo A, Calvache M, Cardona C, Santacoloma C, Monsalve ML (2011). Ch. 3 volcanic eruption and lahars. In: Breitkreuz C, Gursky HJ (eds) Geo-risk management–a German Latin American approach, pp 69–80

Pulgarín BE, Cardona CA, Agudelo AD, Santacoloma CR, Calvache MA, Murcia CA, Cuéllar MA, Medina EN, Balanta RE, Calderón YO, Leiva ÓM (2015) Erupciones Recientes del Volcán Nevado del Huila: Lahares Asociados y Cambios Morfológicos del Glaciar. Servicio Geológico Colombiano, Boletin Geológico 43:75–87

Punongbayan RS, Newhall CG, Bautista MLP, Garcia D, Harlow DH, Hoblitt RP, Sabit JP, Solidum RU (1996) Eruption hazard assessments and warnings. In: Newhall CG, Punongbayan R (eds) Fire and mud: eruptions and lahars of Mount Pinatubo, Philippines (p 1126). Philippine Institute of Volcanology and Seismology, Quezon City, pp 67–99

Santacoloma CC, Londono JM, Cardona CE (2011) Recent Eruptions of Nevado del Huila Volcano, Colombia after 500 years resting. In: AGU fall meeting abstracts, vol 1, p 2675

Scarth A (2002) La Catastrophe: Mount Pelée and the destruction of Saint-Pierre. Dunedin Academic Press Ltd, Martinique

Schwaiger HF, Denlinger RP, Mastin LG (2012) Ash3d: a finite-volume, conservative numerical model for ash transport and tephra deposition. J Geophys Res: Solid Earth 117(B4). doi:10.1029/2011JB008968

Sobradelo R, Bartolini S, Martí J (2014) HASSET: a probability event tree tool to evaluate future volcanic scenarios using Bayesian inference. Bull Volcanol 76(1):770. doi:10.1007/s00445-013-0770-x

Stone J, Barclay J, Simmons P, Cole PD, Loughlin SC, Ramón P, Mothes P (2014) Risk reduction through community-based monitoring: the vigías of Tungurahua, Ecuador. J of Appl Volcanol 3(1):11. doi:10.1186/s13617-014-0011-9

Surono, Jousset P, Pallister J, Boichu M, Buongiorno MF, Budi-Santoso A, Costa F, Andreastuti S, Prata F, Schneider D, and Clarisse L, Oppenheimer H (2012) The 2010 explosive eruption of Java's Merapi volcano —a '100-year'event. J Volcanol Geotherm Res 241–242. doi:10.1016/j.jvolgeores.2012.06.018

Swanson DA, Fiske RS, Rose TR, Kenedi CL (1983) Predicting eruptions at Mount St. Helens, June 1980 through December 1982. Science 221:1369–1375

Tayag J, Insauriga S, Ringor A, Belo M (1996) People's response to eruption warning: the Pinatubo experience, 1991–1992. Fire and mud. Eruptions and Lahars of Mount Pinatubo, Philippines, University of Washington Press, 13

Tilling RI (1989) Volcanic hazards and their mitigation: progress and problems. Rev Geophys 27(2):237–269. doi:10.1029/RG027i002p00237

Tilling RI (2009) El Chichón's "surprise" eruption in 1982: lessons for reducing volcano risk. Geofísica Internacional 48(1):3–19

UN ISDR (2015) Sendai framework for disaster risk reduction 2015–2030, retrieved from: http://www.preventionweb.net/files/43291_sendaiframeworkfordrren.pdf [accessed 14/02/17]

UN ISDR PPEW (2006) Global Survey of Early Warning Systems: An assessment of capacities, gaps and opportunities toward building a comprehensive global early warning system for all natural hazards. In: Platform for the Promotion of Early Warning (UNISDR—PPEW) (ed.). UN, p 46

United Nations, International Strategy for Disaster Reduction (UN-ISDR). Hyogo framework for action 2005–2015: building the resilience of nations and communities to disasters. In: World conference on disaster reduction, Kobe, Japan, January 2005. Accessed 12 May 2017

United Nations Office for Disaster Risk Reduction (UN-ISDR) (2015) Sendai framework for disaster risk reduction 2015–2030. United Nations—Headquarters (UN) pp. 32 Retrieved from http://www.wcdrr.org/preparatory/post2015. Accessed 12 May 2017

Voight B (1990) The 1985 Nevado-Del-Ruiz volcano catastrophe—anatomy and retrospection. J Volcanol Geoth Res 42:151–188. doi:10.1016/0377-0273(90)90075-Q

Voight B, Hoblitt RP, Clarke AB, Lockhart AB, Miller AD, Lynch L, McMahon J (1988) Remarkable cyclic ground deformation monitored in real-time on Montserrat, and its use in eruption forecasting. Geophys Res Lett 25:3405–3408

White R, McCausland W (2016) Volcano-tectonic earthquakes: a new tool for estimating intrusive volumes and forecasting eruptions. J Volcanol Geoth Res 309:139–155. doi:10.1016/j.jvolgeores.2015.10.020

Winson AE, Costa F, Newhall CG, Woo G (2014) An analysis of the issuance of volcanic alert levels during volcanic crises. J Appl Volcanol 3(1):1. doi:10.1186/s13617-014-0014-6

Yokoyama I, Tilling RI, and Scarpa R (1984) International mobile early-warning system (s) for volcanic eruptions and related seismic activities: report of an UNESCO/UNEP sponsored preparatory study in 1982–84. UNESCO

Communication Demands of Volcanic Ashfall Events

Carol Stewart, Thomas M. Wilson,
Victoria Sword-Daniels, Kristi L. Wallace,
Christina R. Magill, Claire J. Horwell, Graham S. Leonard
and Peter J. Baxter

Abstract

Volcanic ash is generated in explosive volcanic eruptions, dispersed by prevailing winds and may be deposited onto communities hundreds or even thousands of kilometres away. The wide geographic reach of ashfalls makes them the volcanic hazard most likely to affect the greatest numbers of people. However, forecasting how much ash will fall, where, and with what characteristics, is a major challenge. Varying social contexts, ashfall characteristics, and eruption durations create unique challenges in determining impacts, which are wide-ranging and often poorly understood. Consequently, a suite of communication strategies must be applied across a variety of different settings. Broadly speaking, the level of impact depends upon the amount of ash deposited and its characteristics (hazard), as well as the numbers and distribution of people and assets (exposure), and the ability of people and assets to cope with the ashfall (resilience and/or vulnerability). Greater knowledge of the likely impact can support mitigation actions, crisis planning, and emergency management activities. Careful, considered, and well-planned communication prior to, and during,

C. Stewart (✉)
Joint Centre for Disaster Research, GNS
Science/Massey University Wellington Campus,
Wellington, New Zealand
e-mail: c.stewart1@massey.ac.nz

T.M. Wilson
Department of Geological Sciences, University of
Canterbury, Christchurch, New Zealand

V. Sword-Daniels
Department of Civil, Environmental and Geomatic
Engineering, University College London, London,
UK

K.L. Wallace
Alaska Volcano Observatory, U.S. Geological
Survey, Anchorage, AK, USA

C.R. Magill
Risk Frontiers, Macquarie University, Sydney,
Australia

C.J. Horwell
Department of Earth Sciences, Institute of Hazard,
Risk and Resilience, Durham University, Durham,
UK

G.S. Leonard
GNS Science, Lower Hutt, New Zealand

P.J. Baxter
Institute of Public Health, Cambridge University,
Cambridge, UK

Advs in Volcanology (2018) 23–49
https://doi.org/10.1007/11157_2016_19

a volcanic ashfall crisis can substantially reduce physical, economic and psychosocial impacts. We describe the factors contributing to the complex communication environment associated with ashfall hazards, describe currently available information products and tools, and reflect on lessons from a range of case-study ashfall events. We discuss currently-available communication tools for the key sectors of public health, agriculture and critical infrastructure, and information demands created by ash clean-up operations. We conclude with reflections on the particular challenges posed by long-term eruptions and implications for recovery after ashfall.

Keywords

Volcanic ashfalls · Societal impacts · Information demands · Information resources

1 Introduction

All explosive volcanic eruptions generate tephra, fragments of glass, rock, and minerals that are produced when magma or vent material is explosively disintegrated. Volcanic ash (tephra < 2 mm diameter) is then convected upwards within the eruption plume and carried downwind, falling out of suspension and potentially affecting communities and farmland across hundreds, or even thousands, of square kilometres. Ashfall is the most widespread and frequent of the hazards posed by volcanic eruptions. Although ashfalls rarely endanger human life directly, disruption and damage to buildings, critical infrastructure services, aviation and primary production can lead to substantial societal impacts and costs, even at deposit thicknesses of only a few millimetres (Table 1; Fig. 1). Impacts vary with proximity to the volcano, how much ash has been deposited, physical and chemical properties of the ash, characteristics of the receiving environment (such as climate and land use) and adaptive capacity of the affected communities (Fig. 1; Wilson et al. 2012). Ashfall impacts are more complex and multi-faceted than for any other volcanic hazards (Jenkins et al. 2015).

Even with small eruptions generating minor quantities of ash, information demands may be heavy and complex. A recent example is the small, but locally high profile, 6 August 2012 eruption of Tongariro volcano, New Zealand. Despite its small size, following this eruption there was intense demand for information from the public, media, and government agencies on questions such as: Was this event a precursor to larger scale activity? What hazards were expected? Was the ashfall hazardous? (Leonard et al. 2014). Similarly, in Alaska, eruptions occur on average one to two times per year, ashfall deposits are typically only a few mm thick on populated areas, and impacts are considered more disruptive than catastrophic. Yet the demand for information is high. During recent eruptions in Cook Inlet, Alaska, the Alaska Volcano Observatory website received as many as 30 million page views in a single month, up to 3000 emails, and thousands of phone calls seeking information throughout the crises (Fig. 2; Adleman et al. 2010; Schaefer et al. 2011).

In this chapter, we describe the factors contributing to the complex communication environment associated with ashfall hazards, describe currently available information products and tools, and reflect on lessons learned from a range of case-study events. We discuss in more detail: ash hazard assessment tools; communication tools available for the key sectors of public health, agriculture, and critical infrastructure; and

Table 1 Volcanic ash impacts on society (adapted from GAR 2015 report: Brown et al. 2014)

Sector	Impacts	Example/photo
Public health	Exposure during an ashfall may not often endanger human life directly, except where thick accumulations cause structural damage (e.g., roof collapse) or when reduced visibility or slippery roads cause traffic accidents. However, very fine ash as PM2.5 and PM10 is a health hazard when it is readily suspended in the air by wind and traffic (Carlsen et al. 2012a; Wilson et al. 2012). Short-term effects commonly include irritation of the eyes and lung airways, and exacerbation of pre-existing asthma and chronic lung diseases (Horwell and Baxter 2006; see also www.ivhhn.org). The presence of respirable crystalline silica in some eruptions will cause much concern over the risk of silicosis, a chronic lung disease which is entirely preventable by adequate measures to reduce exposure in prolonged crises (e.g., Montserrat, 1995–2010). Affected communities can also experience psychological stress from disruption of livelihoods and other social impacts (Carlsen et al. 2012a, b).	**Caption** Windy conditions in Jacobacci, Argentina on 9 September 2011 leading to high levels of fine airborne ash due to remobilisation of fall deposits from June 2011 eruption of Cordón Caulle. *Credit* J. Mellado
Critical infrastructure	Damage and disruption to critical infrastructure services from ashfall impacts can substantially affect normal functioning of societies. Electricity networks are vulnerable, mainly due to ash contamination causing flashover and failure of insulators (Wilson et al. 2012). Ash can also disrupt transportation networks through reduced visibility and traction; and be washed into drainage systems. Wastewater treatment systems that have an initial mechanical pre-screening step are particularly vulnerable to damage if ash-laden sewage arrives at the plant. Suspended ash may also cause damage to water treatment plants if it enters through intakes or by direct fallout (e.g. onto open sand filter beds). In addition to direct impacts, system interdependence is a problem. For example, air- or water-handling systems may become blocked by ash leading to overheating or failure of dependent systems. Specific impacts depend strongly on network or system design, ashfall volume and characteristics, and the effectiveness of any applied mitigation strategies (Wilson et al. 2012, 2014).	**Caption** Suspended ash in waste water caused accelerated wear to pumping station impellors in Bariloche waste-water network, Argentina, following the 2011 eruption of Cordón Caulle. *Credit* C. Stewart
Agriculture	Fertile volcanic soils commonly host farming operations. Impacts will be dependent on how much ash has been deposited, characteristics of the ash, characteristics of the receiving environment, style, intensity and practises of the exposed farm, time of year (as it will determine climate and agricultural activities), and risk management actions taken by the farmer and supporting agencies (Wilson et al. 2011a). Ashfall can contaminate and (if sufficient deposition) bury pastures resulting in reduced availability of feed; contaminate, (if thick enough) lodge and bury horticultural crops, reducing yields and quality; cause adverse effects on livestock health by contaminating feed and (more rarely) cause toxicity hazards; contaminate and disrupt agricultural water supplies; abrade and corrode farm vehicles, machinery and infrastructure increasing maintenance costs; and cause disruption to essential services, such as power supplies, transportation and communication systems.	**Caption** Chillis damaged by acidic surface coating during the Merapi 2006 eruption, Indonesia. *Credit* G. Kaye

(continued)

Table 1 (continued)

Sector	Impacts	Example/photo
	Ashfalls can be beneficial or detrimental to soil depending on the characteristics of the ash (particularly with respect to its soluble salt burden, which can add plant growth nutrients to pastoral systems). The time of year in the agricultural production cycle strongly determines the level of impact (Cook et al. 1981). For example, ripe crops are usually ash tolerant, but are vulnerable to pollination disruption and contamination when close to harvest. Under very thin ashfall (<1 mm) crops and pastures can suffer from acid damage or shading from light; as ashfall depths increase these effects intensify and loading damage may occur. Thick ashfalls (>100 mm) typically require soil rehabilitation, e.g. thorough mixing or removal, to restore agricultural production (Wilson et al. 2011a; 2015). For livestock, ashfall may cause starvation (damaged or smothered feed), dehydration (water sources clogged with ash), deaths from ingesting ash along with feed, and (more rarely) acute or chronic fluorosis if ash contains moderate to high levels of bioaccessible fluoride (Cronin et al. 2003).	
Buildings	The load associated with an ashfall can cause the collapse of roofing material (e.g. sheet roofs), the supporting structure (e.g. rafters or walls) or both and, under great enough loads (> > 100 mm), the entire building may collapse (Blong 1984; Spence et al. 2005). Non-engineered, long-span and low-pitched roofs are particularly vulnerable to collapse, potentially under thicknesses of around 100 mm or less. Under thinner ashfall (< 100 mm), structural damage is unlikely although non-structural elements such as gutters and overhangs may suffer damage (Wilson et al. 2015). Wetted ash is up to twice as dense as dry ash thus loading is correspondingly higher. Building components and contents may also be damaged from ashfall due to ash infiltration into interiors, with associated abrasion and corrosion.	**Caption** Volcanic ash cleaned off a hospital roof in Heimaey following 1973 Eldfell eruption, Iceland (tractor for scale). *Credit* G. Oskarsson
Economy	Economic losses may arise from damage to physical assets, e.g. buildings, or reductions in production, e.g. agricultural or industrial output. Most economic activities will be impacted, even indirectly, under relatively thin (< 10 mm) ashfall, for example through disruptions to critical infrastructure. Losses may also result from precautionary risk management activities, e.g. business closures or evacuations. During or after an ashfall, clean-up from roads, properties, and airports is often necessary to restore functionality. Large volumes of ash require time-consuming, costly and resource-intensive efforts (Wilson et al. 2012).	**Caption** 20–30 mm of volcanic ash covering aeroplanes during the 2011 Cordón Caulle eruption, Chile. *Credit* Bariloche Airport

information demands created by ash clean-up operations. Impacts of airborne ash on aviation are covered elsewhere in this volume. We conclude with reflections on the particular challenges posed by long-term eruptions and implications for recovery after ashfall.

Increasing attention is being paid to the human health, environmental and aviation

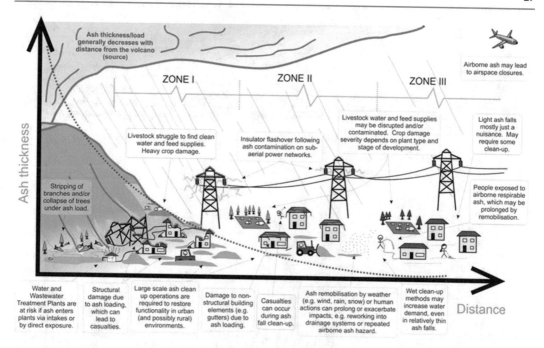

Fig. 1 Schematic of some ashfall impacts with distance from a volcano. This schematic diagram assumes a large explosive eruption with significant ashfall thicknesses in the proximal zone and is intended to be illustrative rather than literal. Three main zones of ashfall impact are defined: (1) Destructive and potentially life-threatening (Zone I); (2) Moderately damaging and/or disruptive (Zone II); (3) Mildly disruptive and/or a nuisance (Zone III). From Brown et al. (2014)

hazards of resuspension and dispersal of ash from fallout deposits (Folch et al. 2014; Wilson et al. 2011b; Hadley et al. 2004). We acknowledge the communication challenges associated with resuspension events, but consider them outside the scope of this chapter.

As a caveat, we note that we, the authors, are all based in countries with advanced economies, and thus our perspective—informed by our own experiences—may be less applicable in dissimilar countries.

2 The Complex Communication Environment Associated with Ashfalls

2.1 Disaster Risk Reduction Context

Empowering society to utilise scientific and technological advances to reduce the impacts of disasters is a well-established challenge (Alexander 2007; Few and Barclay 2011; McBean 2012; Mileti 1999; Cutter et al. 2015). Both the UNISDR Sendai Framework for Action (SFA) and Integrated Research on Disaster Risk (IRDR) programs call for more integration of research with the needs of policy and decision makers (ICSU 2008; UNISDR 2015). Few and Barclay (2011) also stress the need to promote integrated, inter-disciplinary approaches, strengthen two-way links between science providers and end-users, and support more effective research/end-user partnerships.

Because of the low recurrence rates of eruptions at many of the world's volcanoes, ashfalls can be rare events, even in volcanically-active regions. Wilson et al. (2014) note that the rarity of volcanic events can result in low risk awareness, particularly during periods of quiescence. Furthermore, even if knowledge of proximity to volcanic hazards and susceptibility to their consequences is reasonable, this does not ensure that mitigative actions will be taken, and preparedness

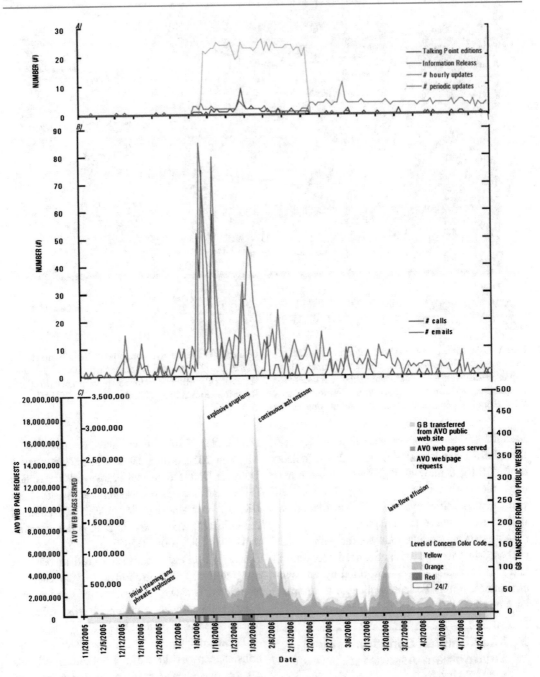

Fig. 2 *Top* Daily totals of information items produced during the 2005–6 unrest and eruption at Augustine volcano. *Middle* Daily totals of recorded phone calls and emails received. *Lower* AVO Website statistics of gigabytes transferred, webpage served and webpage requests. Reproduced from Adleman et al. (2010)

levels often remain low in proximal regions, even in developed countries (Paton et al. 2008). For risk communication, simply providing information often fails to change risk perception or motivate volcanic hazard preparedness, implying that more engaged and appropriate strategies are required. Thus, more participatory processes, whereby stakeholders (e.g. communities and organisations) actively participate as legitimate partners, are recommended (Covello and Allen 1988; Paton et al. 2005; Twigg 2007).

2.2 Complex Communication Environment

Effective management of volcanic ashfall risk requires effective communication between a range of groups and individuals during crisis and non-crisis periods (Höppner et al. 2010). Some countries have coordinating structures which aid information sharing to enhance decision-making during these periods. A broad and evolving array of communication channels may be utilised. Communication between parties is ideally two-way; however, specific ashfall hazard, risk and management information needs to be generated and communicated by expert groups for stakeholders to make risk management decisions, often under urgency. Ideally this evolves into discussions as experts tailor communications to the evolving risk and social context with, for example, the media, public, critical infrastructure and other businesses providing vital situational awareness to emergency managers, and useful data to scientists.

Volcano-specific agencies and emergency managers need to work closely as a team. This multi-agency group must conduct pre-planning and joint exercises. Several communication products can and should be pre-prepared, including contingency messaging for the various possible outcomes of ash characterisation, for example in the event of high levels of crystalline silica in respirable size fractions (see Sect. 4.1.1). Other products should have a pre-planned format and framework but need to be completed dynamically in response to the specific event, such as ashfall forecast maps. As many communication channels as possible should be two-way, allowing for dialogue rather than just provision of information. Ashfall mapping, collection, and testing are substantial activities that require rapid, widespread collaboration and are necessary to inform critical communication messages. An idealised representation of the flow of communication between key actors during a volcanic ashfall crisis illustrates the complex relationships that emerge amongst organisations, processes and communication products (Fig. 3). For example, the provision of authoritative health advice to the public requires wide cooperation between organisations; integration with ash collection and analysis processes; and alignment with other communication products, all at the same time. While these three elements could be illustrated separately, the cross-dependencies would be lost. Figure 3 is adapted from an earlier version developed by Paton et al. (1999), who noted that information management during an eruption is highly complex, owing to the rarity of these events, the complexity of hazard effects and the diversity of agencies involved.

A diverse range of stakeholders have information needs that evolve throughout ashfall crises (Wilson et al. 2012). These are summarised in Table 2 for the following groups: general public, media, emergency management and emergency services, local government, public health agencies, utility managers, farmers and agricultural agencies and private businesses. Experience has shown that information demands are most intense in the following areas:

- Effects on public health from inhaling or ingesting ash (e.g., Horwell and Baxter 2006);
- Potential of ashfall to contaminate water supplies and food chains (e.g., EFSA 2010);
- Impacts of ashfall on agriculture and rural communities (e.g., Wilson et al. 2011a, b);
- Ash clean-up and disposal methods (e.g., Wilson et al. 2012).

Risks to public and animal health are typically considered most urgent by both the public and public health authorities, although often the public concern outweighs the actual risk and the role of the agencies is to allay that concern with event-specific and science-based information. For example, following the April 2010 eruption of Eyjafjallajökull volcano, Iceland, and the subsequent transport of an extensive ash plume over Europe, the European Food Safety

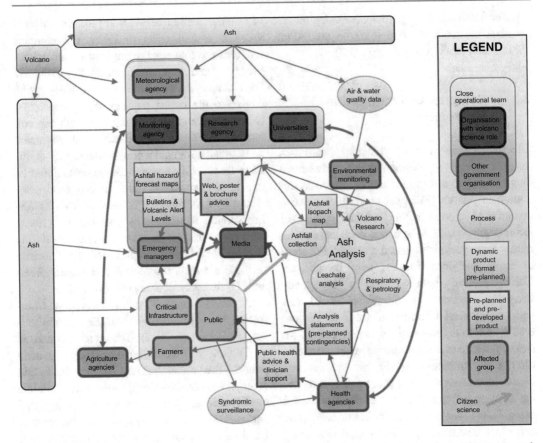

Fig. 3 Idealised flow of communication between key participants during a volcanic ashfall crisis illustrating the complex relationships that emerge amongst organisations, processes and communication products (after Paton et al. 1999)

Authority (EFSA) undertook an urgent assessment of risks for public and animal health (EFSA 2010). Information was urgently sought on questions such as the composition of the ash falling across Europe, with particular concern expressed about the fluoride content of the ash; important pathways of dietary exposure; recommendations for further data collection and comments on the effectiveness of mitigation methods.

3 Tools for Ash Hazard Characterisation and Dissemination

A range of products exists to meet the information demands of stakeholders. Some products are for an international audience and some have been produced according to local (domestic) needs. The need for the products evolves with changing risk and social context before, during and after an ashfall. We summarise, in general terms, some of these evolving needs in Table 2. Explanations about the deployment of specific tools throughout an event are provided in Table 3.

Communication tools and resources can be used during crisis and non-crisis times to contribute to societal resilience[1] to ashfall events. Effective communications summarise hazards and impacts, recommended preparedness, and

---

[1]*Resilience*: The ability of a system, community or society exposed to hazards to resist, absorb, accommodate to and recover from the effects of a hazard in a timely and efficient manner, including through the preservation and restoration of its essential basic structures and functions. http://www.unisdr.org/we/inform/terminology.

Table 2 Evolution of information demands throughout an ashfall crisis/event, by sector

Groups	Typical Information Demands/Questions			
	Quiescence[a]	Before ashfall (volcanic unrest)	During ashfall	After ashfall
All (including the public)	*Typically minimal interest* • If eruption occurs, how much ash will be received and what will the effects be?	• Will the ash be harmful to people? To animals? • Where is ash likely to fall? • How much ash is likely to fall at my location? • When will ashfall start? • When will ashfall stop? • What will be the impacts? • What can be done to prepare (especially for health)? • How should buildings and services be protected from ash ingress?	• Will the ash be harmful to people? To animals? • What protective measures can I take? • How much ash will fall? • When will the ashfall stop? • How should buildings and services be protected from ash ingress?	• What are the longer term health effects? • Will more ash fall? • How and when should ash be cleaned up? • How and where should ash be disposed of? • Can ash be added to gardens?
Media	See 'All'	See 'All' Questions follow public interest in eruption and are (ideally) guided by scientific communiques. • What can people do to prepare (especially for health)?	See 'All' Questions follow public interest in eruption and are (ideally) guided by scientific communiques. • Where has ash fallen and where will it fall in the future?	See 'All' Questions follow public interest in eruption and are (ideally) guided by scientific communiques. • What is the likelihood of more ashfall? Where would it fall?
Emergency Managers and Emergency Services	See 'All' • What is the risk of ashfall (function of likelihood and consequences) as part of risk assessment planning? • Information sources for hazard, impacts and mitigation	See 'All' Require broad overview of how to manage ash risk across all sectors. • How to access most up to date scientific information on eruption and ashfall crisis • How to prepare, respond, remediate and recover from ash impacts	See 'All' Require broad overview of how to manage ash risk across all sectors. • How to access most up to date scientific information on eruption and ashfall crisis • How to prepare, respond, remediate and recover from ash impacts	See 'All' Require broad overview of how to manage ash risk across all sectors. • How to access most up to date scientific information on eruption and ashfall crisis • How to respond, remediate and recover from ash impacts • What was learnt from this event?

(continued)

Table 2 (continued)

Groups	Typical Information Demands/Questions			
	Quiescence[a]	Before ashfall (volcanic unrest)	During ashfall	After ashfall
Utility Managers	See 'All' • What is the risk of ashfall (function of likelihood and consequences) as part of risk assessment planning • Sector specific hazard, impact and risk management information and what are the information sources	See 'All' • Sector specific hazard, impact and risk management information and what are the information sources	See 'All' • Sector specific impact and risk management information • Engineering characteristics of ash	See 'All' • Sector specific impact and risk management information • Engineering characteristics of ash • Sector specific best-practise clean-up methods • What was learnt from this event?
Farmers and Agricultural Agencies	See 'All' • Sector specific hazard, impact and risk management information and what are the information sources • Agriculturally relevant characteristics of ash	See 'All' • Sector specific hazard, impact and risk management information and what are the information sources • What are the implications of ashfall for food chains? • What are the agriculturally relevant characteristics of ash • What are the ash remediation strategies	See 'All' • Sector specific hazard, impact and risk management information and what are the information sources • What are the implications of ashfall for food chains? • What are the agriculturally relevant characteristics of ash • What are the ash remediation strategies	See 'All' • Sector specific hazard, impact and risk management information and what are the information sources • What are the implications of ashfall for food chains? • What are the agriculturally relevant characteristics of ash • What are the ash remediation strategies
Public Health Agencies	See 'All' • What is the risk of ashfall (function of likelihood and consequences) as part of risk assessment planning • What are information sources for hazard, impacts and mitigation	See 'All' • Health specific hazard, impact and risk management information and what are the information sources • What are the short and long term health relevant characteristics of the ash *Will be looking to inform standard public health messaging and modify if required*	See 'All' • Health specific hazard, impact and risk management information and what are the information sources • What are the short and long term health relevant characteristics of the ash *Will be looking to inform standard public health messaging and modify if required*	See 'All' • What are the short and long term health relevant characteristics of the ash • What was learnt from this event?

(continued)

Table 2 (continued)

| Groups | Typical Information Demands/Questions | | | |
	Quiescence[a]	Before ashfall (volcanic unrest)	During ashfall	After ashfall
Private Business	See 'all' • Some businesses will undertake specific ash risk business continuity planning	See 'all' • Business specific hazard, impact and risk management information and what are the information sources	See 'all' • Business specific hazard, impact and risk management information and what are the information sources	See 'all' • Business specific hazard, impact and risk management information and what are the information sources

[a]Level of interest is strongly context-dependent and may be influenced by high-profile eruptions at other volcanoes, proximity to a volcano, previous experiences, etc.

Table 3 Evolution of information products and activities throughout an ashfall event[a]

Quiescent phase	Pre-event phase	During eruption	Post-eruption
• Background hazard maps • Public hazard and risk education and outreach (e.g. information resources, public talks) • Sector-specific impact, mitigation and preparedness resources • Sector-specific hazard and risk information (e.g. volcano science advisory groups, volcanic risk professional courses, engagement with industry/sector groups) • Development and exercising of communication protocols, structures and guidelines	• Preparation of event-specific hazard maps • Deployment of ashfall forecast maps • Enhanced public hazard and risk education • Dissemination of sector-specific resources (e.g. ashfall preparedness posters for utilities) • Dissemination of sector-specific hazard and risk information • Optimisation of communication protocols, structures and guidelines	• Preparation of dynamic crisis hazard maps (iterative process) • Ashfall forecasts (modelled) • Ashfall maps (mapped and modelled) • Consistent public messaging on ashfall preparedness and impact advice • Syndromic surveillance for health intelligence • Ash analyses for: – Eruption forecasting – Health hazard assessment – Agricultural hazard assessment – Engineering hazard assessment (e.g. resistivity characteristics)	• Ongoing communication about risks of ashfall e.g. health, agriculture, etc. • Consistent public messaging on ashfall response and recovery advice • Sharing of lessons learned and revision/optimisation of existing products as required • Calibration of numerical hazard models with event data • Continued syndromic surveillance • Updating of hazard maps

[a]Evaluation and review may be necessary as needs of community evolve

response actions, over a variety of user-preferred platforms. Various media products have been developed for communicating ashfall hazard, risk and impacts, including hazard maps, traditional static media such as posters and brochures, and online resources. Websites have found considerable favour over the past decade, including global resources such as the website of the International Volcanic Health Hazards Network (www.ivhhn.org) and the U.S. Geological

Survey-hosted ash impacts and mitigation website http://volcanoes.usgs.gov/ash. Rapidly-emerging technologies include passive and active provision of information on social media and mobile phone applications (apps) (Leonard et al. 2014).

3.1 Hazard Maps (Background and Crisis)

Hazard maps are a common component of volcanic warnings. Maps can broadly be grouped into (a) background maps prepared in quiescent times, covering the range of possible future events based on past events and/or geological studies and (b) crisis maps for use during a specific event. Maps can also be grouped into those focussed on proximal hazards, generally with some implication for life safety near a volcano, or maps of more distal, far-reaching hazards, primarily ashfall. In addition, hazard maps may depict a single hazard (e.g., ashfall) or multiple hazards emanating from the volcano (including pyroclastic flows, lava flows and lahars).

Prior to a crisis, hazard maps are a tool for education and planning, providing information on areas most likely to be impacted by ashfall, and the probable accumulation of ash deposits. Hazard maps may be combined with spatial exposure and vulnerability information to estimate building and infrastructure damage, evacuation needs, likely transport and utility disruptions, and clean-up requirements. During a crisis, hazard maps are a valuable communication tool used to complement broadcasted alert levels.

Hazard maps for individual volcanic centres are often based on the extent of past eruptive deposits with local topography and environmental factors taken into account. Numerical modelling is often incorporated to help understand the uncertainties surrounding future activity and is particularly important in assessing ashfall hazard, as variations in wind conditions must be considered in conjunction with potential eruption scenarios. At a regional scale, aggregated multi-volcano probabilistic approaches can enable the long-term estimation of ashfall hazard at any particular location. For example, Jenkins et al. (2015) present global and regional maps of probabilistic ashfall hazard which show average recurrence intervals for ashfalls exceeding 1 mm (chosen as a threshold that may cause concern for aviation and critical infrastructure). These authors also presented a detailed local assessment for the municipality of Naples, Italy, merging probabilistic ashfall hazards from both Vesuvius and Campi Flegrei to generate a hazard map for ashfall loading on structures (in units of kPa).

Although every region is unique, crisis hazard maps in support of ashfall communication should contain: version, date, period of validity, impact information or links/references to get impact and mitigation information, reference to any other map types (e.g. background probabilistic), north arrow and scale, legend, and disclaimers as needed (e.g. to clarify that ashfall maps are not flight level forecasts). The triggers for revised versions, the revision process and timeframes should be considered.

If no hazard map exists, we recommend eight key areas for consideration: (1) audience; (2) purpose (e.g. life safety, disruption to infrastructure); (3) timeframe (background probabilistic versus crisis); (4) spatial scale (regional, whole volcano, vent/microzone); (5) organisations and their roles with procedures for discussion and ratification; (6) key messages from emergency managers; (7) hazards and zone styles to be depicted; (8) geological, historical and/or computer-modelled input data to be used. These topics should be considered in approximately this order.

3.2 Ash Forecasting Products

The ability to forecast where and when ashfall will occur is an essential step towards estimating potential consequences and providing useful warnings to stakeholders. Monitoring agencies and emergency managers aim to deliver warnings and forecasts of impending ashfalls to at-risk communities and organisations. Volcanic ashfall forecast products have been developed by several volcano monitoring agencies (e.g., USGS,

USA; JMA, Japan; GNS Science, New Zealand). Typically, these forecast products are updated regularly leading up to and throughout an eruption, and inform which areas are likely to be impacted by ash and how much ash is forecast to accumulate. More advanced models inform forecast ashfall arrival time and ashfall duration. The forecasts can provide useful warnings to exposed stakeholders (e.g., emergency managers, public health authorities, critical infrastructure, general public, etc.). Products may be in graphical, animated graphical, numeric or text formats, but a graphical map product is most common. Generally, a graphical map product is the most easily understood, particularly if it is from a perspective rather than plan view. This information is ideally released alongside advice about what people should do before, during and after ashfall and may be paired with volcano alert bulletins.

In New Zealand, for example, basic ashfall prediction maps are automatically pre-prepared three times per day for all frequently active New Zealand volcanoes, and are available for rapid deployment within a Volcanic Alert Bulletin in an eruption event or a period of unrest. Nine scenarios are pre-calculated each time, representing combinations of three height scenarios and three volume scenarios. These maps show model results computed using the Ashfall programme (Hurst 1994) and are based on wind models supplied by New Zealand's MetService. An example of the automatically-generated map for 1800 h on 9 November 2015, for the scenario of a 1 km^3 volume eruption and 20 km plume height, and incorporating current weather conditions, is shown as Fig. 4. Figure 5 shows an example of a map that was released by the volcano monitoring agency GNS Science on 13 August 2012, following the 6 August 2012 eruption at Te Maari vent (Leonard et al. 2014). This day was forecast to have little low-elevation wind and the most-likely eruption scenario was small volume and low plume height, thus the predicted ashfall extent was localised and centred on Tongariro. While these maps were not a major communication tool during this event, as the probabilities of a larger event remained low, they would have become more important had the activity escalated (Leonard et al. 2014).

An important distinction is that ashfall prediction maps are not relevant to flight level forecasts, which are issued by Volcanic Ash Advisory Centres (VAACs). Whilst beyond the scope of a chapter on ashfall hazard communication, the International Civil Aviation Organization (ICAO) has undertaken substantial work in management and communication of ash cloud hazard for aviation, through the International Airways Volcano Watch system (IAVW). There are nine Volcanic Ash Advisory Centres (VAAC) throughout the world tasked with monitoring volcanic ash plumes within their assigned airspace. Analyses are made public in the form of Volcanic Ash Advisories (VAA) and often incorporate the results of computer simulation models called Volcanic Ash Transport and Dispersion (VATD) to analyse the extent, height and concentration of ash particles in the atmosphere for aviation safety.

A number of issues need to be considered when developing ashfall forecasts to allow broad utility and understanding:

- Forecast dissemination: Forecasts need to be actively and passively disseminated to appropriate stakeholders in an appropriate format and in a timely manner.

  – Where possible dissemination pathways should be established pre-eruption and allowing the forecast product to be made widely available.
  – Uncertainty of input parameters, such as eruption plume height and eruption duration, can limit accuracy of the modelled output, and updating these parameters based on observation during an eruption may delay forecast output. Time spent collecting more accurate input parameters and calibration information needs to be balanced with delivering a timely forecast product. Some agencies deal with this challenge by generating pre-eruption and

Fig. 4 Example of automatically-generated map for Ruapehu volcano for 1800 Monday 9 November 2015, showing predicted ashfall extent for one of nine pre-calculated scenarios (1 km³ eruption volume, 20 km plume height)

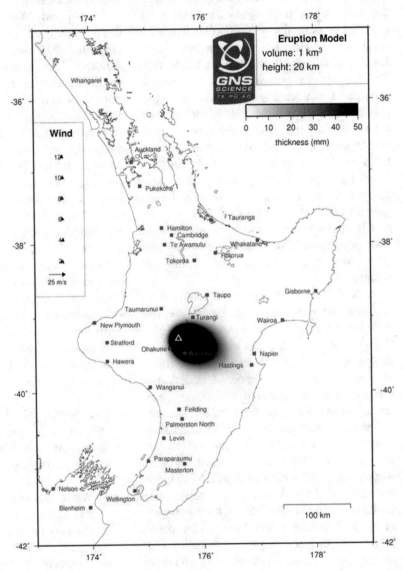

syn-eruption forecasts, with each forecast utilising improved eruption and wind input information. Post-eruption simulations may involve calibration with observed ash accumulation data.

- Hazard intensity measure: Stakeholders may require different hazard intensity measures (HIMs). For example, ash

loading (kg/m²) is critically important for impacts such as roof collapse and loading onto pastures, whereas ground-level airborne particle concentrations (µg/m³) are more directly relevant to assessing exposure to respirable ash, and visibility. Some users may require multiple HIMs. For example, both airborne particle

Fig. 5 Ashfall prediction map released with Volcanic Alert Bulletin TON-2012/17 (Geonet 2012) on 13 August 2012. The most likely eruption that might occur was small, and there was little wind that day, so the predicted ashfall extent was localised and centred on Tongariro

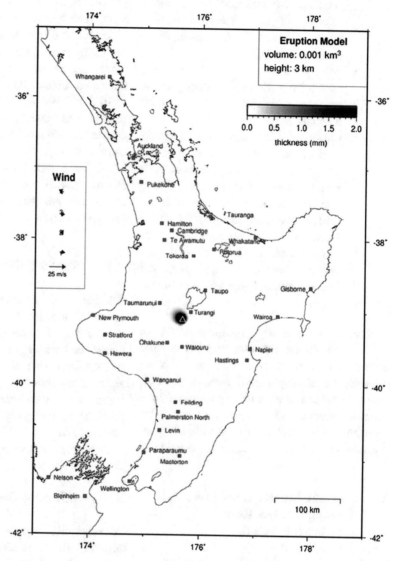

PREDICTED ASHFALL AREA
For a Tongariro eruption at 0600 Monday 13 August 2012

concentrations and ashfall loading may be relevant to the management of road networks through impacts on visibility and traction, respectively.

- Ashfall model uncertainty: Uncertainty associated with eruption parameters and climatic conditions, and simplifications applied in numerical simulation, make it challenging to forecast ash dispersal accurately, especially in near real-time. Therefore such forecasts nearly always have some degree of uncertainty attached to them, which can be challenging to communicate to end-user recipients.

- Relating ashfall hazard to consequences: The numerical models increasingly used to produce both deterministic and probabilistic ashfall hazard forecasts usually do not relate the predicted ash accumulation to potential consequences. However, this information is essential for stakeholders to make meaning of the forecasts and ultimately improve risk management decision making.

- Advice: As with any warning product, ashfall forecasts should either provide or direct recipients to advice so they may take appropriate action.
- Cartography: Not all users have a good level of map literacy, thus other forms of communication may be more suitable for some end users in addition to graphical products. Thompson et al. (2015) have noted that map properties (such as colour schemes and data classification schemes chosen) can influence how users engage with and interpret probabilistic volcanic hazard maps.

Many of these issues are dependent on the requirements of the end user and the specific context within which the warning is being received. Developing ashfall forecast products with stakeholders, along with regular review, can optimise communications. This process is also supported by research which relates ashfall quantity, subsequent effects, and appropriate action.

3.3 Public Involvement in Ashfall Mapping: The Role of Citizen Science

First-hand observers of ashfall are among the best sources of information because their reports can include details about the timing, amount and nature of ashfalls over vast geographic areas, and they can provide physical samples for detailed characterization. Local residents may be best placed to make observations before ash is removed, remobilised, or compacted. For decades, Alaskans have reported ashfall by telephone, email, web, mail, and social media campaigns (Adleman et al. 2010) to the Alaska Volcano Observatory, as a result of a long-running two-way communication effort by

AVO. A web-enabled database, "Is Ash Falling?" collects ashfall observations and encourages sample collections from the public (Wallace et al. 2015). This tool will soon be operational at other U.S. volcano observatories. It is open-source, and can easily be exported and modified for use at other observatories or agencies that collect information on ashfall around the world.

In the United Kingdom, citizen science-based methods were integrated into a suite of methods used to quantify ash deposition from the May 2011 eruption of Grímsvötn, Iceland (Stevenson et al. 2013). The British Geological Survey in Ecuador, Bernard (2013) has suggested a design for a home-made ash meter, constructed from simple, low-cost materials, to improve field data collection.

3.4 Media Releases

Scientists and emergency managers regularly release information to the media in the form of structured media releases. These are often timed to include new warnings or forecasts or are triggered by significant events. The most effective media agencies are those that already understand their importance as a communication device prior to a crisis, have relationships and trust developed with officials, and who feel empowered as part of the crisis-management team or process.

3.5 Informal Communication

A substantial proportion of communications between all groups takes the form of telephone calls, emails and face-to-face meetings. These are often not considered as formal communication devices, but they may constitute a large proportion of the time and effort of communicating during a crisis. Ideally these should be linked to the other types of communication and incorporate reference to warnings, hazard maps, and other supporting resources (e.g., preparedness resources). We also note that agencies must have an authoritative, and preferably interactive, presence on social media channels or else misinformed members of the community may occupy this space.

3.6 Standard Protocols for Determining Hazardous Characteristics of Ash

As part of the immediate emergency response, there should be rapid dissemination of information about the physical and chemical properties of the ash and its hazardous potential. Volcanic ash can be highly variable in its characteristics, both among and within eruptions. Therefore, it is necessary to assess the hazardous characteristics of ashfall *specifically* for each eruption, and with sufficient sampling to capture within-eruption spatial and temporal variability.

Specific protocols to assess hazardous characteristics of ash have been developed by the IVHHN and are described further in the following sections. These protocols are intended for use by scientists who then communicate their findings to public health and agricultural agencies, who may then modify their standard public advice messages as required. For example, after the 6 August 2012 eruption of Tongariro volcano, health and agricultural agencies were strongly interested in the levels of available fluorine (F) in the ashfall, because of reported livestock deaths from fluorosis following the 1995–1996 eruptions of Ruapehu volcano (Cronin et al. 2003). Expedited analyses of the available F content of the ash enabled distribution of results to public health officials by 10 August 2012. While the F content of the ash was moderate, the hazard to human and animal health was limited by the small volume of ash produced (Cronin et al. 2014).

3.6.1 Protocol for Assessment of Respiratory Health Hazards

A protocol for analysis of bulk ash samples for respiratory health hazard assessment (introduced in Damby et al. 2013) has been developed by the International Volcanic Health Hazard Network (IVHHN) and can be downloaded from www.ivhhn.org. The initial (rapid analysis) phase of this protocol involves particle size analysis to determine the proportion of respirable size fractions in each sample. Samples containing <1 %

(by volume) <4 μm or <2 % <10 μm are not considered respirable and do not require further analysis. 'Respirable' samples may require more detailed characterisation (e.g., crystalline silica content for non-basaltic ash), particularly if there is significant or prolonged public exposure to airborne ash (e.g., long-duration eruptions or resuspended ash), to ascertain long-term health hazards. Important health-relevant characteristics of volcanic ashfall include particle size distribution (Horwell 2007), crystalline silica content (Le Blond et al. 2009), and particle surface reactivity (Horwell et al. 2007).

3.6.2 Protocol for Assessment of Hazards from Leachable Elements

Freshly-erupted ash may contain a range of potentially toxic soluble elements such as fluorine, which may be released either rapidly or more slowly upon contact with water or body fluids. A protocol to assess the leachable element content of fresh volcanic ashfall has been developed by the IVHHN (Stewart et al. 2013). The methods include a general purpose deionised water leach, relevant to assessing impacts on drinking water supplies, livestock drinking water, fish hatcheries, and availability of soluble elements for plant uptake; and a gastric leach for a more realistic assessment of the hazards of ash ingestion for livestock.

4 Sector-Specific Considerations for Communication of Ashfall Hazards and Risks

4.1 Public Health

There are wide differences among the responses in high- and low-income countries to the hazards of volcanic ashfall, as reflected in their infrastructure, transport and communication systems. From the health standpoint, low-income countries (where many active volcanoes are located) may have different epidemiological profiles to those of advanced economies with divergent health concerns to match.

Typical public concerns about the health impacts of ashfall (see Table 1) include the effects of inhaling ash; the potential for long-term effects; and the effects on vulnerable groups (Horwell and Baxter 2006). Most concern revolves around vulnerable groups within the population: children, the elderly and those with pre-existing health problems such as cardiovascular and respiratory diseases.

The World Health Organization currently recommends that communities stay indoors during ashfall and wear light-weight, disposable face masks should they go outside. However, staying indoors is impractical during long-duration events and there is currently little evidence that lightweight masks, such as surgical masks, are effective at blocking the inhalation of respirable ash particles (although an IVHHN study is underway). The IVHHN has produced a pamphlet on "The Health Hazards of Volcanic Ash: A Guide for the Public" (downloadable from www.ivhhn.org). This internationally-ratified pamphlet provides *generally applicable* advice for the public, and is available in nine languages, and is supported by a second pamphlet on how to prepare for ashfall, "Guidelines on Preparedness

Fig. 6 Civil defence advice for ashfall, Sistema Nacional de Protección Civil, Colima, México. *Source*: Dr Maria Aurora Armienta, UNAM, México City, México

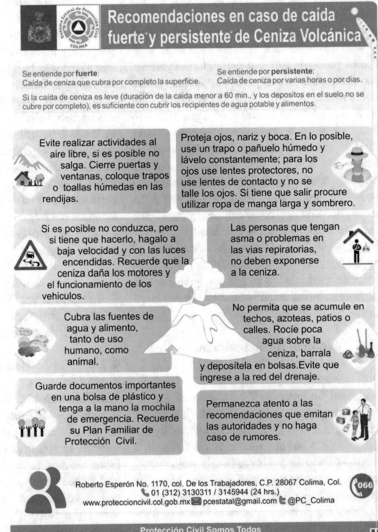

Before, During and After Ashfall", aimed both at the public and emergency managers. Many countries have also developed their own civil defence advice, typically addressing topics such as covering open water supplies, protecting human and animal health, and cleaning up private property (e.g., Fig. 6).

Another common concern is risks to drinking-water supplies, livestock and crops contaminated by ashfall as fresh ash can carry a soluble salt burden that is readily released on contact with water (Stewart et al. 2006). The leachate protocol, described in Sect. 3.6.2, addresses these concerns.

Finally, we note that social and economic disruption resulting from volcanic activity may cause psychological stress that may outweigh physical impacts, particularly for long-lived eruptions. Avery (2004) notes (in relation to the long-lived volcanic crisis on Montserrat) that the social and economic disruption has had a far more profound influence on the health of the ~4500 residents of Montserrat than any purely physical effects related to ash inhalation.

4.1.1 Crystalline Silica

The most hazardous eruptions are those generating fine-grained ash with a high content of *free crystalline silica*, as this mineral has the potential to cause silicosis (a chronic lung disease resulting in scarring damage to the lungs and impairment of their function). Silicosis is primarily an occupational disease associated with occupations such as stone-cutting, tunnel building, and quarrying. To date, no cases of silicosis have been attributed to exposure to volcanic ash, although this may be due to the relatively small population affected.

Rapid determination of quantities (wt%) of free crystalline silica in bulk ash samples after ashfall, using reliable methods, is important (e.g., Damby et al. 2013). Particular care must be taken by agencies conducting and reporting on analyses to avoid any confusion between free crystalline silica (where the individual minerals cristobalite, quartz and tridymite are quantified) and total silica content (commonly used to quantify the bulk composition of ash). Within days of the 1980 eruption of Mt St Helens, there were reports in the media that the Mt St Helens ash contained 60 % or more free crystalline silica—far greater than the actual 3–7 % in the sub-10 µm size fraction (Mount St. Helens Technical Information Network 1980). This misinformation occurred because of a misunderstanding of the difference between free and total silica, and difficulties interpreting the X-ray diffraction pattern due to overlapping feldspar peaks.

In the event of prolonged population exposure to airborne respirable ashfall with a substantial crystalline silica content (in particular, if the eruption is long-lived or ash is being continuously remobilised by wind) it may be necessary for public health officials to conduct more detailed studies on population exposure by using cyclone air samplers to collect samples of airborne respirable dust. The results can then be compared to occupational and environmental exposure limits (Searl et al. 2002).

The groups most heavily exposed are outdoor workers who have to conduct their jobs while exposed to ash (Searl et al. 2002). They include police and traffic controllers, rescuers, emergency staff in utility companies, road and repair workers, clean-up crews, and farmers, who will need specific health messages and advice on personal protective equipment and occupational health risk assessments. There are occupational exposure limits for respirable crystalline silica and to adhere to these will require occupational health and safety input to monitor exposure of workers and to show legal compliance. For the general public the most appropriate exposure limits for health risk assessment are those for particulate matter (see Sect. 4.1.2). Neither of these enforceable sets of limits were designed for volcanic eruptions and so are unrealistic except as guides for communicating potential health risks; specialist advice will be needed for every new eruption, taking into account local circumstances, as was applied after Mount St Helens in 1980 and the volcanic crisis on Montserrat in 1995 onwards (Baxter et al. 2014).

4.1.2 Particulate Matter

In 2013, a review by the World Health Organization concluded that inhalation of any particulate matter sub-2.5 μm diameter (known as $PM_{2.5}$) may impact chronic and acute morbidity and mortality in relation to a range of diseases including cardiovascular and respiratory diseases (World Health Organization 2013). In the USA and European Union countries, there are legal standards on ambient air quality and established air monitoring networks, together with general awareness about the health effects of low levels of air pollutants from sources such as traffic emissions.

Major concerns exist about the health impacts from breathing in air containing elevated levels of respirable ash particles (of non-specific composition), especially in children, and the measures needed to prevent such high exposure. A significant problem after explosive eruptions in dry or semi-arid regions, or during unseasonal droughts, is the resuspension of ash deposits by wind and traffic, leading to exceedances of daily PM_{10} and $PM_{2.5}$ air quality targets by at least one order of magnitude until rain helps to clear the air and consolidate the material, which can be exceedingly fine (including sub-micron particles). The consolidated deposits in inhabited areas should be removed to prevent remobilisation. Strategies such as placing restrictions on vehicle speeds and dampening ash deposits with water may be helpful (Wilson et al. 2013).

Health conditions like asthma and chronic obstructive pulmonary disease are common in the general population, and symptoms of these are likely to be aggravated by exposure to ash. Patients with pre-existing health problems may need to discuss with their physicians the wisdom of moving away from badly affected areas until air quality improves. Public health officials and physicians will need to become well-versed in the acute and chronic health issues surrounding ambient $PM_{2.5}$ in particular. These are complicated for non-specialists to grasp. A further challenge is the development of expertise in communicating the potential health risks associated with exposure to levels of $PM_{2.5}$ that are considerably higher than typical ambient levels in regulated urban environments. Syndromic surveillance (where real-time data are collected from existing public health networks used to monitor the outbreaks of disease) may be useful in communicating the need for health protection strategies where impacts (such as an increase in asthma cases) are recorded (Elliot et al. 2010).

4.2 Agriculture

Impacts of ashfall on agricultural depend on a complex array of factors (Table 1), as well as the inherent vulnerability of the exposed farming systems, on scales ranging from regional (e.g., related to climate) to individual farm-scale (e.g., availability of shelter and supplementary feed). While certain impacts tend to be commonly observed, others may be more site or eruption specific. Thus, in addition to generic impact and mitigation advice, more tailored mitigation strategies may be required.

The assessment of the potential for ashfall to contaminate food chains, as required by modern agricultural production and food safety regulations, is critical. This is essential information for a wide range of stakeholders, from farmers who need to manage and minimise impacts, to food safety organisations. Considerable anxiety can be created for farmers, agricultural markets, and consumers if this issue is not managed and communicated effectively. For example, during the 2010 Eyjafjallajökull, Iceland, eruption, the European Commission asked the European Food Safety Authority (EFSA) to assess the possible short-term threats to food safety in the European Union (EU) from ashfall. The EFSA had no prior information on this hazard and so had to rapidly review and compile scientific information for its assessment (EFSA 2010). No ashfall composition information was available at the time to guide their review. The ESFA identified fluoride as the main component that could pose a short-term risk to food and feed safety, although the risk was assessed as negligible given the very small quantities of ashfall on mainland Europe.

Public anxiety around this issue was considerable, requiring rapid and authoritative communication of the risk to reassure consumers and agricultural markets.

Information demands from farmers, agricultural support organisations, media and other key stakeholders before, during and after an ash-generating eruption can be considerable and diverse, and typically evolve as the risk context changes. Topics on which information is sought include all aspects of volcanic activity, ashfall hazard, likely impacts, and recommended mitigation actions.

From our experience conducting post-event interviews with farmers and farming support organisations, we identify the following information demands that commonly arise before, during, and after ashfall:

1. Will I receive ashfall, and if so, how much and when?
2. What impacts will it have on my farming operations (including effects on pasture, soil, crops, livestock, and farm infrastructure?
3. When will hazard characterisation of the ash be completed? (e.g., characterisation of the environmentally available elements)
4. What actions can I take to mitigate potential consequences before, during, and after ashfall?
5. What support is available? (including sources of advice and direct financial assistance)

In our experience, pre-existing and regularly maintained relationships, protocols, and information resources can greatly ease communication and management demands in a crisis.

The U.S. Geological Survey hosts an ash impacts website, delivering information on ashfall impacts and mitigation for the agricultural sector (U.S. Geological Survey 2015). However, we note that case studies on tropical agricultural systems are limited. Country-specific information resources have been developed for New Zealand (MPI 2012).

4.3 Infrastructure

Ashfalls of just a few millimetres can be damaging and disruptive to critical infrastructure services (also known as 'utilities' in some countries), such as electricity generation, transmission and distribution networks, drinking-water and wastewater treatment plants, roads, airports and communication networks (Wilson et al. 2012). Additionally, disruption of service delivery can have cascading impacts on wider society. Specific impacts of ashfall vary considerably, depending on factors such as plant or network design, ashfall characteristics (e.g., loading, grain-size, composition and levels of leachable elements), and environmental conditions before and after the ashfall (Wilson et al. 2011a, b). Evidence is growing that a range of preparedness and mitigation strategies can reduce ashfall impacts for critical infrastructure organisations (Wilson et al. 2012, 2014).

Volcanic eruptions that produce heavy ashfall are, in general, infrequent and somewhat exotic occurrences and consequently, in many parts of the world, infrastructure managers may not have devoted serious consideration to management of a volcanic crisis. Therefore, during non-crisis periods, risk communication activities should be primarily concerned with volcanic ashfall hazard and impact awareness and education, and making utility companies aware of where information and expertise resides. This incorporates hazard, impact and risk assessment, vulnerability analysis, and formal and informal network building (Daly and Johnston 2015). During crisis periods, provision of specialist, sector-specific impact information is essential to enable rapid decision making in order to minimise consequences. In both instances, preparation of pre-prepared information resources has been beneficial (Leonard et al. 2014). Ideally, a collaborative, participatory process develops these resources for reach region (Twigg 2007).

A successful example of a collaborative process is the creation of a suite of ten posters designed to improve preparedness of critical infrastructure organisations for volcanic ashfall hazards (Wilson et al. 2014; see download link provided in Sect. 4.4). Key features of this process were: (1) a partnership between critical infrastructure managers and other relevant government agencies with volcanic impact scientists, including extensive consultation and review phases; and (2) translation of volcanic impact research into practical management tools. Whilst these posters have been developed specifically for use in New Zealand, the authors propose that these posters are widely applicable for improving resilience to volcanic hazards in other settings (Wilson et al. 2014).

4.4 Clean-up

The removal of ash from urban areas is vital for recovery. However, clean-up operations are more complex than just removal; the ash also needs to be disposed of and stabilised to avoid future problems from remobilisation. Areas exposed to ash hazards should have clean-up plans in place beforehand, covering the following aspects:

- Personnel and equipment requirements, including mutual support agreements for ash clean-up as part of regional emergency management contingency planning.
- Provisions for management of health and safety risks.
- An incident management system/database to manage the clean-up operation.
- Identification of potential disposal sites.
- Strategies for stabilisation of deposits.

Volunteers commonly assist with clean-up operations following an ashfall. Volunteer labour can significantly speed up these operations, but requires effective management and integration with professional crews. An effective communication strategy should include regular briefings of volunteers, liaison officers and health and safety support (Wilson et al. 2014). Clear and ongoing communication with the public during clean-up operations aids efficiency, public trust and goodwill. Guidance on appropriate clean-up methods aids effectiveness, and the coordinated clean-up of neighbourhoods will optimise use of resources and reduce recontamination of cleaned sections.

An example of the value of having pre-existing plans in place, and then communicating them clearly to the public, comes from the May 2010 eruption of Pacaya volcano, Guatemala, which deposited an estimated 11,350,000 m^3 of medium to coarse basaltic ash on Guatemala City, covering approximately 2100 km of roads to depths of 20–30 mm (Wardman et al. 2012). The municipality of Guatemala City utilised a pre-existing emergency plan originally devised for clearing earthquake debris (as a local response to the devastating earthquakes in Haiti and Chile earlier in 2010). An important factor in the success of this clean-up was clear communication with the public. The public were instructed to clear ash from their own properties (roofs and yards), collect it in sacks and to pile the sacks on the street frontage or take them to designated collection points. Sacks were obtained from local sugar and cement companies (Director of Public Works, Municipality of Guatemala City; 2010, pers. comm.). Streets were cleaned with street sweepers or manually, and the ash loaded onto lorries with small excavators. While there were some ongoing problems with flooding caused by ash ingress into storm drains, the main transport routes in Guatemala City (which generates 70 % of the GNP of Guatemala) were cleared within days and the city returned rapidly to its pre-existing level of functionality.

Lessons from this and other eruptions are summarised on the poster "Volcanic Ashfall: Advice for Urban Cleanup Operations" (Auckland Lifelines 2014).

5 Ongoing Communication Demands: Managing Long-Duration Eruptions

In some cases volcanic activity is not confined to a short period of time, but may continue to threaten populations for many years. Some current examples of long-duration and/or ongoing eruptions include: Sakurajima, Japan (intermittent since 1955); Rabaul, Papua New Guinea (intermittent since 1994); Merapi, Indonesia (events every few years since the turn of the 20th century); Soufrière Hills, Montserrat (1995 to present); and Tungurahua, Ecuador (1999 to present). Long-duration eruptions generate hazards of varying intensity over time, where more frequent hazards include ashfalls, gases and acid rain. These hazards can generate widespread losses across societies (Table 1). In long-duration eruptions, this may undermine resilience in the long-term as losses are often not accounted for by governments and businesses, and become absorbed by households and communities. The recurrent nature of the hazards creates challenges for recovery (Sword-Daniels et al. 2014). The complex range of impacts and losses for infrastructure and societies, their cumulative nature, and their long-term manifestations are not well known (Sword-Daniels et al. 2014; Tobin and Whiteford 2004). In general, there are few studies to inform appropriate communication and management strategies and long-term mitigation options for long-duration eruptions.

At some frequently active volcanoes, communication strategies have been developed between disaster managers and communities, but because hazards may vary over time, challenges in communication can arise when the type of hazard changes or is unforeseen (De Bélizal et al. 2012). In many long-duration eruptions, the type of activity can suddenly switch from effusive (dome-building) to explosive, with each presenting entirely different hazards and impacts for the affected communities. For long-duration eruptions, communication strategies, therefore, need to be flexible under changing hazard conditions, must reach and meet the needs of a diverse range of stakeholders and residents during hazard events, and become established such that they can be quickly enacted even after periods of quiescence.

In Montserrat, West Indies, the onset of a long-duration eruption in 1995 (ongoing at the time of writing) of the Soufrière Hills volcano prompted the creation of an exclusion zone in 1996, and relocation of the population further

from the volcano. Despite this, ongoing ashfalls, acid rain and gases intermittently affected populated areas of this small island (e.g. from November 2009 to February 2010), and continued for prolonged periods of time (Wadge et al. 2014). Communication strategies for managing ashfalls have developed and improved over time, creating both formal (often broadcast via radio) and informal local information networks. These provide information about which areas of the island are affected by ashfalls and any temporarily affected infrastructure and services; and advice for residents about protective actions for public health and safety. In particular, dome-forming eruptions, such as Soufrière Hills, create ash containing abundant crystalline silica which has the potential to cause diseases such as silicosis (Baxter et al. 1999, see Sect. 4.1.1). Thus, monitoring and reporting on the crystalline silica content (to government agencies) allowed informed decision-making on population exposure, and was an important part of hazard communication during this eruption (Baxter et al. 2014).

6 Communication Demands During Recovery

Each recovery context is unique, depending on the level of impact (where different impacts are experienced by different groups), available resources, and the social, political and economic context (Smith and Birkland 2012; Tierney and Oliver-Smith 2012). Recovery plans should ideally be in place before a hazard event so that all stakeholders share a common understanding and expectations of the recovery process (Phillips 2009). Tools and strategies that promote community engagement and participation are essential in order to account for multiple perspectives, the needs of different groups, and to guide the recovery process. Effective communication requires clarity and transparency in decision-making during all stages of the process.

In the early stages of recovery after an ashfall event, information and communication should focus on providing emergency assistance (where necessary), undertaking damage assessments, ashfall clean-up activities, restoring the function of infrastructure and services, access to livelihoods, and providing psychosocial support. Rapid responses may reduce longer-term impacts.

In the longer-term, tools and strategies need to transition to become focused on any changes that can be made to increase resilience. Aspects that may be considered include: livelihood diversity, possible adaptations, improvements in reconstruction techniques, land-use planning for future development, ensuring social wellbeing and social security mechanisms, the preservation of culture, and strategies for long-term economic stability.

7 Lessons

Lessons from volcanic ashfall events point to the following key considerations for effective communication:

1. Consistent messages must be delivered from different official agencies wherever possible. This may be fostered through regular inter-agency meetings and structures (e.g., Leonard et al. 2014; Madden et al. 2014) and requires a high level of situation awareness and information sharing.
2. Messages need to be repeated periodically during a prolonged event.
3. Planning needs to allow for time-varying messages. Messages are often evolving, with more data becoming available over time.
4. Agency jurisdictions—over who is authorised to issue different types of messages—need to be discussed and formalised before crises. Usually scientists give information on the volcano status and emergency managers give messages on public safety and instructions to evacuate. However, this needs to be formalised (e.g., Madden et al. 2014).
5. Key messages should be pre-planned wherever possible to ensure complete coverage of essential advice and to reduce workload during crisis periods (e.g., standard public

health messaging). However, there needs to be flexibility in line with the evolving situation.
6. Volcanic ashfall hazard awareness should start with sector-specific background information delivered during quiescent times.
7. Information needs before, during, and after ashfall events vary for different audiences; thus pre-planned messages and resources should be developed and tested with diverse audiences in mind.

Acknowledgments For constructive review comments, we thank Bill McGuire and an anonymous reviewer; and Cheryl Cameron of the State of Alaska Division of Geological and Geophysical Surveys and Game McGimsey of the U.S. Geological Survey. We also thank Emma Hudson-Doyle for helpful discussions.

References

Adleman JN, Cameron CE, Snedigar SF, Neal CA, Wallace KL (2010) Public outreach and communications of the Alaska Volcano Observatory during the 2005–2006 eruption of Augustine volcano. In: Power JA, Coombs ML, Freymueller JT (eds) The 2006 eruption of Augustine Volcano, Alaska. U.S. Geological Survey Professional Paper, pp 631–644

Alexander D (2007) Making research on geological hazards relevant to stakeholders' needs. Quatern Int 171–172:186–192

Auckland Lifelines (2014) Volcanic ashfall: advice for urban clean-up operations. Poster downloadable from: http://www.aelg.org.nz/document-library/volcanic-ash-impacts

Avery G (2004) Inhaling volcanic ash on Montserrat. Occup Environ Med 61:184

Baxter PJ, Bonadonna C, Dupree R, Hards VL, Kohn SC, Murphy MD, Nichols A, Nicholson RA, Norton G, Searl A, Sparks RSJ, Vickers BP (1999) Cristobalite in volcanic ash of the Soufrière Hills Volcano, Montserrat, British West Indies. Science 283:1142–1145

Baxter PJ, Searl A, Cowie HA, Jarvis D, Horwell CJ (2014) Evaluating the respiratory health risks of volcanic ash at the eruption of the Soufrière Hills Volcano, Montserrat, 1995–2010. In: Wadge G, Robertson R, Voight B (eds) The eruption of Soufrière Hills Volcano, Montserrat from 2000 to 2010. Memoir of the Geological Society of London

Bernard B (2013) Homemade ashmeter: a low-cost, high efficiency solution to improve tephra field-data collection for contemporary explosive eruptions. J Appl Volcanol 2:1–9

Blong RJ (1984) Volcanic hazards: a sourcebook on the effects of eruptions. Academic Press, North Ryde, New South Wales, Australia

Brown SK, Loughlin SC, Sparks RSJ, Vye-Brown C (2014) Global volcanic hazards and risk: technical background paper for the UN-ISDR global assessment report on disaster risk reduction 2015, global volcano model and the international association of volcanology and chemistry of the earth's interior (IAVCEI)

Carlsen HK, Gislason T, Benediktsdottir B, Kolbeinsson TB, Hauksdottir A, Thorsteinsson T, Briem H (2012a) A survey of early health effects of the Eyjafjallajökull 2010 eruption in Iceland: a population-based study. BMJ Open 2(2):e000343

Carlsen HK, Hauksdottir A, Valdimarsdottir UA, Gíslason T, Einarsdottir G, Runolfsson H, Briem H, Finnbjornsdottir RG, Gudmundsson S, Kolbeinsson TB, Thorsteinsson T, Pétursdóttir G (2012b) Health effects following the Eyjafjallajökull volcanic eruption: a cohort study. BMJ Open 2(6): e001851

Cook RJ, Barron JC, Papendick RI, Williams GJ (1981) Impact on agriculture of the Mount St. Helens eruptions. Science 211:16–22

Covello VT, Allen F (1988) Seven cardinal rules of risk communication U.S. Environmental Protection Agency, Washington, D.C.

Cronin SJ, Neall VE, Lecointre JA, Hedley MJ, Loganathan P (2003) Environmental hazards of fluoride in volcanic ash: a case study from Ruapehu volcano, New Zealand. J Volcanol Geotherm Res 121:271–291

Cronin SJ, Stewart C, Zernack AV, Brenna M, Procter JN, Pardo N, Christenson B, Wilson TM, Stewart RB, Irwin M (2014) Volcanic ash leachate compositions and assessment of health and agricultural hazards from 20102 hydrothermal eruptions, Tongariro, New Zealand. J Volcanol Geotherm Res 286:233–247

Cutter SL, Ismail-Zadeh A, Alcántara-Ayala I, Altan O, Baker DN, Briceno S, Gupta H, Holloway A, Johnston DM, McBean GA, Ogawa Y, Paton D, Porio E, Silbereisen RK, Takeuchi K, Valsecchi GB, Vogel C, Wu G (2015) Pool knowledge to stem losses from disasters. Nature 522(7556):277–279

Daly M, Johnston DM (2015) The genesis of volcanic risk assessment for the Auckland engineering lifelines project: 1996–2000. Journal of Applied Volcanology 4:7

Damby DE, Horwell CJ, Baxter PJ, Delmelle P, Donaldson K, Dunster C, Fubini B, Murphy F, Nattrass C, Sweeney S, Tetley T, Tomatis M (2013) The respiratory health hazard of tephra from the 2010 Centennial eruption of Merapi with implications for occupational mining of deposits. J Volcanol Geotherm Res 261:376–387

De Bélizal É, Lavigne F, Gaillard JC, Grancher D, Pratomo I, Komorowski J-C (2012) The 2007 eruption of Kelut volcano (East Java, Indonesia): phenomenology, crisis management and social response. Geomorphology 136(1):165–175

Director of Public Works, Municipality of Guatemala City; pers. comm. Guatemala City, 23 September 2010

EFSA (2010) Statement of EFSA on the possible risks for public and animal health from the contamination of the feed and food chain due to possible ashfall following the eruption of the Eyjafjallajokull volcano in Iceland: urgent advice. EFSA J 8(4):1593

Elliot AJ, Singh N, Loveridge P, Harcourt S, Smith S, Pnaiser R, Kavanagh K, Robertson C, Ramsay CN, McMenamin J, Kibble A, Murray V, Ibbotson S, Catchpole M, McCloskey B, Smith GE (2010) Syndromic surveillance to assess the potential public health impact of the Icelandic volcanic ash plume across the United Kingdom, April 2010. Eur Surveill 15(23):6–9

Few R, Barclay J (2011) Societal impacts of natural hazards, a review of international research funding. UK Collaborative on Development Sciences, London

Folch A, Mingari L, Osores MS, Collini E (2014) Modeling volcanic ash resuspension—application to the 14–18 October 2011 outbreak episode in central Patagonia, Argentina. Nat Hazards Earth Syst Sci 14:119–133

Geonet (2012) Volcanic alert bulletin TON-2012/17-Tongariro Volcano. Available at: http://info.geonet.org.nz/pages/viewpage.action?pageId=2195927

Hadley D, Hufford GL, Simpson JJ (2004) Resuspension of relic volcanic ash and dust from Katmai: still an aviation hazard. Weather Forecasting 19:829–840

Höppner C, Bründl M, Bucheker M (2010) Risk communication and natural hazards. CapHaz WP5 Report, Swiss Federal Research Institute WSL

Horwell CJ (2007) Grain size analysis of volcanic ash for the rapid assessment of respiratory health hazard. J Environ Monitor 9(10):1107–1115

Horwell CJ, Baxter PJ (2006) The respiratory health hazards of volcanic ash: a review for volcanic risk mitigation. Bull Volc 69(1):1–24

Horwell CJ, Fenoglio I, Fubini B (2007) Iron-induced hydroxyl radical generation from basaltic volcanic ash. Earth Plan Sci Lett 261(3–4):662–669

Hurst AW (1994) ASHFALL—a computer programme for estimating volcanic ash fallout: report and users guide. GNS Science, New Zealand

ICSU (2008) A science plan for integrated research on disaster risk. http://www.irdrinternational.org/2012/12/29/irdr-science-plan/

Jenkins SF, Wilson T, Magill C, Miller V, Stewart C, Blong R, Marzocchi W, Boulton M, Bonadonna C, Costa A (2015) Volcanic ash fall hazard and risk. In: Loughlin SC, Sparks RSJ, Brown SK, Jenkins SF, Vye-Brown C (eds) Global volcanic hazards and risk. Cambridge University Press, Cambridge

Le Blond JS, Cressey G, Horwell CJ, Williamson BJ (2009) A rapid method for quantifying single mineral phases in heterogeneous natural dust using X-ray diffraction. Powder Diffr 24:17–23

Leonard GS, Stewart C, Wilson TM, Procter JN, Scott BJ, Keys HJ, Jolly GE, Wardman JB, Cronin SJ,

McBride SK (2014) Integrating multidisciplinary science, modelling and impact data into evolving, syn-event volcanic hazard mapping and communication: a case study from the 2012 Tongariro eruption crisis, New Zealand. J Volcanol Geotherm Res 286:208–232

Madden J, Power J, VanPeursem KW, Holt G, Devaris A, Brennell J, Hartig L (2014) Alaska interagency operating plan for volcanic ash episodes

McBean GA (2012) Integrating disaster risk reduction towards sustainable development. Current Opinion in Environmental Sustainability 4:122–127

Mileti D (1999) Disasters by design. Joseph Henry Press, Washington, DC

Mount St. Helens Technical Information Network (1980) Bulletin #13 research into the free crystalline silica content of Mount St. Helens ash. Federal Coordinating Office, FEMA. Downloadable from: http://volcanoes. usgs.gov/ash/health/bulletin13_msh.html

MPI. 2012. Volcanic eruption: impacts and hazard mitigation for New Zealand's primary production industries. Downloadable from: https://www.mpi. govt.nz/document-vault/137

Paton D, Johnston DM, Houghton B, Flin R, Ronan K, Scott B (1999) Managing natural hazard consequences: planning for information management and decision-making. J Am Soc Profess Emerg Plan 6:37–47

Paton D, Smith L, Daly M, Johnston DM (2008) Risk perception and volcanic hazard mitigation: Individual and social perspectives. J Volcanol Geoth Res 172:179–188

Paton D, Smith L, Johnston DM (2005) When good intentions turn bad: promoting natural hazard preparedness. Austr J Emerg Manag 20(1):25–30

Phillips BD (2009) Disaster Recovery. CRC Press, Boca Raton

Schaefer JR, Bull K, Cameron C, Coombs M, Dieffenbach A, Lopez T, McNutt S, Neal C, Payne A, Power J, Schneider D, Scott W, Snedigar S, Thompson G, Wallace K, Waythomas C, Webley P, Werner C (2011) The 2009 eruption of Redoubt Volcano, Alaska, Alaska Division of Geological and Geophysical Surveys

Searl A, Nicholl A, Baxter PJ (2002) Assessment of the exposure of islanders to ash from the Soufriere Hills volcano, Montserrat, West Indies. Occup Environ Med 59:523–531

Smith G, Birkland T (2012) Building a theory of recovery: institutional dimensions. Int J Mass Emerg Disasters 30(2):147–170

Spence R, Kelman I, Baxter PJ, Zuccaro G, Petrazzuoli S (2005) Residential building and occupant vulnerability to ashfall. Nat Hazards Earth Syst Sci 5:477–494

Stevenson JA, Loughlin SC, Font A, Fuller GW, MacLeod A, Oliver IW, Jackson B, Horwell CJ, Thordarson T, Dawson I, Willis P (2013) UK monitoring and deposition of tephra from the May 2011 eruption of Grímsvötn. Iceland. Journal of Applied Volcanology 2:3

Stewart C, Horwell CJ, Plumlee G, Cronin S, Delmelle P, Baxter PJ, Calkins J, Damby DE, Morman S, Oppenheimer C (2013) Protocol for analysis of volcanic ash samples for assessment of hazards from leachable elements. IAVCEI Commissions joint report: international volcanic health hazard network and cities and volcanoes. Protocol downloadable from: www.ivhhn.org

Stewart C, Johnston DM, Leonard G, Horwell CJ, Thordarsson T, Cronin S (2006) Contamination of water supplies by volcanic ashfall: a literature review and simple impact modelling. J Volcanol Geotherm Res 158:296–306

Sword-Daniels V, Wilson TM, Sargeant S, Rossetto T, Twigg J, Johnston DM, Loughlin SC, Cole PD (2014) Consequences of long-term volcanic activity for essential services in Montserrat: challenges, adaptations and resilience. In: Wadge G, Robertson REA, Voight B (eds) The eruption of Soufriere Hills Volcano, Montserrat from 2000 to 2010. Geological Society of London Memoirs, London, pp 471–488

Thompson M-A, Lindsay JM, Gaillard J-C (2015) The influence of probabilistic volcanic hazard map properties on hazard communication. Journal of Applied Volcanology 4:6

Tierney K, Oliver-Smith A (2012) Social dimensions of disaster recovery. Int J Mass Emerg Disasters 30 (2):123–146

Tobin GA, Whiteford LM (2004) Chronic hazards: health impacts associated with ongoing ashfalls around Mt. Tungurahua in Ecuador. Papers of the applied geography conferences vol 27, pp 84–93

Twigg J (2007) Characteristics of a disaster-resilient community: a guidance note. DFID Disaster Risk Reduction Interagency Coordination Group

UNISDR (United Nations International Strategy for Disaster Reduction) (2015) Sendai framework for disaster risk reduction 2015–2030. UNISDR, Geneva

US Geological Survey (2015) Volcanic ash impacts and mitigation website. Accessible at: http://volcanoes. usgs.gov/ash

Wadge G, Voight B, Sparks RSJ, Cole PD, Loughlin SC, Robertson REA (2014) An overview of the eruption of Soufrière Hills Volcano, Montserrat from 2000 to 2010. Geological Society, London, Memoirs 39:1–40

Wallace K, Snedigar S, Cameron C (2015) 'Is Ash Falling?', an online ashfall reporting tool in support of improved ashfall warnings and investigations of ashfall processes. J Appl Volcanol 4:8

Wardman J, Sword-Daniels V, Stewart C, Wilson TM (2012) Impact assessment of the May 2010 eruption of Pacaya volcano, Guatemala. GNS Science Report 2012/09, New Zealand

Wilson TM, Cole J, Cronin S, Stewart C, Johnston DM (2011a) Impacts on agriculture following the 1991 eruption of Vulcan Hudson, Patagonia: lessons for recovery. Nat Hazard Rev 57(2):185–212

Wilson TM, Cole JW, Stewart C, Cronin SJ, Johnston DM (2011b) Ash storms: Impacts of wind-remobilised volcanic ash on rural communities and agriculture following the 1991 Hudson eruption, southern

Patagonia, Chile. Bull Volcanol 73:223–239. doi:10.1007/s00445-010-0396-1

Wilson TM, Jenkins S, Stewart C (2015) Impacts from volcanic ashfall. In: P. Papale (ed) Hazard, risks and disasters: volume 2 volcanic hazards, risks and disasters. Elsevier Inc., Amsterdam

Wilson TM, Stewart C, Bickerton H, Baxter PJ, Outes V, Villarosa G, Rovere E (2013) Impacts of the June 2011 Puyehue-Cordón Caulle volcanic complex eruption on urban infrastructure, agriculture and public health. GNS Science Report 2012/20, New Zealand

Wilson TM, Stewart C, Sword-Daniels V, Leonard GS, Johnston DM, Cole JW, Wardman JB, Wilson G, Barnard ST (2012) Volcanic ash impacts on critical infrastructure. Phys Chem Earth, Parts A/B/C 45–46:5–23

Wilson TM, Stewart C, Wardman JB, Wilson G, Johnston DM, Hill D, Hampton SJ, Villemure M, McBride S, Leonard GS, Daly M, Deligne NI, Roberts L (2014) Volcanic ashfall preparedness poster series: a collaborative process for reducing the vulnerability of critical infrastructure. J Appl Volcanol 3(10):1–25

World Health Organization (2013) Review of evidence on health aspects of air pollution—REVIHAAP project: final technical report. WHO European Centre for Environment and Health, Bonn

Volcanic Ash and Aviation—The Challenges of Real-Time, Global Communication of a Natural Hazard

Peter Lechner, Andrew Tupper, Marianne Guffanti, Sue Loughlin and Tom Casadevall

Abstract

More than 30 years after the first major aircraft encounters with volcanic ash over Indonesia in 1982, it remains challenging to inform aircraft in flight of the exact location of potentially dangerous ash clouds on their flight path, particularly shortly after the eruption has occurred. The difficulties include reliably forecasting and detecting the onset of significant explosive eruptions on a global basis, observing the dispersal of eruption clouds in real time, capturing their complex structure and constituents in atmospheric transport models, describing these observations and modelling results in a manner suitable for aviation users, delivering timely warning messages to the cockpit, flight planners and air traffic management systems, and the need for scientific development in order to undertake operational enhancements. The framework under which these issues are managed is the International Airways Volcano Watch (IAVW), administered by the International Civil Aviation Organization (ICAO). ICAO outlines in its standards and recommended practices (International Civil Aviation Organization 2014a, b) the basic volcanic monitoring and communication that is necessary at volcano observatories in Member States (countries). However, not all volcanoes are monitored and not all countries with volcanoes have mandated volcano observatories or equivalents. To add to the efforts of volcano observatories, a system of Meteorological Watch Offices, Air Traffic Management Area Control Centres, and nine specialist Volcanic Ash Advisory Centres (VAACs) are

P. Lechner (✉)
New Zealand Civil Aviation Authority, Wellington, New Zealand
e-mail: peter.lechner@caa.govt.nz

A. Tupper
Australian Bureau of Meteorology, Melbourne, Australia
e-mail: a.tupper@bom.gov.au

M. Guffanti · T. Casadevall
U.S. Geological Survey, Reston, USA
e-mail: guffanti@usgs.gov

T. Casadevall
e-mail: tcasadevall@usgs.gov

S. Loughlin
British Geological Survey, Nottingham, UK
e-mail: sclou@bgs.ac.uk

Advs in Volcanology (2018) 51–64
https://doi.org/10.1007/11157_2016_49
© The Author(s) 2017
Published Online: 04 June 2017

responsible for observing, analysing, forecasting and communicating the aviation hazard (airborne ash), using agreed techniques and messages in defined formats. Continuous improvement of the IAVW framework is overseen by expert groups representing the operators of the system, the user community, and the science community. The IAVW represents a unique marriage of two scientific disciplines, volcanology and meteorology, with the aviation user community. There have been many multifaceted volcanic eruptions in complex meteorological conditions during the history of the IAVW. Each new eruption brings new insights into how the warning system can be improved, and each reinforces the lessons that have gone before. The management of these events has improved greatly since the major ash encounters in the 1980s, but discontinuities in the warning and communications system still occur. A good example is a 2014 ash encounter over Indonesia following the eruption of Kelut where the warnings did not reach the aircraft crew. Other events present enormous management challenges—for example the 2010 Eyjafjallajökull eruption in Iceland was, overall, less hazardous than many less publicised eruptions, but numerous small to moderate explosions over several weeks produced widespread disruption and a large economic impact. At the time of writing, while there has been hundreds of millions of US dollars in damage to aircraft from encounters with ash, there have been no fatalities resulting from aviation incidents in, or proximal to volcanic ash cloud. This reflects, at least in part, the hard work done in putting together a global warning system—although to some extent it also reflects a measure of good statistical fortune. In order to minimise the risk of aircraft encounters with volcanic ash clouds, the global effort continues. The future priorities for the IAVW are strongly focused on enhancing communication before, and at the very onset of a volcanic ash-producing event (typically the more dangerous stage), together with improved downstream information and warning systems to help reduce the economic impact of eruptions on aviation.

Keywords

Volcanic ash · Aviation · Hazard · Global communication

1 Introduction

Since the advent of the jet age in the 1960s, there has been a significant and continuing growth in air travel with ever increasing densities of high technology aircraft in limited available civil airspace. Over the same period, the correspondingly increased probability and potentially dire consequence of aircraft encounters with volcanic clouds has become clearly apparent.

Damage to aircraft from volcanic ash cloud encounters can be immediate and long term (Casadevall 1993). As aerospace technology develops, jet-turbine running temperatures have increased markedly seeking increasing thrust and economy. Modern high-bypass jet-turbine engines run at temperatures in excess of the melting point of many minerals and silicates. Similarly, the fine tolerances of airframe fabrication, and electrical, hydraulic, and navigation systems can all be compromised by the nature, density, and size of volcanic ash particles. The accretion of volcanic ash silicates on turbine engine blades can, and has, resulted in engine stalling and inability to restart. The accretion or incidence of volcanic ash silicates on and in the airframe can lead to critical interruption of electrical and hydraulic aircraft systems. Even marginal encounters with low density volcanic ash cloud results in accelerated wear and tear on aircraft and engines (International Civil Aviation Organization 2007, Manual on Volcanic Ash, Radioactive Material and Toxic Chemical Clouds, Doc 9691). Any aircraft encounter with a volcanic ash cloud therefore carries both a safety and an economic consequence.

From 1953 to 2009, there have been over 129 reported incidents of aircraft encountering volcanic ash (Guffanti et al. 2010); 79 of these resulted in some physical damage to the aircraft. Of these damaging encounters, 26 can be considered severe, including nine incidents that resulted in loss of in-flight power in one or more engines. Some of the latter have been widely documented, such as the first "all engines out" encounter by a Boeing B747 in 1982 with ash near Indonesia from Galunggung volcano, and the Boeing B747 encounter in 1989 with ash from Redoubt volcano over Alaska (Miller and Casadevall 2000). In contrast some encounters have received little public attention, such as an all-engines failure in a Gulfstream II survey aircraft in 2006 over Papua New Guinea due to an encounter with ash from Manam volcano (Tupper et al. 2007a, b).

From 1953 to 2014, eruptions from 40 volcanoes located in 16 countries have caused damaging encounters of aircraft with ash clouds (Fig. 1). While the most damaging encounters have occurred within 24 h of eruption onset and/or within 1000 km of the source, less safety-significant but still economically damaging encounters have occurred at greater distances and extended times (Guffanti et al. 2010).

The potential risk arising from such encounters has often been highlighted by the international

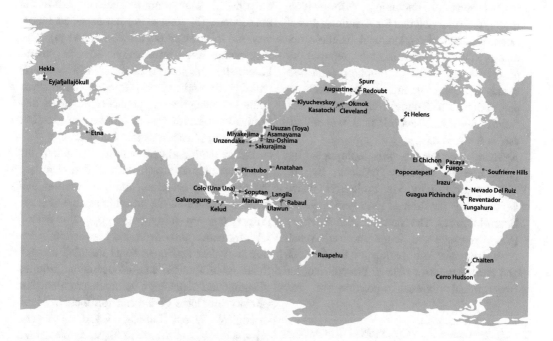

Fig. 1 Map of source volcanoes responsible for damaging encounters of aircraft with ash clouds (modified from Table 7 of Guffanti et al. 2010)

civil aviation community as a priority area in need of systematic global mitigation and further development of risk reduction measures (International Civil Aviation Organization 2012).

2 International Airways Volcano Watch

In response to the demand for globally co-ordinated mitigation of the volcanic ash risks to aviation, the IAVW was established in 1987 by ICAO in close co-ordination with the World Meteorological Organization. Since that time a collaborative approach, led by the IAVW, has matured into a comprehensive worldwide monitoring and notification system (International Civil Aviation Organization 2014a, b).

The IAVW system is an operational programme binding on all ICAO member States (countries) through the Chicago Convention.[1] The system is made up of three main components:

1. Observing component—this comprises existing international ground-based monitoring and observations (including VONA—Volcano Observatory Notice for Aviation), global satellite based detection and in-flight air reports (VAR—volcanic ash reports) to observe/detect volcanic eruptions and ash clouds and pass the information quickly to appropriate Air Traffic Management Area Control Centres, Meteorological Watch Offices, and VAACs.

2. Advisory component—this comprises the production of advisory products by the VAACs for use by Meteorological Watch Offices and air traffic management Area Control Centres. The Volcanic Ash Advisory (VAA) message and its graphical equivalent (VAG) contain information on the position and current eruptive state of the volcano, the current and expected position of any

associated volcanic ash cloud, along with relevant contextual information on plume height, observation sources and expectation for the timing of next issue.

3. Warning component—this provides the necessary warnings to aircraft and air traffic management through two types of messages: SIGnificant METeorological information about aviation weather hazards (SIGMETs) that are issued by Meteorological Watch Offices, and NOTices to Air Men (NOTAMs) for changes in airspace status that are issued by Area Control Centres.

The SIGMETs and NOTAMs are based on advisory information supplied by nine designated VAACs, whose aggregate areas of responsibility cover most of the globe. The VAACs in the IAVW system are: Anchorage, Buenos Aires, Darwin, London, Montreal, Tokyo, Toulouse, Washington and Wellington. The approximate VAAC areas of responsibility are shown in Fig. 2.

IAVW services can also be categorised in four areas: (1) monitoring information on the threat, onset, cessation, scale and characteristics of an eruption, (2) monitoring the volcanic ash in the atmosphere, (3) forecasting the expected trajectory and location of the ash cloud, and (4) communicating the information to the users. Essentially, the success of the IAVW system is entirely dependent on requisite information gathering, analyses and prediction, targeted dissemination of information, and the procedural or automatic application of that information.

Before and during a volcanic eruption, the co-ordination and flow of information regarding the (potential) eruption, and the location and forecast position of the volcanic ash cloud is the primary concern. It involves co-operation among all information providers, and between information providers and operational decision makers. Such co-ordination and co-operation requires planning and preparation before an eruption. The primary providers of information include Meteorological Watch Offices, VAACs, volcano observatories, and aircraft in flight, supplemented by information from the research and broader communities.

---

[1]The Chicago Convention on International Civil Aviation was signed in 1944. Standards and procedures for safe and economic international aviation are set out in detail in 19 Annexes to the Convention.

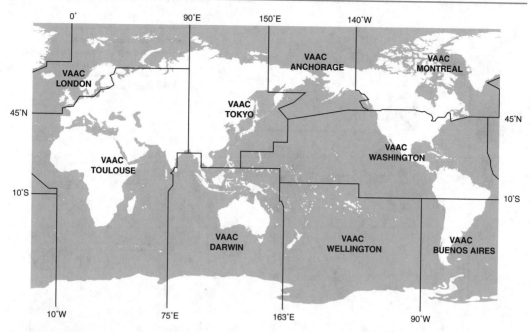

Fig. 2 Areas of responsibility of the nine volcanic ash advisory centres

Users of information (operational decision makers) are Air Traffic Management Systems including Aeronautical Information Services, Air Traffic Control, and Air Traffic Flow Management, private and commercial flight crews, and airline operations centres. Regulatory co-operation between civil aviation authorities and aircraft operators using the information provided is essential for support of the pre-flight planning process, and the in-flight and post-flight decision-making processes; all as part of overall safety risk mitigation.[2]

The lines of communication and responsibility in the IAVW are shown in Fig. 3.

3 Volcano Monitoring

Volcano observatories are loosely organised under the banner of the World Organization of Volcano Observatories, a commission of the International Association of Volcanology and the Earth's Interior, itself a member association of

the International Union of Geodesy and Geophysics. Not all volcanoes are monitored and not all countries have volcano observatories.

Volcano observatory staff can detect volcanic unrest, provide eruption forecasts, identify the onset of an eruption, and advise on the evolution and end of an eruption. Ideally these volcano observatories provide guidance on the changing eruption characteristics through time such as plume heights, altitudes of dispersing ash layers in the atmosphere, likely particle size distribution (post initial eruption) and possible mass eruption rates that can be used in numerical dispersion and transport models. Many observatories may analyse eruption products providing information on composition of ash and also gas emissions that impact on aircraft systems. Volcano observatories typically also hold information on past eruptions of a given volcano so they are able to provide likely eruption scenarios and a range of likely eruption parameters, such as possible ash ejection heights, before an eruption occurs. They are also responsible for monitoring ground hazards such as ash fall and volcanic gas dispersal.

Volcano observatories build long-term relationships with civil protection/defence

---

[2]Refer to; ICAO Doc 9974 Flight Safety and Volcanic Ash.

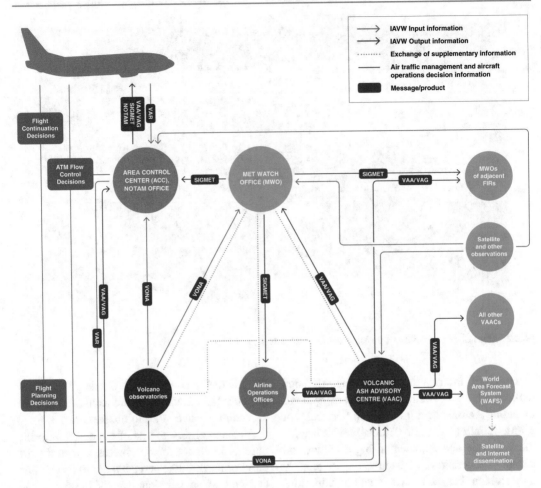

Fig. 3 IAVW system elements and relationships

organisations, local and national authorities, and communities that live around volcanoes. Relationships between VAACs and Meteorological Watch Offices are similarly long-term and strong, given that many observatories run weather stations and have strong links with local meteorologists due to the need to forecast rain-induced lahars (volcanic mud flows), volcanic ash, and gas dispersal. The ascent of magma towards the Earth's surface before an eruption typically generates physical signals that can be detected if appropriate volcano monitoring is in place, thereby allowing eruption forecasting and early warning. Pre-eruptive signals (volcanic unrest) may be detected using a variety of methods, including, but not limited to: volcanic earthquake monitoring using seismometers, ground

deformation measurements and observations of hydrologic activity change, gas emissions change monitoring, and steam explosion observations.

The status activity of a volcano is best communicated in a succinct manner to inform decision makers. To assist with this, the international aviation community has established a four-level colour code chart for quick reference to indicate the general state of a given volcano (Table 1). The colour code identifies the state of the volcano (i.e. unrest vs. eruption) but is not intended to represent the status of distal ash in the atmosphere (Guffanti and Miller 2013) or to represent risk to aviation or to people and assets on the ground.

While the international community has developed the colour-code system, it should be noted that, for various reasons, these codes are

Table 1 ICAO Aviation Colour-Code

Colour-code	Status of volcano activity
GREEN	Volcano is in normal, non-eruptive state, *or, after a change from a higher alert level*: Volcanic activity considered to have ceased, and volcano reverted to its normal, non-eruptive state.
YELLOW	Volcano is experiencing signs of elevated unrest above known background levels *or, after a change from higher alert level*: Volcanic activity has decreased significantly but continues to be closely monitored for possible renewed increase.
ORANGE	Volcano is exhibiting heightened unrest with increased likelihood of eruption *or,* Volcanic eruption is underway with no or minor ash emission *[specify ash-plume height if possible].*
RED	Eruption is forecasted to be imminent with significant emission of ash into the atmosphere likely, *or,* Eruption is underway with significant emission of ash into the atmosphere *[specify ash-plume height if possible].*

not assigned to all volcanoes. While an international standard, the colour code is currently only formally used by the United States, Russia, New Zealand, and Iceland. The reasons for this vary, but most States not using the colour code indicate difficulties in using international systems in parallel with their own locally accepted and appropriate alert levels for ground hazards. It is also recognised that different colours are associated with adverse situations in different cultures and ethnicities.

In 2008, the IAVW Operations Group introduced a new message format to assist volcanologists in the timely provision of information on the state of a volcano to support the issue of volcanic ash advisories by VAACs, the issue of SIGMET information by Meteorological Watch Offices, and the issue of NOTAM for volcanic ash by Air Traffic (Control) Services. This especially formatted message is referred to as Volcano Observatory Notice for Aviation or VONA (International Civil Aviation Organization 2014a, b). The VONA (or something similar based on discussion and agreement on a case by case basis between the specific agencies involved) is expected to be issued by an observatory when the aviation colour code changes (up or down) or within a colour code level when an ash producing event or other significant change in volcanic behaviour occurs. Two-way discussions are essential between volcano observatories and aviation information providers about the observations and information needed during eruption, formats required, challenges and

limitations, as well as an explanation as to how the information will be used and who will receive outputs. For example, the Icelandic Met Office (Iceland's volcano observatory) and the UK Met Office (through the London VAAC) have established a specific format co-designed to suit volcano observatory operating capacities, VAAC needs, and reflect joint experience (Webster et al. 2012). The VONA is a good starting point for such discussions.

4 The Challenges

Introducing and continually improving high technology systems to mitigate safety and economic risk from natural events inevitably bring great challenges. For the volcanic risk to aviation these challenges include the detection of volcanic ash cloud, forecasting its dispersion, and the timely and targeted communication of this information, along with system improvement that is well informed by developing scientific understanding.

4.1 Ash-Cloud Detection and Forecasts

Today's volcanic ash cloud forecasts, provided by the VAACs, are basic textual and graphical information produced using the output from atmospheric dispersion and transport models. Most of the numerical ash dispersal forecast models utilised by VAACs comprise a meteorological model including wind speed and direction, into which volcanic ash is introduced specifying input parameters related to the volcanic source (Eruption Source Parameters). Eruption Source Parameters may include vent location, plume height, eruption duration or start/stop time, mass eruption rate, particle size distribution, vertical distribution of mass with height above the vent and distal fine ash fraction (Mastin et al. 2009). Uncertainty in any of the various source parameters can result in large errors in the resultant volcanic ash cloud forecasts (Webster et al. 2012). Sensitivity analysis can identify the most critical parameters and demonstrate the range of outcomes under different conditions of uncertainty.

Meteorological forecasters evaluate the model outputs before issuing and during the validity of VAA messages. That analysis includes real-time verification of the ash cloud model output against a range of observational resources, principally remote sensing by satellite but also including reports from aircraft and increasingly ground-based sensing such as LIDAR. Post-eruption, model predictions in the distal environment can be compared with observational datasets to examine overall model performance (e.g. Webster et al. 2012).

The current two primary volcanic ash forecast products are the VAA and the SIGMET. The VAACs provide VAA in a text and graphic-based format (VAG) that sets out an analysis of the current position of the ash cloud, and a six, 12 and 18-h forecast location of the ash cloud, setting out position, altitude and thickness using aviation flight level nomenclature. Work has been undertaken informally at each VAAC to provide forecast location of ash cloud out to 24 h. This may become a standard time-step in the future. Meteorological Watch Offices issue volcanic ash cloud SIGMETs based on the guidance provided by the associated VAAC in their respective VAA and VAG products. These SIGMETs are valid for up to 6 h and describe the current and expected location of the ash cloud within the Flight Information Region or area of responsibility of the Meteorological Watch Offices.

As a supplementary service, at time of writing, the European and North Atlantic regions use forecast ash cloud concentration charts issued alongside official VAAC products. Such charts, depicting forecast ash concentration were first provided to users in April 2010 in response to the Eyjafjallajökull volcanic event. It is important to note that there are currently no globally agreed standards and procedures for the production, provision, and use of concentration charts (Guffanti and Tupper 2015).

4.2 Communications

In elementary terms, the IAVW system is required to provide volcanic ash cloud information to airline operators and Air Traffic Management system providers who then pass the information to airline dispatchers and pilots. Figure 3 depicts the information flow following a volcanic eruption and identifies participants in the provision of volcanic ash cloud information.

In practice, and despite some excellent initiatives to improve it, communication can fail at any stage. For many significant aviation encounters, aircraft crew members had no knowledge of the eruption encountered despite it being evident to people on the ground—this was the case recently with an aircraft experiencing a damaging encounter with ash from Kelut, Indonesia, 6 h after the 13 February 2014 eruption (airline sources, unpublished communication, 2014). The worst known example occurred in 1991 when there were at-least sixteen in-flight encounters with volcanic ash from Pinatubo, in the Philippines. These encounters occurred despite extensive information being available. Casadevall et al. (1996) noted that the response within the Philippines was relatively effective, but the international response was not, as summarised:

> ...information and warnings about the hazard of volcanic ash either did not reach appropriate officials in time to prevent these encounters or that those pilots, dispatchers, and air traffic controllers who received this information were not sufficiently educated about the volcanic ash hazard to know what steps to take to avoid ash clouds... the key to

communicating information about volcanic eruptions in a timely and readily understandable form is to involve all interested groups (geologists, meteorologists, pilots, and air traffic controllers) in the development of information and to streamline the distribution of this information between essential parties....

Other documented examples include the Manam eruptions during 2004–05 in Papua New Guinea where a large number of pilot reports of volcanic activity collected in flight were not passed on outside the airline involved, regardless of international requirements (Tupper et al. 2007a, b). Conversations with the air traffic management community have also indicated that air traffic controllers are often too busy to pass on messages that they believe have a lower priority than managing the separation of aircraft (Tupper, personal communication, 2014).

When communications are working well, initial reports of volcanic ash can result in useful information being delivered to the end user. In most cases, information about a volcanic ash cloud will be provided to the pilot, either in flight, or during pre-flight planning, in the form of SIGMETs, NOTAMs, reports from pilots, or VAA/VAG. Each of these products is distinct in format and content, but all can provide information regarding the location of volcanic ash cloud. It is critically evident that all of these

products must be consistent in their overall message. When the situation is changing rapidly, that can be extremely challenging.

The 18 August 2000 eruption of Miyakejima, Japan, illustrated this point (Tupper 2012). At least four non-Japanese aircraft encountered the cloud, with two sustaining significant damage. The eruption was sudden, but there was very strong awareness amongst domestic and some foreign airlines of the potential for activity at the volcano. The eruption was well observed, and the speed of response by Japanese authorities was exceptional. Nevertheless, there were some minor communication issues at several stages in the warning chain, resulting in inconsistencies in the information available, particularly during the rapidly developing early stages of the eruption.

To illustrate the potential differences of estimated volcanic ash cloud height in various real-time warnings, Fig. 4 sets out the ash cloud heights stated in VAA, SIGMET, and NOTAM, with respect to their issue time and validity, against the post eruption evaluation of the approximate real volcanic ash cloud height for the 2000 Miyakejima event. The times and approximate altitudes of four confirmed aircraft encounters with the cloud are shown. During the critical first half hour of the eruption, the VAA and then consequent warnings responded to

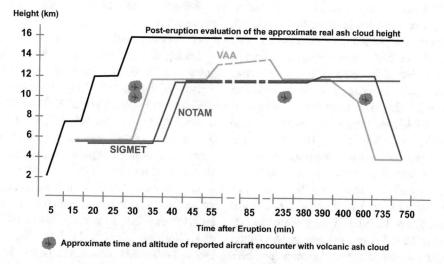

Fig. 4 Schematic showing volcanic ash cloud height as set out in various real-time warnings along with approximate real ash cloud height for the 2000 Miyakejima event, over approximate time after eruption. After Tupper et al. (2004)

multiple, and in some cases time-lagged observations and information loops. The two early encounters occurred before a concurrent height of the eruption was reflected in the warnings. The actual warning response was relatively good for this eruption, but the schematic illustrates the complexity of messaging, in a fast changing environment, particularly with multiple warning types.

Getting the official communication for warnings right can be made easier, but can also be complicated by non-official communications. In recent years, the rise of social media, enhanced remote communications, and omnipresent digital photography has meant that unofficial eruption and hazard notifications have become almost expected. Operational centres can and do use this to their advantage, particularly for early alerting. However observers can be mistaken—for example during the eruption at Bardarbunga Iceland in 2014 there were *Twitter* reports of an eruption ash cloud based on web cam pictures but it was in fact a dust storm from a nearby sandur plain. Another downside is the amount of 'chatter' and the potential for conflicting messaging. Nevertheless, the necessity for public engagement during an event has also risen. The relative level of safety risk of events is also not necessarily reflected in the attention that particular eruptions get in public.

As a result of the avalanche of non-official communications during volcanic events, VAACs and Meteorological Watch Offices endeavour to authenticate all incoming information to establish the reliability and weighting of such information. For example, in 2010, earth scientists and atmospheric scientists in Iceland and the UK enhanced their relationships in a number of ways, including through visits between operational institutions (VAAC at the UK Met Office and Iceland's volcano observatory, the Icelandic Met Office) to better understand processes and working practices used by the other organisation. In parallel, civil protection authorities in the UK sought information and advice about impacts to the UK through UK national research institutions, who in turn consulted Icelandic scientists including the Icelandic Met Office.

In order to support both aviation and civil protection sectors and to facilitate strategic science, a memorandum of understanding was established between the UK and Iceland to facilitate the flow of information between nations, and to enable wider management of the impacts of cross-border hazards and co-ordination of distal observations of volcanic ash cloud. This memorandum of understanding now underpins long-term productive cross-disciplinary research and relationships. The Icelandic Met Office with the National Commissioner of Icelandic Police (Iceland's Civil Protection) continue to make a great deal of data and information available in close to real time during volcanic unrest and eruptions (including that on social media) to enhance communication across sectors.

4.3 Science Challenges

Operational enhancements will continue to need wide scientific development work and expansion of the understanding of the full volcanic ash hazard and risk to aviation.

The central theme of scientific concern is how to accurately determine the constituents (solid particles, gases, and aerosols), density, and three-dimensional shape of a volcanic cloud at particular times and locations. Understanding engine and airframe tolerances to ash ingestion and gas effects will better inform the operational risk management of airlines.

Reducing uncertainties in ash reporting and plume modelling is expected to eventually provide critical warning system enhancements in the future.

During volcanic eruptions, a number of toxic gases may be emitted in addition to ash; these include sulphur dioxide (SO_2), hydrogen fluoride (HF), and hydrogen sulphide (H_2S) amongst many others. Each of these gases has different atmospheric dispersion properties, and so gas clouds may be found coincident or separate from volcanic ash clouds. Of these gases, SO_2 is of particular importance as it may be emitted in large quantities and potentially has significant health effects, as well as longer term effects on aircraft. Further engineering and science work is needed to fully understand this area and reflect any advances in the IAVW system.

In pursuing these objectives the aviation community has been well supported for many years by the science community, including the World Meteorological Organization and the International Union of Geodesy and Geophysics, member associations, and many dedicated individuals.

Particular support in contributing to and co-ordinating these scientific endeavours in support of the ICAO IAVW will continue to be provided by the WMO Sponsored Volcanic Ash Advisory Group and VAAC Best Practices workshops. Supplementing this, the periodic WMO sponsored volcanic ash science meeting is expected to provide the academic forum for reporting of developments and scientific collaboration.

In supporting this growing area of work, future science investment will be essential to continue developing the IAVW.

5 Warning System Enhancements

The eruption of Iceland's Eyjafjallajökull volcano during March to May 2010 demonstrated again the vulnerability of aviation to volcanic eruptions. According to an analysis by Oxford Economics (2015), more than 100,000 commercial flights were cancelled during the Eyjafjallajökull volcano eruptive phase, and US $4.7 billion in global GDP was lost. The report estimated the gross loss to airlines worldwide at US$2.6 billion due to this single volcanic event.

Later, in 2010 and in response to the Eyjafjallajökull episode (International Civil Aviation

Organization 2013a), ICAO established an International Volcanic Ash Task Force as a multidisciplinary global group to further develop and co-ordinate work related to volcanic ash. Before it concluded its work in 2012, it addressed issues related to air traffic management, aircraft airworthiness, aeronautical meteorology, and volcanological and atmospheric sciences. The Task Force identified further work to be undertaken, by existing bodies such as the IAVW Operations Group and collaborative best practice development amongst the nine VAACs, co-ordinated by the WMO (Guffanti and Tupper 2015).

Also of significance, over the last decade ICAO has gradually developed and begun implementing a Global Air Navigation Plan (ICAO 2013b Doc 9750, 2013–2028) as an overarching air navigation framework, including key civil aviation policy principles to assist regions and States with the preparation of their Regional and State air navigation plans.

The objective of the Global Air Navigation Plan is to increase airspace capacity and improve efficiency of the global civil aviation system while improving, or at least maintaining safety. The Plan includes an upgrade framework, and guidelines for associated technology development covering communications, surveillance, navigation, information management, meteorology and avionics.

The Plan reflects all of the science, communications and operational recommendations of the International Volcanic Ash Task Force 4th

Table 2 Enhancement of Volcanic Ash Risk Mitigation—Excerpts from the ICAO Global Air Navigation Plan (Abridged) with supporting and progress commentary

Global air navigation plan module	Commentary	Progress
Completion by 2018		
Implement widely collaborative processes	Acceptable developments should take into account the needs of all directly involved	Common views have been established on the collaborative treatment of volcanic ash cloud extending across different Flight Information Regions and VAAC areas
Increase the provision of Volcano Observatory Notices to Aviation	Not all State Volcano Observatories are issuing VONA	State Volcano Observatories have been encouraged to issue the VONA. The number of States doing so is increasing
Develop information confidence levels	This work responds to a request from the International Air Transport Association (IATA)	VAAC provider States are actively developing confidence level concepts from both a science and operational standpoint

(continued)

Table 2 (continued)

Global air navigation plan module	Commentary	Progress
Completion by 2023		
Enhance the provision of SIGMETs	A large volcanic ash cloud over congested, multi-State areas can result in multiple SIGMET information messages, all being in effect at the same time	Work is underway better support MWO responsibilities to issue SIGMET Work is also underway to develop a regional approach to the issue of SIGMET
Transition to all-digital format	Volcanic cloud information needs to be provided in a digital form to support ingestion directly into flight planning and flight management systems	A large area of work is under way as part of the ICAO Meteorology Panel set of projects to support the development of data-centric meteorological information in GML/XML form
Increase Volcanic Advisory message frequency and time steps	Operators need frequent updates of volcanic ash information especially in congested airspace and around constrained airports	Dynamic provision of forecast data is being considered as part of the overall move to data-centric meteorological products
Continued improvements in observing networks	Reliable and granular observation of meteorological phenomena including volcanic ash is pivotal in improving forecast products	The expansion of ground-based networks, satellite platforms and sensors, and airborne sampling will continue building on existing accomplishments
Completion by 2028		
Implementation of Now-casts	Aircraft operating at up to 1000 km/h need to know the current location of a volcanic ash cloud at any given time	It is expected that a three-dimensional representation of the current or near-current volcanic ash boundaries could eventually be made available and extracted by the user as required
Implementation of Probabilistic forecasts	Current volcanic ash forecasts are deterministic forecasts. They are a *yes/no* forecast, with respect to the depiction of the airspace impacted by discernible[a] volcanic ash	Probabilistic forecasts will provide decision makers with an assessment of all the likelihoods of risk of occurrence exceeding a defined magnitude
Completion after 2028		
Other contaminant forecasts	There is a need to expand the warning services to other toxic emissions from volcanic eruptions	This issue is currently being studied by both ICAO and WMO experts
Trajectory based operations	Integration of volcanic cloud now-casts and forecasts, combined with the use of probabilistic forecasts to address uncertainty, is expected to significantly reduce the effects of volcanic cloud on air traffic flow	The meteorological and air traffic management communities are starting to work more closely on this objective
Development of Index levels for aircraft ash tolerances	Aircraft operators increasingly need quantitative volcanic ash forecasts to take advantage of yet to be specified aircraft and engine limits.	The development of a volcanic ash index for ash/gas tolerances of various types of engine/aircraft combinations is in the very initial stages of engineering review and concept design.
Airborne detection equipment	A few basic systems to alert pilots to the distal presence of volcanic ash are under evaluation	To allow operators to take advantage of tactical on-board volcanic ash detection equipment, new air traffic management processes will need to be developed

[a]*Discernible* ash is defined as "volcanic ash detected by defined impacts on/in aircraft or by agreed in-situ and/or remote-sensing techniques"; *Visible* ash is defined as "volcanic ash observed by the human eye" and not defined quantitatively by the observer

Meeting (2012) to further develop and co-ordinate work related to volcanic ash risk mitigation. The main approaches from these initiatives to be implemented by the IAVW Operations Group are set out chronologically in Table 2.

In essence, the Plan calls for the enhancement of collaborative processes in the observation and provision of information, better land and aircraft-based volcano and ash observations, the introduction of confidence levels to forecast information, increased frequency of information, introduction of probabilistic forecasts, and the introduction and use of data on aircraft ash tolerances.

6 Conclusion

There is no doubt that future volcanic eruptions, coupled with certain meteorological conditions, have the potential to cause significant disruptions to air transport (Sammonds et al. 2011).

The ongoing development of the IAVW system continues to reveal significant challenges, some of which may remain unresolved. Enacting a fit-for-purpose warning network that brings volcanic hazard warnings into the aircraft cockpit requires the bridging of gaps between two sciences (volcanology and meteorology) in order to understand the hazard, to knit the operational parts of those sciences together in a single warning system, and then to connect with operations in the time and resource-critical aviation industry.

Fortunately, and arguably due in large part to the IAVW system, there have been no fatalities associated with aircraft operations proximal to volcanic ash clouds. However, where the eruption is forecast, warned for, resulting volcanic ash clouds tracked, and with communications procedures in place and followed, experience shows that aircraft are still not always able to avoid volcanic ash clouds. Naturally, where science or communications cannot provide usable information, the operational risk rises.

The objective remains to provide increasingly granular and robust information that will allow aircraft to operate, safely and economically, proximal to volcanic ash in the atmosphere. While much has been achieved, there is more to do, in procedures, science, engineering, and in practical communications. Without good warning system communications, fully informed by the social sciences that assist in the 'uptake' of the message, and by robust, reliable operational practices, the fruit of science and policy development will remain compromised.

Lastly, because the advent of the IAVW has brought the meteorological and aviation communities much closer to the volcanological community, there is an exciting opportunity to bring potentially useful practices further into the combined geophysical hazards space. For example, volcanic tsunami, lahar warnings, ash fall, and even rainfall induced lava dome collapses, are all areas where the two fields will need to work together well to produce enhanced warning and communication services.

References

Casadevall TJ (1993) Volcanic hazards and aviation safety: lessons of the past decade. Flight safety foundation, flight safety digest, May 1993

Casadevall TJ, Delos Reyes PJ, Schneider DJ (1996) The 1991 Pinatubo eruptions and their effects on aircraft operations. In: Newhall CG, Punongbayan RS (eds) Fire and mud: eruptions and lahars of Mount Pinatubo, Philippines. Philippines Institute of Volcanology and Seismology & University of Washington Press, Quezon City & Seattle, pp 625–636

Guffanti M, Miller TP (2013) A volcanic activity alert-level system for aviation—review of its development and application in Alaska. Nat Hazards 69:1519–1533. doi:10.1007/s11069-013-0761-4

Guffanti M, Tupper A (2015) Volcanic ash hazards and aviation risk. In: Pappale P (ed) Volcanic hazards, risks and disasters. Elsevier, Amsterdam, pp 87–108

Guffanti M, Casadevall T, Budding K (2010) Encounters of aircraft with volcanic ash clouds: a compilation of known incidents, 1953–2009. USGS data series 545

ICAO (2012) International Volcanic Ash Task Force, 4th meeting Montreal, Canada, June (2012), meeting report

International Civil Aviation Organization (2007) Doc 9691—manual on volcanic ash, radioactive material and toxic chemical clouds

International Civil Aviation Organization (2010) Journal, vol 65, No 4. Eyjafjallajoull's Aftermath

International Civil Aviation Organization (2012) Doc 9974, Flight safety and volcanic ash. Available via http://www.icao.int/publications/Documents/9974_en.pdf. Accessed 1 Dec 2014

International Civil Aviation Organization (2013a) Journal, vol 68, No 1. Heeding Eyjafjallajoull's Lessons

International Civil Aviation Organization (2013b) Doc 9750, 4th edn. 2013–2028 Global Air Navigation Plan

International Civil Aviation Organization (2014a) International airways volcano watch operations group eighth meeting Melbourne, Australia, February (2014) Roadmap for international airways volcano watch (IAVW) in Support of International Air Navigation

International Civil Aviation Organization (2014b) Doc 9766—Handbook on the international airways volcano watch (IAVW). Available via http://www.icao.int/safety/meteorology/iavwopsg/Pages/default.aspx. Accessed 1 Dec 2014

Mastin LG, Guffanti M, Servranckx R, Webley P, Barsotti S, Dean K, Durant A, Ewert JW, Neri A, Rose WI, Schneider D, Siebert L, Stunder B, Swanson G, Tupper A, Volentik A, Waythomas CF (2009) A multidisciplinary effort to assign realistic source parameters to models of volcanic ash-cloud transport and dispersion during eruptions. J Volcanol Geotherm Res 186(1–2): 10–21. doi:10.1016/j.jvolgeores.2009.01.008

Miller TP, Casadevall TJ (2000) Volcanic ash hazards to aviation. In: Sigurdsson H (ed) Encyclopedia of volcanoes. Academic Press, San Diego, pp 915–930

Oxford Economics—the economics of air travel restrictions due to volcanic ash. Prepared for AirBus Industries. https://www.oxfordeconomics.com/my-oxford/projects/129051. Accessed 1 June 2015

Sammonds, McGuire, Edwards (2011) Volcanic hazard from Iceland—analysis and implications of the Eyjafjallajoulls eruption, 2011 UCL Institute for Risk and Disaster Management. Operations Group

Tupper A (2012) Managing diffuse eruption clouds—the experiences of 2010–11 and the results of the IVATF. 28th congress of the aeronautical sciences

Tupper A, Kamada Y, Todo N, Miller E (2004) Aircraft encounters from the 18 August 2000 eruption at Miyakejima, Japan. In: Second international conference on volcanic ash and aviation safety. Office of the federal coordinator for meteorological services and supporting research, Alexandria, Virginia, USA, pp 1:5–1:9

Tupper A, Guffanti M, Rose B, Patia H, Richards M, Carn S (2007a) The "Gulfstream incident"; Twin-engined flame-out over the Papua New Guinea highlands. In: 4th international work-shop on volcanic ash, Rotorua, New Zealand, 26–30 Mar 2007 [Proceedings]: World Meteorological Organization, Commission for Aeronautical Meteorology, 6 pp. http://www.caem.wmo.int/moodle/course/info.php?id=27. Accessed 17 June 2015

Tupper A, Itikarai I, Richards MS, Prata F, Carn S, Rosenfeld D (2007b) Facing the challenges of the international airways volcano watch: the 2004/05 eruptions of Manam, Papua New Guinea. Weather Forecast 22(1):175–191

Webster HN et al (2012) Operational prediction of ash concentrations in the distal volcanic cloud from the 2010 Eyjafjallajökull eruption. J Geophys Res 117: D00U08. doi:10.1029/2011JD016790

Volcanic Gases: Silent Killers

Marie Edmonds, John Grattan and Sabina Michnowicz

Abstract

Volcanic gases are insidious and often overlooked hazards. The effects of volcanic gases on life may be direct, such as asphyxiation, respiratory diseases and skin burns; or indirect, e.g. regional famine caused by the cooling that results from the presence of sulfate aerosols injected into the stratosphere during explosive eruptions. Although accounting for fewer fatalities overall than some other forms of volcanic hazards, history has shown that volcanic gases are implicated frequently in small-scale fatal events in diverse volcanic and geothermal regions. In order to mitigate risks due to volcanic gases, we must identify the challenges. The first relates to the difficulty of monitoring and hazard communication: gas concentrations may be elevated over large areas and may change rapidly with time. Developing alert and early warning systems that will be communicated in a timely fashion to the population is logistically difficult. The second challenge focuses on education and understanding risk. An effective response to warnings requires an educated population and a balanced weighing of conflicting cultural beliefs or economic interests with risk. In the case of gas hazards, this may also mean having the correct personal protection equipment, knowing where to go in case of evacuation and being aware of increased risk under certain sets of meteorological conditions. In this chapter we review several classes of gas hazard, the risks associated with them, potential risk mitigation strategies and ways of communicating risk. We discuss carbon dioxide flows and accumulations, including lake overturn events which have accounted for the greatest

M. Edmonds (✉)
Earth Sciences, University of Cambridge, Downing Street, Cambridge, Cambridgeshire, UK
e-mail: me201@cam.ac.uk

J. Grattan
Aberystwyth University, Aberystwyth, UK

S. Michnowicz
University College London, London, UK

Advs in Volcanology (2018) 65–83
https://doi.org/10.1007/11157_2015_14
© The Author(s) 2015
Published Online: 26 March 2017

number of direct fatalities, the hazards arising from the injection of sulfate aerosol into the troposphere and into the stratosphere. A significant hazard facing the UK and northern Europe is a "Laki"-style eruption in Iceland, which will be associated with increased risk of respiratory illness and mortality due to poor air quality when gases and aerosols are dispersed over Europe. We discuss strategies for preparing for a future Laki style event and implications for society.

Volcanic gases have claimed directly the lives of >2000 people over the past 600 years (Auker et al. 2013). Millions more people have been impacted by volcanic gas, with effects ranging from respiratory irritation to neurological impacts, to crop failure and famine. Gas hazards contrast markedly with other volcanic hazards such as lahar, pyroclastic flows and ash fall; they are silent and invisible killers often prevailing over large areas of complex terrain. Volcanic gases may accumulate far from their source and flow down valleys as a gravity flow, engulfing and asphyxiating people as they sleep. Sometimes the hazard is visible in the form of a condensing plume emanating from a vent, with acidic gases capable of corroding buildings and aircraft, damaging crops and causing respiratory disease and skin burns. The trajectory and dispersal of such a plume is subject to local meteorology. The plume or gas cloud must be detected and tracked by sophisticated instrumentation. Designing a warning system that works in real time whilst incorporating both measurements and models tests the ingenuity of personnel at volcano observatories and meteorological agencies. Yet these hazard-warning systems are necessary if people are to live at close quarters with degassing volcanoes. The dissemination and communication of warnings associated with gas hazards requires effective alerts and systems in place to ensure that the warning gets to the part of the population at risk. The population must react to the warning in a way that mitigates risk; this is only possible if sufficient understanding of the hazard exists. The insidious hazard of volcanic gases is often poorly understood and overlooked. In this chapter, we review the challenges associated with monitoring, detecting and communicating gas hazards and managing risk associated with gases. We start by reviewing the types of hazard.

1 Volcanic Gases, Insidious Hazards

A single event dominates the inventory of deaths due to volcanic gases: in August 1986 Lake Nyos (Cameroon, Africa) emitted a dense cloud of carbon dioxide (CO_2) gas in the middle of the night, which rapidly flowed down surrounding valleys, suffocating immediately 1700 sleeping people up to 20 km away from the lake (Kling et al. 1987). Many other deaths have occurred as a result of people encountering accumulations of CO_2 or hydrogen sulfide (H_2S) gases in low-lying areas or in the form of flows and clouds. In a recent analysis volcanic gas inundation was recognized as the second most common cause of death in the most frequent, fatal volcanic events (Auker et al. 2013). The key characteristic of this hazard is that usually there is no warning and no visible sign of it. Gas concentrations may creep up unnoticed until it too late, or a sudden inundation may leave no time for escape (Fig. 1).

Fatalities arising from the secondary effects of volcanic gases run into the millions over historical times (Rampino et al. 1988). Large explosive eruptions inject SO_2 directly into the stratosphere, which transforms rapidly (within hours to days) to sulfate aerosol (Robock 2000). The aerosol scatters and reflects incoming visible and UV radiation from the sun, causing tropospheric cooling over the lifetime of the aerosol (typically

(a)   **(b)**

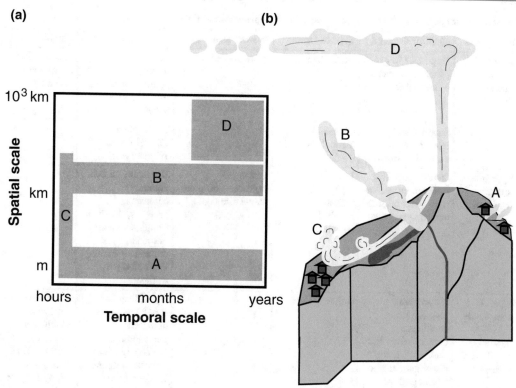

Fig. 1 Cartoon to show the range of gas hazards and the scale of their impacts. **a** Diffuse degassing through fractures and faults. These gases are sourced from deep magma reservoirs. They may persist for long periods between and during eruptions. They typically affect local areas only but present significant hazards to people when gases accumulate in basements and topographic lows. **b** Acidic tropospheric plumes from active volcanic vents contain SO_2 and halogen gases. They lead to pervasive vog (sulfate aerosol) that may cause or exacerbate respiratory diseases. They may persist for many years during non-eruptive activity at some volcanoes and the plumes are dispersed over 10 s of km. **c** Sudden flows of cold CO_2-rich gases occur as a consequence of lake overturn or phreatic explosions. They may last only minutes but may travel many 10 s of km in that time, flowing close to the ground with lethal concentrations of CO_2. **d** Large explosive eruptions inject SO_2 directly into the upper troposphere or stratosphere. The resulting sulfate aerosol has potential to cause significant regional and/or global environmental and climatic effects that may lead to cooling and crop failure, acid rain, increased mortality and crop failure over years timescales

a few years Fig. 1). Volcanic cooling has caused crop failure and famine for many years after large eruptions. Some recent eruptions (e.g. Pinatubo, Philippines, 1991 and El Chichon, Mexico, 1982) have allowed direct measurement of the reduction in direct radiative flux into the troposphere, total aerosol optical depth and tropospheric temperature (Dutton and Christy 1992), which validated predictions of the effects of stratospheric sulfate aerosol on climate. Large historic eruptions such as that of Tambora Volcano in 1815 (Indonesia) were associated with global cooling, leading to famine, social unrest and epidemic typhus, leading to the "Year Without a Summer" (Oppenheimer 2003). A dramatic European example is the Laki (Iceland) eruption of 1783, which was followed by several years of crop failure and cold winters, resulting in the deaths of >10,000, ~20 % of the Icelandic population (Grattan et al. 2003; Thordarson and Self 2003).

Another class of volcanic gas hazards is generally non-fatal, but gives rise to or exacerbates significant chronic and acute health conditions (Table 1). Persistent gas plumes at low levels in the atmosphere are common at many volcanoes

Table 1 Health effects of volcanic gases (Hansell and Oppenheimer 2004)

Gas species	Mode of dispersal	Type of hazard	In what quantity?	Acute effects	Chronic effects
Sulfur dioxide, sulfate aerosol	Tropospheric gas plumes from vents or lava lakes	Acidic irritant	More than a few Mt	Upper airway irritation, pulmonary edema, nose, throat, skin irritation	Exacerbation of respiratory disease
	Stratospheric injection during explosive eruption	Climate-forcing, particularly in tropics		Tropospheric cooling lasting 10^0–10^1 years	
Hydrogen sulfide	Diffuse degassing from the ground or from vents prior to or during eruptions	Irritant, asphyxiant, inhibitor of metabolic enzymes	Prolonged exposure >50 ppm may cause death	Headache, nausea, vomiting, confusion, paralysis, diarrhea. Cough, shortness of breath, pulmonary edema. Eye and throat irritation	
Fluoride compounds (HF, fluoride dissolved in water)	Tropospheric plumes during eruptions. Groundwaters and acid rain (through dissolution and/or leaching of ash particles)	Acidic irritant		Hypocalcemia, coughing, bronchitis, pneumonitis, pulmonary edema. Nausea, vomiting. Eye and throat irritation. Slow healing skin burns	Permanent lung injury. Mottling or pitting of dental enamel. Osteoporosis, kyphosis spine
Chloride compounds (HCl, other chlorides in gaseous and aqueous form)	Tropospheric plumes during eruptions. Groundwaters and acid rain. Plumes arising from the contact of lava and seawater	Acidic irritant		Coughing, bronchitis, pneumonitis, pulmonary edema. Eye and throat irritation	Permanent lung injury
Carbon dioxide	Diffuse/vent degassing pre- or syn-eruption. Overturn CO_2-saturated lakes	Inert asphyxiant		Asphyxia, collapse	Paralysis, neurological damage
Carbon monoxide	Diffuse/vent degassing between or prior to eruptions	Noxious asphyxiant, binds to haemoglobin		Collapse, coma	Paralysis, neurological damage
Metals e.g. mercury Hg	Tropospheric plumes during eruptions, groundwater and diffuse degassing	Oxidant irritant		Bronchitis, pneumonitis, pulmonary edema. Neurotoxicity	Neurotoxicity

worldwide. These plumes may be rich in sulfate aerosol, generating a pervasive, choking haze. At Kīlauea Volcano, Hawai'i (Fig. 2), studies have shown a link between incidences of plume inundation and asthma attacks in children (Longo et al. 2010a). These plumes give rise to acid rain and

Fig. 2 Volcanic plume from the summit of Kīlauea Volcano, Hawai'i. This plume contains acid gases and condensed water droplets, conducive to the formation of "vog" (volcanic smog, or sulfate aerosol). *Photograph credit* United States Geological Survey

their corrosive properties (arising from not just the SO_2 but also the acid halogen gases HCl and HF) leads to the damage of buildings, vehicles and infrastructure. These plumes may persist for decades or longer (Fig. 1), making them a significant health hazard (Delmelle et al. 2002). In other areas, interception of magmatic gases by groundwater aquifers may lead to contamination of water supplies that are tapped by springs. In East Africa, for example, the high concentrations of fluorine in the spring water, once dissolved in magmas many kilometres below, have caused widespread dental fluorosis (D'Alessandro 2006).

What are volcanic gases? Volcanic gases are mixtures of volatile compounds released from the ground's surface or directly from volcanic vents, into the atmosphere. They are generated when magmas exsolve volatiles at low pressures during their ascent to the surface and eruption.

Volcanic gases may precede the arrival of lava at the surface by several weeks or even months. In some cases, persistent and diffuse emissions of gases may take place continuously between eruptions, even when the eruptions occur very infrequently. The gases have different compositions depending on: tectonic setting, how close to the surface the degassing magma is stored and whether the fluids are interacting with a wet hydrothermal system prior to reaching the atmosphere (Giggenbach 1996). The gases that typically emanate from deep magma intrusions between and prior to eruptions are dominantly carbon dioxide (CO_2) and hydrogen sulfide (H_2S). When magma reaches the surface, the gas composition becomes dominated by the more melt-soluble components: water (which may make up >85 % by volume of the gas mixture), with lesser amounts of CO_2 and SO_2 (which make up 2–10 %), halogen gases hydrogen fluoride (HF) and hydrogen chloride (HCl), and carbon monoxide (CO) and other minor components. If the gases interact with a hydrothermal system the acid gases SO_2 and HCl are removed, or "scrubbed" (Symonds et al. 2001); this is typical of the early stages of an eruption, or of "failed" eruptions (Werner et al. 2011). The components of volcanic gases that are of greatest concern for health are (Table 1), primarily CO_2, SO_2, H_2S, HCl, HF and metals such as mercury (Pyle and Mather 2003) and short-lived radioactive isotopes such as radon (Baxter et al. 1999). These gases and aerosols are of course also produced in many industrial settings and the risk of accidents in these settings has prompted most of the studies on their effects on health. Some gases undergo chemical reactions in the plume, resulting in secondary products that can cause health and environmental effects. Sulfur dioxide reacts with water to form sulfuric acid aerosol droplets that leads to acid rain in the troposphere (Mather et al. 2003). When injected into the stratosphere, the aerosols may reflect and absorb radiation from the sun, resulting in the cooling of the Earth's surface for up to a few years for the largest eruptions over the past few decades, perhaps longer for larger classes of historic eruptions (Robock 2000).

There are multiple factors governing the magnitude of the volcanic gas health hazard and consequently, risk: the concentrations of gases (a function of both gas flux and composition), the mode of delivery to the atmosphere (e.g. from a point-source or over large areas; tropospheric or stratospheric) and the longevity or duration of the event. Monitoring networks should fulfill several functions in order to produce a realistic picture of the hazard: instrumentation coverage, precision (both spatial and temporal) and timeliness are critical. Once the hazard is identified and assessed, the nature of it must be communicated effectively to the communities at risk via an alert or warning system. The reaction and response of the community to the risk communication must be appropriate and prompt, otherwise delays in evacuations and other risk mitigation procedures might occur. Preparing for future events requires an understanding of the hazard and its recurrence interval, robust monitoring networks and alarm systems, sophisticated models to simulate possible outcomes and risk mitigation plans to reduce or prevent fatalities. Whilst this sequence is well-developed for a subset of hazards in some localities, such as lahar, ash fall and lava flow inundation, there are very few examples of successful alert systems for gas hazards and even fewer that have been tested in extremely hazardous scenarios which might allow us to evaluate the effectiveness of hazard communication and risk mitigation. Challenges specific to gas hazards relate to: (1) the difficulty of achieving adequate coverage with regard to monitoring (e.g. gas concentrations may be low across most of an area, but there may be localized regions of high concentrations, so dense networks of instrumentation are required); (2) developing alert and early warning systems that will be communicated in a timely fashion to the population. Gas hazards may develop rapidly and be highly dispersed, making communication of warnings problematic. (3) Ensuring that an educated population will respond in a timely and appropriate way. An amenable response to warnings or evacuation orders requires an educated population and a balanced weighing of conflicting cultural beliefs or economic interests with risk. In the case of gas hazards, this may also mean having the correct personal protection equipment, such as gas masks; knowing where to go in case of evacuation (e.g. high ground); and being aware of increased risk under certain sets of meteorological conditions (e.g. on still days with no wind). Different hazards require vastly different responses. Large eruptions which inject gas (and ash, see Chap. XXX) into the upper atmosphere for example, give rise to regional, or global hazards that have their own unique set of challenges that focus on dealing with both immediate health effects and longer term impacts (social and economic) resulting from climate forcing. In this chapter we review some key case studies and discuss the monitoring, alert and risk mitigation schemes that were in place or could be implemented for future events. We discuss the particular challenges inherent in dealing with gas hazards on all temporal and spatial scales and suggest profitable approaches for future development.

2 Developing Risk Mitigation Strategies for CO_2 Flows and Accumulations

Over the course of a decade beginning in 1979, our understanding of gas hazards was to take a dramatic turn. Events served as a stark reminder that volcanic gas hazards were capable of causing significant loss of life. Hazards from atmospheric CO_2 are usually limited, because atmospheric dispersion tends to dilute volcanic or hydrothermal gas emissions to the extent that concentrations become non-lethal rapidly away from a vent or degassing area. If however, geological, geographical, hydrological or meteorological factors bring about the accumulation of CO_2, or its concentration into a flow, the effects are life-threatening. Within the Dieng Volcanic Complex in central Java, on 20 February 1979, a sequence of earthquakes was followed by a phreatic eruption and sudden release of CO_2 (Allard et al. 1989; Le Guern et al. 1982). The area was known for its hydrothermal manifestations, with boiling mud pools, hot springs and areas of tree kill indicative of CO_2; local people are aware of "death valleys" in which vegetation is dead up to a certain level on the valley walls, and animals are often killed. People lived (and still do) in the low areas adjacent to grabens and phreatic craters known to have been sites of explosions and gas emissions in the past. After three large earthquakes between 2 and 4 a.m., a phreatic explosion at 5:15 was associated with the ejection of large blocks and a lahar that reached the outskirts of the village Kepucukan (Allard et al. 1989). Frightened by the activity, people attempted to escape from the village, walking west along the road to Batur, another village just 2 km away. Halfway there,

142 people were engulfed in "gas sheets" that emanated from the erupting crater, which killed them instantly. Gas emissions, dominated by CO_2, continued for another 8 months (Allard et al. 1989) and may have reached a total volume of 0.1 km^3 (Allard et al. 1989).

Today, more than 500,000 people live in an area at high risk of hazardous CO_2 flows in Dieng caldera. Gas emission events occur frequently, heralded by seismicity (every few years with large events every few decades). A recent survey showed that 42 % of the people are aware of the risk of "poisonous gas" but only 16 % link this hazard to volcanic activity (Lavigne et al. 2008). Most people show a reluctance to accept the risk and a greater reluctance to leave the area due to a combination of religious and cultural beliefs (the area has been a sacred Hindu site since the 7th century) and economic factors (Dieng is agriculturally rich and in addition attracts many tourists). Farmers work within metres of dangerous mofettes (cold CO_2-producing fumaroles) and mark them with mounds of earth. Villages are situated at the mouths of valleys that connect phreatic craters on high ground with the caldera floor and which channel cold CO_2 flows (Fig. 3). Monitoring the hazards is therefore of utmost importance and takes place using a network of in situ logging geochemical sensors and seismometers, maintained by the Indonesian volcanological agencies. Monitoring is not easy: the sensors are difficult to maintain, have short lifetimes and do not have the spatial coverage required to monitor all of the gas-producing vents and areas. Since 1979, there have been six phreatic eruptions accompanied by elevated CO_2 emissions. Degassing crises in 2011 and in 2013, however, were successfully managed using the existing system, with CO_2 concentration levels used to assign alert levels. Gas emission forced the evacuation of 1200 residents following a phreatic eruption at Timbang crater on 29 May 2011, and people were advised to remain at least 1 km away from the crater, where dead birds and animals were found (Global Volcanism Program Report 2011). An improved network of telemetered arrays of sensors, webcams and linked siren warning systems for the surrounding villages was approved for USAID/USGS funding in 2013. For future events, it is widely assumed that phreatic eruptions will be preceded by significant seismicity (Le Guern et al. 1982). Evacuations of far larger areas will be necessary to protect the population from the gas hazard and Early Warning Systems are needed to communicate encroaching hazards.

It was not until 1986 that the wider public was exposed to the idea of volcanic gas hazards, when the 8th largest volcanic disaster in historical times occurred near to Lake Nyos in Cameroon. A landslide triggered the overturn of a density-stratified lake, within which CO_2 had concentrated in its lower levels. The sudden depressurization of the lake water upon overturn caused an outpouring of CO_2 from the lake and into a valley, killing 1746 people by asphyxiation, up to 25 km from the lake, as well as thousands of cattle (Kling et al. 1987). Around 15,000 people fled the area and survived but developed respiratory problems, lesions and

paralysis as a result of their exposure to the gas cloud (Baxter et al. 1989). There were no monitoring systems in place, no warning system and no assessment of risk before the event; scientists had no idea that this kind of event was possible prior to 1986.

It transpired, from isotopic analysis of the CO_2, that the gas had a magmatic origin, and had entered the lake from fault systems channeling gases from deep in the crust, derived ultimately from the mantle (Kling et al. 1987). There was no direct volcanic activity associated with the disaster. Gas sensor networks linked to siren systems were immediately set up at the edges of the lake and at the heads of the valleys to warn of future gas flow events. A unique hazard mitigation system was set up in 1999, funded by the United States and supplemented by the governments of Cameroon, France and Japan, with the aim of artificially degassing Lake Nyos by decompressing deep lake waters using three pipes, which work in a self-sustaining way,

Fig. 3 Condensed steam and CO_2 accumulating in a valley close to Timbang Crater, Dieng Plateau, Indonesia in 2011. Note the dead vegetation below the level of the gas as a result of the high CO_2 concentrations. *Photograph credit* Andy Rosati, Volcano Discovery

initially pumping deep water towards the surface but thereafter driven by the degassing of CO_2 (KIing et al. 1994). The scheme has reduced gas pressures in the lake substantially, reducing the risk of future overturn and gas flow events, which would otherwise have occurred every few decades. A new hazard has been identified however, in the shape of a weak dam, raising the possibility that dam breach and removal of water from Lake Nyos could be a potential future trigger for a gas emission event, regardless of the degassing pipes. Added to this is the increasing risk to people, as they gradually resettle the area.

The Lake Nyos event was not unique; two years before the disaster a similar limnic eruption occurred at Lake Monoun, killing 38 people. Other lakes are associated with significant risks of similar events: at Lake Kivu, on the border of the Democratic Republic of Congo and Rwanda, recent measurements have shown that ~ 300 km^3 of CO_2 (at standard temperature and pressure) are present in the lake's permanently stratified deep water (Schmid et al. 2005). Release of these gases by limnic overturn would have deadly consequences for the two million people living along the lake shore. It has been suggested that limnic eruptions in the Holocene have been responsible for local extinction events (Haberyan and Hecky 1987). Elsewhere, limnic eruptions have been implicated in the deaths of a wide range of Eocene vertebrates, which were subsequently preserved to an exceptional degree, at the Messel Pit (Germany), which was, in Eocene times, a crater lake over a maar (Franzen and Köster 1994). Limnic eruptions remain, however, a rare, if extremely hazardous, event.

Outstanding questions are those concerning how to mitigate hazard and manage early warning systems and how to reduce risk associated with these silent, yet deadly hazards. Considerable interest in modeling gas flow over topography has arisen from recent developments in CO_2 transport as a supercritical fluid through long-range pipelines for carbon sequestration (Duncan and Wang 2014). The possibility of a breach in a pipeline and associated gas flow has prompted investment in gas hazard assessment. At Mefite D'Ansanto in central Italy, a near-pure

CO_2 gas flows down a channel at a rate of ~ 1000 tonnes per day (Chiodini et al. 2010). The flow reaches a height (defined by a gas concentration of 5 vol%) of 3 m above the valley floor (far higher than a typical human). Using measurements of CO_2 concentration at various heights and distances in the valley to constrain the model and a local wind field, a gas transport model (TWODEE-2; Folch et al. 2009) was used to simulate the gas flow and to predict the zones of potential hazard for humans in terms of dangerous (>5 vol%), very dangerous (>10 vol%) and lethal (>15 vol%) concentrations, which has been used successfully for risk mitigation in the area. Gas transport models will have great utility in areas subject to dense, cold gas flows and are relatively inexpensive to implement, given appropriate constraints and calibrations provided by field measurements. Their unique advantage is that they provide a means to convert discrete measurements of gas concentrations using sensors into a fully 3-D continuous model of gas concentration and hazard that can be straightforwardly incorporated into warning systems.

The gas flows described above are extreme; there are numerous examples of smaller scale gas accumulation hazards that have caused loss of life. These kinds of manifestations have been shown to be the most frequently associated with deaths in the record (Auker et al. 2013) and as such, require robust monitoring, alert systems and risk assessment. Areas of tree kill and asphyxiated animals were reported at Mammoth Mountain, inside Long Valley Caldera, beginning in 1990 and caused by the diffuse emission of CO_2 over 0.5 km^2 that reached up to 1200 tons/day at its peak (Farrar et al. 1995), following a swarm of earthquakes and an intrusion in 1989. The emissions have caused fatalities: in 2006 three ski patrollers died after falling close to a fumarole. The gas hazards occur in a recreational area visited by 1.3 million skiers in the winter and 1.5 million hikers in the summer. Monitoring has been undertaken since 1990 in the form of campaign-style measurements using soil gas chamber spectrometers, and then through three permanently installed soil gas instruments, operated and monitored by United States

Geological Survey scientists (Gerlach et al. 2001). Risk mitigation measures include the posting of signs in prominent areas warning of the hazards associated with gas accumulations in topographic lows. For this lower level of hazard, this communication method is effective and has resulted in a largely safe enjoyment of the area by a largely educated public, despite the gas emissions.

In the Azores, in the mid-Atlantic, the situation is rather more precarious. On Sao Miguel Island, villages are situated within the Furnas volcanic caldera (Baxter et al. 1999; Viveiros et al. 2010). This is the site of numerous gas manifestations such as boiling fumaroles, diffuse emissions and cold CO_2-rich springs. It is an area popular with tourists, who enjoy the thermal spas. Up to 98 % of the houses, however, are situated over CO_2 degassing sites (Viveiros et al. 2010). A study in 1999, which has been repeated many times subsequently, showed that lethal concentrations of CO_2 (>15 vol%) existed in non-ventilated confined spaces in the houses (Baxter et al. 1999). There have been no confirmed cases of deaths in the area from CO_2 asphyxia but there exist frequent anecdotal records of people being "overcome" by gases (Baxter et al. 1999). No formal early warning or alert system exists, but there are soil gas flux spectrometers and soil temperature sensors located in the village that telemeter data back to the Azores Monitoring Centre for Volcanology and Geothermal Energy in real time. A survey of the population of the village of Furnas carried out in 1999 showed that, astonishingly, not a single one of 50 random adult respondents had any knowledge about the existence of gas hazards in the area. Upon closer questioning of the wider population only a very small fraction, mainly civil defense and medical workers, were aware of the hazard (Dibben and Chester 1999). This shows a profound lack of education of the general population by the scientific establishment at the time of the survey. Whilst a more recent survey has not been carried out, it is likely that this has improved in recent years with the enhancement of monitoring and the responsibility to safeguard tourists. But this

situation raises some thorny issues concerned with risk mitigation (Dibben and Chester 1999). Highlighting the most vulnerable areas in the village is likely to reduce the value of property in those areas and so the public will likely be averse to accepting such information. Gas hazard alerts might affect tourism and hence the economic status of the area. Building regulations to prevent the build up of CO_2 in basements might be harder for the poor to comply with, resulting in a socially divisive vulnerability structure. Lastly, installation of a high spatial coverage, precise and reliable monitoring and early warning system might lead the population to believe that they are no longer threatened, encouraging risky behaviors.

3 Monitoring and Communicating "Vog" Hazards

When magma is close to the Earth's surface (and when the gases do not interact with extensive wet hydrothermal systems), the gas hazards fall into a different category to those described above. In this case, acidic gases such as sulfur dioxide, hydrogen chlorine and hydrogen fluoride become important hazards. Active volcanism is therefore associated with thick plumes containing a mixture of these acid gases, as well as water, CO_2 and minor carbon monoxide (CO) and hydrogen sulfide (H_2S). Under these conditions, volcanic smog or "vog" may cause acute respiratory difficulties and skin, noise and throat irritation. Vog, which is made up of sulfate aerosol particles, has been linked to asthma and other respiratory diseases (Hansell and Oppenheimer 2004). Some volcanoes degas prodigious fluxes of gases quasi-continuously. Mount Etna, in Italy, for example, produces several thousand tons of SO_2 and significant quantities of other acidic gases every day and activity has persisted at this level for decades (Allard et al. 1991). Other prodigious producers of tropospheric volcanic gas plumes are Nyiragongo (Democratic Republic of Congo), Ambrym (Vanuatu), Kīlauea (USA), Erebus (Antarctica), Masaya (Nicaragua), Erta Ale (Ethiopia) and Villarica (Chile). Some of these volcanoes are sparsely populated; others have major urban centres within range of their plumes.

Kīlauea Volcano, Hawai'i, has been in continuous eruption since 1983. At Kīlauea, magma is outgassing at both the summit (since 2008) and from eruption sites and active lava fields on the east rift zone (Longo et al. 2010a), giving rise to multiple sources of gases. The emissions affect not only the 2 million visitors to Hawai'i Volcanoes National Park every year, but also wider areas of Big Island and the other Hawaiian islands via dispersal by the trade winds (Fig. 4). It has been shown that indoor SO_2 concentrations regularly exceed the World Health Organisation guidelines in the affected areas of Big Island (Longo et al. 2010b) and that during periods of enhanced volcanic outgassing there are synchronous increases in the occurrence of acute respiratory conditions requiring treatment on the island (Longo et al. 2010a). In response to the clear need for a system of monitoring and early warning, SO_2 concentration sensor data from inside the park and around the island are combined with SO_2 emission rates and a model for plume dispersion to produce a vog model that forecasts air quality for the Hawaiian Islands (Fig. 5). These warnings have proven to be a very successful way of mitigating risks due to vog; statistical analysis has shown that the predictions lie within one standard deviation of the data for forecasts up to 24 h ahead (Reikard 2012). Advice to residents to minimize their exposure to vog once a forecast or warning for high aerosol concentrations has been issued include closing windows and doors, limiting outdoor activities and exertion and having medications on hand. Communication of vog

Fig. 4 Hawaiian Islands, December 3, 2008, showing a pervasive tropospheric vog plume carried westwards from Kīlauea Volcano by the Trade winds. Image acquired by the Moderate Resolution Imaging Spectroradiometer (*MODIS*) on NASA's Aqua satellite

Island of Hawaii SO₂

Concentration (PPM) averaged between   0 m and   100 m
Integrated from 0400 14 Oct to 0500 14 Oct 14 (HST)
SO2 Release started at 0200 14 Oct 14 (HST)

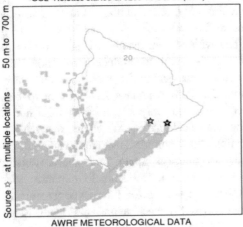

AWRF METEOROLOGICAL DATA

Island of Hawaii Sulfate Aerosols

Concentration (ug/m3) averaged between   0 m and   100 m
Integrated from 0400 14 Oct to 0500 14 Oct 14 (HST)
SO4 Release started at 0200 14 Oct 14 (HST)

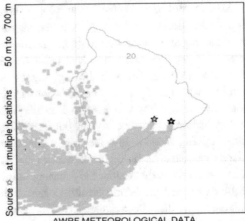

AWRF METEOROLOGICAL DATA

Fig. 5 Model to forecast "vog" and communicate vog hazard warnings for the Hawaiian Islands. The model uses estimates of volcanic gas emissions along with forecast winds to predict the concentrations of sulfur dioxide gas (SO₂, *left*) and sulfate aerosol particles (SO₄, *right*) downwind of the ongoing Kīlauea Volcano eruption. Images from the Vog Measurement and Prediction Website (*VMAP*; http://weather.hawaii.edu/vmap), hosted by the School of Ocean and Earth Science and Technology, University of Hawai'i at Manoa

warnings takes place via the web, radio, field units and road signs. This style of monitoring, modeling, forecasting, warning and communication might profitably be applied to many other volcanic centres facing similar tropospheric volcanic aerosol pollution in the future.

4   The Great Dry Fog: Preparing for a Future Laki-Style Event

The Laki (Lakigigar) eruption 1783–1784 is known to be the largest air pollution incident in recorded history and its effects were felt throughout the northern hemisphere (Grattan 1998). Activity in this area of southern Iceland began in mid-May 1783 with weak earthquakes which intensified into June. On the 8th of June, the 27 km long fissure opened up with more than 140 vents (Thordarson and Hoskuldsson 2002; Thordarson et al. 1996). The eruption pumped 100 million tonnes of SO₂ into the westerly jet stream, producing sulfur-rich plumes that were dispersed eastwards over the Eurasian continent and north to the Arctic. The reaction of SO₂ with

atmospheric vapour produced 200 million tonnes of sulfate aerosol, of which 175 million tonnes were removed during the summer and autumn of 1783 via subsiding air masses within high pressure systems (Thordarson and Hoskuldsson 2002; Thordarson and Self 2003). At its peak, this mechanism may have been delivering up to six million tonnes of sulfate aerosol to the boundary layer of the atmosphere over Europe each day (Stothers 1996). The explosive activity from the eruption produced a tephra layer that covered over 8000 km² and is estimated to have produced 12 km³ of tholeiitic lava flows. Ten eruption episodes occurred during the first five months of activity at Laki, each with a few days of explosive eruptions followed by a longer phase of lava emissions. Volcanic activity began to decrease in December 1783 and ceased on the 7th of February 1784 (Steingrímsson 1998; Thordarson and Hoskuldsson 2002; Thordarson and Self 2003).

The consequences of the eruption were catastrophic. In Iceland, acid rains destroyed grazing and more than half of the livestock died from starvation or in combination with skeletal fluorosis (bone deformation resulting from the ingestion of high levels of fluorine) precipitated from erupted fluorine gases. More than a quarter of Iceland's population subsequently died from starvation and the survivors suffered from growths, scurvy, dysentery, and ailments of the heart and lungs (Steingrímsson 1998). The aerosol produced in the atmosphere resulted in a "dry fog" which hung over Britain, Scandinavia, France, Belgium, the Netherlands, Germany and Italy during the summer of 1783, affecting human health and withering vegetation (Durand and Grattan 2001). The aerosol also caused severe climatic perturbations. In the UK, August temperatures in 1783 were 2.5–3 °C higher than the decadal average, creating the hottest summer on record for 200 years. A bitterly cold winter followed, with temperatures 2 °C below average (Luterbacher et al. 2004). Coincidentally, in England, the death rate doubled during July 1783–June 1784 with 30,000 additional deaths recorded (Federation of Family History Societies 2010; Grattan et al. 2007; Witham and Oppenheimer 2004b). This period is classified as a 'mortality crisis' because the annual national mortality rate was 10–20 % above the 51-year moving mean (Wrigley and Schofield 1989). Two discrete periods of crisis mortality occurred: August–September 1783 and January–February 1784, which in combination accounted for around 20,000 additional deaths, with the East of England the most affected region (Witham and Oppenheimer 2004a). Crisis years are not unusual however, during the period 1541–1870 there were 22 crises where the death rate was 20–30 % higher, which is greater than the 1783–84 crisis of 16.7 % (Grattan et al. 2003). Whilst it is difficult to prove a direct causal link between the eruption and the mortality crisis, the connection between temperature extremes and mortality of the elderly or vulnerable is well established (Keatinge and Donaldson 2004; Kovats 2008; Royal Society 2014; Wilkinson et al. 2004). The effects of the Laki volcanic cloud are implicated in the climatic anomalies of 1783–4 and it is therefore likely that the Laki Craters eruption did contribute to the crises (Grattan et al. 2003; Witham and Oppenheimer 2004a).

Current levels of particulate air pollution in many parts of the UK exert considerable impact upon public health (Public Health England 2014). Epidemiological studies have linked premature mortality with exposure to air pollution, particularly to particles smaller than 2.5 μm in diameter (PM2.5) (Pope and Dockery 2006). During a 14 day period in March and April 2014, air pollution was 'very high' (based on government monitoring of PM10 and PM2.5) across the UK, which resulted in 3500 additional healthcare visits for acute respiratory symptoms and approximately 500 for severe asthma (Smith et al. 2015). The air pollution episode was due to anticyclonic atmospheric conditions which brought together local air pollution emissions, pollution from continental Europe and dust transported atmospherically from the Sahara (Smith et al. 2015). Air pollution levels resulting solely from local emissions also regularly breach European Union directives; NO_2 is of particular concern and in April 2015 the UK Supreme Court ruled that the government must submit new air quality plans to the European Commission by the end of the calendar year (Supreme Court Press Office 2015).

Given that air pollution in parts of the UK is regularly at (or in breach of) permissible levels, even a modest-sized eruption in Iceland could push UK cities over the threshold into very high levels of pollution. Over the last 1130 years, there have been four fissure eruptions in Iceland that caused environmental and climatic perturbation, of which Laki was the second largest and the occurrence of a contemporary Laki-style eruption poses a serious threat to the health of European populations. The need for preparedness for such an event was raised by a Geological Society working group in 2005 (Sparks et al. 2005) and subsequently added to the National Risk Register of Civil Emergencies (Loughlin et al. 2014).

Recent modelling of likely excess mortality resulting from a modern Laki reveals that a similar-sized eruption would produce, on average, 120 % more PM2.5 over background levels, which would result in 142,000 additional deaths, an increase of 3.5 % in the mortality rate

(Schmidt et al. 2011). This rate of mortality is much lower than actually occurred during the 1780s, which could be due to several factors, including the assumption that modern populations are more resilient to air pollution and environmental stress (which may not be the case), and that the concentration response functions in the model do not account for all adverse health effects (i.e. asthma caused by elevated SO_2) (Schmidt et al. 2011).

The link between elevated mortality and extremes of temperature is also well-established and therefore volcanically-induced anomalous weather could also contribute to a post-eruptive death toll. The European heatwave of 2003 was a three week period of abnormally hot weather which resulted in over 52,000 deaths across Europe with cities particularly affected (Royal Society 2014). There were over 14,800 fatalities in France, with excess mortality greater than 78 % in Paris, Dijon, Poitiers, Le Mans and Lyon. In the UK there were 2091 fatalities of which 616 occurred in London alone (Kovats and Kristie 2006; Royal Society 2014). There was a resultant increase in heat health warning systems across Europe (heat surveillance systems with associated risk warnings and awareness raising) with 16 active by 2006, which resulted in a reduction in the mortality following the 2006 heatwave (Royal Society 2014). The World Health Organisation's EuroHEAT project researches heat health effects in European cities, preparedness and public health system responses. It has highlighted that the health burdens fall disproportionately on those living in urban areas, particularly if they are also physiologically susceptible, socio-economically disadvantaged and live in degraded environments; a variety of practical measures to increase resilience have been suggested alongside legislation, national plans and social capital-building (World Health Organization 2007).

A future eruption similar to Laki would likely be forecast days to weeks in advance using the sophisticated volcano monitoring networks that are in place (Sigmundsson et al. 2014). The eruption itself would likely be accompanied by prolonged high fluxes of gases and ash,

producing an aerosol-laden plume in the troposphere, as observed in recent Icelandic eruptions. During some prolonged or particularly intense periods of eruption the plume may even reach the stratosphere (Thordarson and Self 2003). The plume will be modified physically and chemically as it moves away from the vent. Dispersal largely depends on wind direction and shear, meteorological conditions, synoptic-scale features (Dacre et al. 2013) and the stability of the atmosphere. Reactions take place in the gas phase and on the surfaces of ash and aerosol particles, where SO_2 is transformed to sulfate aerosol as well as other chemical reactions involving halogen radicals and ozone and NO_x species (von Glasow et al. 2009). Chemical transformations of the plume will depend on the availability of surfaces for reactions and will be affected by particle aggregation and sedimentation. The lifetime of sulphate aerosols and SO_2 in the troposphere depends on altitude and season and is of the order of 5–10 days at the low altitudes between UK and Iceland (Stevenson et al. 2003). The source parameters and associated uncertainties for modelling of a Laki eruption scenario were developed by the British Geological Survey who determined that once an eruption was underway and assuming the least favourable meteorological conditions for the UK (a strong north-westerly wind), there would be a minimum lead time of approximately six hours (Loughlin et al. 2013). A sustained supply of gas and aerosol from the source and unfavourable meteorology might maintain long-term (months) direct impacts in the UK (Loughlin et al. 2014).

Most of the risks associated with the eruption could be mitigated, given sufficient time to prepare for them, but there is work to be done in preparing guidelines to deal with hazards such as acid rain, increased levels of atmospheric pollutants, contaminated water, and the effect of aerosol on aviation (Loughlin et al. 2014). An effective response to an impending crisis will also require a much better understanding of plume chemistry and dispersion and its effects on the environment and on climate; there is a clear need to make these a research priority. Tracking volcanic clouds using satellites is now possible for eruptions in most parts of the world (Fig. 6), but there is clearly scope to improve coverage in both time and space (including depth resolution in the atmosphere). Air quality monitoring networks would require augmentation and coordination to be used as input to forecasting models.

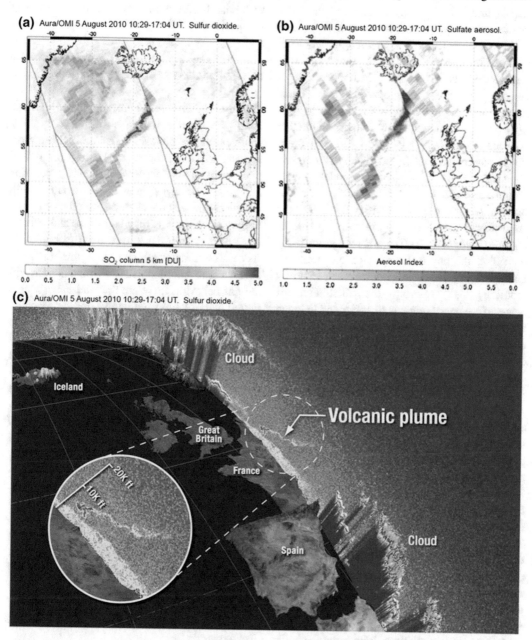

Fig. 6 Risk mitigation during a future large eruption in Iceland will depend on effective monitoring and hazard forecasting, which will be possible with a new generation of satellite-based sensors e.g. ESA's Sentinal 5 Precursor mission. Here we show data from existing satellite-based sensors. The OMI instrument on Nasa's Aura satellite can image the spatial distribution (in x-y) of **a** sulfur dioxide and **b** sulfate aerosol in the atmosphere from volcanic eruptions. These simultaneous traces were recorded on 8 May 2010 during the Eyjafjallajökull eruption (NASA). **c** on April 17, 2010, during the same eruption, NASA's Cloud-Aerosol Lidar and Infrared Pathfinder Satellite Observations (CALIPSO) satellite captured this image of the Eyjafjallajökull Volcano ash and aerosol cloud, providing a vertical profile of a slice of the atmosphere

There are many examples of smaller scale gas and aerosol monitoring and alert systems that have been successful (e.g. Kīlauea, USA; Mijakejima, Japan), but there are particular challenges applying these kinds of strategies to large regions potentially to include the whole of northern Europe. A major breakthrough has been the development of sophisticated modelling of aerosol formation, transport and loss. Early models used Global Circulation Models to simulate aerosol formation and its effects on climate (Chenet et al. 2005; Highwood and Stevenson 2003) but it was recognised that fully coupled chemistry and microphysics models were required in order to simulate aerosol size distributions (Schmidt et al. 2010). Recently, the atmospheric chemistry and meteorology model NAME (Jones et al. 2007) has shown promise for modelling the physical dispersion and transformation of volcanic SO_2 to aerosol. Current modelling is exploring the likelihood of near-surface concentrations of sulfur and halogen species exceeding health thresholds and the effects of acid deposition on ecosystems (Witham et al. 2014). Whilst these models are sophisticated, it is important to note that all models inherently involve uncertainties; particularly significant here are the estimated volcanic ash emission rates (Witham et al. 2012). A striking new finding from modelling the effects of tropospheric SO_2 emissions from the 2014 Holuhraun eruption has been that the sulfate aerosol increases the albedo of liquid clouds, causing a radiative forcing that might have been observable, had the eruption continued into summer 2015 (Gettelman et al. 2015). Radiative forcing of this magnitude is sufficient to cause changes in atmospheric circulation and might be a feasible mechanism to explain the far-reaching climatic effects of the 1783 Laki eruption (Gettelman et al. 2015). Understanding how dominantly tropospheric SO_2 emissions from large Icelandic flood basalt eruptions may affect climate and ultimately European air quality is a critical component of mitigating risk from a future eruption. The recent eruptions of Eyjafjallajökull (2010), Grímsvötn (2011) and Holuhraun (2014) illustrate well that Icelandic eruptions have

potential to disrupt aviation, our economy and air quality; the impacts of an even larger future eruption will undoubtedly extend into the realms of human health, agriculture and the structure of our society.

5   Perspectives for the Future

We have shown that the hazards due to volcanic gases are diverse in terms of not only their chemical nature but also their impacts. Monitoring and modeling the hazards, producing effective warning or forecast systems and risk mitigation strategies are all associated with unique challenges not shared with other volcanic hazards. Gas hazards may be diffuse and affect a large area. While there have been examples of successful monitoring strategies that integrate observations into sophisticated models describing gas behavior, these are few and far between. Future work requires innovative and far-reaching solutions to these monitoring challenges that can be applied in developing countries with minimal maintenance. Arguably the greatest strides are being made in modelling, with sophisticated models that couple chemistry with particle microphysics showing great promise as a monitoring and risk mitigation tool when combined with high quality ground- and satellite-based observations of volcanic emissions. Overcoming the challenges associated with educating populations with regard to gas hazards and maintaining effective communications is critical for future risk mitigation. Our greatest challenge may be a future large fissure eruption in Iceland, which may have significant consequences for air quality, our economy and environment in Europe and in North America.

References

Allard P, Carbonnelle J, Dajlevic D, Le Bronec J, Morel P, Robe M, Maurenas J, Faivre-Pierret R, Martin D, Sabroux J (1991) Eruptive and diffuse emissions of CO2 from Mount Etna. Nature 351 (6325):387–391

Allard P, Dajlevic D, Delarue C (1989) Origin of carbon dioxide emanation from the 1979 Dieng eruption, Indonesia: implications for the origin of the 1986 Nyos catastrophe. J Volcanol Geoth Res 39(2):195–206

Auker MR, Sparks RSJ, Siebert L, Crosweller HS, Ewert J (2013) A statistical analysis of the global historical volcanic fatalities record. J Appl Volcanol 2(1):1–24

Baxter PJ, Baubron J-C, Coutinho R (1999) Health hazards and disaster potential of ground gas emissions at Furnas volcano, Sao Miguel, Azores. J Volcanol Geoth Res 92(1):95–106

Baxter PJ, Kapila M, Mfonfu D (1989) Lake Nyos disaster, Cameroon, 1986: the medical effects of large scale emission of carbon dioxide?: BMJ. Br Med J 298 (6685):1437

Chenet A-L, Fluteau F, Courtillot V (2005) Modelling massive sulphate aerosol pollution, following the large 1783 Laki basaltic eruption. Earth Planet Sci Lett 236 (3):721–731

Chiodini G, Granieri D, Avino R, Caliro S, Costa A, Minopoli C, Vilardo G (2010) Non-volcanic CO_2 Earth degassing: case of Mefite d'Ansanto (southern Apennines), Italy. Geophys Res Lett, 37(11)

D'Alessandro W (2006) Human fluorosis related to volcanic activity: a review. Environ Toxicol 1:21–30

Dacre H, Grant A, Johnson B (2013) Aircraft observations and model simulations of concentration and particle size distribution in the Eyjafjallajökull volcanic ash cloud. Atmos Chem Phys 13(3):1277–1291

Delmelle P, Stix J, Baxter P, Garcia-Alvarez J, Barquero J (2002) Atmospheric dispersion, environmental effects and potential health hazard associated with the low-altitude gas plume of Masaya volcano, Nicaragua. Bull Volcanol 64(6):423–434

Dibben C, Chester DK (1999) Human vulnerability in volcanic environments: the case of Furnas, Sao Miguel, Azores. J Volcanol Geoth Res 92(1):133–150

Duncan IJ, Wang H (2014) Estimating the likelihood of pipeline failure in CO_2 transmission pipelines: new insights on risks of carbon capture and storage. Int J Greenhouse Gas Control 21:49–60

Durand M, Grattan J (2001) Effects of volcanic air pollution on health. Lancet 357(9251):164

Dutton EG, Christy JR (1992) Solar radiative forcing at selected locations and evidence for global lower tropospheric cooling following the eruptions of El Chichón and Pinatubo. Geophys Res Lett 19(23):2313–2316

Farrar C, Sorey M, Evans W, Howle J, Kerr B, Kennedy BM, King C-Y, Southon J (1995) Forest-killing diffuse CO_2 emission at Mammoth Mountain as a sign of magmatic unrest. Nature 376 (6542):675–678

Federation of Family History Societies (2010) National Burial Index

Franzen J, Köster A (1994) Die eozänen Tiere von Messel —ertrunken, erstickt oder vergiftet. Nat Mus 124 (3):91–97

Gerlach T, Doukas M, McGee K, Kessler R (2001) Soil efflux and total emission rates of magmatic CO_2 at the Horseshoe Lake tree kill, Mammoth Mountain, California, 1995–1999. Chem Geol 177(1):101–116

Gettelman A, Schmidt A, Kristjánsson JE (2015) Icelandic volcanic emissions and climate. Nat Geosci 8 (4):243

Giggenbach W (1996) Chemical composition of volcanic gases, monitoring and mitigation of volcano hazards. Springer, Berlin, pp 221–256

Global Volcanism Program (2011) Report on Dieng Volcanic Complex (Indonesia). In: Sennert, SK (ed.), Weekly Volcanic Activity Report, (25 May-31 May 2011). Smithsonian Institution and US Geological Survey

Grattan J (1998) The distal impact of Icelandic volcanic gases and aerosols in Europe: a review of the 1783 Laki Fissure eruption and environmental vulnerability in the late 20th century. In Maund JG, Eddleston M (eds) Geohazards in engineering geology, pp 97–103

Grattan J, Durand M, Taylor S (2003) Illness and elevated human mortality in Europe coincident with the Laki Fissure eruption. Geological Society, vol 213, no 1. Special Publications, London, pp 401–414

Grattan J, Michnowicz S, Rabartin R (2007) The long shadow: understanding the influence of the Laki fissure eruption on human mortality in Europe. Living Under Shadow Cult Impacts Volcanic Eruptions, pp 153–175

Haberyan KA, Hecky RE (1987) The late Pleistocene and Holocene stratigraphy and paleolimnology of Lakes Kivu and Tanganyika. Palaeogeogr Palaeoclimatol Palaeoecol 61:169–197

Hansell A, Oppenheimer C (2004) Health hazards from volcanic gases: a systematic literature review. Arch Environ Health Int J 59(12):628–639

Highwood E-J, Stevenson D (2003) Atmospheric impact of the 1783–1784 Laki Eruption: part II climatic effect of sulphate aerosol. Atmos Chem Phys 3(4):1177–1189

Jones A, Thomson D, Hort M, Devenish B (2007) The UK Met Office's next-generation atmospheric dispersion model, NAME III, air pollution modeling and its application XVII. Springer, US, pp 580–589

Keatinge WR, Donaldson GC (2004) Winter mortality in elderly people in Britain—action on outdoor cold stress is needed to reduce winter mortality. Br Med J 329(7472):976

Kling GW, Evans WC, Tuttle ML, Tanyileke G (1994) Degassing of Lake Nyos. Nature 368(6470): 405–406

Kling GW, Clark MA, Wagner GN, Compton HR, Humphrey AM, Devine JD, Evans WC, Lockwood JP, Tuttle ML, Koenigsberg EJ (1987) The 1986 Lake Nyos gas disaster in Cameroon, West Africa. Science 236(4798): 169–175

Kovats RS, Kristie LE (2006) Heatwaves and public health in Europe. Eur J Public Health 16(6):592–599

Kovats S (2008) Health effects of climate change in the UK 2008: an update of the Department of Health Report 2001/2002. In: Health DO, Agency HP (eds) Crown Copyright, p 113

Lavigne F, De Coster B, Juvin N, Flohic F, Gaillard J-C, Texier P, Morin J, Sartohadi J (2008) People's behaviour in the face of volcanic hazards: perspectives from Javanese communities, Indonesia. J Volcanol Geoth Res 172(3):273–287

Le Guern F, Tazieff H, Pierret RF (1982) An example of health hazard: people killed by gas during a phreatic eruption: Dieng Plateau (Java, Indonesia), February 20th 1979. Bull Volcanologique 45(2):153–156

Longo BM, Yang W, Green JB, Crosby FL, Crosby VL (2010a) Acute health effects associated with exposure to volcanic air pollution (vog) from increased activity at Kilauea Volcano in 2008. J Toxicol Environ Health Part A 73(20):1370–1381

Longo BM, Yang W, Green JB, Longo AA, Harris M, Bibilone R (2010b) An indoor air quality assessment for vulnerable populations exposed to volcanic vog from Kilauea Volcano. Family Community Health 33 (1):21–31

Loughlin SC, Aspinall WA, Vye-Brown C, Baxter PJ, Braban CF, Hort M, Schmidt A, Thordarson T, Witham C (2013) Large magnitude fissure eruptions in Iceland: source characterisation. British Geological Survey

Loughlin SC, Aspinall WP, Vye-Brown C, Baxter PJ, Braban CF, Hort M, Schmidt A, Thordarson T, Witham C (2014) Large-magnitude fissure eruptions in Iceland: source characterisation British Geological Survey Open File Report, v. OR/12/098, p. 123

Luterbacher J, Dietrich D, Xoplaki E, Grosjean M, Wanner H (2004) European seasonal and annual temperature variability, trends, and extremes since 1500. Science 303(5663):1499–1503

Mather T, Pyle D, Oppenheimer C (2003) Tropospheric volcanic aerosol. Volcanism Earth's Atmos 139:189–212

Oppenheimer C (2003) Climatic, environmental and human consequences of the largest known historic eruption: Tambora volcano (Indonesia) 1815. Prog Phys Geogr 27(2):230–259

Pope CA, Dockery DW (2006) Health effects of fine particulate air pollution: lines that connect. J Air Waste Manag Assoc 56(6):709–742

Public Health England (2014) Estimating local mortality burdens associated with particulate air pollution. Crown Copyright, Oxfordshire, p 46

Pyle DM, Mather TA (2003) The importance of volcanic emissions for the global atmospheric mercury cycle. Atmos Environ 37(36):5115–5124

Rampino MR, Self S, Stothers RB (1988) Volcanic winters. Annu Rev Earth Planet Sci 16:73–99

Reikard G (2012) Forecasting volcanic air pollution in Hawaii: tests of time series models. Atmos Environ 60:593–600

Robock A (2000) Volcanic eruptions and climate. Rev Geophys 38(2): 191–219

Royal Society (2014) Resilience to Extreme Weather

Schmid M, Halbwachs M, Wehrli B, Wüest A (2005) Weak mixing in Lake Kivu: new insights indicate increasing risk of uncontrolled gas eruption. Geochem Geophys Geosyst 6(7): Q07009

Schmidt A, Carslaw K, Mann G, Wilson M, Breider T, Pickering S, Thordarson T (2010) The impact of the 1783–1784 AD Laki eruption on global aerosol formation processes and cloud condensation nuclei. Atmos Chem Phys 10(13):6025–6041

Schmidt A, Ostro B, Carslaw KS, Wilson M, Thordarson T, Mann GW, Simmons AJ (2011) Excess mortality in Europe following a future Laki-style Icelandic eruption. Proc Natl Acad Sci 108 (38):15710–15715

Sigmundsson F, Hooper A, Hreinsdóttir S, Vogfjörd KS, Ófeigsson BG, Heimisson ER, Dumont S, Parks M, Spaans K, Gudmundsson GB (2014) Segmented lateral dyke growth in a rifting event at Bar [eth] arbunga volcanic system. Nature, Iceland

Smith GE, Bawa Z, Macklin Y, Morbey R, Dobney A, Vardoulakis S, Elliot AJ (2015) Using real-time syndromic surveillance systems to help explore the acute impact of the air pollution incident of March/April 2014 in England. Environ Res 136:500–504

Sparks S, Self S, Grattan J, Oppenheimer C, Pyle D, Rymer H (2005) Super-eruptions: global effects and future threats. Report of a Geological Society of London Working Group

Steingrímsson J (1998) Fires of the Earth. Nordic Volcanological Institute, 95 p

Stevenson D, Johnson C, Highwood E, Gauci V, Collins W, Derwent R (2003) Atmospheric impact of the 1783–1784 Laki eruption: part I chemistry modelling. Atmos Chem Phys 3(3):487–507

Stothers RB (1996) The great dry fog of 1783. Clim Change 32(1):79–89

Supreme Court Press Office (2015) R (on the application of ClientEarth) (Appellant) v Secretary of State for the Environment, Food and Rural Affairs (Respondent) [2015] UKSC 28: UK

Symonds R, Gerlach T, Reed M (2001) Magmatic gas scrubbing: implications for volcano monitoring. J Volcanol Geoth Res 108(1):303–341

Thordarson T, Hoskuldsson A (2002) Iceland. Terra Publishing, Classical Geology in Europe, 224 p

Thordarson T, Self S (2003) Atmospheric and environmental effects of the 1783–1784 Laki eruption: a review and reassessment. J Geophys Res Atmos 108 (D1): 29

Thordarson T, Self S, Oskarsson N, Hulsebosch T (1996) Sulfur, chlorine, and fluorine degassing and atmospheric loading by the 1783-1784 AD Laki (Skaftar fires) eruption in Iceland. Bull Volcanol 58(2–3):205–225

Viveiros F, Cardellini C, Ferreira T, Caliro S, Chiodini G, Silva C (2010) Soil CO_2 emissions at Furnas volcano, São Miguel Island, Azores archipelago: Volcano monitoring perspectives, geomorphologic studies, and land use planning application. J Geophys Res Solid Earth (1978–2012) 115(B12)

von Glasow R, Bobrowski N, Kern C (2009) The effects of volcanic eruptions on atmospheric chemistry. Chem Geol 263(1):131–142

Werner CA, Doukas MP, Kelly PJ (2011) Gas emissions from failed and actual eruptions from Cook Inlet Volcanoes, Alaska, 1989–2006. Bull Volcanol 73 (2):155–173

Wilkinson P, Pattenden S, Armstrong B, Fletcher A, Kovats RS, Mangtani P, McMichael AJ (2004) Vulnerability to winter mortality in elderly people in Britain: population based study. Bmj 329(7467): 647

Witham C, Felton C, Daud S, Aspinall W, Braban C, Loughlin S, Hort M, Schmidt A, Vieno M (2014) UK hazard assessment for a Laki-type volcanic eruption. EGU General Assembly 2014, vol 16

Witham C, Hort M, Thomson D, Leadbetter S, Devenish B, Webster H (2012) The current volcanic ash modelling setup at the London VAAC. UK Meteorological Office Internal Report

Witham CS, Oppenheimer C (2004a) Mortality in England during the 1783-4 Laki Craters eruption. Bull Volcanol 67(1):15–26

Witham CS, Oppenheimer C (2004b) Mortality in England during the 1783–4 Laki Craters eruption. Bull Volcanol 67(1):15–26

World Health Organization (2007) Improving public health responses to extreme weather/heat-waves—EuroHEAT Meeting Report Bonn, Germany, 22–23 Mar 2007

Wrigley EA, Schofield RS (1989) The population history of England 1541–1871. Cambridge University Press, Cambridge

Active Hydrothermal Features as Tourist Attractions

Patricia Erfurt-Cooper

Abstract

Tourists are looking increasingly for adventurous experiences by exploring unusual and interesting landscapes. Active volcanic and hydrothermal landscapes and their remarkable manifestations of geysers, fumaroles and boiling mud ponds are some of the surface features that fascinate visitors of National Parks, Geoparks and World Heritage areas worldwide. The uniqueness of hydrothermal activity based on volcanism has provided popular tourist attractions in many countries for several thousand years. The Romans for example have used hydrothermal springs on the Italian island Ischia and visited the Campi Flegrei for recreational purposes. In Iceland the original Geysir already attracted international visitors over 150 years ago, who came to observe this spectacular hydrothermal phenomenon. In Greece and Turkey volcanic hot springs have historically provided attractive destinations, as well as in New Zealand, Japan and the Americas. The fact that locations with hydrothermal activity based on active volcanism have acquired various forms of protected site status, adds a further dimension to their attraction and demonstrates a significant contribution to sustainable and nature based tourism. Countries such as Iceland, New Zealand and Japan have a long tradition of using hydrothermal activity in its various forms to offer tourists a unique natural experience. These environments however are also known for their unpredictable and potentially hostile nature, as the use of hydrothermal features as a natural resource for tourism does harbour certain risks with the potential to affect human health and safety. Hydrothermal systems have erupted in the past, thereby causing the destruction of their immediate environment. Depending on the level of magnitude explosions of super heated water and steam mixed with fractured rocks and hot mud can be violent enough to create craters varying in size from a few metres

P. Erfurt-Cooper (✉)
GEOTOURISM Australia, Canberra ACT, Australia
e-mail: Patricia.Erfurt@my.jcu.edu.au

Advs in Volcanology (2018) 85–105
https://doi.org/10.1007/11157_2016_33

to several hundred metres in diameter. Apart from unexpected eruptions of hydrothermal vents with the potential to cause thermal burns, further risk factors include seismic activity such as earthquakes, lethal gas emissions of hydrogen sulphide (H2S) as well as ground instability through hydrothermal alteration. While it is essential to prevent injuries to tourists the management of hydrothermal hazards remains problematic. Precursory signs are not well understood by the general public and the communication of imminent danger is frequently unachievable. As a consequence serious thought needs to be given to the risk factors and the potential danger of areas in the proximity of active hydrothermal manifestations such as extreme hot springs and geysers. To improve the safety standards in hydrothermal landscapes that are used as main features in tourism, strategic guidelines for best practice management must cover ALL active volcanic and hydrothermal areas. This chapter looks at management issues at hydrothermal destinations with special consideration of areas where these unique features are integrated as tourist attractions. Examples from destinations traditionally based on active volcanic and hydrothermal phenomena are presented as case studies to highlight the risk management processes in individual countries. Potential hazards in volcanic and hydrothermal areas are assessed with a focus on the prevention of accidents and injuries to tourists.

Keywords

Hydrothermal activity · Protected site status · Risk management · Sustainable tourism · Volcanic environments

1   Introduction

1.1   Visitor Safety in Hydrothermal Environments

Tourists are looking increasingly for adventurous experiences by exploring unusual and interesting destinations. Active volcanic and hydrothermal landforms and their remarkable manifestations of geysers, fumaroles and boiling mud ponds are some of the surface features that fascinate tourists worldwide. The uniqueness of hydrothermal activity based on volcanism has provided popular tourist attractions in many countries for several thousand years. The Romans for example used hydrothermal springs on the Italian island Ischia and visited Campi Flegrei for recreational purposes. In Iceland the original Geysir attracted international visitors over 150 years ago who came to observe this spectacular hydrothermal phenomenon. In many other countries worldwide (Greece, Turkey, New Zealand, Japan, China and the Americas) hydrothermal activity in its various forms has historically provided attractive destinations.

The fact that locations with hydrothermal activity, commonly based on active volcanism, have acquired various forms of protected site status (e.g. National Parks, Geoparks and World Heritage Areas) adds a further dimension to their attraction and demonstrates a significant contribution to sustainable and nature based tourism. However, these environments are also known for their unpredictable and potentially hostile nature and the use of hydrothermal features as a natural resource for tourism does have certain risks with the potential to affect human health and safety.

This chapter examines the communication of hazards and the potential risks associated with sites where hydrothermal features are major tourist attractions. Examples from tourist destinations based on active hydrothermal phenomena are presented as brief case studies to highlight the importance of hazard and risk communication. To prevent unnecessary exposure to hazards, which can escalate into a crisis situation, visitors of hydrothermal attractions must be made aware of potential hazards that could carry the risk of personal injury or death. As many hydrothermal areas are located in close proximity to active volcanic systems, this chapter occasionally refers to both environments and their correlated hazards.

1.2 Definitions of Hazard, Risk and Vulnerability

First of all, the actual meaning of the common language terms hazard, risk and vulnerability needs to be clarified in relation to the subject matter of this chapter. While the term hazard is often used as a synonym for danger and/or risk, a hazard is scientifically defined as the *probability* of a natural event occurring as well as being a potential source of vulnerability (exposure to danger). The term vulnerability refers to the susceptibility and inability of humans or physical structures to withstand the impacts of natural hazards. Vulnerability in hydrothermal areas based on volcanic activity takes into account the real possibility of causing injury, damage and loss of life (Aspinall and Blong 2015; Barclay et al. 2015; Jolly and De La Cruz 2015; McGuire et al. 2009; UNISIDR 2016). Vulnerability can be a consequence of either being unsuspecting of potential risks or ignoring these while visiting sometimes remote or unsafe areas without suitable defence structures or shelters. Taking a risk in these environments therefore can result in vulnerability due to exposure of hazards and includes the probability of being harmed in the process. This can be based on a lack of awareness about the potential risks and/or a lack of appropriate hazard communication.

Potential hazards in areas of hydrothermal activity (Table 7.1) are generally assessed with the main focus on minimising the risk of accidents and injuries. The process of risk identification recognises potential hazards as well as any potential vulnerability from the damaging effects of a hazard (Aspinall and Blong 2015; UNISDR 2014). Also, the risk to visitors is considered to increase with extended time spent in an active hydrothermal area (Bratton et al. 2013). It is therefore highly recommended that all visitors of active hydrothermal environments are aware of the particular hazards and the potential risks in these areas.

In this regard, a crisis can be defined as an unstable and hazardous situation of increased danger that has reached a critical phase (De La Cruz-Reyna et al. 2000). Communication of scientific advice in a crisis situation must clearly reflect the level of danger to raise awareness about the real hazard level and to avoid misunderstanding as to the likely consequences (Jolly and De La Cruz 2015).

Table 7.1 Examples of the most common hydrothermal hazards

Potential hazards in active hydrothermal areas	
Seismic activity—unexpected earthquakes	Sudden change of location of hot water source
Toxic fumes and gas emissions	Sudden change of flow rates, direction and currents of hot water source
Unstable ground	Thermal burns from extreme hot springs
Unexpected hydrothermal eruptions	Health hazards from thermophilic microbes and bacteria
Hydrothermal steam discharge	
Sudden change of water temperature	

1.3 Hydrothermal or Geothermal?

Now and again there appears to be confusion between the terms hydrothermal and geothermal. The term hydrothermal refers to hot water and is derived from the Greek meaning of *hydros* for water and *thermos* for heat. Likewise, the term geothermal has its origin in the Greek language with the prefix *geo* referring to earth. To clarify the difference between hydrothermal and geothermal Heasler et al. (2009) describe hydrothermal as a subset of geothermal, whereby geothermal refers to any system that transfers heat from the interior of the earth to the surface involving water, both as a liquid and steam (Keary 1996). All hydrothermal systems related to volcanism are based on a geothermal heat source in the form of an active magma chamber or a cooling magma body (Hochstein and Browne 2000). Consequently, water emerging from a hydrothermal vent is correctly termed geothermal water, but this term can also refer to water heated by convective circulation deep underground without the proximity of a magma body (Heasler et al. 2009). The actual process of heat transfer involves the circulation of groundwater from a subterranean reservoir to the surface. Here, individual hydrothermal features emerge in the form of hot springs and geysers, or in the case of subaqueous hydrothermal vents in close proximity to an underlying magma body as the emission of superheated mineral rich solutions, known as Black or White Smokers.

The temperature range of such hydrothermal systems is estimated to be typically between 50 °C and up to over 400 °C in deeper reservoirs (Haase et al. 2009). According to Heasler et al. (2009) hydrothermal systems present a continuum of resource temperatures that is relatively open-ended. In comparison the temperature range of natural hot springs utilised for health and recreational facilities lies generally between 37 °C and the boiling point of water at sea level (100 °C). These natural hot springs, independent of whether their origin is volcanic or non-volcanic, are frequently referred to as either geothermal or just as thermal springs (Erfurt-Cooper 2012).

2 The Challenges of Hydrothermal Tourist Sites

2.1 Direct Use of Hot Springs as Tourist Attraction

Hydrothermal features play an important role in tourism and are a favourite with people who are looking for unusual natural experiences with a touch of adventure (Tables 7.2 and 7.3). Many hydrothermal areas are marketed as family friendly must-see destinations, offering a once-in-a-lifetime occasion to encounter the raw power of nature. Another reason for people to visit active areas is that features may be ephemeral and may become inaccessible or disappear altogether due to earthquakes or volcanic activity (e.g. Valley of the Geysers in Kamchatka—mudflow in 2007).

Hot springs, geysers, boiling lakes, bubbling mud pools and hissing steam vents are common in countries with active as well as dormant volcanism (e.g. New Zealand, Japan, Iceland (Case Study 7.1), the Americas, Africa and China). Located in protected areas such as national parks or geoparks can be an advantage in controlling access to hazardous sites, although providing safety is a definite challenge, which depends on many variables including language barriers. For example, the family affected in Case Study 7.2 did not speak English, which may have contributed to the horrific accident.

Table 7.2 Examples of hydrothermal features based on volcanism that are used as tourist attractions worldwide

Hydrothermal Features used as Tourist Attractions	
Fumaroles/steam vents	Boiling mud pots
Geysers	Explosion craters
Hot lakes	Sinter terraces
Hot rivers and streams	Hot springs (for bathing)
Hot waterfalls	Hot springs (for cooking)
Mud volcanoes	Geothermal power stations

Table 7.3 Hydrothermal features play an important role in the marketing of tourist destinations

The role of hydrothermal features in tourism	
• Unique selling point for destination development • Value adding when combined with other recreational facilities • Significant resource for Geotourism with opportunities to learn about hydrothermal features and their geological heritage	• Integration into Health & Wellness tourism by utilising geothermal spring water • Sustainable development based on the use of renewable energy • Economic benefits through the use of geothermal energy for local infrastructure

Case Study 7.1: Iceland

Volcanic and hydrothermal tourist attractions are an important part of the visitor experience in Iceland. The geysers at Haukadalur and the hot springs at Landmannalaugar and Hveragerði are commonly included in trip agendas. The Blue Lagoon, not far from the capital Reykjavik and the Mývatn Nature Baths in Iceland's north-east are unique bathing pools fed by excess geothermal water from the neighbouring power stations Svartsengi and Krafla. In fact, tourists can visit most of the Icelandic geothermal power plants and learn about the generation of clean renewable energy. "With more than 100,000 visitors a year, the geothermal power plants and related installations in Iceland are one of the top tourist destinations in Iceland" (Think Geoenergy 2015).

Case Study 7.2: Rotorua—New Zealand

While tens of thousands of people safely visit New Zealand's geothermal parks each year, tragic accidents can happen. In 2010 a ten-year old boy died after falling into one of Rotorua's hot water pools (NZ Herald 2013). According to eye witnesses the boy had burns from his head to his feet and was flown to a hospital in Auckland to be treated in intensive care, but later died (BBC News 2010). Following the accident, Rotorua District Council reviewed the park's safety and added 60 new warning signs as well as additional fencing. However, according to council officials, visitors are frustrated when there are too many fences, and climb over for a better view (TVNZ 2012). An inquest later found that the boy had climbed a wall and fell into one of the hot pools, suffering burns to almost 100% of his body.

Geothermal water is used worldwide as a renewable resource for generating energy, commercial, agricultural and industrial purposes, space heating, bathing and rehabilitation as well as for drinking and cooking. At Yellowstone (USA) Native Americans have historically used hydrothermal features for food preparation. Today geothermal cuisine is regularly used as a tourist attraction at many active volcanic and hydrothermal destinations. In New Zealand tourists can visit natural cooking pools such as *Ngāraratuatara* (Rotorua) and observe ancient *Māori* cooking techniques. In Iceland geothermal or "geyser cooking" is attractive to tourists, with a restaurant in *Hveragerði* specialised in geothermal cuisine using steam from volcanic activity. On the volcanic islands of the Azores one of the most remarkable tourist attractions of *Furnas* (San Miguel) is geothermal cooking, offering typical Azorean cuisine at many local restaurants. Hot spring cooking is equally popular in Japan with *jigoku-mushi* one of many sought after dishes prepared using geothermal steam. Apart from these well-known examples, many countries worldwide, including Kenya, the Philippines, Mexico and Indonesia, use geothermal water for cooking.

On account of their proximity to active volcanism, the management of tourist areas with underlying hydrothermal systems is therefore not an easy task. Given the potential hazards of active environments, both volcanic and hydrothermal, it is essential that sufficient warnings and safety guidelines are available and emergency procedures and crisis response resources have been prepared along with appropriate channels for effective communication.

2.2 Potential Hazards—Beauty or Beast?

Although hydrothermal areas are attractive to visitors, there are a number of inherent hazards and risks. Depending on the magnitude, explosions of super-heated water and steam mixed with fractured rocks and hot mud can be violent enough to create craters varying in size from a few metres to several hundred metres in diameter. Apart from the unexpected eruptions of hydrothermal vents with the potential to cause thermal burns, further risk factors include seismic activity such as earthquakes and lethal gas emissions, as well as ground instability through hydrothermal alteration.

Considering the number of hydrothermal hazards (Table 7.1) it becomes clear that there is indeed a possibility for accident and injury, despite the fact that fumaroles, geysers and bubbling mud pools present such a picturesque photo opportunity. Adding to the list of potential problems is the frequent underestimation of the safety risk from nearby active volcanoes, where circumstances can quickly change in case of unexpected eruptions. Failing to seek information about current activity levels prior to visiting such areas, or not following warnings, can lead to serious injuries or death as many people are not aware of the various hazards they may encounter in these environments.

While life-threatening hydrothermal eruptions are relatively rare, toxic fumes and gas emissions are rather common in active areas. These naturally occurring gases are emitted from volcanic craters and fumaroles or diffused through the soil (Hansell et al. 2006). Toxic gas emissions can also occur in the absence of eruptive activity. Some of the most common gases include

Hydrogen Sulphide (H_2S), Carbon Dioxide (CO_2) and Sulphur Dioxide (SO_2). All of these are dangerous to human health with H_2S causing instant death at extreme levels (Case Study 7.3).

Case Study 7.3: Akita, Japan

In December 2005 a family of four tragically died near a hot spring resort from hydrogen sulphide poisoning. While playing two young boys tried to retrieve their Frisbee from a snow covered hollow, unaware that the depression in the ground contained a lethal concentration of H_2S. When both children suddenly collapsed their mother tried to save them, but also died instantly after inhaling the toxic gas. When searching for his family the father discovered them lying on the ground and also entered the hollow. He initially survived, but passed away a day later in hospital (Japan Times 2005).

Being denser than air H_2S can accumulate in low lying areas such as hollows and depressions in the landscape and remain trapped if not dispersed by wind (USGS[1] 2014, Whittlesey 2014; Williams-Jones and Rymer 2000). Although H_2S at low levels has a distinctive odour often described as rotten egg smell, at higher concentrations this gas cannot be detected through smell which means there is no warning.

In Hawai'i tourists regularly visit the Hawai'i Volcanoes National Park to observe the glowing lava flows. A special attraction is the area where the lava flows enter the ocean, instantly boiling the seawater and turning it into vapour (Hansell et al. 2006; Heggie et al. 2010; Williams-Jones and Rymer 2000). This chemical interaction between molten lava and sea water creates a white plume known as lava haze or *LAZE* and is frequently mistaken for a 'harmless' steam cloud. This plume however contains a mixture of hydrochloric acid (HCl) and concentrated seawater with up to 2.3 times average salinity and with pH levels as low as 1.5–2.0 (Heggie et al. 2010; USGS[1] 2014). As a precaution the USGS information website (USGS[2] 2014) clearly advises people not to stand beneath the volcanic laze plume or downwind of it because hydrochloric acid is toxic and can cause irritation of the throat, lungs, eyes, and nose. In fact, volcanic laze is dangerous enough to kill (Case Study 7.4).

Case Study 7.4: Clouds that can kill— Acidic LAZE plumes

During November 2000 in the Hawai'i Volcanoes National Park two people were caught in a volcanic laze plume near the point where lava flows enter the ocean and died as a result of pulmonary oedema caused by inhalation of volcanic laze. Conditions near this point involved the threat of exposure to dense hydrochloric acid mist, which subsequently engulfed the victims in an extremely hot and acidic cloud. Nevertheless, the area can be accessed without restrictions, although warning signs and safety instructions should never be ignored (Heggie et al. 2010).

Hydrothermal areas for example in Japan, New Zealand or at Yellowstone warn visitors about the risk of encountering unstable ground, as hydrothermal features may only be covered by a thin crust that easily breaks underfoot, and may cause thermal burns (Case Study 7.5; Fig. 7.1).

Unexpected hydrothermal eruptions can become a serious danger also, causing impact injuries and burns from scalding steam emissions, hot mud and ejected rocks. Hydrothermal steam discharge from fumarolic vents can suddenly increase without warning, or steam plumes can change their direction with the wind, which can result in thermal burns as well as respiratory problems. Likewise, a sudden change in the water temperature of hot streams, rivers and lakes, where geothermal spring water mixes with cooler water, can result in serious injuries if people take a soak in what they initially perceive as warm water.

Case Study 7.5: New Zealand— Geothermal Wonderlands

New Zealand's hydrothermal sites provide walkways for tourists in potentially hazardous areas. Boiling lakes, geysers, mud pools, and especially areas of unstable ground are fenced off to reduce the risk for accidents and injury due to underlying geothermal activity. This practice protects both the tourists and the mineral deposits which are part of the attraction of hot springs (Roscoe 2010). However, there is still a need for multilingual warning signs at hydrothermal destinations to increase awareness about potential hazards and visitor safety.

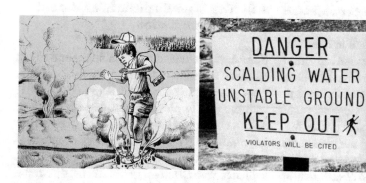
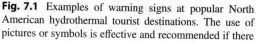

Fig. 7.1 Examples of warning signs at popular North American hydrothermal tourist destinations. The use of pictures or symbols is effective and recommended if there is the lack of warnings in several languages (Compare Figs. 7.2, 7.3, and 7.4). *Source* Public Domain

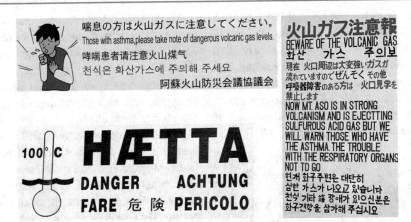

Fig. 7.2 Examples of warning signs in Japan and Iceland. The effort to communicate hazards to international visitors is obvious and commendable, while at the same time adding urgency to the warnings. *Source* Patricia Erfurt-Cooper

Other risks involve the unexpected change of location of the hot water source, especially if these are located below the surface of a stream of lake. The sudden change of flow rates, direction and currents of the hot water source is another hazard and a risk that people frequently underestimate. The temperature of hydrothermal features such as erupting geysers can also be underestimated from a distance, which can result in thermal burns on approach.

Another important health hazard involves disease-causing organisms such as thermophilic microbes and bacteria. *Legionella* bacteria and *Naegleria fowleri* have been identified at thermal pools used for recreation (Fig. 7.3) with some hydrothermal locations reportedly having problems with *Primary Amoebic Meningoencephalitis* (PAM), a rare but life-threatening infection caused by the organism *Naegleria fowleri* (Erfurt-Cooper and Cooper 2009). While casual exposure via the skin does not result in infection, the inhalation of contaminated water can cause serious problems, as the pathogenic amoebae migrate up the sinuses and surrounding tissue to the brain (Barnett et al. 1996). A common hot pool safety warning in New Zealand advises that *"When swimming in natural hot pools, where the water comes out of the ground, keep your head above water because there is a small risk of contracting an illness called amoebic meningitis. While very rare, this illness is serious"*.

Public bathing facilities in the form of hot spring pools are often also accessible in hydrothermal areas. Bathing accidents are not unusual at natural hot springs, thermal health spas and geothermally heated communal pools. Being generally careless or consuming alcohol prior to using hot spring pools shows a lack of common sense and ignorance of safety advice, but is unfortunately all too common. In general, developed hot spring spas and pools advise visitors on the risk of excessive soaking in hot water when suffering from certain health conditions.

3 Communicating, Forecasting and Managing Natural Hazards— A Mission Impossible?

3.1 The Main Challenges of Hazard Communication

Hazard education and risk communication are an essential foundation for effective risk management (Leonard et al. 2008; Paton et al. 2001) with the aim to prevent harm through injury or death, while crisis communication focuses on informing the public during a crisis event (Leonard et al. 2008; Steelman and McCaffrey 2013; Williams and Olaniran 1998). To maximise the effectiveness of warnings, signage plays a critical role in increasing hazard awareness of the public

WARNING

DO NOT ALLOW WATER TO ENTER YOUR NOSE

NAEGLERIA FOWLERI

AN AMOEBA COMMON TO THERMAL POOLS MAY
ENTER CAUSING A RARE INFECTION AND DEATH

AVISO

NO DEJAR QUES SE META AGUA POR LA NARIZ

NAEGLERIA FOWLERI

ES UNA AMIBA QUE SE ENCUENTRA EN
MANANTIALES CALIDOS Y PUEDE SER
CAUSA DE UNA INFECCION GRAVE O LA
MUERTE SI SE ENTRA AGUA POR LA NARIZ

Fig. 7.3 Bilingual sign at a thermal pool warning visitors of the risk of contracting amoebic meningitis. *Source* Public Domain

(Dengler 2005; Leonard et al. 2008). Visitors of hydrothermal and volcanic environments may be aware of potential hazards, but once an emergency situation develops, reaching people in remote locations can be fraught with problems. Apart from the geophysical hazards, a number of additional challenges at hydrothermal tourist sites also have to be considered (Table 7.4).

While effective communication starts with essential warning signage in hazardous areas, warnings, safety instructions and hazard maps, if ignored, can result in the loss of lives. The reliance on smart phones and a potential lack of reception is a further challenge in achieving effective communication. Some of the biggest problems in hazard communication however are language barriers, with more signs displaying multilingual warnings and effective symbols or pictograms needed at many hydrothermal tourist destinations.

Another main difficulty is to keep a record about the exact numbers of people present in an affected area during a crisis situation. Uncertainty over the whereabouts of tourists (hikers, climbers etc.) can cause preventable fatalities if there is insufficient time to reach all affected people under deteriorating conditions. This is complicated by the fact that predictions and/or forecasts of hazardous activity can often only be made based on the frequency and the type of past hazardous activities of a particular area. To realistically forecast future activity and determine which area may be subject to potential threats is only possible if the site is constantly monitored and assessed. Scientists can advise on possible danger zones and create hazard maps in cooperation with local authorities to show unsafe and safe areas, escape routes and shelters, but this is at best only a probability assessment.

Table 7.4 Challenging factors related to hazards in active hydrothermal environments

Potential problems at hydrothermal tourist sites

Challenging factors		Examples/Consequences
Remote areas	Uncertain visitor numbers Uncertain access and escape routes Rescue response delayed or impossible	Valley of Geysers, Kamchatka, Russia
Large areas	Monitoring visitors difficult Emergencies can go unnoticed	Yellowstone, USA (Case Study 7.8)
Large crowds	Crowd control to avoid panic Unknown and/or blocked escape routes	Hydrothermal parks (New Zealand) (Case Study 7.5, 7.7 and 7.9) Yellowstone (USA)
Confined spaces	Blocked escape routes Potential of panic causing injuries	Small hydrothermal parks or Jigoku (Japan) (Case Study 7.6)
Communication barriers	Insufficient warning signs Lack of emergency information Lack of mobile phone reception	Areas with limited or no communication infrastructure, remote areas
Language barriers	Signage does not communicate warnings effectively	Tourist sites only using signs in the local language
Lack of shelters or difficult access to shelters	Shelters not built strong enough to protect from eruption fallout or from toxic gas emissions	
Transport logistics in crisis situations	Access for rescue and transport of injured people in remote locations Treatment and medical care	
Time factor	Sudden onset of crisis	
Political	Access limited or prevented during times of political instability Cross-border disagreements	Erta Ale and Dallol Hydrothermal Field (Ethiopia)
Financial	Lack of funding to implement safety strategies Economic mismanagement	Affects emergency management, strategic planning and rescue response
Topography	Extreme terrain and physical surface features	Area is difficult to negotiate due to environmental factors
Sudden weather changes	Additional natural hazards such as strong wind, rain, snow, temperature drop	Can affect any region Disadvantage for rescue missions
Human resources	Lack of trained staff, unpreparedness Lack of emergency response teams	
Technical problems	Equipment failure, power outage	

Risk management is based on the monitoring of potentially hazardous areas, and restricting access when for example toxic gas emissions reach dangerous levels. Unfortunately, areas at risk quite often lack the necessary funding to install permanent monitoring equipment (Williams-Jones and Rymer 2000). Table 7.4 lists the major factors that can determine success or failure in crisis communication and emergency management.

Case Study 7.6: Japanese Jigoku—Hellish Experience

In the centre of Beppu City on the Japanese island of Kyushu ten small geothermal parks are located. Thousands of tourists visit these parks known as jigoku, which means 'hell' in Japanese, on a daily basis, with many tour groups arriving by bus. There are a variety of geothermal features, including a small turquoise crater lake fed by a permanent geyser (Umi Jigoku or Sea Hell), a steaming red pond known as Blood Hell (Chinoike Jigoku), the Priest Hell with bubbling mud ponds (Onishi Bozo Jigoku), the White Hell ponds (Shiraike Jigoku), and others, all of which are popular tourist sites. For the safety of visitors, the boiling lakes are fenced off and warning signs are located at all hazardous features (Fig. 7.4). Public announcements update visitors on important issues of the individual sites and friendly staff members are always around talking to visitors and guiding them from one geothermal feature to another (e.g. Kamado Jigoku). Safety is rarely an issue as Japanese regulations are very strict. Seismic activity and pressure in the underlying hydrothermal system are constantly monitored due to their location in the volcanic dome complex of Mt Tsurumi, which is classed as an active volcano. Also, hazard communication in Japan has over the past decade been extended to include several languages based on increasing number of foreign tourists. Other 'hellish' locations in Japan include Jigokudani near Noboribetsu on the island Hokkaido and the Unzen Jigoku near Shimabara (Kyushu).

Reaching people in an emergency to rescue them or guide them to a safe area depends on access to the affected site and transport options for evacuation. While a crisis is in progress, weather changes can hinder rescue efforts or make them impossible. For example, during the hydrothermal eruption of Mt Ontake (Japan 2014) rescue workers and Self Defence Force helicopters were carrying injured people to safety, but were battling adverse weather conditions as well as communication problems when trying to locate missing people. This disaster was made worse by not recording visitor numbers as should be done at all active volcanic and hydrothermal areas, although this is difficult logistically.

Further challenges to hazard communication, apart from the already mentioned lack of monitoring facilities at remote or underfunded destinations, include visitors blatantly ignoring warning signs and safety announcements. The co-operation of key stakeholders (e.g. scientists, authorities, tourist organisations) also remains difficult and can result in the lack of sufficient and effective emergency management strategies. Paired with procrastination this can delay the required decision making processes, and the timing of when to warn the public of an imminent danger. This is especially if the crisis situation is exacerbated by unfavourable factors including remoteness, large crowds, an ensuing panic or the threat of bad weather together with low visibility and hostile temperatures.

Fig. 7.4 Warning sign in two languages at a boiling pond at one of Beppu's Jigoku. *Photo* Patricia Erfurt-Cooper

3.2 How Are Hazards and Risks in Hydrothermal Areas Communicated to the Public?

Advising the public of an imminent crisis is generally the responsibility of a local authority, based on the information supplied from scientists monitoring an active area (McGuire et al. 2009), or casual observation. The capacity to communicate safety advice in case of a developing crisis situation depends on a range of factors. Above all the time frame is critical; the sudden onset of a hazardous situation can translate into life-threatening injuries.

Communication of hazards and risks can take place in several ways (Table 7.5). Prior to visiting, people interested in active hydrothermal or volcanic areas certainly have the opportunity to secure information about their chosen destination. While literature about many active environments is available and can be researched at libraries, the most effective way to access up-to-date information is to use reliable online sources for individual destinations (Case Study 7.7) and to check for current conditions and alert levels.

Case Study 7.7: New Zealand

The Waimangu Volcanic Valley is promoted on the internet as "amazing thermal features combined with lovely bush walks" with "hot springs, steaming lakes and colourful mineral deposits" (Waimangu Volcanic Valley 2015). Not promoted is advice on safety and instructions for emergencies. The "All you need to know" section offers "General information" about guided and self-guided tours and refers to "Guide sheets" in nearly a dozen languages. A brochure covering the area describes Waimangu Valley as "the world's youngest geothermal ecosystem, home to many geothermal features of worldwide importance" refers briefly to sustainable management practices including "safe access to the best viewing points". Information related to potential hazards is not always communicated on websites or brochures. However, after contacting the management at Waimangu it was clarified that safety is a very serious issue with strategies in place for all possible events. Great emphasis is placed on

Table 7.5 Availability of information about hydrothermal (and volcanic) hazards before, during and after visiting

Communication channels and their application	Before	During	After
Internet sites of destination (e.g. regular updates, alert levels, webcams)	x		x
Literature (earth science books and journals, guides books, research papers)	x		x
Brochures, fact sheets (park management, tourist offices)	x	x	x
Maps, hazard maps with safety instructions	x	x	
Visitor centres, interpretive centres, science museums	x	x	
Videos (visitor information, safety advice)	x		
Documentaries (related to natural crises)	x		x
Rangers, tour guides	x	x	
Social media (fb, twitter, trip advisor) real time updates	x	x	x
General media (TV, radio)	x		x
Public address systems where installed		x	
Smart phone apps (if any have been developed)	x	x	x
Subscription to text messages, real time updates		x	

trained staff and all visitors are briefed on arrival at Waimangu to stay safe. Some of the guidelines include staying on the paths at all times and following the directional signs. Written interpretations given to visitors contain further safety messages (pers. email communication with H. James CEO Waimangu). Based on further personal experience from previous visits of geothermal destinations in New Zealand staff at visitor centres and rangers/guides are available to advise on safety and potential dangers and are trained to respond to emergencies.

To communicate potential hazards, some National Parks show visitors introductory videos, hand out brochures, explain safety procedures and advise to strictly obey warning signs. The Hawaii Volcanoes National Park shows informational videos about safe conduct near active features prior to entering the Park to warn people in advance about the various hazards they can encounter. These measures cannot be enforced however, with some visitors to hazardous areas choosing to ignore them. Nevertheless, effective hazard communication is possible; an example is to be found at Mt Aso (Kyushu, Japan), where during visits to the summit public announcements in four different languages constantly update tourists about the level of toxic gas emissions from the crater and whether this poses a risk for visitors on the viewing platform and the surrounding walkways. Evacuation of the area is carried out immediately if the wind direction changes and the situation becomes hazardous for visiting tourists.

Another positive example is found in the Yellowstone National Park, USA, where the emphasis is firmly on public safety to avoid accidents and injuries from hydrothermal features, which have led to fatalities in the past (Case Study 7.8).

Case Study 7.8: Yellowstone—10,000 geothermal Features and 3 Million annual Visitors

Yellowstone National Park's chief safety officer says that they "try to educate people starting when they come through the gate" and that it is important for parents to keep a close eye on their children when visiting thermal areas. Prior to visiting the Yellowstone website informs people that "wild animals are not the only dangerous threat in Yellowstone", and that there have been a significant number of "deaths and injuries from geysers and geothermal water" over time. While the geothermal features are known to be hazardous, park management is concerned that visitors and even employees are not aware or ignorant of the potential risks when leaving designated walkways (Yellowstone 2014). According to Whittlesey (2014) there have been 19 confirmed human fatalities in Yellowstone's history as a national park from falling into thermal features including children, adults and even people working in the park. Safety managers at Yellowstone also think that incidents of injuries are higher than reported, because people cannot resist testing the water temperature by putting in their fingers or toes and suffer thermal burns. Warnings related to the dangers of geothermal hazards are clearly communicated on Yellowstone's website and are a good example of informing the public ahead of visiting. Throughout the geothermal areas there are warning signs and rangers are trying their best to keep unwary tourists from endangering themselves (Yellowstone 2014).

Case Study 7.9

The Waikato Regional Council (New Zealand) monitors geothermal sites and develops specific hazard maps for the Waikato region. For visitors to geothermal areas the Waikato Council offers some guidelines on their website about potential hazards and how to avoid them:

- Always check the temperature of the water before putting any part of your body in it. Take care not to fall in the water unless you are sure the temperature is safe.
- Keep your head above the water when bathing in geothermal pools. If you have severe flu-like symptoms within a week after visiting a hot pool, see a doctor immediately to rule out amoebic meningitis.
- Don't drink geothermal water in case it contains the toxic minerals arsenic and mercury.
- Be wary of eating trout caught in geothermal streams and lakes, as these fish may contain high levels of mercury.
- Keep a safe distance away from boiling mud pools, geysers and other areas which may suddenly erupt. Remember, a safe distance may be greater than you think, due to the unpredictable size and frequency of these geothermal features (Waikato Regional Council 2014).

Nevertheless, every country has their own methods and legislations how to deal with complex health and safety tasks emerging during crisis situations in relation to hydrothermal hazards. Based on the type of hazard, appropriate crisis management should include strategies for any level of emergency, including preparations for the evacuation of tourists and host communities if necessary. Hence, courtesy of the advanced media coverage we have come to rely on, hazard and crisis communication has far more opportunities to reach the public than even only one or two decades ago. And an abundance of information is available for those who are willing to do some research before embarking on a trip into a potentially hazardous area. However,

the above listed methods of hazard communication are not always taken advantage of and they may not all be available for a particular site or destination.

4 Hazard and Crisis Communication

4.1 Alerting the Public— Communicating Warnings

As mentioned above, the communication of hazards is legally at the discretion of the official management in charge of public safety. Depending on the hydrothermal activity level, appropriate signage and fencing are essential to warn and protect the public under normal conditions. During a crisis situation accurate and up-to-date information about an imminent danger is one of the key elements of effective communication. Difficulties in translating data from monitoring scientists into relevant facts followed by an appropriate course of action can however affect the successful management of an emerging crisis situation (Jolly and De La Cruz 2015; Gregg et al. 2015; McGuire et al. 2009).

In the build-up to a crisis warnings are disseminated through local media outlets (TV, radio, newspaper, website updates). Prior to and during a crisis emergency advice and directions can be communicated through rangers and tour guides, assisted by hazard maps and fact sheets. To communicate alert levels in real time colour coded warning lights and public announcement systems are suitable methods to reach tourists in large or remote areas with volcanic and hydrothermal activity. Finally, depending on the actual area, rescue workers and emergency personnel on site should be available to assist the public. Mobile phone message subscriptions and digital Apps are increasingly playing a part in reaching people prior and during emergencies, which shows that text messages carrying information about geo-hazards can be communicated to registered users in real time. To avoid misinterpretations Ghosh et al. (2012) note that "*a generalized system that could be deployed for any geo-hazard across any region*" should be developed. However, as many hydrothermal tourist destinations are located near active volcanoes, alert level systems as described in detail by several scientists (Fearnley et al. 2012; Gregg et al. 2015; Jolly and De La Cruz 2015; McNutt 2015; Williams-Jones and Rymer 2015) could possibly be modified and implemented at hydrothermal sites where required.

4.2 The Main Stakeholders and Their Responsibilities

As with volcanic environments planning for hydrothermal areas includes consultation and discussion between all stakeholders (Gregg et al. 2015). To communicate potential hazards to stakeholders (Connor et al. 2015), prior education through ranger talks, videos, brochures, and the internet can prepare visitors for the need of hazard and crisis communication in hydrothermal areas by raising their awareness. At the first signs of an emerging crisis situation, effective and reliable communication must be established between the stakeholders to develop a strong working relationship to cope with the crisis as it unfolds. While the most important stakeholder group in terms of tourism are the **visiting tourists**, three main stakeholder groups are identified by McGuire et al. (2009): the monitoring scientists, the emergency managers, and the media. However, local authorities and resident communities also constitute important stakeholder groups at hydrothermal tourist destinations. Heath et al. (2009) acknowledge that stakeholder partnerships, which include the public are the key to effective hazard and crisis communication.

Monitoring scientists are responsible for detecting early warning signs and assessing activity levels to provide information and guidance in an emergency situation. The role of emergency management committees is to determine an appropriate response strategy based on such scientific data, to develop hazard maps and risk management strategies and to take a pro-active role in educating the public about the nature of the emergency situation (Gregg et al. 2015; McGuire et al. 2009). To prevent

misunderstandings that may result in misinterpretation and delay in the decision making process in a crisis situation, it is critical that the communication between scientists and all other stakeholders is clear and unambiguous (Doyle et al. 2014). The media then should focus only on information, which is specific to the situation and present warning messages in a form that is clearly understood by everybody. To convey information effectively all irrelevant data should be avoided in media releases as it can confuse the public (Leonard et al. 2008; McGuire et al. 2009; Sorensen 2000). Warning messages should be consistent and designed to include those with poor literacy, language problems or disabilities effectively in the information stream. Pictures, drawings and video footage are useful ways to communicate with international visitors, and at the same time avoid confusion with terminology (Erfurt-Cooper and Cooper 2010, Erfurt-Cooper 2014; McGuire et al. 2009).

Scientists sometimes may be reluctant to communicate scientific information to other stakeholders based on their experience of misinterpretation or selective use of their data (Boykoff 2008). While there have been advances in the field of communicating geo-hazards, much needs to be done to improve the engagement between all stakeholders (monitoring scientists, politicians, government agencies, emergency managers, representatives of the media, and the public) to ensure seamless co-operation and effective communication (Liverman 2010; McGuire et al. 2009).

5 Hazard Management

5.1 Why Are People Reluctant to Respond to Warnings?

An emergency situation in areas of hydrothermal activity can be caused by a natural event and affect many people, but very little can be done to prevent it from happening. Disaster preparedness can offset some of the hazards in this situation, but not all. However, another type of emergency is caused by humans who are frequently injured as a result of being careless, and this situation can be avoided. In the Yellowstone National Park for example accidents and injuries are nearly always due to visitors being irresponsible in thermal areas with a number of reasons for injuries identified by Whittlesey (2014):

- Walking in off-limit areas
- Walking in darkness
- Losing balance
- Being intoxicated
- Being distracted
- Over-confidence
- Ignoring warnings
- Careless running

Whittlesey (2014) rightly points out that *"a balance is needed between adequate warnings and basic responsibility from the visitor"*. Individuals, who choose not to respond to hazard warnings or worse, in a crisis situation may not do so because (a) they are unfamiliar with the hazard, or (b) they think they can avoid the hazard altogether, while (c) others believe it could be a false alarm, and (d) if not, somebody will come to their rescue.

One of the key factors influencing the decisions of individual people in these situations is their personal risk perception, which can range from being overcautious to ignoring a potential risk, or to the complete denial of any danger. In addition, apart from a possible uncertainty about the risk, a negative attitude towards authorities could be another reason for ignoring safety advice. Furthermore, tourists frequently overestimate their personal ability to cope with dangerous situations while at the same time underestimating the actual risk and their own vulnerability. Here it would be advisable to provide all visitors of active hydrothermal environments with detailed safety guidelines, which they should refer to before and during their visit.

Virtual reality is in this day and age a valuable tool for travel planning and the internet offers many sources to assist detailed research of planned destinations including webcams, videos and computer simulations. While cyber visits to extreme landscapes and hazardous areas may lack the actual risk, they can help visitors and host communities to understand the potential difficulties as well as the risks that can be encountered in the real world (Erfurt-Cooper and Cooper 2009). However, despite the advent of real-time internet resources, the task of hazard and risk management in volcanic and hydrothermal environments remains extremely challenging, as varying degrees of potential danger from hydrothermal activity as well as correlated volcanic and seismic events generate different types of hazards (Erfurt-Cooper and Cooper 2009).

5.2 Why Are Authorities Reluctant to Announce Evacuations?

Not being aware of the potential hazards in an active area prior to visiting could mean the difference between safety and injury. So, if this is true, and if maintaining a high level of communication means that public warnings are the responsibility of the authorities (McGuire et al. 2009), why are some authorities reluctant to announce evacuations? As suggested by Francis and Oppenheimer (2004), all details of risk management strategies need to be planned and in place, including successful evacuation, transport and medical care. However, this may not be common knowledge, as for the tourism industry disasters are generally bad for business. Thus, if warnings are given too early, they might be ignored by the public, or in the case of repeated warnings without anything happening they may also be ignored (the cry wolf syndrome). Nevertheless, to avoid endangering the public, good management strategies are required to assist with hazard communication, crisis planning and to prevent emergency situations getting out of hand.

When it comes to the drastic measure of calling an evacuation, authorities may be reluctant to do so for several reasons. Uncertainty about the actual risk combined with a lack of scientific knowledge and experience on the part of some stakeholders can make the decision making process difficult. In the case of insufficiently trained staff and/or in the absence of monitoring facilities the reliable assessment of imminent danger can be impossible. Emergency managers and response teams can also have their tasks complicated by unfavourable logistics based on remote terrain, weather, lack of time and/or inadequate strategies. When local authorities are incapable of dealing with emergencies, external rescue response teams may have to be called in to assist, which again may delay the decision to evacuate. Occasionally political disagreements such as cross border problems can come into play, presenting another reason for making hazard communication and crisis management including successful evacuation unworkable.

To generate effective and accurate communication before, during and after a crisis therefore remains a challenge, because it depends on numerous variables. More to the point, hazard communication is mainly focussed on volcanic environments with hydrothermal areas in need of higher levels of targeted research to improve hazard communication and crisis management. If reliable information is not available or is only partial, it becomes obvious how delays in effective crisis response can occur.

6 Conclusion

Health and safety issues play a major role in the tourism industry based on hydrothermal and volcanic resources (Erfurt-Cooper 2008, 2010, 2014). While it is essential to prevent injuries to tourists the management of effectively communicating hydrothermal hazards remains problematic. Precursory signs indicating a dangerous natural event are still not completely understood and the communication of imminent danger is frequently unachievable. As a consequence, serious thought needs to be given to improvements in the communication of risk factors and potential dangers from the proximity of active hydrothermal surface features, including extreme hot springs and geysers, as well as areas affected by hydrothermal alteration resulting in unstable ground properties.

To reduce the risk factor, it is essential to raise visitor awareness about any potential hazards in hydrothermal tourist areas and put management strategies for emergencies in place. Advice for visitors of active hydrothermal environments should include guidance and recommendations how to cope with extreme events in difficult situations (Erfurt-Cooper 2008, 2010, 2014). To improve the safety standards in these areas, strategic guidelines for safe conduct should cover all active hydrothermal areas and must be designed to be understood by every visitor. In seeking to achieve this ideal state, one of the questions arising is whether visitors to these areas seek enough information from available sources prior to their journey (Appendices 1 and 2). However, the overall hazard, risk and crisis communication process for active hydrothermal environments is a problem, which frequently is insufficiently addressed at tourist destinations. While some areas have staken steps to educate visitors about potential dangers the moment they arrive (Yellowstone 2014), other areas do not offer adequate safety information. Here virtual reality can be a valuable tool for travel planning with the internet offering numerous comparable sources (e.g. webcams, videos and computer simulations) to research planned destinations.

This chapter has presented an overview of the most common hazards or risk factors in hydrothermal areas with brief case studies to exemplify different scenarios. To highlight the challenges of risk management and the need for effective communication a number of locations were discussed for their site specific hazards and their potential risks. The literature reviewed for this chapter indicates a scarcity of research for hazard communication in hydrothermal areas and hence information related to this topic remains limited. In conclusion, it is recommended that more studies are undertaken to contribute to the safety of visitors in these environments and that more destination websites include safety advice relating to their hydrothermal tourist attractions.

Appendix 1

Example of safety advice—extract from USGS factsheet (USGS[3] 2007).

U.S. Geological Survey and the U.S. Forest Service—Our Volcanic Public Lands.

Boiling Water at Hot Creek—The Dangerous and Dynamic Thermal Springs in California's Long Valley Caldera.

This Fact Sheet and any updates to it are available online at http://pubs.usgs.gov/fs/2007/3045.

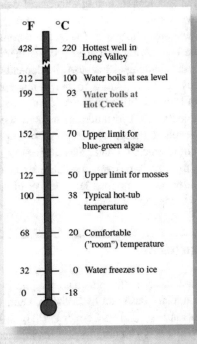

VISITING HOT CREEK SAFELY

Visiting Hot Creek can be an enjoyable and rewarding experience, but you should be aware of the dangers and take them seriously. Know the hazards—boiling or scalding water, steam vents, unstable ground and boulders, hot ground or mud, swiftly flowing water, a stream with unpredictable currents, and water unfit for human consumption.

°F	°C	
428	220	Hottest well in Long Valley
212	100	Water boils at sea level
199	93	Water boils at Hot Creek
152	70	Upper limit for blue-green algae
122	50	Upper limit for mosses
100	38	Typical hot-tub temperature
68	20	Comfortable ("room") temperature
32	0	Water freezes to ice
0	-18	

Follow these tips for your safety:

- *Keep a clear head and be observant— conditions can change quickly.*

- *Keep careful watch on children and pets (always keep pets on a leash).*

- *Follow directions and warnings on signs.*

- *Don't cross over fences or barriers, and stay on walkways.*

How to get there:

The Hot Creek Geologic Site is located northeast of the Mammoth-Yosemite Airport and is accessed from U.S. Highway 395 by the Hot Creek Hatchery Road.

Appendix 2

HAZARDOUS EVENTS AT YELLOWSTONE

Scientists evaluate natural-hazard levels by combining their knowledge of the frequency and the severity of hazardous events. In the Yellowstone region, damaging hydrothermal explosions and earthquakes can occur several times a century. Lava flows and small volcanic eruptions occur only rarely—none in the past 70,000 years. Massive caldera-forming eruptions, though the most potentially devastating of Yellowstone's hazards, are extremely rare—only three have occurred in the past several million years. U.S. Geological Survey, University of Utah, and National Park Service scientists with the Yellowstone Volcano Observatory (YVO) see no evidence that another such cataclysmic eruption will occur at Yellowstone in the foreseeable future. Recurrence intervals of these events are neither regular nor predictable.

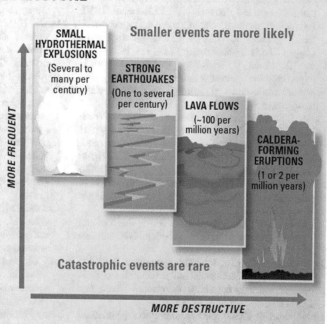

How Dangerous Is Yellowstone?

None of the events described above—cataclysmic caldera-forming eruptions, lava flows, large earthquakes, or major hydrothermal explosions—are common in Yellowstone. Although visitors to Yellowstone National Park may never experience them, some hazardous events are certain to occur in the future. Fortunately, systematic monitoring of Yellowstone's active volcanic and hydrothermal systems, including monitoring of earthquakes and ground deformation, is now carried out routinely by YVO scientists.

This monitoring will allow YVO to alert the public well in advance of any future volcanic changes in the patterns of ongoing seismicity or other indicators of possible geologic unrest are quickly reported to officials responsible for public safety in the National Park Service and other agencies.

Through continuous monitoring and research, YVO is greatly improving understanding of Yellowstone's volcanic, earthquake, and hydrothermal hazards. The work of USGS scientists with YVO is only part of the USGS Volcano Hazards Program's ongoing efforts to protect people's lives and property in all of the volcanic regions of the United States, including California, Hawaii, Alaska, and the Pacific Northwest (Lowenstern et al. 2005).

References

Aspinall W, Blong R (2015) Volcanic risk assessment. In: Sigurdsson H (ed) Encyclopedia of volcanoes, 2nd edn. Elsevier-Academic Press, Amsterdam, pp 1215–1231

Barclay J, Haynes K, Houghton B, Johnston D (2015) Social processes and volcanic risk reduction. In: Sigurdsson H (ed) Encyclopedia of volcanoes, 2nd edn. Elsevier-Academic Press, Amsterdam, pp 1203–1214

Barnett ND, Kaplan AM, Hopkin RJ, Saubolle MA, Rudinsky MF (1996) Primary amoebic meningoencephalitis with Naegleria fowleri: clinical review. Ped Neuro 15:230–234

BBC News (2010) Boy dies after New Zealand geothermal pool fall. 30 December 2010. www.bbc.co.uk/news/world-asia-pacific-12091845. Accessed 10 Aug 2016

Boykoff M (2008) Media and scientific communication: a case of climate change. In: Liverman DGE, Pereira CP, Marker B (eds) Communicating environmental geoscience, vol 305. Geological Society of London Special Publication, pp 11–18

Bratton A, Smith B, McKinley J, Lilley K (2013) Expanding the geoconservation toolbox: integrated hazard management at dynamic geoheritage sites. Geoheritage 5:173–183

Connor C, Bebbington M, Marzocchi W (2015) Probabilistic volcanic hazard assessment. In: Sigurdsson H et al (ed) Encyclopedia of volcanoes, 2nd edn. Elsevier-Academic Press, Amsterdam, pp 897–910

De La Cruz-Reyna S, Meli RP, Quaas RW (2000) Volcanic crises management. In: Sigurdsson H et al (eds) Encyclopedia of volcanoes. Elsevier-Academic Press, Amsterdam

Dengler L (2005) The role of education in the National Tsunami Hazard Mitigation Program. Nat Hazards 35:141–153

Doyle EEH, McClure J, Johnston DM, Paton D (2014) Communicating likelihoods and probabilities in forecasts of volcanic eruptions. J Volcanol Geoth Res 272:1–15

Erfurt-Cooper P (2008) Geotourism: active geothermal and volcanic environments as tourist destinations, presented to The Inaugural Global Geotourism Conference. Perth, Australia, 17–20 August

Erfurt-Cooper P (2010) Introduction to volcano and geothermal tourism. In: Erfurt-Cooper P, Cooper M (eds) Volcano and geothermal tourism: sustainable geo-resources for leisure and recreation. Earthscan, London

Erfurt-Cooper P (2012) An assessment of the role of natural hot and mineral springs in health, wellness and recreational tourism. Dissertation, James Cook University

Erfurt-Cooper P (ed) (2014) Volcanic tourist destinations. Geoheritage, Geoparks and Geotourism Series. Springer, Berlin

Erfurt-Cooper P, Cooper M (2009) Health and wellness tourism: spas and hot springs. Channel View Publications, Bristol, UK

Erfurt-Cooper P, Cooper M (2010) Volcano and geothermal tourism: sustainable geo-resources for leisure and recreation. Earthscan, London

Fearnley CJ, McGuire WJ, Davies G, Twigg J (2012) Standardisation of the USGS volcano alert level system (VALS): analysis and ramifications. Bull Volcanol 74:2023–2036

Francis P, Oppenheimer C (2004) Volcanoes, 2nd edn. Oxford University Press, Oxford

Gregg CE, Houghton B, Ewert JW (2015) Volcano warning systems. In: Sigurdsson H (ed) Encyclopedia of volcanoes, 2nd edn. Elsevier-Academic Press, pp 1173–1185

Ghosh JK, Bhattacharya D, Samadhiya NK, Boccardo P (2012) A generalized geo-hazard warning system. Nat Hazards 64:1273–1289

Haase KM et al. (2009) Fluid compositions and mineralogy of precipitates from Mid Atlantic Ridge hydrothermal vents at 4°48'S. Pangaea

Hansell AL, Horwell CJ, Oppenheimer C (2006) The health hazards of volcanoes and geothermal areas. Occup Environ Med 63:149–156

Heasler HP, Jaworowski C, Foley D (2009) Geothermal systems and monitoring hydrothermal features. In: Young R, Norby L (eds) Geological monitoring: Boulder, Colorado. Geological Society of America, pp 105–140

Heath RL, Jaesub L, Lan N (2009) Crisis and risk approaches to emergency management planning and communication: the role of similarity and sensitivity. J Pub Relat Res 2(2):123–141

Heggie TW, Heggie TM, Heggie TJ (2010) Death by volcanic laze. In: Erfurt-Cooper P, Cooper M (eds) Volcano and geothermal tourism: sustainable geo-resources for leisure and recreation. Earthscan, London

Hochstein MP, Browne PRL (2000) Surface manifestations of geothermal systems with volcanic heat sources. In Sigurdsson H et al (eds) Encyclopedia of volcanoes. Elsevier-Academic Press, Amsterdam

Jolly G, De La Cruz S (2015) Volcanic crisis management. In: Sigurdsson H et al (ed) Encyclopedia of volcanoes, 2nd edn. Elsevier-Academic Press, Amsterdam, pp 1187–1202

Keary P (1996) The new Penguin dictionary of geology. Penguin Books, London

Leonard GS, Johnston DM, Paton D, Christianson A, Becker J, Keys H (2008) Developing effective warning systems: ongoing research at Ruapehu volcano, New Zealand. J Volcanol Geoth Res 172:199–215

Liverman D (2010) Communicating geological hazards: educating, training and assisting geoscientists in communication skills. In: Beer T (ed) Geophysical hazards—minimizing risk, maximizing awareness. International Year of Planet Earth. Springer, The Netherlands, pp 41–55

Lowenstern JB, Christiansen RL, Smith RB, Morgan LA, Heasler H (2005) Steam explosions, earthquakes, and volcanic eruptions—what's in Yellowstone's future? U.S. Geological Survey. http://pubs.usgs.gov/fs/2005/3024/. Accessed 10 Aug 2016

McGuire WJ, Solana MC, Kilburn CRJ, Sanderson D (2009) Improving communication during volcanic crises on small, vulnerable islands. J Volcanol Geoth Res. doi:10.1016/j.jvolgeores.2009.02.019

McNutt S (2015) Eruption response and mitigation. In: Sigurdsson H et al (ed) Encyclopedia of volcanoes, 2nd edn. Elsevier-Academic Press, Amsterdam, pp 1069–1070

NZ Herald (2013) Hot pool death: coroner satisfied with council's actions. www.nzherald.co.nz/nz/news/article.cfm?c_id=1&objectid=10862848. Accessed 10 Aug 2016

Paton D, Millar M, Johnston DM (2001) Community resilience to volcanic hazard consequences. Nat Hazards 24:157–169

Roscoe R (2010) Geothermal parks in New Zealand. In: Erfurt-Cooper P, Cooper M (eds) Volcano and geothermal tourism: sustainable geo-resources for leisure and recreation. Earthscan, London

Sorensen JH (2000) Hazard warning systems: review of 20 years of progress. Nat Hazards Rev 1(2):119–125

Steelman TA, McCaffrey S (2013) Best practices in risk and crisis communication: implications for natural hazards management. Nat Hazards 65:683–705

Think Geoenergy (2015) Geothermal plants among top tourist attractions in Iceland. http://www.think-geoenergy.com/geothermal-plants-among-top-tourist-attractions-in-iceland/. Accessed 10 August 2016

TVNZ (2012) Inquest into boy's death in Rotorua mud pool. 20 Dec 2012. http://tvnz.co.nz/national-news/inquest-into-boy-s-death-in-rotorua-mud-pool-5299381. Accessed 20 Jan 2015

UNISDR (2014) Terminology on DRR. The United Nations Office for Disaster Risk Reduction. www.unisdr.org/we/inform/terminology. Accessed 10 Aug 2016

UNISIDR (2016) Hyogo framework for action (HFA)—building the resilience of nations and communities to disasters. The United Nations Office for Disaster Risk Reduction. www.unisdr.org/we/coordinate/hfa. Accessed 6 Aug 2016

USGS[3] (2007) Boiling water at hot creek—the dangerous and dynamic thermal springs in California's Long Valley Caldera. http://pubs.usgs.gov/fs/2007/3045/. Accessed 10 Aug 2016

USGS[1] (2014) Volcanic gases and their effects. Volcanic gases can be harmful to health, vegetation and infrastructure. Volcano Hazards Program. http://volcanoes.usgs.gov/hazards/gas/. Accessed 10 Aug 2016

USGS[2] (2014) Hydrothermal explosions. Yellowstone volcano observatory. http://volcanoes.usgs.gov/volcanoes/yellowstone/yellowstone_hazard_43.html. Accessed 10 Aug 2016

Waikato Regional Council (2014) Geothermal activity – Regional hazards and emergency management. www.waikatoregion.govt.nz/Services/Regional-services/Regional-hazards-and-emergency-management/Geothermal-activity/. Accessed 10 August 2016

Waimangu Volcanic Valley (2015) Rotorua, New Zealand—amazing geothermal activity. www.waimangu.co.nz. Accessed 10 Aug 2016

Whittlesey LH (2014) Death in Yellowstone: accidents and foolhardiness in the First National Park, 2nd edn. Amazon Digital Services, Inc., Roberts Rinehard

Williams D, Olaniran BA (1998) Expanding the crisis planning function: introducing elements of risk communication to crisis communication practice. Public Relat Rev 24(3):387–400

Williams-Jones G, Rymer H (2000) Hazards of volcanic gases. In: Sigurdsson H et al (ed) Encyclopedia of volcanoes. Elsevier-Academic Press, Amsterdam

Williams-Jones G, Rymer H (2015) Hazards of volcanic gases. In: Sigurdsson H (ed) Encyclopedia of volcanoes, 2nd edn. Elsevier-Academic Press, Amsterdam, pp 985–992

Yellowstone (2014) Geothermal attractions can be dangerous. Deaths and injuries at Yellowstone's geysers and hot springs. www.yellowstonepark.com/2007/01/cautionary-tale/. Accessed 10 Aug 2016

Mapping Hazard Zones, Rapid Warning Communication and Understanding Communities: Primary Ways to Mitigate Pyroclastic Flow Hazard

Franck Lavigne, Julie Morin, Estuning Tyas Wulan Mei, Eliza S. Calder, Muhi Usamah and Ute Nugroho

Abstract

Protection against the consequences of Pyroclastic Density Currents (PDCs) is almost impossible due to their high velocity, temperature, sediment load and mobility. PDCs therefore present a challenge for volcanic crisis management in that specific precautionary actions, essentially evacuations, are required to reduce loss of life. In terms of crisis communication for PDC hazards, there are three challenging questions that arise in terms of reducing risk to life, infrastructure and livelihoods. (1) *How do we accurately communicate the hazardous zones related to potential PDC inundation?* The areas exposed to PDC hazard are difficult to assess and to map. In terms of risk/crisis management, the areas considered at risk are usually those that were affected by PDCs during previous eruptive episodes (decades or centuries ago). In case of "larger-than-normal" eruptions, the underestimation of the hazard zone may lead to refusals to evacuate in the "newly" threatened area. Another difficulty in assessing the PDC hazard zones relate to their transport processes that allow surmounting of the topography and in some cases across the surface of water. Therefore warning systems must be able to cover vast areas in a minimum of time. (2) *How do we efficiently warn people in time?* PDCs are extremely mobile and fast. It is therefore necessary to raise the alert early enough before the onset of the first PDCs. A challenging question in terms of crisis communication is related to the

F. Lavigne (✉)
Laboratoire de Géographie Physique,
Université Paris 1 Panthéon-Sorbonne, Meudon,
France
e-mail: franck.lavigne@univ-paris1.fr

J. Morin
Laboratoire Magmas et Volcans,
Université Clermont Auvergne, CNRS, IRD, OPGC,
Clermont-Ferrand F-63000, France

E.T.W. Mei
Fakultas Geografi, Universitas Gadjah Mada,
Yogyakarta, Indonesia

E.S. Calder
School of GeoSciences, University of Edinburgh,
James Hutton Rd, The Kings Buildings, Edinburgh,
UK

M. Usamah
PYDLOS, Universidad de Cuenca, Cuenca, Ecuador

U. Nugroho
Universitas Padjadjaran, Bandung, Indonesia

Advs in Volcanology (2018) 107–119
https://doi.org/10.1007/11157_2016_34
© The Author(s) 2017
Published Online: 26 May 2017

type of tools used by the local authorities, modern and traditional tools both of which have advantages and disadvantages. (3) *Why are people reluctant to evacuate?* Local inhabitants can be reluctant to evacuate during a crisis if traditional warning signs or signals they are familiar with are lacking, if they don't receive both traditional and official warning, and because they may lose their livelihoods. Thus a deeper understanding of the at-risk communities and efficient dissemination of information are key issues in order to reduce vulnerability in PDC hazard regions.

Keywords

Pyroclastic density currents · Risk communication · Crisis management · Warning · Evacuation · Risk perception

1 Introduction

Pyroclastic Density Currents (PDCs) are rapid flowage phenomena that involve various proportions of volcanic gas and fragmented volcanic rock at high temperatures. PDCs encompass dense pyroclastic flows, which tend to be more topographically controlled, and dilute pyroclastic surges that are less topographically controlled and can surmount topographic obstacles, or travel across the surface of bodies of water. Both dense flows and dilute surges destroy almost everything in their path and therefore protection against the consequences of PDC inundation is almost impossible. In some countries, anti-PDC bunkers have been built in high hazard prone areas to provide a safe shelter to a limited number of people in the situation that they are unable to evacuate on time. It was demonstrated in 2006 on Merapi (Indonesia) that they are not always effective, as two people died trapped in the bunkers where they took refuge (Gertisser et al. 2011). Moreover, hard engineering structures such as SABO dams may actually accentuate the avulsion process of PDCs, e.g. in 2006 and 2010 at Merapi (Lube et al. 2011), or Tungurahua in 2006 (Stone et al. 2014). Thus, PDCs present a challenge for volcanic crisis management in that specific precautionary actions are required to reduce loss of life.

The improvement of crisis management capabilities is based, on one hand, on PDC monitoring and early warning systems as well as robust communications that are not likely to be compromised for example by power failure and, on the other hand, on preparedness of stakeholders and population (MIAVITA Team 2012).

This chapter discusses three challenging questions in reducing the risk associated with PDCs: (1) How should we accurately communicate the hazardous zone related to potential PDC inundation? (2) How should we efficiently warn people in time? (3) Why are people oftentimes reluctant to evacuate and how should we improve the propensity for people to accept and undertake evacuations?

These points are addressed through examples, mostly focusing on Merapi, and other Indonesian volcanoes as well as other volcanoes around the world. This chapter concludes with a discussion on ways to improve volcanic risk management in areas prone to PDC hazards.

2 How Can We Communicate PDC Hazard Zones?

The areas exposed to PDC hazard are difficult to accurately assess and to map. In terms of risk/crisis management, the areas considered at risk are usually those that were affected by PDCs during the last decades or centuries. Scientists in charge of volcano monitoring often use a "reference eruption", the extent of volcanic deposits of which are used to gauge inundation extent for future eruption scenarios in operational hazard maps. For instance, the "danger zones" maps used in Indonesia for emergency planning are provided by the Center of Volcanology and Geological Hazards Mitigation (CVGHM). These maps typically display two zones threatened by PDC hazard: the KRB III (KRB stands for *Kawasan Rawan Bencana* in Indonesian or Hazard Prone Area in English) encompasses areas located close to the summit, frequently affected by dome-collapsed pyroclastic flows, lava flows, rock falls and ejected rock fragments. The KRB II is affected by less frequent and longer runout pyroclastic flows, lahars, volcanic ash fall, and ejected rocks. At Merapi for example, the boundaries of hazard zone III and II were based, until 2010, on the distribution of volcanic products of the largest eruptions of the 20th century. Therefore the maximum distance of the KRB II did not exceed 15 km, which was the approximate maximum extension of the 1930,

1961, and 1969 PDCs. Since its first edition in 1978 (Pardyanto et al. 1978, Fig. 8.1), the volcanic hazards map has been widely disseminated among the communities at risk through the local authorities and non-governmental organizations (NGOs). Although this map was updated following the 2006 eruption (Mei and Lavigne 2012), the contingency plan created in 2009 still did not consider a plinian or subplinian eruption scenario such as the one that occurred in 1872. Several areas affected by the subsequent and devastating 2010 PDCs, the length of which were substantially longer than expected (17 km from the summit), had not been included in the danger zone. As a result 53 people who were resisting evacuation or who were late in the process of evacuating were killed in Bronggang, a village located 13.5 km to the south of Merapi, when dilute surges detached from their parent flows in the adjacent Gendol River and entered the village (Jenkins et al. 2013). Among the survivors, several inhabitants who were unprepared for evacuation took a wrong evacuation route too close to the river (Mei et al. 2013). Since the 2010 eruption, the hazard map has been revised (Fig. 8.1).

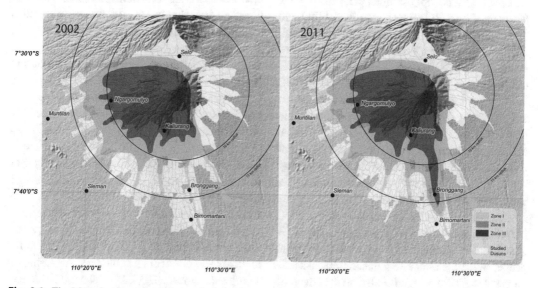

Fig. 8.1 The Merapi volcano hazard map designed by the Indonesian Center of Volcanology and Geological Hazard Mitigation (CVGHM) in 2002 (CVGHM 2002) and after its revision in 2011 (CVGHM 2011)

Although the local authorities are often aware of the worst-case scenario provided by the volcanologists, they cannot use it for contingency planning, i.e. for risk management. Using the worst case scenario for a background hazard map is actually impractical since many existing communities are established on deposits from large eruptions, and an eruption is unlikely to reach worst case without some precursory activity. However, all possibilities should be discussed between volcanologists and authorities well before a crisis, so that outline contingency planning can be made if significant escalation does occur. Communities can live within hazard zones, if they are aware of the threat and there is good planning for evacuations in the event of an escalation. Maps are commonly adjusted as a crisis evolves, as shown in the cases of Merapi. During the 2012 Tongariro eruption crisis in New Zealand, Leonard et al. (2014) highlighted the importance and complementary roles of three map types for communicating volcanic hazard information: background hazard, crisis hazard and ashfall prediction maps. In developing hazard maps there are a range of key points to consider in terms of message, presentation and basis for each map type. For example, perspective view has been shown to increase map readability and public self-location accuracy (Haynes et al. 2007b; Nave et al. 2010). Following Leonard et al. (2014), PDC hazards need careful quantification through modelling. For rapid crisis PDC hazard map zone development the critical factors are (1) having access to and experience in running flow models, (2) having those models tested against the past and expected future parameters of a volcano and (3) having access to the computing resources needed to run enough scenarios in a short (day to days) timeframe.

The main difficulty in communicating PDC hazardous zones occurs when the hazard is almost unknown, or has been forgotten by local people over time after a few generations, such as on the slopes of the Mount Pelée on the island of Martinique before the 1902 eruption (Leone and

Lesales 2009). One day before the total destruction of Saint-Pierre by pyroclastic surges, a scientific expert from mainland France claimed that "Saint-Pierre is not more threatened by the Pelée volcano than Napoli by the Vesuvius volcano". This scientist was not aware that dilute pyroclastic surges could occur at Mount Pelée (such as during the plinian eruption in 1300 AD), since these phenomena were not yet known (Lacroix 1904). Examples as this one are numerous in volcanic areas. For instance, most of the villages of the northern and southeastern coast of Lombok Island (Indonesia) have been built on pumice PDC deposits emplaced during the 1257 AD ultraplinian eruption of the Samalas volcano (Lavigne et al. 2013).

3 How Do We Warn People in Time?

3.1 Difficulties in Providing Timely Warnings

PDCs are extremely mobile, generally travelling at tens to hundreds of km/h. Therefore alerts need to be provided at least several hours before the first PDC occurs. Unfortunately, local populations are not always warned by the authorities before an imminent eruption, for various reasons. In case of gravitational collapses of silicic lava domes, which trigger 'Merapi-type' pyroclastic flows, warning people is not possible until the occurrence of the collapse itself: reliable precursory signals have not yet been identified, as observed on Merapi on 22 November 1994 (Abdurachman et al. 2000), although in some cases, the characteristics of seismic activity can change leading up to a collapse e.g. at Soufriere Hills (Cole et al. 1998). In some cases, the absence of warning may be related to traditional scepticism in technological predictions, when local officials refuse to listen to the scientific forecasts and predictions at the very beginning of a volcanic crisis (IAVCEI Subcommittee for Crisis Protocols 1999). Additional external drivers may prevent an alert by the local authorities (the mayor in many cases), e.g. local elections, as observed at the beginning of the 1902 eruption of Mount Pelée in Martinique (Lacroix 1904).

3.2 Modern Versus Traditional Warning Tools

The type of tools used to communicate warnings by local authorities in times of crisis is critical to ensuring an effective messaging. Modern tools like sirens are increasingly used as "official"

warning systems on active volcanoes. In developing countries, however, the areal distribution of sirens is not homogenous, e.g. around Merapi in Indonesia (Fig. 8.2). Based on a survey carried out among 1969 people in shelters during the 2010 eruption of Merapi, only 16% of the people were warned by sirens before the PDCs totally destroyed the slopes of the volcano (Fig. 8.3, Mei et al. 2013), whereas most people received evacuation alerts directly from the head of village (54%), or from neighbors (11%).

The warning signal may be also transmitted by a mobile system installed on the fire department's vehicles (e.g. in France). In Japan, the

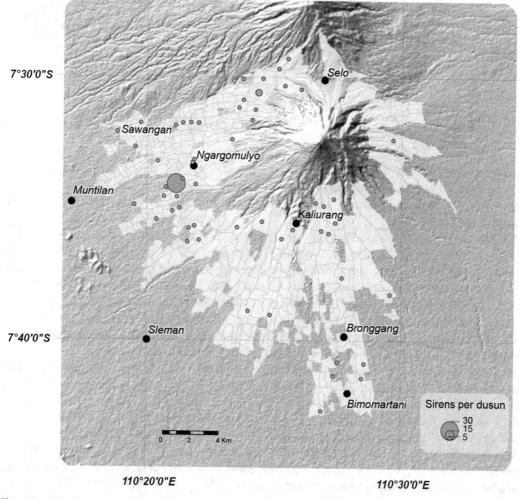

Fig. 8.2 Siren distribution around the flanks of Merapi, Indonesia. *Source* Lavigne et al. (2015), based on Mei et al. (2013)

Fig. 8.3 The source of
warnings during the 2010
volcanic crisis of Merapi,
Indonesia (adapted from
Mei and Lavigne 2013)

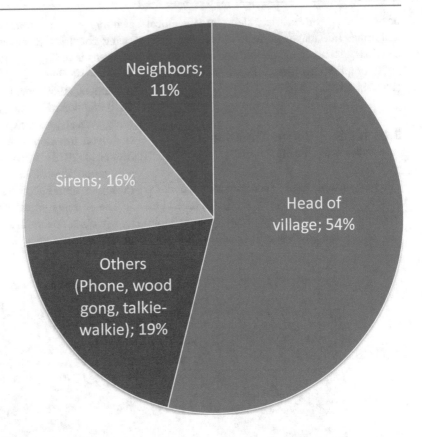

J-Alert system, launched in 2007, aims to allow government officials to address the population directly via loudspeakers, e.g. in case of an eruption alert.

Other modern tools are widely used to warn people of imminent PDCs. Should a volcanic event happen, local people considered at risk may receive a warning message in the form of SMS Text Message and/or Email onto their cell phones, smart phones or other electronic devices like iPads, Laptops, Desktop computers, etc. Usually, this type of warning message would be distributed by Civil Defense Corps and includes all the hazards that could trigger an emergency situation, not only volcanic hazards, e.g. the app provided in Auckland (www. aucklandcivildefence.org.nz/Alerting/Get-the-Applications) or in Hawaii (http://www. hawaiicounty.gov/active-alerts). At Tongariro Volcano in New Zealand, for example, an alert is

provided by the key scientific institution to related agencies (e.g. Civil Defense and Emergency Management) and the media through online bulletins as well as direct communication through its emergency network. This bulletin is also accessible by the public via their network website and social media (Leonard et al. 2014). In Japan, real-time volcanic warning is available to the public on the website (JMA 2015). Beyond volcanic hazards, Japanese agencies send out SMS alerts to all registered mobile phones in the country (Pearson 2015).

Although modern tools are growing, traditional tools are still considered as efficient warning tools by local authorities, especially in remote areas. For example, the Indonesian *kentongan* (bamboo drum: Fig. 8.4a) is traditionally used for warning the public, notably in rural areas or during an emergency period during which electricity might be cut off, meaning that

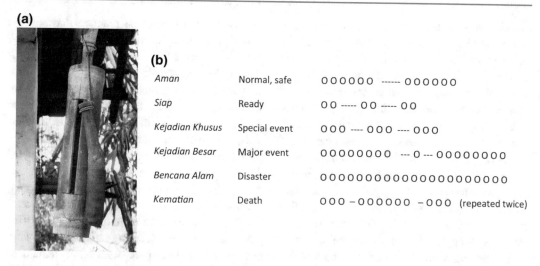

(b)		
Aman	Normal, safe	O O O O O O ------ O O O O O O
Siap	Ready	O O ----- O O ----- O O
Kejadian Khusus	Special event	O O O ---- O O O ---- O O O
Kejadian Besar	Major event	O O O O O O O O --- O --- O O O O O O O O
Bencana Alam	Disaster	O O O O O O O O O O O O O O O O O O O O O O
Kematian	Death	O O O – O O O O O O – O O O (repeated twice)

Fig. 8.4 The use of bamboo drums (kentongan) at Merapi, Indonesia. **a** kentongan at the entrance of a house on the West slope of Merapi (*Photo* F. Lavigne, 2010). **b** Kentongan communication codes. *Source* Lavigne et al. (2015)

modern alert tools relying on electricity power might be dysfunctional. Based on a field survey at Merapi in 2002, over 70% of interviewed villagers thought that the *kentongan* was an efficient warning system (Lavigne et al. 2008). Every kentongan code has its own meaning (Fig. 8.4b). In case of volcanic disaster, the kentongan is beaten repeatedly and continuously with the same tone. It indicates that people should immediately evacuate to a pre-determined location, which is usually a village hall. However, many people among the young generation are not able to interpret the signals anymore. Therefore, the use of this tool is forbidden during Merapi's volcanic crisis in some municipalities or villages, e.g. in Sawangan and Selo on the north slope of the volcano (Mei et al. 2013).

3.3 Official Warning Versus Community-Based Warning

Local communities are still using natural warning signs of various types, as exemplified on the slopes of Merapi: increase in rock fall noise, increase in fumarolic activity from the summit's crater, the descent of monkeys or other wild animals from the hills, ground shaking relating to increased seismic activity, or lightning storms

caused by the emission of ash into the atmosphere. Local inhabitants can be reluctant to evacuate during a crisis if signals they are familiar with are lacking, and if they don't receive both traditional and official warning of a possible eruption (Donovan 2010). Furthermore, some culturally accepted warning signs can create a false sense of security, and it can be a struggle for some to believe those based on scientific monitoring alone. Such problems related to traditional cultural beliefs were reported not only in developing countries, but also in the USA at Mt. Kilauea in Hawaii (Gregg et al. 2004), Mount St. Helens (Greene et al. 1981), and in Italy at Mt. Etna and Mt. Vesuvius (Chester et al. 2008).

The credibility of a given warning and the validity of past warnings and evacuations, both influence the decision to evacuate. Social, economic and political forces may distort risk messages, leading to public reliance upon informal information networks (Haynes et al. 2008), e.g. social networks. Therefore, local organizations play a key role in crisis communication, as exemplified at Merapi by the actions of JalinMerapi (*Jaringan Informasi Lingkar Merapi*, in English *Merapi Circle Information Network*), a local organization supported by several NGOs working around the volcano. This association was established in 2006 by three community-based radio stations. During the emergency response period in 2010, JalinMerapi used various electronic media to quickly and accurately convey important information and data to support the decision making process. JalinMerapi information was transmitted through a website, social networks such as Twitter and Facebook, SMS, radio communications, telephone and through information posters in the field. JalinMerapi was managed by a voluntary network that operated 24 h a day during the 2010 eruption (Mei et al. 2013). Thus, repetition of warnings through different sources of the evacuation command-line increased the chances that people heeded the warning.

Community-based volcano risk communication is also exemplified by the existence of the *los vigias* system in Tungurahua, Ecuador. *Los vigias* literally means watchmen, and comprises organised surveillance of the volcano made up of local community members from different villages situated on the flanks of the volcano. The vigias system has been integrated into the official risk communication of Tungurahua managed by the Volcano Observatory of Tungurahua (Stone et al. 2014).

4   Why Are People Reluctant to Evacuate?

Refusal to evacuate is one of the main issues in volcanic crisis management, as exemplified during the 2010 eruption of Merapi (Fig. 8.5). Evacuations have traditionally been a difficult task to carry out because of people's reluctance to leave their homes and land. Various reasons compound people's reluctance to evacuate in case of a volcanic crisis related to PDC hazard, as exemplified at Merapi (Mei and Lavigne 2013).

First, the principal reason for hesitancy is that some people do not believe that their lives are endangered by PDCs, or that PDCs are likely in that locality. Thus differences in perception of PDC management issues by local communities

and scientists or emergency planners may lead to a disruption of crisis management plans (Johnston and Ronan 2000; Ronan 2013). PDC hazard experience may create an inaccurate localized template for future eruptions, giving local people a false sense of safety (Douglas 1985; Donovan 2010). For instance, despite the efforts of officials, scientists and concerned members of the public of Montserat, about 80 people were in Zones A and B of the Exclusion Zone on 25 June 1997 (Loughlin et al. 2002). Many had become accustomed to the pyroclastic flows and had become overconfident in their own ability to judge the threat by observing repeated flows that had gradually increased runout but remained restricted to valleys. Many people had contingency plans and believed that there would be observable or audible warning signs from the volcano if the activity were to escalate significantly (Loughlin et al. 2002). The feeling of safety is enhanced with the presence of concrete structures like Sabo dams, and by increasing distance of the village from the crater. The feeling of safety felt by the local communities living further than 15 km from Merapi crater in 2010 was enhanced by the extent of the pyroclastic flow hazardous areas delineated by CVGHM, which did not take into account the possibility of a major explosive eruption (Mei et al. 2013). Therefore understanding how people perceive risk has become increasingly important for improving risk communication and reducing risk associated conflicts (Haynes et al. 2008).

Second, it is essential to consider the local and cultural factors in volcanic risk and crisis management (Lavigne et al. 2008). During the 2010 PDC of Merapi, the evacuation refusal of Mbah Marijan (the volcano's gatekeeper or *Juru Kunci*) and his followers led to the deaths of thirty-five people in Kinarhejo, a village located only 5 km from the summit, including the gatekeeper himself. Before this disaster, evacuation refusals along the southern flank of the volcano were mostly conditioned by trust in the gatekeeper and the feeling of being protected by his presence

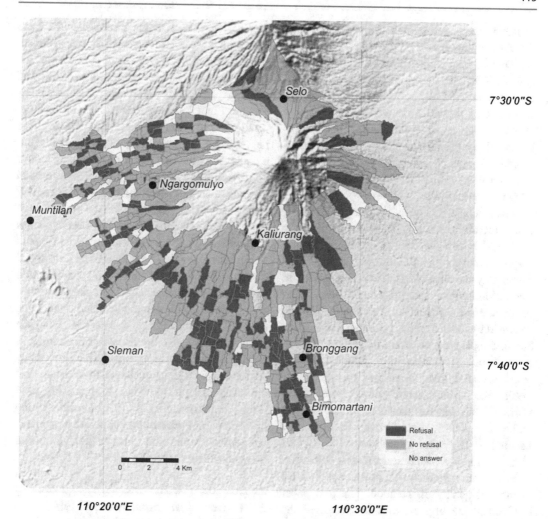

Fig. 8.5 Evacuation refusal during the 2010 eruption of Merapi volcano, Indonesia. Refusal means that at least one person in the village has been identified by the village's chief as being reluctant to evacuate after have received the warning. *Source* Lavigne et al. (2015)

(Mei and Lavigne 2012), even though in 2010 Marijan suggested to people not to follow his decision to stay in the village. Another well-known disaster related to large PDCs was partly due to evacuation refusals for cultural reasons: the 1963 eruption of Mount Agung occurred at the time of a very rare and important Balinese ceremony at Besaki Hindu temple, 7 km away from the crater. The PDCs were therefore interpreted as a punishment from the gods, leading to the death of more than 1000 people. Careful communication and awareness of potential culture clashes might aid

communication within and beyond the scientific community (Donovan and Oppenheimer 2014).

Third, people may be reluctant to evacuate even if they are aware of the danger. Economic pressure may explain people's behaviour during crisis, since they often refuse to evacuate in order to cultivate their crops, take care of their animals and protect their goods. Evacuation can have severe consequences on the economy of a village or a city. During the 2007 volcanic crisis of Kelut volcano in East Java, for instance, 77% of people living in Sugihwaras, a village close to the crater did not pay attention to the warning message issued by the government, and almost a half of the interviewees disregarded the order to evacuate (De Belizal et al. 2012). They chose to stay at home, hiding themselves in their own houses, closing shutters and turning off lights. Almost two third of those interviewed thought it was dangerous to leave their houses and assets behind. They declared that they were afraid of potential looters, a common perception which has been deeply discussed in the literature (e.g. by Quarantelli 1984).

Some people refuse to evacuate until other family members, and also pets, are safe. A study conducted within the community living around Mayon Volcano, the Philippines, reveals how community members were not willing to stay at evacuation centers and preferred to stay with family members should an evacuation warning be issued by the authorities (Usamah and Haynes 2012). People may be reluctant to evacuate due to the sanitary situation in evacuation centres, either actual or due to rumors spread by the media. For instance, some people from Sugihwaras (Kelut) did not evacuate to the shelters during the 2008 volcanic crisis, because they heard that they were insalubrious (De Belizal et al. 2012). The media asserted that infectious diseases were spreading in many camps because of the bad quality of the food. The newspaper condemned the organizations in charge of the evacuation centres. Rumours of such diseases spread quickly and many people believed that problems occurred in every evacuation center. Such rumors have been largely covered by the literature (e.g. Drabek 1999) and may increase people's vulnerability.

5   Building Trust in Hazard and Risk Communication to Ensure Better Responses to Evacuations

Open and transparent communications between the stakeholders before and during a volcanic eruption is a key point in improving crisis management capabilities. In order to enhance these capabilities, it is essential to consider the local and cultural factors in volcanic risk management. As pointed out by Haynes et al. (2008), specific differences between the public, authorities and scientists are often responsible for misunderstandings and misinterpretations of information, resulting in differing perceptions of acceptable risk. Deeper understanding of the at-risk communities is therefore a key issue in order to reduce vulnerability in PDC hazard regions. Information dissemination and education of the people at risk are also key factors in correcting the perception of PDC threats, and therefore in improving crisis communication. Modes of communication should be reviewed regularly in the context of social changes. The need for community participation and involvement in raising PDC hazard awareness is crucial. Risk communication is a dialogue between the communities and people giving the warnings. The take-up of scientific advice is much more efficient when communication of that advice is founded on personal trust rather than written on a report (Haynes et al. 2007a). As a result, communication of PDC hazard directly between scientists and the public is very important, e.g. on Montserrat (Haynes 2005; Donovan and Oppenheimer 2014).

6   Conclusion: Improving Crisis Management Capabilities for PDC's Risk Reduction

Crisis management capabilities may be improved through a set of good practices that are theoretically well-established, but that remain difficult to develop practically by the local stakeholders.

Among the good practices for raising PDC hazard knowledge and public awareness, information related to this specific hazard should be widely disseminated, through the members of hazard mitigation offices from regional to local levels, not only within the PDC hazard zones but also involve villages and cities located tens of kilometres away from the vent. PDC hazard information should be disseminated especially around dormant volcanoes, where volcanic risk perception is usually low.

Video footage is an effective tool for raising PDC hazard knowledge. During the 1991 eruption of Mount Pinatubo (Philippines), the dissemination of a film from the French volcanologists M. and K. Kraft likely saved many thousands of lives. Recently, the World Organization of Volcano Observatories (WOVO) provides video resources through VOLFilm, a Multilingual and multi-platform films database for resilience to risks from volcanic hazards (http://www.wovo.org/ volfilm-multilingual-and-multi-platform-films-for-resilience-to-risks-from-volcanic-hazards.html). Dissemination of information also comprises continuous media slots on PDC risk prevention, preparedness and management taking into account geographical specificities.

Community Based Disaster Risk Reduction (CBDRR) should be considered in integrating top-down and bottom-up approaches and as a channel of information and actions between stakeholders. Indeed it fosters the participation of threatened communities in both the evaluation of risk (including PDC hazards, vulnerability and capacities) and in the ways to reduce it. Community participation and involvement in raising PDC hazard awareness might be accomplished through various ways: socioeconomic factors should be better integrated from daily life to strengthen livelihoods; collaboration should be based on actual collaboration between institutional and upper level stakeholders, local stakeholders, and communities. Several approaches may be taken in order to gain more traditional knowledge of and responses to PDC-related disasters, in the framework of a bottom-up disaster risk reduction programme. Dialogue between the communities and people giving the warnings could be improved through participatory methods, e.g. participatory volcanic hazard mapping, community evacuation simulations, rural appraisal, focused group discussion or participatory three dimension mapping.

Efficient communication between scientific experts on PDC hazard, authorities, the media,

local NGOs, and the population should be enhanced to improve crisis management. Information should be provided to people on time and using simple and clear language, preferably traditional language.

The need for community participation and involvement in raising PDC hazard awareness is crucial, especially among specific stakeholders, e.g. recent immigrants or daily workers coming from outside the PDC hazard zones or women, who usually have a poorer knowledge of hazards than their husband or children. Local and cultural factors should also be considered in risk and crisis management, especially because PDC hazard is often related to local myths. The 2010 Merapi disaster suggests that religion is an essential element of culture and must be carefully considered in the planning process, and not simply dismissed as a symptom of ignorance or superstition. Participatory risk management involving community leaders and their populations is most appropriate to bridge tradition, local realities and the implementation of risk management policies and strategies.

To conclude, CBDRM eventually empowers communities with self-developed and culturally acceptable ways of coping with crises due to PDC hazard.

References

Abdurachman EK, Bourdier JL, Voight B (2000) Nuées ardentes of 22 November 1994 at Merapi volcano, Java, Indonesia. J Volc Geotherm Res 100:345–361

Chester DK, Duncan AM, Dibben C (2008) The importance of religion in shaping volcanic risk perception in Italy, with special reference to Vesuvius and Etna. J Volc Geotherm Res 172:216–228

Cole PD, Calder ES, Druit TH et al (1998) Pyroclastic flows generated by gravitational instability of the 1996–97 lava dome of Soufriere Hills Volcano, Montserrat. Geoph Res Lett 25(18):3425–3428

CVGHM (Center of Volcanology and Geological Hazard Mitigation) (2002) Merapi Volcano Hazard Map. Bandung

CVGHM (Center of Volcanology and Geological Hazard Mitigation), (2011) Revised Merapi Volcano Hazard Map. Bandung

De Belizal E, Lavigne F, Gaillard JC et al (2012) The 2007 eruption of Kelut volcano (East Java, Indonesia): phenomenology, crisis management and social response. Geomorphology 136(1):165–175

Donovan K (2010) Doing social volcanology: exploring volcanic culture in Indonesia. Area 42(1):117–126

Donovan A, Oppenheimer C (2014) Science, policy and place in volcanic disasters: insights from Montserrat. Environ Sci Policy 39:150–161

Douglas M (1985) Risk acceptability according to the social sciences. Russell Sage Foundation, New York

Drabek T (1999) Understanding disaster warning responses. Soc Sci J 36(3):515–523

Gertisser R, Charbonnier SJ, Troll VR et al (2011) Merapi (Java, Indonesia): anatomy of a killer volcano. Geol Today 27(2):57–62

Greene MR, Perry RW, Lindell MK (1981) The March 1980 eruptions of Mt. St. Helens: citizen perceptions of volcano threat. Disasters 5(1):49–66

Gregg CE, Houghton BF, Johnston DM et al (2004) The perception of volcanic risk in Kona communities from Mauna Loa and Hualalai volcanoes, Hawaii. J Volc Geotherm Res 130:179–196

Haynes K (2005) Exploring the communication of risk during a volcanic crisis: a case study of Montserrat. Unpublished Ph.D. Thesis, University of East Anglia, West Indies

Haynes K, Barclay J, Pidgeon N (2007a) The issue of trust and its influence on risk communication during a volcanic crisis. Bull Volc 70(5):605–621

Haynes K, Barclay J, Pidgeon N (2007b) Volcanic hazard communication using maps: an evaluation of their effectiveness. Bull Volcanol 70(5):123–138

Haynes K, Barclay J, Pidgeon N (2008) Whose reality counts? Factors affecting the perception of volcanic risk. J Volc Geotherm Res 172(3–4):259–272

IAVCEI Subcommittee for Crisis Protocols (1999) Professional conduct of scientists during volcanic crises. Bull Volcanol 60:323–334

Jenkins S, Komorowski JC, Baxter P et al (2013) The Merapi 2010 eruption: an interdisciplinary impact assessment methodology for studying pyroclastic density current dynamics. J Volc Geotherm Res 261:316–329

JMA (2015) Volcanic warnings. Japan Meteorological Agency. http://www.jma.go.jp/en/volcano/. Accessed 20 Jan 2015

Johnston D, Ronan K (2000) Risk education and intervention. In: Sigurdsson H, Houghton B, McNutt S et al (eds) Encyclopedia of volcanoes. Academic, San Diego, CA

Lacroix A (1904) La Montagne Pelée et ses éruptions («Mount Pelée and its eruptions» in French). Masson et Cie, Paris

Lavigne F, De Coster B, Juvin N et al (2008) People's behaviour in the face of volcanic hazards; perspectives from Javanese communities, Indonesia. J Volc Geotherm Res 172(3–4):273–287

Lavigne F, Degeai JP, Komorowski JC et al (2013) Source of the Great AD 1257 Mystery Eruption Unveiled, Samalas Volcano, Rinjani Volcanic Complex, Indonesia. Proc Nat Acad Sc USA 110 (42):16742–16747

Lavigne F, Morin J, Surono (eds) (2015) The Atlas of Merapi Volcano, 1st edn. CNRS-CVGHM, Meudon 73 p

Leonard GS, Stewart C, Wilson GS et al (2014) Integrating multidisciplinary science, modelling and impact data into evolving, syn-event volcanic hazard mapping and communication: a case study from the 2012 Tongariro eruption crisis, New Zealand. J Volc Geotherm Res 286:208–232

Leone F, Lesales T (2009) The interest of cartography for a better perception and management of volcanic risk: from scientific to social representations the case of Mt. Pelée volcano, Martinique (Lesser Antilles). J Volc Geotherm Res 186:186–194

Loughlin S, Baxter PJ, Aspinall W et al (2002) Eyewitness accounts of the 25 June 1997 pyroclastic flows and surges at Soufriere Hills Volcano, Montserrat, and implications for disaster mitigation. Geol Soc London Memoirs 21(1):211–230

Lube G, Cronin SJ, Thouret JC et al (2011) Kinematic characteristics of pyroclastic density currents at Merapi and controls on their avulsion from natural and engineered channels. Geol Soc Am Bull 123(5–6):1127–1140

Mei ETW, Lavigne F (2012) Influence of the institutional and socio-economic context for responding to disasters: case study of the 1994 and 2006 eruptions of the Merapi Volcano, Indonesia. Geol Soc London Spec Pub 3:171–186

Mei ETW, Lavigne F (2013) Mass evacuation of the 2010 Merapi eruption. Int J Emerg Manag 9(4):298–311

Mei ETW, Picquout A, Lavigne F et al (2013) Lessons learned from the 2010 evacuations at Merapi volcano. J Volc Geotherm Res 261:348–365

MIAVITA team (2012) Handbook for volcanic risk management—prevention, crisis management, resilience. Bureau de Recherches Géologiques et Minières (BRGM), Orléans

Nave R, Isaia R, Vilardo G et al (2010) Re-assessing volcanic hazard maps for improving volcanic risk communication: application to Stromboli Island, Italy. J Maps 6:260–269

Pardyanto L, Reksowigoro LD, Mitromartono FXS et al (1978) Volcanic hazard map, Merapi volcano, Central Java (1/100 000). Geological Survey of Indonesia, Bandung, II, p 14

Pearson L (2015) Early warning of disasters: facts and figures. http://www.scidev.net/global/communication/feature/early-warning-of-disasters-facts-and-figures-1.html. Accessed 20 Jan 2015

Quarantelli EL (1984) People's reactions to emergency warning. Disaster Research Center, University of Delaware, Delaware

Ronan KR (2013) Education and training for emergency preparedness. Encyclopedia of natural hazards, encyclopedia of earth sciences series: 247–249

Stone J, Barclay J, Simmons P et al (2014) Risk reduction through community-based monitoring: the *vigías* of Tungurahua, Ecuador. J Appl Volc 3:11

Usamah M, Haynes K (2012) An examination of the resettlement program at Mayon Volcano: what can we learn for sustainable volcanic risk reduction? Bull Volc 74(4):839–859

The Communication and Risk Management of Volcanic Ballistic Hazards

R.H. Fitzgerald, B.M. Kennedy, T.M. Wilson, G.S. Leonard, K. Tsunematsu and H. Keys

Abstract

Tourists, hikers, mountaineers, locals and volcanologists frequently visit and reside on and around active volcanoes, where ballistic projectiles are a lethal hazard. The projectiles of lava or solid rock, ranging from a few centimetres to several metres in diameter, are erupted with high kinetic, and sometimes thermal, energy. Impacts from projectiles are amongst the most frequent causes of fatal volcanic incidents and the cause of hundreds of thousands of dollars of damage to buildings, infrastructure and property worldwide. Despite this, the assessment of risk and communication of ballistic hazard has received surprisingly little study. Here, we review the research to date on ballistic distributions, impacts, hazard and risk assessments and maps, and methods of communicating and managing ballistic risk including how these change with a changing risk environment. The review suggests future improvements to the communication and management of ballistic hazard.

Keywords

Volcanic ballistics · Volcanic hazard · Volcanic risk · Risk communication · Risk management

R.H. Fitzgerald (✉) · B.M. Kennedy · T.M. Wilson
Department of Geological Sciences, University of Canterbury, Private Bag 4800, Christchurch 8140, New Zealand
e-mail: rebecca.fitzgerald@pg.canterbury.ac.nz

G.S. Leonard
GNS Science, PO Box 30368, Lower Hutt 5040, New Zealand

K. Tsunematsu
Mt. Fuji Research Institute (MFRI), 5597-1 Kenmarubi Kamiyoshida Fujiyoshida-Shi, Yamanashi 403-0005, Japan

H. Keys
Department of Conservation, PO Box 528, Taupo 3351, New Zealand

Advs in Volcanology (2018) 121–147
https://doi.org/10.1007/11157_2016_35
© The Authors(s) 2017
Published Online: 15 March 2017

1 Introduction

Ballistic projectiles are one potentially lethal and damaging hazard produced in volcanic eruptions. Ballistics are fragments of lava (bombs) or rock (blocks) ejected in explosive eruptions (Fig. 1a, b). Projectiles range from a few centimetres to tens of metres in diameter and separate from the eruptive column to follow nearly parabolic trajectories (Wilson 1972; Fagents and Wilson 1993; Bower and Woods 1996). Their exit velocities can reach hundreds of metres per second and land up to ∼10 km from the vent, although typically within five kilometres (Blong 1984; Alatorre-Ibargüengoitia et al. 2012). Ballistics are associated with all forms of explosive eruptions but are considered major hazards of hydrothermal, phreatic, phreatomagmatic, Strombolian and Vulcanian eruptions, especially those which have little to no precursory signals of volcanic unrest. Managing ballistic hazard and risk on active volcanoes, particularly those permanently occupied or regularly visited, presents considerable challenges: it requires good information and specialist communication strategies around risk mitigation, preparedness, response, and recovery dependent on the state of the volcano, e.g. pre-, during- and post-eruption. In this chapter, we present an overview of volcanic ballistic hazards and impacts and the communication strategies used to manage risk on active volcanoes.

2 Ballistic Hazard and Risk Management

Ballistic projectiles are a risk to life on active volcanoes and can cause substantial damage to exposed infrastructure and the environment due to their high kinetic energy, mass, and often high temperatures (Blong 1984). Volcanic ballistic projectiles are amongst the most frequent causes of fatal incidents on volcanoes, with at least 76 recorded deaths at six volcanoes (Galeras, Yasur, Popocatepetl, Pacaya, Raoul Island and Ontake) since 1993 (Baxter and Gresham 1997; Cole et al. 2006; Alatorre-Ibargüengoitia et al. 2012;

Wardman et al. 2012; Tsunematsu et al. 2016). Many more people have been injured as a result of ballistic impacts, frequently suffering from blunt force trauma (broken bones), lacerations, burns, abrasions and bruising (Blong 1984; Baxter and Gresham 1997). Additionally, damage to buildings (Fig. 1c, e), infrastructure, property and the surrounding environment (Fig. 1d) are also common occurrences from ballistics during explosive eruptions. The high kinetic and thermal energy of ballistics can puncture, dent, melt, burn and knock down structures and their associated systems, such as power supply and telecommunication masts; crater roads; and crush and potentially ignite crops (Booth 1979; Calvari et al. 2006; Pistolesi et al. 2008; Alatorre-Ibargüengoitia et al. 2012; Wardman et al. 2012; Maeno et al. 2013; Fitzgerald et al. 2014; Jenkins et al. 2014). Blong (1981), Pomonis et al. (1999) and Jenkins et al. (2014) estimate a ballistic only needs 400–1000 J of kinetic energy to penetrate a metal sheet roof, far less than the estimated kinetic energy of ballistics ($\sim 10^6$ J) from VEI 2-4 eruptions (Alatorre-Ibargüengoitia et al. 2012).

The distribution (distance from vent, direction, area and density) of ejected ballistics is controlled by the explosivity, type, size and direction of explosive eruptions, and usually creates spatially variable deposits (Gurioli et al. 2013; Breard et al. 2014; Fitzgerald et al. 2014). Generally, the distance travelled and the total area impacted by ballistics increases with increasing explosivity, i.e. particles generally travel further and cover a greater area in Vulcanian eruptions (Nairn and Self 1978; Alatorre-Ibargüengoitia et al. 2012; Maeno et al. 2013) compared with Strombolian eruptions (Harris et al. 2012; Gurioli et al. 2013; Turtle et al. 2016). However, eruptions can be directed, ejecting ballistics at low angles and at distances greater than those from more vertically directed eruptions (Fitzgerald et al. 2014; Tsunematsu et al. 2016). The directionality of these blasts is often unpredictable, and can be influenced by external factors such as landslides (Christiansen 1980; Breard et al. 2014), making it difficult to deterministically forecast future ballistic

Fig. 1 Types of ballistic particles and their impacts: **a** Ballistic bombs from Yasur Volcano, Vanuatu (*Photo credit* Ben Kennedy), **b** Ballistic blocks (1.4 m diameter block) from the August 2012 Upper Te Maari eruption, **c** Damage to a building from ballistics ejected in the 2000 Mt. Usu, Japan eruption, **d** Damage to the environment illustrated by a 4.4 m wide crater from the August 2012 Upper Te Maari, Tongariro eruption, **e** Damage to a hiking hut from 2012 Upper Te Maari ballistics (*Photo credit* Nick Kennedy)

distributions. Mapped deposits from past eruptions are often not symmetrical around the vent, reflecting this directionality (Minakami 1942; Fudali and Melson 1972; Steinberg and Lorenz 1983; Kilgour et al. 2010; Houghton et al. 2011; Gurioli et al. 2013; Fitzgerald et al. 2014), and are sometimes the result of the crater and surrounding topography (Breard et al. 2014; Tsunematsu et al. 2016). Detailed descriptions and maps of ballistic impact distributions are rare, but those published may contain some of the following data: maximum ballistic travel distances (Steinberg and Lorenz 1983; Robertson et al. 1998; Kaneko et al. 2016); the outer edges of a ballistic field (Minakami 1942; Nairn and Self 1978; Yamagishi and Feebrey 1994); and/or maximum particle (Nairn and Self 1978; Steinberg and Lorenz 1983; Robertson et al. 1998;

Swanson et al. 2012) or crater size (Robertson et al. 1998; Maeno et al. 2013; Kaneko et al. 2016). When isopleths of particle size are included these rarely contain individual measurements and may be severely limited by the availability of only specific mapped locations (e.g., Kilgour et al. 2010; Houghton et al. 2011). For this reason, the number of particles, sizes of particles, and spatial density per unit area is rarely reported (only four publications could be found with this level of detail—Pistolesi et al. 2008; Swanson et al. 2012; Gurioli et al. 2013; Kaneko et al. 2016). This leads to a limited understanding of the hazard and risk posed to the area.

Though work has been completed on ballistic hazard (e.g., mapping deposits, better understanding eruption dynamics and the factors that influence ballistic distribution, recording particle velocities, the creation and use of ballistic trajectory models, and the production of hazard maps either focussed solely on ballistics or as an aspect of a multi-hazard map), very little has been focussed on the management of ballistic risk, leaving a large knowledge gap and a need for research in this area. Risk management strategies and mitigation systems are key to protecting life and infrastructure from ballistic hazards (Leonard et al. 2008; Bertolaso et al. 2009; Bird et al. 2010; Jolly et al. 2014b). Table 1 lists some of the strategies and tools used at volcanoes around the world.

Effective communication of ballistic hazard and risk to end-users such as the public, stakeholders in the area and emergency managers underpins effective development and implementation of these risk management strategies. However, ballistic hazard and risk are not and should not be treated the same at all volcanoes. The risk environment (the hazard, the number of people and assets exposed and their associated vulnerability) will determine the strategies, tools and methods of communication, and their relative importance, utilised in the overall risk management strategy. The volcano tourism industry is also growing (Sigurdsson and Lopes-Gautier 1999; Erfurt-Cooper 2011), increasing the number of people exposed to ballistic hazard in proximal areas. In addition, population growth in many volcanic regions means increasing numbers of people are settling closer to and on volcanoes (Small and Naumann 2001; Ewart and Harpel 2004). This creates an increasing demand for ballistic hazard and risk assessments coupled with effective communication strategies to manage ballistic risk at volcanoes. Ballistics are not a hazard in isolation. Their management needs to be integrated with that of other volcanic hazards (especially pyroclastic density currents in terms of near-vent life safety, but also landslides, lahars, lava flows, and volcanic gas emissions/areas of hot ground), and other life safety issues such as severe weather and mountain safety.

3 Assessments of Ballistic Hazard and Risk

Successful management of the risk from ballistic hazards typically requires first assessing the level of risk. This may range from the simple recognition that ballistics may endanger people or their activities on a volcano through to a sophisticated quantitative hazard or risk assessment (e.g. Alatorre-Ibargüengoitia et al. 2012; Jolly et al. 2014b). Ballistic hazard assessments determine the likelihood of ballistic-producing eruptions and the areas that may be impacted (Thouret et al. 2000; Alatorre-Ibargüengoitia et al. 2012). Risk assessments estimate the likelihood of consequences (i.e. death, injury, damage) from exposure to ballistics, typically with an associated probability of occurrence (Blong 1996). Once the level of risk has been assessed it can be used as the robust basis for risk management strategies, such as exclusion zones, hazard/risk maps and signs, and land-use planning. Ideal assessments involve a number of steps including: (1) a review of the eruption history of the volcano to determine past eruption frequencies and magnitudes, thus informing future eruption probabilities; (2) field mapping, remote sensing and/or review of past reports and literature to determine the nature and extent of past ballistic distributions; (3) utilising ballistic trajectory models to explore possible future distributions and areas of hazard; (4) identifying exposed assets in the area such as humans (visitors and inhabitants) and infrastructure; and (5) estimating their vulnerability to the hazard i.e. likelihood of fatality or damage (Nadim 2013). Assessments are ideally probabilistic, providing spatially varying probabilities of occurrence and damage from a range of scenarios varying in frequency and magnitude, and accounting for model and input parameter uncertainty. They should be constantly refined and improved as new information becomes available.

A hazard map is a primary tool used to present hazard and risk information (Sparks et al. 2013). Zonation is generally used as a means to

Table 1 Risk management and communication strategies with selected example volcanoes where they have been employed

Risk management strategy	Description	Selected examples of volcanoes where strategy has been used	References
Hazard and risk assessments	Hazard assessments determine the likelihoods of hazard producing events and the areas that may be impacted. These can be expanded to risk assessments to determine the likelihood of consequences to people and/or other societal assets. They underpin and inform other risk management strategies	Tongariro Mt. Fuji Popocatepetl El Chichon	Mount Fuji Disaster Prevention Council (2004), Alatorre-Ibargüengoitia et al. (2012), Jolly et al. (2014b), Alatorre-Ibargüengoitia et al. (2016)
Hazard and risk maps	Identify zones of relative hazard and/or risk, typically in a two dimensional representation	Popocatepetl Mt. Fuji	Mount Fuji Disaster Prevention Council (2004), Alatorre-Ibargüengoitia et al. (2012)
Volcano monitoring and research	Systems deployed on and around a volcano to monitor volcanic activity, and indicate when the volcano is in unrest or eruption. Research is also conducted on the eruptive behaviour (e.g. magnitude and style) and eruption frequency of a volcano	Mt. Etna Tongariro Sakurajima Mt. Ontake	https://www.geonet.org.nz/volcano/info/tongariro; http://www.ct.ingv.it/en/mappa-stazioni.html; JMA (2013a, b)
Real-time warning systems	To monitor and detect a hazardous event (e.g. eruption) and communicate a warning to those potentially exposed.	Ruapehu	Leonard et al. (2008), Keys and Green (2010)
Volcanic alert levels, bulletins and media advisories	Formal communications from a volcano observatory which communicate changes in volcano behaviour, notify emergency managers and the population of an eruption and advise on mitigation.	Yasur Tongariro	https://www.geonet.org.nz/volcano/info/tongariro; http://www.geohazards.gov.vu/
Emergency response plans	Plan for directing response actions which aim to reduce the impacts of an eruption. Plans are best executed with training and exercises in their use	Ruapehu Sakurajima	Leonard et al. (2008); http://www.city.tarumizu.lg.jp/kikikanri/kurashi/bosai/bosai/taisaku/sakurajima.html
Rescue services	Deploy in emergencies to provide aid to affected persons and properties, e.g. Search and Rescue, police, ambulance, and fire services	Ontake Ruapehu	Kilgour et al. (2010), The Japan Times 27/9/2015
Land use planning	Policy and regulations used to minimise or exclude the development of settlement and construction of high-value assets in hazard zones.	Usu Volcano Tongariro Ruapehu	Becker et al. (2010), Keys and Green (2010)

(continued)

Table 1 (continued)

Risk management strategy	Description	Selected examples of volcanoes where strategy has been used	References
Construction of protective shelters	Structures designed to withstand specific hazards e.g. ballistic shelters in high-risk areas	Sakurajima Aso Stromboli	Bertolaso et al. (2009), Erfurt-Cooper (2010)
Exclusion or restriction zones	Area restrictions commonly used temporarily during eruptions or unrest. Permanent zones may be required when risk is sufficiently high or frequent	Sakurajima Stromboli	Bertolaso et al. (2009), Kagoshima City (2010)
Stakeholder engagement	Involvement of stakeholders (e.g. people and organisations potentially affected by an eruption) in planning and activities to manage the risk. This is essential for effective hazard and risk communication, and to establish appropriate and acceptable risk management strategies	Tongariro Sakurajima	Williams and Keys (2013), Jolly et al. (2014b), http://www.data.jma.go.jp/svd/vois/data/fukuoka/506_Sakurajima/506_bousai.html
Hazard and risk education resources	Education resources which aid communication of hazard and risk information, often developed specifically for non-expert stakeholders. These can include pamphlets and brochures, websites, warning signs, videos, and public talks and meetings.	Auckland Volcanic Field, Tongariro	DOC (2012), Wilson et al. (2014)

distinguish areas of hazard, exposure, vulnerability and risk (Sparks et al. 2013). Ballistic hazard map zones may be classified by maximum travel distance of particles (either any size or a specific sized particle; Alatorre-Ibargüengoitia et al. 2012), number of ballistic impacts per unit area (Gurioli et al. 2013), probability of a specific size of ballistics reaching a given area (Artunduaga and Jimenez 1997), or probability of a specific consequence occurring e.g. death, injury, damage (Fitzgerald et al. 2014). A good example of a ballistic hazard map that follows the best-practice steps above was created by Alatorre-Ibargüengoitia et al. (2012) of Popocatepetl Volcano, Mexico. In this example, eruption history and frequency of occurrence are used to define three eruption scenarios (High: VEI 2–3 (as they are more frequent), Intermediate: 4, and Low: 5 (though an eruption of this

size would affect more people and impact a larger area, it has a much lower likelihood of occurring). The maximum travel distance of ballistic projectiles from each scenario (based on field and model distributions) is then used to define the extent of the hazard zones. Additionally, the map identifies nearby towns and roads exposed to ballistic hazard.

In many instances, it may not be possible or warranted to complete all of the steps involved in an ideal risk assessment. For example, Gareloi Volcano, Alaska is located on an uninhabited island, thus a detailed ballistic hazard assessment was not the priority of initial hazard assessments. Coombs et al. (2008) explore the eruptive history of Gareloi Volcano, though eruption frequency is only narrowed down to one eruption every 20–50 years and is not broken down into eruption magnitudes. Ballistic hazard is confined to one hazard zone (a 5 km concentric radius around the vent), whose extent is based on Blong's (1996) assessment that ballistics generally do not travel further than 5 km from vent. It is also mentioned that recent ballistic distributions have not travelled further than several hundred metres from vent. Neither a deterministic or probabilistic approach was taken, instead a value was adopted from other eruptions around the world.

Very few studies exist on ballistic risk or vulnerability. We summarise the three that could be found. Booth (1979) presents an example of a volcanic risk map for the La Primavera Volcanic Complex, Mexico. Though ballistics are included, they are not ascribed a probability of occurrence, instead, one zone at risk of ballistic fall is defined by the maximum travel distance for ballistics up to 0.1 m in diameter. The equation that Booth used to calculate risk includes probability of occurrence, indicating that eruption frequency has been examined; however, neither the probability used nor the description of prior eruptive history are provided in the publication. Thus, though an end-product of a risk map is produced, the process itself is not documented. Pomonis et al. (1999) utilise the Blong (1981) impact energy thresholds for roof perforation to assess building vulnerability from an eruption of Furnas Volcano, the Azores. Two risk zones are assigned (moderate and high) based on the statement that ballistics generally land within 5 km of the vent, but sometimes up to 10 km. The study only considers one eruption (the last major eruption), thus is lacking eruption frequency and magnitude, and does not provide any probabilities of building damage occurring. Building vulnerability to ballistic impact has been assessed by Jenkins et al. (2014) for Kanlaon and Fogo volcanoes (Philippines and Cape Verde, respectively) using estimates of energy required to penetrate roof materials by Blong (1981) and Pomonis et al. (1999). This study, however, focussed only on the vulnerability of the built environment and did not include an overall assessment of hazard or risk. Eruption frequency and magnitude, the extent of past

ballistic distributions, and modelling of possible future trajectories were not investigated.

Assessments may also vary depending on the state of the volcano. Volcanoes in a state of quiescence allow for (and call for) more in-depth, preferably probabilistic, assessment to be completed, ideally following the steps outlined earlier. However, quiescent volcanoes may not be the primary target for in-depth assessment. Conversely, renewed volcanic activity, especially when unexpected, urgently demands rapid hazard assessments which may, as a result, be too simplistic, overly conservative or lacking sufficient detail to be considered complete. They also need to be focussed on the range of scenarios presenting the risk in that crisis (e.g. from one vent), rather than the entire background risk from that volcano (e.g. from multiple vents). Leonard et al. (2014) describe the process of creating a crisis hazard map for the 2012 Upper Te Maari eruption, comparing this to the existing background hazard map. In the case of a volcano in a state of unrest, assessments may be limited by the availability of safe locations to survey, and this is especially likely once an eruption episode has commenced as evident during the 2012 Upper Te Maari, Tongariro eruptions and assessments presented later. Odbert et al. (2015) have been developing updateable hazard forecast estimates using Bayesian belief networks, which may help to improve rapid hazard assessments in times of crisis.

4 Communication and Risk Management Strategies

Effective communication is essential in managing ballistic hazard and risk (Barclay et al. 2008; Leonard et al. 2014). Science needs to be communicated to decision-makers, stakeholders, and the public and understood and absorbed by them so they can make informed decisions. Similarly, the public, stakeholders, and decision-makers should communicate to scientists what type of information they need to make decisions relevant to their situations. Ballistic communication methods used at volcanoes include hazard and risk assessments, hazard maps, volcano monitoring and research, real-time warning systems, volcanic alert levels; volcano warnings, alert bulletins and communication with agencies; response exercises, education materials, response plans, exclusion and evacuation zones, instructions and signage for what to do in the event of an eruption around the volcano, community engagement, educational materials, and land-use planning and infrastructure design. These methods typically fall under four aspects of emergency management: Mitigation (Reduction), Preparedness, Response and Recovery (UNISDR 2009). Methods must also be integrated with the management of other risks, ideally in one cohesive approach. Ballistic communication strategies will also vary with eruption frequency, the risk context (quiescence or crisis; Fig. 2), whether

volcanoes are frequently visited or inhabited, and the availability of resources. This equally applies to volcanoes at which ballistics are/are not the main hazard.

Effective risk management is built on communication, hazard education and engagement with the at-risk communities (Johnston et al. 1999, 2000; Paton et al. 2001; Twigg 2002; Gregg et al. 2004; Leonard et al. 2008; Dohaney et al. 2015). Appropriate risk management actions by stakeholders, emergency managers and the public require an adequate perception of the risk and the correct actions to take in a crisis, with perception dependent on the hazard information received and exposure to impacts (Johnston et al. 1999; Leonard et al. 2014). Knowledge and understanding of volcanic hazards allows individuals to better decide whether to undertake preparedness and response measures, and if so,

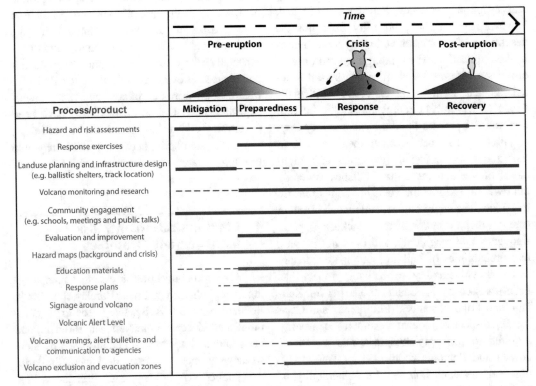

Fig. 2 Various ballistic hazard and risk communication processes (*blue*) and products (*red*) implemented over the changing state of the volcano and the stage of risk or emergency management. The level of activity/importance is indicated by *line style*, with *solid lines* indicating higher use or importance

which are required, thus reducing their vulnerability to the hazard(s) (Siegrist and Cvetkovich 2000; Paton et al. 2008; Bird et al. 2010).

Scientific information can be misunderstood, misrepresented or distorted when passed from scientists to end-users (stakeholders, emergency managers and the public; Barclay et al. 2008). This can occur when end-users do not comprehend or are unaware of the science being presented, the information is not what is actually needed by end-users, the science is communicated poorly to end-users, or there is a lack of trust between groups (Haynes et al. 2007). All groups therefore need to communicate with each other, preferably prior to a volcanic crisis, with communication products tailored to the audience (Haynes et al. 2007; Leonard et al. 2008). Following the 1979 eruption of Mt. Ontake, Japan the National Research Institute for Earth Science and Disaster Prevention in Japan (NIED, though now renamed to National Research Institute for Earth Science and Disaster Resilience) completed a report recommending: regulations on development and land-use, building of ballistic shelters and evacuation facilities, and the development of emergency plans, as an eruption in the summer hiking season would likely result in human casualties (NIED 1980). However, the report may not have been suitable or communicated well to the local municipalities responsible for disaster management as these recommendations were not adopted prior to the 2014 eruption, indicating the need for communication to ensure the information is relevant, understood and acted upon (Barclay et al. 2008; The Japan News, 27/10/2014). Communication delivered jointly by scientists and the local community is also advisable as community members may be better trusted and better communicators to their community than scientists in isolation. Users must be able to trust the source of the information being released as well as how and what is presented (Slovic 2000; Haynes et al. 2008). It is also therefore important for scientists and emergency managers to be honest about what is/is not known to maintain credibility and trust (Lindell 2013).

Best practice suggests the use of multiple sources to disseminate hazard and risk information as preferred forms of media accessed for information vary (Sorensen 2000; Mileti et al. 2004; Haynes et al. 2007; Bird et al. 2010). The public's response to volcanic hazard communication is influenced by the content and attractiveness of the message (which should include a description of the hazard, its impacts, hazard extent, and advice on what to do and when), how comprehensible it is, and the frequency and number of channels the message is received from, as well as the extent of public belief that safety actions are possible and will be effective (Leonard et al. 2008; Sorensen 2013).

4.1 Ballistic Communication Processes and Products in Different Risk Contexts

4.1.1 Volcano Quiescence

Communication and risk management methods vary with changing eruptive states. In times of quiescence focus is placed on risk mitigation and preparedness, with access generally allowed into the hazard zone. In terms of ballistics this includes the completion of ballistic hazard and risk assessments; volcano monitoring and research; land-use and building planning i.e. the building of ballistic shelters capable of withstanding ballistic impacts or the reinforcement of existing structures to specific building standards, and the choice of location for hiking trails, viewing platforms or other visitor facilities; the creation of well distributed hazard maps with instructional text with what to do or where to go in an event of an eruption; and engagement with the local communities including exercises and evaluation (Fig. 2).

Hazard and risk assessments are useful starting points for all communication and management strategies as the nature, extent and consequences of the hazard need to be understood prior to any decisions being made. The assessment should be made available to relevant decision makers, with the authors and science advisors available to advise or answer questions about the assessments. Scientists/authors should always strive to be transparent in their methodology. Transparency builds trust and credibility. It is important that stakeholders know the limitations of the information presented to them and/or informing decisions which affect them. It may not be needed or appropriate for the methods to be presented to the stakeholders in depth but instead it be communicated that they are available if requested. However, it is imperative to think of the risk context when making these decisions, as every situation is different. Methods and assessments should also be made fully available to other scientists so that these methods can be adopted at other volcanoes if chosen, which would increase best-practice and encourage similar and comparable methodologies. These assessments also need to be communicated to the public so that they can make informed decisions about the hazard and risk in the area they choose to enter as well as what steps they need to take to protect themselves.

The main way assessments are communicated is through a map (Haynes et al. 2007). Ballistic hazard maps are rare as they are typically not the only hazard produced in an eruption. Instead ballistics are typically included in 'all-hazard' or 'multi-hazard' maps (Fig. 3) depicting the general hazard for all active vent(s) (Neal et al. 2001; Hadisantono et al. 2002; Mount Fuji Disaster Prevention Council 2004; Kagoshima City 2010; Leonard et al. 2014). Ballistics are usually represented by one hazard zone, often based on the maximum or expected travel distance of a ballistic clast. This is, in part, because the public require concise, easily comprehensible information, rather than being distracted or overloaded with specifics of individual hazards (Haynes et al. 2007; Leonard et al. 2014). An effective hazard map for the public contains clear information on what are the consequences of the hazard(s), where they occur, and what to do (Leonard et al. 2014). For ballistics, impacts may be death or injury; impact locations are usually within 5 km of the vent; and advice may include "if ballistics are landing around you, move out of their oncoming path, seek shelter and make

yourself a small target." Advice on actions to be taken may vary at different volcanoes, although it would be beneficial if messages are consistent across all volcanoes to reinforce actions and increase the likelihood of people following the correct actions. For this to occur, testing of suggested actions would be required to ensure that the safest and most successful measures are being advised. For example, where frequent Strombolian eruptions are the main source of ballistics, it may be possible to watch the low velocity ballistics and move out of their path. However, in many other eruption styles multiple particles may be ejected rapidly toward a person, presenting a situation in which dodging one ballistic may put you in the path of another. It may be more beneficial to make yourself as small a target as possible, seek shelter and use your backpack as a protective shield. Additionally, ballistics may be accompanied by a surge as seen in the 2014 Mt. Ontake (Kaneko et al. 2016; Oikawa et al. 2016) and August 2012 Te Maari eruptions (Breard et al. 2014), inhibiting the ability to see ballistics until it is too late to act.

Map design should also take into account the effect of map properties on communication (understanding/comprehension) such as data classification, basemap or image, colour scheme (e.g. for colour blind readers), content, and key expression (Haynes et al. 2007; Thompson et al. 2015). Haynes et al. (2007) evaluated the effectiveness of volcanic hazard maps as communication tools on Montserrat, West Indies and found that the use of aerial photographs as a basemap improved people's ability to comprehend hazard information compared to traditional contour basemaps. In general, it has been found the public do not comprehend maps well and professional design input guided by iterative evaluation of map comprehension is wise (Haynes et al. 2007; Thompson et al. 2015). In contrast to the public, more specialist stakeholders such as infrastructure managers may require more detailed and hazard specific information about the impacts, location and recommended actions to inform decisions on land-use and building strength e.g. ballistics impacts in zone 1 can be expected to have sufficient energy

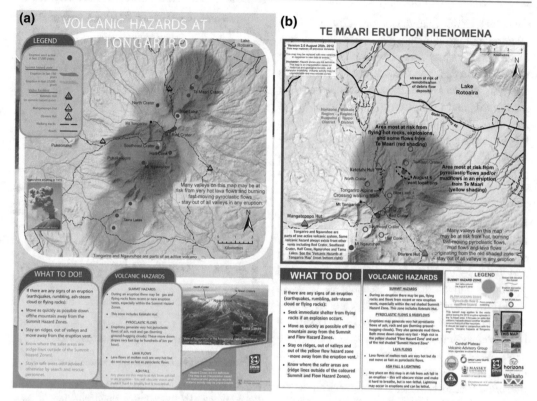

Fig. 3 Volcanic hazard maps of Tongariro volcano, New Zealand: **a** General background hazard map used in quiescent periods (GNS Science 2007), focussed on hazards from events up to a scale that may not have significant precursors to enable warning; **b** Event-specific crisis hazard map following the 2012 eruptions of Upper Te Maari (GNS Science 2012). Note that map A is shown as an *inset* on map B with an explanation as to the complementary but differing nature of the two communication products

to cause severe damage to nearly all types of infrastructure below a certain design standard. Multiple zones of different impact intensity may be shown (e.g. travel distance, density of impacts in an area, size and or energy of expected ballistics in given scenarios). All end-user maps should successfully balance adequate detail and maximum clarity. Hazard maps and additional information should be made available and accessible to the public, and if different maps are made for, or directed to, different audiences their content must be consistent. Public availability may include being posted on signs around the volcanoes entrance(s), in a pamphlet or similar printed media at tourist facilities (e.g. information centres, tourism businesses, hotels, backpackers accommodation, transport operators), and on relevant websites such as volcano

observatories and those charged with managing natural hazards.

Additionally, community engagement and participation in meetings with scientists and managers is encouraged as a means of risk communication, and discussion around management strategies, especially for communities at risk (i.e. tourism providers and those living near or on the volcano) (Cronin et al. 2004; Williams and Keys 2013). Ballistic hazards lend themselves to this type of community engagement because many open system volcanoes that may be constantly erupting but not considered to be in a state of volcanic crisis (e.g. Stromboli, and Yasur) have frequent ballistic-producing eruptions that provide an attraction to tourists and employment for the local community. Ballistics at these constantly erupting volcanoes provide tangible hazards that the community can both relate to and provide valuable observational data on. Meetings should be sufficiently regular to update residents when the status of a volcano is changing and to remind them when necessary of the hazards and risks. Briefing those new to the area, especially the transient visitor, may be the biggest challenge. Engagement allows the community to be prepared in the event of an eruption and to know what to do in the event that they are within hazard areas.

4.1.2 Volcanic Crisis

In a volcanic crisis (when the volcano is showing signs of unrest or is in eruption) communication and emergency management processes and products move toward response (Fig. 2). Real-time warning systems triggered by monitoring equipment, such as the EDS (Eruption Detection System) system installed on Mt. Ruapehu, New Zealand (Leonard et al. 2008), are used to communicate an eruption to those in the immediate vicinity. Wider communication occurs when an event is communicated from monitoring equipment to scientists, then onto emergency managers and decision-makers. Part of this process is the release of alert bulletins/warnings to advise the public of unrest, eruption phenomena, affected areas, and should always include instructions on what to do. Alert bulletins, existing hazard maps and risk and hazard assessments provide emergency managers with information to make decisions on limiting access to parts of the volcano. In the case of ballistics, limits or restrictions on access or development are usually achieved via creation of an exclusion zone, typically 1–4 km in radius (Kagoshima City 2010; Jolly et al. 2014b), or by reducing exposure by limiting the time spent or number of individuals allowed within a zone (Bertolaso et al. 2009).

During the crisis, hazard maps are typically updated and hazard and risk assessments modified. Maps are generally event-specific and only used over a short time-frame, reverting back to the original background hazard maps once the crisis period is over (Leonard et al. 2014; Fig. 3). However, ballistic hazard mapping during a

Fig. 4 Crisis communication sign temporarily used at Ruapehu volcano following a small eruption in 2007, while it was considered there was an elevated risk of further eruptions

crisis can be limited by access restrictions due to the possibility of further eruptions, though as time progresses more detailed mapping is able to be completed (Fitzgerald et al. 2014). The ongoing work by Odbert et al. (2015) in developing a real-time updateable probabilistic risk assessment may prove useful in these situations. The event-specific hazard maps are generally shared around the various media outlets (e.g., television, radio, newspapers, Facebook, Twitter) to inform the public of the updated hazard, as well as through the usual means of communication. They may be augmented by specific life safety signage (e.g. Fig. 4).

Meetings and consultations with local communities, emergency managers and other stakeholders should also occur during and following volcanic crises. The objectives of such meetings are to update communities on the evolving eruptive hazards, build relationships and trust, reduce any miscommunication or misinformation passed along, and to make sure the information being presented is what the end-members need (Barclay et al. 2008; Bertolaso et al. 2009).

Communication of ballistic hazards and risk management vary at frequently erupting volcanoes that commonly enter in and out of crisis, such as Sakurajima in Japan. Access is generally controlled at all times (even during periods of quiescence), sometimes with permanent restriction zones in which nobody is allowed to enter due to the risk of being struck by ballistics (Kagoshima City 2010). In these cases different hazard scenarios may be pre-prepared and communication strategies reused with a population that is well educated about the volcano.

4.2 On-Going Challenges in Ballistic Risk Communication

Many volcanoes are tourist destinations with associated tourist facilities such as ski fields, accommodation and walking tracks (Erfurt-Cooper 2011). One challenge of communicating ballistic risk is to transient populations, especially tourists and other visitors. Tourists spend only a short amount of time in areas (hours to weeks) and often have little knowledge of the hazards or the available protection resources (Murphy and Bayley 1989; Drabek 1995; Burby and Wagner 1996; Bird et al. 2010). They often rely on tourism operators/employees/guides to inform them of volcanic hazards and the correct actions to take in an eruption (Leonard et al. 2008; Bird et al. 2010). This is evident at Yasur Volcano, Vanuatu where guides are frequently relied on to communicate ballistic hazard and safe areas to approach around the volcano, and at Tongariro Volcano, New Zealand where transport operators can give important information to 85% of all those hiking the Tongariro Alpine Crossing (TAC). However, tourism staff may also be somewhat transient, meaning that they may need to be regularly educated, trained or updated on volcanic hazards, appropriate responses and emergency procedures so that they can pass the message down to their patrons (Leonard et al.

2008; Bird et al. 2010; Williams and Keys 2013). Additionally, education material such as pamphlets and hazard maps on volcanic hazards should not only be available at tourism businesses but mechanisms should be in place that ensure that the hazard information is relayed to these transient populations.

Another ongoing challenge in communicating ballistic hazard is the lack of warning time associated with events that have little precursory activity, in which ballistics are typically one of the main hazards. In this scenario volcanic alert levels and bulletins may not be released prior to eruption. Instead, visitors and stakeholders would have to rely on their knowledge of the potential hazards and the response actions to take, especially if there are no real-time warning systems. This places more emphasis and weight on the availability of background hazard maps with messaging covering actions in events up to this size, signage around the volcano (in language(s) appropriate for the audience to comprehend, especially if there is a large proportion of visitors who speak a different language), on pamphlets distributed to businesses and visitors actually reading them, and through communication with their guides. Many visitors to the TAC still assume that they do not need to be concerned because they expect the area to be closed if it is unsafe or to be advised it was unsafe (Keys 2015).

We present the various ballistic risk management and communication approaches taken at four volcanoes: Upper Te Maari, Tongariro Volcanic Complex, New Zealand; Yasur Volcano, Vanuatu; Sakurajima Volcano, Japan and Mt. Ontake, Japan (Table 2). These volcanoes have been chosen for their variation in: frequency of eruption (Sakurajima and Yasur frequently erupt, while Upper Te Maari and Mt. Ontake have longer repose periods), available resources (Yasur has less monitoring equipment and hazard information available than the other three examples), eruptive styles—Yasur predominantly erupts bombs from small Strombolian eruptions; compared with phreatic eruptions from Mt. Ontake and Upper Te Maari and Vulcanian eruptions from Sakurajima that erupt blocks over a larger area, and the similarity in eruptions but with very

Table 2 Comparison of the four case studies and their risk management and communication strategies

	Upper Te Maari	Yasur	Sakurajima	Mt. Ontake
Dominant eruptive style	Hydrothermal	Strombolian	Vulcanian	Phreatic
Recurrence interval	~16 years	Frequently erupting	Frequently erupting	~13 years
Duration of precursory activity	3 weeks of seismicity, 5 min of increasing seismicity			2014: 16 days seismicity, 11 min increasing seismicity and inflation
Number of visitors	100,000 visitors/year	20,000 visitors/year	3,702,000 visitors/year	Hundreds of visitors per day
Hazard map with ballistics	Background and crisis	Yes	Yes	Two background maps
Volcano exclusion and evacuation zones	Yes	Yes	Yes	Yes
Volcanic Alert Levels	Yes	Yes	Yes	Yes
Education material	Yes	Yes	Yes	Yes
Volcano monitoring	Yes	Yes	Yes	Yes
Land-use planning and infrastructure design	Ketetahi Hut (not reinforced). TAC runs through summit hazard zone	No shelters or buildings. Track and viewing platforms along rim of volcano	Concrete shelters around island, evacuation ports	Mountain lodges and shrines (not reinforced). Tracks near active craters
Community engagement	Yes	In progress	Yes	Yes
Response exercises	No	No	Yes	Yes
Evaluation and improvement	Assessments updated post-eruption, creation of crisis hazard map	Currently assessments and maps are being updated	Last update of hazard map was in 2010	New hazard map released November 2015
Signage around volcano	Yes	Yes	Yes	Yes
Volcano warnings, alert bulletins and communication with external agencies	Yes	Yes	Yes	Yes

different consequences between Upper Te Maari and Mt. Ontake. Additionally, all of these volcanoes are relatively accessible and attract large numbers of tourists each year.

5 Case Studies

5.1 2012 Eruptions of Upper Te Maari, Tongariro, New Zealand

On the 6th August 2012, Upper Te Maari Crater, one of the many vents on Tongariro volcano, New Zealand, erupted for the first time in over 100 years (Scott and Potter 2014). The hydrothermal eruption produced multiple pyroclastic surges, an ~8 km high ash plume and ejected thousands of ballistic blocks (Fitzgerald et al. 2014; Lube et al. 2014; Pardo et al. 2014). Blocks were distributed over a 6 km^2 area, affecting ~2.6 km of the popular Tongariro Alpine Crossing (TAC), a walking track frequented by around 100,000 people a year (Fitzgerald et al. 2014). Additionally, Ketetahi Hut, an overnight hut along the TAC, was severely damaged by ballistics. Fortunately, the eruption occurred at night, in winter (the low season) and in bad weather, resulting in no hikers along the TAC or staying at Ketetahi Hut (both around 1.5 km away from the vent and well within the impacted area). A smaller eruption followed on 21 November 2012, though ballistics and pyroclastic surges were confined to within a well posted risk management zone 1 km from the vent and did not affect the TAC.

Ballistics were a known hazard from the active vents of Tongariro, witnessed in the 1974–5 Ngauruhoe eruptions (Nairn and Self 1978). As such they were described on the background hazard map for the volcano (Fig. 3a). The map, published in 2007, consists of a summit hazard zone around each active vent, encompassing gas and ballistics at radii of 2–3 km for different vents based on experience of ballistic ranges in past eruptions at Tongariro National Park. Work is underway to develop ballistic and life safety models to better inform zone radius. Pyroclastic density currents (PDC's) and lava flows are not included in a hazard zone but are mentioned as a possibility in all valleys. Ashfall is stated as a hazard that could occur any place on the map. Text is provided, with instructions including to move quickly down off the mountain and away from summit hazard areas, though ballistics-specific advice was not provided (GNS Science 2007). The background hazard map with associated instructions was permanently posted at the entrances to the walking tracks up the volcano, was available on the GNS and DOC websites as well as on flyers at many of the tourist hubs (Leonard et al. 2008, 2014). The TAC hiking track cuts through most of the summit hazard zones, where access has been open at background levels. One hut, Ketetahi Hut, is located within the summit hazard zone, though is not reinforced to protect against ballistic impact.

Unrest was observed at the volcano up to three weeks before the eruption, initially in the form of increased seismicity and then increased magmatic gas content (Jolly et al. 2014a). In response the Volcanic Alert Level was raised from 0 to 1 (indicating unrest). Seismicity declined in the days prior to eruption and thus the TAC remained open to tourists (Jolly et al. 2014b), with seismicity reoccurring only ~5 min before the event (Jolly et al. 2014a). In the build-up to the eruption, a decision was made to complete response plans and create a crisis hazard map initially for the whole volcanic massif with some focus on the northern flank of Tongariro. However, it was not publically available before the August 6th eruption (Leonard et al. 2014). GNS volcanic alert bulletins were also produced, communicating updates on the precursory phenomena observed at Tongariro (Volcanic Alert Bulletins TON-2012/01–04; Fig. 5e). Meetings and other discussions were held with the local residents and businesses involved with the TAC to discuss the situation and future scenarios. Being wintertime, there was very little use of the track. As there was no one on the hiking trail during the eruption it is difficult to assess the success of the hazard communication strategies, and these strategies would have been different during summer months with heavy track use.

Following the August event, some of the local population evacuated for the night and the TAC was closed for two months due to the risk of further eruption. Within this two-month period an updated hazard and risk assessment was completed (Jolly et al. 2014b). This involved a combination of reviewing the eruptive record to understand eruption frequency and magnitude, and expert elicitation by GNS staff (the institute responsible for monitoring volcanoes and assessing their hazard/risk) working closely with the land manager (Department of Conservation) to produce three possible future eruption scenarios (a 21 November size eruption, a 6 August size eruption, and a magnitude larger eruption) and associated probabilities of these occurring. Probabilities were re-assessed every week immediately after eruption, which was subsequently extended to every month, then every three months as time passed. Hazard extent was considered for ballistics and PDC's for each scenario, exposure time along the impacted area, and the vulnerability (probability of fatality) of an individual to each hazard (using the area of hazard around an individual impact for ballistics, and the presence of a person in the path of a PDC), to calculate the combined risk of fatality for all scenarios (Jolly et al. 2014b).

Initial assessments suggested that ballistics were the main hazard to life from the eruption, though detailed mapping was not able to be carried out until months later when risk levels

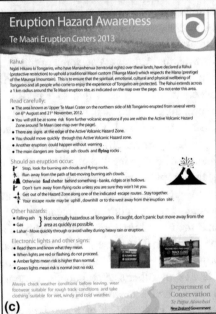

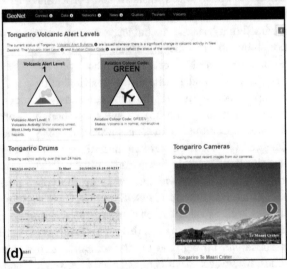

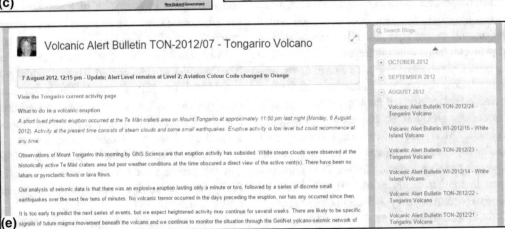

◀ **Fig. 5** Risk communication methods used at Tongariro, New Zealand. **a** Electronic signs communicating risk level and track closure at entrances to the volcano and where it crosses the AVHZ. **b** Signs advising area of increased hazard including a track-specific AVHZ hazard map. **c** Additional information on volcanic hazards at Tongariro (including ballistics), initially handed out to all hikers, provided on Department of Conservation website. **d** GeoNet website showing monitoring data such as Volcanic Alert Level, seismic drums and visuals of the volcano. **e** A Volcanic Alert Bulletin issued on the GeoNet website and distributed to media following the 2012 Upper Te Maari eruption

had decreased (Fitzgerald et al. 2014; Jolly et al. 2014b). The Department of Conservation (DOC), the agency responsible for hazard and risk management at Tongariro, began to implement risk management as part of a recovery programme. The risk assessments by Jolly et al. (2014b) became an important tool for making decisions about reopening. A new, event-specific Te Maari hazard map was created using mapped deposits and the most likely hazard scenarios, in which the main hazard zone was increased to a 3 km radius (choosing the larger potential radius based on historic events) down-slope and deliberately renamed the Active Volcanic Hazard Zone (AVHZ) to distinguish it from the former map (Fig. 3b). It included ballistics, explosions, pyroclastic density currents, lahars, gas and rockfall (Jolly et al. 2014b). The accompanying text to the crisis hazard map was also updated, with a ballistic specific instruction to 'seek immediate shelter from flying rocks if an explosion occurs' (GNS Science 2012). The map was released to the public alongside a Volcanic Alert Bulletin describing the changes made to the map and the source of the data (Volcanic Alert Bulletin TON-2012/23). This was distributed to the media (print, television, web and radio) to inform a wider audience (Leonard et al. 2014). Additionally, the map was posted at either ends of the track and where it crossed the boundaries of the AVHZ. Cordons, initially manned, were established at either ends of the TAC to prevent hikers from entering. Later, the cordon was moved to Emerald Lakes (on the edge of the 3 km Volcanic Hazard Zone) as the track was partially reopened. With declining risk of further eruption (based on the trend of the eruption probability estimates made by GNS to estimate how the expert elicitation might evolve over time), the track was fully opened 5 ½ months after the 21 November eruption.

DOC also published educational information on the eruption hazard at Te Maari including further advice on actions to take in an eruption (Fig. 5c). This included to 'stop, look for flying rocks', to 'find shelter behind something—banks, ridges or in hollows', to not turn away from 'flying rocks unless you are sure they will not hit you' and to 'get out of the Hazard Zone along one of the indicated escape routes' (Department of Conservation 2012). In October 2013 electronic warning signs were installed that informed hikers of the status of the volcano—a red flashing light meant danger-turn back, orange elevated risk and green normal volcanic activity (Jolly et al. 2014b, Fig. 5a). A survey of 203 hikers on the TAC in March–May 2014 indicated that most people saw these signs when activated red and understood the messages irrespective of their native language (Keys 2015). A reinforced public shelter and warden's quarters was one option being considered to replace the damaged Ketetahi Hut. Now the favoured option is to replace it with facilities outside the AVHZ.

5.2 Yasur Volcano, Vanuatu

Yasur Volcano is a frequently erupting basaltic scoria cone located on Tanna Island, Vanuatu (Cronin and Sharp 2002). Strombolian and Vulcanian eruptions have been relatively continuous since 1774 (Eissen et al. 1991). Ballistics are the main hazard produced by these eruptions, responsible for multiple fatalities in the past (Baxter and Gresham 1997). Yasur is one of Vanuatu's main tourist attractions with some twenty thousand people visiting the crater rim each year. The vast majority of people are guided up the volcano by local guides to watch the eruptions occur, with a main viewing area only 150 m from the crater's inner rim. As the majority of people in the area are transient tourists, guides are often relied upon to relay hazard and risk information to their patrons. Volcanic alert levels (VALs) and bulletins are posted on the Vanuatu Meteorology and Geo-Hazards Department (VMGD) website when the behaviour of the volcano changes. These sometimes include hazards maps that provide the locations of where bombs have been observed or are likely to impact, and often caution the public to approach the crater or hazardous areas with care. Maps also urge visitors, tourist agencies and communities to seriously consider the information provided prior to ascending Yasur (Vanuatu Geohazards Observatory 2009). However, advice or instructions are not given for what to do if caught in an area where ballistics are landing. A hazard map is displayed at the carpark before the ascent up the cone, highlighting the 1999 lava bomb impact zone and the observation location for each volcanic alert level—as the alert level increases so does the distance of the observation position from the cone (i.e. restriction zones are emplaced). In addition, visitors to Yasur are warned by a sign to 'Think Safety' before ascending the crater rim, though no further instructions or information is provided. As it is frequently erupting, it is assumed that visitors accept the risk that they are entering into an active volcanic hazard zone.

An updated risk management framework has been developed from 2012 to 2016 including updated bulletins and VALs, background and safety (crisis) hazard maps, and tourist information including education and safety map information. This is associated with an upgrade of Vanuatu's active volcanoes to real-time warning (at the time of writing this included a seismometer and webcam on Yasur and daily OMI satellite monitoring of SO_2 emissions; Vanuatu Geohazards Observatory 2014), supported by the New Zealand Aid Programme and GNS Science in partnership with VMGD. This integrated framework allows for pre-planning of safety zones related to ballistics and other hazards, and integration with warning products such as bulletins, VALs and tourist information. Ballistic zone ranges will initially be based on historic event ranges, but will be updated to include the modelling being developed in New Zealand, once available.

5.3 Sakurajima Volcano, Japan

Another frequently active volcano in which ballistics are a major hazard is Sakurajima Volcano, Japan. Continuous Vulcanian eruptions have occurred since 2009 from the andesitic composite cone (Japan Meteorological Agency 2013b). Sakurajima is constantly monitored by the Sakurajima Volcano Observatory and is considered to be one of the best monitored volcanoes in Japan (GSJ 2013). When activity changes, alert levels are posted on the Japan Meteorological Agency (JMA) website for the public to view. Many people live in close proximity to the volcano ($\sim$4900 within 5 km of the volcano) and millions visit the Kagoshima-Sakurajima area each year (3,702,000 in 2010; Japan Meteorological Agency 2013b), thus JMA and Kagoshima City released a volcanic hazard map with additional information in 2010. This map was distributed to local citizens and posted around the volcano. Three relevant zones are delineated on the map: the first is a 2 km radius (from the active craters) restricted area in which both residents and tourists are restricted from entering at all times; the second is $\sim$3 km away from the active vents showing the area expected to be inundated with volcanic bombs in a 'strong eruption', and lastly a 6 km radius extends around the active vents where 'volcanic rock' is likely to impact from a 'great eruption' (Kagoshima City 2010). Definitions for 'strong eruption' and 'great eruption' are not provided, nor is an explanation of the data that these zones are based on. The hazard map also includes societal components such as important landmarks i.e. schools and the visitor centre, and evacuation buildings and ports. The other half of the map consists of information on precursory phenomena likely to be felt and who to call if detected; how volcanic warnings will be disseminated and the measures needed to be taken; what the five volcanic alert levels are/what activity is expected and the consequent actions needed to be taken; information on major historic eruptions and recent activity; and evacuation procedures. An English version of the map is available in addition to the original in Japanese. This information is also available on the official tourism website of Kagoshima City (http://www.city.kagoshima.lg.jp/soumu/shichoshitu/kokusai/en/emergency/sakurajima.html). Ballistics (called 'cinders') are additionally listed on the site as a possible volcanic hazard accompanied by a description, particle size and travel distance. To prepare for a future eruption from Sakurajima, Tarumizu City (Kagoshima Prefecture) runs an emergency response exercise every year (http://www.city.tarumizu.lg.jp/kikikanri/kurashi/bosai/bosai/taisaku/sakurajima.html).

Three other notable risk communication and mitigation measures have been implemented at Sakurajima. A Volcano Disaster Prevention Council was created as a means of communication to discuss disaster prevention measures between volcanologists, local government, JMA, and other invested agencies (http://www.data.jma.go.jp/svd/vois/data/fukuoka/506_Sakurajima/506_bousai.html). Secondly, signs instructing people on the distance and direction to the nearest eruption safe house and evacuation port have been posted around the volcano. Lastly, concrete roofed shelters have been built around the island to protect visitors from falling ballistics (Erfurt-Cooper 2010).

5.4   2014 Eruption of Mt. Ontake, Japan

Mt. Ontake is a stratovolcano located on the island of Honshu, Japan (Japan Meteorological Agency 2013a). It is not a continuously active volcano with four eruptions (all phreatic) in its historic record (1979, 1991, 2007 and 2014; Japan Meteorological Agency 2013a; Smithsonian Institution 2013).

Mt. Ontake straddles the boundary of two prefectures—Gifu and Nagano, with trails on either side. Both prefectures have developed hazard maps for two eruption scenarios that include ballistics—the first a phreatic eruption similar in size to the 1979 eruption (VEI 2) and the second a larger eruption on the scale of 90,000–20,000 year recurrence interval (Nagano hazard map: http://vivaweb2.bosai.go.jp/v-hazard/L_read/53ontakesan/53ontake_2h03-L.pdf; Gifu hazard map: http://vivaweb2.bosai.go.jp/v-hazard/L_read/53ontakesan/53ontake_2h01-L.pdf). In both maps, ballistic hazard is defined by a 4 km asymmetric zone around an asymmetric vent area encompassing the 1979 vents—the same vents that erupted in the 1991 and 2007 eruptions. The parameter by which the zone is based on is not provided (e.g. maximum travel distance, spatial density of impacts) and no advice accompanies the hazard map, though a residents' handbook was printed that included examples of what ballistics are and how far they can travel. The maps and handbooks are available on the NIED database and the prefectural government websites, though the map is not signposted around the volcano.

Mt. Ontake is constantly monitored by the JMA, with seismometers, GPS stations, tiltmeters, cameras and infrasonic microphones (Japan Meteorological Agency 2013a). In addition, preparedness communication measures also include Volcanic Alert Levels, in place since 2008 (Japan Meteorological Agency 2013a). Similarly to other volcanoes, these VALs range from 1 to 5 and include whether the alert level is a warning or forecast, the target area (e.g. crater area or more distal residential areas), the expected volcanic activity and phenomena with examples of previous cases, actions needed to be taken and also keywords accompanying the level (e.g. level 5 with 'evacuate').

The 27 September 2014 phreatic eruption occurred at lunchtime on a busy autumn day when ~340 hikers were on the mountain (Tsunematsu et al. 2016). Multiple pyroclastic surges were produced, travelling up to 2.5 km from vent, in addition to ballistics that impacted up to 1 km from the vent (Kaneko et al. 2016;

Tsunematsu et al. 2016). Fifty-eight people were killed in the eruption, 55 most likely the result of ballistic trauma relatively close to the summit, with five still missing (as of 24 June 2016; Tsunematsu et al. 2016). An increase in summit seismicity was noted 16 days prior to the eruption resulting in the JMA releasing notices about volcanic activity, though activity was not at levels significant enough to raise the Volcanic Alert Level (there needed to be signs of deformation, which were not recorded until just prior to eruption; The Japan News, 26/10/14; Ui 2015). The eruption was largely unexpected with 11 min of precursory tremor, and uplift detected only seven minutes before the event (Ui 2015). This was a much shorter period of precursory activity than previous eruptions. The 1979 eruption was preceded by earthquake swarms for a year and five months. A month of seismicity was noted prior to the 1991 eruption, increasing in frequency just days before the event. And the 2007 eruption was preceded by inflation and seismicity for three months, accompanied by increasing fumarolic activity the week prior (Japan Meteorological Agency 2013a). Longer periods of precursory activity allow time for warnings to be issued. JMA released warnings prior to the 1991 and 2007 events, although the resulting eruptions were very small, only impacted the immediate area and occurred in winter outside the climbing season (Japan Meteorological Agency 2013a). However, if it had been possible to issue a warning when the precursory activity increased on the day of the 2014 eruption, it is unlikely that it would have resulted in no fatalities. Any evacuation warning prior to an event would need to occur at least an hour before the event and be immediately transmitted to all hikers on the summit area as it takes over an hour for hikers to move out of the ballistic hazard zone. Nonetheless, even a short warning time may have provided more hikers time to get to shelter.

Following the eruption, the Volcanic Alert Level was increased to 3, warning people not to approach the volcano (as access was restricted), and that blocks may be ejected up to 1 km from vent (based on previous eruptions). Signs were posted around the volcano telling people to "keep out" of the restricted area. Search and Rescue teams were deployed to rescue the injured hikers and those that sheltered in the buildings at the summit, and to recover the dead. Those that sheltered in the buildings around the summit survived the 2014 eruption, while many of the fatalities occurred due to hikers choosing to take photos and video of the eruption outside instead of running to the nearest hut. Half of the people autopsied by one doctor were found with cellphones in hand while one person's camera was found with a photo taken 4 min after the eruption occurred (Mainichi Shimbun 10/10/2014). Some then attempted to shelter around the summit shrine which they could not gain access to (the summit shrine is only open from the beginning of July to early September). Fatalities also occurred in exposed areas where there were no buildings in sight to shelter within. Personal safety measures taken by exposed hikers saved lives. This included sheltering behind large rocks, placing backpacks on heads, and wearing hard hats provided inside the mountain huts (NHK 2015).

Numerous risk management and communication tools have since been adopted. Prior to the eruption, Gifu and Nagano prefectures had separate commissions to manage volcanic activity from Mt. Ontake. Following the 2014 eruption they have combined to form one commission for the entire volcano, improving communication between the prefectures and subsequently to the public. The commission, similar to the Sakurajima council, is comprised of volcanologists, local government, JMA and other interested agencies (http://www.pref.nagano.lg.jp/kisochi/kisochi-sei saku/ontakesan/kazanbousaikyougikai.html). The council ran its first eruption evacuation drill on 4th June 2015.

Interviews conducted post-eruption showed that many climbers were unaware of the volcanic activity notices released, while of those that were aware 76% did not consider that they needed to be prepared for an eruption (The Japan News

26/10/2014; Shinano Mainichi Shimbun 2015). JMA subsequently launched a website to provide climbers with its observations of the volcanic activity around Japan, in an attempt to improve communication to climbers. From the 1st April, 2015 the Gifu Prefectural Government made it mandatory for all climbers of Ontake to submit a mountain climbing notification form prior to ascending Mt. Ontake, in an effort to improve knowledge of the number and location of people on the mountain, and to improve communication in times of crisis by recording their emergency contact information (http://www.pref.gifu.lg.jp/English/tourism/mountain/). Kiso, a town in the Nagano Prefecture responsible for one of the mountain trails, has also installed loudspeakers in the mountain cabins prior to easing restrictions in September 2015 (The Japan Times 27/09/2015).

In November 2015, a new hazard map was released by the Ontakesan Volcano Disaster Prevention Council (the combined commission mentioned previously). It provides two ballistic hazard zones—one for a phreatic eruption that extends 2 km from the vent area, and one for a larger magmatic eruption, extending 4 km from the vent area (http://www.city.gero.lg.jp/hazardmap/#12/35.9073/137.5203). The zones are based on research completed for Mt. Fuji on past ballistic distributions from phreatic and magmatic eruptions in Japan and around the world (Mount Fuji Disaster Prevention Council 2004). The asymmetric vent area has also been increased significantly, encompassing 3 km in length and $\sim$2 km in width. In addition, further research has been completed on the ballistic hazard produced in the eruption. Tsunematsu et al. (2016) describe an elongated distribution toward the N-NE resulting from an inclined ejection and topographic controls such as the shape of the valley the vents formed in. The spatial distribution was mapped from aerial photos by Kaneko et al. (2016) and delineated into four zones. The densest zone (A) encompasses areas with impact densities >10 impacts per 5 × 5 m, decreasing in density with distance from the vent to Zone C which has between 0 and 2 impacts per 5 × 5 m.

6 Discussion

6.1 Understand the Context and Assess the Risk

We identify from review of literature and analysis of the four case study volcanoes (Table 2) that understanding the risk context is highly important for effective communication associated with ballistic hazard and risk. Establishing this context and identifying potential risks requires engagement with potential stakeholders, such as those which may be exposed or affected by ballistic, or other, volcanic hazards. Effective communication is an essential component of this. Once these steps are complete, we then suggest that a ballistic risk assessment is undertaken to help underpin effective management and communication of ballistic hazard and risk. Best-practice ballistic risk assessment generally consists of: (1) reviewing the volcano's eruptive history to establish eruption frequency and eruption magnitude; (2) determining the nature and extent of past ballistic distributions; (3) exploring possible future ballistic distributions; (4) identifying assets exposed in the area; and (5) estimating the asset's vulnerability. Once complete, risk can be evaluated and appropriate management and communication strategies implemented. However, we stress that risk assessment alone cannot underpin effective communication of ballistic hazard and risk. But must be carried out in conjunction with the tools and strategies listed in Table 1 and Fig. 2.

It is important to remember that every context is different and what works at one volcano does not necessarily mean it will work or is needed at another. An assessment for a frequently erupting, highly visited volcano where risk management organisations are well resourced will require a different approach compared with an infrequently active, rarely visited volcano in a country where there are few resources available for risk management. The scope and scale of risk management activities should be guided by the risk context, and determine which and how risk management tools and strategies are used.

6.2 Reflections on the Four Case Study Volcanoes

All of the volcanoes studied are capable of sustaining injuries and fatalities from ballistics. The Mt. Ontake 2014 eruption resulted in the most fatalities from any of the case studies, and provides a chance to analyse why this was so with the aim of preventing it from occurring again. Multiple factors contributed to the high fatality rate:

- The eruption happened in peak season when ~340 people were on the mountain.
- Precursory activity only increased 11 min prior to eruption, resulting in an unexpected eruption. This meant no warning was able to be issued to the people on the summit and no closure of the summit prior to the event occurred. Previous eruptions had precursory events that gave more warning of the impending eruption underscoring that past history should not be solely relied on to predict outcomes of future unrest.
- The Alert Level was not raised following increased seismicity beginning 16 days before the eruption. A requirement for this to occur is the presence of ground deformation, which was not recorded until 7 min before the eruption.
- Hikers chose to take images and video of the eruption instead of finding shelter. This decision may have been different had hazard maps been posted around the volcano with instructions on actions to take in an eruption.

Fatalities from ballistics could occur at all of the case study volcanoes. However, a scenario with fatalities on the scale seen at Ontake is unlikely from Sakurajima due to the 2 km restriction zone. Yasur is visited by much fewer tourists than Ontake so it is unlikely to see as many fatalities from one event as occurred at Ontake, although the lack of shelter, lack of hazard advice, and proximity to the vent means that ballistic casualties are still relatively likely at this volcano. Work is ongoing to reduce this risk. The August 2012 eruption of Upper Te Maari is the most comparable to the Ontake eruption as it was largely unheralded and of the same explosivity. If the August 2012 eruption had occurred in peak tourist season, then a similar amount of fatalities as Ontake potentially could have occurred.

6.3 Critical Issues

We identify the following critical issues for contemporary and future communication of volcanic ballistic risk, based on our review of literature and analysis of the four case study volcanoes. We note many of these issues transcend volcanic ballistics to include nearly all volcano types and volcanic hazards:

- What is the most effective way to manage and communicate risk from volcanoes which are (highly) visited and/or settled which experience eruptions with very short and/or no meaningful warnings (e.g. Ontake, Te Maari)? This is a critical issue for managing ballistic risk, as eruptions with longer unrest phases typically allow evacuation of ballistic hazard zones before the eruption.
- What are the most appropriate risk management and communication strategies for volcanoes where ballistic (and other) risk is present which have poorly understood eruptive histories and/or monitoring systems?
- Effective ballistic risk assessment requires greater understanding of (a) the distribution of ballistic from a range of potential eruption styles, (b) the impact of ballistics to people and other societal assets (vulnerability/fragility characteristics), and (c) identification and (crucially) evaluation of what are the most appropriate mitigation actions to reduce ballistic risks before, during and after an eruption.
- Successful management of ballistic risk requires effective engagement (of which communication is a keystone) between authorities responsible for managing risk at volcanoes, those people and organisations who may have economic, cultural and social connections with a volcano, and the scientific community who can help inform hazard and (sometimes) risk considerations. Organisational and governance frameworks to

allow and facilitate this seem to be highly variable globally, but some relatively successful examples do exist (e.g. New Zealand).

- How to manage future risk, particularly for volcanoes where there is significant existing use and/or strong pressure to utilise the resources through tourism (increasing visitor numbers to high risk areas), and agricultural and settlement pressure from population growth.

7 Conclusions

Ballistic projectiles ejected in explosive eruptions present a major proximal hazard to life, infrastructure and the environment. An increasing population living on or close to active volcanoes and a growing volcano tourism industry give rise to an increased number of people exposed to ballistic hazard, presenting a considerable need for detailed ballistic hazard and risk assessments, and specialised communication and management strategies. Recommended strategies would include at least the following:

(1) Hazard and risk assessments (ideally probabilistic) specific to the volcano in question, which include ballistics where appropriate, that are made available to emergency managers and decision makers with authors/ scientists available to answer questions and advise where necessary and practical;

(2) The inclusion of ballistic hazard zones in hazard maps with accompanying advice on what to do. Maps should be updated in a crisis to reflect new information and readily available through a range of media. These maps should continue to be updated after the event when detailed scientific studies are complete;

(3) Volcano monitoring systems to monitor volcanic activity and indicate when a volcano is in unrest;

(4) The use of signage around the volcano to communicate ballistic hazard and risk, integrated with other hazard advice, including warning systems where practical, and with a focus on effectiveness of communication rather than just providing information;

(5) The use of volcanic alert bulletins, media releases or reports to communicate ballistic hazard and risk in crisis phases;

(6) Open, sufficiently frequent communication between scientists, stakeholders, emergency managers and local communities in which updates and training are provided, and informed input made into management and mitigation measures.

These strategies may vary with eruptive state (quiescence or crisis), frequency of eruptions, availability of resources, and whether ballistics are the main hazard at the particular volcano. In addition to the strategies mentioned in this chapter, further work is needed to test and update the advice provided to visitors on the actions to take in a ballistic eruption, in particular personal protective measures. Effort should also be made to provide consistent advice at all volcanoes on the actions to be taken, depending on the volcanic hazards involved. This way the information would be reinforced with visits to different volcanoes and increase the likelihood of visitors acting correctly.

Acknowledgements Funding for this study was provided by DeVoRA (Determining Volcanic Risk in Auckland) and a New Zealand Earthquake Commission (EQC) Biennial Grant (16/727). RHF is also supported by a doctoral scholarship from the Ngāi Tahu Research Centre. We wish to thank Bill McGuire and an anonymous reviewer for their thorough and constructive reviews.

References

Alatorre-Ibargüengoitia MA, Delgado-Granados H, Dingwell DB (2012) Hazard map for volcanic ballistic impacts at Popocatépetl volcano (Mexico). Bull Volc 74(9):2155–2169

Alatorre-Ibargüengoitia MA, Morales-Iglesias H, Ramos-Hernández SG, Jon-Selvas J, Jiménez-Aguilar JM (2016) Hazard zoning for volcanic ballistic impacts at El Chichón Volcano (Mexico). Nat Hazards. doi:10.1007/s11069-016-2152-0

Artunduaga A, Jimenez G (1997) Third version of the hazard map of Galeras Volcano, Colombia. J Volcanol Geoth Res 77:89–100

Barclay J, Haynes K, Mitchell T, Solana C, Teeuw R, Darnell A, Crosweller HS, Cole P, Pyle D, Lowe C, Fearnley C, Kelman I (2008) Framing volcanic risk communication within disaster risk reduction: finding ways for the social and physical sciences to work together. Geol Soc, London, Spec Publ 305(1):163–177

Baxter P, Gresham A (1997) Deaths and injuries in the eruption of Galeras Volcano, Colombia, 14 January 1993. J Volcanol Geoth Res 77:325–338

Becker JS, Saunders WSA, Robertson CM, Leonard GS, Johnston DM (2010) A synthesis of challenges and opportunities for reducing volcanic risk through land use planning in New Zealand. Australas J Disaster Trauma Stud 2010:1

Bertolaso G, De Bernardinis B, Bosi V, Cardaci C, Ciolli S, Colozza R, Cristiani C, Mangione D, Ricciardi A, Rosi M, Scalzo A, Soddu P (2009) Civil protection preparedness and response to the 2007 eruptive crisis of Stromboli volcano, Italy. J Volcanol Geoth Res 182(3–4):269–277

Bird DK, Gisladottir G, Dominey-Howes D (2010) Volcanic risk and tourism in southern Iceland: implications for hazard, risk and emergency response education and training. J Volcanol Geoth Res 189:33–48

Blong RJ (1981) Some effects of tephra falls on buildings. In: Self S, Sparks RSJ (ed) Tephra studies, proceedings NATO Advanced Studies Institute, Laugarvatn and Reykjavik, 18–29 June 1980, pp 405–420

Blong RJ (1984) Volcanic hazards: a sourcebook on the effects of eruptions. Academic Press, Orlando

Blong RJ (1996) Volcanic hazards risk assessment. In: Scarpa R, Tilling RI (eds) Monitoring and mitigation of volcanic hazards. Springer, Berlin, pp 675–698

Booth B (1979) Assessing volcanic risk. J Geol Soc 136(3):331–340

Bower S, Woods A (1996) On the dispersal of clasts from volcanic craters during small explosive eruptions. J Volcanol Geoth Res 73:19–32

Breard ECP, Lube G, Cronin SJ, Fitzgerald R, Kennedy B, Scheu B, Montanaro C, White JDL, Tost M, Procter JN, Moebis A (2014) Using the spatial distribution and lithology of ballistic blocks to interpret eruption sequence and dynamics: August 6 2012 Upper Te Maari eruption, New Zealand. J Volcanol Geoth Res 286:373–386

Burby RJ, Wagner F (1996) Protecting tourists from death and injury in coastal storms. Disasters 20(1):49–60

Calvari S, Spampinato L, Lodato L (2006) The 5 April 2003 vulcanian paroxysmal explosion at Stromboli volcano (Italy) from field observations and thermal data. J Volcanol Geoth Res 149(1–2):160–175

Christiansen RL (1980) Eruption of Mount St. Helens—volcanology. Nature 285:531–533

Cole JW, Cowan HA, Webb TA (2006) The 2006 Raoul Island Eruption—a review of GNS science's actions. GNS Science Report 2006/7 38 p

Coombs ML, McGimsey RG, Browne BL (2008) Preliminary volcano-hazard assessment for Gareloi Volcano, Gareloi Island. Alaska Scientific Investigations Report 2008-5159

Cronin SJ, Sharp DS (2002) Environmental impacts on health from continuous volcanic activity at Yasur (Tanna) and Ambrym, Vanuatu. Int J Environ Health Res 12(2):109–123

Cronin SJ, Gaylord DR, Charley D, Alloway BV, Wallez S, Esau JW (2004) Participatory methods of incorporating scientific with traditional knowledge for volcanic hazard management on Ambae Island, Vanuatu. Bull Volcanol 66(7):652–668

Department of Conservation (2012) Volcanic risk in Tongariro National Park. http://www.doc.govt.nz/parks-and-recreation/places-to-go/central-north-island/places/tongariro-national-park/know-before-you-go/volcanic-risk-in-tongariro-national-park/. Accessed Mar 2015

Dohaney J, Brogt E, Kennedy B, Wilson TM, Lindsay JM (2015) Training in crisis communication and volcanic eruption forecasting: design and evaluation of an authentic role-play simulation. J Appl Volcanol 4:12

Drabek TE (1995) Disaster responses within the tourist industry. Int J Mass Emerg Disasters 13(1):7–23

Eissen JP, Blot C, Louat R (1991) Chronologie de l'activité volcanique historique de l'arc insulaire des Nouvelles-Hébrides de 1595 à 1991. Rapp. Scientifiques Technique, Sci. Terre Geol.-Geophys. – ORSTOM (Noumea) 2

Erfurt-Cooper P (2010) Volcano and geothermal tourism in Kyushu, Japan. Volcano and geothermal tourism: sustainable geo-resources for leisure and recreation, Earthscan, p. 142

Erfurt-Cooper P (2011) Geotourism in volcanic and geothermal environments: playing with fire? Geoheritage 3:187–193

Ewart JW, Harpel CJ (2004) In harm's way: Population and volcanic risk. Geotimes, American Geological Institute. http://www.geotimes.org/apr04/feature_VPI.html. Accessed 15 June 2016

Fagents S, Wilson L (1993) Explosive volcanic eruptions—VII. The ranges of pyroclasts ejected in transient volcanic explosions. Geophys J Int 113:359–370

Fitzgerald RH, Tsunematsu K, Kennedy BM, Breard ECP, Lube G, Wilson TM, Jolly AD, Pawson J, Rosenberg MD, Cronin SJ (2014) The application of a calibrated 3D ballistic trajectory model to ballistic hazard assessments at Upper Te Maari, Tongariro. J Volcanol Geoth Res 286:248–262

Fudali R, Melson W (1972) Ejecta velocities, magma chamber pressure and kinetic energy associated with the 1968 eruption of Arenal volcano. Bull Volc 35:383–401

Geological Survey of Japan (2013) Sakurajima Volcano, 2nd edn. https://gbank.gsj.jp/volcano/Act_Vol/sakurajima/text/eng/exp01-5e.html. Accessed Feb 2015

GNS Science (2007) Volcanic hazards at Tongariro. http://info.geonet.org.nz/download/attachments/8585571/Tongariro_Poster_A4.pdf. Accessed Mar 2015

GNS Science (2012) Te Maari Eruption Phenomena. http://info.geonet.org.nz/download/attachments/8585571/Northern_Tongariro_eruption_phenomena.pdf. Accessed Mar 2015

Gregg CE, Houghton BF, Paton D, Swanson DA, Johnston DM (2004) Community preparedness for lava flows from Mauna Loa and Hualālai volcanoes, Kona, Hawai'i. Bull Volc 66:531–540

Gurioli L, Harris AJL, Colo L, Bernard J, Favalli M, Ripepe M, Andronico D (2013) Classification, landing distribution, and associated flight parameters for a bomb field emplaced during a single major explosion at Stromboli, Italy. Geology 41(5):559–562

Hadisantono RD, Andreastuti MCHSD, Abdurachman EK, Sayudi DS, Nursusanto I, Martono A, Sumpena AD, Muzani M (2002) Peta Kawasan Rawan Bencana Gung Api Merapi, Jawa Tengah dan Daerah Istimewa Yogyakarta scale 1:50 000 Direktorat Vulkanologi dan Mitigasi Bencana Geologi, Bandung

Harris AJL, Ripepe M, Hughes EA (2012) Detailed analysis of particle launch velocities, size distributions and gas densities during normal explosions at Stromboli. J Volcanol Geoth Res 231–232:109–131

Haynes K, Barclay J, Pidgeon N (2007) Volcanic hazard communication using maps: an evaluation of their effectiveness. Bull Volc 70(2):123–138

Haynes K, Barclay J, Pidgeon N (2008) The issue of trust and its influence on risk communication during a volcanic crisis. Bull Volc 70(5):605–621

Houghton BF, Swanson DA, Carey RJ, Rausch J, Sutton AJ (2011) Pigeonholing pyroclasts: Insights from the 19 March 2008 explosive eruption of Kilauea volcano. Geology 39(3):263–266

Japan Meteorological Agency (2013a) 53 Ontakesan. National Catalogue of the active volcanoes in Japan (4th edn). http://www.data.jma.go.jp/svd/vois/data/tokyo/STOCK/souran_eng/souran.htm#kantotyubu. Accessed Nov 2014

Japan Meteorological Agency (2013b) 90 Sakurajima. National Catalogue of the active volcanoes in Japan (4th edn). http://www.data.jma.go.jp/svd/vois/data/tokyo/STOCK/souran_eng/souran.htm#kantotyubu. Accessed Nov 2014

Jenkins SF, Spence RJS, Fonseca JFBD, Solidum RU, Wilson TM (2014) Volcanic risk assessment: quantifying physical vulnerability in the built environment. J Volcanol Geoth Res 276:105–120

Johnston DM, Bebbington MS, Lai CD, Houghton BF, Paton D (1999) Volcanic hazard perceptions: comparative shifts in knowledge and risk. Disaster Prev Manag 8:118–126

Johnston DM, Houghton BF, Neall VE, Ronan KR, Paton D (2000) Impacts of the 1945 and 1995–1996 Ruapehu eruptions, New Zealand: an example of increasing societal vulnerability. Geol Soc Am Bull 112:720–726

Jolly AD, Jousset P, Lyons JJ, Carniel R, Fournier N, Fry B, Miller C (2014a) Seismo-acoustic evidence for an avalanche driven phreatic eruption through a beheaded hydrothermal system: An example from the 2012 Tongariro eruption. J Volcanol Geoth Res 286:331–347

Jolly GE, Keys HJR, Procter JN, Deligne NI (2014b) Overview of the co-ordinated risk-based approach to science and management response and recovery for the 2012 eruptions of Tongariro volcano, New Zealand. J Volcanol Geoth Res 286:184–207

Kagoshima City (2010) Sakurajima Volcano hazard map. http://www.city.kagoshima.lg.jp/soumu/shichos hitu/kokusai/en/emergency/documents/sakurazimahm_eng.pdf. Accessed 19 Oct 2015

Kaneko T, Maeno F, Nakada S (2016) 2014 Mount Ontake eruption: characteristics of the phreatic eruption as inferred from aerial observations. Earth, Planets Space 68:72–82

Keys H (2015) Tongariro Alpine crossing visitors surveyed on effectiveness of new electronic light signs. Tongariro Aug 2015, pp 48–51. www.tongariro.org.nz/tongarirojournals. Accessed 18 Oct 2015

Keys HJR, Green PM (2010) Mitigation of volcanic risks at Mt Ruapehu, New Zealand. In: Malet J-P, Glade T, Casagli N (eds) Proceedings of the mountain risks international conference, Firenze, Italy, CERG, Strasbourg, France, 24–26 Nov 2010, pp. 485–490

Kilgour G, Della Pasqua F, Hodgson KA, Jolly GE (2010) The 25 September 2007 eruption of Mount Ruapehu, New Zealand: directed ballistics, surtseyan jets, and ice-slurry lahars. J Volcanol Geoth Res 191(1–2):1–14

Leonard GS, Johnston DM, Paton D, Christianson A, Becker J, Keys H (2008) Developing effective warning systems: ongoing research at Ruapehu volcano, New Zealand. J Volcanol Geoth Res 172(3–4):199–215

Leonard GS, Stewart C, Wilson TM, Procter JN, Scott BJ, Keys HJ, Jolly GE, Wardman JB, Cronin SJ, McBride SK (2014) Integrating multidisciplinary science, modelling and impact data into evolving, syn-event volcanic hazard mapping and communication: a case study from the 2012 Tongariro eruption crisis, New Zealand. J Volcanol Geoth Res 286:208–232

Lindell MK (2013) Risk perception and communication. In: Bobrowsky PT (ed) Encyclopedia of natural hazards. Springer, Netherlands, pp 870–874

Lube G, Breard ECP, Cronin SJ, Procter JN, Brenna M, Moebis A, Pardo N, Stewart RB, Jolly A, Fournier N (2014) Dynamics of surges generated by hydrothermal blasts during the 6 August 2012 Te Maari eruption, Mt. Tongariro, New Zealand. J Volcanol Geoth Res 286:348–366

Maeno F, Nakada S, Nagai M, Kozono T (2013) Ballistic ejecta and eruption condition of the vulcanian explosion of Shinmoedake volcano, Kyushu, Japan on 1 February, 2011. Earth, Planets Space 65(6):609–621

Mainichi Shimbun 10/10/2014. Ballistic blocks killed 20 people instantly. http://mainichi.jp/select/news/2014 1010k0000m040138000c.html. Accessed 29 June 2016

Mileti D, Nathe S, Gori P, Greene M, Lemersal E (2004) Public hazards communication and education: the state of the art. natural hazards informer, Issue 2. Boulder, p. 13

Minakami T (1942) 5. On the distribution of volcanic ejecta (Part I.): the distributions of volcanic bombs ejected by the recent explosions of Asama. Bull Earthq Res Inst 20:65–92

Mount Fuji Disaster Prevention Council (2004) Report of Mount Fuji Hazard Map Examination Committee (in Japanese). http://www.bousai.go.jp/kazan/fujisan-kyougikai/report/. Accessed 28 Jun 2016

Murphy PE, Bayley R (1989) Tourism and disaster planning. Geogr Rev 79(1):36–46

Nadim F (2013) Hazard. In: Bobrowsky PT (ed) Encyclopedia of natural hazards. Springer, Netherlands, pp 425–426

Nairn IA, Self S (1978) Explosive eruptions and pyroclastic avalanches from Ngauruhoe in February 1975. J Volcanol Geoth Res 3:36–60

Neal CA, McGimsey RG, Miller TP, Riehle JR, Waythomas CF (2001) Preliminary volcano-hazard assessment for Aniakchak Volcano, Alaska. United States Geological Survey Open File Report 00-519, Plate 1

NHK (Japan Broadcasting Corporation) (2015). Ontake: eyewitnesses or eruption. http://www.nhk.or.jp/d-navi/link/ontake2014-en/index.html. Accessed Jul 21 2015

NIED (1980) Field report of the disaster from Ontake 1979 eruption. Natural Disaster Research Report 16, 41 p

Odbert H, Hincks T, Aspinall W (2015) Combining volcano monitoring timeseries analysis with Bayesian Belief Networks to update hazard forecast estimates. EGU General Assembly 2015, 12–17 Apr 2015, Vienna, Austria

Oikawa T, Yoshimoto M, Nakada S, Maeno F, Komori J, Shimano T, Takeshita Y, Ishizuka Y, Ishimine Y (2016) Reconstruction of the 2014 eruption sequence of Ontake Volcano from recorded images and interviews. Earth, Planets Space 68:79

Pardo N, Cronin SJ, Németh K, Brenna M, Schipper CI, Breard E, White JDL, Procter J, Stewart B, Agustin-Flores J, Moebis A, Zernack A, Kereszturi G, Lube G, Auer A, Wallace C (2014) Perils in distinguishing phreatic from phreatomagmatic ash; insights into the eruption mechanisms of the 6 August 2012 Mt. Tongariro eruption, New Zealand. J Volcanol Geoth Res 286:397–414

Paton D, Millar M, Johnston DM (2001) Community resilience to volcanic hazard consequences. Nat Hazards 24:157–169

Paton D, Smith L, Daly M, Johnston D (2008) Risk perception and volcanic hazard mitigation: individual and social perspectives. J Volcanol Geoth Res 172(3–4):179–188

Pistolesi M, Rosi M, Pioli L, Renzulli A, Bertagnini A, Andronico D (2008) The paroxysmal event and its deposits. The Stromboli Volcano: an integrated study of the 2002–2003 eruption. Geophysica, 317–330

Pomonis A, Spence R, Baxter P (1999) Risk assessment of residential buildings for an eruption of Furnas Volcano, Sao Miguel, the Azores. J Volcanol Geotherm Res 92(1–2):107–131

Robertson R, Cole P, Sparks RSJ, Harford C, Lejeune AM, McGuire WJ, Miller AD, Murphy MD, Norton G, Stevens NF, Young SR (1998) The explosive eruption of Soufriere Hills Volcano, Montserrat, West Indies, 17 September, 1996. Geophys Res Lett 25(18):3429–3432

Scott BJ, Potter SH (2014) Aspects of historical eruptive activity and volcanic unrest at Mt. Tongariro, New Zealand: 1846–2013. J Volcanol Geoth Res 286:263–276

Shinano Mainichi Shimbun (2015) Verification of Mount Ontake eruption—living with a volcano. What do we learn from 9.27? The Shinano Mainichi Shimbun Press, Nagano (in Japanese)

Siegrist M, Cvetkovich G (2000) Perception of hazards: the role of social trust and knowledge. Risk Anal 20 (5):713–720

Sigurdsson H, Lopes-Gautier R (1999) Volcanoes and tourism. In: Sigurdsson H, Houghton B, McNutt SR, Rymer H, Stix J (eds) Encyclopedia of volcanoes. Academic Press, Cambridge, pp 1283–1299

Slovic P (2000) Perception of risk. In: Slovic P (ed) The perception of risk. Earthscan, London, pp 220–231

Small C, Naumann T (2001) The global distribution of human population and recent volcanism. Environ Hazards 3:93–109

Smithsonian Institution (2013) Ontakesan bulletin reports, Global Volcanism Program. http://www.volcano.si.edu/volcano.cfm?vn=283040. Accessed Dec 2014

Sorensen JH (2000) Hazard warning systems: review of 20 years of progress. Nat Hazards Rev 1(2):119–125

Sorensen JH (2013) Communicating emergency information. In: Bobrowsky PT (ed) Encyclopedia of natural hazards. Springer, Netherlands, pp 110–112

Sparks RSJ, Aspinall WP, Crosweller HS, Hincks TK (2013) Risk and uncertainty assessment of volcanic hazards. In: Sparks RJS, Hill L (eds) Risk and uncertainty assessment for natural hazards. Cambridge University Press, Cambridge

Steinberg G, Lorenz V (1983) External ballistic of volcanic explosions. Bull Volcanol 46(4):333–348

Swanson DA, Zolkos SP, Haravitch B (2012) Ballistic blocks around Kīlauea Caldera: Their vent locations and number of eruptions in the late 18th century. J Volcanol Geoth Res 231–232:1–11

The Japan News 26/10/2014. Improved steps needed to inform volcano climbers in Japan. http://the-japan-news.com/news/article/0001671312. Accessed 28 Oct 2014

The Japan News 27/10/2014. Mt. Ontake risks reported in 1979. http://the-japan-news.com/news/article/0001673442. Accessed 28 Oct 2014

The Japan Times 27/9/2015. Families of Ontake victims mark first anniversary of deadly eruption. http://www.japantimes.co.jp/news/2015/09/27/national/familieson takevictimsmarkfirstanniversarydeadlyeruption/#.VxRfHDB942w. Accessed 18 Apr 2016

Thompson MA, Lindsay JM, Gaillard JC (2015) The influence of probabilistic volcanic hazard map properties on hazard communication. J Appl Volcanol 4:6

Thouret J-C, Lavigne F, Kelfoun K, Bronto S (2000) Toward a revised hazard assessment at Merapi volcano, Central Java. J Volcanol Geoth Res 100(1–4):479–502

Tsunematsu K, Ishimine Y, Kaneko T, Yoshimoto M, Fujii T, Yamaoka K (2016) Estimation of ballistic block landing energy during 2014 Mount Ontake eruption. Earth, Planets Space 68:88

Turtle EP, Lopes RMC, Lorenz RD, Radebaugh J, Howell RR (2016) Temporal behavior and temperatures of Yasur volcano, Vanuatu from field remote sensing observations, May 2014. J Volcanol Geoth Res. doi:10.1016/j.jvolgeores.2016.02.030

Twigg J (2002) The human factor in early warnings: risk perception and appropriate communications. In: Zschau J, Kuppers AN (eds) Early warning systems for natural disaster reduction. Springer, Berlin, pp 19–26

Ui T (2015) The difficulty of predicting volcanic eruptions and releasing information. Geography 60(5):43–49 (In Japanese)

UNISDR (2009) United Nations International Strategy for Disaster Risk Reduction: UNISDR Terminology on Disaster Risk Reduction (2009). http://www.unisdr.org/eng/terminology/terminology-2009-eng.html. Accessed Jun 2015

Vanuatu Geohazards Observatory (2009) Volcanic Alert Status. http://www.geohazards.gov.vu/index.php/hazards-updated-events/volcano-alert-status. Accessed Mar 2015

Vanuatu Geohazards Observatory (2014) Vanuatu Monitoring Network (2012–2014). http://www.geohazards.gov.vu/index.php/geophysical-monitoring-network/vanuatu-monitoring-network. Accessed Apr 2015

Wardman J, Sword-Daniels V, Stewart C, Wilson T (2012) Impact assessment of the May 2010 eruption of Pacaya volcano, Guatemala. GNS Science Report 2012/09, 90 p

Williams KL, Keys HJR (2013) Reducing volcanic risk on the Tongariro Alpine Crossing. Report of a workshop 24 September 2013. Department of Conservation Tongariro District, 36 p

Wilson L (1972) Explosive volcanic eruptions II. The atmospheric trajectories of pyroclasts. Geophys J Roy Astron Soc 30(1):381–392

Wilson TM, Stewart C, Wardman JB, Wilson G, Johnston DM, Hill D, Hampton SJ, Villemure M, McBride S, Leonard G, Daly M, Deligne N, Roberts L (2014) Volcanic ashfall preparedness poster series: a collaborative process for reducing the vulnerability of critical infrastructure. J Appl Volcanol 3:10

Yamagishi H, Feebrey C (1994) Ballistic ejecta from the 1988–1989 andesitic Vulcanian eruptions of Tokachi-dake volcano, Japan: morphological features and genesis. J Volcanol Geoth Res 59(4):269–278

Imagining the Unimaginable: Communicating Extreme Volcanic Risk

Amy Donovan and Clive Oppenheimer

Abstract

This chapter considers the challenges surrounding the management of extreme volcanic risk. We examine eruption scenarios based on past episodes and assess the key issues that might arise should similar events occur in the future. The nature of such eruptions will entail transboundary and multi-scalar hazards. In a globalised world, the geopolitical and societal issues that are likely to emerge cannot all be predicted, and communication technologies themselves are likely to be affected. We explore two aspects: communication prior to the eruption, and communication during the eruption. To the best of our knowledge, all large eruptions are presaged by sensible phenomena but the enduring challenge for volcanic hazard assessment and risk management will remain the uncertainty surrounding evaluations of the likelihood, timing, nature and magnitude of potentially damaging activity. At present, too, communication of volcanic risk beyond the borders of the country where the volcano is located is generally patchy and unsystematic in most parts of the world (with the exception of the threat of ash clouds to aviation). In the preparatory phase, it is also critical to establish robust communication strategies that are resilient during an eruption. Such strategies would be essential for communicating the availability of supplies, the extent and nature of damage, and the ongoing status of the eruption.

A. Donovan (✉)
Department of Geography, King's College London,
The Strand WC2R 2LS, UK
e-mail: amy.donovan@kcl.ac.uk

C. Oppenheimer
Department of Geography, University of Cambridge,
Downing Place, Cambridge CB2 3EN, UK

Advs in Volcanology (2018) 149–163
https://doi.org/10.1007/11157_2015_16
Published Online: 26 March 2017

1 Introduction: Extreme Eruptions

Large magnitude volcanic eruptions are rare events with typically long return-periods (less frequent than ~1 in 1000 years) at any single volcano (Oppenheimer and Donovan 2015).[1] These range from large magnitude effusive basaltic lava eruptions to explosive super-eruptions. Basaltic lava eruptions, such as the 1783–4 Laki eruptions, are hazardous largely through their emissions of sulphur and halogens to the atmosphere, both locally and via long-range atmospheric transport. Such impacts include problems with air quality, disturbance of terrestrial and marine ecosystems, local contamination of water supplies (e.g. by fluorine leached from ash particles) and climate change. These eruptions may continue episodically for years. In contrast, large magnitude explosive eruptions usually last days or weeks, but can have more prolonged impacts. One estimate for an eruption with 100 times the sulphur yield of Pinatubo suggests that it would impact the climate strongly for approximately ten years (Timmreck et al. 2012). Such an event could have major impacts on food security, energy security and other critical networks.

The challenge of communicating these risks is multi-scalar and multi-dimensional: the potential impacts in the near-field to the far field are diverse, but are linked through the structures that manage them, such as local authorities and nation states. Such events and their impacts can transcend scale and are better defined by relationality: the relationships between groups of scientists with different responsibilities, political/policy groups, populations in different places, and a particular but as yet unidentified volcano. The volcano can be viewed as a geographical anchor in the physical landscape, to which different types of human network are connected. When we discuss communication of extreme volcanic risk, we are effectively examining the nature of the network connections and how they operate. Volcanic risk from large eruptions is fundamentally a systemic risk of low probability but high impact. Communicating such risk requires a very broad approach: the events are potentially global, they would require management by institutions at multiple levels, and they would involve input from a very wide range of experts, stakeholders and citizens (Fig. 1).

Here, we concentrate on aspects of the problem in two different timeframes (pre-eruption and immediate) and two particular but overlapping communication types (those between scientists and policymakers, and between scientists and populations). Initially, we define two scenarios for a large magnitude eruption. We then focus on the nature of systemic risks, and explore generalities of the communication of global systemic risks, focussing on the pre-eruption timeframe (by which we mean the period prior to any detection of anomalous activity). We examine the immediate period, when signals are detected and then an eruption commences. We argue that these two timeframes are not entirely distinct: it is critical that relationships and knowledge exchange takes place pre-eruption in order that it underpins communication in an immediate setting. The subject of this paper is challenging because it deals with scenarios not experienced in recent history, and therefore the discussion is sometimes necessarily speculative. However, we have drawn on examples both from volcanic crises and from the wider risk literature.

2 Volcanic Risk Scenarios

In this section, we outline two particular extreme volcanic risks, and discuss the likely challenges presented by each. This provides the context for the subsequent discussion in the paper.

2.1 Large Magnitude Basaltic Eruptions

The 1783–4 Laki Fissure eruption has been extensively studied (e.g. Thordarson and Self 1993, 2003; Thordarson et al. 1996; Schmidt et al. 2011, 2012; Hartley et al. 2014).

---

[1]In this paper, we focus on the impact of large magnitude eruptions. However, we note that a smaller eruption at the wrong time and in the wrong place could produce an extreme event, and some of the implications discussed in this paper may be relevant to such a scenario.

Fig. 1 Examples of the complex web of ideas, institutions, infrastructure and groups involved in the management of extreme volcanic risk

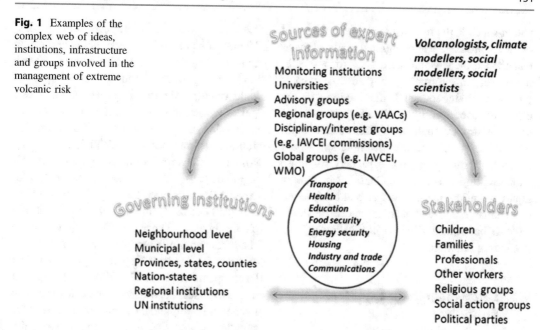

Thordarson and Self (1993) estimated that this eruption produced fire fountains up to 1.4 km high, sourcing plumes that rose into the stratosphere. Schmidt et al. (2011) modelled the potential air quality consequences of a comparable eruption occurring today and suggested that the long-range exposure to volcanogenic particulate it could significantly increase cardiopulmonary mortality across Europe. Thordarson and Self (2003) further implicated the eruption in the hot summer of 1783 and cold winter of 1783–4. Beyond climate impacts, such an eruption in the contemporary world could have a significant impact on aviation—especially if the eruption continued for months or years.

The 2014–15 volcanic eruption in Holuhraun, north of Vatnajökull, in Iceland represents an analogue for such an eruption, albeit at about a tenth of the magnitude scale. Precursory activity to this eruption was roughly two weeks in duration, with evidence for dyke propagation (Sigmundsson et al. 2014) from a central volcano, starting on 16 August 2014 and the first subaerial eruption on 29 August. Based on this scenario, there would be roughly two weeks' warning that magma was rising. As was demonstrated during the precursory activity

however, it is very challenging for volcanologists to forecast what is to the uncertainty in interpreting monitoring data from volcanoes, in general, is a theme that pervades the literature (e.g. Baxter et al. 2008; Marzocchi et al. 2012).

A large magnitude basaltic eruption could produce two primary hazards over large areas—gas hazard and climate forcing. In the near-field, lava flows and tephra could be a problem. Gas hazard would affect the immediate area around the volcano, but its dispersion would be heavily dependent on the height of the plume and the meteorological conditions (e.g. Schmidt 2014). For example, SO_2 from the Holuhraun eruption reached parts of Ireland and Norway (Schmidt et al. 2015; Gettelman et al. 2015). The extent of any climate forcing would be dependent on aerosol formation and transport, at least in the short term, and on the duration, seasonality and latitude of the eruption (e.g. Schmidt et al. 2012). In this scenario, then, the hazard would be spatially specific but long-range and variable. Forecasts of the hazard would be heavily dependent on meteorological data and models and source constraints—such as SO_2 flux measurements and plume height distribution. Management of the hazard would defer to individual nation-states in the first instance, but in the event of more prolonged and regional scale climatic disturbance there would be a need for a collaborative response to manage any adverse impacts on food production and distribution. If the eruptions occur in a populated area and/or a small country, evacuations to other nations might be necessary. Air quality deterioration would affect healthcare provision, especially for those with existing respiratory illness. Air quality issues and airborne ash could also affect aviation at regional scale, and airspace closures would have to be managed reflexively during a long, fluctuating eruption.

2.2 Large Magnitude Explosive Eruptions

Large magnitude explosive eruptions (>M6, according to the scale of Mason et al. (2004)), and super-eruptions (M8) can result in regional to global scale effects on climate (depending on sulphur yield to the atmosphere, location and timing) and regional scale devastation. Climate impacts from a 100× Pinatubo SO_2 release, for example, were modelled by Timmreck et al. (2012), and include decadal-scale global cooling of several degrees. This kind of scenario has major implications for food production globally. It would also affect trade, transportation and communication, particularly close to the source but with ripple-effects worldwide due to the nature of commercial aviation and global markets, and supply and distribution networks. Super-eruptions may even pose an existential risk (e.g. Rampino 2002). The complexities and diversity of the direct and indirect consequences of such large events make it very difficult to assess the risks in a meaningful way, and indeed it could be said that there have been no comprehensive efforts to assess the integrated impacts of a super-eruption on global society. Hereafter, we focus on an M8 scenario similar to that of the Youngest Toba Tuff (YTT) eruption, circa 74 ka BP.

A large magnitude eruption would yield >10 km^3 of tephra in a matter of hours or days. In explosive eruptions of this size, substantial ash clouds are generated and associated plumes would circumnavigate the globe within a few weeks. Pyroclastic flows would likely extend tens of km from the volcano. The estimated minimum tephra mass for the YTT is 2×10^{15} kg (Rose and Chesner 1990): an eruption of this magnitude could affect the continental scale with severe implications for farming and agriculture. A major uncertainty in such an eruption is the volatile budget. There has been considerable debate concerning the sulphur yield of the YTT eruption, for example (e.g. Rampino and Self 1992; Oppenheimer 2002; Williams 2012), and there is even less consensus on the halogen yields of such large eruptions. Cadoux et al. (2015) for example showed that halogen inputs from the large Minoan eruption of Santorini could have had significant impacts on atmospheric ozone. In any case, contamination of water supplies and ecosystems over large areas would likely lead to major food security problems. This could provoke or exacerbate epidemics and social unrest (Oppenheimer and Donovan 2015). While the prospect of such a global hazard can result in a return to environmental determinism (e.g. Rampino 2002), so widely condemned in the disasters literature (e.g. Raleigh et al. 2014), it nevertheless raises the question of global vulnerability and the complexity of networked societies and nations. While many studies have examined the impacts of past large magnitude eruptions (e.g. Oppenheimer 2011), there are no recent analogues to assess the impact of such an eruption on modern globalised society—particularly the impacts on technologies, including communication, transportation and power.

2.3 Volcanic Risk Webs

Approaches to volcanic eruption management have traditionally attempted to follow the so-called linear model, in which scientists produce evidence that they use for a risk assessment, which is then presented to policy-makers who make decisions and then communicate those decisions to the public (e.g. Marzocchi et al. 2012). However, the empirical literature in volcanology and in other fields of environmental policy demonstrates that the linear model is flawed in practice because scientists and policymakers are part of the "public" and make social inferences throughout the process, as well as being affected by a range of political factors (e.g. Owens 2005; Owens et al. 2006; Jasanoff 1990, 2005; Donovan and Oppenheimer 2014). In essence, decision-making is networked and web-like, not linear, and communication and decision-making are not readily disentangled, since decisions will be interrogated by stakeholders. For example, during and after the 2010 eruption of Eyjafjallajökull in Iceland, the UK press attacked the Met Office's handling of the crisis (Harris 2015). Furthermore, large magnitude eruptions will not fit easily into human boundaries—national or institutional. Their management will depend not only on scientific monitoring and information, but also on the complicated networks of food production, security, transportation, electricity, political powers, water supply and communication, for example—and on the connections and dependencies between these networks. Risks of large magnitude volcanic events are best conceptualised in the context of networks or webs of interactions between critical infrastructure, institutions, political powers and the Earth system itself.

In the event of a super-eruption, for example, the impacts on global aviation and even shipping would have major impacts on supply chains of both food and technology (see Fig. 2). This would be significantly compounded by poor harvests in regions affected by volcanically-forced climate change. Increased food prices can trigger or exacerbate civil unrest and public health problems, limiting economic growth (e.g. Benson and Clay 2004). Areas and nations less affected by transport restrictions or more self-sufficient in terms of food requirements would be able to adjust to some extent but a severe economic impact on key banks in, for

example, East Asia, would cause liquidity crises in Europe and the United States.

As the eruption continues, or ash fallout across wide areas becomes re-suspended, much of the globe could have to manage repeated airspace closures and trade disruption. Climate impacts could be felt for years after the eruption (e.g. Timmreck et al. 2012), with poor harvests driving the global economy further into crisis and causing conflicts over resources (e.g. Godfray et al. 2010; see also Gassebner et al. 2010). Communication technologies could be badly affected as ashfall damages infrastructure including telecommunications masts, limiting the scope of cooperative risk management across and within national borders. In this scenario, we can only scope out the potential impacts, but several attempts have been made to rationalise such an extreme event (e.g. Denkenberger and Pearce 2014; Rees 2013). Modelling such events is highly complex, not least because it depends on numerous source factors which have large ranges (such as sulfur load, timing, location) as well as on the fragilities and vulnerabilities of the global food system, transport systems and governance systems. On top of this, uncertainty—both scientific and social—has to be taken into account to produce meaningful results. The problem is transdisciplinary because it requires action by experts (scientists, social scientists), governments and other stakeholders (Fig. 1). The uncertainty is therefore likely to be amplified because of the combination of monitoring methods, models and interpretation across different groups.

A further source of uncertainty is the trans-boundary nature of these large magnitude scenarios. Both of these scenarios pose challenges that transcend national boundaries, but will nevertheless be significantly affected by them, as nations differ in how they manage volcanic risk. At present, the management of volcanic risk is primarily the work and responsibility of individual nation states. Institutional frameworks vary considerably between nations (e.g. Donovan and Oppenheimer 2015a), and are not necessarily readily combined. Volcanoes are inevitably sited in particular places, and if other nations are affected by an eruption, they will be dependent at

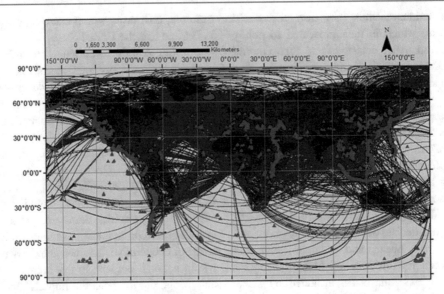

Fig. 2 Global flight routes (*black lines*). Also shown are volcanoes listed in the LaMEVE database (Crosweller et al. 2012) and population density (*green* to *red*)

least in part on the "host nation" for information. This has been demonstrated in recent eruptions in Iceland (Donovan and Oppenheimer 2012), and also in Ethiopia and Eritrea (where the lack of diplomatic relationships caused problems as well; Yirgu et al. 2014). Information about the eruption of Nabro volcano in Eritrea was primarily sourced from satellite data, demonstrating the importance and significance of recent developments in remote sensing of volcanoes (e.g. Biggs et al. 2014).

An additional challenge of transboundary events is the issue of consistency. Nations may vary in their responses to eruptions and in their willingness to issue evacuation orders. One country may evacuate its citizens from the immediate proximity of the volcano and another may not. In the event of threats to air quality from volcanic emissions, one country may provide free masks and another not. These factors are important because of the potential for the situation to be exacerbated by social unrest as citizens in one state feel less well provided-for than those in another. Hence, such an eruption could reverberate through global social networks, amplifying uncertainties and significantly affecting social stability. There are also spatial differences in vulnerability and exposure: societies

will not be affected equally, and this could create significant challenges for security and for the allocation of resources. The Millennium Declaration, for example, states that "global challenges must be managed in a way that distributes the costs and burdens fairly in accordance with basic principles of equity and social justice" (UN General Assembly 2000, p. 1, paragraph 6). This effectively refers to "risk sharing": the principle that risk is reduced for those most affected if it is shared with those who are less affected. This would be challenging in the event of an M8 eruption in which there are global impacts, and requires careful planning at an international level prior to the event.

The probability of either of the scenarios in this paper being realised is very low, if it is based on frequency analysis of past events. Frequency-based probabilistic assessments are commonly used in volcanology as "base-rates" (e.g. Mader et al. 2006; Self 2006). However, as a volcanic crisis unfolds, additional information may become available—such as an increase in seismic activity for example. Many volcanologists would argue that this information suggests that there is an increased probability of an eruption—but a frequency-based analysis cannot incorporate this information as it generally requires higher levels of judgement, and belief-based probabilistic methods may be used (e.g. Bayesian methods, expert elicitation; Marzocchi et al. 2007; Aspinall et al. 2003; Newhall and Pallister 2014). The scenario then becomes a "single event" problem (Gigerenzer 1994). These two scenarios—the longer term risk from large magnitude eruptions and the immediate potential for such an eruption—represent our two time-scales hereafter. Initially, we argue that the longer-term risk from these eruption scenarios requires engagement with policymakers at all levels to ensure that there is awareness of the risk—it has to be on the global agenda as a systemic risk. We then explore some of the implications of this for scientists. Finally we discuss the evolution of a scenario into a single event problem, and the ensuing challenges of communication.

3 Systemic Volcanic Risk: Global Communication Structures and Decision-Making Systems

Ultimately, the risks associated with these scenarios are systemic: they occur at the interface between the human and the physical, and require a holistic approach to risk management. According to Renn and Klinke (2004), "A holistic and systemic concept of risk must expand the scope of risk assessment beyond its two classic components: extent of damage and probability of occurrence." Haldane and May (2011) compare the complex systems of banking to the complexity of ecosystems, for example: there are multiple connections between actors and institutions that are dependent on one another and that transcend scale. Global systemic risk is a direct result of globalisation: it is a networked risk. While networks make risk more manageable in some ways by adding robustness, they can also increase fragility (Beale et al. 2011; Goldin and Vogel 2010), because a break in one part of the network affects the whole network.

The complexity of global networks sits uneasily with the existence of nation states (e.g. Sassen 2006): "bits of territory, authority and

rights" are assembled on multiple scales that transcend the traditional "local to national to global" scalar distinctions. Sassen (2006) argues further that "new types of orderings" are emerging. The global community of volcanologists also exists in this precarious spot between national and global—volcano monitoring scientists work within the institutional structures of governments, yet also participate in a global scientific debate about methods, new interpretations and new data—all of which can feed into their work within a state. This can have the positive effect of adding robustness to risk management within a nation, but can also produce culture clashes between the scientific perception and the local response (e.g. Donovan et al. 2014a, b). Such culture clashes have materialised in connection with outputs from the Intergovernmental Panel on Climate Change (IPCC), for example (e.g. Hulme 2009, 2014), and also in debates in different countries about issues such as genetically-modified crops. The focus in discussions around the IPCC is not merely a result of the evidence and uncertainties, but also a matter of how evidence is presented (Hulme and Mahoney 2010).

Communication of extreme volcanic risk at a global scale is thus immensely complex. It is affected by geography—availability bias, for example, will make some people more readily able to conceive of volcanic impacts than others (Tversky and Kahneman 1974). There are cultural variations not only in the way that risk is perceived, but also in the way that it is managed (e.g. Dake 1992). In the case of extreme volcanic risk that affects multiple nations and cultures, effective communication would have to use a range of media, and would require the involvement of a wide range of actors and institutions with a consistent and culturally sensitive narrative.

3.1 Managing Communication

The volcanic ash advisory centres (VAACs) have some experience in communicating transboundary volcanic hazard from ash plumes. They require governments to monitor their volcanoes and provide information, usually via volcano observatories, when eruptions are imminent or ongoing. Models are run by the VAACs to assess the trajectory of ash clouds and information is provided for use by aviation authorities. It thus addresses a very specific problem with key actors who have some control over operations. However, the systematic allocation of areas of the globe to particular VAACs does ensure that responsibility is clear, and the information is available to those who need it in a straightforward way.

The requirement that there are global systems in place for managing global risks can, however, obfuscate the complexity of the problems at smaller scales. There is, for example, wide variation in the use of colour codes and alert levels between nations (and sometimes within them). This can depend on historical experience and on dominant types of volcanism, for example (e.g. Potter et al. 2014; Fearnley et al. 2012). The use of alert levels is also not very reliable at present (Winson et al. 2014). There are therefore some issues with any potential "global alert level system", and while there is a global aviation colour code system, it is hazard-specific to ash plumes that might affect aircraft.

The absence of any international mechanisms for volcanic risk assessment and communication is problematic for several reasons, not least because it means that any response will be reactive. Goldin and Vogel (2010) note that many of the "obvious" international decision-making institutions (such as the World Bank and United Nations) "are already overloaded": they have been stretched beyond their original remits by globalisation. Global governance itself is highly complex and driven by regulation, but it is also in its infancy, and is beset by issues such as the achievement of global democracy. There are also, importantly, much more pressing issues for global governance to deal with than the small possibility of a large magnitude eruption. Volcanologists are familiar with the difficulties of getting governments to act (e.g. Oppenheimer 2011): volcanoes are not that important. Hence there is no mechanism for

international scientific advice in large magnitude events.

Recent work on volcano early warning systems (e.g. Fearnley et al. 2012; Potter et al. 2014; Winson et al. 2014) has demonstrated some of the challenges of applying such systems in practice. Similarly, work on scientific advisory mechanisms has also demonstrated their social complexity (Donovan and Oppenheimer 2015b; see also for example Hulme 2014 on the IPCC). Nevertheless there is a strong argument that mechanisms for providing global warnings about volcanic activity are needed. It is worth noting that the Sendai Framework for Disaster Risk Reduction 2015–2030 called for a greater role for science (including social, economic, engineering, physical and medical sciences) in disaster risk reduction, and this presents an opportunity for the development of international advisory systems that fully integrate expertise from all of these fields, learning from the experiences of similar bodies such as the IPCC.

Rare events require three things from scientific advisors: imagination, flexibility and rapid response. Even where governments have no interest in preparing, scientists can be ready to offer advice: after all, governments will want it quickly when the need arises! In spite of this, there is a need for at least a mechanism for scientific advice during large magnitude eruptions. The model of the Volcanic Disasters Assistance Program is useful here—it gives precedence to local scientists whilst also providing resources (Pallister 2015). A further consideration, however, is the integration of social sciences throughout the processes of risk assessment and communication. Risk communication in an extreme event would require several characteristics: consistency, transparency and reflexivity. Consistency does not mean that the communication should always say the same thing about the risk; it rather means that messages should be internally consistent and clear about what is and is not known. It can be damaging when one group claims to have better information than another. This is therefore aided by transparency about the information itself and how it is being used, as well as about uncertainty. Reflexivity refers to the subjectivity of those doing the communicating, and how they manage it (Donovan et al. 2014b; see also Alvesson and Sköldberg 2009; Gibbons et al. 1994, for general discussions of reflexivity). This requires a level of personal integrity and self-awareness. It becomes threatened if the communicator has too much emotional investment in the information, for example.

3.2 The Nature of Communication

The systemic nature of these risks requires a long-term scientific and social scientific engagement with decision makers in international, regional, national and local institutions (see Fig. 1). Work in risk communication has demonstrated unequivocally that communication links that are established prior to a crisis are critical in facilitating communication during a crisis (e.g. Donovan and Oppenheimer 2012; Marzocchi et al. 2012; Barclay et al. 2008; Bird et al. 2008; Haynes et al. 2008). One key recommendation is therefore that experts ensure that they are in communication with civil protection organisations even when volcanic activity is low. Such contact does not have to be continuous, but it should be relatively regular (for example, a meeting every six months—though this is likely to vary between institutions and circumstances). Meetings might include simulated drills, discussions of earthquake activity and monitoring data over the last few months, updates on response plans and discussion of new results.

Risk communication has different requirements depending on whether one considers high-frequency low impact events or low-frequency high impact events, especially at the extreme event end of the scale. This requires judgement: one of the problems with extreme events that have low probabilities but high impacts is that they can capture scientists' and journalists' imaginations. As Pidgeon and Fischoff (2011) point out, listening is also a form of communication, and so is silence. Govern-

ments have struggled to identify the "right" time to tell populations about extreme risks, and survey respondents generally want to be told but recognise that the issue is ambiguous (e.g. Eiser et al. 2014; Donovan et al. 2014a, b). In general, low probability risks are communicated if the probability increases. Setting a threshold for this can, however, be challenging because of the fear of "false alarms". Volcanologists have struggled with this balance in the past, and this has fed into studies to find appropriate statistical approaches (e.g. Woo 2008; Aspinall et al. 2003). Successful management depends on the communication not only of risk but of uncertainty. In the presence of high uncertainty, there is no such thing as a false alarm (e.g. Hincks et al. 2014, showed that the evacuations on Guadeloupe in 1976 were justified even though there was no eruption, because there was very high uncertainty). However, high uncertainty regarding a low probability but high impact risk is challenging, and is also a situation in which the precautionary principle can produce paralysis rather than rational decisions (e.g. Sunstein 2005), not least because of the high economic cost of always erring on the side of precaution. With regard to extreme volcanic risk pre-crisis, communication with the public has to be carefully considered and framed. It also has to be transparent: the results of the L'Aquila trial demonstrate the importance of clear and open communication, for example. After the 2009 earthquake, inhabitants felt that they had been misled because they had been explicitly reassured (Alexander 2014), rather than told that the risk was low but not zero.

Framing is a critical aspect of risk communication (e.g. Barclay et al. 2008). A frame is a social construct that allows the interpretation of complex information. Frames can be negotiated through dialogue between all the participants in a conversation to ensure that they are appropriate and effective. It is important not only to state scientific knowledge and uncertainty, but also to engage with the audience. Scientists' perceptions of the audience are also likely to affect their communication (Donovan et al. 2014a) and must be handled carefully. This is a subjective process, but can be rationalised through collaboration and discussion between physical scientists, social scientists, policymakers and the public.

4 Single Event Communication

Once volcanic activity is detected and expert opinion is that the eruption may be very large, the volcanic risk web has to be "activated". Figure 3 is a map of the world showing the location of tweets with the hashtag #Holuhraun from 29 August to 2 September 2014 (there was a small eruption on 29 August, and then the fissure re-opened from 31 August 2014 to 27 February 2015). It demonstrates the velocity with which information about volcanic eruptions travels the web. Secrecy about a potential large eruption would be very difficult to maintain for any length of time. Ultimately, the communication of such risk would be the responsibility of governments, but it would be strongly dependent upon relationships between scientists, social scientists and governments—and the trust that is

placed in them by industry, NGOs and other stakeholders, including the public.

In Donovan et al. (2012), we introduced a modified framework for the management of risk in volcanic crises, based on the Science Studies literature (Wynne 1992; Stirling 2007). A simplified version of this is shown in Fig. 4, demonstrating the transdisciplinary nature of the problem: the types of question that are generated from such a systemic risk are multi-dimensional. There are risks, uncertainties, ambiguities and ignorance that affect multiple connected groups and entities.

The management of such a situation would quickly escalate beyond nations, and require coordination (e.g. by UN institutions). It would require coherence and clarity about the potential impacts of the eruption, and about its management by different groups in the risk web (e.g. distinguishing the responsibilities of government from those of individuals). In this respect, the communication of risk moves far beyond volcanologists—but also requires that volcanologists retain their integrity as sources of expert information. In the next subsection, therefore, we discuss some of the broad issues that affect

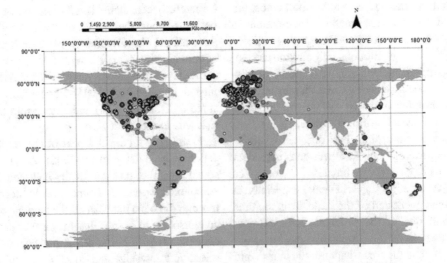

Fig. 3 Tweet map, showing tweets with the hashtag #Holuhraun between 29 August and 2nd September. Colours and size represent the day: *red* 29 August, *orange* 30 August, *yellow* 31st August, *light green* 1 Sept, *dark* *green* 2 Sept. Note that only tweets from users who disclose their geolocation can be mapped; this amounts to ~2500 tweets, of a total of ~5600

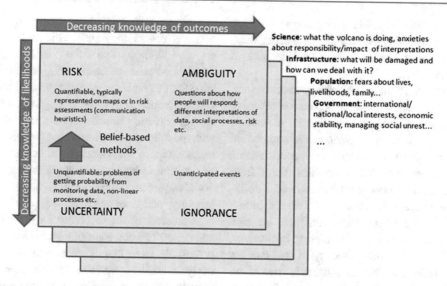

Fig. 4 Different aspects of uncertainty around volcanoes, modified from Stirling (2007), Wynne (1992) and Donovan et al. (2012). Each category represents a different aspect of uncertainty that will be represented in each part of a network, including scientific institutions, infrastructure (transport, energy, agriculture etc), government institutions and different social groups

volcanologists involved in expert advice. This affects both long-term and single event communication: we suggest that the communication of extreme volcanic risk, though conceptually broken into these two aspects, actually depends on their integration. The management of an extreme volcanic event is likely to be dependent upon communication networks that are instituted prior to the detection of activity.

4.1 Professionalising Volcanology

Some of the issues that have been raised in this chapter speak directly to IAVCEI's Crisis Protocols (Newhall et al. 1999). Recent events, such as the trial of six seismologists in L'Aquila, Italy (e.g. Marzocchi 2012; Alexander 2014) have led to renewed calls for protocols and guidance for volcanologists who are involved in policy advice (e.g. Aspinall 2011). Volcanology in certain circumstances comes to resemble a formal profession (such as medicine or law; Baxter et al. 2008). In light of social media developments as well as an increasingly litigious world, the importance of individual responsibility cannot be

ignored in the context of global risk. Figure 3, for example, shows that the interest in the Holuhraun eruption was global and not scaled. In Iceland, researchers were asked not to tweet photos of themselves that were inconsistent with doing scientific work: the government was under pressure to allow tourists into the restricted area, which was a flood plain that would be affected rapidly in the event of subglacial eruption. The additional requirement to monitor social media sites may be an important consideration for future planning by volcano monitoring institutions. The interest and availability of information about volcanic activity—without any quality controls—is therefore challenging to manage. Communication technologies such as Twitter and Facebook are double-edged: they can both complicate and aid disaster management. Furthermore, they may require additional resources from responsible institutions in the event of a crisis. Where there is potential for a large magnitude eruption—for example, an unrest episode at a known supervolcano—it is likely that media attention would be significant and that volcanologists around the world might be asked to comment on a volcano about which they know relatively little. The likely involvement of the global volcanology community in a large magnitude event necessitate a brief consideration of these issues here.

The implications of this for researchers are, ultimately, value-driven: they require reflexive management of subjective tendencies to want to be involved in every volcanic crisis, and careful consideration of comments on social media, following professional guidelines (e.g. from IAVCEI). Awareness that there may be a political context that individuals at a distance cannot see is also important. However, there are also implications for volcano monitoring institutions, particularly concerning the need to monitor social media but also its potential. Bird et al. (2008) demonstrated that the level of information provided by the Icelandic Meteorological Office on its website was appreciated by the public. Donovan and Oppenheimer (2012) found that scientists interviewed in Iceland felt that the availability of data had led to the public becoming, over time, able to use that information effectively. There are considerable advantages to a long-term dialogue with the public that familiarises people with the kinds of information available in a crisis—though there may be disadvantages and understandable insecurities in making data too readily available, not least the potential for misinterpretation (Donovan and Oppenheimer 2015a, b). Again, this issue is magnified in the case of supervolcanoes, as data indicating an increase in activity might cause widespread concern.

A further uncertainty in the present case—large magnitude eruptions—is that the source volcano may not have received much attention, compared with frequently active systems such as Etna. There may well be a trade-off between the involvement of scientists who have experience in policy advice but no local knowledge, and the involvement of, for example, a single local scientist who happened to do their PhD studies on that volcano, but is considerably more familiar with petrological methods than with providing hazard advice. Again, reflexivity is an approach that may aid in this situation.

Comparisons have been made in the past between volcanology and medicine (e.g. Baxter et al. 2008), and the introduction of "evidence-based volcanology" (Aspinall et al. 2003) builds on ideas from the medical literature. One difference is that volcanologists do not take the Hippocratic Oath and are not trained to deal with people or indeed to refer patients to colleagues whose specialties might be more appropriate. The culture of academia in particular is not always conducive to effective delegation or humility, and academics are not trained for the high pressure of responsibility for matters of life and death. The detection of signals that suggest a potentially large magnitude eruption at a previously unstudied volcano would have major impacts on the expectations that the public have of volcanologists, and on the pressure and responsibility faced by experts. Academic training, at present, rarely includes mandatory training in science communication for policy, or indeed in statistics for risk assessment—but volcanology is slowly turning into a profession, and this requires some adjustments in pedagogy.

5 Conclusions

This chapter has addressed some of the issues around the communication of extreme volcanic risk. We have demonstrated that such a risk is fundamentally systemic, not local, and it transcends scale. It is a networked risk that can be visualised as a web of connections between governments, institutions, infrastructures, industry, experts and populations, and "grounded" geographically through an as-yet-unspecified volcano. Such an event would reverberate through the risk web, and requires transdisciplinary collaboration. It is therefore important that communication pathways are established prior to any eruption. Communication about extreme volcanic risk prior to the detection of any activity will dictate the nature of communication when activity is detected. It should be characterised by four ideas:

- Appropriate framing: this requires careful consideration of the social context of the scientific information and also the potential impacts that it might have on the population.
- Intersubjective validation: this represents the importance of some level of scientific consensus—peer review—so that messages are clear.
- Dialogue and listening: communication is a two-way process, and this means that concerns from policymakers and populations have to be taken seriously and addressed
- Reflexive adaptation: in a rapidly developing situation, it is important that individual fears, assumptions and anxieties are acknowledged and addressed within the communication process.

These concepts are suggested as a means of structuring discussions between social and physical scientists, policymakers, officials and populations: they require transdisciplinary discussions. In the event of an imminent large-magnitude eruption, there will not be time to build relationships from scratch, nor to debate the appropriate role of different stakeholders and experts. Experts and civil institutions have to work together both to establish relationships and to assess existing communication technologies for their resilience to large magnitude hazards. However, much of this is required at a political level. The management of the high uncertainty in volcanic crises, the communication challenges and the need for engagement over the long term suggest that volcanology is becoming a profession in which agreed standards for practice are required.

References

Alvesson M, Sköldberg K (2009) Reflexive methodology: New vistas for qualitative research. Sage, Beverly Hills

Alexander DE (2014) Communicating earthquake risk to the public: the trial of the "L'Aquila Seven". Nat Hazards 72(2):1159–1173

Aspinall WP, Woo G, Voight B, Baxter PJ (2003) Evidence-based volcanology: application to eruption crises. J Volcanol Geoth Res 128(1):273–285

Aspinall W (2011) Check your legal position before advising others. Nature 477:251

Barclay J, Haynes K, Mitchell T, Solana C, Teeuw R, Darnell A, Kelman I et al (2008) Framing volcanic risk communication within disaster risk reduction: finding ways for the social and physical sciences to work together. Geol Soc Lond Spec Pub 305(1):163–177

Baxter PJ, Aspinall WP, Neri A, Zuccaro G, Spence RJS, Cioni R, Woo G (2008) Emergency planning and mitigation at Vesuvius: a new evidence-based approach. J Volcanol Geoth Res 178(3):454–473

Beale N, Rand DG, Battey H, Croxson K, May RM, Nowak MA (2011) Individual versus systemic risk and the Regulator's Dilemma. Proc Natl Acad Sci 108 (31):12647–12652

Benson C, Clay EJ (2004) Understanding the economic and financial impacts of natural disasters. World Bank Publications

Biggs J, Ebmeier SK, Aspinall WP, Lu Z, Pritchard ME, Sparks RSJ, Mather TA (2014) Global link between deformation and volcanic eruption quantified by satellite imagery. Nat Commun 5

Bird D, Roberts MJ, Dominey-Howes D (2008) Usage of an early warning and information system web-site for real-time seismicity in Iceland. Nat Hazards 47(1):75–94

Cadoux A, Scaillet B, Bekki S, Oppenheimer C, Druitt TH (2015) Stratospheric Ozone destruction by the Bronze-Age Minoan eruption (Santorini Volcano, Greece). Scientific reports, 5

Crosweller HS, Arora B, Brown SK, Cottrell E, Deligne NI, Guerrero NO, Venzke E et al (2012) Global database on large magnitude explosive volcanic eruptions (LaMEVE). J Appl Volcanol 1(1):1–13

Dake K (1992) Myths of nature: culture and the social construction of risk. J Soc Issues 48(4):21–37

Denkenberger D, Pearce JM (2014) Feeding everyone no matter what: managing food security after global catastrophe. Academic Press, London

Donovan A, Oppenheimer C (2015a) At the mercy of the mountain? Field stations and the culture of volcanology. Environ Plan A Abstr 47(1):156–171

Donovan AR, Oppenheimer C (2015b) Modelling risk and risking models: The diffusive boundary between science and policy in volcanic risk management. Geoforum 58:153–165

Donovan A, Oppenheimer C (2012) Governing the lithosphere: insights from Eyjafjallajökull concerning the role of scientists in supporting decision-making on active volcanoes. J Geophys Res Solid Earth (1978–2012):117(B3)

Donovan A, Oppenheimer C (2014) Extreme volcanism: disaster risks and societal implications. Extreme Nat Hazards, Disaster Risks Soc Implic 1:29

Donovan A, Oppenheimer C, Bravo M (2012) Science at the policy interface: volcano-monitoring technologies and volcanic hazard management. Bull Volcanol 74 (5):1005–1022

Donovan A, Eiser JR, Sparks RSJ (2014a) Scientists' views about lay perceptions of volcanic hazard and risk. J Appl Volcanol 3(1):1–14

Donovan A, Oppenheimer C, Bravo M (2014b) Reflexive volcanology: 15 years of communicating risk and uncertainty in scientific advice on Montserrat. Geol Soc Lond Mem 39(1):457–470

Eiser JR, Donovan A, Sparks RSJ (2014) Risk perceptions and trust following the 2010 and 2011 Icelandic Volcanic Ash Crises. Risk Anal

Fearnley CJ, McGuire WJ, Davies G, Twigg J (2012) Standardisation of the USGS volcano alert level system (VALS): analysis and ramifications. Bull Volcanol 74(9):2023–2036

Gassebner M, Keck A, Teh R (2010) Shaken, not stirred: the impact of disasters on international trade. Rev Int Econ 18(2):351–368

Gettelman A, Schmidt A, Kristjánsson JE (2015) Icelandic volcanic emissions and climate. Nat Geosci

Gibbons M, Limoges C, Nowotny H, Schwartzman S, Scott P, Trow M (1994) The new production of knowledge: the dynamics of science and research in contemporary societies. Sage, Beverly Hills

Gigerenzer G (1994) Why the distinction between single-event probabilities and frequencies is important for psychology (and vice versa). In: Wright G, Ayton P (eds) Subjective probability. John Wiley, New York, pp 129–161

Godfray HCJ, Beddington JR, Crute IR, Haddad L, LawrenceD Muir JF, Toulmin C (2010) Food security: the challenge of feeding 9 billion people. Science 327 (5967):812–818

Goldin I, Vogel T (2010) Global governance and systemic risk in the 21st century: lessons from the financial crisis. Global Policy 1(1):4–15

Haldane AG, May RM (2011) Systemic risk in banking ecosystems. Nature 469(7330):351–355

Harris A (2015) Forecast communication through the media Part 1: Framing the forecaster. Bull Volcanol 77:29

Hartley ME, Maclennan J, Edmonds M, Thordarson T (2014) Reconstructing the deep CO_2 degassing behaviour of large basaltic fissure eruptions. Earth and Planetary Science Letters 393:120–131

Haynes K, Barclay J, Pidgeon N (2008) The issue of trust and its influence on risk communication during a volcanic crisis. Bull Volcanol 70(5):605–621

Hincks TK, Komorowski JC, Sparks SR, Aspinall WP (2014) Retrospective analysis of uncertain eruption precursors at La Soufrière volcano, Guadeloupe, 1975–77: volcanic hazard assessment using a Bayesian Belief Network approach. J Appl Volcanol 3(1):1–26

Hulme M (2009) Why we disagree about climate change: understanding controversy, inaction and opportunity. Cambridge University Press, Cambridge

Hulme M (2014) Behind the curve: science and the politics of global warming. Clim Change 126(3–4):273–278

Hulme M, Mahoney M (2010) Climate change: what do we know about the IPCC? Progress Phys Geogr

Jasanoff S (1990) The fifth branch: science advisers as policymakers. Harvard University Press, Harvard

Jasanoff S (2005) Designs on nature: science and democracy in Europe and the United States. Princeton University Press, Princeton

Mader HM, Coles SG, Connor CB, Connor LJ (ed) (2006) Statistics in volcanology. Geol Soc Lond

Marzocchi W (2012) Putting science on trial. Physics World

Marzocchi W, Newhall C, Woo G (2012) The scientific management of volcanic crises. J Volcanol Geoth Res 247:181–189

Marzocchi W, Sandri L, Selva J (2007) BET_EF: a probabilistic tool for long-and short-term eruption forecasting. Bull Volcanol 70(5):623–632

Mason BG, Pyle DM, Oppenheimer C (2004) The size and frequency of the largest explosive eruptions on Earth. Bull Volcanol 66(8):735–748

Newhall CG, Pallister JS (2014) Using multiple data sets to populate probabilistic volcanic event trees. Volc Hazards Risks Disasters 203

Newhall C, (IAVCEI Subcommittee on Crisis Protocols) et al (1999) Professional conduct of scientists during volcanic crises. Bull Volc 60(5):323–334

Oppenheimer C, Donovan AR (2015) On the nature and consequences of super-eruptions. In: Schmidt A et al (eds) Volcanism and global environmental change. Cambridge University Press, Cambridge

Oppenheimer C (2002) Limited global change due to the largest known quaternary eruption, Toba≈ 74kyr BP? Quatern Sci Rev 21(14):1593–1609

Oppenheimer C (2011) Eruptions that shook the world. Cambridge University Press, Cambridge

Owens S (2005) Making a difference? Some perspectives on environmental research and policy. Trans Inst Br Geogr 30(3):287–292

Owens S, Petts J, Bulkeley H (2006) Boundary work: knowledge, policy, and the urban environment. Environ Plan C 24(5):633

Pallister J (2015) Volcano disaster assistance program: preventing volcanic crises from becoming disasters and advancing science diplomacy. Global Volc Hazards Risk 379

Pidgeon N, Fischhoff B (2011) The role of social and decision sciences in communicating uncertain climate risks. Nat Clim Change 1(1):35–41

Potter SH, Jolly GE, Neall VE, Johnston DM, Scott BJ (2014) Communicating the status of volcanic activity: revising New Zealand's volcanic alert level system. J Appl Volcanol 3(1):1–16

Raleigh C, Linke A, O'Loughlin J (2014) Extreme temperatures and violence. Nat Clim Change 4 (2):76–77

Rampino MR (2002) Super-eruptions as a threat to civilizations on earth-like planets. Icarus 156:562–569

Rampino MR, Self S (1992) Volcanic winter and accelerated glaciation following the Toba super-eruption. Nature 359(6390):50–52

Rees M (2013) Denial of catastrophic risks. Science 339 (6124):1123

Renn O, Klinke A (2004) Systemic risks: a new challenge for risk management. EMBO Rep 5(S1):S41–S46

Rose WI, Chesner CA (1990) Worldwide dispersal of ash and gases from earth's largest known eruption: Toba, Sumatra, 75 ka. Palaeogeogr Palaeoclimatol Palaeoecol 89(3):269–275

Sassen S (2006) Territory, authority, rights: from medieval to global assemblages, vol 7. Princeton University Press, Princeton

Schmidt A, Leadbetter S, Theys N, Carboni E, Witham CS, Stevenson JA, Shepherd J et al (2015) Satellite detection, long-range transport and air quality impacts of volcanic sulfur dioxide from the 2014–15 flood lava eruption at Bárðarbunga (Iceland). J Geophys Res Atmos

Schmidt A, Thordarson T, Oman LD, Robock A, Self S (2012) Climatic impact of the long-lasting 1783 Laki eruption: inapplicability of mass-independent sulfur isotopic composition measurements. J Geophys Res Atmos (1984–2012):117(D23)

Schmidt A, Ostro B, Carslaw KS, Wilson M, Thordarson T, Mann GW, Simmons AJ (2011) Excess mortality in Europe following a future Laki-style Icelandic eruption. Proc Natl Acad Sci 108 (38):15710–15715

Schmidt A (2014) Volcanic gas and aerosol hazards from a future Laki-type eruption in Iceland. Volc Hazards Risks Disasters 377

Self S (2006) The effects and consequences of very large explosive volcanic eruptions. Philos Trans R Soc A: Math Phys Eng Sci 364(1845):2073–2097

Sigmundsson F, Hooper A, Hreinsdóttir S, Vogfjörd KS, Ófeigsson BG, Heimisson ER, Eibl EP et al (2014) Segmented lateral dyke growth in a rifting event at Bar [eth] arbunga volcanic system, Iceland. Nature

Stirling A (2007) Risk, precaution and science: towards a more constructive policy debate. EMBO Rep 8 (4):309–315

Sunstein CR (2005) Laws of fear: beyond the precautionary principle (vol 6). Cambridge University Press, Cambridge

Thordarson T, Self S (1993) The Laki (Skaftár Fires) and Grímsvötn eruptions in 1783–1785. Bull Volcanol 55 (4):233–263

Thordarson T, Self S (2003) Atmospheric and environmental effects of the 1783–1784 Laki eruption: a review and reassessment. J Geophys Res: Atmos (1984–2012):108(D1), AAC-7

Thordarson T, Self S, Oskarsson N, Hulsebosch T (1996) Sulfur, chlorine, and fluorine degassing and atmospheric loading by the 1783–1784 AD Laki (Skaftár Fires) eruption in Iceland. Bull Volcanol 58(2–3):205–225

Timmreck C, Graf HF, Zanchettin D, Hagemann S, Kleinen T, Krüger K (2012) Climate response to the Toba super-eruption: regional changes. Quatern Int 258:30–44

Tversky A, Kahneman D (1974) Judgment under uncertainty: heuristics and biases. Science 185(4157):1124–1131

UN General Assembly (2000) United Nations Millennium Declaration. Resolution A/55/L.2

Williams M (2012) Did the 73 ka Toba super-eruption have an enduring effect? Insights from genetics, prehistoric archaeology, pollen analysis, stable isotope geochemistry, geomorphology, ice cores, and climate models. Quatern Int 269:87–93

Winson AE, Costa F, Newhall CG, Woo G (2014) An analysis of the issuance of volcanic alert levels during volcanic crises. J Appl Volcanol 3(1):1–12

Woo G (2008) Probabilistic criteria for volcano evacuation decisions. Nat Hazards 45(1):87–97

Wynne B (1992) Uncertainty and environmental learning: reconceiving science and policy in the preventive paradigm. Glob Environ Change 2(2):111–127

Yirgu G, Ferguson DJ, Barnie TD, Oppenheimer C (2014) Recent volcanic eruptions in the Afar rift, northeastern Africa, and implications for volcanic risk management in the region. Extreme Nat Hazards Disaster Risks Soc Implic 1:200

Part One Summary: Adapting Warnings for Volcanic Hazards

William J. McGuire and Carina J. Fearnley

Of all the geophysical threats, volcanic activity is unique in having a particularly large and diverse portfolio of associated phenomena capable of causing death and injury, societal and economic disruption and damage to population centres and attendant infrastructure. Potentially hazardous phenomena as wide-ranging as ash, noxious gases, lava flows, pyroclastic density currents and tsunamis differ in terms of nature, predictability, scale, extent, impact and perception. As such, a 'one size fits all' approach does not provide the most effective means of addressing the communication of volcanic hazards, and while general principles apply, warnings that seek to manage and mitigate the effects of individual hazardous phenomena need to be adapted or tailored.

The chapters that form Part One of this volume demonstrate how this approach may be utilised successfully to tackle a variety of specific hazards, ranging from those that apply in the immediate vicinity of a volcano, notably ballistic ejecta, hydrothermal events, and pyroclastic density currents, to those that have more widespread ramifications, most especially tephra and the manifold consequences of extreme volcanic events.

To date, little consideration has been given to the hazard arising from the ejection of ballistic material from explosive vents, perhaps because the numbers of people exposed tend to be small and death and injury tolls have been limited. As Fitzgerald et al. observes in "The Communication and Risk Management of Volcanic Ballistic Hazards", casualties of ballistic events are largely restricted to tourists, hikers, locals and volcanologists, who visit and linger in the immediate vicinity of active vents. Nonetheless, notes the author, ballistic ejecta can be hurled to distances of several kilometres from source, and is capable of causing significant damage to property and infrastructure within range. Fitzgerald reviews the current state of thinking on the ballistic hazard and the methodologies used to communicate and manage associated risk, and addresses the potential for developing new tools and management approaches.

Closely linked to ballistic hazard is the general threat to life presented by areas of active hydrothermal activity, principally because explosions at hydrothermal vents often generate ballistic ejecta. The hazards characterising hydrothermal fields are, however, more diverse, and include geysers, fumaroles and pools of boiling mud and water. In "Active Hydrothermal Features as Tourist Attractions", Erfurt-Cooper highlights the fact that such fields constitute environments, the unpredictability and potential

W.J. McGuire
Department of Earth Sciences, UCL Hazard Centre,
University College London,
London WC1E 6BT, UK
e-mail: w.mcguire@ucl.ac.uk

C.J. Fearnley (✉)
Department of Science and Technology,
University College London, Gower Street,
London WC1E 6BT, UK
e-mail: c.fearnley@ucl.ac.uk

Advs in Volcanology (2018) 165–167
https://doi.org/10.1007/11157_2017_24
Published Online: 11 November 2017

hostility of which are rarely fully appreciated by visitors. This, the author observes, is making the management of hydrothermal hazards increasingly problematic as tourist numbers increase. Erfurt-Cooper focuses attention, in particular, on hydrothermal areas that are also significant tourist attractions; evaluating potential hazards in the context of reducing accidents and limiting injuries to visitors.

The destructive and lethal potential of pyroclastic density currents (PDCs) is well known, as are the issues that make successful mitigation problematic; high velocities; extreme temperatures; sediment and debris load; and mobility. In "Mapping Hazard Zones, Rapid Warning Communication and Understanding Communities: Primary Ways to Mitigate Pyroclastic Flow Hazard", Lavigne et al. discuss how communication of a hazard that has, effectively, no warning time, might best be approached. They note that, given the properties of PDCs, the only useful approach is a precautionary one centred around building understanding in at-risk communities, hazard zonation, and rapid and effective warning. Challenges abound, however, which Lavigne and his colleagues analyse, concluding that efficient dissemination of information and an improved understanding of how at-risk communities are likely to respond, are key to mitigating the effects of PDCs on populated areas close to active volcanoes.

Volcanic ash and gases provide the means whereby the reach of an erupting volcano may be extended over many thousands of square kilometres and, in the most extreme cases, across the globe. The communication demands of such spatially extensive hazards are very different from those relating to hazards confined to a volcano and its immediate environs. As Stewart et al., note in "Communication Demands of Volcanic Ashfall Events", forecasting when, where and how much ash will fall during an eruption, constitutes a major challenge that requires a suite of communication strategies rather than a single solution. Stewart and her co-authors examine the factors that contribute to the complex 'communication environment' that characterises ashfall hazard and review the

various communication tools and methodologies available across a range of sectors, including public health, agriculture and critical infrastructure. They also highlight the peculiar challenges presented by long duration, ash-producing, eruptions, and during the clean-up and recovery following ashfall events.

The major threat that widespread clouds of volcanic ash present to the aviation industry, has been highlighted in the last six years by disruptive eruptions at Eyjafjallajökull (2010) and Grimsvötn (2011)—both in Iceland—and the Chilean volcanoes, Puyehue-cordón Caulle (2011) and Calbuco (2011). In addition, there have been numerous examples of encounters between individual aircraft and ash clouds in the vicinity of erupting volcanoes that are close to flight paths. In "Volcanic Ash and Aviation—The Challenges of Real-Time, Global Communication of a Natural Hazard", Peter Lechner et al., focus on the range of issues that arise due to interactions between ash and aircraft, noting that informing aircraft in-flight of the exact distribution of potentially dangerous ash clouds in the vicinity remains a challenge. The authors present a picture of how the aviation industry currently deals with ash clouds, within the framework of the International Airways Volcano Watch (IAVW), observing that management of ash cloud events has improved considerably since the 1980s. Lechner and his co-authors highlight the import role international collaboration has played in developing a global warning system, but recognise that there is still work to do, notably in improving communication prior to, and at the onset of, an ash-producing event.

In the manner of ash, volcanic gases can have serious consequences close to a volcano, but also half a world away. As Edmonds et al., note in "Volcanic Gases: Silent Killers", proximally, the effects may include asphyxiation, respiratory diseases and skin burns while, further afield, widespread famine may result from the climate-modifying ramifications of sulphate aerosols injected into the stratosphere. The authors identify a number of key challenges related to tackling the volcanic gas threat, notably difficulties in monitoring gas concentrations and communicating the results to those who need to know, and improving understanding of the gas risk amongst at-risk populations. Edmonds and her colleagues examine a range of scenarios, including carbon dioxide release from lake overturn events and the large-scale loading of the troposphere and stratosphere with volcanic sulphate. In the latter context, there is a particular focus on the 1783 Laki (Iceland) eruption, and its impact on climate and health across the UK and Europe, and how a future eruption of this type might be managed.

It seems fitting that, in the final chapter of Part One, "Imagining the Unimaginable: Communicating Extreme Volcanic Risk" Donovan and Oppenheimer consider how we might tackle an eruption large enough to qualify as a global geophysical event. Such eruptions recognise no boundaries and have the potential to impact in some way upon many, if not all, nation states. As such, the problems of communicating the associated hazards are unique. Donovan and Oppenheimer focus on communication both prior to and during such an eruption. While noting that no volcano erupts on such a scale without warning signs, the authors caution that great uncertainties in likelihood, timing, nature and magnitude of extreme volcanic events would, nonetheless, provide a major challenge to effective communication of the hazard and risk, as would problems arising in seeking to provide clear messages beyond the borders of the country that hosts the volcano. Donovan and Oppenheimer also highlight the need to ensure that communication strategies developed prior to the start of the eruption are robust enough to survive through the course of the eruptive event, so as to provide for reporting of the status of the eruption, the degree and nature of damage and disruption, and any logistical requirements.

Part Two
Observing Volcanic Crises
Gill Jolly and Carina J. Fearnley

Volcanic Unrest and Hazard Communication in Long Valley Volcanic Region, California

David P. Hill[ID], Margaret T. Mangan
and Stephen R. McNutt

Abstract

The onset of volcanic unrest in Long Valley Caldera, California, in 1980 and the subsequent fluctuations in unrest levels through May 2016 illustrate: (1) the evolving relations between scientists monitoring the unrest and studying the underlying tectonic/magmatic processes and their implications for geologic hazards, and (2) the challenges in communicating the significance of the hazards to the public and civil authorities in a mountain resort setting. Circumstances special to this case include (1) the sensitivity of an isolated resort area to media hype of potential high-impact volcanic and earthquake hazards and its impact on potential recreational visitors and the local economy, (2) a small permanent population (~ 8000), which facilitates face-to-face communication between scientists monitoring the hazard, civil authorities, and the public, and (3) the relatively frequent turnover of people in positions of civil authority, which requires a continuing education effort on the nature of caldera unrest and related hazards. Because of delays associated with communication protocols between the State and Federal governments during the onset of unrest, local civil authorities and the public first learned that the U.S. Geological Survey was about to release a notice of potential volcanic hazards associated with earthquake activity and 25-cm uplift of the resurgent dome in the center of the caldera through an article in the Los Angeles Times published in May 1982. The immediate reaction was outrage and denial. Gradual acceptance that the hazard was real required over a decade of frequent meetings between scientists and civil authorities together with

D.P. Hill (✉) · M.T. Mangan
U.S. Geological Survey, California Volcano
Observatory, Menlo Park, CA 94025, USA
e-mail: hill@usgs.gov

S.R. McNutt
School of Geosciences, University of South Florida,
Tampa, FL 33620, USA

Advs in Volcanology (2018) 171–187
https://doi.org/10.1007/11157_2016_32
© The Author(s) 2017
Published Online: 26 March 2017

public presentations underscored by frequently felt earthquakes and the onset of magmatic CO_2 emissions in 1990 following a 11-month long earthquake swarm beneath Mammoth Mountain on the southwest rim of the caldera. Four fatalities, one on 24 May 1998 and three on 6 April 2006, underscored the hazard posed by the CO_2 emissions. Initial response plans developed by county and state agencies in response to the volcanic unrest began with "The Mono County Volcano Contingency Plan" and "Plan Caldera" by the California Office of Emergency Services in 1982–84. They subsequently became integrated in the regularly updated County Emergency Operation Plan. The alert level system employed by the USGS also evolved from the three-level "Notice-Watch-Warning" system of the early 1980s through a five level color-code to the current "Normal-Advisory-Watch-Warning" ground-based system in conjunction with the international 4-level aviation color-code for volcanic ash hazards. Field trips led by the scientists proved to be a particularly effective means of acquainting local residents and officials with the geologically active environment in which they reside. Relative caldera quiescence from 2000 through 2011 required continued efforts to remind an evolving population that the hazards posed by the 1980–2000 unrest persisted. Renewed uplift of the resurgent dome from 2011 to 2014 was accompanied by an increase in low-level earthquake activity in the caldera and beneath Mammoth Mountain and continues through May 2016. As unrest levels continue to wax and wane, so will the communication challenges.

1 Geologic Setting and Background

Long Valley caldera is a 15- by 30-km oval-shaped topographic depression located midway between Mono Lake and the town of Bishop in east-central California (Fig. 1). It is nestled against the western escarpment of the large graben formed by the Sierra Nevada on the west and the White Mountains on the east. This impressive eastern Sierra landscape has developed over the past three million years as a result of repeated slip on the range-front normal faults and persistent volcanism. The dominant volcanic event in the area was the massive eruption of the Bishop Tuff during the collapse of Long Valley caldera 760,000 years ago. This caldera-forming eruption spewed some 600 km^3 of rhyolitic ash across much of the western United States. Frequent, smaller eruptions have continued within the caldera and along the adjacent Mono-Inyo

volcanic chain right up to the recent past, the most recent of which was a small eruption from the north side of Paoha Island in the middle of Mono Lake some 250 years ago (Bailey 2004; Hildreth 2004).

From a purely scientific viewpoint, the Long Valley Caldera-Mono Craters volcanic field is a natural laboratory for studying the interaction between active tectonic and magmatic processes in a transtensional continental regime. In the presence of people and their societal accoutrements, however, active geologic processes pose geologic hazards. The earthquake hazard in this area is emphasized by range-front faults with Holocene offsets (Chen et al. 2014), the great, ($M_W \sim 7.8$) Owens Valley earthquake of 1872, together with the series of $M > 7$ earthquakes this century that ruptured much the Eastern California-Central Nevada Seismic Belt (Wallace 1981). The volcanic hazard is reflected in the recurrence of small to moderate volcanic

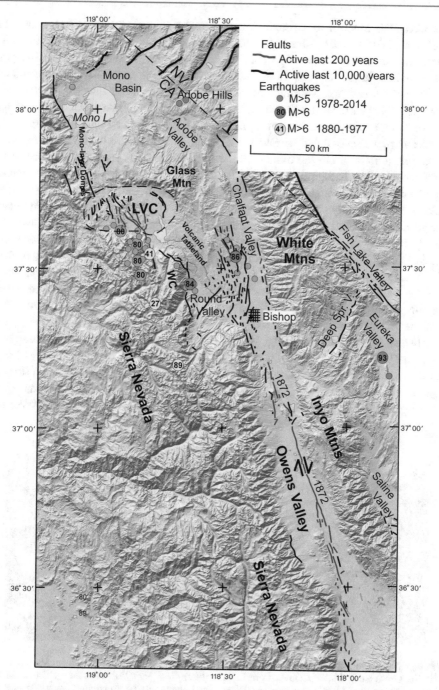

Fig. 1 Shaded relief map showing regional setting of Long Valley Caldera (LVC) along the eastern escarpment of the Sierra Nevada with M ≥ 6 earthquake epicenters (*yellow circles* for 1880–1977 and *orange circles* for 1978–2014) and M ≥ 5 earthquake epicenters for 1978–2014 (*small orange circles*) where numbers are last two digits of the year. *Black lines* are faults with Holocene offset, *red lines* are faults with offsets in the last 200 years. Note the surface rupture of the M 7.8, 1872 Owens Valley earthquake, which was dominantly right-lateral, strike-slip. Updated from Fig. 3 in Hill (2006) (courtesy of The Geological Society, London)

eruptions along the Mono-Inyo volcanic chain over the past 50,000 years. Over the past 5000 years, for example, some 20 small to moderate eruptions have occurred from vents scattered along the Mono-Inyo volcanic chain at intervals of 200–700 years. Most were explosive, rhyolitic eruptions accompanied by ash clouds and occasional pyroclastic flows; a few were effusive basaltic eruptions. On the basis of this 5000-year record, the background (unconditional) probability of an eruption from somewhere along the Mono-Inyo chain is roughly 0.5% per year. For perspective, this is comparable to the probability of a M ~ 8 earthquake along the San Andreas fault or an eruption from some of the major Cascade volcanoes. It is thus typical of the background probability for geologic hazards throughout much of the active margin of the North American Plate. The particular challenge in the case of Long Valley caldera lies in (1) making a meaningful assessment of the probability gain (conditional probability) for an eruption as this large magmatic complex displays varying levels of seismicity, ground deformation, and magmatic gas

emissions, and (2) effectively communicating this and associated uncertainties to civil authorities, the public, and the media. The problem is one of reliable eruption forecasting and real-time, probabilistic hazard assessment.

Eastern California is an important recreation area heavily used by much of urban California as well as visitors from elsewhere in the country and the world. Mammoth Mountain volcano, which last erupted ~55,000 ybp, stands on the southwestern rim of Long Valley caldera (Hildreth et al. 2014). It hosts one of the largest ski areas in the country. The resort town of Mammoth Lakes at the base of Mammoth Mountain sits within the southwestern corner of the caldera (Figs. 2 and 3). Mammoth Lakes, with a permanent population of ~8000 and a temporary population that swells to over 40,000 during major ski weekends, is a year-round, destination resort providing ready access to the adjacent high Sierra as well as mountain biking, fishing, and golf during summer months. This does not mix easily with talk of geologic hazards, particularly high-profile hazards posed by volcanoes and earthquakes (Hill 1998).

(a)

(b)

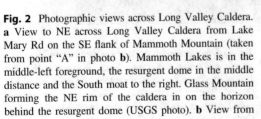

Fig. 2 Photographic views across Long Valley Caldera. **a** View to NE across Long Valley Caldera from Lake Mary Rd on the SE flank of Mammoth Mountain (taken from point "A" in photo **b**). Mammoth Lakes is in the middle-left foreground, the resurgent dome in the middle distance and the South moat to the right. Glass Mountain forming the NE rim of the caldera in on the horizon behind the resurgent dome (USGS photo). **b** View from the resurgent dome westward to Mammoth Mountain (taken from point "B" in photo **a**). The 40 MW geothermal plant is in the foreground, Hwy 395 cuts across the middle distance, and Hwy 203 extends toward Mammoth Lakes at the base of Mammoth Mountain. High peaks of the Sierra Nevada form the distant horizon to the right of Mammoth Mountain (USGS photo)

2 Hazard Communication (and Miscommunication) During Two Decades of Strong Volcanic Unrest (1978–2000)

A $M_L = 5.8$ earthquake on October 4, 1978, located beneath Wheeler Crest 14 km southeast of Long Valley caldera (roughly midway between Bishop and Mammoth Lakes) marked the onset of the extended episode of unrest in the caldera and vicinity that continues today (Figs. 3 and 4; Hill 2006; Shelly and Hill 2011; Lewicki et al. 2014). Over the following year and a half, seismic activity in the form of $M > 3$ and occasional $M > 4$ earthquakes

gradually migrated northwestward and the southern margin of the caldera. Then, on May 25, 1980, just seven days after the catastrophic eruption of Mount St. Helens, three M ~ 6 earthquakes shook the southern margin of the caldera accompanied by a rich aftershock sequence. By the morning of May 27, the Director of the U.S. Geological Survey (USGS) had released a formal "Hazard Watch" (Table 2a) noting the possibility of additional M ~ 6 earthquakes in the area. Just hours later, a fourth M ~ 6 earthquake shook the area—a successful "short-term" forecast!

That summer, Savage and Clark (1982) re-leveled a section of Highway 395 through the

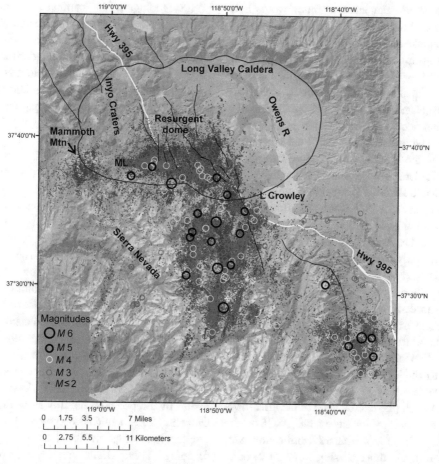

Fig. 3 Map of epicenters for earthquakes with magnitudes $1.5 \geq M \leq 6.5$ from 1987 to 2014 scaled by symbol size. Epicenters for the 1978–1983 interval (prior to installation of the locally dense seismic network in 1984) are from the standard NCSN catalog and include only M $\geq$ 3 earthquakes. Epicenters for M $\geq$ 1 earthquakes the 1984–2014 interval are from the double-difference catalog of Waldhauser (2009). *ML* indicates the town of Mammoth Lakes. Background shaded relief and seismicity thanks to Stuart Wilkinson

area affected to document the co-seismic displacement expected from this series of four M ~ 6 earthquakes. What they found instead, was a broad, dome-shaped uplift of the caldera floor (the resurgent dome). Monuments near the center of the resurgent dome were 25 cm higher in the summer of 1980 than they had been in 1975. Measurements of a trilateration network spanning the area made in 1979 suggested that most, if not all of this deformation developed sometime between the summer of 1979 and the summer of 1980. With fresh images of Mount St. Helens in mind, it required no great leap to recognize a volcanic signature in this combination of strong earthquake swarm activity and ground deformation. Still, if the activity had died away as with most aftershock sequences, attention would have soon focused elsewhere.

Earthquake activity continued, however, with frequent swarms that included locally felt earthquakes (M ~ 3–5 events), rapid-fire bursts of small earthquakes (spasmodic bursts), which are often associated with active volcanoes, and evidence that focal depths appeared to be getting shallower with time (Ryall and Ryall 1983). Roy Bailey of the USGS, who produced the Long Valley Caldera geologic map (Bailey 1989) and was Volcano Hazards Program Coordinator, had initial responsibility for communicating the significance of this activity to local authorities. In discussions between scientists in the USGS, the California Division of Mines and Geology (CDMG), and Alan Ryall at the University of Nevada, Reno over the winter of 1981 and a meeting in early May 1982, a consensus developed that we had an obligation to inform local civil authorities of our concerns about the volcanic nature of this activity. The result was a memo to the Chief Geologist in USGS headquarters dated May 17 recommending a volcanic Hazard Watch (the middle level of the three-level hazard terminology in use by the USGS at the time; see Table 2a). Hazard notification was top-down at the time, passing from the Federal government (USGS) to the State of California Governor's Office of Emergency Services (CalOES) and from there, on to the counties and cities affected. A dialogue between USGS headquarters and CalOES on the precise wording of the alert ensued. Meanwhile, a science reporter for the Los Angeles Times got wind that something was up. An article announcing that the USGS was about to release a "Notice of Potential Volcanic Hazards" appeared in the Los Angeles Times on the morning of May 25, 1982, just two days before the Memorial Day weekend and short circuiting the official release of the "Notice of Potential Volcanic Hazards" (the lowest level in the Notice/Watch/Warning hazard terminology) scheduled for the next day: The local response was one of outrage, anger, and disbelief (what volcano?!) exacerbated by inflammatory headlines and news stories about "brewing lava eruptions" and a town in denial. Geologists, and USGS geologists in particular, immediately became persona non grata in Mammoth Lakes and Mono County—an attitude that only gradually mellowed over the years. Under the best of circumstances, communicating information on a newly recognized hazard is tricky—this serves as an outstanding example of the wrong way to start such a dialogue.

Roy Bailey, Coordinator of the Volcano Hazards Program, oversaw USGS monitoring efforts in the caldera through the summer and fall. In December 1982 the USGS instituted the Long Valley Monitoring Project with David Hill as Chief Scientist in charge of coordinating research, monitoring, and hazard communication efforts in Long Valley Caldera. The California Division of Mines and Geology (CDMG) assumed a comparable responsibility in 1982, and purchased an automatic earthquake recording system and eventually hired a volcano seismologist, Stephen McNutt in spring 1984. In an effort to clarify the significance of the Notice of Potential Volcanic Hazards, Miller et al. (1982) published USGS circular 877 describing the nature of the potential hazards from future volcanic eruptions, and CDMG prepared a series of eruption scenarios as the basis for response planning. These documents together with the persistence of ongoing unrest stimulated disaster preparedness efforts by State, County, and Federal land management agencies in the area that included establishing Incident Command

Table 1 Key to events noted in Fig. 4

Number	Date	Event
1	4 Oct. 1978	Wheeler Crest M 5.8 earthquake
2	25–27 May 1980	Three M 6 earthquakes on 25th. USGS releases Hazard Watch for possible additional M ~6 earthquakes on the 27th; a fourth M 6 earthquake four hours later (a "successful" forecast)
3	1980–1981	Savage and Clark (1982) re-level Hwy 395 to discover a 25-cm uplift of the resurgent dome. Ongoing swarm activity in the caldera
4	25–26 May 1982	L.A. Times article announcing planned release of USGS Notice of Potential Volcanic Hazards on 25th. Official release of USGS Notice of Potential Volcanic Hazard on the 26th
5	December 1982	D. Hill appointed USGS Scientist-in Charge (SIC) of the Long Valley project (LVO). CDMG established comparable role
6	7–14 Jan. 1983	Strong south moat swarm including two M 5 earthquakes
7	31 Aug. 1983	Hill and Filson attend "1000-year lunch" with business leaders
8	Oct 1983	Dedication of the "Mammoth Scenic Loop", a 2nd road into town
9	Sept. 1983	USGS Director replaces Notice-Watch-Warning alert system by a single-level Hazard-Warning system
10	Aug. 1984 Oct. 1984	The Town of Mammoth Lakes incorporates S. McNutt joins CDMG as volcano seismologist
11	23 Nov. 1984	A M 6.1 earthquake in Round Valley
12	21 July 1986	A M 5.9 foreshock followed 12 h later by the M 6.4 Chalfant Valley earthquake
13	Aug. 1987	Hill and Bailey LVC field trip for Dr. Graham and Andrea Lawrence
14	May 1989 to March 1990	11-month long earthquake swarm beneath Mammoth Mtn. Onset of deep LP (volcanic) earthquakes and magmatic CO_2 emissions from flanks of Mammoth Mtn
15	17–19 May 1990	Town and County officials attend USGS workshop on the 10th anniversary of the May 1990 Mount St. Helens eruption
16	1991	LVO adopts alphabetic A-E, alert-level system
17	1991–1997	Recurring earthquake swarms and continued resurgent dome inflation
18	June 1997	LVO replaces alphabetic alert-level system with a 5-level color code
19	Nov. 1997–Jan. 1998	Earthquake swarm activity including M 4.8 and 4.9 earthquakes accompanied by elevated resurgent dome inflation
20	28 June–4 July 1998	Mammoth Lakes City Manager accompanies Hill to 1st IAVCEI Cities on Volcanoes meeting in Rome and Naples, Italy.
21	1998–1999	M 5.1 earthquakes on 8/6/98 and 14/7/98 near the south-east margin of the caldera and a M 5.6 earthquake on 15/5/99 in Sierra Nevada block 5 km S. of caldera
22	Oct. 2006	USGS adopts a uniform ground/aircraft hazard warning system for all five volcano observatories
23	March 2008	Mammoth Mountain Ski Area opens the "Top of the Mountain" display illustrating the geologic and volcanic history of the area
24	Jan. 2009	M. Mangan becomes LVO SIC
25	2011–2014	Inflation of the resurgent dome resumes and continues at a rate of 2 cm/year through 2014 accompanied by low-level swarm activity under Mammoth Mountain, the caldera, and the Sierra Nevada block to the south
26	Feb. 2012	The California Volcano Observatory (CalVO) is established with responsibility for all California volcanoes including Long Valley
27	July 2014	Joint USGS-CGS Earthquake Hazard Scenarios report released (Chen et al. 2014)

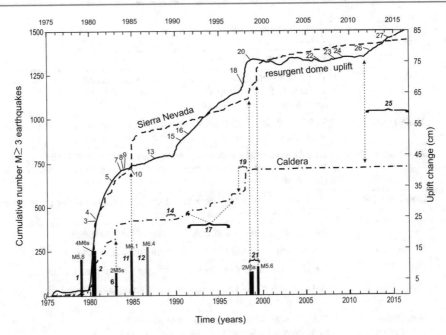

Fig. 4 Time history of resurgent dome uplift (*solid line*) and the cumulative number of M ≥ 3 earthquakes in Long Valley Caldera (*dot-dashed line*) and the Sierra Nevada block (*dashed line*) from 1978 through May 2016. *Solid line* represents uplift of the center of the resurgent dome with numbers tied to numbered events in Table 1. Heavy vertical bars mark occurrences of M ≥ 6 earthquakes (including their M ≤ 5 aftershocks) with length proportional to magnitude. *Horizontal brackets* indicate periods of earthquake swarm activity. *Dashed arrows* point to associated rate changes in cumulative number of earthquakes The *grey bar* is the M = 6.4 Chalfant Valley earthquake and its aftershocks (see Fig. 1), which do not contribute to the cumulative number count for Sierra Nevada earthquakes. *Bold italic* numbers are linked to numbered events in Table 1

(ICS) and Unified Command (UCS) systems as well as emergency response plans including "Plan Caldera" of the California OES and the "Basic Emergency Plan" of Mono County (Mader et al. 1987).

As low-level earthquake swarm activity continued through the summer and fall of 1982, however, a series of public meetings did little to mitigate the simmering anger of much of the public and business community. Then, on the afternoon January 7, 1983, activity abruptly resumed with an intense earthquake swarm in the south moat of the caldera that included two M = 5.3 earthquakes accompanied by nearly constant felt shaking from frequent M3 to 4 earthquakes over the next several weeks (Fig. 4). This was an El Nino winter and the snow was piled high along the roads within Mammoth Lakes and along the only paved road connecting the town to Highway 395. At his own initiative,

the Chairman of the Mono County Board of Supervisors, ordered second (dirt) road plowed. He also initiated steps to have this road widened and paved to provide an alternate way out of town. This was not a popular decision, in part, because it carried an implicit acknowledgment that there might actually be a volcanic hazard. Setting some sort of record from inception to completion, the newly paved "escape route" was formally dedicated in October of 1983 as the "Mammoth Scenic Loop". The Chairman and a second member of the board who had been pro-active in support of the new road and other mitigation issues (including monthly public updates by the USGS on the evolving caldera unrest) were voted out of office in a special recall election over the following summer.

Re-leveling of the deformation network along Highway 395 during the summer of 1983 showed that the resurgent dome had been uplifted by an additional 7 cm over the winter. Following the intense January 1983 swarm, both the earthquake activity and deformation rates within the caldera activity began a gradual decline that persisted through the remainder of the 1980s.

Resentment of the media attention attracted by the "Notice of Potential Volcanic Hazards" was most acutely expressed by the real estate and business communities. Other agencies with responsibilities in the area, including the U.S. Forest Service, the Bureau of Land Management, National Park Service, and Mammoth Mountain Ski Area together with a number of residents were supportive of efforts to monitor the activity and communicate its significance. The Ski Area, for example, instigated a series of day-long geology field trips for the public and local officials led by Hill and Bailey that proved to be an especially effective means of explaining the active geologic setting of Long Valley caldera and Mono Craters area, and Hill provided a series of invited public lectures on the significance of the ongoing activity in the early 1980s.

On August 31, 1983, John Filson (then Chief of the USGS Office of Earthquakes, Volcanoes, and Engineering) and Hill attended a luncheon meeting in Mammoth Lakes with local business leaders that was organized by the president of the Chamber of Commerce. John Filson recalls that occasion as the "1000-year lunch." One of the people at the 1000-year lunch was an influential resident rumored to have drafted several critical letters to the USGS Director on behalf of the local business community. He had earned BS and Ph.D. degrees in electrical engineering from prestigious universities, and at the time was working as a consultant for the aerospace industry from his home in Mammoth Lakes. As it turns out, this same individual, Dr. William Graham, became President Ronald Reagan's Science Advisor from 1986 to 1989. Needless to say, scientists monitoring Long Valley caldera developed a bit of angst when they learned that the President's science advisor and the resident from the 1000-year-lunch were one and the same person! In August 1987, Bailey and Hill had the opportunity to take the Science Advisor and

Andrea Mead Lawrence, an influential member of the Mono County Board of Supervisors (also a 1952 Olympic gold medalist in Alpine skiing), on a field trip around the caldera. In the end, both were instrumental in easing relations between scientists and the local business community as they came to understand the nature of hazard.

The continued existence of the "Notice of Potential Volcanic Hazards" was a thorn in the side of the Mammoth Lakes business community. The Federal Register (v. 24, no. 7, 1977) defined this notice as "*Information on the location and possible magnitude of a potentially hazardous geologic condition. However, available evidence is insufficient to suggest that a hazardous event is imminent or evidence has not been developed to determine the time of occurrence*". Given this definition and the geologic history of the caldera and its unrest, USGS scientists had no basis for rescinding the notice. In September, 1983, the Director of the USGS announced that the three-level Notice/Watch/Warning hazard notification system was being replaced by a one-level "hazard warning" system, which would be used only when the situation required a "near-term" (hours to days) public response (Table 2a). This change in hazard communication terminology became official on 24 January 1984 (Federal Register, v. 49, no. 21, pp. 3838–3839, January 31, 1984). By default, this change removed the "Notice of Potential Volcanic Hazard" for Long Valley caldera. In August 1984, the town of Mammoth Lakes was formally incorporated, thereby requiring the scientific community to interact with a new civil authority. While activity remained low within the caldera following the January 1983 swarm, this was not the case for seismic activity outside the caldera. On November 23, 1984, the $M_L = 6.1$ Round Valley earthquake and its many aftershocks (located midway between the southeastern caldera boundary and Bishop) shook the region (Figs. 1 and 4). The area again was repeatedly shaken from late July through mid August, 1986, by the rich foreshock and aftershock sequence associated with the $M_W = 6.4$ Chalfant Valley mainshock of July 21, 1986 located 20 km southeast of the caldera. Each of these earthquake sequences generated a

Table 2 Evolution of USGS hazard warning systems (HWS) in abbreviated form

(a) USGS-wide systems for hazard statements released by the Director

1977–1983, *Federal Register*, v. 42, no. 70, 1977

Terminology	Significance[a]
NOTICE OF POTENTIAL HAZARD	Info on potential hazard but insufficient data to time of occurrence
HAZARD WATCH	Info on potentially catastrophic event within months to years
HAZARD WARNING	Info on time, location, and magnitude of a potentially disastrous geologic event

11 October 1983, *Federal Register*, v. 48, n. 197, 1983

HAZARD WARNING	Info on a potential geologic hazard posing a significant threat requiring a timely response

(b) Local hazard level systems for the Long Valley Caldera volcanic field with authority for releasing hazard statements delegated to the LVO Scientist-in-Charge

17 April 1991 Alphabetic system[b] (Hill et al. 1991)

STATUS	USGS RESPONSE	ACTIVITY LEVEL	RECURRENCE
N	Normal monitoring	Background	–
E STATUS	Notify responsible personal as appropriate	Weak unrest	Weeks
D STATUS	Notify mid-level USGS personnel, OES, CDMG, USFS	Moderate unrest	Weeks–months
C STATUS	Notify USGS Office Chief, OES headquarters, State Geologist	Strong unrest	Months–years
B ALERT	Alert USGS Director, trigger an EVENT RESPONSE	Intense unrest	Years–decades
A ALERT	Issue HAZARD WARNING	Eruption likely	Decades–centuries

(c) June 1997 LVO Color-code system[c] (Hill et al. 2002)

GREEN (NORMAL)	Normal operations with information calls as appropriate	Background to / Strong unrest	Most of the time / Months to years
YELLOW (WATCH)	Full call-down Event response	Intense unrest	Years to decades
ORANGE (WARNING)	Full call-down Event response (if not in place)	Accelerating unrest eruption likely	Decades to centuries
RED (ERUPTION)	Full call-down Event response (if not in place)	Minor to moderate eruption with possible increase in intensity	Centuries to millennia

Table 2 (continued)

(**d**) 2006 A uniform ground-based and aviation warning system adopted for all USGS volcano observatories (Gardner and Guffanti 2006). Authority for releasing hazard statements delegated to the respective Observatory Scientists-in-Charge

Ground-based system for USGS Volcano Observatories[d]

Terminology	Description
NORMAL	Non-eruptive, background activity levels
ADVISORY	Activity levels above background, or ongoing eruptive activity declining
WATCH	Heightened or escalating unrest, or ongoing eruption poses limited hazard
WARNING	Hazardous eruption imminent or underway

Aviation color code used by USGS Volcano Observatories[d]

Color	Description
GREEN	Volcano is in a non-eruptive state
YELLOW	Volcano is showing elevated unrest levels
ORANGE	Volcano is exhibiting escalating unrest: potential eruption time-frame uncertain, or minor eruption underway with no or minor ash emissions
RED	Eruption is imminent or underway with significant ash emissions high into the atmosphere

[a]No stand-down criteria specified
[b]Stand-down criteria based on current ACTIVITY LEVEL
[c]Stand-down criteria based on current ACTIVITY LEVEL
[d]Stand-down criteria based on current activity levels noted under DESCRIPTION

flurry of news stories on seismic activity and volcanic unrest in the Mammoth Lakes area, further aggravating the local business leaders over negative publicity.

Activity returned to the vicinity of the caldera with the onset of a persistent earthquake swarm beneath Mammoth Mountain that began in early May and continued into March 1990 (Hill et al. 1990; Langbein et al. 1993; Cramer and McNutt 1997). Only a handful of earthquakes in this swarm had magnitudes as large as M = 3, but the activity included numerous spasmodic bursts, which are commonly associated with active volcanic systems. The swarm was accompanied by the onset of magmatic CO_2 emissions around the flank of the mountain and a marked increase of magmatic Helium ratios to values as high as $R/R_A \sim 7$, where $R = (^3He/^4He)$ is the ratio of light to heavy Helium isotopes measured in fumarole gasses and R_A is the isotope ratio in the atmosphere. Taken together, these observations suggest an intrusion of magmatic fluids into the shallow crust beneath Mammoth Mountain (Hill and Prejean 2005). In an effort to keep local civil authorities apprised of the situation, Hill called the City Manager for Mammoth Lakes several times a week with updates on evolving activity. At one point in a conversation with Steve McNutt, the City Manager asked in apparent exasperation if we couldn't provide him with some sort of written criteria for how seriously he should regard the varying levels of activity. His request led to development a response plan that included an alphabetic scheme of five "alert levels" (E through A in ascending order of concern) modeled after that used for the Parkfield earthquake prediction experiment (Table 2b; Hill et al. 1991). This alphabetic system included criteria for stepping down from an elevated alert level—an element that was absent in the old "Notice-Watch-Warning" system. With an important addition, the USGS Director ceded local authority to the Chief Scientist of the Long Valley monitoring project to communicate changes in alert level directly to local officials thus avoiding delays associated with high-level, inter-agency discussions.

In an effort to help local civil authorities better appreciate the issues involved with volcanic hazards, the USGS invited Mono County and Mammoth Lakes civil authorities to attend a conference commemorating the tenth anniversary of the May 1980 Mount St. Helens eruption. This conference, which was held adjacent to Mount St. Helens in Kelso Washington on May 17–19, 1990, included presentations by local civil authorities as well as scientists who were directly involved in responding to this catastrophic eruption. The occasion turned out to be enormously informative for all concerned.

Meanwhile, frequent trilateration measurements with the 2-color geodimeter showed that deformation rates across the resurgent dome began to increase substantially in October of 1989 (Langbein et al. 1993). Three months later, earthquake swarm activity resumed in the south moat of caldera and continued to wax and wane through mid-1997, accompanied by relatively steady uplift of the resurgent dome at a rate of 2- to 3-cm/year (Fig. 4; Langbein et al. 1993). Other developments that persisted through the 1990s involved high concentrations of magmatic CO_2 in the soil around Mammoth Mountain and growing number of long-period (LP) volcanic earthquakes occurring at depths of 10–20 km beneath the southwest flank of Mammoth Mountain (Pitt and Hill 1994). In May 28, 1998, a cross-country skier died of CO_2 asphyxiation after falling into a collapsed snow cave filled with 70% CO_2 at the base of Mammoth Mountain (Hill 2000) Both the CO_2 emissions and LP earthquakes were coincident with the 1989 Mammoth Mountain swarm, and together they serve as a reminder that this 11,000-foot high volcano, which last erupted 55,000 years ago, is not extinct.

The increased activity within the caldera provided ample opportunity to exercise the "alert-level" system. For a while, it seemed to work just fine as we moved back and forth between the lower "status levels" from E through C. The new system included a matrix, rather than top-down, system for information flow in which alerts are passed to State, County, and City authorities simultaneously. Flaws in this alphabetic scheme began to emerge, however, as the media added their own twist on the status levels. On a slow news day, for example, a swarm including a couple of M $\sim$ 3, locally felt earthquakes corresponding to a "D-status" under this scheme would get reported in headlines as a "D-level volcano alert" (the latter two words added by the reporter or headline writer). Of course, most of the public didn't understand what a "D-level alert" meant—except that is sounded serious. The result was exaggerated concern in the public (should I cancel my planned vacation to Mammoth?), and renewed frustration in the business community over negative "volcano" publicity (see Mader et al. 1987).

The USGS in consultation with Mono County, Mammoth Lakes, the California State Geologist and head of CDMG, the California Earthquake Prediction Evaluation Council (CEPEC), and the California Office of Emergency Services (CalOES), began discussing ways to improve this system by making it less susceptible to misinterpretation by both the media and the public at large. The result was a four-level color scheme, with the three lower status levels (E, D, and C) grouped under "condition GREEN" (no immediate risk), replaced level B by "condition YELLOW" (watch), and replaced level A by "condition ORANGE" (warning). A final color, "condition RED" was added to indicate an eruption was actually under way. Mono County suggested that we incorporate shapes with this scheme. The result associated GREEN with a circle, YELLOW with a square, and ORANGE with a diamond (skiers will recognize these as the shapes indicating beginner through advanced slopes). A triangle (the shape of a volcano) was chosen to represent RED. This new system was formally adopted in

June 1997, shortly before the onset of a sustained period of escalating unrest in the caldera (Table 2c; Fig. 4; Hill et al. 2002).

In April–May of 1997, the two-color geodimeter data again showed hints of an increasing horizontal deformation rate across the resurgent dome (Langbein 2003). By early July, earthquake swarm activity in the caldera picked up with an increasing frequency and intensity that persisted through the remainder of the year and into January 1998 (Fig. 4). The peak in this activity from mid-November through early January 1998 included nine earthquakes with magnitudes of M = 4.0 or greater accompanied by thousands of smaller events. The three largest earthquakes had magnitudes of M = 4.8–4.9. Resurgent dome deformation escalated through the second half of 1997, reaching a peak uplift rate of over 2 cm/month in mid-November. By the end of the year, the center of the resurgent dome was approximately 10 cm higher than in early May. Mammoth Mountain earthquake swarm activity also increased in September through December 1997. Under the new color-code notification system, the condition remained GREEN through this extended period of elevated activity. During particularly strong swarm sequences on November 22 and 30, however, activity came extremely close to meeting the guidelines for a condition YELLOW (it would have been level C under the old system—see Table 2). The Long Valley caldera web page, on which most of the monitoring data are available in real-time together with frequent written updates on the current condition and its significance, was receiving tens of thousands of "hits" a day. It also attracted lots of email with messages ranging from "why aren't we at YELLOW yet?", "you're a pawn of the Realtors", "what are you covering up?", "I'm moving to Maine!" to "thanks for an excellent job of keeping us informed and keeping things in perspective". Both the seismicity and deformation rates within the caldera gradually slowed through the first half of 1998. Six months later, however, Mammoth Lakes and vicinity was again shaken by three M 5 earthquakes in the Sierra Nevada just south of the caldera—a pair of M 5.1 events

on 8 June and 14 July 1998, respectively, and a M 5.6 event on 15 May 1999. The abundant aftershocks to these earthquakes included a number of M > 3 earthquakes, which were felt locally. On August 6, 1999, the Long Valley monitoring project was officially established as the Long Valley Observatory (LVO) with Hill as Scientist-in-Charge (SIC).

After nearly two decades of intense, episodic unrest, the Mammoth Lakes Town Council agreed to support travel for the Town Manager to attend the first Cities on Volcanoes (COV) meeting held in Naples and Rome from 28 June to 4 July 1998. This was the first of an ongoing series of international meetings under the IAVCEI (International Association of Volcanology and Chemistry of the Earth's Interior) umbrella bringing together scientists and civil authorities for discussion on volcanic hazards and mitigation. Having the Mammoth Lakes Town Manager at the COV marked an important milestone in the efforts of scientists to grow community awareness.

3 Maintaining Community Awareness and Preparedness During Low-Level Volcanic Unrest (2001–May 2016)

Since the turn of the century, unrest has diminished in Long Valley Caldera (Fig. 4; Wilkinson et al. 2014). The communication challenge has thus become one of maintaining public awareness and preparedness when there are few signs of volcanic unrest recognized by the public. Two notable exceptions during the relative quiescence from 2000 to 2011 included: (1) the tragic death of three skiers on April 6, 2006, as they fell into a CO_2—filled snow cave that had developed over the fumarole on the upper flank of Mammoth Mountain, and (2) the temporary onset of geyser activity in Hot Creek (a popular bathing area) in 2006. Instrumentally detected low-level unrest between 2000 and 2011 included additional inflation of the resurgent dome by 3–4 cm in 2002 accompanied by minor seismicity (Feng and Newman 2009). In 2006, 2008, and 2009 short-lived swarms of earthquakes, all too small to be felt, occurred at depths at 25–30 km below Mammoth Mountain (Shelly and Hill 2011). An increase in shallower earthquakes, again too small to be felt, followed within a few months of each swarm and magmatic CO_2 emission increased in tandem at the surface (Lewicki et al. 2014). Scientists relate these phenomena to deep intrusion and degassing of basaltic magma in the lower crust followed by upward migration of CO_2-rich fluids through fracture development. To the non-scientist, however, the significance of "stealth" magma supply events is hard to appreciate. Few earthquakes were felt, and the uptick in magmatic gas could be measured, but not seen. Years of low-level volcanic unrest require a sustained effort to educate a growing population base and a changing cast of elected officials, land managers, and civil authorities.

Effective hazard education requires persistence and strategic networking within at-risk communities. From the beginning, USGS scientists have demonstrated their commitment to hazard mitigation in the Long Valley region by attending quarterly meetings of the Mono and Inyo County Office of Emergency Services, Unified Command System (UCS). Through participation in the UCS scientists have learned to speak the language of emergency response teams and have become integral members. The standalone "Plan Caldera" developed in the 1980s during the height of caldera unrest has evolved into the Mono and Inyo county Emergency Operations Plan, which includes regularly updated contributions written by scientists. USGS scientists have provided face-to-face volcano hazard training to citizens in the local Community Emergency Response Team (CERT), a volunteer disaster response organization sponsored by the Mammoth Lakes Police Department, Mammoth Fire District and Mono County Public Health, as well as, specialized trainings for civil authorities through a partnership with the Federal Emergency Management Agency—National Disaster Preparedness Training Center and the University of Hawaii. In collaboration with USGS scientists, the USFS Ranger Districts in Mammoth Lakes and Mono Lake visitors centers established an information display on the active geology of the region, and Mammoth Mountain Ski Area developed a "Top of the Mountain" display taking advantage of a spectacular overview of regional geology.

Sustaining a proactive hazard education program in the Long Valley region is time-consuming and costly, but necessary. In 2009 this responsibility fell to Margaret Mangan as the newly appointed LVO SIC as Hill stepped aside to become Scientist Emeritus. The necessity of continuing education is underscored by renewal of caldera inflation, which began in early 2011, and has continued through early 2017 (Montgomery-Brown et al. 2015). The time-averaged uplift rate from 2011 through 2014 was about 2 cm/year, comparable to that of the mid-1990s. Seismicity within the caldera increased as well with modest earthquake swarms in the caldera in June and July 2014 followed by the most energetic swarm since 1999 in September 2014 with eight earthquakes of magnitudes M ∼ 3.0–3.8 that were felt in the town of Mammoth Lakes. Just prior to this swarm, USGS and CGS scientists had briefed the Mono and Inyo County Unified Commands on earthquake hazards in the region in preparation for release of a joint report describing scenario earthquake hazards for the Long Valley—Mono Lake area (Chen et al. 2014). Meanwhile, Mammoth Mountain has continued to produce episodic, low-level swarm activity (Shelly et al. 2015).

Periodic evaluation of the effectiveness of hazard communication is also necessary, and new methodologies must be developed to meet modern needs. Several years after the introduction of Long Valley's four-level color-code warning system, for example, the USGS Volcano Hazards Program recognized the need to establish a national alert-notification system, one that covers all US volcanoes and that distinguishes between ground-based and atmospheric hazards (Gardner and Guffanti 2006; Fearnley et al. 2012). By 2006, two new communication vehicles where in use—Volcano Alert Notifications (VANs) and Volcano Observatory Notices for Aviation (VONAs), both of which use four-tiered threat level systems (Table 2d).

VANs use Normal-Advisory-Watch-Warning to specify increasing threat on the ground (e.g., lahars, lava flows) and VONAs use Green-Yellow-Orange-Red to specify increasing threat to aviation (ash clouds and volcanic aerosols). The dual ground-level and aircraft-ash systems recognize that a low-level effusive eruption might pose a serious local hazard (Warning) without posing a serious ash hazard for aircraft at 30,000 feet. In keeping with communications in the digital age, an email-based Volcano Notification Service (VNS) automatically delivers VANs, VONAs, and other volcano information to all subscribers who register online at http://volcanoes.usgs.gov/vns (Accessed 13 February 2017).

In 2012, the USGS Director announced establishment of the California Volcano Observatory (CalVO), giving the former LVO the responsibility to monitor Long Valley caldera, as well as the other young and restless volcanoes in the state (Stovall et al. 2014). As the director explained "By uniting the research, monitoring, and hazard assessment for all of the volcanoes that pose a threat to the residents of California, CalVO will provide improved hazard information products to the public and decision makers alike."

4   Conclusions

Applying scientific research to understand the processes driving magmatic unrest in a large caldera and communicating the significance of this research in terms of volcanic hazards represents a microcosm of the challenges that much of science faces today as we grapple with the societal relevance of scientific research. Long Valley caldera unrest has provided a rich experience in terms of both scientific return and the challenges in trying to effectively communicate socially useful information on natural hazards to the public.

We have come a long way since the highly charged relations between geologists and the business community of the early 1980s. Those involved on the scientific side learned to be much more effective at presenting messages on activity

levels and their significance in terms that can be understood by the general public, and the residents of Mono County and eastern California have gained a much better appreciation for the geologically active environment in which they live. The following three points are important in this regard:

1. We must continue to do the best science possible as we track the ongoing activity. Understanding the processes driving the unrest is key to understanding the nature of the hazards posed by the unrest and their likely evolution in terms of long-term forecasts and short-term predictions. This is similar to medicine, for which understanding the processes that cause disease provides the best guidance for treatment. Ultimately, our credibility with the public rests on a foundation of sound science.

2. Establishing and maintaining an effective and credible working relation with civil authorities and the public requires constant attention and a major commitment in time. Long-term continuity with scientific personnel and a stable policy with clearly defined lines of responsibility are important elements in this process. The tenures of Hill as LVO SIC from 1982 to 2009, Mangan as LVO SIC from 2009 to 2011 and CalVO SIC from 2012 to present have provided continuity of scientific and hazard information spanning 4 Mammoth Lakes Town Managers, 3 Mono County Sheriffs, 5 State Geologists, and 6 California Office of Emergency Services Directors.

3. Finally, although the "geologists not welcome" signs have long since disappeared from motel and restaurant windows in Mammoth Lakes, it's important to keep in mind that a protracted crisis and an evolving resort population will severely test the good will and trust scientists worked so hard to establish since the early 1980s. The threat of an impending volcanic eruption inevitably presents a narrow path and an element of uncertainty to a successful response. On one side, an over-anxious response will exacerbate the "false alarm" problem (one, or at most

two, "needless" evacuations, for example, may well destroy our credibility and effectiveness as scientists). On the other side, an overly conservative response (no need to worry yet) may lead to serious casualties and fatalities if an explosive eruption does develop. The 1991 eruption of Mount Pinatubo in the Philippines stands as an example of the narrow path to success in this business with timely evacuations implemented less than a week in advance of the most hazardous, explosive phase of the eruption. This places a premium on good science, experience "under fire", cool heads, and, yes, a little luck.

Additional information sources

Those not familiar with Long Valley caldera and its activity can find maps, background information, diagrams, and plots summarizing current activity under Long Valley caldera on the California Volcano Observatory web site at: http://volcanoes.usgs.gov/volcanoes/long_valley/ (Accessed on 15 January 2015)

Acknowledgements We are grateful to J. Lewicki and J. W. Cole for constructive reviews and to C. Jones for helping resolve inconsistencies of the dates of several events in the early 1980s.

References

Bailey RA (1989) Geologic map of Long Valley Caldera, Mono Inyo Craters volcanic chain, and vicinity, eastern California, U.S. Geological Survey Miscellaneous Investigation Series I-1933, I:62,5000, p 11

Bailey RA (2004) Eruptive history and chemical evolution of the precaldera and postcaldera basaltic-dacite sequences, Long Valley, California: implications for magma sources, current magmatic unrest, and future volcanism, U.S. Geol Surv Prof Paper 1692, 76p

Chen R, Branum DM, Wills CJ, and Hill DP (2014) Scenario earthquake hazards for the Long Valley Caldera-Mono Lake Area, East-central California, U. S. Geological Survey Open-File Report 2014–1045; California Geological Survey Special Report 233, 92p

Cramer CH, McNutt SR (1997) Spectral analysis of earthquakes in the 1989 mammoth mountain swarm near Long Valley, California. Bull Seismol Soc Am 87:1454–1462

Fearnley CJ, McGuire WJ, Davies G, Twigg J (2012) Standardisation of the USGS Volcano Alert Level

System (VALS): analysis and ramifications, Bull Volcanol. doi:10.1007/s00445-012-0645-6

Feng L, Newman AV (2009) Constraints on continued episodic inflation at Long Valley Caldera, based on seismic and geodetic observations. J Geophys Res 114 (B06403). doi:10.1029/2008JB006240

Gardner CA, Guffanti MC (2006) U.S. Geological Survey's alert-notification system for volcanic activity, USGS Fact Sheet 2006–3139, 4p. http://pubs.usgs.gov/fs/2006/3139/

Hildreth W (2004) Volcanological perspectives on Long Valley, Mammoth Mountain, and Mono Craters: several contiguous but discrete systems. J Vol Geotherm Res 136:169–198

Hildreth W, Fierstein J, Champion D, Calvert A (2014) Mammoth mountain and its mafic periphery—a late quaternary volcanic field in eastern California. Geosphere 10(6):1315–1365

Hill DP (1998) Seismological Society of America meeting —presidential address: science, geological hazards, and the public in a restless Caldera. Seismol Res Lett 69(5):400–404

Hill PM (2000) Possible asphyxiation from carbon dioxide of a cross-country skier in eastern California: a deadly volcanic hazard. Wilderness Environ Med 11:192–195

Hill DP (2006) Unrest in Long Valley Caldera, California, 1978–2004. In: Mechanisms of activity and unrest at large calderas. In: Troise C, De Natale G, Kilburn CRJ (eds) Geological society special publication, vol 269. The Geological Society, London, pp 1–24

Hill DP, Prejean S (2005) Magmatic unrest beneath Mammoth Mountain, California. J Vol Geotherm Res 146:257–283

Hill DP, Ellsworth WL, Johnston MSJ, Langbein JO, Oppenheimer DH, Pitt AM, Reasenberg PS, Sorey ML, McNutt SR (1990) The 1989 earthquake swarm beneath Mammoth Mountain, California: an initial look at the 4 May through 30 September activity. Bull Seismol Soc Am 80:325–339

Hill DP, Johnston MJS, Langbein JO, McNutt SR, Miller CD, Mortensen CE, Pitt AM, Rojstaczer S (1991) Response plans for volcanic hazards in the Long Valley Caldera and Mono Craters Area, California, U. S. Geol Survey Open File Report 84-500, 64p

Hill DP, Dzurisin DE, WL, Endo ET, Galloway DL, Gerlach TM, Johnston MJS, Langbein J, McGee KA, Miller CD, Oppenheimer D, Sorey M (2002) Response plan for volcano hazards in the Long Valley Caldera and Mono Craters Region California, U.S. Geol Surv Bull 2185, 57 p

Langbein JO (2003) Deformation of the Long Valley Caldera, California: inferences from measurements from 1988 to 2001. J Volcanol Geotherm Res 127:247–267

Langbein JO, Hill DP, Parker TN, Wilkinson SK (1993) An episode of re-inflation of the Long Valley Caldera, eastern California: 1989–1991. J Geophys Res 98:15851–15870

Lewicki JL, Hilley GE, Shelly DR, King JC, McGeehin JP, Mangan M, Evans WC (2014) Crustal migration of CO_2-rich magmatic fluids recorded by tree-ring radiocarbon and seismicity at Mammoth Mountain, CA, USA, Earth Planet Sci. Lett 390:52–58

Mader GG, Blair ML, Olson RA (1987) Living with a volcanic threat: response to volcanic hazards, Long Valley, California, William Spangle, and Associates. Portola Valley, CA. 105 p

Miller DC, Mullineaux DR, Crandall DW, and Bailey RA (1982) Potential hazards from future volcanic eruptions in the Long Valley-Mono Lake Area, east-central California and southwest Nevada—a preliminary assessment, U.S. Geological Survey Circular 877, 10 p

Montgomery-Brown EK, Wicks CW, Cervelli PF, Langbein JO, Svarc JL, Shelly DR, Hill DP, Lisowski M (2015) Renewed inflation of Long Valley Caldera, California (2011–2014) Geophys. Res Lett 42:5250–5257. doi:10.1002/2015GL064338

Pitt AM, Hill DP (1994) Long-period earthquakes in the Long Valley Caldera region, eastern California. Geophys Res Lett 21(16):1679–1682

Ryall AS II, Ryall F (1983) Spasmodic tremor and possible magma injection in Long Valley Caldera, eastern California. Science 219:1432–1433

Savage JC, Clark MM (1982) Magmatic resurgence in Long Valley Caldera, California, possible cause of the 1980 Mammoth Lakes earthquakes. Science 217:531–533

Shelly DR, Hill DP (2011) Migrating swarms of brittle-failure earthquakes in the lower crust beneath Mammoth Mountain, California. Geophys Res Lett 38. doi:10.1029/2011/GL049336

Shelly DR, Taira TA, Prejean SG, Hill DP, Dreager DS (2015) Fluid-faulting interactions: fracture-mesh and fault-valve behavior in the February 2014 Mammoth Mountain, California, earthquake swarm. Geophys Res Lett 42(14):5803–5812

Stovall WK, Marcaida MM, Mangan MT (2014) The California volcano observatory—monitoring the state's restless volcanoes, USGS fact sheet 2014–3120

Waldhauser F (2009) Near-real-time double-difference event location using long-term seismic archives, with application to Northern California. Bull. Seismol Soc Am 99:2736–2748. http://ddrt.ldeo.columbia.edu/DDRT/index.html. Accessed 15 Jan 2015

Wallace RE (1981) Active faults, paleoseismology, and earthquake hazards in the western United States, In: Earthquake prediction—an international review, Maurice Ewing Series 4, American Geophysical Union, Washington, D.C., pp 209–215

Wilkinson S, Hill DP, Langbein JO, Lisowski M, Mangan M (2014) Long Valley Caldera 2003 through 2014: overview of low level unrest in the past decade, USGS Open-File report 2014–1222, 1 sheet. http://pubs.usgs.gov/of/2014/1222. Accessed 15 Dec 2014

Volcanic Hazard Communication at Pinatubo from 1991 to 2015

Chris Newhall⊙ and Renato U. Solidum

Abstract

When Pinatubo re-awakened in early 1991, very few people within the vicinity were familiar with volcanic hazards, and even fewer believed that Pinatubo could impact them. Scientists knew more, but were still struggling to answer:

- How often and how explosively did Pinatubo erupt, and when was its most recent eruption?
- What precursors could be expected in advance of a very large (VEI ≥ 6) explosive eruption?
- What was happening beneath Pinatubo that was driving 1991 unrest?

To reach an exceptionally diverse audience and to counter widespread scepticism, scientists tried a whole package of communication measures, including simplified alert levels; a "worst case" hazard map; a probability tree; personalized briefings for local and national government officials, military and civil defense officials, nuns, and the news media; use of a IAVCEI video on volcanic hazards on broadcast TV and in briefings; volcanology tutorials for school teachers; talks on the mountain with villagers and anti-government guerrillas; and beer and hotdogs too. Forecasts were just-in-time and generally correct about what areas would be at risk. Overall, pre-eruption communication achieved its goal of getting people out of harm's way. Three lessons stand out: use simple, multipronged communications, especially video; include worst case scenarios in your warnings, together with estimated probabilities thereof; and be willing, as scientists

C. Newhall (✉)
Mirisbiris Garden and Nature Center, Santo
Domingo, Albay, Philippines
e-mail: cgnewhall@gmail.com

R.U. Solidum
Philippine Institute of Volcanology and Seismology
(PHIVOLCS), Quezon City, Philippines

Advs in Volcanology (2018) 189–203
https://doi.org/10.1007/11157_2016_43
Published Online: 15 March 2017

and decision makers, to recommend evacuations even if uncertainty is still high and there is still a chance of false alarm. For more than a decade after the 1991 eruption, rain-induced lahars threatened even more people and more infrastructure than the eruption itself. Several groups of scientists and engineers worked on the lahar threat, each coming up with slightly different long-term assessments that appeared to the public as bickering or incompetence. Scientists' credibility was seriously diminished. Decisions of what lahar-mitigation projects to build—including a succession of inadequate ones—were influenced less by science and more by public pressure, pragmatism, back-room politics, and profit. Short-term or immediate lahar warnings were communicated by scientists and by police-manned watch points. The scientific warnings were technically superior but the police warnings had greater credibility, as they were from familiar sources and easily understood. Communication of hazard information at Pinatubo saved many lives, and we are proud and privileged to have been part of preventing a much worse disaster. However, margins of safety were narrow and some deaths that did occur could have been prevented by better communication.

1   Introduction

The 1991–92 eruptions of Mount Pinatubo, Philippines, would affect a population in Central Luzon that was unfamiliar with and initially sceptical about volcanic hazards. That population was exceptionally diverse, including indigenous Aeta people on the volcano itself, helped by trusted nuns and pastors, and sharing the volcano with a small but influential band of guerrillas of the New Peoples' Army. Around the volcano were nearly a million lowland Filipinos in several large cities and towns of three provinces, and two large American military bases that were more like America than the Philippines. And in the skies above and around Pinatubo, commercial as well as military aviation had to be alerted.

Following small phreatic explosions on April 2, 1991, a team of Filipino and American scientists were trying urgently to decipher the history of the volcano and the unrest. What little we knew of Pinatubo's geologic history indicated that a major explosive eruption was possible, and geophysical and geochemical monitoring—just started—was indicating continuing but (as-yet) not escalating unrest. It was not clear that the

volcano WOULD erupt, but it was clear that IF IT DID, the eruption was likely to be large and explosive.

Facing strong scepticism from officials and the public unfamiliar with volcanic threats, and with no indication of how much time remained before an eruption could occur, we had to quickly find ways to teach about volcanoes and overcome that scepticism. If unrest escalated, officials would have to be convinced to evacuate large numbers of people to safety. Unrest did escalate, and most of those at high risk were successfully evacuated.

After the large eruption of June 15, 1991, focus shifted to the hazard of rain- (and lake-) related lahars. Though there was more time for a lahar information campaign, the long (>10 year) duration complicated the effort. Again, large numbers of people were at risk, and nearly all were moved to safety.

Although communications and mitigation were broadly successful, we stress that the successes were often "just in time" and "just barely enough." Most of the deaths that did occur could have been prevented, and many more deaths could have occurred had the eruption or the lahars been slightly sooner or larger. We tell this story in the hope that it will help colleagues in similar situations in the future add margins of safety.

2 Audiences: Who Needed to Learn About Volcanic and Lahar Hazard?

Audiences for our hazard communications were primarily decision-makers, community opinion leaders, and the news media. They in turn conveyed hazard messages and suggested actions to the public. Given that the total population at risk was roughly 1 million, the task was large and messages would need to be disseminated through many levels and in many ways. Scientists communicated directly with hundreds of officials at various levels, and they in turn reached out to the million.

In the Philippines, there is a well-established hierarchy of government and NGO bodies charged with disaster risk mitigation. At the top is the National Disaster Risk Reduction and Management Council (NDRRMC), known at that time as the National Disaster Coordinating Council, (NDCC), which includes top civilian, military, NGO officials, and scientists. The executive arm of the NDRRMC is the national Office of Civil Defense (OCD), led at the time by Engineer Fortunato Dejoras. Similar crisis coordinating functions are replicated at the regional (RDRRMC), provincial (PDRRMC), city or municipal (MDRRMC), and barangay (village) level, and execution is by local offices of OCD and corresponding offices of other government agencies and NGOs. Needs are assessed from the bottom or middle up; decisions are made mostly at the regional or provincial level, or higher if necessary, and passed down to local levels. In the case of volcanic hazards, assessments and recommendations are made by the Director of the Philippine Institute of Volcanology and Seismology (PHIVOLCS), or his/her designated representative.

At and around Pinatubo, virtually no one in this hierarchy had experience with volcanic hazards, so all needed urgent education. The sole exceptions were geoscientists from PHIVOLCS, joined by colleagues from the US Geological Survey (USGS) and a few university-based geoscientists.

News media had perhaps slightly more experience with volcanic hazards, but not much. There are relatively few science reporters in the Philippines, and some of the best reporters turned out to be generalist regional reporters who familiarized themselves with the issues better than national reporters who were pulled in many more directions.

Before the climactic eruption of June 15, 1991, nearly all communications with the preceding groups were led by the late Dr. Raymundo Punongbayan, then-director of PHIVOLCS. Ray was a talented communicator who quickly earned the trust of those he briefed. Briefings were held for the NDCC (in a meeting led by then-Defense Secretary and later President Fidel Ramos), for the Governors and staff of the three provinces that adjoin at the summit of Pinatubo (Zambales, Pampanga, and Tarlac), and for the Mayor of Angeles City. The Mayor of Olongapo City had the benefit of personal advice from several university-based scientists, including Dr. Kelvin Rodolfo and a young PHIVOLCS scientist, here the 2nd author. Other mayors joined the provincial level meetings. Dr. Punongbayan also developed a good rapport with news media, and he tapped them to help disseminate information. Other PHIVOLCS and USGS scientists were largely free to concentrate on field work and data interpretation, undistracted by media because Dr. Punongbayan handled their requests.

The American military bases were self-contained, with their own command structure, logistics, hospital, schools, businesses, and the like. Although the primary mission of the USGS scientists was to work with long-time PHIVOLCS colleagues in support of the PHIVOLCS mission, the USGS scientists were also granted access to logistical resources of the US military and, in return, kept the US military informed of developments in parallel with communications to the NDCC. Because the US Air Force was preparing a contingency plan in case evacuation became necessary, it sought scientific advice not only for commanders but also for enlisted personnel on whom base operations depended, for teachers, and for hospital staff. An initial attempt to also teach schoolchildren about the hazard had to be abandoned—for lack of time—in favour of a single-day briefing for all science teachers of Clark Air Base.

3 Pre-eruption Messages

One might liken the scientific effort—and the communication effort—to running a race against an unknown competitor, Pinatubo. At the time of the first phreatic explosions on April 2, 1991, no instruments monitored the volcano and there were no background data of monitoring. Fortunately, there was some prior geologic knowledge of Pinatubo, from a site safety study for a nearby nuclear power plant and from exploration for geothermal power, which told of what HAD happened before and COULD happen again, but what officials wanted was information about what WOULD happen. Furthermore, even worldwide, there was no information about the expectable precursors of a VEI 6 eruption, which we could see from the geology was a strong possibility.

Scientists, by training, are conservative in what they say. Data must be ample and convincing, and uncertainties should be low before a paper is published or advice is given. However, that conservatism must sometimes be overridden in times of volcanic crises.

In chronologic order, pre-eruption messages evolved through the following sequence:

- Pinatubo is a volcano, and is restless, so here is a primer on volcanic hazards (April–May).
- Near the volcano, the only way to protect yourself is to evacuate before an eruption. We may recommend evacuations from within 10, 15, 20 km radius (later, adding 30 and 40 km radius). (With radii adjusted for each specific volcano, this is a standard PHIVOLCS message in the face of any volcanic crisis) (April–June).

- We are not yet sure whether Pinatubo WILL erupt, but if it does, the eruption will be big and bad (late April).
- Evidence (as of May 13) shows that rising magma is causing the unrest.
- Even though this volcano is new to us too, we will try to raise alerts progressively if an eruption is approaching, giving you days up to two weeks of advance warning and, later, final notice hours or a day before an eruption (May 13).
- In return, we ask you (officials and the public) to understand that there might be a false alarm or two along the way. Please bear with us.
- Many areas have been swept by lethal volcanic flows in the past, or affected by ashfall, and could be similarly affected by a new eruption (hazard map, May 23).
- The volcano MAY erupt within 2 weeks (June 5).
- The volcano MAY erupt within 24 h (June 7). All within a 20 km radius of the summit should evacuate.
- The volcano has started to erupt, though only with a lava dome (June 7).
- The volcano has started explosive eruptions (June 9).
- Even though what you saw yesterday and today was impressive, the big one is still to come (June 12, 13, 14). On June 14, all those within 30 km radius of the summit should evacuate.
- A typhoon will arrive on June 15 (message from the weather bureau PAGASA).
- A massive eruption is in progress (June 15). Those within 40 km radius of the summit should evacuate, a recommendation easier made than followed.

Our message about possible false alarms was given in briefings to Governors and other decision-makers. Volcanologists have been notably averse to false alarms ever since the 1976 crisis at Soufrière Guadeloupe (Fiske 1984) and initially, we worried about this at Pinatubo as well. But we were reassured in our briefing in Zambales province when one of the attendees, a nun working with indigenous Aetas, told us they would happily accept premature warnings. They were more concerned that warnings might be too late. Not all would have been so forgiving, but we took the nun's words to heart. Social science research on earthquake and weather hazards recognizes some so-called "cry-wolf" reduction in scientific credibility (e.g., Atwood and Major 1998), but less than often presumed (Dow and Cutter 1998; Barnes et al. 2007 and references therein).

One communication that was, in retrospect, underemphasized, warned of ash in areas far from the volcano. Although warning was technically made, it was not emphasized sufficiently, especially in light of the typhoon rain that almost doubled the weight of ash on roofs and led to most of the eruption-related deaths. Another result of this under-emphasis was failure to warn aircraft outside the Philippine Flight Information Region (FIR), mostly over Indochina. We thought we had the aviation hazard covered through the Philippine FIR, and didn't even think of FIR's beyond the Philippines.

Additional details of eruption warnings, including dates and text of alert levels and recommendations for evacuation radii, may be found in Punongbayan et al. (1996) and Tayag et al. (1996).

4　How Were the Warnings Prepared and Presented?

4.1　Briefings and Video

By mid-May, as soon as we were reasonably sure that magma was rising and we had drafts of warning materials, the Office of Civil Defense arranged for the Director of PHIVOLCS (Dr. Punongbayan) to brief the Governors of each affected province (Pampanga, Zambales, and Tarlac), and the Mayor of Angeles City. The Governors, in turn, arranged for attendance of town mayors, captains of barangays closest to the volcano, and representatives of NGOs and news media. The general format of each meeting included an introduction by the Governor or Vice-Governor, followed by presentation of what

scientists knew about Pinatubo in general and about the hazard facing each province, and showing and discussion of a rough cut of the IAVCEI video "Understanding Volcanic Hazards" prepared by Maurice Krafft and others. Finally, there would be open discussion of general mitigation steps, though most discussion of specific mitigation steps came after these briefings. In most of these briefings, attention levels were high and many good questions were asked; in only one, that for the Mayor of Angeles City, was there official disinterest or hostility.

The IAVCEI video deserves special mention. Shortly after the terrible mudflow (lahar) disaster at Nevado del Ruiz Volcano (Colombia) in 1985, volcanologists agonized over how to prevent such disasters in the future. Because people of Armero and other towns at risk seemed not to have understood the seriousness of warnings they received, they did not walk to safety even though they could have done so. Apparently, they did not understand that a "flujo de lodo" (literally, a flow of mud) could in fact be a huge wall of mud, sand, boulders, trees, and more. Accordingly, Maurice Krafft and others set out to make a video that showed, in starkly graphic ways, the nature of each volcanic hazard, how far and fast it travels, and what it does when it hits houses and people. The ad-hoc steering committee for this video debated whether to include dead bodies—lest audiences find it too hard to watch—but in the end decided for inclusion to shock audiences into attention. Pinatubo was the next big event to threaten a population unfamiliar with volcanic hazards and, fortunately, scientists had by then a rough cut of Maurice's video, sufficient for public screening. This video was highly effective. Many who simply hadn't grasped the threat before seeing this video became converts soon after. The old saying, "A picture is worth 1000 words" might be re-written to say a "A video is worth 1,000,000 words!" Interestingly, we found no great revulsion from scenes of dead bodies—perhaps TV programming had already made this an everyday sight. But a short clip of a young girl trapped and shivering in the Armero lahar deposit drew audible gasps of horror in every briefing. It was VERY effective.

Unaware yet of its powerful reception at Pinatubo, Maurice Krafft and his wife Katia were dissatisfied with the video's footage of pyroclastic flows (much better footage is available now), so enroute to Pinatubo they stopped at Unzen to get better footage. They misjudged the threat and were themselves killed by a pyroclastic flow. That sad irony further increased the impact of the video as it was shown at Pinatubo.

Although technically the video was copyrighted by IAVCEI, we decided in the interest of time and the spirit in which the video was made to show it on broadcast TV and to distribute copies freely at each briefing. No doubt many copies of copies were also made. Today, one might post a video on YouTube; at that time, none of today's social media were available. Readers of this paper wishing to order a re-mastered DVD of this video, and a sequel, may order it at http://www.volcanovideo.com/p1IAVCEI.html.

In addition to briefings at the provincial level, Dr. Punongbayan also gave briefings to key national leaders. In one, he briefed then-President Corazon Aquino. In another, led by then-Secretary of National Defense Fidel Ramos and held at Camp Aguinaldo, he briefed assembled Cabinet members and other key officials. We recall a multi-tasking, busy Gen. Ramos calling a timeout after this briefing, and remarking to Dr. Punongbayan that he heard the message loud and clear, and half-joked that the Philippine government had better hurry up the renegotiation of the RP-US military bases agreement before there was nothing left to negotiate about.

While Dr. Punongbayan handled briefings at the national and provincial levels and for the media, several other members of the joint PHIVOLCS-USGS team provided briefings at lower levels. One PHIVOLCS team was based in villages on the NW flank of Pinatubo and quickly developed trust and provided information on the hazard. Field parties often encountered and stopped to talk with other villagers. One time, a group approached our helicopter while we were installing a seismic station. Other times, we stopped in roadside villages on our way to study outcrops, and the requisite courtesy call on a village captain would inevitably and fruitfully turn into quite a long discussion. In one village, we met with both indigenous Aetas and rebel guerrillas of the New People's Army. In another village from which the volcano could not even be seen, we didn't find the village captain so we spoke instead with his wife, who seemed not to understand at all, yet we know that residents of this village did ultimately evacuate to safety. These encounters and outreach were by no means comprehensive. A few PHIVOLCS scientists started systematic outreach in towns around the foot of the volcano, but owing to limits in time and transport, most of our outreach was bootlegged onto our field work.

Those of us on Clark Air Base at the East foot of Pinatubo also provided near-daily briefings for military officers and officials from nearby towns. We spoke to several classes of schoolchildren but quickly realized that the only way we could reach larger numbers would be to teach the teachers, especially science teachers. Inside Clark Air Base, a session for science teachers was quickly arranged; outside the base, we connected with only a few teachers, far from all.

Also within Clark Air Base, we gave briefings to individual units including the hospital and, at the suggestion of the Deputy Base Commander, to a level of staff called "chief master sergeants." The "chiefs" were eminently practical—the real "doers" of the base. Our makeshift Pinatubo Volcano Observatory (PVO), in a crowded apartment for enlisted personnel, became a hub for curious and concerned daily visitors, including the base meteorologists, officers, and even their wives. On one occasion when activity ramped up, we summoned top officers of Clark and Subic to PVO. Their arrival by helicopter and thence in cars bearing the flags of admirals and generals drew quite a lot of attention and gossip throughout the base. Because we relied initially on the base weather office for access to a fax machine, US Air Force weathermen (and by extension, their superior officers) were also privy to fax communications between scientific team at Clark and Director Punongbayan in Manila. One fax from Director Punongbayan, about alert levels if we recall, arrived at PVO with yellow highlights! Even though the Base Commander wanted to keep our work quiet, there was really no way to do so, and occasional breaches were actually quite useful! Finally, in late May, the US Air Force command realized it needed to run an interview with a scientist on the base TV station, but the interview was tightly scripted and the pre-scripted wrap-up by the interviewer was much more reassuring than the interview itself.

4.2 Alert Levels

We soon recognized that, given the large, diverse audience and widespread unfamiliarity with the threat, all of our warnings would need to be very simple. One such simplification was definition of numbered alert levels from 0 (no unrest) to 5 (large explosive eruption in progress) (for details, see Punongbayan et al. 1996; Tayag et al. 1996). These were patterned, from vague and stressed memory, on alert levels first introduced at Rabaul (Papua New Guinea) and later adapted for eruptions in Alaska (USA) (see brief histories of alert levels in Fearnley et al. 2012; Fearnley 2013; Winson et al. 2014). Levels 3 and 4 of the Pinatubo scheme anticipated forecast time windows (2 weeks and 24 h) within which an eruption might occur, and each level had an interpretation of activity. We intended that Civil Defense could design and key their mitigation actions to these alert levels. Probably, they would have done so had not events developed so rapidly in early June that –in effect—recommended evacuation radii from the Director of PHIVOLCS pre-empted plans that Civil Defense was still preparing. In recent years there has been much re-examination of alert schemes and debate of whether they should include forecasts and formal linkage to responses, making scientists de-facto decision makers. Given the extreme urgency at Pinatubo, we think both the forecast and the de-facto decision-making role of PHIVOLCS were necessary, though we acknowledge that these matters should be discussed and agreed elsewhere on a country-by-country basis.

PHIVOLCS and civil defense leaders still link alert levels and responses, but in recognition of uncertainties and differences between volcanoes, PHIVOLCS has made the forecast windows less precise.

The general upward progression of levels resembled a familiar 3-level alert used for typhoons in the Philippines, but had the opposite sense to a countdown of alerts used by the US military. Fortunately, the US military agreed to use our scheme rather than their own.

We did note one misunderstanding of the wording on the alert levels. For alert levels 3 and 4, the wording stated that "an eruption was possible within (a specified timeframe, 2 weeks or 24 h)." Strictly speaking, we meant that we could no longer guarantee that an eruption would not occur within that period; however, the simpler reading of "could occur within that time-frame" would have been an acceptable simplification. In Pilipino, that would have been stated as "ma-aaring mangyari sa loob ng (dala-wang lingo o 24 oras)." Perhaps since we wrote it only in English, many misread it to mean that "an eruption WILL occur within 2 weeks or 24 h" or, worse yet, "an eruption WILL occur in exactly 2 weeks or 24 h." If differences between the terms "could occur" versus "will occur", and between "in" versus "within" might be misunderstood, use the local language to clarify! No serious harm was done, as those making final evacuation recommendations understood the terms as intended.

Our alert level scheme also included guidance for step down, with built-in delays to guard against premature lowering of alert levels. We think such guidance for step-downs is helpful, partly to guard against sudden decreases in activity that are "calm before the storm," and partly to allow orderly stepdown, without embarrassment, should unrest truly stop. Many magma intrusions fail to reach the surface and moveable alert levels (up and down) are designed as an alternative to forecasts that may prove wrong. In the case of Pinatubo, we didn't have occasion to use the stepdown until well after the climactic eruption.

4.3 Pre-eruption Hazard Map

Hazard maps are the geoscientist's standard response to the question, "What areas are at risk?" Our pre-eruption map was based mainly on quick reconnaissance of the maximum extent of pyroclastic flows from past eruptions of Pinatubo. In most areas this was immediately obvious both on the ground and on aerial photos. Our geological team doing field work compared notes with Director Punongbayan who was interpreting aerial photos from his office in Quezon City. In most cases, the field exposures (in canyon walls) extended slightly beyond the distinctive, dissected topography of old pyroclastic flows; in a few cases, e.g., near Barangay Pasbul, of Floridablanca town, suspect topography was found in aerial photos and later confirmed by ground visit. The aerial photos we had were well out of date, and we note with pleasure how much better it is to have modern satellite coverage with current roads and quarries.

We sketched the outline of prehistoric pyroclastic flows onto a single sheet of paper on which we had also traced main highways and towns. For safety, we added 0.5–1.0 km of buffer zone around the known pyroclastic flow extent. In our original sketch we showed that ash could fall anywhere on the area of the map; in a cleaned-up sketch, the most likely directions of ashfall were shown. Potential lahars were shown only as hash marks down the main river channels—a simplification that would be greatly expanded after the eruption. Without calling it as such, we intended this map to reflect our "worst-case" scenario, VEI 6 eruptions from Pinatubo. Fortunately, the actual reach of pyroclastic flows almost perfectly matched the hazard zones on our map, and did not reach as far as what we later discovered was the reach of an even larger prehistoric eruption. These sketched maps were shown, copied and distributed at all briefings after May 23 and also reproduced in one or more national newspapers.

In retrospect, we can see that our hazard map —while useful—was difficult for some audiences to understand. Many otherwise well-educated people are unable to read maps, and maps in standard plan view were an even more difficult abstraction for those less educated. Haynes et al. (2007) and Leone and Lesales (2009) offer excellent suggestions on how to make hazard maps more readily understandable, e.g., by use of 3D visualization.

4.4 A Probability Tree

Newhall and Hoblitt (2002) and Newhall and Pallister (2015) describe relatively simple ways to estimate probabilities of volcanic events and their consequences, and thereby help officials and those at risk to decide on what risks to take. At Mount St. Helens, calculations were carried all the way to annual risk of death, which allowed loggers and others to compare volcanic risk to more familiar occupational and lifestyle risks.

The first probability tree at Pinatubo (May 17) considered just one scenario, of pyroclastic flows to the East and onto Clark Air Base. We carefully did not exaggerate any hazard or risk, but we wanted to make sure that officials understood that hazards and risk were, in fact, unacceptably high. Our estimate of a 3% chance of pyroclastic flows reaching Clark and killing thousands within the coming months was immediately understood by US Air Force officers to be unacceptable. The Director of the national Office of Civil Defense, Engr. Dejoras, saw similarly unacceptable risk for the civilian population all around Pinatubo. Social science research has reaffirmed the pitfalls of using ambiguous adjectives like "high" or "low" to describe hazard and risk, and the advantages of quantifying those hazards (Doyle et al. 2014 and references therein). Once quantified, volcanic risk can be compared to more familiar risks and to levels of risk judged acceptable under the circumstances.

Public discussion of the probability tree at Pinatubo might have helped in our general education campaign, but probability is a difficult concept for many non-technical persons so for want of time we discussed it only with those who already understood probabilities. When we did discuss probabilities, we usually spoke in terms of "percent chance" or "odds" rather than strict decimal probability numbers, as "odds" and frequency expressions are more easily understood (Gigerenzer and Edwards 2003; Leclerc and Joslyn 2012; Henrich et al. 2015). We did not carry estimation of hazard all the way on to estimation of risk, so there was no quantification of how much risk could be reduced by various mitigation options. Pyroclastic flows are so lethal that vulnerabilities would have been nearly 100% and without evacuation, exposure would have been 100% too. Discussion moved quickly to plans for evacuations should an eruption become imminent.

We did prepare one update to this tree, on June 10, but by that time evacuation of Clark Air Base and nearby areas was already well underway and the new probabilities—much higher of course—were effectively moot.

In neither tree did we estimate uncertainty of our probability estimates, but in briefings about the first tree, we did indicate that uncertainty was at least plus or minus one order of magnitude. Simply using probabilities already indicated uncertainty about what would transpire. Indeed, in every episode of volcanic unrest there is a range of possible outcomes, including the null event of "no eruption." Use of probabilities, frequencies, or odds (especially, when combined with alert levels) allows scientists to bypass the as-yet unreachable goal of making very specific deterministic predictions of what will occur. Yes, officials and the news media will ask for such predictions, but we believe that it is scientifically more correct and educationally more useful to indicate the range of possible scenarios and to discuss how the probabilities of each scenario can be estimated and can change. Interestingly, recent research by Leclerc and Joslyn (2015) found that including probabilities in frost forecasts increased recipients' willingness to accept false alarms.

4.5　Recommendations for Evacuation

In the Philippines, PHIVOLCS typically recommends that a certain radius around a volcano be declared a "permanent danger zone" and that residence in this zone be forbidden. Outside that zone, restrictions or measures such as evacuations are at the discretion of local government officials. As Pinatubo had not erupted in historical time, no permanent danger zone had been designated, nor had there been any resolutions within provincial or municipal governments about when and where evacuations might be ordered. The Director of PHIVOLCS, with blessing from the national Office of Civil Defense, took on the responsibility for recommending radii of evacuation, and the Philippine Army together with the Departments of Public Works and Social Welfare and Development implemented those recommendations. Although an early, limited evacuation (in early April) proved to be porous, later evacuations in June had the advantage of visible eruptions and were relatively smooth and effective.

Although the boundaries of hazard zones are naturally irregular, based on topography, Director Punongbayan and Director Dejoras decided that recommended evacuations would be circles with radii of 10 and 20 km (later, including 30 and 40 km). Circles drawn around the outer boundaries of the mapped hazard inevitably include some areas of relative high ground and safety, but it was judged to be simpler and more effective to base evacuations on simple circles rather than on the boundaries of hazard maps that some might not understand. Details are given in Tayag et al. (1996).

4.6　Personal Communications

Most of our communications with officials were of the formal types listed above—briefings, alert levels, hazard map, a probability tree, and recommended radii of evacuation. However, we found that informal, personal communications were sometimes just as effective as the formal ones, if not more effective. We already mentioned the gut-level emotional impact elicited by film clips of the young girl trapped and shivering in lahar deposit in Armero, Colombia.

A different and very effective form of personal communication was movement of scientists themselves to safer, fall-back positions—in the northwest, from Sitios (hamlets) of Tarao and Yamot to Barangay Poonbato, and on the east, from the center of Clark Air Base to the far eastern edge of the air base. The move was for our own safety, but we realized as we planned it that our action would be a strong message for others. As we hoped, local officials and residents took note and took our warnings more seriously thereafter.

Yet another form of personal communication was developing simple friendships and trust with those at risk. Several members of the PHI-VOLCS team on the northwest side of Pinatubo, led by Julio Sabit, developed quick rapport with local residents. Eating together, drinking together, and sharing family histories and aspirations are wonderful ways to build the trust that becomes so essential when urgent warnings must be issued. A similar experience on the east side was generated accidentally on May 18. The scientific team was exhausted and stressed, so we called a time-out for a BBQ, inviting Air Force officers to join. Over hot dogs and beer, the officers discovered to their surprise that we scientists were just normal people, with families of our own just like them. Probably, scientists had the same revelation about the military officers. Up until that time we had regarded each other with some puzzlement and caution; after the BBQ and beer, things lightened up and more trust was evident.

5 Post-eruption Lahar Messages

As soon as the climactic eruption occurred, it was obvious to scientists that the big threat in coming years would be from rain-induced lahars (Janda et al. 1996). We didn't know exactly what percentage of the fresh deposit would be washed into the lowlands in lahars (and in normal muddy streamflow)—estimates ranged from around 15% to around 50%. Even 15% of the new deposit would be enormous and would more than fill river channels; 50% would bury huge areas of farmland and towns under several meters of sediment. Even on the back of an envelope,

scientists could see that these lahars would be bigger than anything they had ever seen or imagined. In contrast, most people at risk in the lowlands were blissfully oblivious to the lahar threat, grateful that they had survived the eruption. Even engineers and officials had a hard time envisioning the scale of the impending lahars.

Our messages for long-range and short-range lahar hazard included:

- There is an enormous amount of loose sand and pumice on the volcano that will be carried into the lowlands in coming months and years. (unquantified, no maps yet)
- Depending on assumptions, large areas including many towns may be buried, and here (on hazard maps) are the areas at high and lesser risk.
- If you build dikes to contain the sediment, most of these will fill and can breach if you don't build them big enough and strong enough.
- Some towns are not going to survive unscathed and may need to be sacrificed. People from those towns will need to be resettled in other places for the foreseeable future.
- Immediate warnings of lahars, e.g., "A (small/large) lahar has formed in the Sacobia River and will reach populated areas by (specified time)".
- Because lahars can go overbank and even breach protective dikes, populations still remaining at risk from these lahars should evacuate immediately.

In the late 1990s and in 2000–2001, two more messages were added:

- As the caldera lake fills, it will eventually overtop and may pose a severe lahar threat to Botolan (1998–2001).
- The threat of a breakout lahar still remains, because we didn't succeed to induce a rapid scouring/ breach of the loose material that forms a dam at the head of the Maraunot/Balin Baquero/Bucao river system (late 2001).

6 How Were Lahar Warnings Prepared and Presented?

6.1 Briefings and Video

Briefings followed more or less the same pattern as during pre-eruption time, though more were held at the regional level than before the eruption. The Regional Disaster Coordinating Council (RDCC 3) assumed a greater role for lahars than it had before the eruption. More players were involved as well—with notably increased involvement by the Department of Public Works and Highways (DPWH); the Department of Social Welfare and Development (DSWD) (for emergency relief and more permanent resettlement of those displaced by lahars), and the new Mount Pinatubo Commission (a mechanism to coordinate funding and response, but relying heavily on DPWH, DSWD, and their contractors).

The Krafft video was still used occasionally, but was increasingly supplanted by live and taped coverage of actual Pinatubo lahars, beamed over broadcast TV. Scientists and at least two television networks also prepared their own video documentaries on Pinatubo lahars.

6.2 Hazard Maps

Everyone wanted to know if his or her town would be hit by lahars. Engineers and planners also needed to know the likely volumes of sediment that would move into the lowlands, for planning engineering structures and for debating the relative merits of trying to control the sediment versus simply relocating communities and letting the sediment flow.

The earliest hazard maps were prepared by PHIVOLCS (Punongbayan et al. 1991) by the Pinatubo Lahar Hazards Taskforce (PLHT, a cooperative effort of the Mines and Geosciences Bureau, Univ. of the Philippines, Univ. of Illinois at Chicago, and PHIVOLCS) (1991a, b), and by the Bureau of Soils and Water Management (1991). Revised maps were prepared by Pierson et al. (1992), PHIVOLCS (1992, 1994); and the Zambales Lahar Scientific Monitoring Group

(ZLSMG) (1993, 1994). ZLSMG was the successor to PLHT, comprised of university-based scientists including some PHIVOLCS scientists on study leave. By mutual agreement with PHIVOLCS, the ZLSMG handled most of the lahar study on the west side of Pinatubo (1992, 1994).

Although there were minor differences between the various maps, most of the maps were broadly similar and confusion between maps seemed not to be a serious problem. Greater confusion may have arisen over the use of lines rather than gradations. Lines are satisfying, but they inevitably give a sense of more certainty than actually exists. Lahar hazard in lowland areas is gradational, without sharp boundaries.

Later, PHIVOLCS was given an additional duty of certifying whether specific land parcels (e.g., those for new construction and bank loans) were "safe" or "unsafe" from lahar. In general, that meant simply locating the parcel of land on the published hazard map and certifying that it was inside or outside hazard zones. Though unstated, delineation of "safe" zones implies a choice of how low a hazard must be in order to call the zone safe. Everything was changing too fast for us to identify an "X-year floodplain" (one event in X years) but qualitatively, areas outside the hazard zones were judged to have "very or extremely low" probability of being inundated.

In general, the hazard maps served dual purposes of letting communities know their (qualitative) chances of being buried in the coming years, and letting engineers and planners design appropriate responses. To be sure, some mitigation measures were technically inadequate or even foolish (e.g., construction in 1992 of a new school in Sta. Barbara, Bacolor, which would soon be buried by more lahars), but these instances were not for want of good scientific information. More likely, they were driven by inattention, bureaucracy, politics, or profit.

6.3 Short-Term, Immediate Lahar Warnings

Four systems were used for lahar warnings. One, installed by DPWH with information fed by radio telemetry directly to the RDCC at Camp Olivas, San Fernando, used trip wires and rain gauges, placed near the foot of a pyroclastic fan and the head of the corresponding alluvial apron. They didn't last long, with the trip wires almost immediately tripped or stolen. A second, installed by PHIVOLCS and the USGS, used rain gauges high in the watershed and acoustic flow sensors (inexpensive, high-frequency exploration seismometers) lower on the pyroclastic fans but still above the alluvial fans. Data were telemetered to PVO where they were interpreted 24/7, and warnings were relayed to RDCC by telephone. Sometimes, PHIVOLCS observers would also report from watchpoints but these were not an essential part of PHIVOLCS' warnings. A third was direct observations of lahars by scientists from the ZLSMG, from watchpoints they manned in Dalanaoan, San Marcelino, and Malumboy, Botolan. Dalanaoan was midway down the alluvial fan of the Marella/Sto. Tomas River but still upstream from populated areas; Malumboy was well down the Bucao River but still 11 km upstream from Botolan. Results were sent in real-time to authorities. The fourth system also used direct observations of lahars, by policemen posted near the heads of several alluvial fans who then radioed reports to the RDCC.

Of the four systems, the PHIVOLCS and ZLSMG systems were scientifically superior, giving early and semi-quantitative information about both the scale and the travel speed of lahars, and the benefit of scientific interpretation. In retrospect, the PHIVOLCS system would have been better accepted if one of its scientists with good communication skills had spent more time at the RDCC. The police reports were less accurate than those from scientists but had the advantage of being simple and from familiar sources.

Because everyday thundershowers at Pinatubo are very localized, and neither the national meteorological service (PAGASA) nor the military had modern Doppler radar, meteorologists did not play as great a role at Pinatubo as they would today. However, they did make major contributions by warning of incoming typhoons

that invariably generated lahars across the entire volcano.

In general, short-term lahar warnings did reach those at risk and saved many hundreds or even thousands of lives. Regrettably, they didn't manage to save all. We recall one instance in which PHIVOLCS had strong evidence for a lahar headed for Dolores, Mabalacat, and relayed its warning, but local officials deferred to the police system which did not sound the same alarm. Approximately 100 died as a result. In another instance, all systems warned of a major lahar headed for Bacolor town (by then, largely evacuated) and its satellite barangay, Cabalantian (not evacuated). Unfortunately, a large dike gave a false sense of security until it breached, and as many as 400 of those who did not evacuate perished.

6.4 Probabilities of Lahars

No effort was made to estimate either long-term or short-term probabilities of lahars. Most of the effort toward quantification was focused on estimating rates of sediment transport, both in lahars and in normal muddy streamflow.

6.5 Personal Factors

Ironically, the extra time afforded by lahars (relative to the pre-eruption period) created extra difficulties in communication. There were more end-users to be informed, and more scientists providing the information. In addition to the main players mentioned above, there were also engineering consultants from many countries. Scientists operating on limited (sometimes, shoestring) budgets anticipated the magnitude of the lahar hazard while engineers turned that information into lucrative contracts for sediment control and reconstruction.

There were, we must admit, some unfortunate clashes between scientists which contributed to the loss of scientific credibility. One area of seeming disagreement was on the efficacy of engineering works. Without going into more details, suffice it to say that both real and imagined differences got translated, on front pages and lead stories, into a personalized competition of scientific expertise. Scientists' credibility would have been higher, and the public served better, had these differences been resolved behind closed doors, and officials and the public been given consensus statements.

7 Lessons to Remember

Pinatubo was a pressure-cooker for scientists, demanding warnings well before full data sets could be collected. Many times, we had to suppress our scientific instinct to say, "Wait, let me collect more data." Information had to be given immediately. Timely communication of often-uncertain hazards information, pre- and post-eruption, saved thousands, perhaps even tens of thousands of lives.

Below, we list some lessons that we ourselves learned, and that we commend to readers:

- Start your communications immediately, preferably long before unrest begins. If that is not possible, then begin teaching about the volcano at least start as soon as unrest is noticed. Do NOT wait until you are sure about what the volcano will produce. Communicating well with all parties takes time, and volcanic crises can develop so quickly that there might not be enough time for communications if you wait until you know more.
- Make a checklist of what data must be gathered and analysed, and what information must be communicated to whom. It would be easy, in the rush and stress of unrest, to forget one thing or another. Do not think of this as an affront to your professional experience; rather, think of it as a pilot thinks of his or her checklist—as an extra safety measure.
- Even if the most immediate pre-eruption concern is for pyroclastic flows, don't forget to warn thoroughly of ash, in the air and on roofs. It can be life-threatening, even far from the volcano.

- Expect scepticism, especially where a volcano has long been dormant. This will make your communication job more difficult, but also more essential. Consider using a variety of approaches and tools (e.g., briefings, video, hazard maps, alert levels, probability trees, and personal touches) to reach both sceptics and converts. Different tools will be required for different audiences. Video is especially effective.
- Help officials to understand that every volcanic crisis has several possible outcomes, and that their relative likelihoods may change with time as new geologic and monitoring data are collected. Response plans should be flexible enough to account for several different scenarios.
- Don't be afraid to give warnings based on your best current data, even if you know those data are woefully inadequate. Some will argue that giving interpretations and warnings based on incomplete data may be worse than no warnings at all. We respectfully disagree, as we think it is scientists' societal duty to give the best warnings possible AT ALL STAGES of a crisis, even early on while uncertainties remain high. If Director Punongbayan had not been willing to risk his reputation and perhaps even his job to give warnings when he did, even though uncertainties remained high, we doubt that the evacuations would have been as successful as they were.
- Don't be afraid to include a "worst case scenario" among various scenarios, and put it in context by estimating relative probabilities of various scenarios. In the case of Pinatubo, we saw so much geologic evidence for VEI 6 eruptions and so little evidence for smaller eruptions that we put a high probability on the "worst-case" VEI 6 scenario.
- Similarly, don't be overly afraid of a false alarm. To be sure, officials and the public do have limited tolerance for false alarms (e.g., Atwood and Major 1998), but it is not zero, especially if you explain all of the possible scenarios and uncertainties. Use of alert levels and probability trees reduces the likelihood of

false alarms. Consider also a social contract with officials and those at risk: If they want to be sure of warning, they must accept some risk of a false alarm.

- After an eruption that produces a large volume of ash and other pyroclastic debris, expect lahars over an extended period. With more time and players, communications will get more complicated, and may require new, proactive communication strategies to keep scientific advice in the forefront of further planning.
- Both before and after an eruption, strive for scientific coordination of all messages BEFORE they go to officials and the public.
- Recognize the importance of trust and personal connections between scientists, officials, and the news media. Sometimes, these are as important or more important than the formal warnings.

We close this chapter, proud of the successes but also conscious of the near-misses and failures. Communication of hazards information at Pinatubo was complicated, but also absolutely essential to mitigation of what would have otherwise been a much worse disaster.

References

Atwood LE, Major AM (1998) Exploring the 'cry wolf' hypothesis. Int J Mass Emerg Disasters 16(3):279–302

Barnes LR, Gruntfest EC, Hayden MH, Schultz DM, Benight C (2007) False alarms and close calls: a conceptual model of warning accuracy. Weather Forecast 22:1140–1147 and corrigendum 24:1452–1454

Bureau of Soils and Water Management (1991) Mudflow and siltation risk map (as of Oct 1991). Bureau of Soils and Water Management. Scale 1:200,000

Dow K, Cutter SL (1998) Crying wolf: repeat responses to hurricane evacuation orders. Coast Manag 26:237–252. doi:10.1080/08920759809362356

Doyle EEH, McClure J, Johnston DM, Paton D (2014) Communicating likelihoods and probabilities in forecasts of volcanic eruptions. J Volcanol Geoth Res 272:1–15

Fearnley CJ (2013) Assigning a volcano alert level; negotiating uncertainty, risk, and complexity in decision making process. Environ Plan A 45:1891–1911

Fearnley CJ, McGuire W, Davies G, Twigg J (2012) Standardisation of the USGS volcano alert level system (VALS): analysis and ramifications. Bull Volc 74:2023–2036

Fiske RS (1984) Volcanologists, journalists, and the concerned local public: a tale of two crises in the eastern Caribbean. In: Explosive Volcanism: interception, evolution, and hazard. National Academy Press, Washington, DC, pp 170–176

Gigerenzer G, Edwards A (2003) Simple tools for understanding risks: from innumeracy to insight. Br Med J 327:741–744

Haynes K, Barclay J, Pidgeon N (2007) Volcanic hazard communication using maps: an evaluation of their effectiveness. Bull Volc 70:123–138

Henrich L, McClure J, Crozier M (2015) Effects of risk framing on earthquake risk perception: life-time frequencies enhance recognition of the risk. Int J Disaster Risk Reduct 13:145–150

Janda RJ, Daag AS, Delos Reyes PJ, Newhall CG, Pierson TC, Punongbayan RS, Rodolfo KS, Solidum RU, Umbal JV (1996) Assessment and response to lahar hazard around Mount Pinatubo, 1991–1993. In: Newhall CG, Punongbayan RS (eds) Fire and mud: eruptions and lahars of Mount Pinatubo, Philippines: Quezon City, Philippine Institute of Volcanology and Seismology, and Seattle, University of Washington Press, pp 107–139

LeClerc J, Joslyn S (2012) Odds ratio forecasts increase precautionary action for extreme weather events. Weather Clim Soc 4(4):263–270. doi:10.1175/WCAS-D-12-00013.1

LeClerc J, Joslyn S (2015) The cry wolf effect and weather-related decision making. Risk Anal 35:385–395

Leone F, Lesales T (2009) The interest of cartography for a better perception and management of volcanic risk: from scientific to social representations: the case of Mt. Pelée volcano, Martinique (Lesser Antilles). J Volcanol Geoth Res 186:186–194

Newhall CG, Hoblitt RP (2002) Constructing event trees for volcanic crises. Bull Volc 64:3–20

Newhall CG, Pallister JS (2015) Using multiple data sets to populate probabilistic volcanic event trees. In: Papale P (volume ed) Volcanic hazards, risks, and disasters, v. 2 of Hazards, disaster & risks series. Elsevier, pp 202–232

Philippine Institute of Volcanology and Seismology (1992) Pinatubo Volcanic Lahar Hazards Map, as of August 1992. Prepared by PHIVOLCS; produced by National Economic Development Authority, NEDA. Scale 1:200,000 (updated after each major lahar)

Philippine Institute of Volcanology and Seismology (1994) Pinatubo Volcanic and Flood Hazards Map, as of December 1994. Prepared by PHIVOLCS; produced by National Economic Development Authority, NEDA. Scale 1:200,000

Pierson TC, Janda RJ, Umbal JV, Daag AS (1992) Immediate and long-term hazards from lahars and excess sedimentation in rivers draining Mt. Pinatubo,

Philippines. US Geological Survey, Water-Resources Investigations Report 92-4039, 35 pp (see Plate 1)

Pinatubo Lahar Hazards Taskforce (PLHT; compilers, Paladio ML, Umbal JV) (1991a) Lahar hazard map of the Bucao River system, Pinatubo Volcano, Luzon, Philippines (as of October 11, 1991). Scale 1:50,000

Pinatubo Lahar Hazards Taskforce (PLHT; compilers, Umbal JV, Rodolfo KS) (1991b) Lahar hazard map of the Sto. Tomas River system, Pinatubo Volcano, Luzon, Philippines (as of August 23, 1991). Scale 1:50,000

Punongbayan RS, Besana GM, Daligdig JA, Torres RC, Daag AS, Rimando RE (1991) Mudflow Hazard Map, as of 01 October 1991. PHIVOLCS, printed and published by the National Mapping and Resource Inventory Authority, NAMRIA. Scale 1:100,000

Punongbayan RS, Bautista MLP, Harlow DH, Newhall CG, Hoblitt RP (1996) Pre-eruption hazard assessments and warnings. In: Newhall CG, Punongbayan RS (eds) Fire and mud: eruptions and lahars of Mount Pinatubo, Philippines: Quezon City, Philippine Institute of Volcanology and Seismology, and Seattle, University of Washington Press, pp 67–85

Tayag J, Insauriga S, Ringor A, Belo M (1996) People's response to eruption warning: The Pinatubo experience, 1991–92. In: Newhall CG, Punongbayan RS (eds) Fire and mud: eruptions and lahars of Mount Pinatubo, Philippines: Quezon City, Philippine Institute of Volcanology and Seismology, and Seattle, University of Washington Press, pp 87–106

Winson AEG, Costa F, Newhall CG, Woo G (2014) An analysis of the issuance of volcanic alert levels during volcanic crises. J Appl Volcanol 3(1):1–12

Zambales Lahar Scientific Monitoring Group (ZLSMG; compilers Alonso RA, Rodolfo KS, Remotigue CT, Umbal JV, Jalique-Calamanan V) (1994) Lahar Hazard Map of Western Mount Pinatubo, Zambales Province (as of July 1, 1994, revised from Pinatubo Lahar Hazards Taskforce, 1991, and ZLSMG maps of 1 July 1993, 21 September 1993, 8 October 1993,for Sto. Tomas and Bucao Rivers). Scale 1:50,000

Zambales Lahar Scientific Monitoring Group (ZLSMG; compilers Alonso RA, Rodolfo KS, Remotigue CT) (1993) Lahar Hazard Map of the Santo Tomas-Pamatawan Plain, Southern Zambales Province, as of September 1, 1993). Scale 1:50,000

Instrumental Volcano Surveillance and Community Awareness in the Lead-Up to the 1994 Eruptions at Rabaul, Papua New Guinea

Chris McKee, Ima Itikarai and Hugh Davies

Abstract

Instrumental volcano surveillance and community awareness played key roles in preparing for the outbreak of the 1994 VEI 4 volcanic eruptions at Rabaul (pop. 17,000). The eruptions were preceded by 23 years of fluctuating unrest involving swarms of caldera earthquakes (max M_L 5.2) and co-seismic uplift of parts of the floor of Rabaul Caldera. Eruption contingency planning was formally driven by government authorities and involved all sections of the community. Community awareness of the volcanic threat was enhanced by the dissemination of relevant information by the Public Information Unit of the East New Britain Provincial Government and reached a peak in the mid-1980s at the time of a large increase in the strength and frequency of earthquake activity (between August 1983 and July 1985). However, the intensity of the unrest declined after July 1985 and another 9 years elapsed before a new and dramatically stronger phase of unrest took place. The strong and sustained earthquake activity on 18 September 1994, together with marked co-seismic uplift that took place that night, was the final episode of volcanic unrest prior to the outbreak of eruptions on the morning of 19 September 1994. Memories and stories of the seismic prelude to the previous eruptions, in 1937, are reported to have been a major influence on community response

C. McKee (✉)
Port Moresby Geophysical Observatory,
P.O. Box 323, Port Moresby, NCD,
Papua New Guinea
e-mail: chris_mckee@mineral.gov.pg

I. Itikarai
Rabaul Volcanological Observatory, P.O. Box 386,
Rabaul, ENBP, Papua New Guinea

H. Davies
Earth Sciences, University of Papua New Guinea,
P.O. Box 414 University NCD, Port Moresby, Papua
New Guinea

Advs in Volcanology (2018) 205–233
https://doi.org/10.1007/11157_2017_4

to the seismicity on 18 September 1994. The evacuation of all areas within the caldera proceeded efficiently from late afternoon of 18 September until the early hours of 19 September. These areas were almost deserted when the eruptions started at two vents, Tavurvur and Vulcan, on opposite sides of the caldera at 0606 and 0717 LT respectively on the morning of 19 September 1994. Ten deaths in the first six weeks of eruptive activity were volcano-related. Damage inflicted by the eruptions was severe. About 70% of Rabaul Town was destroyed by tephra fall from Tavurvur, and several villages were obliterated by pyroclastic flows and heavy tephra fall from Vulcan. The 23 years of precursory activity and the events around the start of the 1994 eruptions delivered a number of important lessons in the fields of volcano surveillance, communications and disaster management. Perhaps the most important lessons of all are that co-existence with active and potentially active volcanoes requires (i) open and effective lines of communication between volcano scientists, government officials, town authorities and the general public, facilitated by designated public information officers, and (ii) the establishment and frequent exercising of eruption contingency plans.

1   Introduction

Rabaul is the name of both a complex volcanic system and a town at the northeastern tip of the Gazelle Peninsula, New Britain Island, Papua New Guinea (Fig. 1). Since the latest caldera-forming eruption at Rabaul, at about 1400 BP (Heming 1974; Walker et al. 1981), small cones have grown at five vent areas within Rabaul Caldera. The recorded eruption history of the Rabaul system spans a period of about 250 years and includes six eruptive episodes that took place at intervals of 24–59 years: 1767, 1791, 1850s, 1878, 1937 and 1994 (McKee et al. 2016). The most active vents during the historical period were those in the Vulcan area and at Tavurvur, on the western and eastern sides of Rabaul Caldera respectively (Fig. 1). These centres were simultaneously active in 1878, 1937 and 1994.

The town of Rabaul is one of very few towns or cities worldwide that has been built inside the caldera of an active volcanic system. The town was established within Rabaul Caldera by the German Administration of New Guinea in the period 1904–1910 (Johnson and Threlfall 1985), about 30 years after the 1878 eruptions (Brown 1878;

Johnson et al. 1981). The early German capital had been at Kokopo about 20 km southeast of Rabaul (Fig. 1), outside the caldera, but the seat of government was shifted to Rabaul in 1910. The attraction of the new site was the sheltered deep-water harbour. The tenure of Rabaul as the capital of German New Guinea was short-lived. In 1914, at the beginning of the First World War, Germany lost its New Guinea territory to an invasion of Australian forces. After World War I, administration of Rabaul, and indeed of New Guinea, became the responsibility of Australia.

The disastrous eruptions of May 1937, during which more than 500 lives were lost (Fisher 1939; Johnson and Threlfall 1985), led to broad-ranging investigations by the Australian Administration concerning the suitability of Rabaul as a capital city. There was disagreement about the future of the town. However, it was proposed that Rabaul could remain as the capital provided that a volcanological observatory was established to monitor the volcanoes and to warn of future eruptions. Routine volcanological observations at Rabaul commenced in 1938, and in 1940 the first Rabaul Volcanological Observatory (RVO) was established, at a site on the

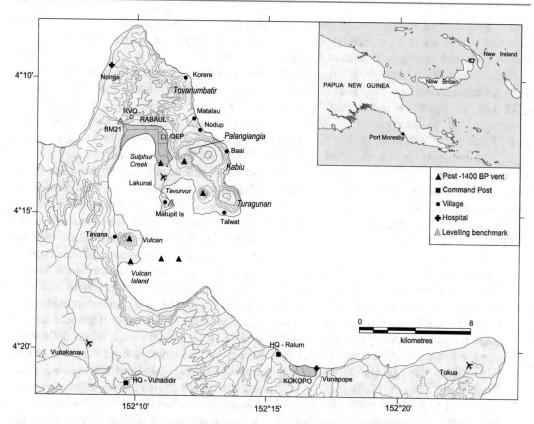

Fig. 1 The northeastern tip of the Gazelle Peninsula, New Britain Island, showing the major stratovolcanoes, post-1400 BP vents within Rabaul Caldera, the towns Rabaul and Kokopo, RVO, villages mentioned in the text, emergency command posts, hospitals, airports and the emergency assembly area, Queen Elizabeth Park (QEP). Surveying benchmarks on the northwestern caldera rim (BM 21) and at Matupit Island mark the ends of the main levelling line. Inset map shows the islands of New Britain and New Ireland, and the capital city Port Moresby on mainland Papua New Guinea

northern rim of the caldera. The same site is occupied by the current RVO (Fig. 1). Following a resurgence of eruptive activity in June 1941 (Fisher 1976), a decision was made by the Administration in September 1941 to shift the capital to Lae on the New Guinea mainland, thus ending a period of 31 years in which Rabaul was a colonial capital.

Apart from suffering the effects of volcanic eruptions in the period 1937–1941, the town was totally destroyed by bombing during World War II, and sustained about 70% destruction by volcanic activity in 1994 (Blong and McKee 1995; Davies 1995). Intermittent eruptions between 1994 and 2014 occasionally made life at Rabaul unpleasant.

Before the outbreak of the 1994 eruptions, the population of the town was about 17,000. A much larger population of about 100,000 lived in the satellite towns, villages and plantations of the Rabaul-Kokopo area. The urban and, to a lesser extent, plantation communities contained a mixture of short- and long-term residents, while residents of the villages could be regarded as constituting traditional communities. Apart from Rabaul Town the most-threatened population centres within the caldera are at the villages Tavana (near Vulcan), Talwat (near Tavurvur) and Matupit Island (between Vulcan and Tavurvur), as shown in Fig. 1. These villages, and indeed all areas within Rabaul Caldera, are frighteningly close to sources of volcanic threat.

The frequency of eruptions at Rabaul in the historical period, about 2 per century, has conditioned the traditional communities to the local volcanic threats. Local knowledge, particularly from the experience of the 1937 eruptions, coupled with the results of instrumental volcano monitoring and the long period of precursory activity (starting in 1971) resulted in a high level of volcano awareness prior to the 1994 eruptive outbreak. This awareness was an important element in the response to the volcanic unrest that preceded the 1994 eruptions.

The eruptive period that started at Rabaul in 1994 is important both as:

(i) a destructive event at a densely populated caldera volcano with a recent history of eruptions and caldera unrest—a case commonly identified as being a particular challenge for both monitoring and for warning communication, as in the context of Campi Flegrei, Italy, and Masaya, Nicaragua (e.g. Newhall and Dzurisin 1988).

(ii) an unusual case in which self-warning and evacuation of communities occurred in parallel with (and in some instances, in advance of) the formal volcano alert levels.

2 Precursory Activity (I): 1971–1985

2.1 The Nature of Volcanic Unrest at Rabaul—Characteristics of Rabaul Volcanic Crises

Historical records suggest that unrest between volcanic eruptions at Rabaul may be common, manifest as ground deformation and earthquakes. Uplift was reported to have occurred in the Sulphur Creek to Matupit Island area (see Fig. 1) in the lead-up to an eruption at Sulphur Creek in about 1850 (Brown 1878; Boegershauser 1937; Fisher 1939). Immediately prior to the 1878 eruptions that created Vulcan Island and formed a new crater at Tavurvur, small rocky islets near the site of the Vulcan Island eruption were raised

about 1 m and massive uplift of $\approx$6 m occurred along the foreshore at the southern foot of Tavurvur (Brown 1878; Johnson et al. 1981). Frequent local earthquakes preceded the eruptions, and tsunami activity was reported but its timing is unclear (Brown 1878; Fisher 1939). An account of earthquake activity at Matupit Island in 1891 by Edward Hernsheim (Sack and Clark 1983) refers to continual earth tremors and subterranean rumblings heard by residents of the island (including Hernsheim) that became so pronounced towards the end of 1891 that an eruption was considered imminent and partial evacuation of the island took place. Another phase of unrest, involving strong local seismicity and marked ground deformation in the Matupit Island area, occurred between 1916 and 1919 (Fisher 1939). The next eruptive activity eventuated in 1937, 46 years after Hernsheim's report and 21 years after the start of the 1916–1919 unrest, but not before a brief period of intense unrest, involving strong and sustained seismicity and emergence of near-shore areas, that commenced 27 h before the onset of the eruptions at Vulcan and Tavurvur (Fisher 1939; Johnson and Threlfall 1985).

Seismic unrest within the Rabaul volcanic system primarily takes the form of volcano-tectonic (VT) earthquakes. VT earthquakes are high frequency events similar to tectonic earthquakes but are associated with a volcano and are therefore assumed to represent failure (fault movements) resulting from strain induced by volcanic processes (Lahr et al. 1994; Chouet et al. 1994). The VT earthquakes at Rabaul occur as discrete events and in clusters or swarms. The largest non-eruptive unrest events are termed "volcanic crises".

Volcanic crises at Rabaul are characterized by swarms of $\geq$150 to thousands of VT earthquakes that occur within periods of a few hours (Mori et al. 1986, 1989). The threshold of 150 was based, somewhat arbitrarily, on the experience of the 1983–1985 "Crisis Period" (see below). The strongest earthquakes usually take place at or near the beginning of a crisis. The largest recorded event in any crisis was a M_L 5.2

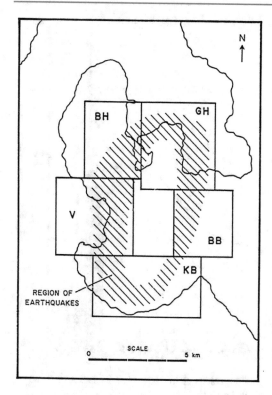

Fig. 2 Regions within the Rabaul Caldera seismic zone. *BH* Beehives; *GH* Greet Harbour; *V* Vulcan; *BB* Blanche Bay; *KB* Karavia Bay

Fig. 1) has amounted to as much as 100 mm for an individual crisis. However, the greatest deformation is usually off-shore and is not readily measurable (Greene et al. 1986; McKee et al. 1989). The focus of the ground deformation is usually in the central part of the caldera, within the region bounded by the zone of caldera seismicity. However, the deformation may extend beyond the zone of seismicity.

2.2 Volcanic Crises in the Period 1971–1985

Following the eruptions of 1937–1943, volcanic unrest at Rabaul was at low levels until 1971. As far as is known, seismicity within the caldera from 1943 to 1971 was weak and consisted of occasional, isolated VT earthquakes of relatively small magnitude. Rates of ground deformation were also low as indicated by minimal elevation changes of a benchmark installed at Matupit Island in 1949 (Lauer, unpublished data). This benchmark is different to the one occupied in levelling surveys starting in 1973.

1. *1971–1983*

The relative calm of the period following the eruptions of 1937–1943 was disturbed in November 1971 by the first recorded swarm of caldera earthquakes (Cooke 1977). Compared with many of the seismic swarms and crises in the following two decades, the crisis of November 1971 was unremarkable. During the period November 1971–July 1983, there was a trend of increasing numbers of earthquakes in the seismic swarms and crises, from <200 in November 1971 to $\approx$1200 in January 1982 (Fig. 3). Earthquake magnitudes also increased —the strongest earthquake was a M_L 5.2 event in the swarm of October 1980, and a slightly weaker event, M_L 5.1, was recorded in the swarm of March 1982 (Table 1). M_L determinations were made at RVO using an electronically-simulatedWood-Anderson seismograph. For

earthquake in October 1980. Seismicity of individual crises is usually confined to specific regions within the caldera seismic zone (Fig. 2), which is believed to include the fault system that bounds the youngest caldera (Mori et al. 1986, 1989). Caldera earthquakes take place at depths of 0–5 km, although most events are shallow, 1–2 km. In contrast, tectonic earthquakes of the northern Gazelle Peninsula region have a depth range of between about 10 and 280 km (Ripper et al. 1996; Ghasemi et al. 2016).

The ground deformation that takes place during volcanic crises is typically, but not always, inflationary, involving shallow-focussed uplift and tilting (McKee et al. 1984, 1989). The deformation occurs rapidly in direct association with the strongest earthquakes. Measured uplift at Matupit Island relative to a bench mark on the northwestern rim of the caldera (BM21, see

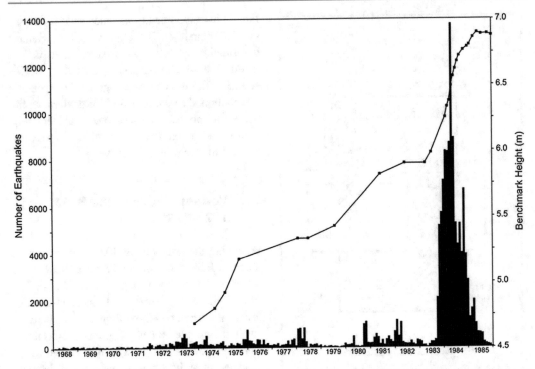

Fig. 3 Monthly caldera earthquake totals 1968–1985 and levelling results 1973–1985. Note increased frequency of levelling surveys starting from late-1983, during the early part of the Crisis Period

comparison the M_L 5.2 event of 28 October 1980 was assigned M_L 5.0 by Port Moresby Geophysical Observatory and m_b 4.9 by the National Earthquake Information Centre's PDE catalogue.

The spatial distribution of seismicity at Rabaul was discerned for the first time during this period due to improvements to both the local seismic network and to earthquake location techniques. The pattern of surface-projected seismicity was first represented as a "D"-shaped zone (Cooke 1977), and subsequently as two inward-facing arcuate zones, on the eastern and western sides of the caldera (Almond and McKee 1982). Later analysis suggested that the seismicity was related to a ring fault system (Mori and McKee 1987; Itikarai 2008) elongated north–south. The distribution of caldera seismicity for this period (1971–1983) is shown in Fig. 4.

Ground deformation monitoring during this period comprised levelling surveys, tiltmeter and dry tilt measurements, and strandline measurements. The results of these surveys indicated a generally steady rate of uplift of about 110 mm/year in the central part of the caldera (Fig. 3). Gravity measurements reflected this uplift (McKee et al. 1989).

2. *1983–85: Crisis Period*

The period September 1983–July 1985 was one of sustained intensified activity at Rabaul. At the time it was generally considered that the increased activity could be the prelude to an eruption (McKee et al. 1984). The period was characterized by frequent major crises (many hundreds of VT earthquakes) and a large number of minor crises (a few hundred VT earthquakes), and therefore was termed the "Crisis Period" by Mori et al. (1986). This period was immediately heralded by a generally steady build-up in background seismicity from a "normal" level of 5–10 caldera earthquakes per day in August 1983 to 50–80 per day before the first crisis in September 1983 (Mori et al. 1986).

The first crisis of this period took place on 19 September 1983, in the eastern part of the caldera seismic zone (Blanche Bay, Fig. 2) and included

Table 1 Seismic crises: October 1980–May 1985 (adapted from Mori et al. 1989)

Date	Location	No. Events (4 or more stations)	Largest event (M_L)
28 Oct. 80	**Vulcan**	**—*[1]**	**5.2**
4 Mar. 82	**Blanche Bay**	**—*[1]**	**5.1**
19 Sep. 83	Blanche Bay	210	4.2
15 Oct. 83	Greet Harbour/Beehives	334	3.6
28 Oct. 83	Greet Harbour	195	4.0
26 Nov. 83	Beehives/Vulcan	212	2.9
15 Jan. 84	Greet Harbour/Beehives	342	4.6
13 Feb. 84	Greet Habour	164	2.6
18 Feb. 84	Karavia Bay/Vulcan	212	4.3
27 Feb. 84	Blanche Bay	159	3.0
3 Mar. 84	**Blanche Bay**	**484**	**5.1**
17 Mar. 84	Vulcan	219	2.1
25 Mar. 84	Beehives/Vulcan	398	3.5
11 Apr. 84	Greet Harbour	217	2.0
13 Apr. 84	Greet Harbour	212	2.2
20 Apr. 84	Greet Harbour	302	3.0
21 Apr. 84	Greet Harbour	287*[2]	3.7
22 Apr. 84	**Greet Harbour**	**485**	**4.8**
4 May 84	Beehives/Vulcan	168	3.5
29 May 84	Beehives/Vulcan	323	3.6
13 Jul. 84	Vulcan	151	3.5
2 Aug. 84	Beehives	242	3.4
17 Oct. 84	**Greet Harbour**	**233*[2]**	**4.9**
18 Oct. 84	Blanche Bay	158	3.8
26 Oct. 84	Vulcan	209	3.2
3 Mar. 85	Beehives	333	3.6
10 May 85	Vulcan	—*[3]	3.7

Bold signifies stronger activity

M_L determined at RVO using electronically-simulatedWood-Anderson seismograph

The event counts are for a 24 h period following the start of the crisis

*[1]Event count not available

a M_L 4.2 earthquake. During the following few months the seismicity increased and by January 1984 the monthly total of caldera earthquakes had climbed to about 8300. In February and March 1984 the monthly total numbers of earthquakes remained about the same as in January. The crisis on 3 March 1984 was one of the strongest of the entire Crisis Period and included the largest earthquake, a M_L 5.1 event in the Greet Harbour area (Fig. 2). Re-occupation of the level line indicated that the southern tip of Matupit Island was rising at about 35 mm/month.

The activity reached a peak in April 1984 (Fig. 3). There were 5 crises in that month, all of which took place in the northeastern part of the caldera seismic zone (Greet Harbour, Fig. 2).

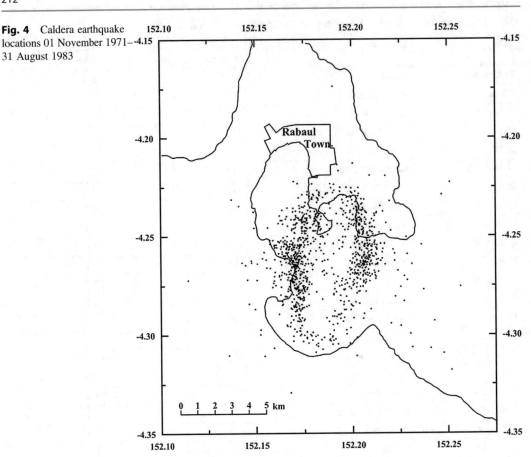

Fig. 4 Caldera earthquake locations 01 November 1971–31 August 1983

Crises took place on 3 consecutive days, 20th, 21st and 22nd of April, and the crisis on 22 April included a M_L 4.8 earthquake. The total number of earthquakes in April 1984 was about 14,000. Ground deformation was greatest in the central-northeastern part of the caldera. As much as 80 mm of uplift was recorded.

From May 1984 the activity declined. However, crises continued to take place occasionally, including one on 17 October 1984 during which the second strongest earthquake of the Crisis Period, a M_L 4.9 event, was recorded from the Greet Harbour area (Fig. 2). By the end of July 1985, seismic activity had returned to pre-Crisis levels. Also, there was a generally steady decline in the rate of uplift of the southern tip of Matupit Island, dropping to only a few mm/month by July 1985.

Altogether, there were 25 major crises during the period September 1983–May 1985 (Mori et al. 1986; Table 1). The intervals between the major crises ranged between 1 and 128 days. The seismicity shifted from one region to another of the caldera seismic zone during the Crisis Period but without showing any systematic patterns. The most active regions were those in the east to northeast (Blanche Bay, Greet Harbour; Fig. 2) and those in the west to northwest (Vulcan, Beehives; Fig. 2). Earthquake locations for the entire Crisis Period, September 1983–July 1985, are shown in Fig. 5. The greatest measured uplift for the 23 months of the Crisis Period was 767 mm, at a benchmark at the southeastern end of Matupit Island (McKee et al. 1989).

Fig. 5 Caldera earthquake locations for the Crisis Period, 01 September 1983–31 July 1985

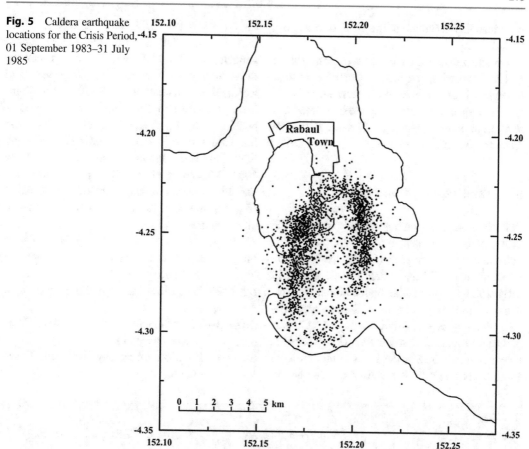

3 Impact of Precursory Activity (I): Volcano Monitoring, Contingency Planning and Public Awareness, 1983–1985

3.1 Volcano Monitoring

Prior to the 1983–85 Crisis Period the volcano monitoring system at Rabaul was unable to provide the necessary monitoring information promptly. Seismic data acquisition relied on manual timing of earthquakes from a microfilm recorder (Develocorder) and off-site data processing using the computing services of an accounting company. Thus, the determination of earthquake locations could take hours to days. Ground deformation monitoring relied mostly on levelling which was out-sourced to surveying staff from Department of Lands. Surveys were conducted at intervals ranging from 3 to 27 months.

Levelling surveys were more frequent from November 1983, at intervals of less than 1 month to about 3 months depending on events, and the surveys continued to be conducted by Lands Department staff. During a visit to Rabaul by Norman Banks of the United States Geological Survey (USGS) in December 1983, an electronic distance measuring (EDM) network was established and EDM monitoring commenced. In recognition of the need for a range of in-house ground deformation monitoring capabilities, a new position at RVO, Principal Surveyor, was created in 1985. The position was occupied from July 1985 and allowed RVO much greater control over the deformation monitoring program. With assistance from the Volcano Disaster Assistance Program (VDAP) of the USGS, telemetering electronic tiltmeters and tide gauges were installed at Rabaul in the wake of the Crisis Period.

On-site seismic data processing commenced in 1985 following the recruitment of new seismological staff equipped with personal computers. Earthquake location processing time was reduced to minutes, although the same microfilm recorder continued in service.

3.2 Contingency Planning

Concern about the growing volcanic unrest in the early 1980s prompted the development of a volcano contingency plan (VCP). Instigated by RVO, the first VCP for Rabaul was developed in early 1983 by provincial and national government authorities guided by UN-sponsored disaster planning expert, Brian Ward. The VCP was part of a Provincial Disaster Plan (PDP) for East New Britain, revised in 1985 by Captain Dwayne Hunt, RANR, and further revised in 1987 and in early 1994 by the Provincial Disaster Committee (PDC). The PDP set out the organizational structure of the PDC which was to include Sub-Committees for Rescue, Transport and Requisition, Evacuation and Welfare, and Command, Control and Communications. The composition and responsibilities of each of the Sub-Committees were specified in the PDP. Volcanic hazards assessments (McKee 1981; McKee et al. 1985) were the basis for definition of areas of danger and of relative safety (Fig. 6), and for the development of a 4-stage alert system:

Stage 1—Eruption in years to months. Normal preparedness.

Stage 2—Eruption within months. Prepare safe areas.

Stage 3—Eruption within weeks or days. Voluntary evacuation.

Stage 4—Eruption within days or hours. Evacuation.

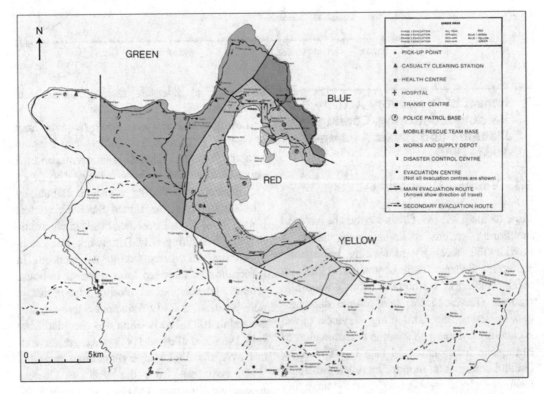

Fig. 6 Part of a poster showing hazard/danger zones and (relatively) safe areas; developed in 1983–1984 during the early part of the Crisis Period

The main objective of the VCP was the re-location of about 60,000 people from the danger areas to areas of relative safety. Duplication of principal services, including alternative airport, hospital and communications, would support the planned re-location. In considerable detail, the plan defined the roles of National and Provincial Disaster Organizations and the actions of 12 Government Departments and other organizations, including: Civil Aviation, Education, Electricity, Health, Harbours and Marine, Police, Communications, Works and Supply and Red Cross. It is remarkable that the major churches were not involved initially. Several exercises of the contingency plan were conducted over the following decade. Typically, the exercises concentrated on communications between government agencies, and did not involve businesses or the general community of Rabaul.

Wide acceptance that Rabaul was heading towards an eruption prompted the development of eight new Acts of disaster-related legislation, which were passed by the National Parliament in the first two weeks of March 1984 (Davies 1995). The eight Acts were:

Disaster Management
Emergency General Powers
Emergency (Defence Force)
Emergency (Register Evacuees)
Emergency (Requisition)
Emergency (Requisition Compensation)
Emergency (Third Party)
Emergency (Workers Compensation)

These actions demonstrated that concerns about an impending eruption at Rabaul had reached to the highest levels of public service in PNG.

Significant progress was made as a result of the Crisis Period in preparing for duplication of services and infrastructure in safe areas, notably:

1. Re-opening of a former airstrip (Vunakanau, a WW2 airstrip).
2. Roads to and water supplies in the safe areas were improved.

3. A safe area headquarters command post was established (at Vunadidir).
4. Equipment and medical supplies were moved from Nonga Hospital, near Rabaul, to Vunapope Hospital, near Kokopo.
5. Construction began on a wharf near Kokopo.
6. Construction started on a new airport (Tokua, a WW2 airstrip).

It was considered that logistical arrangements for RVO staff and civil defence officials must be established well in advance of any volcanic emergency for the following:

– Food
– Water
– Accommodation
– Care of families
– Transport
– Security
– Emergency electricity supplies
– Rostering of staff
– Protection of property
– Emergency communications

3.3 Establishment of Public Information Unit

The marked increase in volcanic unrest in August–September 1983 led to the establishment of a Public Information Unit (PIU) to disseminate information about the unrest. The declaration of Stage 2 volcanic alert in late-October 1983 accentuated the public need for volcano information, but because of increased workloads in the volcano monitoring program volcanologists were unable to maintain the flow of information to the public. Some public distrust and disquiet developed with fears that important information was being withheld. These fears were fanned by irresponsible and over-dramatic stories in the press, some of which stressed the possibility of a catastrophic eruption.

The PIU was established by the East New Britain Provincial Government in early-February

1984 (Davies 1995). At the suggestion of RVO the PDC requested the services of an Australian government geologist (Davies) to lead the unit. Davies had served in PNG for some years and was familiar with Rabaul, and was known to the RVO staff.

3.4 Three Initiatives of PIU

1. *Make all information available to the public*

As a first step towards confronting the problem of community distrust it was agreed among the volcanologists that all information should be made publicly available. Information released in the first week of March 1984 included plots showing the dramatic increases in seismic activity and in rates of ground deformation and explained the logic behind the declaration of a high-risk zone in the central–eastern part of the caldera, defined by ground deformation. A process was established for producing regular press releases: volcanologists presented information, PIU asked questions, and the resulting draft press release was vetted by volcanologists before release. The press releases were made available to all sections of media. Emphasis was placed on radio broadcasting as this was deemed the most effective method of reaching the largest audience.

2. *Town meetings and meetings with special interest groups*

Meetings involved:

Chamber of Commerce
School Principals
Staff and Students of the most-affected schools
Church Officials
Insurance Company Representatives
Defence Force Officers
Harbours Board Management

The meetings provided valuable opportunities for the PIU to deliver information directly and, at the same time, to learn the concerns of the community.

The most-frequently-asked questions were:

- Would there be adequate warning?
- Might the situation jump from Stage 2 to Stage 4, without an intervening time in Stage 3?
- What was the best estimate of the size of the eruption?

Davies recalls that he was tempted to respond to questioners with reassuring words but realised that the success of his mission depended on retaining the trust of the people and for this reason it was necessary to be absolutely honest. He was pleased to find that the public response was not one of alarm but of calm, albeit concerned, acceptance. As had been observed by the celebrated volcanologist Gordon A. Macdonald: "People show a great ability to face dangerous situations with equanimity if they understand the situation. It is the unknown or the not understood danger that terrifies" (United Nations 1977).

3. *Local newspaper*

The national daily newspaper Post-Courier, printed in Port Moresby, proved to be an unsatisfactory conduit for conveying volcano information to the public. This problem was met face-on when local businessman, Joe Speccatori, established and produced Rabaul's own newspaper, the Rabaul Gourier (a pun based on the Melanesian Pidgin word "guria", meaning earthquake, and the name of the national newspaper). This provided up to date accurate information and was immensely popular during the worst of the crisis.

4 Precursory Activity (II): 1985–1994

4.1 Fluctuating Activity: August 1985 to Mid September 1994

For about 3 years following the 1983–85 Crisis Period the level of unrest at Rabaul was low. Seismicity was generally weak, interrupted occasionally by earthquake swarms. Subsidence

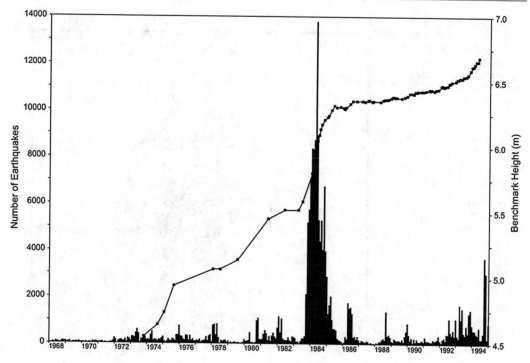

Fig. 7   Monthly caldera earthquake totals 1968–1994 and levelling results 1973–1994

was recorded near the margin of the youngest caldera during this interval, but the central part of the caldera floor continued to rise, albeit slowly (Fig. 7).

A M_L 2.8 caldera earthquake on 10 April 1988 and associated inflation of the caldera floor marked the beginning of a period of 4 years of moderate activity. Swarms of caldera earthquakes became more frequent with the strongest earthquake being a M_L 3.5 event in July 1990. Measured uplift at the Matupit Island benchmark during this interval amounted to about 150 mm, giving a rate of about 40 mm/year. This rate is less than half the uplift rate of the pre-Crisis Period.

An unusual swarm of earthquakes that started on 2 May 1992 marked the beginning of a prolonged period of increased activity that continued until December 1993 (Fig. 7). The seismicity in early May 1992 took place about 2 km outside the northern part of the caldera seismic zone

(Fig. 8), under the eastern fringe of Rabaul Town, and consisted of more than 300 high frequency earthquakes, the largest of which was M_L 4.2 (Table 2). The locations of these earthquakes appeared to define a northeast-trending zone and led to the term "Northeast (NE) earthquakes" (Itikarai 2008). The significance of this seismicity remains uncertain, although a connection between NE earthquakes and phases of eruptive activity in the post-1994 period has been suggested (Itikarai 2008).

Caldera seismicity and rates of ground deformation began to increase after May 1992. A succession of caldera earthquake swarms was recorded, and one crisis took place, in May 1993 (Table 2). The crisis of 20 May 1993 was the first since the end of the 1983–85 Crisis Period, and included four M_L 3.8 earthquakes in the Greet Harbour area (Fig. 2). Data from the tide gauge network indicated uplift of 50–60 mm in the area of inferred maximum deformation, in the

Fig. 8 Caldera earthquake
locations for the period 01
August 1985–17 September
1994. Note the "northeast
earthquake" zone near the
eastern fringe of Rabaul Town

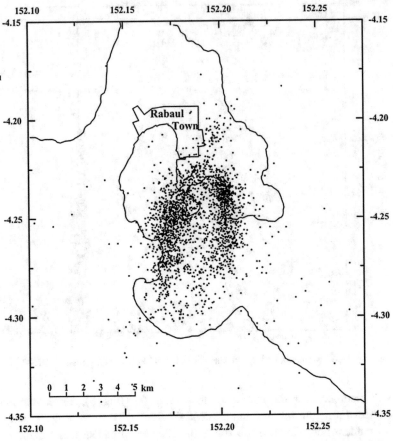
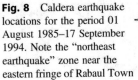

Table 2 Larger Rabaul
earthquakes: 2 May 1992–
18 September 1994
(adapted from Stewart and
Itilkarai, unpublished data)

Date	Time (LT)	Location	Magnitude (M_L)
02 May 92	0714	Namanula	4.2
20 May 93	0643	Greet Harbour	3.8
25 May 94	1043	Beehives	3.3
25 May 94	1058	Vulcan	3.0
18 Sep. 94	**0251**	**Greet Harbour**	**4.9**
18 Sep. 94	**0251**	**Vulcan**	**5.1**
18 Sep. 94	1203	Vulcan	4.3
18 Sep. 94	1709	Vulcan	4.4
18 Sep. 94	2042	Vulcan	4.4
18 Sep. 94	2054	Vulcan	4.5
18 Sep. 94	2307	Vulcan	4.1
18 Sep. 94	2323	Vulcan	4.1

Bold signifies stronger activity

eastern-central part of the caldera. Tide gauge data confirmed that the deformation coincided with the seismic crisis.

Through most of 1994 until the day preceding the outbreak of eruptions a trend of generally declining seismicity was recorded (Fig. 7). There were many small earthquake swarms, but two stronger swarms took place on 25 May 1994 involving M_L 3.3 and 3.0 earthquakes in the Beehives and Vulcan areas respectively (Fig. 2, Table 2). Unusual "hybrid" earthquakes having high frequency onset and low frequency codas were recorded in the aftermath of the swarms of 25 May. The significance of those events was uncertain at that time, but it is now known that hybrid earthquakes of this type are common at other volcanoes with active hydrothermal systems and can be interpreted as episodes of fracture creating permeability that allows movement of hydrothermal fluids, or alternatively as episodes of fracture associated with propagation of magma-filled dykes (Faria and Fonseca 2014). Seismic swarms ceased after a relatively small swarm on 19 July 1994. In late August 1994, several hundred discrete earthquakes were recorded from near Tavurvur, and one hybrid earthquake was recorded. Earthquake locations for the period 01 August 1985–17 September 1994 are shown in Fig. 8.

Elevation changes within the caldera were moderate in the period January to mid-September 1994. A rise of about 100 mm was recorded at benchmarks at the southern end of Matupit Island during this period. The implied uplift rate is slightly greater than the average long-term rate of the pre-Crisis Period. However, in detail uplift rates declined at mid-year and there appeared to be slight subsidence in July. The decline in uplift rates may have coincided with the generation of the hybrid earthquakes, and may reflect depressurization of the hydrothermal system at Tavurvur.

Although the events of the period May 1992 to mid-September 1994 were reported to authorities, the significance of the NE earthquakes, the hybrid earthquakes, and the decline in uplift rates in mid-1994 was not appreciated at that time. In addition, maintenance and equipment problems led to a lack of telemetered data from tiltmeters and tide gauges by late 1993 which rendered the reliable interpretation of available ground deformation data more difficult. These deficiencies conspired to impede the generation of warning messages from RVO.

4.2 The Ultimate Crisis of 18–19 September, 1994

1. Seismicity

The 1994 eruptive outbreak at Rabaul was immediately preceded by the "Ultimate Crisis", 27 h of vigorous and fluctuating seismicity and remarkable ground deformation (Blong and McKee 1995). The timeframe and nature of these events in 1994 are eerily similar to the events that immediately preceded the 1937 eruptions (Fisher 1939). The seismicity of the Ultimate Crisis was initiated by two strong caldera earthquakes at 0251 on 18 September (Fig. 9, Table 2). The first earthquake was a M_L 4.9 event located near Tavurvur. The second earthquake was larger (M_L 5.1), and is believed to have originated from beneath the southern part of the Vulcan headland (Stewart and Itikarai, unpublished data). A small tsunami was generated by this earthquake and had greatest impact in the southern part of the caldera (Nishimura et al. 2005; Stewart and Itikarai, unpublished data).

The early part of this intense seismicity resembled a typical crisis. However, by early afternoon of the 18th the sustained nature of this activity caused concern at RVO and in some communities, notably those on the western side of the caldera and at Matupit Island. After about 14 h of seismic swarm activity, at about 1700 on the 18th, the seismicity intensified and it became clear to many people that conditions were unusual. Numerous felt earthquakes, many with magnitudes equal to or greater than M_L 3.5, occurred between 1700 on the 18th and 0300 on the 19th. The felt seismicity tapered off towards dawn on the 19th.

Most of the earthquakes were located in the west-northwestern part of the caldera seismic

Fig. 9 Caldera earthquake locations 18–19 September 1994

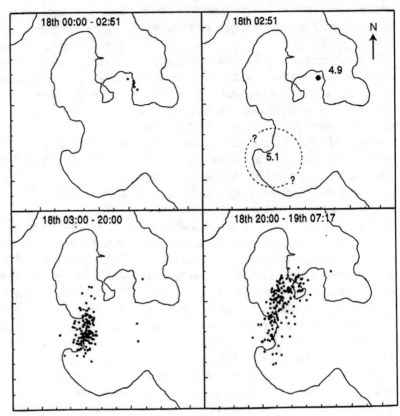

zone, i.e. the area between the Vulcan headland and Matupit Island, although the initial focus of the seismicity, following the main (M_L 4.9 and 5.1) events, was under the Vulcan headland (Fig. 9).

2. *Ground Deformation*

The new telemetering electric tiltmeters and tide gauges that had been deployed as a result of the 1983–85 Crisis Period had become unoperational by 1994 because of maintenance and funding difficulties. Remarkable ground deformation started on the night of the 18th or early next morning, but was not detected until dawn, at about 0515 on the 19th. Uplift had occurred in the Vulcan area where a tide gauge pylon was almost fully exposed, indicating uplift of about 6 m (Fig. 10). The western and southern coasts of Matupit Island had also been raised and the southern shoreline had migrated about 70 m

south. The estimated uplift of the western and southern parts of Matupit Island was about 2 m. There was little or no evidence of elevation changes at Tavurvur. The overall pattern of the ground deformation appeared to be east-hingedup-tilting of the caldera block within the caldera seismic zone.

4.3 Outbreak of Twin Eruptions: 19 September 1994

New eruptions at Rabaul began just after 0600 on 19 September 1994, ending a repose period of about 51 years. Following the pattern of the previous two eruptive episodes, in 1878 (Brown 1878; Johnson et al. 1981) and 1937–43 (Fisher 1939; Johnson and Threlfall 1985), there were almost simultaneous outbursts on opposite sides of the caldera as Tavurvur and Vulcan began erupting at 0606 and 0717, respectively.

Fig. 10 Newly emerged sea floor at the eastern foot of Vulcan as seen at 6:15 a.m. on 19 September 1994. The near total exposure of the tide gauge pylon shown here in the small embayment indicates about 6 m of uplift in this area. Photo courtesy of N. Lauer, formerly of RVO

5 Responses to the Ultimate Crisis and Outbreak of Eruptions

5.1 Timeline of Events and Responses, 18–19 September 1994

Responses to the physical (volcano-related) events in the immediate pre-eruption period were reported by Davies (1994, 1995) and are summarized in Table 3. The responses include the actions by RVO, by the PDC and by communities resident within the caldera. The responses are discussed in the following sections.

5.2 Response of Government Authorities

As earthquake activity intensified during the afternoon of 18 September it became apparent that this activity was more than a typical crisis. A recommendation to raise the alert to Stage 2 was made by RVO to the PDC at about 6 p.m. on 18 September. The raising of the alert level was broadcast by local radio and spread by word of mouth.

An exodus of people from villages near Vulcan started in late afternoon. By dusk people started to evacuate Matupit Island and the southern town area. Those who did not have transport were advised to congregate at the Queen Elizabeth Park (QEP) sports field (Fig. 1) in the Rabaul Town area (as stipulated in the VCP).

At around 11 p.m. the PDC began the evacuation of about 5000 people who had gathered at QEP. All available transport worked through the night until the early hours of the next day to ferry people out of the caldera.

At about midnight RVO advised the PDC that "Rabaul Volcano was on an irreversible course towards an eruption" and that a recommendation for declaration of a Stage 3 Alert was imminent. By 2 a.m. on 19 September that recommendation

Table 3 Events and responses timeline, 18–19 September 1994

Time	Event	RVO	PDC (NDES)	Communities
18 Sept				
0251	M_L 4.9, 5.1 earthquakes, tsunami	Processed and analysed eq. data		Most people awakened by eq. shaking
0800–1200	Sustained moderate-strong seismicity	Processed eq. data Investigated eq., tsunami impact		
1200–1700	Sustained strong seismicity	Processed eq. data		Growing concern with continuing seismicity
1700	Seismicity intensifies	Processed eq. data		Mobilization of communities on W side of caldera (Tavana, Valaur)
1800	Sustained intensified seismicity	Communication to PDC to increase alert to Stage 2	Stage 2 declared, broadcast on local radio	Evacuation of Matupit Is., southern part of Rabaul Town Assembly of residents at QEP
2300	"		Directed evacuation of people gathered at QEP	Evacuation of most communities resident within caldera
2400	"	Communication to PDC that eruption anticipated	Conveyed message that eruption anticipated	Continued evacuation
19 Sept				
0200	"	Communication to PDC and NDES to increase alert to Stage 3 Arranged for aerial inspection at 0600	Continued to convey message that eruption anticipated	Evacuation of all communities resident within caldera completed
0300			Stage 3 declared (NDES), broadcast on Radio Australia	
0515	Emergence of seafloor around Vulcan and Matupit Is first observed	Notified PDC that eruption imminent		
0606	Outbreak of eruption at Tavurvur	Observed outbreak during aerial inspection		
0700			Stages 3 and 4 declared locally, and request for declaration of State of Emergency (subsequently declared)	
0717	Outbreak of eruption at Vulcan	Observed outbreak from RVO		

had been faxed from RVO to the PDC and to the National Disaster and Emergency Services (NDES) headquarters in Port Moresby. The Director of NDES (Leith Anderson) subsequently notified the Port Moresby-based representative of Radio Australia, Sean Dorney, of this development which was then broadcast (at about 3 a.m.).

Following the advice from RVO that an eruption was inevitable members of the PDC advised clubs, hotels, boarding schools and others that people should move out from the caldera. This was done by phone calls and by Police and Red Cross officials moving through the streets advising evacuation.

The local radio station remained on air until 12:30 a.m. on 19 September, and resumed broadcasting at 5:30 a.m. until the eruption began. Through this time the station continued to broadcast the message that Rabaul was on Stage 2 Alert and that people should remain in their homes until further notice. This anomaly arose because the PDC had decided to delay declaring Stage 3 Alert until first light. The PDC reasoned that the evacuation was proceeding smoothly during the night and it was felt that announcement of a heightened state of alert might cause panic and unnecessary havoc on the roads.

Rabaul Town and all villages within the caldera were evacuated by the early hours of 19 September. The evacuation was conducted smoothly and the entire process was completed within 12 h. However, some residents of Talwat village, just outside the eastern margin of the

caldera, did not join the mass exodus from Rabaul on 18–19 September. Likewise, most residents of villages east and northeast of Rabaul Town (Baii, Nodup, Matalau, Korere etc., see Fig. 1) did not evacuate at this time.

At 7 a.m. on 19 September the PDC declared simultaneously Stage 3 and Stage 4 Alerts, and requested the declaration of a State of Emergency. By that time Tavurvur's eruption had already started and within about 1.5 h a dense emission plume had engulfed Rabaul Town (Fig. 11). The State of Emergency was approved by the Prime Minister and declared by the Director NDES (Leith Anderson). Anderson was appointed Controller of the Emergency and the Chairman of PDC, Ellison Kaivovo, was appointed Deputy Controller.

A command post for emergency operations was established at Ralum, near Kokopo, on the second morning of the eruptions. The original plan for command post operations to be headquartered at Vunadidir was abandoned when that area received tephra fall from the Vulcan

Fig. 11 Tavurvur emission plume engulfing Rabaul Town at about 7.25 a.m. on 19 September 1994, as viewed from RVO. Photo courtesy of N. Lauer, formerly of RVO

emissions. The Ralum location for the command post probably was a better option, being more conveniently-located for a range of services.

Telephone communications in the region were seriously disrupted during the first few days of the eruption. Lightning strikes in the mid-late afternoon of 19 September damaged the telephone system at RVO. The loss of the Rabaul Telephone Exchange on the morning of 20 September, due to collapse of the roof of the building under a heavy load of wet volcanic ash, caused a temporary break in communications in the Rabaul area. Earlier on the same day operation of the Kokopo Telephone Exchange was halted by an electrical power outage.

Loss of volcano-monitoring capability at RVO began on the first day of the eruptions. Seismic monitoring equipment and telemetry links were damaged by heavy falls of tephra and by lightning strikes. On the morning of 20 September heavy rainfall overwhelmed drainage from the roof of the main RVO building and incursion of muddy water into the roof space resulted in water damage inside the building and distraction from eruption monitoring activities.

The PIU played an important part in conveying volcano-related information from RVO to the PDC, to the displaced people in care centres (Fig. 12) and to the news media. Led by H. Davies, who travelled from Port Moresby to

Fig. 12 Locations of care centres for evacuees

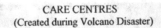

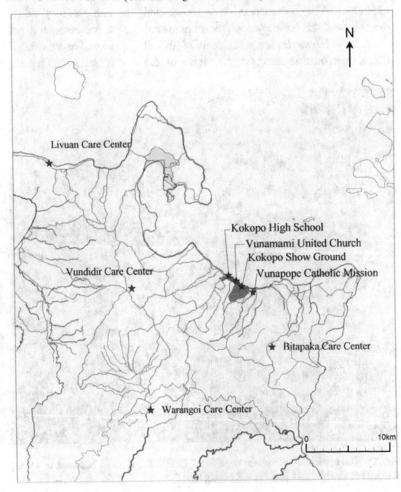

Kokopo on 23 September, the PIU was assisted by local resident geologist David Lindley (Lindley 1995) and occasionally by volcanologists from RVO.

5.3 Community Response

The responses of the traditional communities to the volcanic events of 1994 were shaped by experience and knowledge. Memories and stories of the events preceding the 1937 eruptions and of the approximately 500 fatalities near Vulcan during that eruption are still strong within these communities. The 1937 eruptions were preceded by 27 h of intense local earthquake activity, accompanied by dramatic late-stage emergence of the seafloor in the Vulcan and Matupit Island areas (Fisher 1939; Johnson and Threlfall 1985). The timeframe and the nature of the geological activity on 18–19 September 1994 were very similar to the events of 1937. However, the general community response to the pre-eruption geological activity in 1994 probably was based only on the earthquake activity, as also suggested by Neumann (1996) from anthropological research, as the massive uplift in the Vulcan and Matupit Island areas took place at night and was not observed until first light on 19 September. By that time nearly all of the people normally resident within and near the caldera had evacuated.

Another element of the general community response in 1994 was the conditioning through 23 years of precursory activity. The attention of the community to the volcanic threat became highly focussed during the mid-1980s Crisis Period (McKee et al. 1985). Information provided by the PIU and other agencies prompted individuals, family units and larger groups to make their own volcano contingency plans in the light of the broader official volcano contingency plan. However, the long delay (10 years) between the peak of the Crisis Period and outbreak of eruptions may have caused some complacency within the community about the lingering volcanic threat. Additionally, changes to the migrant and expatriate communities would have weakened the focus on preparedness for eruptive activity.

It was reported by Neumann (1996) that a group of people from Valaur Village (near Tavana village) visited RVO on the morning of 18 September anxious about the seismic activity. The village group was allegedly told by RVO staff "not to worry" about the earthquakes. The details of that exchange cannot be verified, however at that early stage of the Ultimate Crisis a judgement on the significance of the seismicity would have been difficult to make. Similar activity had been recorded previously (see Table 1), notably September–October 1980 (including an M_L 5.2 earthquake), March 1982 (including an M_L 5.1 earthquake), March 1984 (including an M_L 5.1 earthquake), April 1984 (three consecutive days of crisis activity including an M_L 4.8 earthquake) and October 1984 (including an M_L 4.9 earthquake).

Differences between the activity of 18–19 September 1994 and crises of the 1980s became clearer through the afternoon of the 18th when felt earthquake activity continued and appeared to be strengthening. A series of magnitude 4 earthquakes from the Vulcan area began at about midday on the 18th, with more frequent such earthquakes occurring after about 2000 LT (Table 2).

The rapid evacuation of areas near Vulcan in the afternoon of the 18th appears to have been a reasoned and conscious response to a situation perceived to be different to those experienced during the crises of the 1980s, and bearing similarities to the earthquake activity prior to the 1937 eruptions as remembered by village elders. A similar response may have occurred at Matupit Island. As the spontaneous evacuation gathered pace some people may have been influenced by the actions of those around them to join the exodus. Such behaviour would have been further fuelled by the sight and sound of aircraft taking off from Rabaul airport (Lakunai; Fig. 1) during the hours of darkness on 18–19 September. Those actions may have provided powerful communications to observers, as a case of "actions speaking louder than words". This phenomenon was evident from interviews with survivors of the 2011 Tohoku tsunami (Ando et al. 2013) which showed that many people began to evacuate coastal towns before the tsunami arrived, simply because they saw other people evacuating.

There was a sharp contrast between the actions and reactions of village communities east and northeast of Rabaul Town and those resident within the caldera. The village communities east and northeast of Rabaul Town may have heeded the local radio broadcast messages to stay put. Also, the lower intensity of felt earthquake activity in these areas may have been a factor in prompting these communities to not evacuate during the afternoon or night of the 18th. Evacuation of these communities and the remaining residents of Talwat took place on 20–21 September, assisted by a number of ships (Fig. 13).

Fig. 13 Evacuation of some residents of Talwat village on 20 September 1994, assisted by MV *Madang Coast*. A dense, convoluting emission plume rises from Tavurvur in the background. Further in the background is the stratovolcano Kabiu. Photo courtesy of South Pacific Post Courier

One very disturbing response to the Ultimate Crisis and outbreak of eruptions in 1994 was looting. The looting started in Rabaul Town on the night of 18 September as the evacuation proceeded. During the early part of the eruption looting became widespread as residents of villages outside of the caldera took advantage of unoccupied houses and the lack of strong action on the part of police to secure the town. Check points were established at the main roads but there was little effort to stop people from going into the town area or to stop trucks laden with looted goods from leaving.

5.4 Outcomes

The eruption at Vulcan was a powerful, short-lived plinian event (Fig. 14), producing about 260×10^6 m^3 of tephra mostly in the first few days of activity, and concluding on 2 October 1994 (Blong and McKee 1995). Tavurvur's eruption was less powerful (vulcanian) but persistent, continuing for years. The output rate at Tavurvur peaked during the first days of activity when about 40×10^6 m^3 of tephra was erupted.

The death toll from the 1994 eruptions was relatively small, officially 10. The 10 eruption-related deaths comprised 3 due to tephra fall and associated asphyxia, 1 due to lightning, 3 from road trauma, 2 from desertion (ill and incapacitated persons), and 1 from drowning in a flash flood 6 weeks after the start of the eruptions (Dent et al. 1995). This small death toll is in stark contrast to 1937 when as many as 500 people perished, mostly caught in pyroclastic flows and heavy tephra fall in the Vulcan area (Fisher 1939). As mentioned, the events of 1937 and the recognition of the on-going volcanic threat led to the establishment of RVO in 1940. The instrumental volcano surveillance that ensued helped to better inform the residents of the Rabaul area of the local volcanic threat and led to the detailed planning that helped to manage the long precursory period and the inevitable eruptions in 1994.

Fig. 14 Vulcan's eruption column and plume as seen from the Space Shuttle at about 7 a.m. on 20 September 1994. At this time the eruption column was rising to at least 18 km (astronaut's estimate) and feeding a broad, wedge-shaped emission plume the axis of which was oriented southwest. In this view from the east-northeast, the eastern coast of New Ireland can be seen beneath the extensive white cloud cover in the foreground, and part of the eastern coast of the Gazelle Peninsula is visible in the left middleground. Photo courtesy of NASA

A UNDHA mission to Rabaul in February 1995 (Tomblin and Chung 1995) reported that "the Rabaul Emergency Plan had in most respects worked very well". Analysis of the events and responses of the precursory period and of the first few months of the eruptive period identified problems with telecommunications as one of 22 issues "which could be improved for the future". This issue is discussed further in the next Sect. 6.

While responses to the events of 18–19 September 1994 were not ideal in some instances, the general recognition of the impending eruptive threat, the implementation of emergency plans, the willingness of most people to evacuate and the successful evacuation of areas of greatest danger resulted in avoidance of a repeat of the outcome of Vulcan's 1937 eruption and of the sort of tragedy that befell the town of Armero during the 1985 Nevado del Ruiz volcano catastrophe. At Armero, the loss of >22,000 lives was attributed mainly to "cumulative human error—misjudgement, indecision and bureaucratic shortsightedness" (Voight 1990, p. 383).

An irony of the 1994 eruption is the extent of damage to Rabaul Town (70% destroyed) from the relatively small volume of Tavurvur tephra emissions (40×10^6 m^3). The output from Vulcan was much greater (260×10^6 m^3) but its impact was mostly in the vicinity of Vulcan itself. Several (evacuated) villages were obliterated by the pyroclastic flows and heavy tephra fall from Vulcan. The damage from Tavurvur tephra was maximized by the close proximity of Rabaul Town to Tavurvur and by the prevailing winds which carried the tephra directly towards the town. Such seemingly disproportionate impact was also evident at (the snow and ice covered) Nevado del Ruiz Volcano in 1985 where $<5 \times 10^6$ m^3 of magma ejected as pyroclastic flows was able to generate about 60×10^6 m^3 of lahars from 10 to 20×10^6 m^3 of melt water (Calvache 1990; Pierson et al. 1990).

Real-time monitoring capability for both earthquake locations and seismic amplitudes (RSAM-Murray and Endo 1989) was deployed at Rabaul in October 1994 by a team from VDAP-USGS, replacing the damaged and seriously weakened existing seismic network. At the same time new electronic tiltmeters were also installed. High resolution GPS monitoring commenced in 1999 with sensors deployed at Matupit Island and at RVO. All of the immediate efforts to restore and upgrade volcano-monitoring facilities at Rabaul focussed on data transmission to RVO. Subsequently, plans were developed for a re-location of the volcano-monitoring hub to Kokopo, and in 2016 the first element (accommodation block) of a re-located volcano observatory had been completed.

In general, the scientists who visited Rabaul during the eruptions followed the principles of not engaging in discussions with local media and individuals, and releasing information only through official channels i.e. RVO and PIU. However, in one case a visiting scientist caused significant anxiety through public discussions of a personal prediction of increased volcanic activity during the second week of the eruptions.

Negative outcomes of the eruptions include: losses of lives, homes, property, businesses and livelihoods. Many people suffered significant financial loss. Some of these losses were a result of the inability to get adequate insurance because premiums had been raised due to the 1983–85 Crisis Period. The displacement of 105,000 people during the first weeks of the eruptions created a huge logistical problem of providing food, water and shelter in care centres. The number of displaced people was swelled by the unnecessary movement of some people from safe area locations to care centres. Probably this was driven by fear of the unknown while care centres offered the perception of "safety in numbers", the provision of food and water, and access to information about the unfolding drama. For some, the displacement and disruption continued for many months. The short- and long-term psychological effects of the trauma of displacement and disruption were largely not addressed officially.

6 Lessons

The 1994 eruptions at Rabaul, which destroyed much of the town and several villages (Fig. 15), were remarkable for the very small loss of life. This outcome resulted from timely evacuations from areas most impacted by volcanic activity. Nevertheless, the rapid development of the late-stage precursory activity caught most residents by surprise, allowing insufficient time to pack up properly, with the result that many small businesses are no more and many people lost all possessions.

Experiences during the long period of precursory activity and during the 1994 eruptions, and comparisons with experiences during the 1937 eruptions and earlier activity, have framed the following lessons:

1. *Volcanic crises and unrest are common at Rabaul*

Historical records suggest that volcanic crises and unrest are common between eruptions at Rabaul. Most, if not all, of the volcanic crises at Rabaul follow a similar pattern. The earthquakes of crises are of high frequency type and the strongest events occur near the beginning of a crisis to be followed by a sequence of smaller earthquakes. Typically, uplift of parts of the caldera floor occurs during crises in association with the earthquake activity. Greater and more rapid uplift accompanies the sustained strong seismicity that occurs immediately before an eruption outbreak. Tsunami are generated by the stronger crises, particularly those immediately preceding eruptions.

2. *An individual crisis may or may not lead directly to an eruption.*

In terms of maximum earthquake magnitude, there was nothing to distinguish the crisis that immediately preceded the 1994 eruption from earlier crises. Caldera earthquakes of magnitude (M_L) 4.9 or greater occurred on at least 4 occasions (Mori et al. 1989) before the magnitude (M_L) 4.9 and 5.1 events of the Ultimate Crisis in

Fig. 15 Aerial view of Rabaul Town from the northwest in early January 1996 showing destruction of the southern and central parts of the town area. A small eruption cloud stands over Tavurvur. The neighbouring stratovolcanoes Turagunan, Kabiu and Palangiangia are looming threats—all were active in the Middle-Late Holocene. Photo courtesy of Gazelle Restoration Authority, Kokopo, East New Britain

1994 (Table 1). These events were: October 1980—M_L 5.2, March 1982—M_L 5.1, March 1984—M_L 5.1, October 1984—M_L 4.9. However, the sustained high rate of felt earthquake occurrence in 27 h periods preceding the 1937 and 1994 eruptions was distinctive compared with other crises.

3. *Late stage precursory activity can develop rapidly*

Late stage precursory activity developed rapidly in 1994: only 27 h separated the start of the final crisis and the outbreak of eruptions. There was a re-intensification of seismicity 14–16 h after the initial strong earthquakes. The resurgence of seismicity halfway through the immediate pre-eruption crisis was associated with the rapid uplift of parts of the caldera floor that began less than 12 h before the start of the 1994 eruptions. Memories and stories of a similar pattern of late-stage precursory activity that preceded the 1937 outbreak influenced the responses of traditional communities to the events of 1994.

4. *Eruptions can occur at more than one vent*

In the three most recent eruptive periods, 1878, 1937 and 1994, two vents on opposite sides of Rabaul Caldera erupted simultaneously. This pattern of activity has implications concerning the stability of Rabaul Volcano. Simultaneous activation of different parts of the volcanic system could lead to major de-stabilization, culminating in events of significantly larger scale.

5. *Seismic data acquisition needs to be automated*

A microfilm and paper chart seismic recording system was deployed through most of the 23 years of precursory activity and for about 10 days into the 1994 eruptions. Processing of data in this system involved manual timing of earthquakes and manual entry of data into a computer for locating the earthquakes. This system provided no facility for monitoring seismic amplitudes. The system was adequate for routine operations in non-eruptive periods when the most intense activity was isolated crises, but could not cope with a sustained high-intensity crisis or eruptions outbreak. Real-time monitoring capability for both earthquake locations and seismic amplitudes is essential.

6. *Continuous real-time telemetered ground deformation monitoring is required*

Remarkable ground deformation occurred on the night before the 1994 eruptive outbreak but was not observed. Telemetering electronic tiltmeters and tide gauges had been deployed at Rabaul as a result of the 1983–85 Crisis Period but had become unoperational by 1994 because of maintenance and funding difficulties. If operational, these instruments may have provided an earlier warning that an eruption was imminent.

7. *Clear established links with media are needed*

The experience of the Crisis Period in 1983–85 demonstrated the advantages of having a Public Information Unit to disseminate information from scientists and civil defence officials and to handle all media enquiries about volcanic activity. This allowed the scientists to concentrate on volcano monitoring. Nevertheless, it is desirable for scientists to communicate directly with media, if the circumstances permit.

8. *Strong communication links between scientists and civil defence agencies are needed*

Serious communications difficulties between scientists and civil defence officials were encountered during the first dew days of the 1994 eruptions because of the loss of telephone services. In order to ensure effective and reliable official communications it is necessary to have multiple communications links, including radios and both mobile and landline telephones. Protection and duplication of communications infrastructure, such as telephone exchanges, is essential.

9. *Logistical arrangements for volcano monitoring teams and civil defence personnel need careful planning*

At Rabaul in 1994 volcanologists and disaster co-ordinators were also victims of the eruptions and had to deal with a number of logistical issues, including care of families, accommodation, security, protection of property and emergency communications. This had a heavy impact on the ability of these officials to function effectively during the early days of the 1994 eruptions.

10. *Protection of Property*

One disturbing aspect of both the 1937 and 1994 eruptions was looting. In 1994 looting started on the night before the eruptions began, when there was a mass evacuation of Rabaul. In 1937 vigilante groups patrolled the town and severely punished looters. This did not happen in 1994 and during the first few days and weeks of the 1994 eruptions looting took place on a large scale, essentially unchallenged. The looting was a serious distraction for RVO staff and emergency officials who were residents of Rabaul Town and whose homes and possessions were targeted by looters.

7 Concluding Remarks

1. Living near active and potentially-active volcanoes requires open and effective lines of communication between volcano specialists, national and provincial government officials, town authorities and the general public. Communication between all of these sectors is vital at all stages of volcano surveillance, eruption contingency planning, town planning and disaster management. The creation of the PIU by the East New Britain Provincial Government in 1984 assisted greatly in the dissemination of volcano-related information.

2. The availability of volcano contingency plans provides great support during crisis periods. Involvement of all sectors of threatened communities in the creation of these plans is desirable. Frequent rehearsals and reviews of the plans maintains a high level of alertness.

3. A majority of the various communities of the Rabaul area coped well with the prospect of impending eruption once they were provided with relevant information. This information coupled with local knowledge and lessons from previous eruptions, particularly that of 1937, resulted in a high level of volcano awareness that led to decisive actions to evacuate the most-threatened areas during the immediate pre-eruption period. For some, "actions may have spoken louder than words", as some residents may have been influenced to evacuate by the movements of others around them, resulting in a mass exodus of people from within the caldera.

Acknowledgements The manuscript benefitted greatly from the detailed reviews by Professor Emeritus Russell Blong and Dr. Simon Day. Marissa Sari Egara of the Port Moresby Geophysical Observatory helped with word processing of the manuscript. Sonick Taguse of Papua New Guinea's Mineral Resources Authority helped with preparation of some of the line diagrams. COM and II publish with the permission of the Secretary, Mr. Shadrach Himata, Department of Mineral Policy and Geohazards Management, Papua New Guinea.

References

Almond RA, McKee CO (1982) Location of volcano-tectonic earthquakes within the Rabaul Caldera. Geological Survey of Papua New Guinea Report 82/19, 10 p

Ando M, Ishida M, Hayashi Y, Mizuki C, Nishikawa Y, Tu Y (2013) Interviewing insights regarding the fatalities inflicted by the 2011 Great East Japan Earthquake. Nat Hazards Earth Syst Sci 13(9):2173–2187

Blong R, McKee CO (1995) The Rabaul eruption: destruction of a town. Natural Hazards Research Centre, Macquarie University

Boegershauser G (1937) Eruption of a volcano at Rabaul. Rabaul Times 35:15 (6 Aug 1937)

Brown G (1878) Journal of the Rev. G. Brown 1860–1902 (11 volumes). Mitchell Library, Sydney, Australia

Calvache VML (1990) Pyroclastic deposits of the November 13, 1985 eruption of Nevado del Ruiz volcano, Colombia. In: Williams SN (ed) Nevado del Ruiz Volcano, Colombia, I. J Volcanol Geotherm Res 41:67–78

Chouet BA, Page RA, Stephens CD, Lahr JC, Power JA (1994) Precursory swarms of long-period events at Redoubt Volcano (1989–1990) Alaska: their origin and use as a forecasting tool. J Volcanol Geoth Res 62:95–135

Cooke RJS (1977) Rabaul Volcanogical Observatory and geophysical surveillance of the Rabaul Volcano. Aust Phys Feb 1977, pp 27–30

Davies H (1994) The 1994 eruption of Rabaul Volcano—the events of 18–19 September 1994. Report prepared for ENB Provincial Disaster Committee (unpublished)

Davies H (1995) The 1994 eruption of Rabaul volcano—a case study in disaster management. Report prepared for the UNDP Office Port Moresby (unpublished), 35 p

Dent AW, Davies G, Barret P, de Saint Ours PJA (1995) The 1994 eruption of the Rabaul volcano, Papua New Guinea: injuries sustained and medical response. Med J Aust 163(4/18):635–639

Faria B, Fonseca JFBD (2014) Investigating volcanic hazard in Cape Verde Islands through geophysical monitoring: network description and first results. Nat Hazards Earth Syst Sci 14:485–499

Fisher NH (1939) Geology and vulcanology of Blanche Bay, and the surrounding area, New Britain. Territ N Guin Geolog Bull 1, 68 p

Fisher NH (1976) 1941–42 eruption of Tavurvur Volcano, Rabaul, Papua New Guinea. In: Johnson RW (ed) Volcanism in Australasia. Elsevier, Amsterdam, pp 201–210

Ghasemi H, McKee C, Leonard M, Cummins P, Moihoi M, Spiliopoulos S, Taranu F, Buri E (2016) Probablistic seismic hazard map of Papua New Guinea. National Hazards. doi:10.1007/s11069-015-2116-8

Greene HG, Tiffin DL, McKee CO (1986) Structural deformation and sedimentation in an active caldera. J Volcanol Geotherm Res 30:327–356

Heming RF (1974) Geology and petrology of Rabaul Caldera, Papua New Guinea. Geol Soc Am Bull 85:1253–1264

Itikarai I (2008) The 3-D structure and earthquake locations at Rabaul Caldera, Papua New Guinea, M. Phil. Thesis. The Australian National University

Johnson RW, Threlfall NA (1985) Volcano Town—the 1937–43 eruptions at Rabaul. Robert Brown and Associates, Bathurst

Johnson RW, Everingham IB, Cooke RJS (1981) Submarine volcanic eruptions in Papua New Guinea: 1878 activity of Vulcan (Rabaul) and other examples. In: Johnson RW (ed) Cooke–Ravian volume of volcanological papers. Geol Surv Papua New Guinea Memoir 10:167–179

Lahr JC, Chouet BA, Stephens CD, Power JA, Page RA (1994) Earthquake classification, location and error analysis procedures for a volcanic sequence: application to 1989–1990 eruptions of Redoubt Volcano, Alaska. J Volcanol Geotherm Res 62:137–151

Lindley ID (1995) The 1994–1995 Rabaul volcanic eruptions: human aspects. R Soc NSW Bull 188 (Part 1), 189 (Part 2) and 190 (Part 3)

McKee CO (1981) Recent eruptive history of the Rabaul volcanoes, present volcanic conditions and potential hazards from future eruptions. Geological Survey of PNG Report 81/5

McKee CO, Mori J, Talai B (1989) Microgravity changes and ground deformation at Rabaul Caldera, 1973–1985. In: Latter JH (ed) IAVCEI proceedings in volcanology, vol 1. Volcanic hazards. Springer, Berlin, pp 399–428

McKee CO, Patia H, Kuduon J (2016) Recent eruption history at Rabaul: volcanism since the 7th century AD caldera-forming eruption. Geohazards Management Division Report 2016/01

McKee CO, Lowenstein PL, de Saint Ours P, Talai B, Itikarai I, Mori JJ (1984) Seismic and ground deformation crises at Rabaul Caldera: prelude to an eruption? Bull Volcanol 47–2:397–411

McKee CO, Johnson RW, Lowenstein PL, Riley SJ, Blong RJ, de Saint Ours P, Talai B (1985) Rabaul Caldera Papua New Guinea: volcanic hazards, surveillance and eruption contingency planning. J Volcanol Geotherm Res 23:195–237

Mori J, McKee CO (1987) Outward-dipping ring-fault structure at Rabaul Caldera as shown by earthquake locations. Science 235:193–195

Mori J, McKee C, Itikarai I, Lowenstein PL, Talai B (1986) Account and interpretation of the seismicity during the Rabaul Seismo-deformational crisis September 1983 to July 1985. Geological Survey of Papua New Guinea Report 86/26

Mori J, McKee CO, Itikarai I, Lowenstein PL, de Saint Ours P, Talai B (1989) Earthquakes of the Rabaul seismo-deformational crisis September 1983 to July 1985: Seismicity on a caldera ring fault. In: Latter JH (ed) IAVCEI proceedings in volcanology, vol 1. Volcanic hazards. Springer, Berlin, pp 429–462

Murray TL, Endo ET (1989) A real-time seismic amplitude measurement system (RSAM). U.S. Geological Survey Open File Report 89–684, 25 p

Neumann K (1996) The 1994 volcanic disaster in East New Britain and its aftermath: comments and observations. Department of Archaeology and Anthropology, Australian National University ISBN 0 642 25077 4, 30 p

Newhall CG, Dzurisin D (1988) Historical unrest at large calderas of the world. US Geol Surv Bull 1855 2:1108

Nishimura Y, Nakagawa M, Kuduon J, Wukawa J (2005) Timing and scale of tsunamis caused by the 1994 Rabaul Eruption, East New Britain, Papua New Guinea. In: Satake K (ed) Tsunamis: case studies and recent developments. Springer, Dordrecht, pp 43–56

Pierson TC, Janda RJ, Thouret J-C, Borrero CA (1990) Perturbation and melting of snow and ice by the 13 November 1985 eruption of Nevado del Ruiz, Colombia and consequent mobilization, flow and deposition of lahars. In: Williams SN (ed) Nevado del Ruiz Volcano, Colombia, I. J Volcanol Geotherm Res 41:17–66

Ripper ID, Letz H, Anton L (1996) Seismicity and seismotectonics of Papua New Guinea presented in earthquake depth zones. Geological Survey of Papua New Guinea Report 96/9

Sack P, Clark D (1983) Eduard Hernsheim South Sea Merchant. Institute of Papua New Guinea Studies, 230 p

Stewart RC, Itikarai I (in prep.) Seismic activity associated with the 1994–95 eruption of Rabaul Volcano

Tomblin J, Chung J (1995) Papua New Guinea, analysis of lessons learnt from Rabaul volcanic eruptions and programming for disaster mitigation activities in other parts of the country. UNDHA Report on Mission from 17 to 26 February 1995

United Nations (1977) Disaster prevention and mitigation, vol 1: volcanological aspects

Voight B (1990) The 1985 Nevado del Ruiz volcano catastrophe: anatomy and retrospection. J Volcanol Geotherm Res 42:151–188

Walker GPL, Heming RF, Sprod TJ, Walker HR (1981) Latest major eruptions of Rabaul Volcano. In: Johnson RW (ed) Cooke-Ravian volume of volcanological papers. Geol Surv Papua New Guinea Memoir 10:181–193

Challenges in Responding to a Sustained, Continuing Volcanic Crisis: The Case of Popocatépetl Volcano, Mexico, 1994-Present

Servando De la Cruz-Reyna, Robert I. Tilling and Carlos Valdés-González

Abstract

Popocatépetl Volcano, located in the central Trans-Mexican Volcanic Belt, is surrounded by a densely populated region with more than 20 million people. During the past 23,000 years, this volcano has produced eruptions ranging widely in size and style, including Plinian events and massive sector collapses. However, the historical activity of Popocatépetl, recorded in detail since 1500, consists of only nineteen small to moderate eruptions, several similar in style to the current eruptive episode (1994-present). After nearly 70 years of quiescence since its eruptions in the mid-1920s, Popocatépetl reawakened in December 21, 1994. This eruptive activity, which is still ongoing, has been characterized by a succession of lava dome growth-and-destruction episodes: pulses of effusive and moderately explosive activity alternating with periods of almost total quiescence. This pattern appears to be characteristic of all historical eruptions, several of which lasted for decades, with interspersed lull periods that in some cases make it difficult to identify the end of the eruptive episodes. In this chapter, we discuss the problems and challenges posed by a prolonged, low-level volcanic crisis (or "semi-crisis") of variable intensity that has lasted for more than 20 years, without showing any signs of coming to an end. Paradoxically, this still-continuing crisis has spawned two opposite developments: (1) during periods of little visible activity, people dwelling near the volcano become somewhat

S. De la Cruz-Reyna (✉)
Instituto de Geofísica, Universidad Nacional
Autónoma de México UNAM, CDMX, Mexico
e-mail: sdelacruzr@gmail.com

R.I. Tilling
Volcano Science Center, U.S. Geological Survey,
Menlo Park, USA

C. Valdés-González
Centro Nacional de Prevención de Desastres
CENAPRED, CDMX, Mexico

Advs in Volcanology (2018) 235–252
https://doi.org/10.1007/11157_2016_37
Published Online: 15 March 2017

apathetic and indifferent; but (2) during times of easily observed visible activity, awareness of changes at the volcano—and their hazardous implications—is rapidly and greatly enhanced by the common use of social media by people.

1 Introduction

Disasters can occur when society fails to identify and foresee the potentially hazardous manifestations of a natural phenomenon. However, disasters may also occur if society fails to adopt adequate measures to reduce the risks—to people, property, and infrastructure—posed by hazardous phenomena even if recognized in advance. Responding effectively to hazards is a process as complex as is the fabric of society itself, as each hazardous phenomenon has a variety of destructive manifestations, and each may affect different sectors of society in particular ways. This is especially true when dealing with volcanic crises. As has been long recognized (e.g., Fiske 1984; Peterson 1986, 1988; Tilling 1989; Voight 1990; Peterson and Tilling 1993; Haynes et al. 2008; Solana et al. 2008; Fearnley 2013), the process of crisis response entails close interaction between three main entities: (1) The scientists studying the hazardous phenomena and their potential social outcomes; (2) the authorities in charge of public safety and infrastructure; and (3) the affected populace. The wide spectrum of backgrounds and attitudes of all the involved stakeholders during such interaction, together with the vague or imprecise information generally available during the crisis, often combine to hinder effective communications among the entities involved. Poor communications in turn complicate the perception of the risk, a factor likely to increase societal vulnerability. Thus, during an evolving crisis, it is critical to develop a perception of risk as uniform as possible among all stakeholders—no easy task when the affected population is measured in millions. To achieve this goal requires searching for communication tools that can describe—as

simply as possible—the relations between the level of threat posed by the volcano, and the level of response of the authorities and the affected public. In the case of Popocatépetl, the Civil Protection of Mexico addressed this challenge by developing and implementing the *Volcanic Traffic Light Alert System* (VTLAS). Distinct from other volcano alert systems (VALs)—typically referenced to the activity of the volcano—used to communicate warning information from scientists to civil authorities managing volcanic hazards (Fearnley 2013; Potter et al. 2014), the VTLAS scheme (discussed in detail below) additionally was intended to reduce the possibility of ambiguous interpretations of intermediate alert levels by the large populations at risk. This additional component marks a significant advance in the management of volcanic crises in Mexico (De la Cruz-Reyna and Tilling 2008).

Before proceeding further, we should summarize how risk is managed by the National Civil Protection System of México (SINAPROC), which was created in 1986 after the catastrophic disaster caused by a M 8.1 earthquake on September 19, 1985. The executive body of the SINAPROC at the federal level is the General Coordination, housed within the Ministry of the Interior (Secretaría de Gobernación). The General Coordination is supported by four agencies: (1) the National Direction of Civil Protection, an operational body in charge of implementing the preventive and relief actions; (2) the General Direction of Integral Risk Management, which provides the funding for prevention and emergency actions; (3) the General Direction of Interlinking and Regulations, which coordinates the different government levels involved with civil protection; and (4) the National Center for Disaster Prevention (CENAPRED), a technical body created in September 19, 1988, with substantial technical and generous financial support from the government of Japan. The mission of CENAPRED is to promote the applications of science and technology for the prevention and mitigation of disasters, to train and inform professionals and technicians on these subjects, and to disseminate the necessary information for preparedness and self-protection.

CENAPRED also acts as an active interface between the operative, decision-making authorities of the SINAPROC and the academic scientific community. In conducting its work, CENAPRED utilizes four advisory scientific committees on topics relevant for disaster prevention, composed of prominent, experienced Mexican scientists in the areas of Earth sciences, hydro-meteorological sciences, social sciences, and chemical and industrial hazards. There are also ad-hoc sub-committees, as is the case of the Advisory Committee for Popocatépetl Volcano, on which several international volcanologists—especially from the U.S. Geological Survey

(USGS)—have actively participated. This advisory sub-committee will be herein referred as the Popocatépetl Scientific Committee (PSC).

Popocatépetl has remained persistently active for over 20 years, thereby creating a long-standing volcanic crisis that has imposed additional difficulties in the management of the volcanic risk. CENAPRED has been and remains in charge of the monitoring of Popocatépetl volcano, and it also continues to host and coordinate PSC sessions as needed.

2 Popocatépetl Volcano: Geologic Setting and Eruptive History

Popocatépetl Volcano (19.02°N, 98.62°W), which lies within the central Trans-Mexican Volcanic Belt (Fig. 1), is located about 60 km SE of México City and 45 km W of Puebla City.

These two populations centers, combined with other nearby cities within a 100-km radius around the volcano, contain a total population of

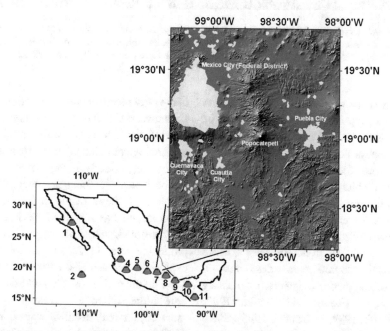

Fig. 1 Sketch map showing the location of Popocatépetl Volcano and other historically active volcanoes of Mexico: *1* Tres Vírgenes; *2* Evermann (Socorro); *3* Ceboruco; *4* Colima; *5* Parícutin; *6* Xitle; *7* Popocatépetl; *8* Pico de Orizaba (Citlaltépetl); *9* San Martín Tuxtla; *10* El Chichón; *11* Tacaná. The inset shows the distribution of cities and major towns (*yellow*) within 100 km of Popocatépetl's active crater. The cities of Mexico City, Puebla, Cuernavaca, Cuautla and others located within the inset have a combined population over 20 million

Fig. 2 Popocatépetl (*right*) and Iztaccíhuatl Volcanoes: Above, "Camino a Chalco con los volcanes" painting by José María Velasco, 1891 (Museo Nacional de Arte, MUNAL, Mexico). *Below*, view of these two prominent peaks through helicopter windshield (*Photograph* by S. De la Cruz-Reyna in September 2002)

over 20 million people. Rising 5454 meters above sea level, the conical volcanic edifice of Popocatépetl is topped by a 600 × 800 m elliptical summit crater. Popocatépetl is the youngest peak within the Sierra Nevada, a volcanic mountain range which extends in a roughly N–S direction. To the north of Popocatépetl, the volcanic complex Iztaccíhuatl complements the iconic volcanic landscape of the region (Fig. 2).

Geological evidence indicates that a large eruption about 23,000 years destroyed a pre-existing volcanic edifice, generating massive debris avalanches (Robin and Boudal 1987; Boudal and Robin 1989; Siebe et al. 1995). Since then, Popocatépetl's eruptive history has been characterized by at least seven major explosive eruptions and many smaller eruptions that have produced large volumes of ash and pumice. Three

of the most recent explosive eruptions (ca. 3000 B. C., between 800 and 200 B.C., and ca. A.D. 800) affected human settlements, as indicated by archaeological remains buried by ashfall deposits and pottery shards incorporated by mudflows (Siebe et al. 1996; Siebe and Macías 2004). After the last of these major eruptions, activity at Popocatépetl has remained moderate for nearly 1200 years. Batches of magma were extruded, producing lava domes and associated moderate explosions and ashfalls. Eyewitness reports since 1354 (in the native Nahuatl and Spanish language translations) describe episodes of activity, while more recent and detailed written reports since 1500 document that about 16 small and 3 moderate eruptive episodes have occurred within the past 500 years, some of them probably involving dome growth-and-destruction processes similar to those of the current, ongoing activity (De la Cruz-Reyna et al. 1995; De la Cruz-Reyna and Tilling 2008).

3 Ongoing Unrest, Eruptive Activity, and Volcanic Crisis

Management of risk posed by volcanic unrest requires a comprehensive understanding of the natural phenomenon. In this regard, a sustained activity makes it particularly difficult to forecast the future activity and its consequences, because the commonly employed methodologies to recognize and assess the relevance of precursors of increased activity are obscured by the persistent low-to-medium level of activity. It is thus important to define and establish the context of the unrest to develop decision-making criteria.

Before the current activity, there was little public awareness outside the scientific community about Popocatépetl being an active and potentially hazardous volcano. The previous eruption of Popocatépetl was not a major event. It began in 1919, and the available descriptions of that activity (Friedländer 1921; Waitz 1921; Dr. Atl 1939) indicate that it consisted of a succession of dome emplacements and destructions, similar to the current eruptive episode, probably lasting until 1927. Then, after nearly 70 years of quiescence (except for a minor fumarolic event in 1947), Popocatépetl volcano reawakened in 1993 with increased fumarolic and seismic activity (De la Cruz-Reyna et al. 2008). By October 1994, this unrest further escalated, culminating with a series of moderately large phreatic explosions at the crater during the early hours of 21 December 1994. These explosions produced ashfalls on several towns to the east and northeast of the volcano, including the large city of Puebla.

At the time of the explosions, glaciers with an estimated total area of 0.54 km^2 blanketed the northern flank of the cone, below the crater (Delgado-Granados 1997; Huggel and Delgado-Granados 2000). With the vivid memories of the 1985 Nevado del Ruiz disaster in the minds of authorities and scientists, nearly 25,000 people living in some of the most vulnerable

towns located along the likely paths of pyroclastic flows and lahars were evacuated in the afternoon of 21 December as a precautionary measure. A week later, the eruptive activity decreased and its largely phreatic nature became better understood, the evacuated residents were allowed to return home. Ash emissions or protracted explosions consisting mostly of gas and steam with relatively low concentrations of ash and a characteristic emerging seismic signal were referred to as "exhalations." This type of relatively low-level activity persisted through 1995 and into early 1996, with decreasing intensity (De la Cruz-Reyna and Siebe 1997).

About a year later, seismicity and exhalation activity increased again, and on 26 March 1996 a lava dome was first observed growing on the crater floor. This dome was partially destroyed by an explosion on April 30, 1996, which propelled ejecta several kilometers into the sky and hot debris as far as 4 km and caused the only reported fatalities to date directly related to the Popocatépetl activity. Despite public warnings not to enter the 12 km-radius restricted area around the mountain, five members of a sports club climbed to the summit crater rim to obtain good images and videos of the activity. These climbers were struck and killed by incandescent fragments during their descent, a few hundreds of meters downslope from the crater, as evidenced by the images recovered from their cameras. Dome-building activity resumed mid-March 1997 (GVN 1996, 1998). These 1996–1997 events marked the beginning of a series of dome growth-and-destruction cycles that have continued up until this writing. Although most of the explosions have been moderate, some of them have been large enough to produce pyroclastic flows (in 2001) and lahars (in 1997 and 2001; Capra et al. 2004).

In summary, the current eruptive episode to date has consisted of a succession of moderate-size eruptions, characteristic of Popocatépetl's activity since the 14th century (De la Cruz-Reyna et al. 1995). Nonetheless, given the huge population in the region potentially at risk, together with concerns about possible escalation of the ongoing eruptive activity to Plinian phases, or even worse a major sector collapse, the management of the "volcanic crisis" at Popocatépetl—already persisting for 20-plus years—remains a major challenge for volcanologists, national and local Civil Protection authorities, and the threatened populations (De la Cruz-Reyna and Siebe 1997). From experience gained through 1997, involved scientists and decision-making authorities became increasingly convinced of the need to attain a more appropriate and uniform perception of the changing risk among all the stakeholders. Accordingly, in 1998 the Volcanic Traffic Light Alert System (VTLAS) was developed as a risk- communication protocol, hazards-warning system and response scheme specifically designed to manage the ongoing volcanic crisis at Popocatépetl more effectively (Fig. 3). A detailed description of VTLAS and its use are given by De la Cruz-Reyna and Tilling (2008), but we summarize below three salient points:

1. A level of activity of the volcano is defined by the PSC and translated into the most likely scenarios, describing them in specific terms, including time scales, names of threatened areas, etc. In general terms, these sets of scenarios may be grouped according to seven levels of response of the SINAPROC, which in turn are managed as phases within each of the Traffic light colors: two for Green, three for Yellow, and two for Red.
2. SINAPROC authorities translate the level of volcanic hazard defined by the PSC into one of three alert levels for the population (not of the volcano) that leave no room for uncertainty: **Green**, everything is fine; **Yellow**, you must be aware of the hazard and pay attention to any official announcements; and **Red**, you must leave the area according to the instructions given by the authorities.
3. All decisions involving mitigative actions are undertaken by the Civil Protection authorities according to the selected phase within the color level. It is important to emphasize that, in Mexico, the management of risks associated to natural phenomena is by law a

responsibility assumed by the three levels of government: federal, state, and municipal. The Scientific Committees are officially appointed advisory groups of "more than 10 but less than 15 experts in the subject, which can emit opinions and recommendations about the origin, evolution and consequences of hazardous phenomena, aimed to technically induce decision making for prevention and mitigation to the population...", as stated by the Organization and Operation Manual of the National System of Civil Protection within the General Law of Civil Protection.

4 Evolution of the Activity Influences Public Perception of Hazards

Popocatépetl's historical volcanic activity has been dominated by an irregular, yet sustained dome-emplacement and destruction processes, which in turn have resulted in some significant fluctuations in the level of the summit crater floor. The floor, which was about 277 m deep in 1906 (Friedländer 1921), was raised almost 100 meters because of the accumulation of dome-forming products during the 1919–1927 eruptive episode (Gómez-Vázquez et al. 2016) The episode beginning in 1994 further accumulated enough material to raise the level of the crater floor to near the crater rim after the peak of activity in 2000–2001, thereby reducing the capacity of the crater walls to contain or reduce the range of potential pyroclastic flows. In fact, under these conditions an explosion occurring on January 22, 2001 produced pyroclastic flows that traveled about 4.5 km from the crater, affecting the glacier on its northeast slopes and triggering a lahar that reached the outer limits of Xalitzintla, one of the nearest villages (Sheridan et al. 2001; Capra et al. 2004; Macías and Siebe 2005). However, the general level of activity, as measured by exhalation-explosion events and dome-emplacement rates, significantly decreased after 2003, as the rate of dome destruction exceeded the rate of lava emplacement, causing

Fig. 3 A poster designed for public offices showing the VTLAS

some reversals in the rate of volcanic material accumulation within the crater, and a significant increase of its maximum depth (Gómez-Vázquez et al. 2016). As of March 2015, the summit crater was again refilling with volcanic debris (Fig. 4), but so far, no other pyroclastic flows have been observed.

Figure 5 shows the evolution of seismicity during the period 1996–2014, as indicated by the fluctuations in total seismic energy release, a basic volcano-monitoring parameter expressed by RSAM values and their cumulative curve. RSAM (**R**eal-time **S**eismic **A**mplitude **M**easurement) is a system that continuously samples the

Fig. 4 View of Popocatépetl's summit crater on 3 March 2015 showing that volcanic materials filling the crater had reached within 30 m below the lowest part of the rim (*photograph* downloaded from https:// www.whatsapp.com/). While providing useful information about the state of the volcano, this *photograph* was taken by a member of a group of adolescents who had climbed the summit area without authorization (see text for further discussion)

absolute amplitude of the seismometer signals (Murray and Endo 1989; Endo and Murray 1991). From RSAM data, the cumulative seismic energy released by the volcano provides a very good proxy of its vigor of eruptive activity, as it reflects the total seismic energy release from all sources: VTs, explosions, exhalations, tremors, etc. Figure 5 exhibits that the overall level of activity varied little between 1996 and early 2000, with a weak increasing trend from the onset of the activity before sharply increasing in late 2000 and the start of 2001. Afterwards, the

level of activity gradually decreased keeping a relatively low level until a new marked increase in RSAM counts in mid-2010. During the period 2003–2010, however, far fewer reports of visible volcanic manifestations—such as observed explosions, exhalations or lava accumulation— induced many people to believe that the eruptive activity perhaps was approaching its end, despite the seismic-monitoring data indicating otherwise. Expectedly, public interest in the volcano waned and perception of its hazards slowly started to dissipate. Yet, volcano monitoring and the

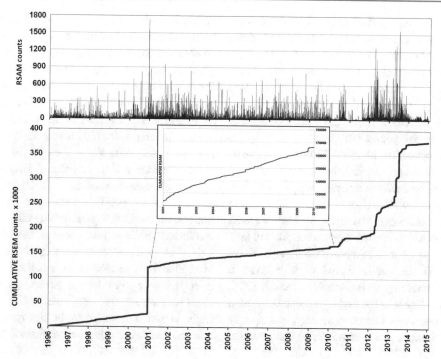

Fig. 5 Evolution of seismicity during the period 1996–2015, as recorded by the Chiquipixtle seismic station (PPX), located to the WSW of the crater, at a height of 3980 m. asl. *Above*, each vertical bar represents a ten-minute RSAM average of the seismic signal amplitude sampled at a rate of 100 samples per second. *Below*, cumulative RSEM (sum of successive squared RSAM values, proportional to the total seismic energy detected by the monitoring station). In December 2000, the activity increased sharply, including recorded manifestations such as saturating harmonic tremors (see Fig. 6). This surge in seismic activity prompted a precautionary evacuation of towns exposed to pyroclastic and lahar flows. The seismic energy release since 2010 has been actually larger, but at a lower rates. The inset in the center shows a zoom of the waning RSEM in the period 2001–2010

information about the volcano activity were never scaled down, as the daily reports on the volcano condition and the status of the VTLAS, which remained in Yellow, were published every day by CENAPRED in its website. Nevertheless, only when activity manifestations such as explosions, exhalations or persistent fumaroles become visible, the media massively reproduces such reports. The proliferation of social networks and fixed webcam sites, such as http://www.webcamsdemexico.com/ (with more than half a million followers), has made possible wide dissemination of information about the occurrences of even minor events, thereby improving in this way the persistence of awareness.

5 Development of Risk—Mitigation Strategies Since 1994

Perhaps the main challenge in managing the response to the ongoing volcanic activity of Popocatépetl has been posed by its rather pedestrian, anti-climactic character, particularly during the 2003–2010 period. The initially impressive phreatic eruption of 1994 that sharply contrasted with the quietness of the previous 70 years prompted the frenzied making of a Popocatépetl's volcanic hazards map (Macías et al. 1995) in only a few months under high-stress conditions. This map—the first such for Popocatépetl specifically intended for use by civil authorities—was made with a general consensus of the involved Mexican and U.S. experts. The high-urgency 1994–95 scientific and governmental response, which also included the development of the VTLAS, then gradually declined until the onset of effusive magmatic activity in 1996. The VTLAS was set at Yellow for the population at the moment of its implementation in mid-1995. The slow decline of the ash–emission events, the direct source of public and authority awareness, was not much in agreement with the data from the volcano-monitoring instruments. The VTLAS thus remained in Yellow even when the volcano appeared to be in a relative state of rest. News media and the public started to joke about a "busted" traffic light. However, in March 1996, the emplacement of the first lava dome confirmed an ongoing level of eruptive activity, and thereby rekindled public and media interest and concern. Unlike the initial 1994 episode, the now much-improved monitoring data allowed a better understanding of the 1996 activity. Because the character of the dome -emplacement processes was effusive and confined within the summit crater, the PSC continued to recommend maintaining the VTLAS in Yellow, as the probabilities of pyroclastic flows or lahars were still low. It was at that time that some members of a sport clubs climbed to the crater rim. Tragically, this imprudent action resulted in the above-mentioned casualties caused by the first dome-destruction explosion, again rekindling the interest of public and media.

Dome-emplacement and destruction activity continued in the ensuing years with a somewhat increasing trend, but without exceeding the levels set by the scenarios marked by the VTLAS, so it continued in condition Yellow. However, the dome-destruction explosions in 1997, and particularly the 13 km-high ash column of 30 June that caused ashfalls in Mexico City and impelled closing its airport for 12 h, prompted changing the VTLAS to Red for a few hours. However, no evacuations of populations were ordered, thereby generating some confusion among people and authorities. Studies at other volcanoes (Solana et al. 2008) indicate that, although civil

authorities are aware of the volcanic hazards, their understanding of how to respond during an emergency can be incomplete, and that understanding how people perceive risk is important for improving risk communication and reducing risk-associated conflicts (Haynes et al. 2008).

At this stage, the need of an embedded scale within the three-color alert levels designating the alert level of authorities became immediately evident; see De la Cruz-Reyna and Tilling (2008) for a detailed account of the VTLAS levels. The quick return to condition Yellow, as no evacuations were needed, again prompted the news media and public to joke about a "busted" traffic light. Public discussions on this subject, however, ultimately proved to be beneficial, because it helped to convey to the general public and many authorities that the color of the Traffic Light is not a description of the state of the volcano, but rather it is a description of the threat on people and thus reflects the state of awareness of individuals. Hence, the VTLAS remained in Yellow, although the phase, i.e., the level of alert for Civil Protection authorities, has changed several times.

The VTLAS was only temporarily set again in condition Red at the peak of intensity (i.e., rate of seismic energy release) of the entire eruptive episode during December 15–19, 2000. Unlike the 1997 event, the management of the December 2000 eruption was more efficient as lessons were learned from the 1997 experience. The colors and phases of the VTLAS, and a safety radius of 12 km around the crater, were then clearly defined by the PSC. This radius narrowly excluded the closest towns to the volcano. The Red-VTLAS was set on the basis of a 24-h forecast made by the PSC from the large amplitude of the volcanic tremor signals, and using results from a load-and-discharge time-predictable model consisting of a succession of episodes of variable intensity, in which the seismic energy released by a high intensity episode of activity is followed by a lower-intensity period with a duration proportional to the previous energy drop (see for example De la Cruz-Reyna 1991). The exclusion radius was then extended to 13 km, and an evacuation of the towns within that radius was undertaken

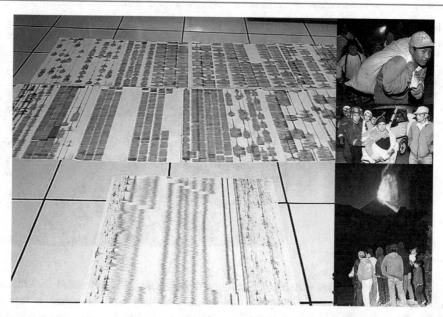

Fig. 6 *Left* Seismograms of the December 2000 activity (*back*), showing highly saturated recordings, compared with seismicity recorded for the largest previous eruptive event in June 1997 (*front*), which caused ashfall on Mexico City, obtained from the same CENAPRED monitoring station. *Right* Images of the evacuation and of evacuees watching the eruption from a safe distance (*photographs* courtesy of Associated Press)

(Fig. 6). This precautionary action was taken because the dome-destroying eruptions ejected large amounts of hot debris onto the glacial ice that could induce localized melting, such that even minor pyroclastic flows could produce powerful lahars. One of the vexing difficulties in the management of the December 2000 response was a marked overreaction of authorities in some small towns slightly beyond the exclusion radius. Concerned that some nearby towns were being evacuated, these officials decided to order immediate evacuations, without waiting for confirmation of the National Security Committee. No criteria for the return of these people to their homes were defined at the time, and some evacuees remained up to 10 days in temporary shelters. The problems of overreaction and the lack of clear "return" criteria to lower the level of alert have not yet been solved (De la Cruz-Reyna et al. 2000).

The December 2000–January 2001 explosive activity marked a watershed in the evolution of the ongoing volcanic crisis at Popocatépetl. Before 2001, the accumulation rate of dome lavas and debris exceeded the rate of removal by explosive activity. After a period of irregular activity lasting until 2003, that trend slowly reversed, and the main crater slowly began to recover some of its former capacity (Gómez-Vázquez et al. 2016). After 2003, a lower lava emplacement and explosion rates (shown as a diminishing slope of the cumulative RSEM counts in the inset of Fig. 5) prompted a reduction of the VTLAS phase from Y-3 to Y-2, and then to Y-1 in 2004, still maintaining however condition Yellow on affected populations. The slightly increased seismic activity in late 2005 and 2006 raised the VTLAS phase back to Y-2. Overall, between 2005 and 2009, the rate of dome-lava growth did not exceed the rate of debris removal by explosions and exhalations, so that the crater continued to deepen slightly. Explosive activity gradually increased again in mid-2010 and continued with minor fluctuations through 2011. On 20 November 2011, a powerful explosion ejected large ballistic blocks to distances of 4 km; this explosion also generated a shock wave that was felt by some people as far away as 10 km from the volcano, but luckily with no damaging consequences.

Fig. 7 The results of a telephone survey of 800 people, conducted on 19 April 2012, to gauge the general public's perceptions of the ongoing, but generally low level, activity of Popocatépetl. (Image from http://kaleydoscopio.mx/)

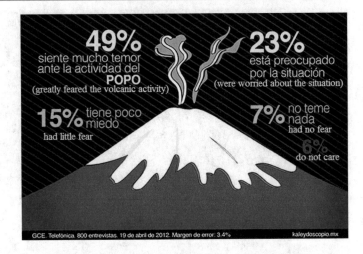

Since 2001, dome building, exhalations, and explosions have continued sporadically to the date of this writing (July 2016). Accordingly, this persistent, though irregular, activity has necessitated that the VTLAS remains for long times in condition Yellow, leading to some wearisomeness and complacency among the populace and some municipal authorities of towns near the volcano during protracted lull periods. However, no complacency existed among authorities at the federal level. During periods of relative inactivity, the PSC and the CP authorities discussed in depth the pros and cons of making the VTLAS more dynamic, particularly lowering it to Green during relative quiescences. However, after much debate, a strong argument finally gained consensus: Green conditions would immediately allow the occupation and/or reoccupation of previously restricted areas close to the limit of the National Park, and well within the exclusion radius of 11 km. Then, should a new episode of more intense explosive activity arise, it would be much more difficult to evacuate people than it would be had the VTLAS remained in Yellow. Thus, it was decided to retain the current protocols, until there was solid evidence that the now two-decade-long eruptive episode had completely finished and the volcano had re-entered another long repose period.

The continuing, persistent moderate level activity of the volcano, together with the need to keep an intermediate level of alert among people and authorities, has generated an awkward situation that may well be called a "semi-crisis," which appears to be taken less seriously than a full-blown crisis among authorities and communities surrounding the volcano. The ambivalence of how this "semi-crisis" is viewed may be gleaned from the results of a telephone survey conducted in April 2012. Of the 800 people surveyed, 49% greatly feared the volcanic activity, 23% were worried about the "situation", but the remaining 28% had little or no fear, or had no opinion (Fig. 7). Of continuing concern to the authorities, some people clearly do not perceive the volcano as a risk, as evidenced, for example, by the actions of a group of adolescents in March 2015. This group—numbering about 15 and all under the age of 18—had climbed to Popocatépetl's summit crater five times, against government restrictions and apparently oblivious to dangers from possible volcanic activity during their ascents (https://www.youtube.com/watch?v=higZ7j98Og8).

6 Scientific Strategies and Scientific Challenges

The main function of the Popocatépetl Scientific Committee (PSC) is to assess the hazards related to the activity of Popocatépetl volcano. From its earliest sessions in 1995, a methodology that proved to be efficient and practical was adopted.

Sessions are chaired by a moderator, usually the director of CENAPRED, or a high authority from the National Autonomous University of Mexico (UNAM) or from the Ministry or Interior, and usually limited to 2 h, unless a special situation requires more time. The sessions are split into three parts. In the initial one, the monitoring groups describe the recent observations, always in the same order: seismic, geodetic, visual, geochemical, and others (Fig. 8). Questions are allowed, but no discussions are permitted until the exposition of all the data is complete. In the second part, the members of the PSC discuss the observations and propose possible explanations of the observed data. This discussion is steered towards interpretations for which consensus seem to exist. Non-consensual matters are set aside to be discussed later elsewhere. In the third part, the PSC proposes the most likely scenarios based on the consensus of the previous discussions. When a small number of likely scenarios is agreed upon, the PSC summarizes the likely scenarios and makes recommendations to the Civil Protection (CP) authority based, on all previous deliberations. This method, which is similar to the differential diagnosis used in medical science, has proved effective to make diagnostics and prognostics of the volcano activity, and to present them to CP as a list of the possible scenarios in a descending order of likelihood. We re-emphasize that the PSC **does not** set the level of the VTLAS, which is done by the CP authorities based on the PSC recommendations.

The main difficulties faced by the scientists of the PSC may be summarized in two different realms: firstly, the scientific and technical one related to understanding of volcanic processes and, secondly, the operational aspects in effectively communicating hazards information to the CP authorities, news media, and the affected populace. With regard to the former, apart from the typical instrumental and technological inadequacies and limitations in the amount and

Fig. 8 A session of the PSC in progress in 2014, at which the relevance of changes in seismic spectrograms related to observed visible signs of volcanic activity is being discussed. The inset shows a session in 2015 at which the main subject was the policies to deal with people disregarding the security radius to climb Popocatépetl Volcano. All sessions of the PSC are recorded

quality of monitoring data, an important issue has been the evolution of the precursors' meaning and possible implications. Specifically, what do the variations in precursory activity portend what the volcano might do next? During the initial years of the crisis at Popocatépetl, acquired experience allowed identification of clear precursors to explosions, such as harmonic tremors; repetitive short-duration, low-amplitude LP events (drumbeats); accelerated rate of RSAM counts, etc. However, and with no clear watershed (although it may be related to the change of eruptive regime in 2001–2003), in several instances such seismic signals were not necessarily followed by explosions. On the other hand, cumulative volcano-tectonic (VT) energy release, rate of dome growth, and VT appearing in specific locations seemed to have become more relevant precursors.

With regard to effective communications of hazards information, in addition to the general factors considered previously discussed, an additional recurrent hindrance during the Popocatépetl crisis has been the frequent changes of decision-making authorities over time scales shorter than the duration of the volcanic activity. Federal and state governments change every 6 years, so that within each period of administration the responsible CP authorities with whom the scientists interact may be replaced more than once. Thus, it is always necessary to train the new authorities in the communication process. Fortunately, in all cases, the communication of volcanic risk based on likely scenarios has made it possible to deal with this problem relatively easily.

Although the basic tools for scientific assessment of the hazards and systematic monitoring of the activity have not always been sufficient, CENAPRED has made, and is making, a major effort in maintaining the highest technological standards in the volcano-monitoring networks. In addition, a new volcanic-hazards map by a team of volcanologists at the Instituto de Geofísica, and other UNAM institutes has replaced the current one (Macías et al. 1995), which was prepared under rushed conditions with minimal

data available at the time. The new hazards map considers a wealth of new geological information collated over the past 2 decades (in particular, extent of lahar inundation areas, magnitude and timing of past Plinian eruptions, ash dispersal data, etc.). The updated hazards map and assessment were released by CENAPRED in (2016). Participants in the current map include some of the authors of the 1995 map: Siebe, Macías, Capra, Delgado, and Martin del Pozzo, plus their associates (e.g., postdocs, students). The CP system provided a special fund (FOPREDEN. Disaster Prevention Fund) to finish the new map, although much of the new scientific data were obtained with research projects financed by CONACYT (Mexican National Science Council) and DGAPA (The UNAM fund to support major research projects).

7 Concluding Remarks and Future Challenges

The historical activity of Popocatépetl has been characterized by eruptive episodes of activity similar to the current one (1994-present). The relatively low magnitudes of the eruptions reported since the year 1500 have not left enough geological evidence to assess their evolution with time. Nonetheless, historical accounts (detailed in De la Cruz-Reyna et al. 1995, and summarized in De la Cruz-Reyna and Tilling 2008) suggest that the described events may correspond to the most visible manifestations (i.e., dome-destruction explosions) within long-duration episodes of successive dome emplacement and destruction. Without further evidence other than the general, and sometimes vague, wording of the reports, the events listed in De la Cruz-Reyna and Tilling (2008) may be grouped into about six of such episodes, as shown in Table 1.

No statistical analysis of Table 1 is attempted because of the subjective grouping of episodes. However, some hints may be gleaned from the inspection of the information summarized in the table. The current eruptive episode is not anomalously long, and within the range of

Table 1 Grouping of the 19 known historical eruptions of Popocatépetl volcano (De la Cruz-Reyna and Tilling 2008) into six long-duration episodes involving successive dome-emplacements and dome-destruction explosions, interspersed with long periods of quiescence

Period	Estimated minimum duration (y)	Estimated lapse until next episode (y)	VEI range of eruptions in the period
1512–1548	36	23	2–3
1571–1592	21	50	2
1642–1665	23	32	2–3
1697–1720	23	199	1–2
1919–1927	8	67	2
1994-present	>20	?	2–3

durations of previous historical episodes. Moreover, the expected lapse to a future episode does not seem to be correlated with the duration of the previous episode or the duration of the previous lapse. Unfortunately, the historical information precludes any reliable means to forecast the precise time and explosivity of the next eruptive episode after the current one ends. The possibility that the next episode could involve a destructive Plinian phase cannot be ruled out; Plinian eruptions have occurred in the geologic past, with a mean return period of about 1500 years (Siebe et al 1996; Mendoza-Rosas and De la Cruz-Reyna 2008). A formidable challenge for scientists and the civil authorities in anticipating Popocatépetl's future behavior is to address this question: if the current already-long episode continues as a semi-crisis for a lot more time, what are the best options to counteract the wearisomeness and indifference of some of the stakeholders? There are no easy answers. Suggestions have been made to replace the VTLAS by another alert system. Apart from the confusion that a new, untested system may induce in the large population involved, it may not help necessarily solve the problem, as the dullness of the situation is not caused by the alert system, but instead by the pauses characterized by the absence of major explosions or any other visible expressions of increased volcanic vigor. A positive aspect is that when a highly visible event such as a moderate explosion occurs on a clear day, the public's volcano awareness seems to recover almost instantly, as has been the case

since 2011. Figure 9 illustrates the positive reaction of the news media and the CP preparations in response to a raise in the alert level of the VTLAS to Yellow Phase 3.

Identifying the end of the current, relatively minor eruptive episode or the possible precursors of a much more explosive activity poses other major challenges. In particular, as seismic records constitute the dominant volcano-monitoring data for Popocatépetl and most other active volcanoes, it is crucial to better understand the empirical relationships between the seismic signals recorded since 1994 and the nature and vigor of the observed volcanic activity preceding, during, and following the seismicity. Physical models then can be developed to explain the source of the diverse seismic signals. Such method is inherently non-unique, and in some cases the ambiguity of the possible causes may lead to inadequate or even incorrect assessment of the hazards. Reduction of such levels of non-uniqueness based on integral analysis of different types of geophysical and geochemical data is a critical need to be fulfilled in the future with additional data and more diagnostic analytical methodologies. In the meanwhile, however, some pragmatic actions must be implemented. That it is why the Popocatépetl Scientific Committee, in attempting to reduce such a complex problem, has strongly emphasized the consensual approach in its deliberations. In a broader context, experience gained over recent decades at volcanoes worldwide indicates a sobering reality: despite the considerable advances in volcano-monitoring techniques, except for

Fig. 9 Front pages of the Puebla City news media announcing the raise in alert level to *Yellow Phase* 3 in the VTLAS. This phase requires having emergency shelters ready (*upper right*), and CP personnel mobilized to carry out a possible evacuation (*lower right*)

very rare exceptions, current state-of-the-art volcanology still lacks a routine, reliable capability to always correctly interpret a volcano's precursory signals and to accurately predict the outcomes of volcano unrest (e.g., Tilling 2014).

Maintaining a continuous flow of up-to-date information to the public about volcanic activity, hazards, and risk reduction seems to be the best and most practical solution to minimize the weariness and indifference that at times develop among the authorities and populations at risk during lulls of visible activity. In this regard, the daily posting of the activity reports of Popocatépetl on the CENAPED website (http://www.cenapred.gob.mx/reportesVolcan/BuscarReportesVolcan?opt Busqueda=1) apparently has been quite effective. This, together with the exponentially growing influence of fixed webcam sites and social media networks, has greatly increased public awareness of the occurrence of even minor events at Popocatépetl and other volcanoes in Mexico or elsewhere. Nonetheless, we recognize that the influence of user-generated social media reporting needs to be regarded with caution, because there is no assurance of the accuracy of the content of the transmitted information. Although most reports and comments diffused via social media are generally informative, at times scientifically unsupported remarks and predictions are also included, thereby contributing to possible confusion and generating a negative impact on the general awareness. To deal with this problem, SINAPROC opened a twitter account to spread reliable information. Dealing with modern-day modes of information dissemination poses another major challenge for all those involved in the management of volcanic risk.

Acknowledgements This research has been partially supported by the UNAM-DGAPA-PAPIIT project IN-106312. We wish to express our appreciation to Jose Luis Macías and an anonymous reviewer for helpful constructive reviews of an earlier version of this paper.

References

Boudal C, Robin C (1989) Volcán Popocatépetl: recent eruptive history, and potential hazards and risks in future eruptions. In: Latter JH (ed) Volcanic hazards. IAVCEI proceedings in volcanology. Springer, Berlin, pp 110–128

Capra L, Poblete MA, Alvarado R (2004) The 1997 and 2001 lahars of Popocatépetl volcano (Central Mexico): textural and sedimentological constraints on their origin and hazards. J Volcanol Geoth Res 131 (3):351–369

CENAPRED (2016) http://www.cenapred.gob.mx/es/Publicaciones/archivos/357-CARTELMAPASDEPELIGRO SDELVOLCNPOPOCATPETL.PDF; http://www.cena pred.gob.mx/es/Publicaciones/archivos/270-INFOGRA FAPOPOCATPETL-FLUJOSPIROCLSTICOS.PDF; http://www.cenapred.gob.mx/es/Publicaciones/archivos /270-INFOGRAFAPOPOCATPETL-LAHARES.PDF; http://www.cenapred.gob.mx/es/Publicaciones/archivos /270-INFOGRAFAPOPOCATPETL-DISPERSINDEC ENIZA.PDF; http://www.cenapred.gob.mx/es/Publica ciones/archivos/270-INFOGRAFAPOPOCATPETL-A VALANCHAS.PDF; http://www.cenapred.gob.mx/es/ Publicaciones/archivos/270-INFOGRAFAPOPOCATP ETL-CADADEBALSTICOS.PDF; http://www.cenapr ed.gob.mx/es/Publicaciones/archivos/270-INFOGRAF APOPOCATPETL-LAVAS.PDF; http://www.cenapr ed.gob.mx/es/Publicaciones/archivos/270-INFOGRAFA POPOCATPETL-SEMFORODEALERTAVOLCNICA. PDF

De la Cruz-Reyna S (1991) Poisson-distributed patterns of explosive eruptive activity. Bull Volcanol 54:57–67

De la Cruz-Reyna S, Siebe C (1997) The giant Popocatépetl stirs. Nature 388(6639):227

De la Cruz-Reyna S, Quezada JL, Peña C, Zepeda O, Sánchez T (1995) Historia de la actividad del Popocatépetl (1354–1995). Volcán Popocatépetl, Estudios Realizados Durante la Crisis de 1994–1995. CENAPRED-UNAM, México DF, pp 3–22

De la Cruz-Reyna S, Meli RP, Quaas RW (2000) Volcanic crises management. Encyclopedia of Volcanoes. Academic Press, San Diego, pp 1199–1214

De la Cruz-Reyna S, Tilling RI (2008) Scientific and public responses to the ongoing volcanic crisis at Popocatépetl Volcano, Mexico: importance of an effective hazards-warning system. J Volcanol Geoth Res 170:121–134. doi:10.1016/j.jvolgeores.2007.09.002

De la Cruz-Reyna S, Yokoyama I, Martínez-Bringas A, Ramos E (2008) Precursory Seismicity of the 1994 Eruption of Popocatépetl Volcano, Central Mexico. Bull Volcanol 70(6):753–767. doi:10.1007/s00445-008-0195-0

Delgado-Granados H (1997) The glaciers of Popocatépetl volcano (Mexico): Changes and causes. Quatern Int 43–44:53–60. doi:10.1016/S1040-6182(97)00020-7

Dr Atl (Gerardo Murillo) (1939) Volcanes de México, vol I. La Actividad del Popocatépetl, Editorial Polis, México

Endo ET, Murray TL (1991) Real-time seismic amplitude measurement system (RSAM): a volcano monitoring and prediction tool. Bull Volc 53:533–545

Fearnley CJ (2013) Assigning a volcano alert level: negotiating uncertainty, risk, and complexity in decision-making processes. Environ Plann A 45:1891–1911

Fiske RS (1984) Volcanologists, journalists, and the concerned local public: a tale of two crises in the eastern Caribbean. Explosive volcanism: interception, evolution, and hazard. National Academy Press, Washington, DC, pp 170–176

Friedländer I (1921) La erupción del Popocatépetl. Memorias Sociedad Científica Antonio Alzate 40:219–227

Gómez-Vázquez A, De la Cruz-Reyna S, Mendoza-Rosas AT (2016) The ongoing dome emplacement and destruction cyclic process at Popocatépetl volcano, Central Mexico. Bull Volcanol 78(9):1–15. doi:10.1007/s00445-016-1054-z

GVN (1996) Global Volcanism Network, Smithsonian Institute Bulletin 21(8)

GVN (1998) Global Volcanism Network, Smithsonian Institute Bulletin 23(2)

Haynes K, Barclay J, Pidgeon N (2008) Whose reality counts? Factors affecting the perception of volcanic risk. J Volcanol Geoth Res 172:259–272

Huggel C, Delgado-Granados H (2000) Glacier monitoring at Popocatépetl Volcano, México: glacier shrinkage and possible causes. In: Hegg C, Vonder Muehll D (eds) Beiträge zur Geomorphologie. Proceedings Fachtagung der Schweizerischen Geomorphologischen Gesellschaft, Bramois, WSL Birmensdorf, pp 97–106

Macías JL, Carrasco-Nuñez G, Delgado-Granados H, Martin de Pozzo AL, Siebe C, Hoblitt RP, Sheridan FM, Tilling RI (1995) Mapa de peligros del Volcán Popocatépetl: Instituto de Geofísica, UNAM, México, DF (scale 1: 250,000)

Macías JL, Siebe C (2005) Popocatépetl's crater filled to the brim: significance for hazard evaluation. J Volcanol Geoth Res 141:327–330

Mendoza-Rosas AT, De la Cruz-Reyna S (2008) A statistical method linking geological and historical eruption time series for volcanic hazard estimations: applications to active polygenetic volcanoes. J Volcanol Geoth Res 176:277–290. doi:10.1016/j.jvolgeores.2008.04.005

Murray TL, Endo ET (1989) A real-time seismic amplitude measurement system (RSAM). US Geol Surv Open-File Rep 89–684:1–21

Peterson DW (1986) Volcanoes: Tectonic setting and impact on society. In: Active tectonics, Geophysics Study Committee, National Research Council: Washington, DC, National Academy Press, pp 231–246

Peterson DW (1988) Volcanic hazards and public response. J Geophys Res 93:4161–4170

Peterson DW, Tilling RI (1993) Interactions between scientists, civil authorities, and the public at hazardous volcanoes. In: Kilburn CRJ, Luongo G (eds) Monitoring active lavas. UCL Press, London, pp 339–365

Potter SH, Jolly GE, Neall VE, Johnson DM, Scott BJ (2014) Communicating the status of volcanic activity: revising New Zealand's volcanic alert level system. J Appl Volcanol 3:13. doi:10.1186/s13617-014-0013-7

Robin C, Boudal C (1987) A gigantic Bezymianny-type event at the beginning of modern volcano Popocatépetl. J Volcanol Geotherm Res 31:115–130

Sheridan MF, Hubbard B, Bursik MI, Siebe C, Abrams M, Macías JL, Delgado GH (2001) Short-term potential volcanic hazards at Popocatépetl, Mexico. EOS Trans Am Geophys Union 82(16):187–189

Siebe C, Abrams M, Macías JL (1995) Derrumbes Gigantes, Depósitos de Avalancha de Escombros y Edad del Actual Cono del Volcán Popocatépetl. Volcán Popocatépetl, Estudios Realizados Durante la Crisis de 1994–1995. CENAPRED-UNAM, México, pp 195–220

Siebe C, Abrams M, Macías JL, Obenholzner J (1996) Repeated volcanic disasters in pre-Hispanic time at Popocatépetl, Central Mexico. Past key to the future? Geology 24:399–402

Siebe C, Macías JL (2004) Volcanic hazards in the Mexico City metropolitan area from eruptions of Popocatépetl, Nevado de Toluca, and Jocotitlán stratovolcanoes and monogenetic scoria cones in the Sierra Chichinautzin. Volcanic Field guide, Penrose Conf. Neogene-Quaternary Continental margin Volcanism. Geol Soc Am, pp 1–77

Solana MC, Kilburn CRJ, Rolandi G (2008) Communicating eruption and hazard forecasts on Vesuvius, Southern Italy. J Volcanol Geoth Res 172:308–314

Tilling RI (1989) Volcanic hazards and their mitigation—Progress and problems. Rev Geophys 27:237–269

Tilling RI (2014) Volcano hazards and early warning. In: Meyers RA (ed) Encyclopedia of complexity and systems science. Springer, New York, 19 pp. doi:10.1007/978-3-642-27737-5_581-2

Voight B (1990) The 1985 Nevado del Ruiz volcano catastrophe: anatomy and retrospection. J Volcanol Geoth Res 44(3):349–386

Waitz P (1921) La nueva actividad y el estado actual del Popocatépetl. Memorias Sociedad Científica Antonio Alzate 37:295–313

Organisational Response to the 2007 Ruapehu Crater Lake Dam-Break Lahar in New Zealand: Use of Communication in Creating an Effective Response

Julia S. Becker[iD], Graham S. Leonard, Sally H. Potter, Maureen A. Coomer, Douglas Paton, Kim C. Wright and David M. Johnston

Abstract

When Mt. Ruapehu erupted in 1995–1996 in New Zealand, a tephra barrier was created alongside Crater Lake on the top of Mt. Ruapehu. This barrier acted as a dam, with Crater Lake rising behind it over time. In 2007 the lake breached the dam and a lahar occurred down the Whangaehu Valley and across the volcano's broad alluvial ring-plain. Given the lahar history from Ruapehu, the risk from the 2007 event was identified beforehand and steps taken to reduce the risks to life and infrastructure. An early warning system was set up to notify when the dam had broken and the lahar had occurred. In combination with the warning system, physical works to mitigate the risk were put in place. A planning group was also formed and emergency management plans were put in place to respond to the risk. To assess the effectiveness of planning for and responding to the lahar, semi-structured interviews were undertaken with personnel from key organisations both before and after the lahar event. This chapter discusses the findings from the interviews in the context of communication, and highlights how good communication contributed to an effective emergency management response. As the potential for a lahar was identifiable, approximately 10 years of lead-up time was available to

J.S. Becker (✉) · G.S. Leonard · S.H. Potter
M.A. Coomer · D.M. Johnston
Joint Centre for Disaster Research, GNS Science,
P.O. Box 30368, Lower Hutt, New Zealand
e-mail: j.becker@gns.cri.nz

G.S. Leonard
e-mail: g.leonard@gns.cri.nz

S.H. Potter
e-mail: s.potter@gns.cri.nz

M.A. Coomer
e-mail: m.coomer@gns.cri.nz

D.M. Johnston
e-mail: david.johnston@gns.cri.nz

D.M. Johnston
Joint Centre for Disaster Research, School of
Psychology, Massey University, P.O. Box 756,
Wellington 6140, New Zealand

D. Paton
School of Psychological and Clinical Sciences,
Charles Darwin University, Darwin, NT 0909,
Australia
e-mail: douglas.paton@cdu.edu.au

K.C. Wright
Ministry of Civil Defence & Emergency
Management, Wellington, New Zealand
e-mail: Kim.Wright@dpmc.govt.nz

Advs in Volcanology (2018) 253–269
https://doi.org/10.1007/11157_2016_38

install warning system hardware, implement physical mitigation measures, create emergency management plans, and practice exercises for the lahar. The planning and exercising developed effective internal communications, engendered relationships, and moved individuals towards a shared mental model of how a respond to the event. Consequently, the response played out largely as planned with only minor communication issues occurring on the day of the lahar. The minor communication issues were due to strong personal connections leading to at least one incidence where the plan was bypassed. Communication levels during the lahar event itself were also different from that experienced in exercises, and in some instances communication was seen to increase almost three-fold. This increase in level of communication, led to some difficulty in getting through to the main Incident Control Point. A final thought regarding public communication prior to the event was that more effort could have been given to developing and integrating public information about the lahar, to allow for ease of understanding about the event and integration of information across agencies.

Keywords

Volcano · Lahar · Communication · Emergency management planning · Mt. Ruapehu · New Zealand

1 Introduction

1.1 Background

In 1995 and 1996 Mt. Ruapehu in New Zealand underwent a series of eruptions that initiated approximately 30 eruption-induced lahars, and several ash falls over parts of New Zealand's North Island (Galley et al. 2004; Cronin et al. 1997; Johnston et al. 1999). A tephra layer was also deposited at the head of the Whangaehu Valley on the Ruapehu Crater Lake outflow channel, creating a natural dam to the lake (Manville et al. 2007). Over time it was anticipated that the water in the Crater Lake would rise behind the dam to a volume of 7–9 million cubic metres, and eventually breach it, resulting in a dam-break lahar (Keys and Green 2008). This lahar would travel down the Whangaehu Valley, and despite much of the area being quite remote, people and infrastructure were potentially in

danger (Fig. 1). In addition to many lahars caused by other factors (such as eruptions and rain mobilisation; Keys and Green 2008), previous dam-break lahars are known to have occurred from Ruapehu, including a devastating lahar in 1953. In that instance, the dam-break lahar travelled down the valley and seriously damaged a rail bridge at the location of Tangiwai (Board of Inquiry 1954). Soon after, a passenger train hurtled into the lahar-flooded Whangaehu River and 151 people died. Given the consequences of the 1953 event, and the anticipation that a new dam-break lahar could occur, it was considered imperative that something be done to reduce the risk from a potential future event. Options included removing the tephra dam through to planning an emergency management response to a break-out lahar. The dam-break lahar eventually occurred on 18 March 2007. This chapter describes the mitigation and planning actions that were taken to reduce the risk, and based on a

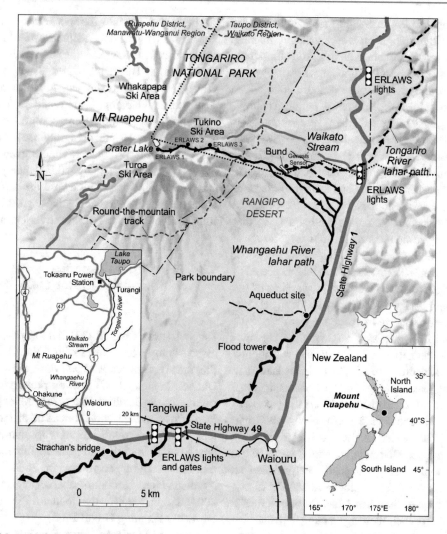

Fig. 1 Map of Mt. Ruapehu Crater Lake, showing potential lahar path, and locations of significance (adapted from Keys and Green 2008)

series of interviews with those involved in the planning process, outlines the role of communication in creating an effective response.

1.2 What to Do About an Anticipated Lahar?

A number of studies were conducted to assess the hazard and risk of a dam-break lahar at Ruapehu (e.g., Hancox et al. 1997, 1998; Taig 2002) and to identify options for reducing the risk (DOC 1999). While some strongly advocated for intervention at the tephra dam, e.g., digging a

tunnel through the barrier to remove the potential for catastrophic collapse (Hancox et al. 1997), it was decided that no engineering intervention should take place (Keys 2007b). Instead, a mixture of warning systems, mitigation works and emergency planning was instigated to deal with the dam-break lahar issue (Keys and Green 2002, 2008; Norton 2002; Massey et al. 2009).

To mitigate the risk of a lahar, a 'bund' (levee) was built in early 2002 at the spill-over point of the Whangaehu River into the Waikato Stream (Galley et al. 2004). The objective of this was to stop a lahar from getting into the catchment of the Tongariro River and thus protecting infrastructure (i.e. Tongariro Power Station) and people downstream. Other mitigation works included making changes to infrastructure at risk from the lahar, such as strengthening and raising the road bridge at Tangiwai by two metres (Wakelin 2006).

At about the same time as the bund was built a warning system called the Eastern Ruapehu Lahar Warning System (ERLAWS) was set up. ERLAWS (a system of sensors and preventative mechanisms) was designed to sense vibrations when the lahar had been triggered and send electronic warnings out to people (e.g., via pager, computers, electronic road signs), so that emergency responders could initiate a response (Keys 2009; Leonard et al. 2008; Wakelin 2006; Massey et al. 2009). The system was expected to give a maximum of one hour of warning of the impending lahar at State Highway One (SH1) on the Desert Road, and two hours warning at the Tangiwai Bridge (Wakelin 2006).

While the ERLAWS in particular was a good way of detecting when a Crater Lake lahar had been triggered, the system was also subject to being set off accidentally and creating a 'false alarm'. This could particularly occur when rain storm events caused heightened stream flow.

An Eruption Detection warning System (EDS) was also present on Ruapehu and provided an additional avenue for warning of lahars. The EDS is triggered by earthquakes when an eruption of a certain size occurs, allowing warning for lahars that are potentially generated by volcanic eruptions (Leonard et al. 2008). Such lahars usually occur from hot volcanic material falling on snow and melting the snow to generate a lahar which could then impact on the local ski fields. However, an eruption could also potentially generate a Crater Lake break-out lahar if a dam has blocked the lake, and thus the EDS provided another warning option for the dam-break lahar situation.

Another element of managing a Ruapehu Crater Lake dam-break lahar was that of emergency planning. It was decided that response plans should be prepared to deal with a dam-break lahar if, and when, it occurred (Keys 2009). The region affected by a potential lahar was divided into two parts: a Northern part which would only be affected if a lahar were to overtop

the bund and get into the Waikato Stream and Tongariro River; and a Southern part which was the main area that would be affected if a lahar moved from Ruapehu Crater Lake via the Whangaehu River to the sea. A Northern Ruapehu Lahar Planning Group and a Southern Ruapehu Lahar Planning Group were formed to develop plans on an effective response for before, during and after a dam-break lahar event. Both groups were convened by 2003 (Galley et al. 2004). The following section describes the nature of the planning undertaken by these groups in the anticipation of responding effectively to a lahar.

1.3 Planning a Response to a Ruapehu Crater Lake Lahar

Taupo District Council (TDC) took on primary responsibility for developing the Northern Emergency Management Plan, while the Ruapehu District Council (RDC) took on primary responsibility for the Southern Emergency Management Plan. Development and refinement of the plans took place in collaboration with many other organisations with responsibilities (Table 1), with the final sign-off occurring in 2004 for the Northern Plan (Northern Lahar Planning Group 2004) and in 2005 for the Southern Plan (Southern Lahar Planning Group 2005).

A central Incident Control Point (ICP) in the town of Ohakune was identified in the plan. The ICP was where the majority of responders would meet to coordinate response to the lahar. There were also a number of agencies that would operate remotely from other Emergency Operations Centres (EOCs) and from out in the field where the lahar was occurring.

The plans outlined the actions to be undertaken before (relating to readiness for the event, and reducing the risk), during (response phase) and after (recovery phase) the lahar took place, and the organisations responsible for those actions. As part of the readiness and reduction phase, warning levels numbered from 1 to 5 were assigned for different levels of the Crater Lake (Table 2). These warning levels were based on historic records of

Table 1 Main organisations involved in the Northern and Southern Ruapehu Lahar Planning Groups

Organisations	Northern plan	Southern plan	Responsibilities
NZ Army	X	X	• Provide staff and resources as required
Department of Conservation (DOC)	X	X	• Respond to pager activation and notify DOC staff • Implement Tongariro National Park Lahar response plan • Provide advice on lahar status and on-going support to emergency operations centre (EOC)/Incident control point (ICP) • Provide specialist personnel • Follow own contingency plans
Environment Waikato (Regional Council)	X		• Activate group EOC • Notify incident controller • Provide regional liaison officer to ICP • Monitor situation • Coordinate regional response as required
New Zealand Fire Service	X	X	• Provide staff as required to ICP and roadblocks • Provide staff to assist with evacuations
Genesis Energy (Power company)	X	X	• Follow own contingency plans • Provide engineering advice to ICP
Good Health Wanganui (Wanganui District Health Board (DHB))		X	• Establish liaison with St John ambulance • Implement plans • Alert other District Health Boards • Health advisor for EOC provided • Health media liaison and spokesperson • Casualty documentation process estimated • Coordination with police • DHB and St John develop transport plan • DHB coordinating health response
Horizons Manawatu-Wanganui (Regional Council), Lakes DHB, Waikato DHB, Bay of Plenty DHB		X	• Activate Group EOC • Notify district councils, Incident Controller, EOC staff, National Crisis Management Centre (NCMC) • Coordinate response across region • Monitor situation and update NCMC • Coordinate regional support • SRLPG[1]
GNS Science (GNS-Earth science research agency)		X	• Notify police communications of lahar status • Monitor lahar status • Provide advice to EOC
Justice Department	X	X	• Respond to pager
Ministry of Civil Defence & Emergency Management (MCDEM)	X	X	• Notify MCDEM staff • Provide national liaison officer to ICP • Monitor situation • Coordinate national response to lahar as required • SRLPG
Opus Consultants	X	X	• Contractors for infrastructure assessment

(continued)

Table 1 (continued)

Organisations	Northern plan	Southern plan	Responsibilities
New Zealand Police	X	X	• Staff road blocks • Assist with evacuations • Initiate Coordinated Incident Management System (CIMS) structure[2] • Activate fire siren • Provide staff to ICP/EOC • Direct agencies as required • Implement traffic management plan • Check on state of bund • Contact transit to open road • Initiate CIMS structure • Provide staff to check river • Coordinate the use of aerial assistance in upper mountain and river areas to make public safe • Coordinate Search and Rescue personnel • SRLPG
Rangitikei and Wanganui District Council		X	• Activate lahar response plan • Advise local school • Advise Chief Executive Officer and Roading Manager • Commence telephone tree call out • Provide staff as required
Ruapehu District Council (RDC)	X	X	• Notify RDC staff • Activate siren if not working • Provide staff to EOC/ICP • Initiate telephone tree calls • Monitor RDC assets • Advise recovery manager • Maintain EOC • Provide on-going info to EOC • Bridge inspection • Contact transit to open road • SRLPG
Search and Rescue (SAR)	X	X	• Activate if required
Taupo District Council (TDC)	X		• Establish EOC • Monitor TDC assets • Resource to the EOC as required
Transpower (Owns and operates New Zealand's electricity grid)	X	X	• Implement agency lahar response plan • Implement Transpower contingency plan • Monitor situation • Provide engineering advice to EOC
Transit New Zealand[3] (Responsible for the New Zealand State Highway network)	X	X	• Notify staff and contractors • Activate agency lahar response plan • Deploy contractors to snow huts • Activate/check variable message signs • Contractors to close snow gates • Provide engineering advice to ICP • Provide contractors to road blocks • Engineering inspection of Transit assets • Implement traffic management plan

(continued)

Table 1 (continued)

Organisations	Northern plan	Southern plan	Responsibilities
Tranz Rail/Ontrack[4] (rail network maintainer and operator)	X	X	• Notify Police communications of activation of Tranz Rail lahar warning system • Cease train traffic • Isolate power at Tangiwai rail bridge and rail crossing • Activate agency lahar response plan • Provide advice to ICP as required • All train traffic to remain stopped until directed by Incident Controller
Winstone Pulp International (local business in the lahar path)		X	• Implement lahar response plan • Alert personnel and evacuate 'at risk sites' • Advise ICP when evacuation complete
Works Consultancy		X	• Contractors for infrastructure assessment

Adapted from Galley et al. (2004), Northern Lahar Planning Group (2004), Southern Lahar Planning Group (2005)

[1]*SRLPG* agencies specifically named as part of the Southern Group in the group plan

[2]CIMS is a scalable command and control structure widely used in emergency management in New Zealand and in many other countries under various incident management or coordination system names

[3]From 2008 Transit New Zealand has been operating as the New Zealand Transport Agency

[4]From 1995 the organisation responsible for maintaining New Zealand's rail network was named Tranz Rail. The name was changed to Ontrack in 2004. Since 2008 the rail network has operated under the new name of KiwiRail

previous lahars and the damage they caused (Galley et al. 2004). At Crater Lake Warning Level 1, when the lake was at a low level, trained staff were expected to be able to "initiate a response" within 30 min (Northern Lahar Planning Group 2004; Southern Lahar Planning Group 2005). In contrast, at Warning Level 4, where the lake level was higher, trained staff were expected to initiate a response within 5 min. As well as using the warning levels for the plan, public information was also provided around the current level of the lake and what this meant in terms of the likelihood and risk of a lahar occurring (e.g. leaflets, websites, free-phone) (Fig. 2).

For the response phase, the Plans outlined actions that each organisation should take for time periods following the activation of ERLAWS (that is, after 5, 30, 60, and 90 min from when a lahar warning had been triggered). For example, in the first 5 min following ERLAWS activation and pager receipt, common activities that agencies listed in the Plans had to do included responding to the pager, and contacting relevant internal and external personnel to notify them of the activation so they could start responding. Within 30 min, Incident Control Points were

required to be set up, the Coordinated Incident Management System (CIMS) structure initiated, road blocks established, lahar visually confirmed (if possible), key contacts and communication established (with any required information flowing), and individual agency contingency plans implemented. Sixty minutes after the lahar warning was received, much of the response was expected to be in 'monitoring and maintenance mode', where agencies were watching the progress of the lahar, staff were kept at key points as required (or deployed elsewhere if needed), and communication over the state of affairs was continuing. After 90 min the watching brief was expected to continue, and response was then to move toward the recovery phase.

After an official commencement of the recovery phase, the plans outlined the need for briefings, implementation of traffic management plans, provision of staff as required, engineering inspections of key infrastructure, and coordination of media and communication. These activities and monitoring of the situation continued for two hours, after which it was expected that debriefing and media activities would be occurring.

Table 2 Crater Lake warning levels

Level of readiness	Lake level (msl)	%	Explanation	Actions	Indicative time for lake to rise to next alert Level in summer
Normal	Below 2527	<95	Base level of readiness as per normal civil defence planning	Planning, preparation, and training	
Level 1	2526.5	95	Critical trigger point, 3 m below the new rock overflow level. *Waves caused by eruptions and small landslides could overtop barrier but the probability of a small collapse lahar caused by a resulting barrier failure is extremely low*	Planning completed. Full response capability available and ready. Response within 30 min	
Level 1b	2529.5	100	Lake full to the buried rock rim outlet level and the base of the tephra dam. Probability of barrier failure at this level is still very low		1–6 months to fill from Alert Level 1b to 2
Level 2	2533	108	Sudden collapse could produce a lahar equivalent to the 1975 event. *This is the largest historic lahar that has passed under the Tangiwai bridges (without causing damage) and down the Tongariro River* Conditional probability of barrier failure at this level is 1–2%	Response within 20 min	0.7–1.9 months to fill from Level 2 to 3, or 7.8 months to drop down to Level 2 from Level 3, depending on infill rates. *This large variation is due to the possibility of the filling spanning fast and slow filling rates, and seepage. Slow fill rates will probably result in net drops in lake level above about 2532 m*
Level 3	2535	113	Equivalent to a large moderately fast lahar. Conditional probability of barrier failure at this level is 5–10%	Response within 10 min	0.4–0.6 months to fill from level 3 to 3b, or 3.2 months to drop down to level 3 from 3b
Level 3b	2536	115	Conditional probability is 50–60%		0.2–0.3 months to fill from level 3b to 4, or 1.1 months to drop down to level 3b from 4
Level 4	2536.5	116	Equivalent to a large, fast lahar. Conditional probability is 90%	Response within 5 min	0.2–0.3 months to fill from level 4 to 5, or 0.7 months to drop down to level 4 from level 5
Level 5	2536.9	117	Lake at top of the tephra dam. Conditional probability is 100%		

Warning levels were associated with actions in the Northern and Southern Emergency Management Plans. Warning levels were also used to communicate to the public about the height of the lake in relation to the tephra dam and crater rim, and the rate at which the lake was filling with water (from Galley et al. 2004; Massey et al. 2009)

The Plans also included procedures for false alarms generated by ERLAWS. Only the Incident Controller could declare a false alarm based on either expert advice from the Duty Scientist monitoring the lahar, visual confirmation that there was no lahar, or if the time estimated for the lahar to reach Tangiwai had passed (Northern Lahar Planning Group 2004; Southern Lahar Planning Group 2005). False alarms provided the opportunity to practice for an effective response.

Individual agencies also developed their own emergency management plans (e.g. the Central District Police Lahar Response Plan), which were consistent with both the Northern and Southern Plans.

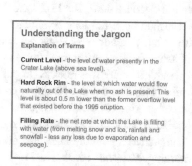

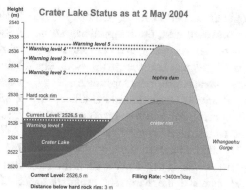

Fig. 2 Public information about the current level of the lake (e.g. for May 2004) and what this meant in terms of the likelihood and risk of a lahar occurring (replicated from a leaflet provided by the Wanganui-Rangitikei Emergency Management Committee 2004)

A key part of planning for the lahar included practicing the response using the Plans. A training and exercise schedule was devised and included in the Plans. Some practice was also obtained when ERLAWS was activated accidentally (e.g. by bad weather) or when the EDS detected a small volcanic eruption was in progress in 2006 (Mordret et al. 2010). In this instance, responders received a message on their pagers saying that a "LAHAR [was] POSSIBLE", and they then knew they needed to start the process of stepping up to respond to a potential event.

1.4 Communication

Communication related to the lahar event needed to be considered in three distinct contexts. First, *internal communication* needed to occur between those involved in planning for and responding to the lahar. Inter- and intra-agency communication about the lahar issue developed over a period of 10 years from the occurrence of the 1995–1996 Ruapehu eruptions through to the 2007 lahar. At first, communication was constrained by issues including awareness of the historic Tangiwai disaster; the range and desirability of options available to treat the lahar risk, and differences in opinion over how that risk should be managed (DOC 1999; Keys 2009); a strong desire from local iwi to let the lahar occur as a natural process; politics over who should be planning for and responding to the risk (Dittmer 2008; Keys 2009); and the quality of initial emergency plans created by RDC for responding to the lahar (Dittmer 2008). Such conflicts raised issues for effective communication around planning. Agencies often disagreed about the best approach to take, and who should be taking it.

The Minister of Conservation made a final decision about management of the lahar in 2001, guided by the Assessment of Environmental Effects report which outlined various treatment options (DOC 1999) and the debate which had occurred around the issue. The Minister's decision included the establishment of a physical warning system and building of a bund, combined with emergency planning (Dempsey 2002). Once the decision was made, a clear path was defined for the context in which future communication would take place. From this point onwards, inter and intra-agency communication developed as emergency management planning for the lahar evolved.

The second primary communication area was *the external communication of public information* about the status of the Crater Lake levels and response to the lahar event. A Lahar Information Management group, comprising the lead agencies in management of the event, was set up to address media communications related to the lahar (Southern Lahar Planning Group 2005). A lahar information plan, outlining management of information before, during and after the lahar, was included as part of the main Southern Emergency Management Plan appendices.

The third area of communication related to *technical capacity*. An effective response to the lahar relied on technical support, the primary one being timely communication about an impending lahar by ERLAWS. Other aspects of technical importance included the hardware required to report to others verbally and visually about the nature and progress of the lahar. It was imperative that these technical aspects were able to operate when the lahar occurred so that responders could communicate effectively.

The organisations involved in the planning process came from different geographic and functional jurisdictions (Leonard et al. 2005). Many agencies had not worked together in the past. This meant that diverse world views and expectations were likely to exist amongst responders, and there was a need to construct a shared mental model over response and communications (Paton et al. 1999). Galley et al. (2004) conducted an analysis of the planning process up to 2003 and made some observations on a variety of issues, many of which pertained to communication. They suggested that clarification of communication roles, lines of communication, and methods of communication should be undertaken to assist with creating a shared mental model. They recommended improving internal communication by continuing to conduct multi-organisation planning meetings, education, training and exercises. There was a need to ensure the technical ERLAWS system was robust and had adequate 'check-back systems'. Methods were also required for updating new information into plans and processes.

2   The Lahar Event

On Sunday 18 March 2007, after a prolonged period of wet weather, the tephra barrier began to breach (Massey et al. 2009). Around 10:06 on Sunday morning slumping started to occur at the Crater Lake tephra dam. By 11:20 the dam was fully breached by rising lake levels due to heavy rainfall, and a lahar was flowing down the Whangaehu Valley. Although the volume of water was high, water was released in pulses over a 45 min period, and the event was considered 'moderate' in size (Wakelin 2007).

The ERLAWS sensor at site 1 on the dam (see Fig. 1) signalled a possible lahar just after 10:00 that morning. Between 10:06 and 11:42 all three ERLAWS sensors (Sites 1–3) and the Genesis sensor (Site 4) were triggered as the lahar travelled down the Valley. The alarms sent data to Tokananu Power Station and warning messages to police, rail and road authorities, and to infrastructure agencies. Alarms also initiated the closure of automatic barrier arms and flashing lights and signs on the State Highways (Keys 2007a).

By 11:00 am the Ohakune Incident Control Point (ICP) was activated, and responders had gathered at the ICP. Police had stopped rail traffic at Waiouru, farmers downstream were notified of the lahar, and signs and road blocks on State Highway 1 and State Highway 49 (SH49) were in place. At 11:28 the Whangaehu Valley telephone tree was activated, notifying valley residents of an impending lahar. By 11:35 the DOC lahar duty staff and response plan were activated and media releases were being prepared (Keys 2007a).

By around 11:44 the lahar had reached the bund built to prevent overflow from the lahar into the Tongariro River, but did not breach it. By 11:50 a helicopter had taken response staff to the area for visual confirmation of the lahar but bad weather and visibility kept observers from viewing the Crater Lake (Wakelin 2007).

From 11:50 to 15:30 other personnel who had not received the initial alerts were notified of the lahar including the Minister of Conservation, additional DoC staff, GNS research scientists, infrastructure providers and the media (Keys 2007a). During the day of the lahar event, over 100 media calls were taken by ICP staff. Lahar updates from the Emergency Operations Centre and the Ruapehu Area Manager fed information into media releases which were given to media as per the Emergency Management Plans' procedures (Wakelin 2007).

The lahar reached Tangawai Rail Bridge and peaked at 13:30 and by 15:48 the warning lights on SH1 were turned off and the highway reopened. SH 49 was reopened at 16:06 and the Ohakune Incident Control Point was closed at 18:18 as the lahar passed down the lower Whangaehu River on its way to the sea (Keys 2007a). By the end of the day on 18 March the lahar and the emergency management response had run its course, with no major incidents occurring.

3 Method for Communication Research

As a Crater Lake tephra dam breach could be anticipated beforehand, it provided an excellent opportunity to study communication aspects of the planned emergency response prior to the lahar, and compare that with what happened in an actual event. Twenty interviews were undertaken between March and May 2006 with individuals from organisations involved in the response to the event. The interviews were semi-structured and contained questions about participants' organisations' plans for the wider response (including response to the ERLAWS warning system, roles and responsibilities, information management, communication, and training) and their expectations for responding to a lahar event. The interviews were repeated between April and June 2007, after the lahar had taken place. The same questions were asked, but this time participants were asked to reflect on how the actual response had gone compared with the anticipated response. All interviews were digitally taped and transcribed, and entered into the software package ATLAS.ti. Coding of the interviews was undertaken and key themes were extracted as described by Braun and Clarke (2006). The themes are organised under the following headings, which were identified as important in the discussion above:

1. Internal communication (in the planning and response process)
2. External communication
3. Technical capacity

4 Results and Discussion: Communication for the Lahar Event

The following section discusses the key themes related to communication that were identified during analysis of the data.

4.1 Internal Communication

4.1.1 Internal Communication During the Planning Process

Two types of internal communication were identified from the interviews. The first related to internal communication during the planning process. It included communication between agencies as the planning process took place, as well as communication within agencies. When the first set of interviews was undertaken in 2006, participants reported that communication during the planning process had been good. This is likely due to the 10-year timeframe it had taken for communication to develop. Tasks such as plan development, meetings, training, exercising and practices all contributed to better communication between agencies over time. These tasks contributed twofold: they helped build relationships between personnel; and they helped develop agreement over roles, responsibilities, and protocols for responding to the lahar, which could subsequently be embedded in a plan. An interviewee from Ontrack (responsible for the rail network) reflected on how communication had improved over time, based on the activities and training that people had undertaken together:

> We've got better at communicating in the last three or four years. We use text messaging and conferencing a lot more and so we are actually going, shifting more to CIMS models now than the ad hoc system we used to work, because we all did CIMS training last year and so we therefore tend to work to a CIMS model.

On discussion about planning for future hazard events, an interviewee from the Police noted how relationships had been built and communication had improved, which would be useful for the future:

I think what we would [use from the lahar planning] is communications with people like [GNS Science], DOC, Horizons, some of those relationships will last a long time …and actually put us in good stead…in years to come.

People's expectations in 2006 of how communication would occur during a future lahar event seemed to predominantly correspond with the details in the plans that had been developed. Interviewees noted that communications, including roles and contact information, were "all laid out" in the plan (as quoted by one DOC interviewee). People's knowledge showed that those involved in the future response had a good understanding of how communication was likely to unfold, and their roles and responsibilities. Two people did anticipate that communication during an actual response may go wider than what the plan detailed, and suspected that some people may approach them personally to seek information.

Interviewees commented on the challenge of keeping momentum and interest in the planning process in the lead up to the lahar, due to the long time frame. As one interviewee from Genesis stated, "…it is just a challenge of keeping that [focus] going". They noted the need to keep information updated and circulated to internal staff and external responders so that people knew the status of the Crater Lake and any new readiness or response procedures that were put in place.

4.1.2 Internal Communication During the Response

The second type of internal communication related to anticipated and actual communication during the response. In general, according to all of the interviewees communication during the response worked well. People followed procedures that had been agreed upon, written down, and practiced. The relationships that had been developed were found to have largely assisted with response-related communication. People had noted prior to the lahar in 2006 that they sometimes had difficulty in getting hold of others during exercises, but that by following the protocols set down in the plans they were able to make contact with an alternate person.

There were a few instances where a breakdown in communication protocols occurred. In one instance, this was due to the relationships that had been developed during the planning process. Such a strong set of relationships were formed that one of the actions regarding confirmation of lahar activation in the planned response was undertaken out of sequence. A call was made by a key responder to another agency to confirm the lahar had occurred, but this person was not officially supposed to do it. Given the occurrence of this out-of-sequence action, the response was delayed slightly until the proper process was reinstated and the response moved forward as per the plan.

Additionally, some participants had expectations that they would receive calls from one of the primary responders to confirm the lahar activation (even though the plan had not outlined this as an action), but this was not the case. This created a period of anxiety and uncertainty as agencies waited for the official word to reach them that a lahar had indeed occurred.

Interviewees reported that multiple and diverse technical sources of communication during the response were useful, as both primary modes of communication, and as a back-up. Types of communication documented in both the Northern and Southern Emergency Management Plans included VHF radio, radio telephone (RT), landline and mobile telephones (for regular communications, telephone trees to notify local residents of the lahar, etc.), satellite phone, fax (e.g., for situation reports), email, conference calling, and web-cameras showing the river. Participants reported finding a common radio connection useful in aiding connections, as multiple people could sit in and listen on the channel for updates. For instance, in 2006 the Emergency Manager from Taupo District spoke about how they could "turn on our radios and do a listening brief because we can tune into DOC's radios, we can tune into the police radios and just see what's going on". A particular point was also made about the usefulness of having visual information through pictures and video footage of the lahar for understanding of how the event was progressing, and for having a common point of discussion between agencies that may not be sitting in the same room, but were able to see the same imagery. An emergency manager from Horizons Regional Council reflected on this point by saying:

> I think one of the biggest things, from a management point of view, was the webcam. That was a huge bonus. In fact, far more useful than I had ever expected it to be. I was talking on the phone to [...] the National Controller, and we were just chatting about how things were going and what information we were getting and all that sort of carry on and I said "We're projecting our webcam image up onto the wall" and he said "Yeah, we're doing the same." Which is good. That's what it was there for and we were making sure that the page was refreshing itself regularly and then [the National Controller] said, "Oh, refresh your page." So we refreshed the page and it had gone from the base of the tower to being over the top of the tower within a few minutes and immediately we knew what was going on, the guys in Wellington knew what was going on, we knew roughly how big it was, and from a management point of view that was really beneficial, being able to see it...

The Horizons participant also remarked that methods of communication such as radio, teleconferencing and web-cam were useful for understanding how the event was progressing in "real-time" and that the response needed to evolve to match its progress.

Videoconferencing between the different agencies was not undertaken for this event, but it was noted that for future events video-conferencing would be useful and might add an additional aspect to the response. As a Horizons emergency manager stated, "you look into a room and immediately get a feel for how things are going".

One of the 'surprises' faced by responders was that the amount of communication differed during the actual 2007 lahar event in comparison with the exercises and practices. An interviewee from the Police suggested that communications were of greater number during the event itself —"they tripled"—and that he was called by a variety of people both from official agencies (e.g., other Police, Ohakune ICP, Genesis), and unofficial sources (e.g., the public ringing to enquire about traffic disruption on SH1). The

emergency manager from Wanganui District also said that his cell phone range constantly from agencies who rang him personally. The increase in communication also relates to the fact that relationships were so good between responders, that they would make contact with each other to find out more information, even if it wasn't specifically part of the plan. Conversely, one DOC interviewee located in the ICP said that she did "very little communication, because everyone had everything under control".

Frequent updates of information were praised by responders, who found the frequency useful for understanding what was going on, and responding. An interviewee from Ontrack gives one example of this:

> The NCM [Network Control Manager] sent text messages out in a real timely manner right through the whole event. It was really good. You didn't actually have to ring the Network Control Manager up because the text messages were coming oh, round about every thirty to forty minutes [...] telling you what's going on.

4.2 External Communication

4.2.1 Communication of Public Information Prior to the Lahar

Prior to the lahar occurring information was provided to the public about the potential lahar and its management. Media interest was high, demonstrated by a number of stories on varying aspects of the lahar (Dittmer 2008). While there was a good effort made to provide timely, relevant and clear information about the potential lahar, sometimes confusion surrounded the information that was presented. The planning group attempted to be proactive in addressing confusion.

For example, some participants perceived that there was confusion amongst the public about the information provided on the warning levels for the lahar, and that more explanation was required. Additionally, as the warning levels changed (i.e. from 1 to 2) messages required updating and clarification. The Horizons emergency manager explained that in response to these communication needs, the 0800-freephone message was re-written to outline the current level of the lake, the likelihood and risk of the lahar occurring, and stated some actions that people should do. Messages for subsequent warning levels were prepared in advance in case they needed to change at short notice.

Interviewees also reported public confusion around some of the messages that were used on the lahar signs placed on SH1 to warn travellers about the potential for a lahar to occur. The Police noted that people did not understand the word 'lahar' that was used on the signs, so the terminology was changed to 'flash flooding' instead.

It was suggested by one of the interviewees that while the Lahar Information Group was active in both the readiness and the response phases, it could have directed more of its efforts to providing public information during the readiness phase. This would have allowed the provision of public information prior to the event, and efforts to provide information by various agencies (e.g., DOC, GNS Science, MCDEM, Horizons, etc.) could have been more integrated. It was suggested by one participant that a specific pre-lahar communication plan could have been developed to assist in readiness education.

4.2.2 Communication of Public Information During Response to the Lahar

Provision of public information about the lahar during the lahar response was generally effective, likely due to the pre-planning that had occurred to coordinate a media response. There was some discussion following the response that some of the media releases were a bit slow to be produced and disseminated to the public. This was thought to be because the person in charge of the media releases was unable to get to the ICP quickly, and another staff member was required to take over issuing media releases. A DOC interviewee suggested that the delay was related to the person

who took over the media releases being "swamped" or overloaded with work and that it was a "staffing-level thing".

4.3 Technical Capacity

Prior to the Ruapehu lahar, interviewees predominantly reported that they had worked to develop a robust and integrated technical system of communication. ERLAWS needed to be reliable, and report as few false alarms as possible. For the most part this was the case, and it was estimated that there was one false activation per month (pers. comm. Keys 2007c). For the false alarms that did occur, protocols were put in place to identify the false alarm and shut down the response. These procedures were agreed on in the Plans, and practiced by the responders. If false alarms started to become too frequent, (e.g., in April 2006 there were four in one month) this was reviewed and new procedures put in place (pers. comm. Keys 2007c).

It was deemed essential that ERLAWS was linked into response systems of multiple organisations. It was also important that organisations were able to communicate together on compatible equipment. An Ontrack employee described how they ensured that the ERLAWS system was consistent with the other systems the Railways used to keep people safe. In another example, an interviewee from Genesis (2006) remarked:

> All DOC ERLAWS systems go through [...] the Genesis handling systems because we have all the communications that work on the mountain, so it made sense to do that. So obviously, we work really closely with DOC to make sure that their ERLAWS systems which run through ours are robust...

Technical communications worked well on the day of the lahar. ERLAWS was activated as expected, and everyone was able to communicate via the various technical means outlined in the planning and used in the exercises. There were only a couple of reports of responders having difficulty getting through to the ICP. These were mostly related to "black spots" in the operation of equipment (i.e. remote areas where equipment could not pick up signals), and the fact that the networks were busy. For example, a participant from Ontrack describes their communication difficulties with the ICP:

> Initially, we had some problems trying to get through to the ICP centre at Ohakune. However, when we did get through they were really good. They gave us a cell phone number to get round the telephone numbers and I suggested to the Network Control Manager next time it happens, we've got a fax number for them, fax them instead of a fax saying 'please ring' and to make the contact, because it's important that we make that link.

This demonstrates how communication issues were solved by users being able to use different modes of communication to get around the problem, highlighting the importance of multiple modes of communication.

5 Conclusions

This chapter has provided a summary of the organisational communication that occurred before, during and after the 2007 Ruapehu Crater Lake break-out lahar. It is evident that effective communication before and during an event, can assist an effective response. The Ruapehu Crater Lake lahar had approximately 10 years of lead-in time before the event occurred. During this time, agencies were able to agree on an approach to manage the lahar risk, develop a technical warning system, implement mitigation measures, start a planning process, develop plans, hold exercises, and practice responding to false alarms. Being involved in many of these activities helped build a shared mental model of how a diverse collection of agencies were going to respond to the lahar (Galley et al. 2004; Paton et al. 1999). Exercises and practices also contributed to testing whether technical aspects of the warning system were robust, and allowing changes to be made to plans when necessary. The on-going communication and cooperation between agencies developed robust relationships that enabled people to work together in an effective manner. Given these conditions, the eventual response to the 2007 lahar was very effective, and communication reportedly worked well.

Only a few minor communication issues were reported.

- First, it was suggested that more effort could have been given to developing and integrating public information about the lahar prior to the event.
- Second, a number of participants in the study reported having trouble getting through to the ICP due to communications being very busy, but this problem was averted by responders making use of the diversity of communication modes and networks to find alternative paths to the ICP.
- Third, it was identified that communication levels during the lahar event itself were different from exercises or practices, and in some instances communication was seen to increase almost three-fold. This phenomenon should be highlighted for those planning for future events so that they have realistic expectations of what a response will be like.
- Finally, the development of relationships was extremely important in facilitating communication, planning, and an effective response. However, this produced some downfalls. Personal relationships affected the response, with at least one person bypassing communications actions in the plan in favour of communicating with someone else not noted in the plan, resulting in a brief communication breakdown. This minor issue should also be considered when undertaking planning for future events.

The findings from this research on the 2007 Ruapehu lahar support current literature on good practice for effective warnings. Leonard et al. (2008) suggest that there are 5 key components of an effective warning system including:

1. Early warning system hardware
2. Planning
3. Co-operation, discussions, and communication

4. Education and participation
5. Exercises.

These five components are supported and informed by 'Research and science advice' and 'Effectiveness evaluation'. Although this chapter has focused on the key role of Co-operation, discussions and communication, it is evident that the actual response to the Ruapehu lahar contains most of the elements required for a warning to be effective. More development could perhaps have been undertaken in the areas of 'Education and participation' and 'Effectiveness evaluation, but other areas were robustly handled. This may explain why the response to the Ruapehu lahar was very effective.

The 2007 Ruapehu lahar was a unique event in that the nature of a potential future lahar could be reasonably well anticipated. The Crater Lake took over 10 years to fill to the point where the lahar occurred, which allowed a long lead in time to build relationships and plan for a response. Not every event will have the luxury of such a lead in time and therefore it is even more imperative that relationship-building, planning and communication takes place before events occur, to allow for an effective response. With respect to future events from Ruapehu and the nearby Central North Island volcanoes, the Central Plateau Volcanic Advisory Group has since been established for this purpose. Meeting two to three times per year, scientists, Civil Defence & Emergency Management personnel, planners, communicators, emergency services and local community representatives discuss plans for the volcanoes of Ruapehu, Ngauruhoe and Tongariro, and build relationships. This structure is likely to have helped in the response to the eruptions of Tongariro in 2012.

Acknowledgements The authors wish to thank all of the participants who talked to us during the course of the research, and shared their experiences about planning for and responding to the 2007 Ruapehu Crater Lake lahar. We also acknowledge the support of the New Zealand Natural Hazards Platform, which provided funding for the research to take place.

References

Board of Inquiry (1954) Tangiwai railway disaster report. Digital reproduction by Transport Accident Investigation Commission (6 September 2001) based on the original publication. Wellington, New Zealand

Braun V, Clarke V (2006) Using thematic analysis in psychology. Qual Res Psychol 3(2):77–101

Cronin SJ, Hodgson KA, Neall VE, Palmer AS, Lecointre JA (1997) 1995 Ruapehu lahars in relation to the late Holocene lahars of Whangaehu River, New Zealand. NZ J Geol Geophys 40(4):507–520

Dempsey B (2002) Planning for a lahar event. Tephra 19:18–19

Department of Conservation (1999) Environmental and risk assessment for the mitigation of the hazards from Ruapehu Crater Lake. Assessment of Environmental Effects Report, Department of Conservation, Turangi

Dittmer M (2008) The Clockwork Lahar. Unpublished Masters thesis in Communication Management. Massey University, Palmerston North, New Zealand, 235 pp

Galley I, Leonard GS, Johnston DM, Balm R, Paton D (2004) The Ruapehu lahar emergency response plan development process: an analysis. Australasian J Disaster Trauma Stud 1

Hancox GT, Nairn IA, Otway PM, Webby G, Perrin ND, Keys JR (1997) Stability assessment of Mt. Ruapehu crater rim following 1995–1996 eruptions. Client Report 43605B Institute of Geological and Nuclear Sciences Limited

Hancox GT, Otway PM, Webby MG (1998) Possible effects of a lahar caused by future collapse of the tephra barrier formed at the Mt. Ruapehu Crater Lake outlet by the 1995–96 eruptions. Client Report 43711B. Institute of Geological and Nuclear Sciences Limited

Johnston DM, Bebbington MS, Lai C-D, Houghton BF, Paton D (1999) Volcanic hazard perceptions: comparative shifts in knowledge and risk. Disaster Prev Manag 8(2):118–126

Keys HJR (2007a) Lahar lahar! Tongariro J Annu 15:22–23

Keys HJR (2007b) Lahars of Ruapehu Volcano, New Zealand: risk mitigation. Ann Glaciol 45:155–162

Keys H (2007c) Discussion about ERLAWS (personal communication)

Keys HJR (2009) Lessons from the warning system and management for Ruapehu's Crater Lake breakout event of 18 March 2007. Paper presented at the symposium of NZ society on large dams, 18 August 2009, Wellington, New Zealand

Keys H, Green P (2002) The crater lake issue—a management dilemma. N Z Alp J 118–120

Keys HJR, Green PM (2008) Ruapehu lahar New Zealand 18 March 2007: lessons for hazard assessment and risk mitigation. J Disaster Res 3(4):284–296

Leonard GS, Johnston DM, Paton D (2005) Developing effective lahar warning systems for Ruapehu. Plann Quart 158:6–9

Leonard GS, Johnston DM, Paton D, Christianson A, Becker JS, Keys H (2008) Developing effective warning systems: ongoing research at Ruapehu volcano, New Zealand. J Volcanol Geoth Res 172 (3/4):199–215. doi:10.1016/j.jvolgeores.2007.12.008

Massey CI, Manville V, Hancox GH, Keys HJ, Lawrence C, McSaveney MJ (2009) Out-burst flood (lahar) triggered by retrogressive landsliding, 18 March 2007 at Mt. Ruapehu, New Zealand—a successful early warning. Landslides 1–13

Manville V, Hodgson KA, Nairn IA (2007) A review of break-out floods from volcanogenic lakes in New Zealand. NZ J Geol Geophys 50:131–150

Mordret A, Jolly AD, Duputel Z, Fournier N (2010) Monitoring of phreatic eruptions using interferometry on retrieved cross-correlation function from ambient seismic noise: results from Mt. Ruapehu, New Zealand. J Volcanol Geoth Res 191(1):46–59

Northern Lahar Planning Group (2004) Ruapehu Lahar: emergency management plan (Northern). Taupo District Council, Taupo, p 49

Norton J (2002) Evaluating the risks and coordinating planning. Tephra 19:20–21

Paton D, Johnston D, Houghton B, Flin R, Ronan K, Scott B (1999) Managing natural hazard consequences: planning for information management and decision making. J Am Soc Prof Emerg Plann VI:37–48

Southern Lahar Planning Group (2005) Ruapehu lahar: emergency management plan (Southern). Ruapehu District Council, Ruapehu

Taig T (2002) Ruapehu lahar residual risk assessment. TTAC Limited, UK. TTAC Limited Report N19. 83 p

Wakelin D (2006) An illustrated guide to Lahars. Tongariro J Ann 14:44–49

Wakelin D (2007) Lahar lahar! Tongariro J Ann 15:17–21

Wanganui Rangitikei Emergency Management Committee (2004) Crater Lake status as at 2 May 2004, Lahar Newsletter for Whangaehu Valley Residents, May 2004, p 4

Crisis Coordination and Communication During the 2010 Eyjafjallajökull Eruption

Deanne K. Bird[iD], Guðrún Jóhannesdóttir,
Víðir Reynisson, Sigrún Karlsdóttir,
Magnús T. Gudmundsson and Guðrún Gísladóttir

Abstract

Eyjafjallajökull became Iceland's most infamous volcano in 2010 when the ash cloud from its summit eruption caused unprecedented disruption to the international aviation industry and considerable challenges to local farming communities and villages. The summit eruption, which began on 14 April 2010, was preceded by a 24-day long effusive flank eruption that produced spectacular fire-fountain activity and lava flows. The 39-day long summit eruption, however, was far more explosive and resulted in medium-sized jökulhlaups to the north, small jökulhlaups and lahars to the south and considerable ash fall to the east and east-southeast of the volcano. As in other crises in Iceland, the Department of Civil Protection and Emergency Management (DCPEM) coordinated efforts and facilitated crisis communication, while collaborating with the Icelandic Meteorological Office, the Institute of Earth Sciences at the University of Iceland and the National Crisis Coordination Centre. The DCPEM's role included providing information to the government and its various agencies and feeding information from scientists to local police officials, civil protection committees and the public. Communication with local residents took place through agencies' websites, the national media and frequent open town hall meetings where representatives of institutions responsible for eruption

The original version of this chapter was revised: An affiliation has been updated. The erratum to this chapter is available at 10.1007/11157_2017_22.

D.K. Bird (✉) · G. Gísladóttir
Institute of Life and Environmental Sciences, University of Iceland, Reykjavík, Iceland
e-mail: deanne.bird@gmail.com

G. Jóhannesdóttir
Department of Civil Protection and Emergency Management, National Commissioner of the Icelandic Police, Reykjavík, Iceland

V. Reynisson
South Iceland Police, Reykjavík, Iceland

S. Karlsdóttir
Icelandic Meteorological Office, Reykjavík, Iceland

M.T. Gudmundsson · G. Gísladóttir
Nordvulk, Institute of Earth Sciences, University of Iceland, Reykjavík, Iceland

Advs in Volcanology (2018) 271–288
https://doi.org/10.1007/11157_2017_6

monitoring, health, safety and livestock handling provided advice. These face-to-face meetings with local residents were critical as ash fall had not affected these areas for over 60 years and plans for dealing with this hazard were not established. This chapter explores these events and in doing so, provides a narrative of crisis coordination and communication in Iceland. The narrative is based on multiple sources, including an analysis of community perspectives of the emergency response and their use and views of the various forms of communication platforms. The chapter also considers the eruptions' impacts at the local level. This exploration reveals that the trust developed through close communication between all involved prior to and during the eruption increased the effectiveness of crisis communication. The experience gained from the Eyjafjallajökull eruption is important for volcanic crisis communication at a local and international level. While the immediate evacuation plans were effective, the ash fall problems illustrated the need for necessary precautions and broadly defined preparedness strategies.

1   Introduction

The Eyjafjallajökull volcano (Fig. 1), which is overlain by a 200 m thick ice-cap bearing the same name, has produced three eruptions since the tenth century: in 1612, from 1821 to 1823, and the recent 2010 events. Past eruptions have produced very fined-grained ash deposits typically found within a 10 km radius from the Eyjafjallajökull crater (Larsen et al. 1999) and only small to medium (3000–30,000 m^3s^{-1}) glacial outburst floods (jökulhlaups) (Guðmundsson et al. 2005).

Initial volcanic risk management plans, developed from as early as 1973 for southern Iceland, did not include response to an Eyjafjallajökull eruption. These plans have, however, undergone revisions since 2002 to rectify this omission due to the realised threat evidenced by continuing magma intrusions in Eyjafjallajökull. The most dangerous aspect of a sub-glacial eruption, based on historic eruptions not least from Katla the volcano underlying Mýrdalsjökull glacier, are jökulhlaups, i.e. massive glacial outburst floods carrying volcanic debris and ice as sub-glacial eruptions melt through the glacier ice. Efforts made in 2003–2006 were therefore

aimed at understanding and mitigating the risk from eruptions in both Mýrdalsjökull (Katla) and Eyjafjallajökull, in relation to reducing the likelihood of accidents and fatalities to jökulhlaups (Guðmundsson and Gylfason 2005). A high degree of volcanic risk awareness leading up to the 2010 eruption and the semi-established communication lines between key stakeholders were partly the outcome of this work.

As the work, however, had focussed on this most dangerous and life-threatening aspect, relatively little attention had been paid to the societal and health effects of weeks of repeated ash fallout on people and the agricultural industry, which is a critical component of the region's economy (Bird and Gísladóttir 2012). Moreover, as the plans were aimed at mitigating jökulhlaup risk, they were developed for a local response only; they did not consider international impacts of an ongoing eruption or a response to deal with the overwhelming international interest in the event (Heiðarsson et al. 2014).

Understandably, local agencies faced significant challenges responding to the 2010 eruptions. Hence, the purpose of this chapter is to provide a narrative of crisis coordination and communication in Iceland through the lens of the

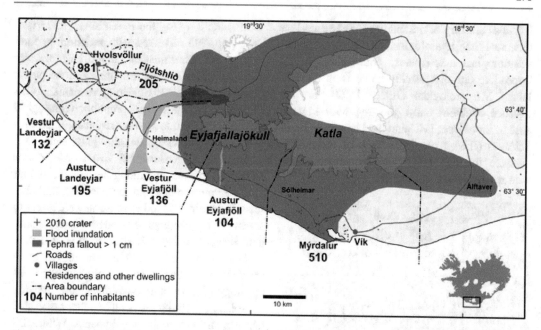

Fig. 1 Eyjafjallajökull, Katla and the surrounding districts. The areas impacted by jökulhlaup (flood) inundation and where cumulated ash fall exceeded 1 cm, are shown. Tephra fallout and flood inundation data derived from Gudmundsson et al. (2012) and Þorkelsson (2012). Population data was sourced from Ísólfur Gylfi Pálmason, the head of the Rangárþing eystra municipality, and Statistics Iceland (2016)

Eyjafjallajökull eruptions. This narrative is developed using multiple sources, including an analysis of community perspectives of the emergency response and their use and views of the various forms of crisis communication. The chapter also considers the eruptions' impacts at the local level.

To provide context, the chapter first describes the roles and responsibilities of local, key agencies involved in civil protection and emergency management in Iceland. This is followed by a description of contributing resources and data collection methods used in this chapter.

2 Civil Protection and Emergency Management in Iceland

The Minister of the Interior is head of civil protection and emergency management in Iceland with the National Commissioner of Police (NCIP) responsible for all issues at the national level (Fig. 2). Sitting within the NCIP is the Department of Civil Protection and Emergency Management (DCPEM), which is responsible for all daily matters including crisis communication, emergency coordination efforts and disaster recovery in relation to all hazards affecting the nation (Almannavarnir 2016a). At the local level, Chiefs of Police (Municipal Authorities) and Civil Protection Committees are responsible for civil protection activities and emergency response plans within their jurisdictions (Johannesdottir 2016).

Representatives from all these levels form the Civil Protection and Security Council, which sets strategies relating to civil protection, security and emergency management for the nation for 3-years at a time. This includes developing and implementing strategies to prevent and/or mitigate physical injury, impacts on public health and damage to the environment and property, in addition to providing emergency relief and assistance (Johannesdottir 2016).

The Civil Protection Act No. 82, 2008 states that NCIP is responsible for assigning alert levels (Text box 1), in collaboration with the relevant Chief of Police, for all natural and man-made hazards (Ministry of the Interior 2008). Based on their real-time hydrological, meteorological and seismological assessments, notifications of an impending eruption usually come from the Icelandic Meteorological Office (IMO). In this instance, DCPEM call a meeting with experts, such as those from IMO and the Institute of Earth Sciences (IES) at the University of Iceland, to evaluate the risk and make recommendations with respect to the alert levels.

Text Box 1: Alert Level Phases, as Assigned by the National Commissioner of the Icelandic Police (Almannavarnir 2016b)

Uncertainty Phase (Óvissustig):
Uncertainty phase/level is characterized by an event which has already started and could lead to a threat to people, communities or the environment. At this stage the collaboration and coordination between the Civil Protection Authorities and stakeholders begins. Monitoring, assessment, research and evaluation of the situation is increased. The event is defined and a hazard assessment is conducted regularly.

Alert Phase (Hættustig):
If a hazard assessment indicates increased threat, immediate measures must be taken to ensure the safety and security of those who are exposed/in the area. This is done by increasing preparedness of the emergency- and security services in the area and by taking preventive measures, such as restrictions, closures, evacuations and relocation of inhabitants. This level is also characterized by public information, advice and warning messages.

Emergency Phase (Neyðarstig):
Emergency phase is characterized by an event which has already begun and could lead, or already has led to, harm to people, communities, properties or the environment. At this stage, immediate measures are taken to ensure security, save lives and prevent casualties, damage and or loss.

In addition to their advisory role with DCPEM, IMO is tasked with monitoring, forecasting and disseminating natural hazard warnings to aviation service providers and the public (Karlsdóttir et al. 2010; Vogfjörd et al. 2005). During the Eyjafjallajökull eruption: hydrological sensors were used to monitor river runoff in terms of chemical composition and jökulhlaup risk; meteorological sensors and visual observations were used to assess lightning hazards, behaviour of the eruption cloud and localised ash fall; and, seismic, strain and GPS sensors were used to assess the geophysical components (Gudmundsson et al. 2010; Karlsdóttir et al. 2010).

The main role of IES is research in earth sciences, including a strong emphasis on volcanology. The institute undertakes core research in volcanic activity as well as the associated hazards and environmental impacts. While IES research includes monitoring via GPS measurements, INSAR, glacier surface surveying and seismic measurements, real-time monitoring remains the responsibility of IMO. As such, IES does not have any statutory obligations with respect to the monitoring and communication of volcanic activity. However, IES is called upon to provide advice to DCPEM and other government agencies prior to and during times of volcanic crises (Þorkelsson 2012).

Also sitting under the NCIP and managed by DCPEM is the National Crisis Coordination Centre (NCCC), comprising staff from NCIP, Emergency Call Centre 112, Icelandic Coast Guard, Icelandic Red Cross, National Health Care System, rescue teams (ICE-SAR), ISAVIA (national airport and air navigation service provider of Iceland) and others (Heiðarsson et al. 2015). Based in Reykjavik, NCCC is responsible for coordinating a national response when the event affects several civil protection districts across the nation.

3 Methods

Multiple sources of data were used to develop this narrative of crisis coordination and communication during the 2010 Eyjafjallajökull

Fig. 2 Iceland's current Civil Protection Structure, as of November 2016 (from Almannavarnir 2016a)

eruptions. Firstly, published reports and academic articles were critically reviewed with reference to the events as they unfolded, with a particular focus on the activities, challenges and achievements of DCPEM, IMO and IES at the local level. Secondly, personal experience of the authors (Jóhannesdóttir, Reynisson, Karlsdóttir, Gudmundsson) who were heavily involved in response and recovery efforts through their positions within DCPEM, IMO and IES provided added detail to that generated from the published reports and academic articles. Thirdly, survey data collected by Gísladóttir and Bird in August 2010 was used to enhance understanding from a community perspective.

The survey incorporated 15 semi-structured interviews with officials and residents alongside a questionnaire disseminated to households living within the vicinity of the Eyjafjallajökull and Katla volcanoes. Out of 61 households, 58 completed the survey giving a response rate of 95%. This included 19 households from Vestur-Eyjafjöll, 26 from Austur-Eyjafjöll, seven from Sólheimar and six from Álftaver (see Fig. 1), covering approximately 141 adults and 38 children. These communities were targeted due to their exposure to volcanic hazards (ash, debris-flows, jökulhlaup, lahars, etc.) during the 2010 eruptions. Every permanent household exposed to volcanic hazards during the 2010 Eyjafjallajökull eruptions was approached in these communities.

The questionnaire and semi-structured interviews asked respondents to detail their: personal experience prior to and during 20 March and 14 April eruptions; affects of the eruptions on individuals, family, property and businesses (agriculture/tourism); and, use of various media sources to access information on Eyjafjallajökull and Katla. Katla was included here as earlier response plans were centred on a Katla eruption rather than a response to an Eyjafjallajökull eruption.

It is beyond the scope of this chapter to present all survey data (some results have been presented by Bird et al. 2011, Bird and Gísladóttir 2012). However, some pertinent points are raised here as they directly relate to crisis coordination and communication during the 2010 events. These points, identified as 'survey' data, are interspersed throughout the following sections. All published reports and academic articles are cited accordingly.

Bringing these sources of information together, the following sections detail the events as they unfolded, with a particular focus on the activities, challenges and achievements of the key agencies and local residents' views of and responses to crisis coordination and communication.

4 Crisis Coordination and Communication During the Eyjafjallajökull Eruptions

4.1 20 March 2010 Events

In light of increased seismic rates in and around Eyjafjallajökull at the start of 2010 (Gudmundsson et al. 2010), the regional Chief of Police and the DCPEM organised emergency management meetings with scientists, local police and rescue teams. These meetings began in February 2010 and included 10 community meetings with residents living in the expected hazard zone around Eyjafjallajökull up to one week prior to 20 March 2010 flank eruption. Evacuation plans in case of jökulhlaup from an Eyjafjallajökull eruption were finalised during this period (Bird et al. 2011).

The initial stages of 20 March 2010 eruption were of very modest magnitude and despite comprehensive monitoring systems, it was first observed that evening by farmers who reported "a fire on top of the mountain" to local police (Bird et al. 2011). Hence, it is not surprising that the survey revealed the majority (50%) of respondents heard about the commencement of the eruption via a family member, friend or neighbour. A further 17% heard from DCPEM while 14% heard of the news over the radio.

As observer reports were received of the initial eruption on 20 March 2010, DCPEM and other authorities began rapid response efforts while IMO and IES scientists monitored real-time data to assess the situation (Þorkelsson 2012). The pre-defined plans for evacuation, based on the hazard assessment from 2005 (see Guðmundsson and Gylfason 2005) were implemented for the first time in the early hours of 21 March. The first evacuation orders were disseminated via an automated phone alert system, supervised by the local Chief of Police (Gudmundsson et al. 2010). This was critical, given that a jökulhlaup from an Eyjafjallajökull eruption has the potential to impact inhabited areas within 1 h (Sigurðsson et al. 2011).

However, some survey respondents indicated that they did not receive the SMS or call to evacuate while others who did receive it chose to ignore it.

> We did not evacuate our home, in these farms here, because there was no risk that the flood would reach the farms, but there was a possibility that we would be cut off. Everyone in this home was pleased with the decision, because it is better to be cut off being at home than to evacuate and unable to return home and attend the livestock.

> …we did not receive any message, neither on the landline nor to the GSM phone. Everyone should have received a message that an eruption had started but we did not receive any for the Fimmvörðuháls eruption or the Eyjafjallajökull eruption.

Nevertheless, approximately 700 residents conformed during the early hours of 21 March and in general, survey respondents were more positive than negative about the coordination of the evacuation (Fig. 3).

The main criticism towards the coordination was in relation to lack of planning, with some residents stating that they were confused about who was to go where.

Fig. 3 Respondents' feelings towards the management of the evacuations in relation to the 20 March and 14 April eruptions

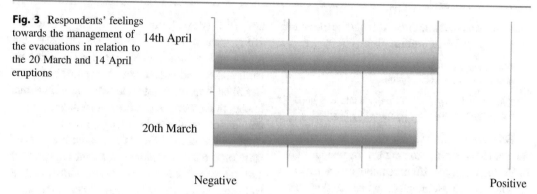

In our area the planning for the evacuation and the evacuation centre was not good, it should have been planned better.

As soon as the eruption site was located in the early morning on the flank of Eyjafjallajökull and not under the ice-cap, residents were allowed to return to their homes.

Although this eruption was relatively small, its impressiveness attracted thousands of onlookers that hiked the 16 km one-way trek with over 1000 m elevation to witness the eruption (Þorkelsson 2012). Continuing for 24 days, this 'tourist eruption' produced spectacular fire fountains and lava flows. As reported by survey interviewees and Þorkelsson (2012), the main task of DCPEM, the police and rescue teams during this time was managing crowd control on site; a task that was made all the more difficult due to the extreme conditions and often very ill-prepared onlookers who lacked suitable clothing for the cold, were exhausted, and sometimes needed assistance with broken down vehicles. Tragically, two people lost their lives while trying to access a suitable viewing point from the north of Eyjafjallajökull. Although tragic, it is surprising that these were the only fatalities considering the adverse conditions and thousands of keen spectators trying to access the eruption site (Donovan and Oppenheimer 2010).

4.2 14 April 2010 Events

On 14 April 2010, after 1–2 days lull in activity, the explosive summit eruption began in the ice-filled caldera of Eyjafjallajökull (Þorkelsson 2012), creating with it new vents under the ice-cap (Sigmundsson et al. 2010). Preparation done in the years prior to the eruption proved crucial to the execution of monitoring, in particular relating to the life-threatening flood hazard. Direct and rapid communication between monitoring scientists at IMO and IES with DCPEM and local police prompted response such as road closures and evacuations in line with the hazard assessment-based pre-defined plans.

Evacuation orders were again disseminated via the automated phone alert system with approximately 800 residents complying before daybreak on 14 April (Gudmundsson et al. 2010). The need for evacuating the region was realised with the impact of jökulhlaups charged with volcanic debris and icebergs, travelling at very high velocities (<20 km/h) (Sigurðsson et al. 2011). These floods destroyed farmlands and caused some damage to roads and infrastructure but no lives were lost (Gislason et al. 2011).

In comparison to the 20 March eruption, the greatest proportion of survey respondents indicated they heard about the possibility of Eyjafjallajökull erupting for a second time from the Chief of Police/police (40%) but received news of the start of the 14 April eruption over the radio (36%). A further 21% stated they heard about the commencement of the eruption via a family member, friend or neighbour. Survey respondents were more positive about the coordination of the evacuations (see Fig. 3), stating that the planning was much improved from the 20 March eruption with better information.

> The evacuation went better because of the experience from previous eruption.

> You never know what will happen and it is good that people are together if something happens.

However, criticism still ensued in relation to the choice of evacuation centre for people living in Vestur-Eyjafjöll. Here, residents were required to evacuate to Heimaland, which is low-lying and considered to be vulnerable to jökulhlaup.

> The planning was in general OK especially for Landeyjar. But the evacuation for Vestur-Eyjafjöll was not OK. It was a terrible mistake to evacuate to Heimaland, that place is not safe.

The potentially life threatening flood hazard subsided after 4–5 days. After that, the ash hazard and possible changes to the activity became more central to emergency response efforts. The eruption was producing very fine-grained ash ejected almost 10 km into the atmosphere (Gudmundsson et al. 2010) with extremely sharp and hard particles, justifying grave concerns for aircraft, as the ash had the ability to cause window and body abrasions as well as melt jet engines (Gislason et al. 2011). The summit eruption also produced lightning with a total of 790 strikes detected (Bennett et al. 2010), gas emissions, heavy sound blasts and lava flows reaching over 3 km down the slopes on the north side (Gudmundsson et al. 2012; Veðurstofa Íslands 2010). Linked to the thick ash layers deposited on the glacier and its foot hills, post-eruptive lahars and debris-flows repeatedly impacted rivers to the south of the crater (Jensen et al. 2013).

Already established communication links, which had been tested through regular exercises with London Volcanic Ash Advisory Centre (VAAC), ISAVIA, EUROCONTROL (European Organisation for the Safety of Air Navigation) and Toulouse VAAC, ensured IMO were better prepared to effectively communicate plume information to relevant international stakeholders. Nonetheless, improvements were made, including 3-hourly reports detailing plume activity for the use of international institutes and organisations and, joint daily reports from IMO

and IES (Karlsdóttir et al. 2010; Þorkelsson 2012), which served the use of DCPEM, the media and general public (see http://en.vedur.is/earthquakes-and-volcanism/articles/nr/1884).

Other improvements included modifications to the IMO website, in order to enhance communication with the broader population. Launched on the second day of the summit eruption, additional pages containing relevant background and overview information were added to the website, alongside IMOs real-time monitoring data (Þorkelsson 2012). These were done 'in order to achieve the goal of being more flexible and communicative to the public, so guaranteeing its considerable educational value and ensuring the public trust in the IMO services' (Heiðarsson et al. 2014, p. 62).

Similarly, IES developed a designated webpage for the Eyjafjallajökull eruption as part of their website (see http://earthice.hi.is/eruption_eyjafjallajokull_2010). Here, the general public along with government officials and agencies were able to access timely data including status reports, satellite images and maps, GPS time series data, chemical composition analyses of rocks and ash, grain size distribution of ash, photos of the eruption, lahar reports and related academic publications. IES considered it essential to make this unpublished primary scientific data open to the public as soon as it was available, due to the international extent of Eyjafjallajökull's impacts (Þorkelsson 2012).

In addition to sharing information via their websites and through frequent meetings, IMO, IES and other institutions, e.g. DCPEM, ISAVIA, the Environmental Agency, Health Authorities, and ministries and embassies, communicated with external agencies and institutes via informal means such as phone conversations and emails (Þorkelsson 2012).

Based on the joint daily reports from IMO and IES, the NCCC media team produced daily (and sometimes multiple daily) reports in Icelandic and English on the eruption and conditions across the nation (Þorkelsson 2012). The first daily report was disseminated on 14 April 2010, with a standardised form of the joint reports adopted some days later, and continuing every day during the summit eruption, after which the less frequently issued reports continued until 9 December 2010 (see http://www.sst.is/displayer.asp?cat_id=413).

In addition to posting on the DCPEM website, the reports were disseminated to government cabinets and ministries, foreign embassies located in Iceland, NGOs including the tourism sector and the Icelandic Red Cross, and media (Þorkelsson 2012). As well as detailing the physical status of the eruption, the reports included impacts on the local environment, impacts on the Icelandic population, response measures being implemented and recommendations for effected populations to consider. These reports were openly discussed with local residents at a Temporary Service Centre established in Heimaland to deal with the crisis, which is in close proximity to Eyjafjallajökull. Residents were encouraged to drop into the Heimaland Temporary Service Centre to attend meetings, ask for information or be provided a meal.

Some survey respondents and interviewees reported that these services were very helpful and trustworthy, while others who did not attend any meetings believed they had missed out on accessing some critical information. A branch of the Temporary Service Centre was later established in Vík in Mýrdal for the population east of the eruption since the main centre in Heimaland was too far for many residents to travel, especially in relation to the ongoing ash fall and resuspension. The DCPEM in conjunction with IMO and IES held meetings critical to local residents in Vík, as well as Heimaland, where concerned citizens were given the opportunity to discuss their worries associated with increasing activity.

Despite the plans and strategies that had been developed to deal with volcanic crises in Iceland, all agencies and organisations involved in disaster risk reduction were faced with an event that they had not previously experienced. Ad hoc procedures were therefore added to the processes already in place in order to deal with unanticipated events as they occurred. One resident stated "[The regional police chief] has done some

very good work in trying to make better plans with the residents." Part of the developing plans included detailed site visits and close collaborations with the tourism industry.

In consideration of the impact the continuing eruption was having on the tourism industry, a special response team was established. Led by the Ministry of Industry and Tourism (now the Ministry of the Interior), this team met every morning during the eruption, and included people from the DCPEM, IMO, IES, tourism operators, airlines and public relations people from the ministries and municipalities. This group played an active role in crisis communication with the responsibility of disseminating information to tourists stranded in Iceland and offering alternative activities to them while they were waiting for flights out of the country.

The role of this group continued well after the eruption was over, with responsibility shifting from crisis communication to the promotion of Iceland through an advertisement campaign entitled 'Inspired by Iceland' in an effort to attract tourists back in the wake of the Eyjafjallajökull eruption. The special response team has also been active during other volcanic crises such as the 2011 Grímsvötn and 2014–15 Bárðarbunga-Holuhraun eruptions.

4.3 Impact on Local Residents

While the evacuation was effective in preventing loss of life and serious injury due to jökulhlaup, there were no immediate plans in place with respect to mitigating the impacts of ash (Bird and Gísladóttir 2012; Heiðarsson et al. 2014). And of all the various volcanogenic hazards impacting local communities, survey respondents and interviewees declared the ash caused the greatest concern. A considerable amount of ash was ejected during the eruption, with large amounts deposited and resuspended causing high levels of localised pollution (Gudmundsson et al. 2012; Thorsteinsson et al. 2012). Many survey respondents and interviewees commented on the resuspended ash, with one stating:

The impacts of the eruption are significant on my farm and at my neighbouring farms. It is very misleading and wrong description that the impacts were confined to Eyjafjöll...We still live with ash storm when the wind is blowing.

Air quality in Iceland, measured against the most commonly used health limit, was exceeded by orders of magnitude in local farming communities and villages during and after the Eyjafjallajökull eruption (Thorsteinsson et al. 2012). The health concerns among local residents (Bird and Gísladóttir 2012) were therefore justified, with resuspended ash proving to be of equal importance to that emitted from the volcano (Thorsteinsson et al. 2012). A study conducted by Carlsen et al. (2012) revealed that residents living in affected communities presented a highly increased prevalence of physical health conditions including tightness in the chest, coughing, phlegm, eye irritations and other respiratory problems.

Gissurardóttir (2015) found that residents with direct experience of and exposure to the Eyjafjallajökull eruption were also at greater risk of psychological morbidity in relation to mental distress and post-traumatic stress disorder. Survey respondents and interviewees noted the ongoing psychological impacts of the eruption and lack of crisis communication regarding health issues:

The ash seems to impact people a lot. They get claustrophobic and become confused and many are not able to make logical decisions.

The individual authorities' approach, such as those concerning health of residents, is reprehensible. Not enough consultation. No understanding on the psychological state of people [who were living under] ash fall and later ash storms for months. During the eruption, the area should have been evacuated – because it was sometimes uninhabitable there – but all tried to survive – many in a state of shock.

Supporting these comments, Gissurardóttir (2015) revealed the following factors as causes of psychological distress:

- Damage to personal property
- Feelings of insecurity during the eruption

- Being required to use protective equipment when working outside during the eruption
- Spending time outdoors in ash fall during the eruption due to work commitments or other duties
- Living in view of the eruption site

The impacted region was, and still is, an important agricultural region, with 15% of all cattle, 6% of all sheep, 17% of all horses and 12% of all dairy production in Iceland in 2010 (Farmers Association of Iceland 2010). Considering that animals are at a great risk of short and long-term mortality when they inhale or ingest fine ash particles (Lebon 2009; Wilson 2009; Wilson et al. 2011), it was understandable that the agricultural industry (Farmers Association of Iceland 2010) and local residents were gravely concerned about the health and wellbeing of their livestock (Bird and Gísladóttir 2012). Moreover, survey interviewees revealed that post-eruptive lahars and debris-flows also caused significant concern to farmers living south of the crater, where sediments were destroying infrastructure and agricultural land.

Thus, accurate, detailed and timely crisis communication was imperative in the lead up to and during the ongoing eruption, not only in relation to jökulhlaup risk but also covering the broad spectrum of impacts and well into the recovery phase.

4.4 Demand for Accurate and Timely Information

The demand for information during the Eyjafjallajökull eruption was unprecedented in Iceland, with communication and media relations suddenly evolving into a major component of the crisis operation (Guðmundsdóttir 2016). Without a doubt, the Eyjafjallajökull eruption dominated international media with the closure of transatlantic and European airspace causing tens of thousands of flights to be cancelled (Harris et al. 2012).

At the local level, the survey data shows that residents were actively using the IMO (Fig. 4)

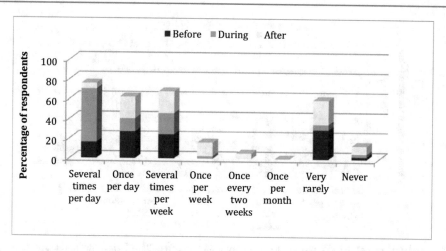

Fig. 4 Respondents' use of the IMO website before, during and after the 2010 Eyjafjallajökull eruptions

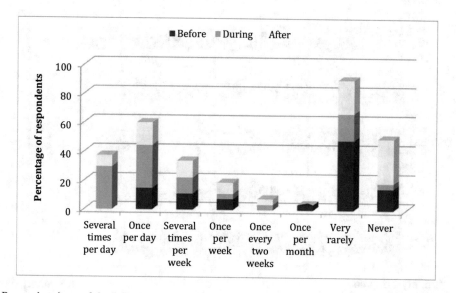

Fig. 5 Respondents' use of the DCPEM website before, during and after the 2010 Eyjafjallajökull eruptions

and DCPEM (Fig. 5) websites up to several times per day during the Eyjafjallajökull eruption. The IES website, which was mostly in English but not streaming real-time data, was not used as actively by survey respondents (Fig. 6); however, it was considered a valuable resource at an international level. In the initial stages, these agencies were not prepared to meet this demand (Heiðarsson et al. 2014).

As Þorkelsson (2012) points out, IMO did not and still does not have a designated press office

that can deal with a huge demand for crisis information yet they were faced with an enormous influx of requests, fielding calls from around 100 international reporters on just day two of the summit eruption. IES and DCPEM designated around 5–7 fulltime staff to work exclusively on dealing with the demand from international media (Þorkelsson 2012).

Due to the demand for information, two media centres were opened under the supervision of DCPEM—one at NCCC in Reykjavík and a

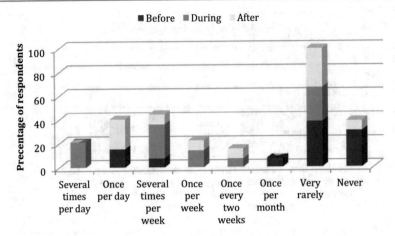

Fig. 6 Respondents' use of the IES website before, during and after the 2010 Eyjafjallajökull eruptions

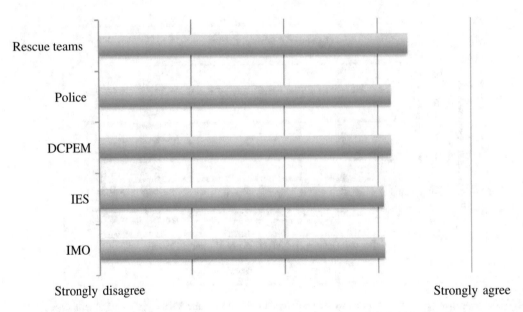

Fig. 7 Respondents views towards the statements 'information released by IMO/IES/DCPEM/police/rescue teams is reliable'

Local Crisis Coordination Centre in Hvolsvöllur (Þorkelsson 2012), which is the closest township to Eyjafjallajökull. With an overall aim to provide information with 'one voice', the teams based at each centre consisted of experts from DCPEM, ISAVIA, the Icelandic Coast Guard, Red Cross, the Icelandic Transport Authority, the Icelandic Ministry for the Environment (currently entitled the Icelandic Ministry for the Environment and Natural Resources), the Icelandic Road and Coastal Administration, ICE-SAR and others. Overall, survey respondents agreed that the information provided by the various agencies was reliable (Fig. 7).

Press conferences were held at both locations to help deal with the enormous interest in the event (Guðmundsdóttir 2016). In addition to distributing status reports, IMO and IES scientists were interviewed about facts and predictions of the continuing eruption, taking valuable time away from research and monitoring. Guðmundsdóttir (2016) reports, however, that many recognised this as an important part of their role as a scientist because media personnel will always be at the heart of the action and if an expert is not available for interview the media will seek information from anyone who is available and willing.

The National Broadcasting Service, 'RÚV' established a crisis-broadcasting studio in the NCCC media centre in Reykjavik enabling them to provide vital information to the Icelandic population (Heiðarsson et al. 2015). Other local news media, such as the 'Fréttablaðið' and 'Morgunblaðið' newspapers and the English newspaper 'The Reykjavik Grapevine' and magazine 'Iceland Review' also broadcasted news of the event via their traditional and online platforms.

Overall, the survey data suggests that residents' more frequently turned to the various media platforms for information (Fig. 8) with television, radio and Internet as the preferred platforms, respectively. However, some respondents also indicated that they accessed information from newspapers, books and information brochures. Once again, increased usage occurred during the eruption with many respondents accessing media sources several times per day (Fig. 9).

Even though survey respondents and interviewees indicated they trusted information from RÚV, they were overall critical of the media during the 2010 eruptions. When local residents were evacuated on 20 March 2010, the media were permitted to access and report from the evacuation zone. This incensed many survey respondents housed in the evacuation centres, where they watched live broadcasts from their evacuated farms, and in some instances, they hadn't had time to lock their front doors. One survey interviewee summed up many people's sentiment with the statement:

> Scientists and media people receive a pass to go into the risk area on their own responsibility. Why are they allowed to go into the risk area when others are not?

Survey interviewees also found it disturbing when the international media portrayed them as helpless victims, dramatizing the summit eruption without any consideration of how people were living their day-to-day lives through the event. For impact, the media sought farmers who were in a state of shock at the realisation of the destructiveness of the heavy ash fall and this angered neighbouring residents.

Fig. 8 Respondents usage of various sources for hazard information

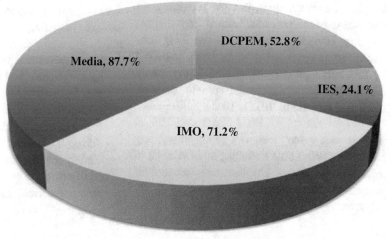

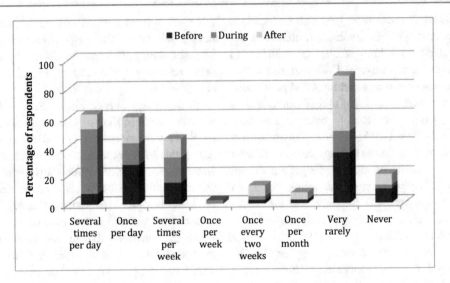

Fig. 9 Respondents' use of media before, during and after the 2010 Eyjafjallajökull eruption

I also found that the media exaggerated when they were broadcasting news from the area. They described it as such that everything here was covered in ash, and the situation in the area disastrous, but it was absolutely not so. A bit "over-dramatizing" if I use slang. And we heard about farmers that were tired of the continuous pressure from the media.

[The media] said that the situation was more severe than it was. The situation was worse in a specific area but not elsewhere. And sometimes the news was erroneous. Media people were too aggressive to the residents, soon after the eruption started.

Residents were also frustrated with the media stating that the Eyjafjallajökull eruptions were "a show and nothing in comparison to what Katla can do". Residents living in close proximity are well aware of the potential hazards from a Katla eruption. However, this was the last thing they wanted to be reminded of while being impacted by ash fall, lightning, lahars and heavy sound blasts from Eyjafjallajökull's summit eruption (Bird and Gísladóttir 2012). To the detriment of the overall aim of DCPEM, IMO and IES to provide 'one voice', Iceland's president publicly stated: "you ain't seen nothing yet" in relation to a Katla eruption (see http://news.bbc.co.uk/2/hi/programmes/newsnight/8631343.stm).

5 Beyond Eyjafjallajökull 2010—Lessons Learnt and Improvements Made in an Unpredictable, Multi-hazard Environment

The enormous workload generated by the Eyjafjallajökull eruptions cannot be denied. Considering Iceland's limited size in terms of manpower and resources, these events placed tremendous pressure on all those working in volcanic crisis management (Guðmundsdóttir 2016; Þorkelsson 2012). Nevertheless, officials and scientists involved in the response made some great achievements under the circumstances, including:

- Pre-eruption risk communication, in particular at the town hall meetings, enabled trust to develop between emergency management officials and the general public, which increased the effectiveness of communication during the crisis.
- Timely evacuation advice and pre-existing trust led to the smooth evacuation of vulnerable populations during the summit eruption.

- Ad hoc modifications to the IMO, IES and DCPEM websites enhanced the availability of information to key stakeholders including the public.
- Good collaborations were maintained with external partners based on already established communication links.
- Establishment of a crisis information centre enabled local residents to access more detailed information and support.
- Establishment of two media centres, with one being in close proximity to the eruption site, met the enormous demands from local and international media and thus relieved pressure on agencies and institutes involved in the response.
- IMO and IES increased the dissemination of primary scientific data on the physical status of the eruption to meet the growing demand.

However, not all of these initiatives worked well for all those involved. While it was considered essential that unpublished primary scientific data was publicly open and accessible for operational use, some foreign researchers viewed this an as opportunity to use the data for their own research agendas and publications (Þorkelsson 2012).

Furthermore, there was some criticism regarding the fact that before the 2010 events, response plans were only in place for jökulhlaup risk while other volcanogenic hazards had not been addressed accordingly (Bird and Gísladóttir 2012). During the eruption, residents were calling for more information about the ash impacts and what these meant to human and animal health in addition to advice on how to mitigate the adverse affects. The need to adopt an all-hazards approach was again apparent during the 2014–15 Bárðarbunga-Holuhraun eruption when poisonous gases were the main cause of health concerns. Since the 2014–15 Bárðarbunga-Holuhraun eruption, NCIP and DCPEM for south Iceland are engaging with local communities in the development of all-hazards crisis management plans. Such engagement invites opportunities to enhance collaboration and communication prior to unrealised events occurring.

Despite these efforts to improve communications and instil learnings internally and externally with key stakeholders (Heiðarsson et al. 2015), more work is needed at the local and national level. For example, as formal guidelines for volcanic crisis communication were not in place prior to March 2010, officials were forced to develop ad hoc strategies during the Eyjafjallajökull eruption. Despite this experience, formal guidelines have not been established.

However, Guðmundsdóttir (2016) postulates a tentative guideline for best practice volcanic crisis communication based on officials' perspectives of the 2010 Eyjafjallajökull, 2011 Grímsvötn and 2014–15 Bárðarbunga-Holuhraun eruptions. These are:

1. Integrating communication into the management process
2. Cooperating at an institutional level
3. Coordinating messages
4. Providing truthful, honest and transparent information
5. Communicating in a proactive way
6. Being accessible and having a good relationship with the media
7. Understanding, informing and cooperating with the audience
8. Improvising if necessary
9. Planning, preparing, documenting the crisis communication

A key underlying theme here is that all officials worked together through mutual respect and understanding, so as to provide crisis communication via 'one voice'. To achieve this, Donovan and Oppenheimer (2012) highlight the importance of building and maintaining relationships well before a volcanic crisis ensues. The hazard and risk assessments, development of response plans and engagement of locals in these efforts in the years prior to the 2010 eruption are an excellent example of that approach.

Heiðarsson et al. (2014) also highlights the importance of the science sector providing clear and consistent messaging that is backed up by all government and non-government stakeholders. This is one of many key lessons that have been documented and investigated through the FUTUREVOLC project. FUTUREVOLC (http://futurevolc.hi.is/), financed by the European Union's Seventh Programme for Research, Technological Development and Demonstration under grant agreement No 308377, was established following the Eyjafjallajökull eruption and has continued while other events and volcanic crises have taken place, e.g. seismic unrest in Bárðarbunga throughout 2014 and the 2014–15 Bárðarbunga-Holuhraun eruption (Heiðarsson et al. 2015).

After the Eyjafjallajökull eruption, IMO and IES began collaborating on a catalogue of Icelandic volcanoes, financed by ICAO. The work was then linked to the FUTUREVOLC project. The aim was to gather all the information electronically in an online catalogue, now available at (http://icelandicvolcanoes.is/), to improve public and official understanding of the current state of Iceland's volcanoes. The catalogue provides descriptions of the geological and tectonic setting of each volcano along with eruption history, characteristics and associated hazards, activity status, monitoring, and possible eruption scenarios. The catalogue has a strong emphasis on ash hazards and impacts, as it is designed with the needs of aviation for rapid information at times of potential crises. It is therefore not comprehensive when it comes to other hazards; e.g. information on flood hazards or air pollution is not extensive, and at present it does not contain information on response or disaster risk reduction plans. The catalogue is written in English, somewhat reducing its potential usefulness for local populations. The aim, however, is to develop an Icelandic version and to include all information and results from the ongoing long-term volcanic risk assessment program for Iceland.

While local residents are, in general, aware of the risks and emergency response procedures associated with a volcanic eruption in southern Iceland (Bird et al. 2009, 2011; Jóhannesdóttir 2005; Jóhannesdóttir and Gísladóttir 2010), tourists are not (Bird et al. 2010; Bird and Gísladóttir 2014). Although volcanic risk awareness among tourists appeared heightened during the Eyjafjallajökull eruption, it did not result in them seeking more information on safety strategies (Lund et al. 2010).

Since the Eyjafjallajökull eruption, tourism has rapidly grown in all regions across Iceland. Tourism currently accounts for more foreign exchange income than any other industry in Iceland, with employment in tourism-related industries increasing more than in any other sector of the economy (Óladóttir 2015). This growth is expected to continue in the coming years.

Increased tourism, however, ultimately leads to increases in the number of tourists exposed to risks associated with volcanic eruptions. Initiatives like the Catalogue of Icelandic Volcanoes and the introduction of real-time alerts in English using social media and text messaging are aimed at enhancing awareness among international visitors to Iceland (Heiðarsson et al. 2015). However, there is currently little evidence as to whether or not these initiatives are successfully inciting risk reduction behaviours.

The experience gained from the Eyjafjallajökull eruption is important and the immediate crisis communication strategies were effective in promoting the successful evacuation of local residents. However, the ash fall problems illustrated the need for necessary precautions and broadly defined preparedness strategies in order to increase the resilience of affected communities. This need for resilience is not confined to the local population, but also includes the ever-increasing number of tourists who visit Iceland and who engage in activities around Iceland's active volcanoes.

Acknowledgements All interviewees and respondents are graciously thanked for their willingness to participate in the survey. Contributions from the Ash Fall Impacts Working Group, Katharine Haynes, Carolina Garcia Londoño and Guðrún Pétursdóttir are appreciated in relation to the development of the questionnaire. The two anonymous reviewers are also thanked for providing invaluable recommendations that helped improve the manuscript from the original. The preparation of this paper has been supported by the Nordic Centre of Excellence for Resilience and Societal Security—NORDRESS, which is funded by the Nordic Societal Security Programme. Þórdís Högnadóttir helped with the preparation of Fig. 1 and Ísólfur Gylfi Pálmason prepared population data for Rangárþing eystra.

References

Almannavarnir (2016a) Civil protection in Iceland. http://www.almannavarnir.is/english/about-the-depart ment-of-civil-protection-and-emergency-management/. Accessed 8 Nov 2016

Almannavarnir (2016b) Emergency response. http://www. almannavarnir.is/english/general-information/emergency-response/. Accessed 8 Nov 2016

Bennett AJ, Odams P, Edwards D, Arason Þ (2010) Monitoring of lightning from the April–May 2010 Eyjafjallajökull volcanic eruption using a very low frequency lightning location network. Environ Res Lett 5:044013

Bird D, Gísladóttir G (2012) Residents' attitudes and behaviour before and after the 2010 Eyjafjallajökull eruptions—a case study from southern Iceland. Bull Volcanol 74:1263–1279

Bird D, Gísladóttir G (2014) Southern Iceland—volcanoes, tourism and volcanic risk reduction. In: Erfurt-Cooper P (ed) Volcanic tourist destinations. Springer, Geoheritage, Geoparks and Geotourism, pp 35–46

Bird DK, Gísladóttir G, Dominey-Howes D (2009) Resident perception of volcanic hazards and evacuation procedures. Nat Hazards Earth Syst Sci 9:251–266

Bird DK, Gísladóttir G, Dominey-Howes D (2010) Volcanic risk and tourism in southern Iceland: implications for hazard, risk and emergency response education and training. J Volcanol Geoth Res 189:33–48

Bird DK, Gísladóttir G, Dominey-Howes D (2011) Different communities, different perspectives: issues affecting residents' response to a volcanic eruption in southern Iceland. Bull Volcanol 73:1209–1227

Carlsen HK, Hauksdottir A, Valdimarsdottir UA, Gíslason T, Einarsdottir G, Runolfsson H, Briem H, Finnbjornsdottir RG, Gudmundsson S, Kolbeinsson TB, Thorsteinsson T, Pétursdóttir G (2012) Health effects following the Eyjafjallajökull volcanic eruption: a cohort study. BMJ Open 2

Donovan A, Oppenheimer C (2012) Governing the lithosphere: insights from Eyjafjallajökull concerning the role of scientists in supporting decision-making on active volcanoes. J Geophys Res 117:B03214

Donovan AR, Oppenheimer C (2010) Commentary: the 2010 Eyjafjallajökull eruption and the reconstruction of geography. Geogr J. doi:10.1111/j.1475-4959. 2010.00379.x

Farmers Association of Iceland (2010) 19 April 2010—announcement from the Farmers Association of Iceland—the volcanic eruption in Iceland and its effects on Icelandic agriculture, Bændasamtök Íslands. http:// www.bondi.is/lisalib/getfile.aspx?itemid=2747. Accessed 19 June 2011

Gislason SR, Hassenkam T, Nedel S, Bovet N, Eiriksdottir ES, Alfredsson HA, Hem CP, Balogh ZI, Dideriksen K, Oskarsson N, Sigfusson B, Larsen G, Stipp SLS (2011) Characterization of Eyjafjallajökull volcanic ash particles and a protocol for rapid risk assessment. Proc Natl Acad Sci U S A 108:7307–7312

Gissurardóttir ÓS (2015) Mental health following the volcanic eruption in Eyjafjallajökull volcano in Iceland in 2010: a population-based study. Master of Public Health Sciences, Centre of Public Health, School of Health Sciences, University of Iceland. Reykjavík, p 45

Guðmundsdóttir BN (2016) Best practices in Icelandic crisis communication during volcanic eruptions: development of a tentative framework. Masters thesis in Environment and Natural Resources, Faculty of Life and Environmental Sciences, School of Engineering and Natural Sciences, University of Iceland. Reykjavik, p 65

Gudmundsson MT, Pedersen R, Vogfjörd K, Thorbjarnardóttir B, Jakobsdóttir S, Roberts MJ (2010) Eruptions of Eyjafjallajökull volcano, Iceland. EOS Trans AGU 91:190–191

Gudmundsson MT, Thordarson T, Höskuldsson Á, Larsen G, Björnsson H, Prata FJ, Oddsson B, Magnússon E, Högnadóttir T, Petersen GN, Hayward CL, Stevenson JA, Jónsdóttir I (2012) Ash generation and distribution from the April–May 2010 eruption of Eyjafjallajökull, Iceland. Sci, Rep, p 2

Guðmundsson MT, Gylfason ÁG (2005) Hættumat vegna eldgosa og hlaupa frá vestanverðum Mýrdalsjökli og Eyjafjallajökli. Ríkislögreglustjórinn and Háskólaútgáfan, Reykjavík, p 210

Guðmundsson MT, Elíasson J, Larsen G, Gylfason ÁG, Einarsson P, Jóhanesson T, Hákonardóttir KM, Torfason H (2005) Yfirlit um hættu vegna eldgosa og hlaupa frá vesturhluta Mýrdalsjökuls og Eyjafjallajökli. In: Guðmundsson MT, Gylfason ÁG (eds) Hættumat vegna eldgosa og hlaupa frá vestanverðum Mýrdalsjökli og Eyjafjallajökli. Ríkislögreglustjórinn and Háskólaútgáfan, Reykjavík, pp 11–44

Harris AJL, Gurioli L, Hughes EE, Lagreulet S (2012) Impact of the Eyjafjallajökull ash cloud: a newspaper perspective. J Geophys Res 117:B00C08

Heiðarsson EP, Loughlin SC, Witham C, Barsotti S (2014) Report on forensic analysis of the Eyjafjallajökull and Grímsvötn communication and risk management response across Europe. European volcanological supersite in Iceland: a monitoring system and network for the future. FutureVolc, Reykjavík, p 198

Heiðarsson EP, Loughlin SC, Witham C, Barsotti S (2015) D3.2—Information for EU-MIC and scenarios for major events. European volcanological supersite in Iceland: a monitoring system and network for the future. FutureVolc, Reykjavík, p 72

Jensen EH, Helgason JK, Einarsson S, Sverrisdottir G, Höskuldsson A, Oddsson B (2013) Lahar, floods and debris flows resulting from the 2010 eruption of Eyjafjallajökull: observations, mapping, and modelling. In: Margottini C, Canuti P, Sassa K (eds) Landslide science and practice, vol 3., Spatial analysis and modellingSpringer, Berlin, Heidelberg, pp 435–440

Johannesdottir G (2016) National risk assessment for Iceland. Executive summary. Department of Civil Protection and Emergency Management, National Commissioner of the Icelandic Police, Reykjavík, p 22

Jóhannesdóttir G, Gísladóttir G (2010) People living under threat of volcanic hazard in southern Iceland: vulnerability and risk perception. Nat Hazards Earth Syst Sci 10:407–420

Jóhannesdóttir G (2005) Við tölum aldrei um Kötlu hér mat íbúa á hættu vegna Kötlugoss. Department of Geology and Geography, University of Iceland, Reykjavík, p 103

Karlsdóttir S, Petersen GN, Björnsson H, Pétursson H, Þorsteinsson H, Arason Þ (2010) Eyjafjallajökull eruption 2010—the role of IMO. http://en.vedur.is/earthquakes-and-volcanism/articles/nr/2072. Accessed 17 July 2016

Larsen G, Dugmore A, Newton A (1999) Geochemistry of historical-age silicic tephras in Iceland. Holocene 9:463–471

Lebon SLG (2009) Volcanic activity and environment: Impacts on agriculture and use of geological data to improve recovery processes. Faculty of Earth Sciences, Masters Thesis in Environmental Sciences and Natural Resources Management, Earth Science Institute, University of Iceland, Reykjavik, ISBN 978-9979-9914-0-3

Lund KA, Benediktsson K, Mustonen TA (2010) The Eyjafjallajökull eruption and tourism: report from a survey in 2010. Icelandic Tourism Research Centre, Reykjavík, p 23

Ministry of the Interior (2008) Civil protection Act No. 82, 12 June 2008. https://eng.innanrikisraduneyti.is/laws-and-regulations/english/civil-protection/. Accessed 22 Sept 2016

Óladóttir OÞ (2015) Tourism in Iceland in figures. Ferðamálastofa, Reykjavík, p 27

Sigmundsson F, Hreinsdóttir S, Hooper A, Árnadóttir T, Pedersen R, Roberts MJ, Óskarsson N, Auriac A, Decriem J, Einarsson P, Geirsson H, Hensch M, Ófeigsson BG, Sturkell E, Sveinbjörnsson H, Feigl KL (2010) Intrusion triggering of the 2010 Eyjafjallajökull explosive eruption. Nature 468:426–432

Sigurðsson O, Sigurðsson G, Björnsson BB, Pagneux EP, Zóphóníasson S, Einarsson B, Þórarinsson Ó, Jóhannesson T (2011) Flood warning system and jökulhlaups—Eyjafjallajökull, Icelandic Meteorlogical Office. http://en.vedur.is/hydrology/articles/nr/2097. Accessed 23 June 2011

Thorsteinsson T, Jóhannsson T, Stohl A, Kristiansen NI (2012) High levels of particulate matter in Iceland due to direct ash emissions by the Eyjafjallajökull eruption and resuspension of deposited ash. J Geophys Res Solid Earth 117

Veðurstofa Íslands (2010) Update on activity: eruption in Eyjafjallajökull, Iceland. http://en.vedur.is/earthquakes-and-volcanism/articles/nr/1884. Accessed 13 July 2011

Vogfjörd KS, Jakobsdóttir SS, Guðmundsson GB, Roberts MJ, Ágústsson K, Arason T, Geirsson H, Karlsdóttir S, Hjaltadóttir S, Ólafsdóttir U, Thorbjarnardóttir B, Skaftadóttir T, Sturkell E, Jónasdóttir EB, Hafsteinsson G, Sveinbjörnsson H, Stefánsson R, Jónsson TV (2005) Forecasting and monitoring a subglacial eruption in Iceland. EOS Trans Am Geophys Union 86:245–252

Wilson T, Cole J, Stewart C, Cronin S, Johnston D (2011) Ash storms: impacts of wind-remobilised volcanic ash on rural communities and agriculture following the 1991 Hudson eruption, southern Patagonia, Chile. Bull Volcanol 73:223–239

Wilson TM (2009) Vulnerability of pastoral farming systems to volcanic ashfall hazards. Doctor of Philosophy in Hazard and Disaster Management, Natural Hazards Research Centre, Department of Geological Sciences, University of Canterbury, Christchurch, p 241

Þorkelsson B (2012) The 2010 Eyjafjallajökull eruption, Iceland. Report to ICAO—June 2012. Icelandic Meteorological Office; Institute of Earth Sciences, University of Iceland; The National Commissioner of the Icelandic Police, Reykjavik, p 206

Supporting the Development of Procedures for Communications During Volcanic Emergencies: Lessons Learnt from the Canary Islands (Spain) and Etna and Stromboli (Italy)

M.C. Solana, S. Calvari, C.R.J. Kilburn, H. Gutierrez, D. Chester and A. Duncan

Abstract

Volcanic crises are complex and especially challenging to manage. Volcanic unrest is characterised by uncertainty about whether an eruption will or will not take place, as well as its possible location, size and evolution. Planning is further complicated by the range of potential hazards and the variety of disciplines involved in forecasting and responding to volcanic emergencies. Effective management is favoured at frequently active volcanoes, owing to the experience gained through the repeated 'testing' of systems of communication. Even when plans have not been officially put in place, the groups involved tend to have an understanding of their roles and responsibilities and those of others. Such experience is rarely available at volcanoes that have been quiescent for several generations. Emergency responses are less effective, not only because of uncertainties about the volcanic system itself, but also because scientists, crisis directors, managers and the public are inexperienced in volcanic unrest. In such situations, tensions and misunderstandings result in poor communication and have the potential to affect decision making

M.C. Solana (✉)
School of Earth and Environmental Sciences, University of Portsmouth, Portsmouth PO1 2UP, UK
e-mail: Carmen.solana@port.ac.uk

M.C. Solana
Instituto Volcanológico de Canarias (INVOLCAN), Antiguo Hotel Taoro, Parque Taoro, 22, 38400 Puerto de La Cruz, Tenerife, Spain

S. Calvari
INGV, Piazza Roma 2, 95125 Catania, Italy

C.R.J. Kilburn
UCL Hazard Centre, Department of Earth Sciences, University College London, Gower Street, London WC1E 6BT, UK

H. Gutierrez
Servicio de Protección Civil Y Atención de Emergencias, Gobierno de Canarias, Carretera de La Esperanza Km 0,8 Edf. Ceplam, 38071 Tenerife, Spain

D. Chester
Department of Geography and Environmental Science, Liverpool Hope University, Hope Park, Liverpool L16 9JD, UK

D. Chester · A. Duncan
Department of Geography and Planning, University of Liverpool, Liverpool L69 3BX, UK

Advs in Volcanology (2018) 289–305
https://doi.org/10.1007/11157_2016_48
© The Author(s) 2017
Published Online: 29 March 2017

and delay vital operations. Here we compare experiences on communicating information during crises on volcanoes reawakening after long repose (El Hierro in the Canary Islands) and in frequent eruption (Etna and Stromboli in Sicily). The results provide a basis for enhancing communication protocols during volcanic emergencies.

Keywords

Communication · Volcano · Emergencies · Canary islands · Etna · Stromboli

1   Introduction

Volcanic crises are complex to manage, owing to uncertainty in the behaviour of volcanoes and of the people and organisations responding to an emergency. Complexity is especially acute at volcanoes showing unrest after long intervals of repose, from several decades to centuries. In such cases, the infrequency of eruptions, and consequent lack of data on the volcanic system, introduces a high level of uncertainty in forecasts of eruptions and their hazards. In addition, few if any of the responding authorities—from monitoring scientists to civil protection agencies and governmental bodies—may have had direct experience of volcanic behaviour. As a result, prepared response plans are either non-existent or tend to be based on generic procedures designed to cover the legal requirements established by the host country for such contingencies. The different authorities may also have been brought together for the first time to address an emergency. In combination, inexperience in the scientific and managerial aspects of a crisis, the absence of a specific response plan and the lack of previous interaction between scientists and emergency managers (and even between scientists of different disciplines) can produce levels of tension among personnel that impair communications between key responders and the quality and timing of the decisions being made (Fiske 1984; Voight 1988a, 1990; Aspinall et al. 2002; Solana and Spiller 2007; Solana et al. 2008; Barclay et al. 2008; McGuire et al. 2009).

The notion to establish professional guidelines for responding to volcanic emergencies has been discussed since at least the 1970s (Tazieff 1977; Bostok 1978; Sigvaldason 1978; Barberi and Gasparini 1979; Fiske 1979; Tomblin 1979). Two decades later, the International Association of Volcanology and Chemistry of the Earth's Interior (IAVCEI) published a set of recommendations for the conduct of scientists during volcanic crises (IAVCEI et al. 1999). The recommendations emphasized that emergency responses are optimized by effective teamwork and the presentation by scientists to non-specialists of a unified and objective evaluation of volcanic unrest and its possible outcomes. Although appealing in theory, the recommendations are not binding and have been followed only erratically in practice. An example is the response to the 2011–1012 eruption of El Hierro, in the Canary Islands, which occurred after more than 200 years of repose. Following the El Hierro crisis, the Spanish Civil Protection agency identified the need to develop protocols to enable better communication of information between scientists and with emergency responders and local governmental agencies and argued that these should be integrated into regional emergency planning. This paper presents the experience gained during the crisis and describes how protocols have been designed and the factors that have hindered their development. It also highlights good practice and the importance of building on the experience gained from communicating information about eruptions at frequently active volcanoes, using emergencies from Etna and Stromboli.

2 Communications During a Volcanic Crisis

Good communication plays a key role in managing a volcanic emergency effectively (Peterson 1988; Tilling 1989; Solana and Spiller 2007; Solana et al. 2008; McGuire et al. 2009; Doyle et al. 2011). Three common obstacles to good communication are: (1) differences in the organisational cultures with which responders are familiar (2) uncertainty in forecasting volcanic behaviour, and (3) inexperience of addressing volcanic unrest. The potential of these factors to diminish efficient communication is rarely appreciated before an emergency begins and this, in turn, favours poor decisions being made under the conditions of high stress during volcanic unrest (McGuire et al. 2009).

2.1 Organisational Cultures

Emergency responses normally require collaboration between academic scientists and the civil authorities. These two groups work in institutions with contrasting organisational cultures. As identified by Handy (1978), academics, scientists and researchers have primarily individualistic personalities. Their association in groups (e.g., universities, research centres or teams) is commonly for personal convenience and to facilitate the advance of individuals in their field of study. Academics do not "willingly take orders [...] or compromise on their own plans" (Handy 1978, p. 39) and are rarely forced to do so by their organisations (Handy 1978). As a result, academic decisions on, for example, the amount and type of information to be communicated during an emergency, normally have to be agreed on an individual basis. This process, although democratic, makes management and decision making through consensus very difficult and time consuming. It is not surprising, therefore, that the key recommendations of IAVCEI et al. (1999)—to value different expertises and approaches equally, to share

information and logistical resources and to work as a team that speaks with a single voice and rewards self-sacrifice—are not always adopted by individuals during emergencies, especially those without previous experience of a volcanic crisis. This was illustrated by an informal survey in 2013 among participants of an international project on volcanic unrest in Europe and Latin-America (Fig. 1). The 21 interviewees represented 12 countries and scientists with different levels of seniority, 15 of which had experience responding or being involved directly or indirectly in 3 or more volcanic crisis. Of these 21 experts, 13 recognised personal ego and individual interests as the main barriers to communications between scientists during volcanic emergencies (Fig. 1).

The individualistic attitudes of many academic scientists contrast sharply with the culture of the civil authorities who are in charge of a crisis. Emergency managers are typically civil servants who work in a hierarchy designed to help their organisation achieve its goals. Individual ideas are rarely expressed except as part of agreed policy (Handy 1978). Managers are thus used to responding to information that has already been agreed by expert advisers and are usually reluctant to engage in evaluating diverging opinions or uncertainty in forecasts of a volcano's behaviour (Solana et al. 2008). Academic debate may therefore be perceived as indecision among experts and so raise questions about the quality of scientific advice being received. As a result, essential information from the advising scientists may not be communicated effectively to the emergency managers.

2.2 Uncertainty

Expressions of uncertainty may be perceived as indecision and so hinder the communication of scientific information. A common example is the delivery of eruption forecasts. Forecasts contain numerous sources of uncertainty because of the natural variations in the behaviour of magmatic

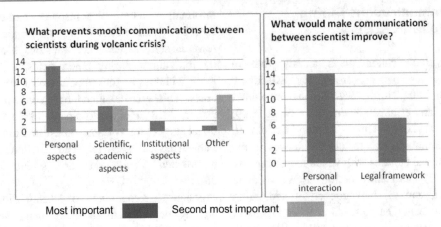

Fig. 1 Results from an informal survey of 21 volcanologists during the "Scientific advice, decision-making and risk communication" Vuelco meeting in 2013 (Solana and Fearnley 2013). *Left*, "Personal aspects" include personality, ego, personal interest, visibility, status. "Scientific/academic aspects" include ownership of data, publishing pressure, career progression. "Institutional aspects" include legal frameworks of institutions, institutional status and agreed responsibilities. "Others" include lack of experience, uncertainty, culture, discipline rivalry. *Right*, Aspects of personal interaction include frequent face to face contact, joint experience in crisis, moderator/expert elicitation methods, grouping all teams together. Legal framework aspects include legalised protocols or rules, pre-agreements for data sharing

systems Marzocchi et al. (2012) and incomplete understanding of a specific volcano (e.g. Doyle et al. 2014) or incomplete data sets from research monitoring networks, especially emergency networks installed after the first signs of unrest. Even when appropriate monitoring is in place, uncertainty remains as to whether unrest will lead to an eruption or intrusion, because both are associated with similar changes in precursory signal (Tilling 1989). Such uncertainty is compounded when monitoring data are contradictory (e.g., increases in geochemical indicators which do not correlate with seismic or deformation data) and can lead to critical delays in forecasting and decision making with fatal consequences, as seen in Mt. Ontake in Japan, where a sudden eruption in September 27, 2014 surprised scientists monitoring the volcano and killed 23 people (Oskin 2014). Eruptions may also occur without detected precursory unrest, such as the so-called "silent" 2004 eruption at Mount Etna in Sicily (Burton et al. 2005) and the "passive" eruptions described for Kilauea in Hawaii (Bell and Kilburn 2011).

To aid the communication of uncertainty, several methodologies have been developed for evaluating the probabilities of an eruption, including Bayesian statistics and expert elicitation (e.g. Newhall and Hoblitt 2002; Aspinall et al. 2003; Marti et al. 2008; Marzocchi et al. 2008; Marzocchi and Bebbington 2012; Sobradelo et al. 2015), as well as the incorporation of deterministic models (Voight 1988b; Kilburn and Voight 1998; De La Cruz-Reyna and Reyes-Davila 2001; Kilburn 2003, 2012). On their own, however, probabilities are open to different interpretations by scientists and civil authorities and agreement on its meaning might not be reached.

As discussed by Doyle et al. (2011), communicating information does not necessarily have to imply that a consensus has been reached or that uncertainty is lacking, but it is imperative that all such issues are conveyed using appropriate language and with an understanding of the scientific culture of groups involved in advising decision makers. Thus, the civil authorities can favour methods other than probabilities for receiving forecasts (Solana et al. 2008), such as preferred time windows (Swanson et al. 1983), precursory scenarios and comparison with uncertainties associated with events that are locally more familiar. Overall, information will generally be more effectively communicated when presented in a form preferred by the recipient rather than imposed by the sender.

2.3 Inexperience

Lack of experience compounds poor communication due to different organisational cultures and the presentation of scientific information. Experience is gained not only through repeated exposure to volcanic unrest, but also by learning from mistakes. Admitting that mistakes have been made tends to be associated with failure and so not to be recorded. Notable exceptions reflecting on the management of volcanic crisis are Voight (1988a, 1990) on Nevado del Ruiz, Fiske (1984) on St. Vincent and Guadeloupe or Aspinall et al. (2002) in Montserrat. Reluctant recording is especially damaging to responses at volcanoes reawakening after long repose where the inherent uncertainty on the behaviour of volcanic system is highest. By virtue of the long repose intervals, few teams or individuals have the opportunity to respond more than once to such an emergency during their professional careers. The benefit of experience strongly depends, therefore, on published accounts of previous events. When accounts focus only on the successful aspects of a response, a false impression may be conveyed that mistakes are rare; and teams may also minimise the role played by good fortune rather than clear judgement Newhall and Punongbayan (1996). As a result, inexperienced teams may undervalue the potential difficulties in responding to an emergency and so unnecessarily repeat mistakes from the past. Global circumstances today are particularly acute, given the gap of nearly 25 years since the last large, VEI 6 eruption (Pinatubo in 1991) and more than 30 years has elapsed since the last volcanic disaster involving tens of thousands of deaths Chester et al. (2000) i.e. the 1984 eruption of Nevado del Ruiz in Colombia (Voight 1990). Based on statistics from the past 200 years (Siebert et al. 2010), at least half the volcanoes in eruption during the next century are expected to reawaken after a repose interval of 100 years or more. The scientists and officials responsible for managing an emergency will have no experience of the particular volcano's behaviour and, possibly, little or no direct experience of responding to volcanic eruptions in general.

The 2011–2012 unrest and eruption of El Hierro illustrates how communication during an emergency can be hampered by inexperience, organisational culture and forecasting uncertainty Marrero et al. (2015), Carracedo et al. (2015). This example is compared below with methods that have been developed at Italy's frequently-erupting volcanoes Etna and Stromboli. Together, the case studies indicate best-practice methods for maintaining good communications when responding to volcanic unrest.

3 Case Studies

3.1 Communications During the 2011–2012 Eruption of El Hierro, Canary Islands

Since the 1990s, the Canary Islands have been recognised as a high-risk volcanic area by the international scientific community. In particular, Mt Teide on Tenerife was selected as a UN-IDNDR decade volcano as well as a EU laboratory volcano. Nevertheless, the long return period of eruptions has encouraged a low perception of volcanic risk among national and local authorities and, hence, a reactive attitude to planning for volcanic emergencies is also evident. An indication of the low priority given to the threat of volcanic activity has been the government's under-investment in a multidisciplinary monitoring system for volcanic activity (Marti et al. 2009). For example, despite a general scientific call for an appropriate monitoring network during the 2004 seismo-volcanic crisis on Tenerife (Salomone 2004), little was invested until the 2011 unrest at El Hierro, when a comprehensive geophysical network was deployed there (López et al. 2012). Initiatives to rationalise and coordinate volcanic research in the region have also lacked economic support. For example, the creation of a comprehensive Volcanological Institute on the Canaries was demanded by the

Spanish Senate in 2005, the Parliament of the Canaries in 2006 and the Spanish Congress in 2009; even so, the only initiative to produce an open and inclusive research group on the islands received only limited funding from the local council in Tenerife and had instead to be created as a commercial entity.

The lack of preparedness was highlighted in 2011, when El Hierro, the westernmost and youngest island in the Canaries, entered eruption in October for the first time in recorded history. It was also the first eruption in the Canaries in 40 years. After four months of low magnitude seismic activity frequently felt by the population, together with enhanced diffuse emissions of CO_2 and H_2S (Pérez et al. 2012) and continuous surface and ground deformation which reached a maximum of 5 cm (López et al. 2012), an eruption was confirmed off the island's southern coast (Fig. 2). Being submarine, the eruption did not cause personal or material damage, but had an important impact on the small local businesses as well as causing anxiety and insecurity to some of the local population, as is clear on the concerns voiced in the media and directly through the digital version of the local newspaper, *Diario del Hierro* (www.diarioelhierro.com).

The legal context of the Canary Islands is important in framing some of the scientific issues that arose during the crisis. Although constitutional relationships between metropolitan Spain and the Canaries (and Spanish law more generally) are complex and beyond the scope of the present paper, a general statement of the key issues is presented next. The Canary Islands are a region of Spain with a special political status which allows them a level of self-determination and government in many areas of policy. The Civil Protection is one of the organisations with relative independence and can plan for and manage local and regional emergencies, though within an overall framework of laws approved by the central Spanish Government. Despite legislation requesting a specific plan for volcanic emergencies in Canaries since 1996, the first protocol to respond to an emergency was created as a response to the seismo-volcanic crisis of 2004 in Tenerife (Salomone 2004; PEVOLCA 2010).

Within this legal framework, in 2008 the Canarian Civil Protection produced a plan for volcanic emergencies known as PEVOLCA, Plan de Emergencias Volcanicas de Canarias (Volcanic Emergencies Plan for the Canaries). The plan was approved in 2010 and established which groups would form the scientific advisory committee and their roles and responsibilities. The original 2004 protocol recommended that the scientific committee be comprised of representatives from most of the local and national scientific institutions involved in volcanic research and monitoring in the Canaries. However, the 2010 PEVOLCA plan instead established a scientific advisory committee consisting of the National Geographical Institute (IGN), the

Fig. 2 The island of El Hierro, its position within the Canary Islands (*inset*) and the location of the offshore 2011 eruption (*star*). The village of La Restinga (pop. 600) is marked with an *open circle*

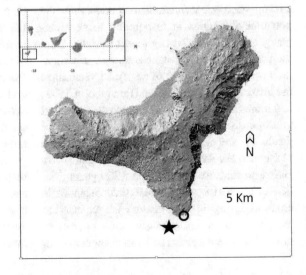

Spanish National Research Centre (CSIC), the National Meteorological Agency (AEMET) and the Civil Protection. This selection created much ill feeling amongst local scientific institutions, which could only participate in the scientific advisory committee meetings by invitation. It also had important implications for the local groups, which, in the absence of a formal role during the emergency, were left in a vulnerable legal position when their staff provided scientific advice and, without specific funding, were placed in financial difficulty when participating in committee meetings.

When the PEVOLCA plan was activated during the 2011–2012 El Hierro emergency, the Civil Protection invited all the principal volcanological research groups to the scientific advisory committee. It soon became apparent that the groups were not readily sharing all the available data, so the Civil Protection authority was eventually forced to act as a mediator. Independently-funded institutions and also groups funded through research projects argued that their data were not public and did not belong to the State. Ill feeling also developed because of the lack of access to resources, such as boats and helicopters for monitoring, and also data: for example, the establishment of an exclusion zone including the eruption site and its surroundings, meant that regular sampling was much easier for members of the Scientific Advisory Committee than it was for other groups. The six-month duration of the crisis also highlighted differences in funding between the formal monitoring group (IGN), that could rely on public funds for their continuous presence on the island and other monitoring groups, which struggled to meet their overheads.

In addition to the PEVOLCA scientists, researchers from other national and international research organisations temporarily joined in monitoring, sampling and research and some occasionally publicized their opinions in the local and national media. Although the majority of opinions were not especially controversial, some of the more speculative comments on the potential evolution of the crisis and the eruption, did create concern within the scientific groups who had followed the crisis from the beginning

as well as some of the local population (see news comments at www.diarioelhierro.com).

3.2 Lessons Learned on Communications During Recent Eruptions of Etna and Stromboli, Italy

Frequently eruptions at a volcano favour the development of reliable monitoring and efficient communications among scientists, the Civil Protection and the public. Etna and Stromboli (Fig. 3a), for example, have permanent, multi-disciplinary monitoring systems which enable precursory signals to be detected and recognised well in advance of an eruption (Bonaccorso et al. 2004; Patané et al. 2004; Puglisi et al. 2004; Martini et al. 2007; Bertolaso et al. 2008a, b; Rizzo et al. 2008; Tarchi et al. 2008; Calvari et al. 2010, 2011; Di Traglia et al. 2014). In addition, the responsibilities of the monitoring scientists have been clearly identified by the Italian Government and Civil Protection authorities. The latter funds monitoring networks and has frequent direct contact with the scientists in charge of coordinating the monitoring (Bertolaso et al. 2008b).

Legal responsibility for volcano and earthquake monitoring in Italy rests with the INGV (Istituto Nazionale di Geofisica e Vulcanologia). The Civil Protection (DPC, Dipartimento della Protezione Civile) funds not just the INGV, but also university teams that may duplicate or complement the data collected by the INGV. The university teams (called Centri di Competenza or Centres of Expertise) share data and ideas with the INGV and DPC during volcanic crises. Although the number of university personnel is much smaller than for the INGV (several tens compared with 800), the duplication of data from different teams allows alternative interpretations that encourages a more complete view of the phenomena being monitored. During the last fifteen years, implementation of this system has produced well-established and efficient management procedures for responding to crises at both Etna and Stromboli (Bertolaso et al. 2008b;

Fig. 3 a Google Map displaying southern Italy, with Stromboli volcano (*red circle*) at the NE end of the Aeolian Archipelago, and Etna volcano just N of Catania, on the East flank of Sicily. The *yellow square* displays the area magnified in **b. b** Google map of the eastern flank of Etna volcano, comprising the summit craters, the Valle del Bove depression with the 1991–1993 lava flow field (*in red*), and the position of Zafferana Etnea town. The *white dotted line* at the end of the lava flow field shows the position of the earth dam built up in January 1992. The *yellow square* displays the area magnified in **c. c** View from East of Etna's eastern flank displaying the Summit Craters, the Valle del Bove, and the position of Zafferana town. Photo by Alfio Amantia, taken from helicopter on 23 December 1991, a few days after the start of the eruption. The *yellow square* displays the area magnified in **d. d** Photo by Alfio Amantia, taken from helicopter on 5 January 1992, showing the earth dam built up at the exit of the Valle del Bove, with several trucks working on its top, and a large lava flow approaching to it

Bonaccorso et al. 2015). Even under such favourable conditions, however, communications during a crisis can be very challenging when a population is directly threatened, as illustrated by the 1991–1993 eruption on Etna (Barberi and Villari 1994; Calvari et al. 1994) and the 2002–2003 activity on Stromboli (Calvari et al. 2005, 2006; Bertolaso et al. 2008a).

When the safety of people is involved, considerable pressure is placed on the scientists from both the population and government agencies, specifically to analyse data and provide information and forecasts quickly, which diminishes both productivity and efficiency. Data sharing and discussion of differing interpretations is of paramount importance, but the scientific community and Civil Protection must provide an agreed interpretation if they are to keep the trust of the population. The 1991–1993 flank eruption at Etna volcano (Fig. 3b) produced the largest compound lava flow field of the last three centuries, with ~250 million m^3 of lava being emplaced along the southern margin of the barren Valle del Bove (VDB) on the eastern flank of the volcano (Calvari et al. 1994; Stevens et al. 1997). From there it approached the village of Zafferana Etnea (Fig. 3b, c), about 9 km from the eruptive fissure (Calvari et al. 1994). The eruption started on 14 December 1991 and, after an initial effusive phase during which the lava rapidly spread within the VDB, the Government decided to build an earthen barrier, 21 m high and 234 m long, across a narrow exit from the valley (Fig. 3b–d) in order to confine the lava and protect Zafferana Etnea (Barberi and Villari 1994). The barrier was completed on January 1st 1992 (Barberi and Villari 1994, Fig. 3d) and worked effectively for three months until April 7th 1992. However, the growth of lava tubes within the lava flow field significantly increased the potential of the lava to spread down slope (Calvari and Pinkerton 1998), and resulted in lava piling up behind and then spilling over the barrier to advance towards Zafferana Etnea (Calvari et al. 1994; Calvari and Pinkerton 1998).

At this stage, the people of Zafferana Etnea were seriously concerned for their safety, because they could observe the lava flows advancing towards their homes, less than 2 km away. Mistrust of the authorities spread among the population, especially when the first attempts to divert the lava (initially carried out at the lower end of the lava flow field) were unsuccessful. This was reinforced by the attention given by the local and national media to differing opinions expressed by scientists and Civil Protection officials over the best solutions to apply to divert the lava. Disagreements were amplified by the media, thus increasing suspicions, doubts and distrust (Barberi and Villari 1994). The population exerted pressure through their elected representatives to halt attempts for changing the course of the lava (e.g., with the use of bombs and destruction of forests) and public trust started to return only after daily meetings between scientists and the population were organised, with additional information being released by radio and television (Barberi et al. 1993; Barberi and Villari 1994; Barberi and Carapezza 2004). At the daily meetings the scientists explained "step by step" and justified what they were doing (Barberi and Villari 1994). Strangely, trust in and esteem for the scientists did not grow even when the safety of Zafferana Etnea was assured by the successful diversion of lava from a location closer to the eruptive fissure, because the attention of the media by then had already moved on to other, more dramatic topics (Barberi and Villari 1994), so highlighting the important role of the media in influencing public perception.

Ten years later, during a major flank eruption in 2002–2003, information to both the media and the population were communicated, not only by radio and television, but also through the internet, where maps, reports, photos and data were continually updated. The collaboration between scientists working for different institutions and universities (even from different countries) proved essential to understand and model eruptive processes, mitigate risks and obtain the best possible results (Bonaccorso et al. 2015). Following this eruption, a key priority was to produce prompt hazard assessment and develop expertise in the modelling of lava flows within the INGV. This was achieved through projects funded by the Italian Government and the DPC.

Since then, communities on Etna have become involved in the monitoring effort by reporting information on any ash fall to the monitoring room of the INGV's Etna Observatory in Catania. They have also been invited to collaborate with ash sampling, thus allowing a unique and extensive sample collection of the several lava fountain episodes spreading ash over a large area (~ 30 km wide and over 200 km long) on and beyond the eastern flank of the volcano (Andronico et al. 2008). This collaboration has increased public understanding of eruptive events and served to increase trust in the work of scientists.

During Etna's 2002–2003 eruption, the INGV had simultaneously to face a volcanic crisis at Stromboli, the most active of the Aeolian Islands about 60 km to the north (Fig. 3a). The start of this eruption was marked by a tsunami that injured three people and caused damage along the coastal area of the island (Tinti et al. 2003). The tsunami was triggered by submarine and subaerial landslides (Bonaccorso et al. 2003) and was soon followed by the opening of an eruptive fissure with lava flowing down the north flank of the island to the sea for several months (Calvari et al. 2005). While lava was still pouring out from the fissure, the summit crater produced a very large explosion (Calvari et al. 2006; Bonaccorso et al. 2012) that damaged houses and buildings in nearby villages and triggered fires on vegetated slopes. The Civil Protection responded rapidly, building an observatory (COA, Centro Operativo Avanzato) where all the monitoring signals from the INGV and other universities were received (Bertolaso et al. 2008a). These monitoring networks were greatly expanded and improved using special financial support from the Civil Protection Authority. Daily meetings were organized between academic and INGV scientists and the Civil Protection to enable data comparison and sharing, and this resulted in a much deeper understanding of the volcanic activity (Calvari et al. 2008). The knowledge of the volcanic system was also greatly improved because of the funds and research projects initiated by the DPC (Bertolaso et al. 2008a). Since the monitoring of Stromboli was shared between INGV and other institutions, such as the Universities of Florence

and Rome (Bertolaso et al. 2008a), the communication of scientific results to the population and the media was strictly controlled by the Civil Protection in order to avoid disseminating divergent or contrasting points of view.

When the eruption started, 400 inhabitants were informed of the possibility of tsunamis, and 330 spontaneously decided to leave the island. Access to the island was temporarily denied, except to monitoring scientists, volcanological guides, The Civil Protection and authorized journalists. Although implemented for safety reasons, the exclusion created ill feelings amongst a team of academic scientists not involved in monitoring, who were prevented access for sampling the erupted products. The Italian Civil Protection addressed this problem by sharing a number of samples between the monitoring and the academic teams. The population returned of their own accord two months later, and within three months from the start of the eruption the island was reopened to tourists.

Initially, a conflict between the local population and the media arose, caused by the fear that dramatic information on the activity of this volcano would scare away tourists and damage the local economy (Bertolaso et al. 2008a). To maintain a high level of awareness of volcanic hazards amongst the local population and to provide guidance on personal safety, the DPC has since annually published and distributed leaflets before the start of the tourist season. However, this basic information has not been welcomed by the whole of the local population, again because of concerns of reducing tourism.

4 General Aspects of the Development of a Protocol for the Communication, Management and Use of Scientific Information During Volcanic Emergencies in Canaries

The experience gained at Etna, Stromboli and El Hierro reinforce the findings from other volcanoes (e.g., Peterson 1988; Tilling 1989; Solana et al. 2008; McGuire et al. 2009; Doyle et al.

2011) that clear procedures for the sharing, discussion and communication of scientific information are crucial for the effective management of volcanic emergencies, especially on volcanoes with long periods of repose, where uncertainty about the volcanic systems are greatest.

In the Canaries, the Civil Protection has proposed that a new Scientific Committee for the Assessment and Surveillance of Volcanic Phenomena (CSEV) be created as an inclusive committee for the study and analysis of seismo-volcanic risk. This committee will be coordinated by the Civil Protection and consist of representatives from each of the groups within the formal advisory committee of PEVOLCA, as well as from each of the local universities and research centres. External advisors may also be appointed at the discretion of the coordinator.

CSEV's aims and objectives are:

- to identify possible precursory phenomena to eruptions in the Canary Islands;
- to assess data obtained from monitoring networks and instruments;
- to produce forecasts of volcanic activity and the consequences for the Civil Protection Authorities;
- to establish a methodology for monitoring *during* volcanic eruptions, for evaluating data and for formulating hypotheses about the likelihood and potential impact of an eruption.

To avoid the difficulties encountered during the El Hierro crisis, CSEV has further established a series of regulations that must be satisfied for participation:

- the requirement to sign an agreement to obey the rules of the committee;
- the use of information and data must be shared by committee members exclusively to advise the Civil Protection (and not for personal or institutional gain);
- the selection of one representative from each group by democratic procedures;
- compulsory attendance at sessions of the committee;

- the need to produce and submit written reports two days before scheduled meetings with information on forecasts, eruption scenarios and their associated probabilities;
- following debate and discussion of data, the need to reach a consensus and produce a written document on the situation or, if a consensus is not reached, the basis for different interpretations should be clearly noted and explained;
- confidentiality and coordination in the dissemination of information about the results and of debates held within meetings.

5 Discussion

The challenges and successes presented in this paper emphasize the varied nature of communication problems that have been encountered in recent crises and the lessons that can be learned. A summary of selected information from these crises is presented in Table 1.

The eruption of El Hierro in 2011 provided the best possible scenario for an eruption from the management point of view. The eruption occurred offshore, meaning that no lives were seriously threatened or property damaged; it affected a sparsely populated area (El Hierro has a population of ca. 10,000, the smallest in the Canary Islands) and hence involved an easily manageable number of people; and it displayed several weeks of increasing precursory activity which permitted emergency arrangements to be put in place. Moreover, a plan to manage volcanic crises was in place before the eruption and therefore, while the scientific management of the crisis presented challenges, the operational management of the crisis ran smoothly. Good fortune certainly helped the smooth operation and, had the 2011 eruption occurred inland, or with brief precursory signals, the consequences might have been more serious.

The difficulties in scientific management stemmed mainly from a non-inclusive policy established by the national government, lack of experience, differences in organisational cultures,

Table 1 Summary of selected characteristics of the case studies, good practice actions and lessons learnt

	El Hierro 2011–2012	Etna 1991–1993	Etna 2002–2003	Stromboli 2002–2003
Previous VE plan	Yes	Yes	Yes	Yes
Previous experience in the volcanic system	No	Yes	Yes	Yes
Understanding of the volcanic system	Low	High	High	High
Direct threat to life/property	No	Yes	No	Yes
Previous experience on VE	Some (1 previous)	High	High	High
Good practice key points	Rapid response to crisis	Coordination of scientific information to the public	Inclusive and open monitoring effort	Inclusive and open monitoring effort
	VE plan revised to incorporate lessons learnt	Post-eruptive needs assessment	Continuous information to the public	Regular (yearly) information campaign
	Ethical behaviour of most scientists directly/indirectly involved	Permits for researchers	Involvement of the public in monitoring	Creation of an in-site inclusive observatory
			Civil protection management of priority research/monitoring funds	Recognition of the roles of other academic researchers
Lessons learnt	More inclusivity in scientific committee	Regular communication with the public	Changeable nature of volcanic threat	Changeable nature of volcanic threat
	Clearer procedures for involvement of scientific groups	Management of scientific disagreements	Importance of duplication in data collection for scientific discussion	Sharing of scientific data and resources (e.g. samples)
	Need of legal frame to cover liability of collaborating institutions/individuals	Improved communication with the media and management of media reports		
	Need of contingency funding for independent advisors/collaborations			

VE Volcanic emergency

lack of formal procedures for the sharing of information and the under-funding of monitoring networks. As demonstrated by the success of analogous procedures at Etna and Stromboli, the CSEV model to produce for the Canary Islands a more inclusive scientific advisory group is a positive step towards establishing a more integrated and collaborative volcanological research community. For successful implementation, the model needs to accommodate four key features.

Financing aspects to ensure that participants are not excluded because of lack of funding, not only for data collection and hazard mitigation, but also for the logistics of attending pre- sin- and post- eruption meetings of the advisory committee. One promising option would be to adapt the

Italian system of providing funds through the national Civil Protection. In the longer term, funding is required to establish permanent and multidisciplinary monitoring networks, including local monitoring centres responsible for specific volcanoes in the Canaries. The centres provide a focus for liaising in particular with local communities and the Civil Protection. It is of course much easier to justify the necessary resources at frequently-erupting volcanoes, such as Etna and Stromboli. However, having at least a contingency for the rapid establishment of a multidisciplinary inclusive institution will encourage the coordination and sharing of data and contribute to the feeling of cohesion and camaraderie amongst the groups involved.

Legal aspects to protect against liability exposure (Marrero et al. 2015). Aspinall and Sparks (2004) describe an example from the United Kingdom Office of Science and Technology, which includes a clause that "appears to indemnify individual members of such a [scientific advisory] Committee" when acting "honestly, reasonably, in good faith and without negligence", although they warn that the phrase "without negligence" is subjective and can be lead to further disputes (quotations from Aspinall and Sparks (2004) pp. 5–6).

Aspects of the format and storage of information, to ensure the effective sharing of data. Examples include:

a. *Type of data that should be shared between scientific groups*, such as real time data, fast data, general data-forecasts, and rock and gas samples.
b. *The content of scientific reports.* On some occasions data would have to be part- processed before being shared in order for non-specialists to understand and be able to use it.
c. *The frequency of delivering information*, to establish how often data are to be received from each group. These timings should be flexible and realistic. For instance, the requirement of the Spanish Civil Protection to receive written forecasts two days in advance of a meeting may be neither feasible nor

helpful, especially if daily meetings are required (following the procedures established at Etna and Stromboli).
d. *The format* preferred by recipients for the communication of forecasts, eruption scenarios and mitigation procedures.
e. *Policies on the storage of information* to decide which agency should be responsible for compiling and storing scientific data, the form of storage (e.g., as an electronic database, or as a website with restricted access), and the length of storage before data are made available for general use.

Procedures for integrating researchers from outside official monitoring groups. Volcanic eruptions provide opportunities for advancing the understanding of volcanic processes and, as identified by IAVCEI et al. (1999), establishing official and legal mechanisms to allow external research groups access to field data can yield insights of potential value to mitigation efforts. It would also reduce the possibility of ill feelings caused by exclusion.

6 Conclusions

Our experience of emergencies at frequently-erupting and long-dormant volcanoes, reinforce the conclusions by IAVCEI et al. (1999) that the management of crises is optimised by officially approving, before an emergency, clear, and legally binding rules and protocols for the communication of scientific information between responding groups. As well as detailing the type, content, amount, format, frequency, storage and use of the information, the protocols should consider aspects such as confidentiality, inclusiveness, ethics, financing and legal aspects such as the liability of scientific groups.

Another important recommendation is to identify a coordinating body outside the monitoring and scientific teams—such as the Civil Protection—to manage discussions, collate forecasts and scenarios and agree a consensus (or the basis for different interpretations) and to ensure that all involved feel that their views and contributions are respected. Recognising the different organisational cultures of the responding groups is vital for a better understanding of each party's needs and limitations and for optimising the design of communication strategies.

Finally, the Italian model demonstrates the benefits of establishing an on-site, inclusive and multidisciplinary institution to produce and coordinate scientific information and to encourage collaboration and camaraderie. It also illustrates the advantages of incorporating external researchers, of sharing resources and of engaging the public to improve the understanding of the volcano in unrest. The allocation of appropriate government funding and resources to all these activities is, of course, key to their success.

Acknowledgements We would like to thank the organisers and participants of the VUELCO meeting in Rome in 2013 for their participation in an informal survey and especially Carina Fearnley for her help with the data collection. We also thank the scientific and civil defence personnel involved in the response to the 2011–2012 El Hierro emergency for sharing their opinions and concerns. A special thank you goes to Roger Jennings for highlighting the institutional and managerial differences between private and public organisations and academia. Finally we would like to thank the anonymous reviewers for their comments and advice.

References

Andronico D, Scollo S, Caruso S, Cristaldi A (2008) The 2002–03 etna explosive activity: tephra dispersal and features of the deposits. J Geophys Res 113:B04209. doi:10.1029/2007JB005126

Aspinall WP, Loughlin SC et al (2002) The Montserrat Volcano observatory; its evolution, organization, role and activities In: Druitt TH, Kokelaar BP (eds) The eruption of Soufriere Hills Volcano, Montserrat from 1995 to 1999; The Geological Society London Memoirs, vol 21, pp 71–91

Aspinall WP, Sparks RSJ (2004) Volcanology and the law. IAVCEI News 1:4–5, 11–12

Aspinall WP, Woo G, Voight B, Baxter PJ (2003) Evidence-based volcanology: application to eruption crisis. J Volcanol Geoth Res 128:273–285

Barberi F, Carapezza ML (2004) The control of lava flows at Mt. Etna. In: Bonaccorso A, Calvari S, Coltelli M, Del Negro C, Falsaperla S (eds) Mt. Etna: volcano laboratory, AGU Geophys Monogr 143:343–369. doi:10.1029/143GM21

Barberi F, Gasparini P (1979) A deontological code for volcanologists? (Letter to the editor). J Volcanol Geotherm Res 6:1–2

Barberi F, Villari L (1994) Volcano Monitoring and civil protection problems during the 1991–1993 Etna eruption. Acta Vulcanol 4:157–165

Barberi F, Carapezza ML, Valenza M, Villari L (1993) The control of lava flow during the 1991–1992. J Volcanol Geotherm Res 56:1–34

Barclay J, Haynes K, Mitchell T, Solana C, Teeuw R, Darnell A, Crosweller HS, Cole P, Pyle D, Lowe CJ, Fearnley C, Kelman I (2008) Framing volcanic risk communication within disaster risk reduction: finding ways for the social and physical sciences to work together. Communicating environmental geoscience. Geol Soc Lond Spec Publ 305:163–177

Bell AF, Kilburn CRJ (2011) Precursors to dyke-fed eruptions at basaltic volcanoes: insights from patterns of volcano-tectonic seismicity at Kilauea volcano, Hawaii. Bull Volcanol 74:325–339

Bertolaso G, De Bernardinis B, Cardaci C, Scalzo A, Rosi M (2008a) Stromboli (2002–2003) crisis management and risk mitigation actions. In: Calvari S, Inguaggiato S, Puglisi G, Ripepe M, Rosi M (eds) The Stromboli volcano, an integrated study of the 2002–2003 eruption. AGU Geophys Monogr 182:373–385. doi:10.1029/143GM30

Bertolaso G, Bonaccorso A, Boschi E (2008b) Scientific community and civil protection synergy during the Stromboli 2002–2003 Eruption. In: Calvari S, Inguaggiato S, Puglisi G, Ripepe M, Rosi M (eds) The Stromboli volcano, an integrated study of the 2002–2003 eruption. AGU Geophys Monogr 182:387–397. doi:10.1029/143GM31

Bonaccorso A, Calvari S, Garfi G, Lodato L, Patané D (2003) December 2002 flank failure and tsunami at Stromboli volcano inferred by volcanological and geophysical observations. Geophys Res Lett 30 (18):1941–1944. doi:10.1029/2003GL017702

Bonaccorso A, Campisi O, Falzone G, Gambino S (2004) Continuous tilt monitoring: lesson learned from 20 years experience at Mt. Etna. In: Bonaccorso A, Calvari S, Coltelli M, Del Negro C, Falsaperla S (eds) Mt. Etna: volcano laboratory. AGU Geophys Monogr 143:307–320. doi:10.1029/143GM19

Bonaccorso A, Calvari S, Linde A, Sacks S, Boschi E (2012) Dynamics of the shallow plumbing system investigated from borehole strainmeters and cameras during the 15 March 2007 Vulcanian paroxysm at Stromboli volcano. Earth Planet Sci Lett 357–358:249–256. doi:10.1016/j.epsl.2012.09.009

Bonaccorso A, Calvari S, Boschi E (2015) Hazard mitigation and crisis management during major flank eruptions at Etna volcano: reporting on real experience. In: Detecting, modelling and responding to effusive eruptions. Geological Society London Special Publications, in print

Bostok D (1978) A deontological code for volcanologists? (Editorial). J Volcanol Geotherm Res 4:1

Burton M, Neri M, Andronico D, Branca S, Caltabiano T, Calvari S, Corsaro RA, Del Carlo P, Lanzafame G, Lodato L, Miraglia L, Muré F, Salerno G, Spampinato L (2005) Etna 2004-05: an archetype for geodynamically-controlled effusive eruptions. Geophys Res Lett 32:L09303. doi:10.1029/2005GL022527

Calvari S, Pinkerton H. (1998). Formation of lava tubes and extensive flow field during the 1991-93 eruption of Mount Etna. J Geophys Res: Solid Earth 103(B11):27291–27301

Calvari S, Coltelli M, Neri M, Pompilio M, Scribano V (1994) The 1991–93 Etna eruption: chronology and lava flow field evolution. Acta Vulcanol 4:1–14

Calvari S, Spampinato L, Lodato L, Harris AJL, Patrick MR, Dehn J, Burton MR, Andronico D (2005) Chronology and complex volcanic processes during the 2002–2003 flank eruption at Stromboli volcano (Italy) reconstructed from direct observations and surveys with a handheld thermal camera. J Geophys Res 110:B02201. doi:10.1029/2004JB003129

Calvari S, Spampinato L, Lodato L (2006) The 5 April 2003 vulcanian paroxysmal explosion at Stromboli volcano (Italy) from field observations and thermal data. J Volcanol Geotherm Res. 149:160–175 doi:10.1016/j.jvolgeores.2005.06.006

Calvari S, Inguaggiato S, Puglisi G, Ripepe M, Rosi M (eds) (2008) The Stromboli volcano, an integrated study of the 2002–2003 eruption. AGU Geophys Monogr 182:390 pp

Calvari S, Lodato L, Steffke A, Cristaldi A, Harris AJL, Spampinato L, Boschi E (2010) The 2007 Stromboli flank eruption: chronology of the events, and effusion rate measurements from thermal images and satellite data. J Geophys Res Solid Earth 115(B4):B04201. doi:10.1029/2009JB006478

Calvari S, Salerno GG, Spampinato L, Gouhier M, La Spina A, Pecora E, Harris AJL, Labazuy P, Biale E, Boschi E (2011) An unloading foam model to constrain Etna's 11–13 January 2011 lava fountaining episode. J Geophys Res 116:B11207. doi:10.1029/2011JB008407

Carracedo JC, Troll VR, Zaczek K, Rodríguez-González A, Soler V, Deegan FM (2015) The 2011–2012 submarine eruption off El Hierro, Canary Islands: New lessons in oceanic island growth and volcanic crisis management, Earth-Science Reviews, vol 150, November 2015, pp. 168–200

Chester DK, Degg M, Duncan AM, Guest JE (2000) The increasing exposure of cities to the effects of volcanic eruptions: a global survey. Environ Hazard 2:89–103

De La Cruz-Reyna S, Reyes-Davila GA (2001) A model to describe precursory material-failure phenomena: applications to short-term forecasting at Colima volcano, Mexico. Bull Volcanol 63:297–308

Di Traglia F, Nolesini T, Intrieri E, Mugnai F, Leva D, Rosi M, Casagli N (2014) Review of ten years of volcano deformations recorded by the ground-based InSAR monitoring system at Stromboli volcano: a tool to mitigate volcano flank dynamics and intense volcanic activity. Earth Sci Rev, in print

Doyle EH, Johnston DM, McClure J, Paton D (2011) The communication of uncertain scientific advice during natural hazards events. N Z J Psych 40(4):39–50

Doyle EH, McClure J, Johnston DM, Paton D (2014) Communicating likelihoods and probabilities in forecasts of volcanic eruptions. J Volcanol Geotherm Res 272:1–15

Fiske RS (1979) A deontological code for volcanologists? (Reply to editorial). J Volcanol Geotherm Res 5:211–212

Fiske RS (1984) Volcanologists, journalists, and the concerned local public: a tale of two crises in the Eastern Caribbean. In: Geophysics Study Committee (ed) Explosive volcanism: interception, evolution and hazard. National Academy Press, Washington, DC, pp 170–176

Handy C (1978) Gods of management, who they are, how they work and why they will fail. Pan LTD ed., 320 pp

IAVCEI, Subcommittee for Crisis Protocols, Newhall C, Aramaki S, Barberi F, Blong R, Calvache M, Cheminee JL, Punongbayan R, Siebe C, Simkin T, Sparks RSJ, Tjetjep W (1999) Professional conduct of scientists during volcanic crises. Bull Volcanol 60:323–334

IGN (s.n.) Vigilancia Volcanica. http://www.ign.es/ign/layoutIn/volcaListadoEstaciones.do?codT=IGN&desT=Estaciones%20s%EDsmicas%20IGN. Accessed Jan 2015

Kilburn CRJ (2003) Multiscale fracturing as a key to forecasting volcanic eruptions. J Volcanol Geotherm Res 125:271–289

Kilburn CRJ (2012) Precursory deformation and fracture before brittle rock failure and potential application to volcanic unrest. J Geophys Res. doi:10.1029/2011JB008703

Kilburn CRJ, Voight B (1998) Slow rock fracture as eruption precursor at Soufriere Hills volcano, Montserrat. Geophys Res Lett 25:3665–3668

López C, Blanco MJ, Abella R, Brenes B et al (2012) Monitoring the volcanic unrest of El Hierro (Canary Islands) before the onset of the 2011–2012 submarine eruption. Geophys Res Lett doi:10.1029/2012GL051846

Marrero JM, Garcia A, Linares A, Berrocoso M, Ortiz R. (2015) Legal framework and scientific responsibilities during volcanic crises: the case of the El Hierro eruption (2011–2014) J Appl Volcanol 4:13. doi:10.1186/s13617-015-0028-8

Marti J, Aspinall WR, Sobradelo R, Felpeto A, Geyer A, Ortiz R, Baxter P, Cole PD, Pacheco J, Blanco MJ, Lopez C (2008) A long-term volcanic hazard event tree for Teide-Pico Viejo stratovolcanoes (Tenerife, Canary Islands). J Volcanol Geotherm Res 178:543–552. doi:10.1016/j.jvolgeores.2008.09.023

Marti J, Ortiz J, Gottsmann J, Garcia A, la Cruz-Reyna De (2009) Chracterising unrest during the reawakening of the central volcanic complex on Tenerife, Canary Islands, 2004–2005, and implications for assessing hazards and risk mitigation. J Volcanol Geotherm Res 182(1):23–33

Martini M, Guidicepetro F, DAuria L, Esposito AM, Caputo T, Curciotti R, De Cesare W, Orazi M, Scarpato G, Caputo A (2007) Seismological monitoring of the February 2007 effusive eruption of the Stromboli volcano. Ann Geophys 50(6):775–788

Marzocchi W, Bebbington MS (2012) Probabilistic eruption forecast at short and long time scales. Bull Volcanol 74(8):1777–1805

Marzocchi W, Sandri L, Selva J (2008) BET_EF: a probabilistic tool for long- and short-term eruption forecasting. Bull Volcanol 70(5):623–632

Marzocchi W, Newhall C, Woo G (2012) The scientific management of volcanic crises. J Volcanol Geotherm Res 247–248:181–189

McGuire WJ, Solana MC, Kilburn CRJ, Sanderson D (2009) Improving communication during volcanic crises on small, vulnerable, islands. J Volcanol Geotherm Res 183:63–75

Newhall CG, Punongbayan RS (1996) The narrow margin of successful volcanic-risk mitigation. In: Scarpa R, Tilling RI (eds) Monitoring and mitigation of volcano hazards, pp 807–838

Newhall C, Hoblitt R (2002) Constructing event trees for volcanic crises. Bull Volcanol 64:3–20. doi:10.1007/s00445010017364: 3. doi:10.1007/s004450100173

Oskin B (2014) http://www.scientificamerican.com/article/how-a-deadly-volcano-erupted-in-japan-without-warning/. Accessed 20 Nov 2014

Patané D, Cocina O, Falsaperla S, Privitera E, Spampinato S (2004) Mt. Etna Volcano: a seismological framework. In: Bonaccorso A, Calvari S, Coltelli M, Del Negro C, Falsaperla S (eds) Mt. Etna: Volcano laboratory. AGU Geophys Monogr 143:147–165. doi:10.1029/143GM10

Pérez NM, Padilla GD, Padrón E, Hernández PA et al (2012) Precursory diffuse CO_2 and H_2S emission signatures of the 2011–2012 El Hierro submarine eruption. Canary I Geophys Res Lett. doi:10.1029/2012GL052410

Peterson DW (1988) Volcanic hazards and public response. J Geophys Res 93:4161–4170

PEVOLCA (2010) Plan Especial de Protección Civil y Atención de Emergencias por riesgo volcánico en la Comunidad Autónoma de Canarias. Boletin Oficial de Canarias 140 (19/07/2010)

Puglisi G, Briole P, Bonforte A (2004) Twelve years of ground deformation studies on Mt. Etna Volcano based on GPS surveys. In: Bonaccorso A, Calvari S, Coltelli M, Del Negro C, Falsaperla S (eds) Mt. Etna: Volcano laboratory. AGU Geophys Monogr 143: 321–341. doi:10.1029/143GM20

Rizzo A, Aiuppa A, Capasso G, Grassa F, Inguaggiato S, Longo M, Carapezza ML (2008) The 5 April 2003 Paroxysm at Stromboli, a review of geochemical observations. In: Calvari S, Inguaggiato S, Puglisi G, Ripepe M, Rosi M (eds) The Stromboli Volcano, an integrated study of the 2002-2003 eruption. AGU Geophys Monogr 182:347–358. doi:10.1029/143GM28

Salomone M (2004) Crisis Volcanica en Tenerife. El Pais, 22 Nov 2004. http://elpais.com/diario/2004/11/22/ultima/1101078001_850215.html. Accessed Jan 2015

Siebert L, Simkin T, Kimberley P (2010). Volcanoes of the world, 3rd edn. Smithsonian Institution & University of California Press, 550 pp

Sigvaldason GE (1978) A deontological code for volcanologists? (Reply to editorial). J Volcanol Geotherm Res 4:I–III

Sobradelo R, Marti J, Kilburn C, Lopez C (2015) Probabilistic approach to decision-making under uncertainty during volcanic crises: retrospective application to El Hierro (Span) 2011 volcanic crisis. Nat Hazards 76(2):979–998

Solana C, Spiller C (2007) Communication between professionals during volcanic emergencies. Eos Trans AGU 88(28)

Solana MC, Kilburn CRJ, Rolandi G (2008) Communicating eruption and hazard forecasts on Vesuvius, Southern Italy. J Volcanol Geotherm Res 172: 308–314

Stevens NF, Murray JB, Wadge G (1997) The volume and shape of the 1991–1993 lava flow field at Mount Etna. Sicily Bull Volcanol 58(6):449–454

Swanson DA, Casadevall TJ, Dzurisin D, Malone SD, Newhall CG, Weaver CS (1983) Predicting eruptions at Mount St. Helens, June 1980 through December 1982. Science 221(4618):1369–1376

Tarchi D, Casagli N, Fortuny-Guasch J, Guerri L, Antonello G, Leva D (2008) Ground deformation from ground-based SAR interferometry. In: Calvari S, Inguaggiato S, Puglisi G, Ripepe M, Rosi M (eds) The Stromboli Volcano, an integrated study of the 2002-2003 eruption. AGU Geophys Monogr 182:359–372. doi:10.1029/143GM29

Tazieff H (1977) La Soufrière, volcanology and forecasting. Nature 269:96–97

Tilling RI (1989) Volcanic hazards and their mitigation: progress and problems. Rev Geophys 27:237–269

Tinti S, Pagnoni G, Zaniboni F (2003) Tsunami generation in Stromboli island and impact on the north-east Tyrrhenian coast. Nat Hazards Earth Syst Sci 3:299–309

Tomblin J (1979) A deontological code for volcanologists? (Reply to Editorial). J Volcanol Geotherm Res 5:213–215

Voight B (1988a) Countdown to catastrophe. Earth Miner Sci 57:17–30

Voight B (1988b) A method for prediction of volcanic eruptions. Nature 332:125–130

Voight B (1990) The 1985 Nevado del Ruiz volcano catastrophe: anatomy and retrospection. J Volcanol Geotherm Res 44(3–4):349–386

Integrating Social and Physical Perspectives of Mitigation Policy and Practice in Indonesia

Supriyati Andreastuti, Agus Budianto and Eko Teguh Paripurno

Abstract

Earthquakes, tsunami, landslide and volcanic eruptions occur frequently in Indonesia. The frequency of events combined with high population and widely varied culture, differing levels of education and knowledge of natural hazards, as well as varied income, combine to give the country a high risk for natural disaster. Communication in hazard zones is affected by a number of factors such as: differing terminology and perceptions of hazards by the public, scientists, and disaster managers; how scientists and emergency managers communicate information; and how effectively the media transfers the information to the public. Communication is also complicated by culture, social factors and a wide variety of local languages. In Indonesia, disaster mitigation efforts at the national level are coordinated by National Disaster Management Agency; whereas, provincial and regional disaster agencies are responsible for managing within their domains and in most cases local authorities are responsible for specific mitigation actions, such as evacuations. Transferring hazard information is an important process in mitigation. In order to obtain efficient communication with the public, trusting relationships between scientists and communities are required. An understanding by scientists and emergency managers of local culture, local languages and people's character facilitates communication and contributes to trust. In addition, the media used for information can contribute significantly to improving communication. Hazard communication also aims to improve the capacity of communities through enhancing their knowledge and strengthening of their mitigation institutions. In hazard zones, effective mitigation requires

S. Andreastuti (✉) · A. Budianto
Center for Volcanology and Geological Hazard
Mitigation, Jl. Diponegoro no 57, Bandung 40122,
Indonesia
e-mail: s.andreastuti@yahoo.com

A. Budianto
e-mail: agusbudianto.vsi@gmail.com

E.T. Paripurno
Universitas Pembangunan Nasional 'Veteran'
Yogyakarta, Jl. SWK 104 (Lingkar Utara),
Condongcatur, Daerah Istimewa, Yogyakarta 55283,
Indonesia
e-mail: paripurno@upnyk.ac.id

Advs in Volcanology (2018) 307–320
https://doi.org/10.1007/11157_2016_36

participation and community empowerment with activities before, during and after disasters. A lesson learned from numerous volcanic eruptions in Indonesia is that each volcano has a different character, based not only the physical characteristics of eruptions but also on geographic, social and cultural features. These features result in different responses of people during crises and they influence the way scientists and government agencies communicate and deal with the process of evacuation and repatriation.

Keywords

Natural hazards · Hazard communication · Information transfer · Lessons learnt · Community capacity

1 Introduction

Indonesia is located between 3 tectonic plates, Indo-Australia, Eurasia and Pacific plates. This plate tectonic configuration exposes the nation to a wide range of geological hazards, i.e. earthquakes, tsunami, landslides and volcano eruptions. According to the Indonesia National Disaster Agency (2014), there are about 200 million people at risk from earthquakes; 4 million at risk from tsunami, 200 million at risk from landslides and 5 million people at risk from volcanic eruptions. With a population of 227,641,326 (Government of Indonesia 2010) and the dense population in hazard zones, risk reduction efforts are a priority of the government.

This chapter will discuss disasters as related to volcanic eruptions. Indonesia has 127 active volcanoes (Fig. 1), 77 are classified Type A, which have experienced one or more eruptions since 1600 AD. Type A are monitoring priority volcanoes. Type B (29 Volcanoes), last erupted before 1600 AD and show evidence of volcanic activity, such as fumaroles or solfatara. Type C (21 volcanoes), do not have any record of historic eruptions, but show fumarole and or sofatara activity. In 2015 (October), there were 18 volcanoes with activity above normal levels, 2 of those were in level 3 (Watch) and 1 in level 4 (Warning).

In hazard mitigation, there are 3 stages; pre-disaster, syn-disaster, and post disaster.

According to Law no 24 (2007) of the Republic Indonesia concerning Disaster Mitigation, efforts of mitigation shall be emphasized in pre-disaster activity programs, such as capacity building to prepare community awareness. Communication with the public is an important part of developing community preparedness in volcanic hazard zones. Our experience in Indonesia shows that disaster mitigation can only be achieved successfully if preparedness is carried out at the community level. During this process, participatory action to empower people is the key to people taking action during crisis according to their preparedness (e.g., Paton and Johnston 2001; Ronan and Johnson 2005).

This paper highlights the process of participatory management of crises and empowerment of responsible media to fill communication gaps between scientist and managers on one side and communities on the other side.

Below we describe methods used in Indonesia and lessons learned in seeking optimal preparedness.

2 Disaster Mitigation in Indonesia

In Indonesia, implementation for disaster mitigation is coordinated by the National Disaster Management Agency (BNPB) and by Provincial and Regional Disaster Management Agencies (BPBD), who are responsible for managing within their domains. The institution responsible

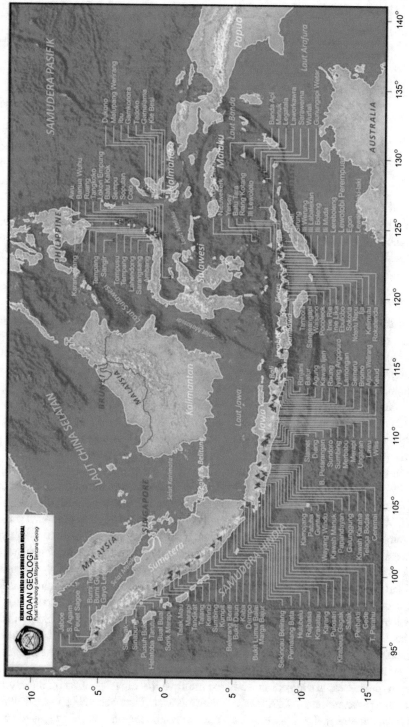

Fig. 1 Distribution of Indonesia volcanoes. Symbols: *red triangles*, Type A; *yellow triangles*, Type B and C

Table 1 Volcano activity in Indonesia

Level of volcanic activity in Indonesia	
Normal level	Visual observations and instrumental records show normal fluctuations, but no change of activity Hazards in the form of poisonous gas may take place near vents according to the volcano's characteristic activity
Waspada level (advisory)	According to visual observations and instrumental records there are indications of increasing of volcanic activity
Siaga level (watch)	According to visual observation and instrumental records there are prominent indications of increasing volcanic activity. Eruptions may take place but do not threaten settlements and/or activities of communities near the volcano.
Awas level (warning)	According to visual observations and instrumental records, there are significant indications of volcanic activity, which are followed eruptions and potentially threaten settlements and or community activities around the volcano

for overall volcano hazard mitigation is the Center for Volcanology and Geological Hazard Mitigation (CVGHM), which uses 4 alert levels (Table 1) to communicate hazards and recommend actions to be taken by the Disaster Management agencies. The alert levels are: Normal, Advisory (Waspada), Watch (Siaga) and Warning (Awas). Characteristics of volcanic activity are defined for each alert level and specific activities of mitigation are linked to the alert levels. For Normal, Advisory and Watch levels these activities include socialization, preparation of contingency plans, simulations (e.g., table top exercises), and evacuation drills. When the highest alert level (Warning) is declared, evacuation of people in a specified threatened area is recommended by CVGHM, and the local authorities take the action to evacuate the people.

During levels Watch and Warning, communication to the public and amongst stakeholders becomes intense and frequent. Communication is accomplished in various ways, including telephone (mobile and land-lines), text messaging, fax, television, radio and radio streaming; the latter is arranged by communities and typically utilizes hand-held citizens-band radios for streaming information.

In order to understand the information and disaster mitigation processes, socialization and simulations (table top exercises, "TTX") are conducted with local disaster mitigation agencies and selected community members (see Fig. 2). Simulations are implemented according to community contingency plans for threatened areas. These exercises help stakeholders and community members understand volcanic hazard information and what to do to respond according to their contingency plans. A wide range of stakeholders are involved, including both local national authorities, such as representatives from agencies involved in public works, social, health, energy and mineral resources (parent agency for CVGHM), and transportation, as well as the Central Bureau of Statistics, Non-Governmental Agencies (e.g., Red Cross), and volunteers.

3   Gaps in Communication

Understanding information flow as a part of early warning systems is essential for dissemination of hazard information to the public. There are a number of factors that can hamper the process of communication, i.e.: culture and language, hazard perception, mandates and policy. According to Damen (1987), culture may be defined as learned and shared human patterns or models for living. Therefore, culture relates to mankind's adaptive mechanisms and includes local beliefs, religion, language, social habits and communication.

In hazard mitigation, a cultural approach is used to improve the capacity of a community to cope with disaster. Donovan (2009) noted that social and economic factors should be considered, such as during the Merapi crisis of 2006. In Indonesia, the culture of a community largely affects

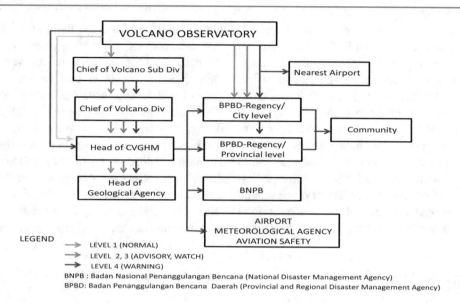

Fig. 2 Information flow during volcanic activity. This diagram illustrates how information flows from CVGHM to Disaster Management Agencies and other related institutions

individuals' perception of hazards. This perception relates strongly to local beliefs, which are typically associated with the community members' experiences during previous eruptions and with physical barriers or obstructions between them and the volcano (Lavigne et al. 2008). An example of a change in physical barriers was evident during the 2006 eruption of Merapi. People on the southern flank of the volcano had long believed that the Gegerbuaya ridge protected them, as it had for decades. The Gegerbuaya was a steep ridge near the summit of Merapi that protected much of the southern flank, as it directed pyroclastic flows to the southwest. However, during the 2006 eruption, the Gegerbuaya was eroded and subsequent pyroclastic flows and surges in 2006 and during the large "100-year" eruption in 2010 had a profound impact not just to the southwest, but also on the southern flank. Unfortunately, the perception of protection of the southern flank persisted long after the "protector" (the Gegerbuaya) had ceased to exist. Consequently, it was important to overcome this long-held misperception through community education during the 2006–2010 time interval.

In most cases, communication from scientists and disaster managers to the public is complicated by uncertainty involved in eruption forecasting and by limitations in understanding by public authorities. This may lead to uncertainty in decision-making and accordingly, in public mitigation actions (Morchio 1993). Consequently, effective two-way communication between parties is important. Simple language to deliver hazard information, supported by a cultural approach is one of the solutions.

Indonesia is a nation of diverse cultures and religions, and as noted by Chester (2005) in such environments religious and community leaders have important roles in disaster management. This is the case in certain areas of Indonesia, where communication between government scientists or disaster managers and society is most effective when done through religious or community leaders. For example, hazard communication during the ongoing Sinabung volcanic crisis in Sumatra has been more effective through the local religion leaders; whereas, communication through other community leaders was effective during recent crises at Merapi and Kelud volcanoes in Java.

The manner of communication, including culturally sensitive approaches, use of local languages, and appropriate means of delivery of information is critical. Indonesia has hundreds of local languages, although Bahasa Indonesia ("language of Indonesia") is the common denominator for hazard communication and in general socialization of hazard information is carried out in Indonesian. However in some places, local languages are still needed for communication and understanding and to be effective. In order to enhance communication and to improve hazard mitigation in this diverse linguistic and cultural nation, CVGHM operates 68 local observatories, staffed by observers from nearby communities who speak local languages.

them to quickly understand the hazard information and respond appropriately during recent crises. In contrast, the communities near Sinabung (which had not erupted in hundreds of years) had less experience, and were thus less aware and less able to respond as effectively during the crisis. This example of a population that is not experienced in volcanic eruptions is by no means unique. For example, Solana et al. (2008) notes that even though local authorities are aware of the hazard at Vesuvius, there is still incomplete understanding among communities as to how to respond during a volcano crisis.

3.1 Hazard Perception

The response of disaster mitigation agencies and communities to anticipate disaster depends on a common understanding of the hazards and an ability to take action during the event. Differences in hazard perception among emergency managers and scientists can lead to different and sometimes confusing and dangerous response actions. Our experience is that such differences may be caused by overlaps in mandate and authority, different levels of knowledge, errors in communication and coordination amongst stakeholders, and by ineffective communication between government authorities and communities. These issues are not unique to Indonesia. Qualitative and quantitative studies by Haynes et al. (2008a) of the factors controlling risk perception during the volcanic crisis on the Caribbean island of Montserrat show that difficulties in communication and understanding of uncertainties pertaining to volcanic risk led to confusion and social, economic and political forces resulted in distorted risk messages.

Hazard perceptions of local governments and communities are also influenced by their experiences during previous disasters. Communities from Kelud and Merapi have had far more experience in dealing with eruptions than the communities around Sinabung. The experience of those near Kelud and Merapi has allowed

3.2 Mandate and Policy

As previously noted, the responsibility for coordination of disaster management in Indonesia lies with the National Disaster Management Agency (BNPB) and with the Provincial and Regional Disaster Management Agencies (BPBD). BNPB has the mandate to coordinate all the stakeholders and to manage the situation during crisis. Therefore, all reports regarding events are firstly delivered to this agency for action. BNPB also has authority to publish policies, guidelines and protocols related to preparedness, mitigation and emergencies regulation. While this is of great value at the national level, a lack of socialization and understanding of the regulation at the local level may result in less than optimum implementation.

Although specific mitigation actions, such as evacuations are carried out by local authorities and managed by BPBD, the overall mandate for the information that results in the mitigation of volcanic hazards in Indonesia is given to CVGHM. This institution is responsible for volcano monitoring, issuing alert levels, and providing recommendations for evacuation to BNPB, BPBD and local governments and for dissemination of hazards information. The protocol for determination of alert levels is established by CVGHM. Problems have arisen in cases when other institutions or individuals have intervened by issuing statements to the mass media, in some cases even causing panic among the public.

4 Problems and Solutions in Communicating Hazards and Achieving Community Preparedness

The aim of the dissemination of knowledge about hazards should be to create awareness among the threatened parties and we suggest that the biggest improvement in communicating hazards in Indonesia will come about through encouraging actors to behave and act equally and in harmony. However, in implementation, we find that there is often a disparity between hazard mitigation actors. Frequently, hazard mitigation institutions attempt to impose their role at levels higher than that of the community. In these cases, key roles of assessment and participatory learning may be considered inappropriate and therefore may not be used. In such cases, dissemination of information tends to be carried out from the top down and centralized, often delaying transmission of critical information and contributing to misunderstanding at the local level. Occasionally, the source of knowledge depends on a single person or group. In other cases during crisis, many groups of people may get involved in dissemination of information without proper knowledge and without adequate coordination with the institution responsible for socialization. We have found that the best result in improving community preparedness is achieved when the style and mode of communication between actors changes gradually. An informal approach is a key factor in good communication, and we find that such an informal mode is an effective means to transfer knowledge through participatory activities during the quiet times between volcanic crises.

4.1 The Role of Media

Involvement of media or mass communication is important to improve public knowledge and response capacity. Access to public media is very much a part of daily public life in Indonesia and at all levels of society. Consequently, it is an effective means for both long-term education and for short-term communication of hazard information. To transfer hazard information to the public several mass media methods are used, namely direct communication through television programs, interviews and articles in newspapers, and webpages. In addition, socialization programs are provided by CVGHM to target priority audiences, utilizing workshops, seminars, exhibitions and formal or informal discussions with the public in the areas of concern.

To enhance the effectiveness of mass media communication, workshops with reporters and scientists on translation of technical data into public language have been carried out. Potential problems include the reporters and the media companies' own interests in how they deliver hazard messages. CVGHM recognizes this need and works with reporters to keep the message appropriate to the scale of the hazards and to make sure that critical public safety messages are effectively communicated. In addition, we recognize value of involving media in observation of capacity building of communities before, during and after disasters. Although the media is typically less attracted in pre-disaster activities, our involvement with news reporters during crises and following disasters helps increase interest in public-interest and capacity-building stories about pre-disaster activities, which might otherwise be neglected by the media.

4.2 Building Trust

To build trust between scientist and communities, equity in level of communication is important, in addition to understanding of local culture, local language and people's character. Sharing information to identify problems will also encourage people be more involved in hazard mitigation. Various targets and conditions of communities during the capacity building process stimulate scientists to be flexible and modify procedures and understanding of hazards. This may include equating perception and

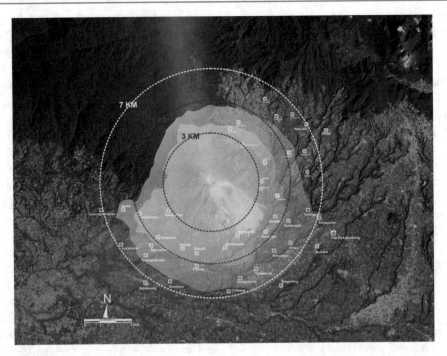

Fig. 3 Map showing direction of pyroclastic flows of Sinabung volcano on 1 February 2014 (*black arrow*) and 5 km exclusion zone (*magenta dashed circle*)

terminology to other more commonly understood threats to communities.

Here we describe how pre-established trust of communities in government (scientists and disaster managers) during the volcano crisis of February 2014 at Kelud resulted in a timely and efficient evacuation. People obtained and distributed information to others by community radio, text messaging and community gatherings. The CVGHM observatory post was one of the main sources of information. New local community leaders emerged from a local disaster preparedness group known as Wajib Latih (Indonesian for "compulsory training"; a group defined in more detail below). Such leaders played an important role in communicating hazard information during the Kelud crisis. The Chief of the District of Ngancar was involved in the distribution of information through radio briefings with the help of local leaders and members of the communities. During the crises, the communities of Kelud were also involved in keeping the public away from exclusion zones.

This case illustrates the independency of communities to take action, and it represents a bottom-up process in mitigation.

In contrast, a low capacity and experience of Sinabung communities resulted in inconsistent responses, such as repeated requests for confirmation of hazard information, attempts to negotiate before taking action, and less consideration in taking risks. A tragic example took place during the Sinabung eruption on 1 February 2014, which resulted in 17 fatalities. As a consequence of ineffective understanding of the risk, these people entered the 5 km exclusion zone (Fig. 3) and were killed by a pyroclastic surge.

Improvements in the capacity of Sinabung communities have resulted from the efforts of staff members of CVGHM conducting socialization work. These improvements are clearly a result of increasing trust of community leaders in the relevant communicator. Such individual trust building is an important way to improve volcanic risk communication (Haynes et al. 2008b).

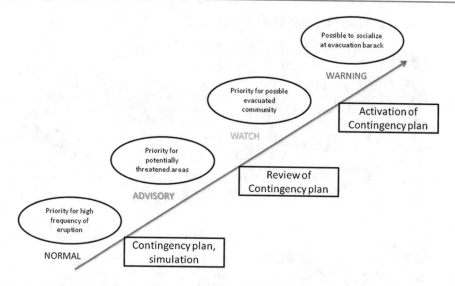

Fig. 4 Capacity building in communities and its implementation during volcano eruption according to alert level

Indonesia has a unique involvement of scientists in mitigation actions related to volcanic eruptions. As a mandate holder, CVGHM is responsible not only for monitoring and volcano hazard evaluation, but also for mitigation of volcanic hazards, as alert levels are directly tied to mitigation actions and areas recommended for evacuation are specified in formal CVGHM notifications. Scientists and decision makers who issue volcano alert levels are in the same institution. Scientists from different institutions may provide input based on research but do not to issue alert levels. Further, decision makers communicate directly with disaster managers, such as BNPB and BPBD and provide specific recommendations regarding mitigation actions. BNPB and BPBD arrange, prepare, and through local authorities enact mitigation plans.

5 Steps of Knowledge Transfer and Communication

In general, there are several steps in delivering information that we use in Indonesia, namely socialization, preparation of contingency plan documents, simulation and evacuation drills.

5.1 Socialization

Socialization is dissemination of hazard information to people at risk. In Indonesia, the level of volcano activity and priority of socialization by CVGHM changes at different alert levels (Fig. 4). At the Normal level, socialization is given to people living close to volcanoes that experience a high frequency of eruptions. At the Advisory level, socialization is carried out with priority to people in the area potentially threatened by hazards. At Watch level, it is carried out in area which will likely be evacuated in the case of a Warning level alert. At the Warning level, additional socialization is conducted in evacuation camps or barracks, if it is needed.

In hazard zones where there is a high frequency of volcanic eruptions, continuity of hazard information is important to maintain awareness of threatened communities. A model of participatory training is most appropriate in these situations. This model aims to implement "training of the trainer" or compulsory training ("Wajib Latih," in the language of Indonesia). Wajib Latih represents the lowest social level of disaster management and mitigation planning. It is a disaster management learning activity that is undertaken to bring

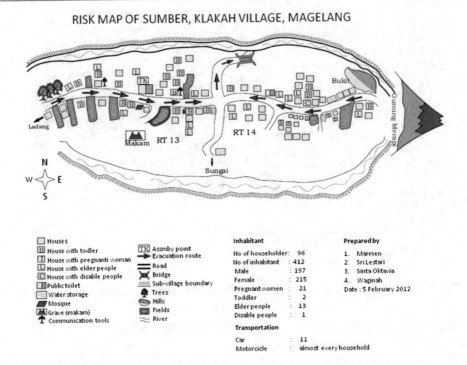

Fig. 5 Village risk map shows total number of inhabitants, resources, locations of vulnerable groups and infrastructure for evacuation

together the instructors, local stakeholders and community leaders. Most activities are conducted by CVGHM and result in the creation of a new local leader. From our experience, the creation and establishment of such a local leader plays an important role in effective community evacuation. Through Wajib Latih, information is shared within communities and by members of communities. In this activity, the group has the task to formulate a contingency plan and SOP of evacuation, including making village risk maps (Fig. 5). Each member of the group is given a specific responsibility and as a result, each acts as a leader of their own task. Because the plan is created by the community itself, it is easily understood and effectively utilized by the community.

The Wajib Latih requires a common perception between government and the community in order to create a sharing environment to find solutions. However, even with this community-based process, this condition is difficult to achieve as even in areas near a volcano, hazard perception may vary from sector to sector.

To a significant degree the success of the method depends on the experience of those involved in dealing with the hazard directly and repeatedly.

5.2 Contingency Plan

According to UNISDR (2009) *"contingency planning is a management process that analyses specific potential events or emerging situations that might threaten society or the environment and establishes arrangements in advance to enable timely, effective and appropriate responses to such events and situations"*.

By Public Law in Indonesia, contingency plans include scenarios and goals, establishment of technical and managerial actions, as well as response plans for mobilization of mutually agreed stakeholders (Government of Indonesia 2007). The formulation of a contingency plan is carried out in two ways, namely community-based and stakeholder-based. Contingency plans are prepared by village (through Wajib Latih groups), and also at regency and national levels (stakeholder-based). At Merapi and Kelud, both of these two approaches were carried out. In preparation of the contingency plans, many stakeholders are involved. Stakeholders involved in Wajib Latih groups include CVGHM, local authorities at village level, community health centers, community preparedness groups, volunteers, and representative community members and leaders. Stakeholder-based contingency plans involve CVGHM, local authorities at regency and provincial levels, local health, public works, social, communication and information, and transportation agencies, community preparedness groups, volunteers, Red Cross/Red Crescent and search and rescue units. During the process of formulating contingency plans, the input of stakeholders is needed to define their appropriate roles. Formulation of the contingency plan includes identification of hazard, threat, and vulnerability, as well as a determination of possible disaster impacts, risk reduction measures, readiness and response mechanisms, and a distribution of tasks, mandates and available resources (Government of Indonesia 2008). Besides providing the contingency plan document and SOP for evacuation, the process also improves communication and coordination of hazard perception amongst stakeholders and community leaders.

The Wajib Latih process is similar to Participatory Rural Appraisal (PRA) methods implemented in the Solomon Islands (Cronin et al. 2004a, b), which involves stakeholders from communities and government to encourage community-based planning. Wajib Latih and PRA emphasize the important of dialogue among stakeholders to integrate various aspects, both social and physical to derive risk assessment and mitigation plans.

Increased capacity of a community requires not just knowledge of the hazard but also effective communication and coordination, in which those responsible for crisis management understand the policy, guidelines and standard operating procedures of the entire process. Therefore, in the preparation of contingency plans it is important to document and empower hazard mitigation institutions (stakeholders) and the

community at risk. Contingency planning also requires understanding of the mandates of each institution and the distribution and coordination of authorities. During contingency planning and through simulation of the plans, needs and gaps are identified and addressed with solutions that will be effective during future crises. Contingency plans are not only useful to identify hazards and vulnerabilities and to enhance understanding of evacuation procedures, but the preparation of these plans also builds a critical communication network and coordination amongst stakeholders, community leaders and members of villages.

As previously noted, preparation of these plans takes place during Normal or Advisory alert levels. In the Watch level, a review of contingency plan is carried out in order to update all data before a real disaster and the contingency plan is changed to an operation plan. At this time and at higher governmental levels (regency/provincial and national levels) an Incident Commander is appointed. The operational plan is then used by the Incident Commander and all stakeholders to guide the response, typically an evacuation (Fig. 4). Activities of communities in each Alert level are shown in Table 2.

5.3 Simulation (Table Top Exercises)

Table top exercises are designed to test the ability of disaster mitigation officials to respond. In this case, the exercise also aims to test and review the procedures set out in the contingency plan. The exercise involves key persons from the various sectors as described in the contingency plan. As used by CVGHM, the nature of the exercise is informal, such that participants may improve the scenarios, share information and experiences, and overall provide input to improve the plan.

5.4 Evacuation Drills

Drills are used to practice evacuations involving communities and all stakeholders. These exercises are carried out in the field, involve

Table 2 Volcano alert levels and community preparedness taken from the Indonesia national standard training guideline for community preparedness to anticipate hazards of volcanic eruption (in press)

Alert level	Activity of community
Normal (normal, level 1)	1. Socialization of volcanic hazard map 2. Understanding of character of volcano hazards 3. Community understanding regarding their settlement within volcanic hazard map 4. Census of inhabitants within hazard zones 5. Inventory of resources within hazard zones 6. Formulation of SOP 7. Preparation of sign and evacuation route 8. Simulation
Advisory (Waspada, level 2)	1. Dissemination of increasing alert level 2. Updating census of inhabitants 3. Updating of vulnerability of inhabitants within hazard zones 4. Intensification of inventory of resources within hazard zones 5. Preparation of equipment and communication system 6. Preparation of evacuation plan 7. Preparation of transportation for evacuation 8. Preparation of evacuation barracks 9. Preparation of logistics 10. Explanation to community 11. Grouping of communities
Watch (Siaga, level 3)	1. Dissemination of increased alert level 2. Sign of alert is ready to be operated 3. Transportation for evacuation is ready to be operated 4. Evacuation barracks are ready to be operated 5. Logistics are ready to be operated 6. Security is ready to be operated 7. SOP is ready to be operated 8. Equipment and communication system is activated 9. Determination of the emergency response command
Warning (Awas, level 4)	1. Dissemination of increased alert level 2. Warning signals are sounded 3. Evacuation order from Incident Commander is executed 4. Activation of SOP 5. Evacuation 6. Activation of evacuation barracks 7. Activation of logistics 8. Activation of security 9. Activation of crisis center

Alert levels are given in English and Indonesian (in parentheses)

vulnerable people and mobilize resources and communities according to scenarios from the contingency plan. Our experience is that review of the contingency plan during evacuation drills substantially improves the plan. For example, risks related to the route of evacuation, location and facilities at evacuation camps or barracks and modes of transportation are all commonly addressed and solutions found.

5.5 Leadership

In disaster management, the role of leadership is important, both for decision makers and local leaders. As previously noted, during the capacity building process of institutions and communities, often a new leader emerges. Such a leader can mobilize disaster management agencies and communities to take action according their capacity and capability to anticipate disaster. The new leader is typically most effective during the evacuation process. In our experience and across a wide variety of Indonesian cultures, we find that a trusted community leader is patient, open-minded, caring, flexible, and a good communicator. Such a leader also has great endurance. The function of a leader in the community is to serve as a role model and initiator; one who encourages people to take proper action according to their capability.

6 Conclusion

Communicating hazard information is a time consuming process, as the interaction between community (social) and scientists (physical perspectives) and disaster management agencies (policy and practice) requires equity in hazard perception. We find that informal approaches are keys to success. An informal nature of communication encourages people to share experience, knowledge and problems without regard to their background differences and is best accomplished through socialization using a participatory knowledge dissemination methods. These activities are carried out through discussion and sharing of experiences among actors, analyzing problems from different perspectives to find solutions, and involvement of various groups to maintain diversity and necessity. The main point of the activity is to identify underlying problems in hazard perception, cultural backgrounds and community characteristics and to harmonize the point of view of both hazard mitigation officials and communities. This is consistent with the conclusion of Howes and Minos-Minopoulos (2004), who pointed out the importance of public perception of hazard, risk and vulnerability in relation to public education programs and disaster management plans.

One of the factors to motivate communities to protect themselves from natural disasters is to encourage people to find out where and how to obtain hazard information and to conduct their own assessments that lead to appropriate actions. This can be achieved by empowerment of communities through participatory learning. Perry and Lindell (2008) proposed that there is correlation between responsibility for self-protection of the community that has had an experience of property damage and information seeking behavior related to protective action. We find this to be the case in Indonesia.

Both Merapi and Kelud communities are experienced the Wajib Latih process. The core activity of Wajib Latih groups is to encourage people to be capable to respond and take action appropriately during crisis. This behavior can be achieved through intensive training and involvement in the preparation of mitigation plans. The training also emphasizes the difference between evacuating and "being evacuated." The first case implies being prepared and actively participating and the second reflects being unprepared and passively participating. The evacuation processes at both Kelud and Merapi communities illustrates the value in preparedness. Even during the short-term crisis of the 2014 Kelud eruption (less than one day of warning), part of the community evacuated themselves before a recommendation for evacuation was issued (i.e., Warning level alert being issued). This condition represents independency of the community to anticipate and take action during a crisis according to their capacity and knowledge.

An important lesson from the experience of disaster mitigation in Indonesia is the necessity to maintain effective communication between scientists and those responsible for mitigation by respecting mandates and authorities for disaster management and by directly involving communities in hazard mitigation. Such effective communication and community involvement is supported by development of policy, strategy and mitigation plans by government which involves public participation.

Acknowledgements We would like to thank to the Director of Center for Volcanology and Geological Hazard Mitigation, Indonesia for providing chance to us to involve in many activities in pre-disaster and during crisis. We are also grateful to John Pallister of the USGS, for provide fruitful comments, and in the preparation and editing of the manuscript.

References

Chester DK (2005) Theology and disaster studies: the need for dialogue. J Volcanol Geotherm Res 146:319–328

Cronin SJ, Gaylord DR, Charley D, Brent V, Alloway BV, Wallez S, Esau JW (2004a) Participatory methods of incorporating scientific with traditional knowledge for volcanic hazard management on Ambae Island, Vanuatu, Bulletin of Volcanology, vol 66, pp 652–668

Cronin SJ, Patterson MG, Taylor PW, Biliki R (2004b) Maximising multi-stakeholder participation in government and community volcanic hazard management programs: a case study from Savo, Solomon Islands. Nat Hazards 33:105–113

Damen L (1987) Culture learning: the fifth dimension in the language classroom, vol 11478. Addison Wesley Publishing Company

Dominey-Howes D, Minos-Minopoulos D (2004) Perception of hazard and Risk on Santorini. J Volcanol Geotherm Res 137:285–310

Donovan K (2009) Doing social volcanology: exploring volcanic culture in Indonesia. Area 42:117–126

Government of Indonesia (2007) Law of Republic Indonesia no 24, 2007 concerning disaster mitigation, published by the Government of Indonesia

Government of Indonesia (2008) Regulation of Head of National Disaster Management Agency No 4, published by the Government of Indonesia

Government of Indonesia (2010) Badan Pusat Statistik (Central Bureau of Statistics). http://www.bps.go.id/tab_sub/view.php?tabel=1&id_subyek=12

Government of Indonesia (2014) Indonesia National Disaster Management Agency, Rencana Nasional Penanggulangan Bencana 2015–2019 (National Disaster Mitigation Plan), published by the Government of Indonesia

Government of Indonesia (2015) Indonesia National Standard, Training guideline for community preparedness against hazards of volcanic eruption (Standar Nasional Indonesia, Panduan Pelatihan Kesiapsiagaan Masyarakat Terhadap Bahaya Erupsi Gunungapi), published by the Government of Indonesia (in press)

Haynes K, Barclay J, Pidgeon N (2008a) Whose reality counts? Factors affecting the perception of volcanic risk. J Volcanol Geotherm Res 172:259–272

Haynes K, Barclay J, Pidgeon N (2008b) The issue of trust and its influence on risk, communication during a volcanic crisis. Bull Volcanol 70:605–621

Lavigne F, de Coster B, Juvin N, Flohic F, Gaillard JC, Texier P, Morin J, Sartohadi J (2008) People's behaviour in the face of volcanic hazard: perspective from Javanese communities, Indonesia. J Volcanol Geotherm Res 172:273–287

Morchio R (1993) The effect of the uncertainty in natural prediction on the user communities. In: Nemec J, Nigg JM, Siggardi F (eds) Prediction and perception of natural hazards. Kluwer, Dordrecht, pp 39–47

Paton D, Johnston D (2001) Disasters and communities: vulnerability, resilience and preparedness. Disaster Prev Manag Int J 10:270–277

Perry RW, Lindell MK (2008) Volcanic risk perception and adjustment in a multi-hazard environment. J Volcanol Geotherm Res 172:170–178

Ronan KR, Johnston DM (2005) Promoting community resilience in disasters; the role for schools, youth, and families. Springer, New York 213 pp

Solana MC, Kilburn CRJ, Rolandi G (2008) Communicating eruption and hazard forecasts on Vesuvius, Southern Italy. J Volcanol Geotherm Res 172:308–314

UNISDR (2009) UNISDR terminology on disaster risk reduction. United Nations International Strategy for Disaster Reduction, Geneva

Social Representation of Human Resettlement Associated with Risk from Volcán de Colima, Mexico

Alicia Cuevas-Muñiz and Juan Carlos Gavilanes-Ruiz

Abstract

This study examines a rural community of several decades of existence called *"La Yerbabuena"*. *La Yerbabuena* belongs to the state of Colima and it is situated on the flanks of the active *Volcán de Colima*; as such, the inhabitants are exposed to high levels of volcanic activity in their daily lives. This community has experienced resettlement on several occasions due to the volcanic risk. The study is based on theories of social representation that deal with how people perceive the events of their daily lives, as well as what happens in their immediate environment, including available information (such as news) and interactions with familiar people. These perceptions are formed partly from personal experience and from information, knowledge, and patterns of thought acquired during a shared tradition, education, and social communication. The social representation of families resettled due to volcanic risk is classified in four categories or assumptions: (a) the volcano represents a potential risk to their lives and possessions; (b) their relocation involved a change in economic, political and cultural factors that impacted on their daily lives; (c) this relocation represented a benefit to their daily lives; and (d) the relocation fractured the social cohesion of the community. Meanwhile, for the families who opposed their own resettlement, social representation was anchored in three aspects: (a) *La Yerbabuena* is not considered a zone of high volcanic risk; instead the resettlement was a *"governmental pretext to expropriate these families of their land and possessions"*; (b) the resettlement was a violation of their human rights, given the harassment they received before and during the resettlement process; and (c) their failure to acknowledge a volcanic risk allowed them to implement strategies of resistant, such as generating discussions and actions appropriate only when the volcano was no threat and became a *"guardian"*. This refers to the old traditional view

A. Cuevas-Muñiz (✉) · J.C. Gavilanes-Ruiz
Universidad de Colima, Colima, Mexico
e-mail: alicia_cuevas@ucol.mx

Advs in Volcanology (2018) 321–334
https://doi.org/10.1007/11157_2017_17

of the volcano as the guardian in which the volcano would protect this community from possible eruptions and it would emit signals in which only these families could perceive and interpret. At the same time, these resistant families depended on the risk management protections implemented by the authorities. Based on this research it is clear that social representation in La Yerbabuena is born from the mental images that both the relocated and the resistant constructed from the sociocultural reality common to all members of the town. It is therefore important that all social actors involved in risk management have an understanding of the culture, risk perception, and forms of social representation of the volcanic risk of the inhabitants of communities high risk zones in order to design plans suitable for prevention. Furthermore, it is critical for the population to have active participation, to facilitate better risk management.

1 La Yerbabuena: A Space of Negotiation and Social Interaction

At the foothills of the Colima volcano, which is located between the states of Colima and Jalisco, there are several rural settlements surrounding this colossus. La Yerbabuena town is located 8 km to the southwest of the volcano's summit. It is an "ejido", an area of communal land used for agriculture on which community members get assigned a parcel but maintain communal ownership of the land. It is officially named "Ex-hacienda San Antonio" and it belongs to the municipality of Comala, which is one of the ten municipalities that constitute the state of Colima.

La Yerbabuena is a young town that for decades struggled with the fair distribution of the land. In 1934 workers of the Hacienda San Antonio and those of the Red Union of Small Farmers, made the first written request of endowment of the agricultural communal land to the Governor of the State of Colima. However, it was not until 1968 that the resolution was approved in favor of the 26 applicants who took possession of 540 ha. The new *ejido* was officially called "Ex Hacienda San Antonio La Yerbabuena." The village was established in 1968 by a group of families from the former hacienda San Antonio and nearby towns such as La Becerrera, Suchitlán, and Cofradía de Suchitlán in Colima; as well as from San José del Carmen, Zapotitlán, and Tamazula in Jalisco (Cuevas 2001).

La Yerbabuena is a traditional social space where power relations are interwoven; social actors perform several traditional acts like maintaining kinship and labor ties as well as negotiating and interacting face to face with the members of the village and some external social groups. In this context, the relocation of a part of the peasantry population formed another community social space with physical-spatial characteristics and habitats different from those of their locality of origin. On February 1999 the government of the state of Colima proposed to relocate La Yerbabuena, but it was not until May 2002 that most of the Yerbabuenians moved to the new settlement (La Nueva Yerbabuena). However, more than 10 families refused to accept the government's conditions and decided to stay at La Yerbabuena. As a result, two groups of settlers formed: those who were "relocated" and, those that were "resistant", building more fractional social relations from that process. The social cohesion of the community was fractionated since that time and it was accentuated by the displacement of part of this rural population.

This study applies the theory of social representation to explore the issues related to the resettlement in the community of La Yerbabuena.

2 Theoretical Foundations of Social Representation

Social representation is defined by Moscovici (1973) p. xiii as:

> Systems of values, ideas and practices with a two-fold function; first, to establish an order which will enable individuals to orientate themselves in their material and social world and to master it; secondly, to enable communication to take place amongst members of a community by providing them with a code for social exchange and a code for naming and classifying unambiguously the various aspects of their world and their individual and group history.

The history of the theory of social representation began with the contributions of Herbart (1825) and French sociologist Emile Durkheim (1893). Herbart developed his concepts from psychosocial situations (understood as the interactions between the individual's mind and its collective environment) and helped to give meaning to the relationship between an individual and society. This is a key factor in the understanding of several situations related to volcanic activity; as an example we have the different cultural roles or perceptions of risk within the village. For his part, Durkheim, speaking of "collective representation" and "cultural factors", proposed the possibility of investigating how this collective representation came to form part of a subjective perspective that positioned an individual against an object and against oneself.

The theory of social representations has been applied in several studies not only related to social psychology but also incorporated into sciences and disciplines as diverse as anthropology, sociology, pedagogy, social work, among others. In fact, recent risk-disaster research includes some areas that have used the theoretical-conceptual tools built by Moscovici (Aparico and Pérez 2014; Bravi 2016).

This study is based on the theory of social representations and its tools used to construct meanings of social actors; in this case, those that resettled and those that resisted. These meanings are socially produced by the actors through culture. Based on them, people organize and give meaning to their experiences and knowledge through narrations, forming beliefs about their own world and themselves. In this way, people legitimize their world from verbal actions and actions based on the relationship between what is done and what is said (Brunner 1992).

According to Moscovici, social representations are a set of concepts, statements, and explanations originated in daily life and the inter-individual communications. In our society, they correspond to the myths and belief systems of the traditional customs. One could say that they are the contemporary view of commonsense (Moscovici 1981, quoted by Perera 1999, p. 10). The relationship established in this work is between the theory of social representations of Moscovici (1969) and the constructivist view of the social actor of Norman Long based on Habermas' concept of "worlds of life" (1987). With the perspective of the social actor we could only illustrate cultural practices and interpretations developed by both relocated and resistant people in the process of permanent resettlement of La Yerbabuena. However, such a perspective would be insufficient to describe how these actors construct their own meanings of displacement. For this reason, from the relocated and resistant world views of life, we explore how these meanings are constructed: first, starting from the theory of social representations, used as an instrument of analysis, since through it we explore how individuals are oriented and act in their relationships with other individuals; and second, on the basis of meanings or understandings about the world that are being created and transformed as the interactions between social actors in the resettlement as it progressed. That is, how the social actors from their feelings, worlds of life, experiences, perceptions and systems of knowledge are building different meanings about the relocation, that in some of them there are similarities, but also differences.

In addition of the perspectives of Long (1998) and Habermas (1989) that enabled us to consider the advancing interactions between social actors as the central object of this study, we based our research on the sole theory of social representation to examine how individuals react to and act in their relations towards other individuals, and on the basis of meanings or intentions how they approach their world views that they are creating or transforming. We considered that these theoretical approaches could enhance our understanding of the role of the resettlement of La Yerbabuena in the social construction of risk.

In people's daily lives, the images that are constructed of a certain social process emerge as a mental elaboration that takes into account the history of the people, the experiences of each person's life, the personal cognitive constructions and above all, their world views of life. In these social processes, there are articulated fields of multiple meanings that are shared by belonging to the same social and cultural space, and having very similar meanings of these processes. Hence each social group tends to make appraisals of reality, which are built from their own experience, but also from the interactions they established with other actors; so, it can be said that the knowledge that is acquired from a social process or a reality corresponds to forms of interpretation of the world that are socially constructed and shared by the members of a group in a given context (Moscovici 1979).

Within social psychology, and according to the approaches of Moscovici (1979) and Jodelet and Tapia (2000), social representation is generated from two phases or processes: first, *"objectification"*, and second, *"anchorage"*. The first consists of transforming an abstract entity or thing into something concrete and material: products of the imagination into physical reality, concepts into forms. Objectification turns a concept into reality, giving an image its corresponding physical counterpart. The outcome is chiefly cognitive. The amount of meanings that a person receives, expresses, and picks up in their daily cycle of interactions may be superabundant. To decrease the separation between the sheer volume of words in circulation and the objects to which they are related, just as one could not speak of "nothing", "linguistic signs" latch onto "material structures": that is to say, the linguistic signs try to attach the word to the thing (Moscovici 1979, p. 75).

Objectification may be defined as an image-forming and representation-structuring process. In this process, the social part translates as the assemblage of knowledge concerning the subject of a representation articulated by a feature of social thought: to make concrete the abstract; to materialize the word. Furthermore, the process of objectification carries within itself two essential operations: (1) naturalization, and (2) classification. In naturalization, social representation is given concrete evidence through conversion into a "common theory" which can categorize autonomous individuals and their behaviors. Classification makes sense of the world around us and introduces a new order that adapts to the existing one mitigating the impact of any new design.

The second phase of social representation is anchorage, which is a process of categorization whereby we classify and give names to things and to people; the integration of the unfamiliar into the familiar. Anchorage denotes the introduction of knowledge into the hierarchy of values, and between transactions that occur in society. In other terms, "through the process of anchorage, society changes a social purpose by an ordering operation, and this purpose is located on a scale of preferences in existing social relationships." (Moscovici 1979, p. 121).

In an artificial way, Moscovici clarifies both processes, arguing that "objectification translates knowledge into the domain of being, and anchorage defines knowledge within the realm of doing" (Moscovici 1979, p. 121); that is to say, just as objectification demonstrates how the elements of knowledge are articulated in a social reality, anchorage makes visible the way in which these elements shape social relationships and also how the elements are expressed.

By Moscovici's definition, social representation arises in four constituent elements. The information which is related to what "I am"; the images which "I see"; the opinions which "I think"; and the attitudes which "I feel". Meanwhile, Jodelet and Tapia (2000) asserts that social representations are areas of knowledge that form part of our personal experiences, but also form information, understanding and patterns of thought that we acquire and share through tradition, education, and social communication. Seen in this way, social representation is directly and exclusively related to common humanity due to the individual reality of all human beings.

3 Methodology

This study adopts a qualitative and cross-sectional methodology guided by the theory of social representations (Moscovici 1979) as a tool for the construction of the image and meanings of the resettlement individuals. This study also analyzes the representation of this social object using two techniques: 'free' (Abric 2001; Borgatti 1996) and 'draw lots or sorting' (Abric 2000; Borgatti 1996), both described further below. These tools explored the meanings attributed both by resistant and resettled individuals to the permanent resettlement. These meanings were constructed by the Yerbabuenians from their own life worlds, experiences, feelings, perceptions, and knowledge of the process of resettlement. The data analysis was carried out using Anthropac 4.9 software (Borgatti 1996), in which a correlation of the knowledge possessed by each informant in relation to the group was calculated to identify the informant's key words; the ones with the highest scores were the people identified to conduct the in-depth interviews.

The studied population were 30 people (randomly selected), amongst them 20 were relocated settlers and 10 were resistant. The study consisted in: (1) exploratory image of the relocation; (2) the dynamics of obtaining the relocation image; and (3) explanatory of the relocation process.

In the image of the resettlement the free association technique was used, as well as the 3-question semi-structured questionnaire instrument called free listing (Abri 2001; Borgatti 1996). Each questionnaire lasted 20 min, and was recorded both in digital audio as well as written hardcopy. The free association technique consists of asking the social actors to name or write all elements that correspond to an inductive term. For example, *to explore word associations relating to the term "relocation" interviewees were asked*: "Please mention all the words that come to mind when you hear the word *resettlement*." These word associations, called 'free listings' are sets of terms used in additional and subsequent data collection tasks, such as pile sorts or indexes or scales. For the free listings, two databases (resettled and resistant) were developed in the Anthropac 4.9 software, frequently used in anthropological studies to study cultural domains. Anthropac 4.9 counts the number of times each descriptor is mentioned by the interviewees, and then organizes the list on a decreasing frequency scale. In this way, it calculates a list of the descriptors, which includes three columns: the frequency, the order in which they were mentioned, and the cultural weight.

The draw lots or sorting approach is characterized by the descriptive phase of the resettlement image. We described how the different forms of social thought are objectified (Moscovici 1979; Jodelet 1984), which included the image and meanings of the resettlement that both the relocated and resistant people have since being displaced. Fifteen descriptors were identified by the Anthropac analysis to apply the draw of lots or sorting to both groups of interviewees (Table 1). The following descriptors were selected:

Table 1 Descriptors selected to apply the draw of lots or sorting

Government	Volcano	Family	Emotional status	Relationship with resettled and resistant
Benefit	Explosion	Houses	Confidence	Resentment
Human rights violation	Danger	Do not sow	Pain	Disunity
Government non-compliance	Risk zone	More expenses	Desperation	Difficulties

4 Construction of Meaning: The Process of Human Resettlement and Volcanic Risk as It Affected the Inhabitants of *La Yerbabuena*

In this section, we present the images and meanings constructed by both the resettled and resistant families of the at-risk community of *La Yerbabuena*, Colima who in beginning of May 2002, the last evacuation began the process of human resettlement.

4.1 Impressions of the Process of Human Relocation

(a) *Impressions of the relocation process formed by resettled habitants of* La Yerbabuena

The impressions of the relocation process for displaced families is divided into five descriptors derived from the *"Free Listings"*: Positive *emotional state; trust on government; benefits to family; volcanic threat;* and *relationship between the resettled and the resistant* (Table 2).

From these descriptors, we can observe that the *La Yerbabuena's* inhabitants have accepted the relocation process, due to the confidence that they had in the authorities (represented by the

highest frequency in the table); moreover, they were amenable to the latter's promises of better services, including health and education, that would benefit all the inhabitants. However, the disadvantages of the relocation process were that it provoked resentment between families and that the resettled families encountered various difficulties in the new settlement. The decision to relocate was not made for the risk that the volcano represented to them, but it was motivated by the promises of the state government, who had assured them that relocation would give them a better quality of life—an assurance that was proved to be false.

(b) *Impressions of the relocation process formed by resistant population of* La Yerbabuena

The impressions of the relocation process formed by the families who resisted resettlement were divided into four descriptors: *authoritative government, family deterioration, negative emotional state, and volcano threat* (Table 3).

These families attributed the relocation process to a decision made by an authoritative government, with hidden intentions, which had nothing to do with their living in a zone of high volcanic risk. They asserted that the displacement of inhabitants was forced because of

Table 2 Descriptors of the impressions of the relocation process held by displaced families

Descriptor	Fr^a	%	Om^b	Cultural weight[c]
Positive emotional state	28	16	4.3	2.018
Trust on government	61	35	2.6	1.707
Benefits to family	47	27	4.7	1.010
Volcanic threat	20	12	2	0.566
Relationship between displaced and resistant families	17	10	10.1	0.309
Total	173	100	23.7	5.61

[a]Frequency (*Fr*) refers to the number of times that each descriptor was mentioned

[b]Order of mention (*Om*) is provided by the subjects according to the order in which the descriptors mentioned the derivatives of the introductory term "relocation". The order of mention "can reveal aspects of the underlying cognitive structure of this domain" (Borgatti 1996, p. 5)

[c]'Cultural weight' is the correlation between frequency and order of mention, and was obtained through the use of Anthropac 4.9 software (Borgatti 1996). It classifies by type in order to select 15 descriptors for the implementation of the *"Drawing Lots"* phase of information collection

Table 3 Descriptors of the impression of relocation formed by resistant families

Descriptor	Fr	%	Om	Cultural weight
Authoritative government	26	41	3.5	1.385
Family deterioration	20	32	3.6	0.949
Negative emotional state	12	19	4.2	0.656
Volcano threat	5	8	3.6	0.111
Total	63	100	14.9	3.101

constant pressures, threats, and harassment from the authorities during the course of their daily lives and during the evacuation processes.[1] They recounted that, for some, resettlement did not affect the dynamics of daily life and that for others, resettlement afforded fewer opportunities of employment and resulting in economic imbalances in households.

As for the families' impressions associated with the volcano, they acknowledged that the local government displayed that *La Yerbabuena* as a zone of high volcanic risk and thus that the volcano presented a threat to the village's inhabitants. However, they declared that the hidden intention of the authorities was to expropriate them of their property and lands.

For both the resettled and the resistant, impressions of the relocation process are 2-fold. The initial impression of the resettled was one of wellbeing and progress, generated by the expectations and promises of the governmental authorities. The ultimate impression of the process formed by the resistant was that the volcanic threat was a pretext of the government to divest them of their goods and to carry out plans for a tourist project in their community.

4.2 Meanings of the Process of Human Relocation

In this section, we explain the phenomenon of resettlement in terms of image and meanings found in the context of everyday lives of social actors of *La Yerbabuena* (both relocated and

resistant), and from their subjective observations. This was achieved through the technique of in-depth interviews.

The results we present are interpretations developed from the theoretical and methodological assumptions of the theory of social representation (Moscovici 1979). These interpretations were elaborated in the analysis of social actors' narratives of their daily lives around the period of displacement of *La Yerbabuena* inhabitants due to volcanic risk. The results are displayed below in two sections: (1) significance of resettlement to relocated families; (2) significance of resettlement to resistant families.

(1) *Significance of resettlement to relocated families:*

This section itself is split into three parts: (a) meanings associated with the resettlement process; (b) impressions of daily life after resettlement; and (c) impressions of volcanic risk.

(a) *Meanings associated with the resettlement process:*

The following meanings, given by social actors in in-depth interviews, are attributed to the economic imbalance encountered when the new settlement's cost of living was higher than in the actors' place of origin. They recounted how previously, in order to provide for their basic necessities, they spent two hundred pesos each week; in the resettlement place (Cofradia de Suchitlán), their expenses doubled. Before resettling, Mrs. Reyna Cervantes along with another female partner was applying for a small loan. With the loan they hoped to buy a *nixtamal* mill with the intention of improving domestic savings by grinding maize and making tortillas.

---

[1]From November 1998 to May 2002, six evacuations of the population of *La Yerbabuena* occurred. The most comprehensive of these were those of May 10th, 1999 and May 18th, 2002.

Once displaced, they continued with their project and, although their incomes were minimal, they were not dissuaded from their objective. *Doña Reyna* said:

> What happened is that here you have to buy everything, and there you didn't. You know that, back there, there is only corn, beans, and sugar—that's all we had; however, here you crave meat… or you want something else. There you only have the produce of the field… here I spend 70 pesos on tortillas alone, and back there I used to make them with two balls of dough, or three balls, each 1.5 kg and people paid 1.50 pesos for each, because I used to grind them myself… for that reason, I think it's better to buy the maize [instead of the tortillas], but only when I have money.[2]

The majority of the inhabitants paid for the transfer of their possessions to the new settlement with their own savings. Furthermore, they enrolled their children in a new school, for which they were required to buy uniforms and school equipment; expenses they did not have encountered in *La Yerbabuena*. For Maria de la Luz Mejia, interviewed at home, the choice to relocate was made because of the constant evacuations of *La Yerbabuena*, and for the hope that resettlement would involve owning a home, as opposed to renting. She commented:

> Beginning with school, there were differences… [in La Yerbabuena] it didn't cost, here it does. Here at the start of the school semester they asked for 50 pesos and over there, deals were made with produce from the allotments… then, my daughter entered kindergarten and it cost 120 pesos for enrollment and breakfast was 5 pesos each week… and now they're asking us to pay for uniform, shoes and extra … over there, you went to school with the little you had but here you can't, you have to go every day with shoes, with these closed shoes or with sneakers and socks… and we had to buy it, and of course it costs money, and if you don't have it then you have to get credit to fulfill it.[3]

Relocated Yerbabuena farmers acquired new transportation costs to go daily to their plots of land. This represented an investment in both time and money, mentioned by Eusebio Montejano.

He was always in favor of resettlement, but several weeks after having been relocated, asserted that: "life was much more expensive [in the new resettlement] and it was very difficult to adapt to this new way of life".

One of the difficulties that resettled families encountered was the diminutive size of their new households. The built area, the distribution of space and the dimensions of individual plots were not equal; in *La Yerbabuena* plots were 30×50 m (1500 m^2), whereas in the new settlement they were 8×25 m (200 m^2), with a built area of 35.86 m^2. All the new houses were equal in extent and distribution, had a living-dining room, a bathroom, a bedroom, kitchen, and an area of 7 m^2 which could be used to plant a vegetable garden, store work tools, or construct another bedroom. The people of *La Yerbabuena* who lived here found it cramped and overcrowded.

The urban style of the houses imposes a new form and style of life on people from a rural environment. Furthermore, as the economy in the rural setting is different from that of the urban, very few of the families are able to construct new and adequate spaces to meet their immediate needs. In the rural environment, families own large plots of land with the understanding that the children will inherit a part of it when they grow up and have their own families. Additionally, it gives them an area to grow vegetables. Hence, leaving these conditions and styles of life for an "urban life" increases a family's vulnerability and limits their access to resources they once had.

The people of *La Yerbabuena* expressed that relocation has brought disunity. Ines Montejano provides one example of family breakup:

> It broke up families as well as people, for example we usually reenact the stages of the cross but this time we did nothing. Here in Cofradía everyone did what he or she wanted. Before the day used to be respected, recently it has not been so, everything has been lost (…) for example in La Yerbabuena we used to have festivals and people were more united (…)[4]

---

[2]Interview with (18RC-M43/04-04). Cofradía de Suchitlán, Colima.

[3]Interview with (6MdLM-Am29/04-04). Cofradía de Suchitlán, Colima.

---

[4]Interview with (16IM-M18/04-04). Cofradía de Suchitlán, Colima.

For other families, the process of relocation was a benefit as it meant they were distanced from the volcanic hazard, as said by Jesús Montejano:

> For me the people who have come have been given a great benefit, as people who had houses made of cardboard now have higher quality homes.[5]

(b) Appearance of daily life in resettlement

Relocation, which shows to have an important difference on your daily life, changes your actions relative to both time and space. One of the most significant disturbances the families experienced is the impact it had on the educational environment. Previously they had a kindergarten with one classroom and one community teacher from CONAFE (Consejo Nacional de Fomento Educativo). At primary level, they had one school with two classrooms for all six grades and one teacher. If one wanted to attend secondary school, they would have to walk to La Becerrera; as mentioned by farmer Jesús Montejano:

> When we lived in La Yerbabuena, I should note, I remember they used to have class three to four times a week and the teacher would have to walk from La Becerrera to La Yerbabuena, and by the time he came in and checked our work he would leave. So what was he teaching us? (...) And here it seems that children have class every day so it is therefore an improvement for them.[6]

Many people considered living in the new homes to mean "a new life" and "a place to live" and furthermore related it with "progress"; this was because many families (especially residents) lived in homes made from flimsy materials. Although the homes were flawed it was a "a new way of life" for the inhabitants; this was expressed by Guadalupe Cueto, municipal commissioner of the new settlement of Corfradía de Suchitlán, and Jesús Montejano, municipal commissioner of La Yerbabuena:

> Because this means we will be starting a new journey, we are starting a new life because we are

beyond what we had and what we were doing, and now we are living in progress (...).[7]

> The most important thing for me is that we are removed from danger, because it was a nuisance that we could be evacuated at any moment, that we could be moved to a shelter, this was a trauma I lived through, although one I was used to, I am now distanced from the danger and that is good.[8]

One of the consequences derived from the resettlement was the differences between the relocated and the resistant, a situation that distanced and broke up family relations. Forced displacement dissolved and dispersed the social ties of the town. However, some villagers consider that the land distribution of the ejido which was configured during the 1960–1970 decade, was not equal nor fair for all, and, as a result, the social fragmentation of the community initiated since that time. Living in small homes unsuitable for large families also caused difficulties in people's daily lives. Ma. De la Luz Mejía, daughter in law of Eusebio Cuellar who was the comisario ejidal (the highest political authority elected by the community) of La Yerbabuena, commented:

> The problem has been around a long time, since they distributed the land. Don Eusebio Cuellar was commissioner at the time, and he moved because the land was being divided, however they gave him a bit more for having fought for everyone, and for this his brother Don Leandro is jealous and it has turned into a rivalry.[9]

(c) The representation of volcanic risk

The scientific community and the government authorities still consider La Yerbabuena as an area of high volcanic risk (Cuevas and Seefoó 2005; Gavilanes-Ruiz et al. 2009; Cuevas and Gavilanes 2013). This perception differs between professionals and the people; therefore, highlighting that risk is not recognized the same way by all involved. It is a social hierarchy where some participants based their opinions on their scientific and technical

---

[5]Interview with (5JM-Eh56/04-04). Cofradía de Suchitlán, Colima.

[6]Interview with (10GC-Ah32/04-04). Cofradía de Suchitlán, Colima.

[7]Interview with (10GC-Ah32/04-04). Cofradía de Suchitlán, Colima.

[8]Interview with (5JM-Eh56/04-04). Cofradía de Suchitlán, Colima.

[9]Interview with (6MdLM-Am29/04-04). Cofradía de Suchitlán, Colima.

knowledge, and others on their experience and their conviviality with the medium.

Relocated families view the volcanic risk from two sides. The first is that the volcano is dangerous for their physical integrity. Others feel that daily coexistence with volcanic activity forms part of their daily lives and geographic location. That is, living on the slopes of the Colima volcano is an everyday experience and not considered dangerous. Jesús Montejano, having lived in the locality for 34 years with no collective memory[10] (Halbwachs 1950; cited in Mendoza 2001) of a volcanic eruption having affected them, says the volcanic risk is not a lived experience for relocated families:

> Look, for me personally it is not dangerous because I have not yet had to experience a large eruption. For all those who say it does nothing we can go 70 years back: the only woman to live there (he refers to La Yerbabuena) has told us she did not live in La Yerbabuena at the time (…) but like I said, one says it does nothing because you've lived here a long time and nothing has happened, but 60 or 70 years is not a long time[11]

(2) *Significance of relocation in resistant families*

This section consists of three parts: (a) significance related to relocation; (b) representation of daily life in the resistant and (c) the representation of the volcanic risk and the volcano.

(a) *Significance related to relocation*

From 2001 when Colima Housing (IVECOL) presented the relocation project to the people of La Yerbabuena many people showed reluctance to relocate, as for them it was only a pretext to

deprive them of their properties and a violation of their human rights. Relocation caused disunity between relocated and resistant families, and an interruption of daily activities such as deprivation to educational services. Antonio Alonso, moral leader of the resistant, states:

> The reason we have been given the volcano is so we can lose (refers to the resistant families) our lives for it, we would be honored to give our lives for this (…) not because someone doesn't impose. We have already stated this publically, we have disclaimed to the government and institutions that are near the volcano, because the right to education is not given and there are 12 children here in the community (…) but we will continue to resist for a few more years.[12]

The displacement caused a shortage of basic products for the resistant families. Shops disappeared and they had to travel to La Becerrera or Comala to buy basic foods. Despite all the drawbacks, none of them are willing to relocate as they argue they had not been requested to do so. The only official petition was realized by the ex-president in 1998 to whom they asked for land to build a house to suit their own needs to inhabit it only during times of volcanic crises without the need for shelters. Antonio Alonso remembers:

> For us the decision we made from the beginning was to stay and that it will be respected. At no point did any of us ask to be relocated. We were also asked at the shelter when the President came to visit; he had very clearly stated that he could provide some ground for everyone to build a home, whether it be made of cardboard or a proper home—everybody needs a home.[13]

The relocation fractured personal relations between people as well as family ties and pre-established social networks. According to the resistant families, the government intended on building a tourist facility in place of the settlement, as stated by Don Eusebio Cuellar:

> The tourist area will be made as it seems one of the Leaño Family wants to by a Cristero Camp (…) and they told me they wanted to build a church and

---

[10]The theoretical antecedents of collective memory can be found in the works of the Frenchman Durkheim, the British Federico Bartlett, the Russian Lev S. Vygotsky and the American George Herbert Mead. However, the concept can be attributed to the French sociologist Maurice Halbwachs. According to this author, collective memory "is the social process of the reconstruction of the past, lived and experienced by a particular group, community, or society" (Mendoza 2001, p. 67).

[11]Interview with (5JM-Eh56/04-04). Cofradía de Suchitlán, Colima.

[12]Interview with (1AA-Ah48/04-04). La Yerbabuena, Colima.

[13]Interview with (1AA-Ah48/04-04). La Yerbabuena, Colima.

cabanas, and so it will be converted into a tourist area.[14]

(b) *Representation of daily life in the resistant group*

There were notable registered changes in the daily lives of the resistant families. The local government suspended various public services, such as public lighting, from the start of the relocation. Later on the scholarship support program for students (Progresa) disappeared; and soon after then closed the school and cancelled all government help programs in the community. Antonio Alonso retells the story:

> From the start the government were attacking us; they took away Progresa (…) they attacked as if we were no longer there (…).[15]

As time passed, the authorities of state were pressuring the resistant group by threatening to demolish their houses by order of the government.[16] Subsequently, they made use of their networks and social capital contacting the Civil Committee of the Zapatista Front of National Liberation (CCFZLN) who support the resistant, as noted by Antonio Alonso:

> Don Rafael, a fellow from here, already had contact with them (refers to the C. C. F. Z. L. N.), that is how we joined forces. Therefore, they said we have to work together, we have to be self-sufficient, we have to find a way to resist, and that is how we came into contact with them.

Gratitude and loyalty for the Zapatista prevailed within the resistant families. With their intervention, they adopted adaptive survival strategies. For example, some farmers did both farm work and hand crafts (dream catchers, bracelets, necklaces, volcanic rock engravings, and papermaking with banana fibers). They

argued that one of the strategies for resisting was knowing the Mexican Constitution to defend themselves against violations of their human rights, as told by Alicia Mejía:

> We rely on the law that is in accordance with the law under the articles, they are supposed to be respected as they are articles of the law and we rely on them to defend ourselves.[17]

The participation of the Zapatista Front was one important factor that contributed to the fracture of the social links between the resistant families, as well as with the displaced families. Some internal disagreements were the result of land conflicts that originated from the foundation of the ejido, others were a consequence of the relocation.

The band "Nativo" joined after the intervention of the Zapatista Front and participated in the first homage to the volcano. At this event, it was proposed to build a temazcal[18] with the objective of obtaining economic resources to cover the costs of the resistant, a proposal that was accepted and consolidated just a few months later. A temazcal was built in one of the houses. The lucky family felt honored but the other families considered that they were victim of some kind of exclusion, which contributed to cause fractures in the social relations.

(c) *The representation of the volcanic risk and the volcano*

For the resistant families, there was no representation of the volcanic risk. Many years of living their daily lives on the slopes of the colossus have allowed them to identify when it going to be dangerous. However, they also denied the

---

[14]Interview with (7EC-H68/04-04). La Yerbabuena, Colima.

[15]Interview with (1AA-Ah48/04-04). La Yerbabuena, Coima.

[16]In the radio show of the former governor Fernando Moreno Peña, "Un Nuevo Colima", he declared the following: "Neither I nor my hand will hesitate to make a decision of that nature"; referring to the displacement of inhabitants.

[17]Interview with (11AM-Am32/04-04). La Yerbabuena, Colima.

[18]Temazcal comes from the Nahuatl language, meaning 'steam house' (Temaz—steam, calli—house). A temazcal is a prehispanic bath that is found throughout Mesoamerican cultures, whose oldest remains are found in archaeological sites in Palenque, México and in Piedras Negras, Guatemala. Historically, temazcals have been used for therapeutical, medicinal, and ceremonial uses; its practice survives in the present day thanks to the oral traditions of diverse indigenous communities in Mexico.

possibility of a major eruption affecting their lives, goods, and property. Although, they relate these events to divine powers, as stated by Maximino Ramírez: "If God wants us to die from the volcano, then we will die happy".

With the help of CCFZLN, the residents organized the first Spring Equinox festival "Atlacohualco", where the environmentalist organization Bios-Iguana, A. C. made their first intervention in the area. From this, the organization joined the resistant movement. Many residents showed an interest in ecology, as demonstrated by Maximino Ramírez:

> So we have confidence that nature is something that can deprive us of life (...). For me it would be very honorable to die due to nature, because I come from nature and there I shall go back, how beautiful it will be, to appreciate that.[19]

The relocation brought about a redefinition of the volcano amongst the resistant. From being a mountain that spews smoke, to being called the "guardian". Some inhabitants claim to have had visions of the volcano and they are sure that the colossus sends signals to "the chosen ones", as they call themselves. It is possible that residents of other volcanic regions such as Popocatépetl and Iztaccíhuatl, respectively called Don Gregorio and Doña Manuela, inspire these visions and forms of redefinition of the volcano. For these residents, the volcanoes are like people who occasionally wander through the village, but their presence, now with the risk of a volcanic eruption, is explained differently from their previous wanderings. Julio Glockner (1996) has called some inhabitants of these communities "tiemperos", "graniceros",[20] or "time workers", who have been sometimes paid by their neighbors to keep the hail stroms (bad for agriculture) away and attract rain.

In the words of Hobsbawn (1983) the visions experienced by the resistant are an "invented

tradition" because they are constructed, formally introduced, and emerge during a short time period. These "invented traditions" are a set of practices normally governed and tacitly accepted as a ritual or something of symbolic nature, which seek to instill values and behavioral norms that imply continuity with the past. One of the principle characteristics is to remain linked in a convenient form to the past—not necessarily the remote past—in which one intends to establish continuity.

For the resistant, the volcano is a being one should respect and learn to live with, because thanks to it, there is life in the location and they can subsist on what Mother Nature has endowed on them.

5 In Conclusion

From the start of the resettlement process the social representation constructed in La Yerbabuena was based on volcanic risk resulting from the processes of interactive significance that the inhabitants of the community developed with other social actors and the interaction amongst themselves. In this significant process, shared values and traditions that form a way of life and history have unified individuals in a cultural belongings and built forms of meaning and explanation from the most mundane and most extraordinary events.

According to the testimonies of the people, we can say that social representation is born from the mental images that both the relocated and the resistant constructed with the information they received of the sociocultural reality common to all members. Social representation represents the organizational forms of symbolic space in which a person develops. The reality appears through social representation and discourses that form the social fabric through which actors related in a particular social space, configure the subjective sense of the spheres of their lives and significance attributed to them and their relationships with others.

Knowledge of how social representation is constructed/built by residents helps link the

---

[19]Interview with (9MR-Ah58/04-04). La Yerbabuena, Colima.

[20]Graniceros and tiemperos are people that have been struck by lightning and survived. The amount of energy that struck them opened energy points within their bodies, which gives people the ability to talk to animals, people, and the Popocatepetl (Anaya 2001).

process of human resettlement and volcanic risk to understand how communities will respond, act, and implement adaptive or resistant strategies at individual, familial, and communal levels during a volcanic emergency. However, despite these representations, volcanic risk managers in the state have implemented mechanisms of information and hierarchical communication for many years while omitting the knowledge and meaning built by the residents; while authorities and scientists implement prevention and mitigation mechanisms based on largely technical and scientific knowledge only.

Undertaking studies on risks and social representations allows us to recognize the ways and processes through which people form and construct their own social reality and bring us closer to the "worldview" they have. The approach to social representations makes it possible to understand the dynamics of social interactions, and to clarify the determinant factors of social practices since representation, discourse, and practice are mutually generated (Abric 1994).

Finally, the lessons learned from this work are that the knowledge of social representations that the social actors constructed regarding volcanic risk and human relocation helped to understand how they will respond, act, and implement strategies, at individual, familiar and collective levels in situations of risk and/or disaster. It is urgent that all social actors involved in risk management have an understanding of the culture, risk perception, and forms of social representation of the volcanic risk of the inhabitants of the communities settled in zones of high risk in order to design plans suitable for prevention and where the population can have an active participation. These measures will improve risk management while strengthening multidisciplinary research.

References

Abric JC (1994) Metodología de recolección de las representaciones sociales. En Practiques sociales et Représentations. Traducción al español por José Dacosta y Fátima Flores (2001). Prácticas Sociales y Representaciones Sociales. Ediciones Coyoacán: México

Abric JC (2001) Prácticas sociales y representaciones. Presses Universitaires de France y Ediciones Coyoacán, S. A. de C. V. México

Anaya Rodríguez Edgar (2001) ¡Feliz cumpleaños Don Gregorio! En: México Desconocido, No. 289. Año XXV marzo 2001:28–37

Aparico AT, Pérez VV (2014) Representaciones sociales del desastre de 1940 en Santa Cruz Pueblo Nuevo, Estado de México. Investigaciones Geográficas, Boletín, núm. 83, Instituto de Geografía, UNAM, México, pp. 89–102

Arce A, Long N (1988) La dinámica de las interfaces de conocimiento entre los burócratas agrarios y los campesinos: un estudio de caso jalisciense. En: Cuadernos. Revista de Ciencias Sociales. CICS. Facultad de Filosofía y Letras. Universidad de Guadalajara. Septiembre-diciembre, No. 8, pp. 3–23

Berger P, Luckmann T (1967) La construcción social de la realidad. Amorrortu, Buenos Aires

Borgatti S (1996) Anthropac 4.0 Methods guide. Analytic Technologies, EE.UU. Natick, MA

Bravi C (2016) Representaciones sociales de la inundación. Del hecho físico a la Mirada social. en: Revista de estudios para el desarrollo social de la comunicación. Universidad de Sevilla, España. No. 13, pp. 133–164

Brunner José Joaquín (1992) América Latina: cultura y modernidad. Grijalbo-Conaculta, México

Cuevas Alicia (2001) El riesgo volcánico como objeto de representación social. Zamora, Mich. El Colegio de Michoacán, A. C, Trabajo de Grado (Maestría), p. 102

Cuevas A, Gavilanes JC (2013) La Historia Oral y la Interdisciplinariedad. Retos y perspectivas. Título del capítulo del libro: La Historia Oral De Una Comunidad Reubicada: Estrategias Adaptativas en los Procesos de Riesgo-Desastre. Año: 2013. País de edición: México. Editorial: Universidad de Colima. No. De edición: 1. Página de inicio: 71. Página final: 102. Tiraje: 1000. ISBN: 9786077010180

Cuevas A, Seefoó JL (2005) Reubicación y desarticulación de La Yerbabuena: Entre el riesgo volcánica y la vulnerabilidad política, Desacatos, No. 19, septiembre-diciembre, pp. 41–70

Durkheim E (1893) The division of Labor in Society: study of the organization of higher societies

Garfinkel H (1967) Studies in Ethnomethodology. Prentice-Hall, Englewood Cliffs, NJ

Gavilanes-Ruiz JC, Cuevas-Muñiz A, Varley N, Gwynne G, Stevenson J, Saucedo-Girón R, Pérez-Pérez A, Aboukhalil M, Cortés-Cortés A (2009) Exploring the factors that influence the percpetion of risk: the caso of Volcan of Colima, México. J Vulcanol Geoth Res (2009)

Glockner Julio (1996) Los volcanes sagrados, mitos y rituales en el Popocatépetl y la Iztaccihualtl. Grijalbo, México DF

Goffman E (1959, 1971) La presentación de la persona en la vida cotidiana. Amorrortu, Buenos Aires

Goffman Ervin (1961) Encounters: two studies in the sociology of interaction. Penguin, Harmondsworth

Habermas Jürgen (1989) Teoría de la acción comunicativa, tomos I y II. Taurus, Buenos Aires

Herbart JF (1825) Psychologie als Wissenschaft. Neu gegründet auf Erfahrung, Metaphysik und Mathematik. Zweiter, analytischer Teil. (Psychology as science: newly founded on experience, metaphysics and mathematics, Second, Synthetic Part). SW VI:1–338

Hobsbawn Eric (1983) Introduction: inventig traditions. The invention of tradition. Cambridge University Press, Cambridge, Nueva Cork y Melbourne, pp. 1–14

Jodelet D (1984) La representación social: Fenómenos, concepto y teoría. En: Serge Moscovici, Psicología Social II. Paídos Barcelona, pp. 469–494

Jodelet D, Tapia AG (2000) Develando la cultura. Estudios en Representaciones Sociales. UNAM, Facultad de Psicología, México, D. F

Long N (1989) Introduction. En: Long N (ed) Encounters at interface: a perspective on social discontinuities in rural development, pp. 1–10

Long N (1993) Introduction. En: Long N, Long A (eds) Battlefields of knowledge: the interlocking of theory and practice in social research and development. Routlrdge, Londres y Nueva York, pp. 3–15

Long N (1996) Globalización y localización: nuevos retos para la investigación rural. En: Huber C (ed) Grammont y Héctor Tejera Gaona (coord.) La sociología rural Mexicana. Frente al nuevo milenio. Vol. I, La inserción de la agricultura mexicana en la economía mundial. INAH, UNAM y Plaza y Valdéz, pp. 35–74

Long N (1998) Cambio rural, neoliberalismo y mercantilización: el valor social desde una perspectiva centrada en el actor. En: Zendejas S, Vries P (eds) Las disputas por el México Rural. Actores y campos sociales, vol. 1. Zamora, Mich. El Colegio de Michoacán, A. C., pp. 45–71

Long N (1999) The multiple optic of interfase analisis. Wageningen University, the Netherlands. UNESCO (manuscrito). http://www.utexas.edu

Long N (2001) Building a conceptual and interpretative framework. En: Norman long development sociology. Actor perspectives. Routledge, London and New York, King United

Mannheim K (1963) Ideology and utopia: an introduction to the Sociology of knowledge. Harcourt Brace and World, New York

Mendoza GJ (2001) "Memoria colectiva". En: Marco A. González Pérez y Jorge Mendoza García (compiladores). Significados colectivos: Procesos y reflexiones teóricas. Tecnológico de Monterrey—CIIACSO, pp. 67–125

Moscovici Serge (1979) El psicoanálisis, su imagen y su público. Presses Universitaires de France, Buenos Aires, Argentina. Huemul SA

Ortoll S (1988, comp.) Colima Textos de su historia, vol. 2. SEP/Instituto de Investigaciones Dr. José María Luis Mora, México

Perera M (1999) A propósito de las representaciones sociales: apuntes teóricos, trayectoria y actualidad. Informe de investigación. La Habana: CIPS. Periódico Oficial del Gobierno del Estado de Colima, 22 de septiembre de 1934

Schutz Alfred (1967) Phenomenology of the social world. Northwestern University Press, Evanston, IL

If I Understand, I Am Understood: Experiences of Volcanic Risk Communication in Colombia

Carolina García and Ricardo Mendez-Fajury

Abstract

In December 1984, after the reactivation of the Nevado del Ruiz Volcano, the Colombian Geological Survey (SGC) began campaigning for the deliver of volcanic risk information. The campaigns, focused mostly on communities located in high volcanic hazard zones, received expert advice and support from national and international volcanologists. Within the context of contrasting, multicultural and multi-ethnic features in Colombia, community reactions to the campaigns have ranged from immediate acceptance of risk, to outright denial and rejection of risk awareness. Religious, political, and philosophical arguments underlie the range of reactions seen in targeted communities. Since December 1984, volcano monitoring has also increased throughout the country. As part of the monitoring strategy, volcanologists have worked on assuring continuous transmission and open access of data to the general public, especially during times of increased volcanic activity. This chapter contains an empirically–based discussion of the measures undertaken by technical volcanologists in Colombia to address volcanic hazard communication for communities located in the hazard zones of Nevado del Ruiz, Nevado del Huila, and the Volcanic Complex Cumbal and Cerro Machin volcanoes. This account is coupled with a review of campaigns described in articles and official reports by the entities in charge of the communication process. The chapter shows how most campaigns focused on delivering technical information to the public. A few cases included inter-agency risk

C. García (✉)
Institución Universitaria Colegio Mayor de
Antioquia, Medellin, Colombia
e-mail: cargalon@gmail.com

R. Mendez-Fajury
Colombian Geological Survey—SGC, Colombia
Volcanological and Seismological Observatory,
Manizales, Colombia
e-mail: ricardomendezfajury@gmail.com

Advs in Volcanology (2018) 335–351
https://doi.org/10.1007/11157_2016_46
Published Online: 05 April 2017

communication campaigns involving social science and participatory activities, within interdisciplinary and participatory educational projects. Further research is necessary in order to analyse the impact of the different communication processes in Colombia. This could provide important feedback to the Colombian volcanological community about how to achieve more effective risk communication campaigns that increase the levels of risk perception and awareness of communities at risk.

Keywords

Risk communication · Multidisciplinary · Active volcanoes of Colombia

Tell me and I will forget. Show me and I will remember. Involve me and I will understand. Step back and I will act.

Old Chinese proverb

1 Introduction

Risk communication aims to increase awareness by encouraging people to adopt preparedness measures that reduce their risk and increase their ability to manage hazard consequences and to make informed and appropriate independent judgments to minimise loss of life and damage to property (Rodriguez et al. 2004; Paton et al. 2008; Haynes et al. 2008). However, providing information about risk is simply not enough, since it is not information by itself that determines whether people act to manage their risk (Paton et al. 2008; Perry and Lindell 2008; Garcia and Fearnley 2012). Rather, decisions to act are determined by how people interpret information in the dynamic context of previous experiences, social relationships, trust and expectations (Perry and Lindell 2008; Haynes et al. 2008). This is especially true when people at risk are in denial about the risk they face or when they expect to be protected by the authorities and emergency personnel, being therefore unprepared to respond appropriately and effectively to warnings. Therefore, risk communication campaigns should intend both to educate and to promote risk reduction (Donovan and Oppenheimer 2014).

Furthermore, the levels of risk perception and awareness are strongly related to the availability, quality and quantity of information, which should be provided at the proper time and should be adapted to the local conditions (Mileti and Sorenson 1990; De Marchi 2007). In this sense, an effective educational or communication campaign needs not just a far-reaching divulgation, but it is fundamental to provide the information in a clear way, using simple language and terminology. To consider local customs and traditions is essential, as well as the real level of perceived risk and the type of information that the population considers more relevant and necessary to improve preparedness.

Several authors agreed that delivering information and disseminating a warning might not be effective at generating an appropriate response unless strongly accompanied by participatory educational campaigns to assure that the warning message is well understood (Paton et al. 2008; IFRC 2009; Bird et al. 2010). Additionally, people are more likely to react appropriately when they have participated in risk education and communication campaigns where there is a process of sharing information among the different stakeholders, among the different actors, a key aspect trust being (Perry and Lindell 2008;

Haynes et al. 2008). Some examples, where sustained prior public education and community preparedness resulted in effective reactions, include the eruption of the Mount Pinatubo Volcano in Philippines 1991 (Punongbayan and Newhall 1998); and, Hurricane Michelle in Cuba 2001 (Wisner 2001).

In Colombia, before the reactivation and subsequent eruption and lahar of Nevado del Ruiz Volcano (NRV) that destroyed the city of Armero in November 1984, volcanology was limited to local studies, especially of geothermal evaluation. It was not only until such reactivation that the study of volcanic hazard and risk assessment began, followed by the dissemination of the results of such studies. Formal volcanic monitoring started after the NRV eruption in November 1985, with the creation of the network of Volcanic and Seismological Observatories of Colombia, based in Manizales (1986), Pasto (1989) and Popayán (1993), the latter being the only volcanological observatory established in Colombia prior to a crisis situation (Agudelo et al. 2012). Regarding volcanic risk communication activities, the ones developed prior to the 1985 Nevado del Ruiz Volcano (NRV) eruption were the first of their kind in Colombia. Since then, the volcanology community has developed regular communication activities, especially during periods of reactivation of any of the fourteen volcanoes considered active in Colombia (Fig. 1). Most of those activities focused on the provision of risk information by technical volcanologists to local communities and authorities. There are however some cases of multi-disciplinary risk communication campaigns.

Colombian experiences on volcanic risk communication provide valuable examples of how effective different communicational approaches are, ranging from provision of scientific information to multidisciplinary and participatory communicational campaigns.

On the other hand, there has been an evolution in the use of technical terminology in Colombia. For example, Monsalve and Méndez (1995) pointed out that at that time of NVR reactivation in 1984 "hazard maps" were incorrectly called "risk maps" by the volcanological community. In

fact, at the beginning of volcanological studies in Colombia the term 'hazard' was not used. The first hazard maps were entitled "Preliminary Potential Volcanic Risk Maps". The term 'hazard' was first used when the volcanic hazard map of NRV was updated and published in October 1986 (Parra et al. 1986). The definitions used in that map were: "volcanic risk: expected consequences on lives and goods in case of a potentially destructive volcanic eruption"; "volcanic hazard: potentially destructive volcanic event which can affect a specific area" (Parra et al. 1986, p. 1). Nowadays, volcanic risk definition is still the same, whereas the definition for volcanic hazard is "the probability of occurrence of a potentially damaging volcanic event within a specific period of time in a given area" (INGEOMINAS 2012, p. 1).

Regarding the legislative framework, Table 1 contains the first legislative instruments on risk management and volcanic risk in Colombia. After that, the Law 1523 of 2012 provided a breakthrough in the legislative framework of Colombia. This law created the National System for Disaster Risk Management, thus evolving from a Prevention and Attention focus into Disaster Risk Management. The law defined Disaster Risk Management as "a social process with the express purpose of contributing to safety, well-being, people's quality of life; and to sustainable development, made up of three main components: risk knowledge, risk reduction and disaster management" (Law 1523 of 2012, article 1). Furthermore, Law 1523 states that risk management is "a development policy indispensable to ensure sustainability, territorial security, collective rights and interests, improvements in the quality of life of populations and communities at risk. Therefore, risk management is inherently related to: safe development planning, sustainable territorial environmental management at all levels of government, and effective people's participation" (Law 1523 of 2012, article 1, paragraph 1).

Besides the above-mentioned laws, the Constitution of Colombia of 1991, has two important articles related to participation and risk management. Article 2 states that some essential

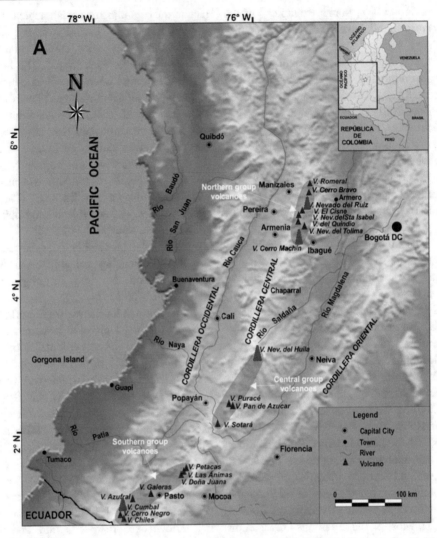

Fig. 1 Location of the active volcanoes of Colombia (Adapted from Ordoñez 2011). The *red triangles* highlight the Nevado del Ruiz volcano, Cerro Machín, Nevado del Huila and Cumbal

purposes of the State include: "to facilitate the participation of all citizens in the decisions that affect them". It further specifies that: "the authorities of the Republic are instituted to protect everyone residing in Colombia: in their life, honour, property, beliefs and other rights and freedoms." In addition, Articles 16, 18, 20 and 28 of the Constitution are related to the right to free development of personality, freedom of conscience, freedom of expression and to the transmission and reception of truthful and impartial information.

This chapter provides an account of four experiences on volcanic risk communication in Colombia, including the campaigns prior to the 1985 eruption of Nevado del Ruiz Volcano (NRV), followed by the attempts to address volcanic hazards communication in the Nevado del Huila volcano, in the Volcanic Complex Cumbal and in the Cerro Machín volcano. In each of the four cases, a description of the reaction by the local communities is discussed. The data presented in this chapter is drawn from archival data, accounts given by the local

Table 1 Legislative background on risk management and volcanic risk in Colombia

Legislative instrument	Implication	Description
Decree—Law 1400 of 1984	Establishment of the first Colombian seismic building code	Generated as a result of the 1983 Popayán earthquake that caused 228 deaths
Decree 3815 of 1985	Assigned to INGEOMINAS research activities on all kinds of geological risks, using volcanological techniques, among others	Following the lahar generated by the NRV eruption of 1985
Law 46 of 1988	Created the national system for prevention and attention of disasters (SNPAD, from its Spanish name)	Following the lahar generated by the NRV eruption of 1985
Decree 919 of 1989	Organized the SNPAD	Established the structure of the SNPAD, including the interagency educational commissions
Decree 98 of 1993	Created the national plan for prevention and attention of disasters	Guidance to develop a national instrument to guide the development of prevention and attention of disasters activities
Law 1523 of 2012	Created the national system for disaster risk management (SNGRD, from its Spanish name)	Following major flooding associated with La Niña of 2010–2011

communities, and the personal experience of the authors via participant observation and engagement.

2 Risk Communication Experiences During Volcanic Emergencies in Colombia

In Colombia, volcanic risk communication activities developed by technical volcanology began prior to the 1985 eruption of Nevado del Ruiz Volcano (NRV). Since then, regular activities for the dissemination of volcanic risk information have been developed, including those describe in the following sections.

2.1 Risk Communication Activities in the Nevado Del Ruiz Volcano (NRV)—States of Caldas and Tolima

After decades of inactivity, reactivation of NVR occurred in 1984, characterized by small fumaroles and earthquakes that generated concern among the population and authorities in the State of Caldas. According to INGEOMINAS (2006)

(National Institute of Mining and Geological Research—nowadays Colombian Geological Survey—SGC), following such reactivation, in January 1985 a new regional body was created and called the Comité de Estudios Vulcanológicos de la Comunidad Caldense (Committee of Volcanological Studies for the Caldas Community). This Regional Committee was coordinated by FIDUCAL (Foundation for Scientific Research and University Development of Caldas) and was made up of by representatives of multiple entities such as several universities, CHEC (Hydroelectric of Caldas S.A.), the Committee of Coffee Growers of Caldas, the financial sector and community in general. The goal of the Regional Committee was mainly to articulate research activities, avoiding duplication of efforts. Some of the functions of this committee included the coordination of technical, logistical and financial support and the generation of the necessary policies needed for the implementation of volcanic seismic instrumentation, development of hazard maps and other research work. The Regional Committee obtained statutory recognition in August 1985 with the Decree 0977 of the State Government of Caldas.

Parallel to the creation of the regional committee, a group of technicians gathered to form a

Local Technical Committee in order to advise to the Mayor of Manizales directly. In the absence of a formal structure to deal with a situation without historical antecedents in Colombia, this Technical Committee was in charge of doing preliminary research, developing community awareness activities, providing assistance to authorities, and giving logistic support for national and international specialized missions, among other tasks. The Local Committee was formally established through Decree 079 signed by the Mayor in March 1985. It should be noted that this local committee was the first technical group in Colombia to formally include community awareness programs among its tasks.

On September 11th, 1985 an eruption characterized by phreatic ash emission generated a small lahar in the Azufrado River and block falls up to 2 km from the crater. Subsequently, on September 17th, a delegation with representatives of the scientific community, the Civic Committee (with delegates from the CHEC, the Civic Corporation of Caldas and the Mayor of Manizales) and the Civil Defense, emphatically demanded from the Ministry of Mines a timely response from the State, warning about the imminent hazard and possible direct consequences of the progress of NRV symptoms (Hall 1990, p. 108; INGEOMINAS 2006, p. 35; Presidency of the Republic of Colombia 1986, p. 46). As a result, the Ministry created a national committee for the scientific investigation of the phenomenon. This national committee was composed of two bodies, the Sub-Committee of Volcanic Monitoring, with geophysics and geochemistry tasks, and the Sub-Committee of the Volcanic Hazards Map, coordinated by INGEOMINAS. The national committee was supported by foreign scientists, in addition to professors and geology students of the University of Caldas.

By that time, due to the lack of solid technical information regarding the possibility of the eruption, the local and national media conducted multiple reports, some of them with partial or confusing information, contributing to confusion and fear in the population of the region,

especially in Manizales (Parra and Cepeda 1990; Villegas 2003; INGEOMINAS 2006).

Regarding the communication activities, the first formal attempt to divulge the scientific findings was during the public "Seminar on Volcanic and Seismic Risks Related to NVR", organized by the National University and CHEC on March 26th 1985. During the seminar, members of the regional Committee coordinated by FIDUCAL presented the findings to the local community, since the main goal of the seminar was to "involve the community" in all the activities and decisions regarding NRV (INGEOMINAS 2006, p. 33).

On September 19th, the Sub-Committee of Volcanic Hazards Map initiated the development of the "Preliminary Map of Potential Volcanic Risk of Nevado del Ruiz Volcano". The 1:100,000 map was finished in on October 7 and contained the "delimitation of potential areas subjected to different risks from volcanic eruptions" (INGEOMINAS 2006, p. 39). The map was delivered to national, regional and local government and Civil Defense authorities and it was published in the national, regional and local press (Hall 1990).

Between October 15th and 20th, 1985, several members of local emergency committees from multiple municipalities, including Armero, received technical advice from representatives of INGEOMINAS, SENA (National Learning Service) and the Institute of Territorial Credit—INSCREDIAL. These institutions developed a massive plan to disseminate the preliminary map of potential volcanic risks of the NRV. In addition to delivering to the population at the hazard zone an illustrative brochure developed by SENA, the representatives gave 87 lectures and open forums on risk prevention to the communities that could be affected in case of a volcanic eruption (INGEOMINAS 2006, p. 59). According to Omar Gomez (verbal communication), Head of Civil Defence at that time, the estimated attendance of this lectures and forums was around 5000 people.

During October, volcanic monitoring continued and scientific delegates gave several lectures in Pereira, Armenia and Manizales to support the organization of Emergency Committees and to improve their knowledge on volcanic phenomena and community preparedness for volcanic disasters (the Presidency of the Republic of Colombia 1986).

Despite the efforts of the scientific community to have a highly accurate volcanic hazards map and disseminate the scientific findings, on 13 November 1985 a small-magnitude eruption of the Nevado del Ruiz Volcano caused lahars that led to the death of about 25,000 people, mainly in Armero and Chichiná municipalities.

It is important to highlight that in Armero, even though the lahars claimed the lives of 22,000 people, about 8000 survived. Many of these survivors evacuated to the nearby hills, previously identified as meeting points. Unfortunately, there are no records of how many people evacuated thanks to the information campaigns developed before the eruption.

In 1985 the Colombian population had no knowledge about previous volcanic disasters and thus there was no experience in risk communication. In consequence, despite the efforts of the scientific community to create risk awareness, the reaction was not effective.

After the 1985 eruption of NRV, there have been varying scenarios in the activity of several volcanoes in Colombia. This variation ranges from mild seismic activity, to eruptive states of different intensity. Consequently, several risk communication activities have been developed, as demonstrated below.

2.2 Risk Communication Activities in the Nevado Del Huila Volcano—the State of Huila

The Nevado del Huila Volcanic Complex (NHVC) commonly known as the Nevado del Huila Volcano, is part of the National Natural Park Nevado del Huila Volcano located between the states of Cauca, Huila and Tolima. It is an active volcanic complex formed by stratovolcanoes and several domes whose maximum height is 5364 m. Its activity has been dominantly

effusive, but in most recent development, it has generated pyroclastic flows produced by collapse and/or explosion of domes (INGEOMINAS 1996, p. 16).

In August 1986 the first evaluation of volcanic hazards of NHVC (Cepeda et al. 1986) included in its results a high risk factor for lahars in areas around Páez River. The recommendations of that work included the promotion of: "permanent school education campaigns about volcanic and other natural phenomena that can affect the population, such as earthquakes, floods, landslides, etc." (Cepeda et al. 1986, p. 83). However, at that time no education campaigns were properly conducted. The only activities in this respect were some lectures given to local authorities and community members in the urban area of Belalcázar municipality and the villages of Wila, Tóez and Ireland. These villages were later greatly affected by flooding and debris associated with the 1994 earthquake and the 2008 lahars.

On June 6th 1994 an earthquake of magnitude 6.8 M_w caused many landslides and consequent flooding of the Paez River, causing more than 1000 deaths. After this disaster, INGEOMINAS (1996) published a second version of the volcanic hazards map. This map included an area of pyroclastic falls and debris avalanches near the crater. The final recommendation of this document was: "the most important measure required is to disseminate the knowledge contained in this report among authorities and inhabitants of the Paez River Basin" (INGEOMINAS 1996, p. 33).

On 19 February and 18 April 2007 and 20 November 2008, the first historically recorded eruptions in the NHVC occurred. There eruptions significantly affected the social fabric and infrastructure in seven municipalities of the states of Cauca and Huila. Afterwards, INGEOMINAS sent to the affected area technicians with broad knowledge of the geology and the hazard of the Nevado del Huila Volcano. These technicians, even if lacking social-science preparation or training, had an extensive experience delivering technical information to potentially affected populations. An intensive communication campaign was then developed between March 6th

Fig. 2 Activities to disseminate volcanic risk information in 2007; addressed to **a** students and teachers of the Normal School of Belalcázar, later destroyed by the 2008 lahar and **b** indigenous governors. *Photo* Ricardo Méndez-Fajury

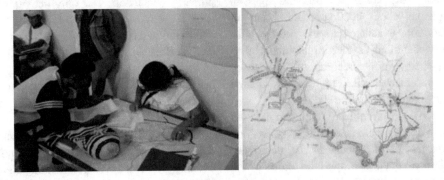

Fig. 3 Development of the map of risk scenarios by the local indigenous community in collaboration with the association of councils Nasa Çxhãçxha. *Photo* Ricardo Méndez-Fajury

and 13th of 2007 addressing about 1000 people from the local community, local emergency committees, students and teachers of the largest school (Escuela Normal) and indigenous governors from the Association of Indigenous Councils Juan Tama and Nasa Çxhã çxha (Fig. 2). Additionally, other lectures were given to Hospital personnel, the NGO Nasa Çxhã çxha and relief organizations like the Red Cross and Civil Defense. Finally, the INGEOMINAS personnel participated in a radio broadcast on the 'community radio station Eucha' whose motto is: "If I understand, I am understood", with simultaneous translation from Spanish to Nasa language, spoken by the indigenous communities of the area.

During the meeting with indigenous governors, a community participant stated that to the indigenous community the volcano was not a threat but a God. Given this fact, the intuitive response of the INGEOMINAS technician who sought an explanation, adapted to the indigenous world-view, was that the God was angry. He explained that if that God was really angry it was equivalent to what technicians call high activity, if the God was more or less angry it would be medium activity and if the God was only little angry it would be low activity and so on. Eventually, if the God gets very angry there could be damage to the high hazard zone. This explanation was welcomed by the indigenous community since it helped them to understand the technical jargon regarding alert levels.

It is important to highlight the work of the Secretariat of Government and the indigenous community of the region in identifying evacuation routes and developing their own map of risk scenarios, for which they had the important support from the Association of Nasa Councils Çxhãçxha and the OSSO Corporation (Fig. 3).

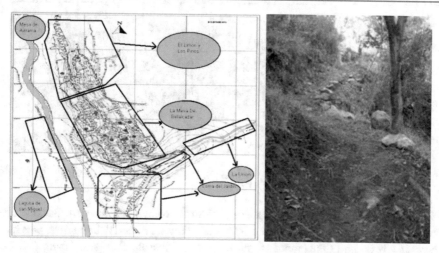

Fig. 4 Evacuation and evacuation paths selected and prepared by the community and local authorities. Map provided by the Government Secretariat Township Belalcázar—Cauca; *Photo* Ricardo Méndez-Fajury

INGEOMINAS also supported the verification of the identified evacuation routes prepared by local authorities and the community (Fig. 4) and suggested extending such routes so the most vulnerable people were able to evacuate in time of emergency. As a result, community and local authorities extended the routes contributing to subsequent successful evacuation.

According to Agudelo et al. (2012), from the moment the Volcanological and Seismological Observatory of Popayán (OVSPo), belonging to INGEOMINAS, reported the increment of activity of the volcano, the local community became aware and participated actively in the preparedness exercises and the monitoring of the volcanic phenomenon. Communities contributed with the worldview of the indigenous communities and their experience of what happened as a result of the 1994 Páez earthquake, thus strengthening the understanding of volcanic phenomena and their response capacity.

With additional support from INGEOMINAS Manizales, the OVSPo continue working with CLOPAD'S, CREPAD'S (today municipal and departmental Risk Management Councils—CMGRD) and communities in urban and rural areas. This work focused on developing contingency plans and providing training in the volcanic area, with the aim of establishing early warning systems for volcanic risk. Additionally,

the OVSPo provided broad and public information on the evolution of the phenomenon through official statements addressed to local and traditional authorities in the area of influence of the volcano, among other joint strategies designed together with the local operational Relief Corps and the community to promote timely and safe evacuation responses (Agudelo et al. 2012).

Besides INGEOMINAS, many organisations participated in the exercise, including the Association of Nasa Çxhãçxha Councils, the OSSO Corporation and the National System for Disaster Prevention and Response (SNPAD), among many others. These organisations contributed to a comprehensive strategic planning exercise for the construction of an Early Warning System—EWS, involving the community in a daily prevention exercise, where each protagonist had a defined role to play in the process, from dialogue and conflict resolution, involving the whole community to reconstruction of the past, definition of the present and planning for the future by the whole community (Peralta 2008).

On 18 April 2007 a lahar that reached speeds close to 80 km/h and average heights up to 13 m, caused damage to the road infrastructure in the area and destroyed six bridges restricting the access to the municipalities of Páez and Belalcázar, whose urban area was also strongly affected (Agudelo et al. 2012). More than

Fig. 5 Municipality of Belalcázar on the River Páez bank, before and after the 2008 Lahar. *Photos* INGEOMINAS

5000 people in the lowlands along the river Páez and Simbola areas, were effectively evacuated to the safe areas in a very short time (about 15 min).

The success of the awareness and preparation activities was achieved for several reasons (Hall 1990; Semana 2008; Efe and Reuters 2008), including the articulated inter-agency work that started after the 2007 reactivation. Another reason was the constant monitoring of permanently available information through online seismograph in the institutional website. In addition, a local newspaper "The Swarm" published regular reports about the volcanic activity. Finally, the constant concern and interest of the community in understanding the volcanic phenomenon and make it part of everyday life contributed greatly to the timely and efficient response.

The risk communication process developed on the NHVC is an excellent example of the high effectiveness of long term multi-agency, interdisciplinary and participatory work. This is evidenced in the big difference between what happened in the 1994 Páez Earthquake when 1100 lives were lost, and the evacuation in April 2007 and November 2008 when there were "only" 10 deaths (El Tiempo 2008), even though the 2008 event was more intense than the one of 1994 (Fig. 5).

2.3 Risk Communication Activities in the Volcanic Complex Cumbal—State of Nariño

The Cumbal Volcanic Complex—CVC is composed of two active volcanoes: Cumbal and Mundo Nuevo. Its frequent hazards are mainly lava flows, ash and debris flows, plus hidroclastic and pyroclastic falls and lahars (Monsalve and Méndez 1988). The lava flows covers a 3.5 km radius from the crater and the hazard by pyroclastic flows goes up to 10 km radius from the crater.

From February 1988 on there were a series of reports from the inhabitants of the Cumbal municipality, concerning changes in the normal activity of the Volcán Cumbal, such as noise and changes in the fumaroles. For this reason, the authorities of the state of Nariño requested that the Government draw up a map showing the vulnerable sites in case of a possible volcanic eruption (Monsalve and Méndez 1988), since they feared a similar tragedy to the one in Armero in 1985. The result was the Preliminary Map of Potential Volcanic Hazard Complex Cumbal (Monsalve and Méndez 1988) in which it was shown that the population of Cumbal is located close to the pyroclastic flows hazard zone.

During the development of the hazard map two technical talks were given to the community of two villages in rural Cumbal to explain the mapping process. Later, during a fieldwork session, locals reacted aggressively and forced the withdrawal of the SGC personnel, claiming that scientists were "perverting the volcano" and stealing from it (sampling). What is more, the community did not believe that the volcano was under study, and insisted that mining practices were being carried out instead, which they deduced from the name of the institution: INGEOMINAS (National Institute of Mining

Geological Research). This, coupled with the armed conflict in the area, led to the interruption of the field work. Fortunately this did not stop the completion of the hazard map and its technical report finished by the end of 1988. Soon after the hazard map was finished, three talks were given to the community and local authorities of Cumbal and two talks were given to the regional authority in Pasto.

Over the next 25 years, permanent volcanic telemetric monitoring continued and the volcano did not show any significant changes in activity, until 2011 when the associated seismic activity increased significantly. As a result, the Colombian Geological Survey—SGC, began a process of updating the hazard map.

Activities for updating the hazard map include field work and communication activities to inform the community about the work performed by the SGC. Even though the updated map was about to be completed at the time of preparation of this chapter, it is important to note that great difficulty has been encountered in developing both communication activities talks and fieldwork, as the community, mostly indigenous, has repeatedly rejected the work regarding volcanic phenomena of the SGC, claiming that the SGC works constantly for the multinational mining companies (which is emphatically rejected by local communities).

The SGC repeatedly tried to talk to the Community Action Council, but failed because of constant evasive tactics. There were even threats made to the SGC technicians, which claimed that SGC was entering by force (despite constant attempts to disseminate the scientific information and to explain to the community that the scientific work regarding volcanic phenomena was completely unrelated to mining practices). To resolve conflicts with the community, the SGC has requested the support of local, regional and national authorities, who have been taking action, although, without significant results up to now.

SGC issued on its website weekly bulletins that described the changes in the volcanic activity of the Cumbal Volcanic Complex, as well as the description of communication activities undertaken with communities and authorities. Prominent among these bulletins was the report of an institutional meeting held on 27 March 2014, between the SGC and the Ministry of the Interior called in order to analyse the situation regarding compliance of SGC missionary activity in CVC area, since SGC had not been able to enter the top of the volcano for almost a year because of opposition from some indigenous communities), (Boletín semanal de actividad del Complejo Volcánico Cumbal—Weekly Bulletin Cumbal activity Volcanic Complex 1 April 2014). Later, with the support of the Bureau of Indian Rom and Minorities Affairs of the Ministry of the Interior, between 4 and 6 August 2014, meetings were arranged with indigenous authorities and community members of the Indigenous Cumbal Reserve. The meetings, addressing the problem of admission to the upper Cumbal volcano, established some short-term strategies to move forward in this complex situation (Boletín semanal de actividad del Complejo Volcánico Cumbal, 12 August 2014).

Currently, fluctuations in seismic activity continue as well as gas emissions. The SGC continues with the agenda of proposed activities with the support of the Bureau of Indian Rom and Minorities Affairs and the participation of indigenous communities and members of the Indigenous Reserve Cumbal (Boletín semanal de actividad del Complejo Volcánico Cumbal, 30 September 2014).

2.4 Risk Communication Activities in the Cerro Machín Volcano—State of Tolima

The Cerro Machin Volcano is a highly volatile tuff-ring active volcano, reaching a Volcanic Explosive Index—VEI 5 with eruptions dominantly magmatic with highly destructive pyroclastic surges, whose deposits are distributed mainly in about a 12 km radius of the crater (Cepeda et al. 1996).

A group of scientists from INGEOMINAS initiated the study of the Machin volcano in 1988 led by geologist Luis Armando Murcia (RIP).

Following the death by natural causes of Murcia, the study of the volcano was temporarily stopped because the information gathered was partially lost. In 1995 work on the volcano resumed and the results were compiled in an internal document describing the volcanic deposits and different scenarios of volcanic activity, without a hazard analysis (Cepeda et al. 1996). Among its conclusions this document recommends "to disseminate the study results among government authorities, planners, investors and the general public, using methodologies of formal and non-formal education, with emphasis on disaster prevention and knowledge of the physical environment" (Cepeda et al. 1996, p. 45). Despite this recommendation, initiatives to disseminate volcanic risk information were not immediately undertaken.

In 1999 the hazard assessment of the Cerro Machin volcano began and some informative talks were given to advise the community and local authorities about the conducted scientific work. As a result, some relief agencies carried on dissemination activities such as the one shown in Fig. 6, in addition to handing out over a few days, flyers at tollbooths to inform tourists that they were entering into an active volcanic area.

The results of the studies initiated in 1999 were compiled into a document and a hazard map

(Méndez et al. 2002). The document states that the volcano had produced six eruptive periods during the Holocene (four Plinian episodes and two dome collapses), the last one about 800 years ago. It has also produced domes, eruption columns of over 20 km above the crater, pyroclastic flows and surges and large volumes of lahar deposits, (corresponding to debris and hyperconcentrated dacitic flows that cover an area of around 1000 km^2 to the East, in the valley of the Magdalena River). Currently, nearly one million people live in the area of influence of past eruptions where strategic plans for the national economy are being developed. The area also includes a key viaduct for 50% of domestic and international trade and for the communication between the central and southwest part of the country, including some coffee regions. The main vulnerable population centres are Ibagué, Armenia, Girardot, Calarcá, Cajamarca, Espinal, Flanders and Guamo (Méndez et al. 2002). It is important to clarify that the resulting hazard map does not present hazard levels, but represents the different volcanic products expected in areas of influence.

Regarding the initiatives to disseminate volcanic risk information, Méndez et al. (2002, p. 60) recommended: "The evaluation results should be considered for the development of

Fig. 6 Public poster announcing different risk management activities carried out by the Colombian Red Cross and German Red Cross in 1999. *Photo* Ricardo Méndez-Fajury

local, regional and national contingency plans... In order to achieve this, it is necessary to develop risk assessment... In this process the participation of the stakeholders located in hazardous zones is absolutely necessary."

As a result, in 2004, after the formalization of the document, multiple communication activities began in the area of influence. Since then there have been more than 1000 lectures on Cerro Machin addressed to the exposed population, relief agencies, student community and local, regional and national authorities (Fig. 7).

The talks given by the SGC were completely technical and included the location and characteristics of the volcano, an account of historical eruptions, with comparison to similar eruptions, for example Pinatubo in Philippines.

Among the experiences of volcanic communication there are a couple of interesting anecdotes. On Friday 28 December 2007 at 7:30 pm, a large earthquakes swarm associated with the volcano occurred. A strong earthquake was felt in the cities of Ibagué, Armenia and Pereira, more than 60 km away from the volcano. Following the instructions of the National Risk Management System, the emergency alarm protocol was activated, which is to alert system members about the variation of the volcanic activity so it would not take them unaware in case of a more significant change. The event coincided with Christmas celebrations, the Innocents Day and the taking of office of a new local Mayor (public officials had finished their contract that day at 6 pm). Because of this

situation, many members of the emergency phone teams relinquished their responsibilities and passed them on to the next in the emergency phone chain claiming they thought the call was a joke, that they were celebrating, or just that it was not his/her responsibility.

Another anecdote worthy of mention is that on November 8th and 9th, 2008 when a large swarm of earthquakes, including two earthquakes of magnitude 4.6 activated again the emergency phone chain. When calling the people located on the volcano, they reported that the earthquake had created panic in the Machin community, so they asked the SGC if it was necessary to evacuate. The SGC representative responded that he could not give the evacuation order, since it was the responsibility of the Local and Regional Committees for Risk Management (which SGC subsequently called). The community decided not to wait and started the evacuation to Ibague, even before receiving the official order, showing high risk awareness. Parallel to this, heavy rains generated large landslides blocking the roads to Ibague and Armenia, forcing people on the road, mostly tourists, to go to the town of Cajamarca. Tourists felt seismic activity and when asked what was happening, the local community, informed them of the existence of an active volcano nearby, (without further explanation), causing panic among tourists who had no idea of what was happening.

Currently, the only information on the Machin volcano available for tourists, is a big public board next to the road to Cajamarca. The board

Fig. 7 Lectures given in 2004 in the Municipality of Guamo and Toche village in the Municipality of Ibagué. *Photo* Ricardo Méndez-Fajury

Fig. 8 Public board next to the road to Cajamarca welcoming visitors to Cerro Machin volcano. *Photo* John Jairo Sánchez

simply welcomes visitors to the Cerro Machín volcano (Fig. 8).

Tourists and other itinerant populations are highly vulnerable during volcanic crisis due to their lack of hazard knowledge. Therefore communication about the specific volcanic hazard and risk to the travelling population is necessary as well as training of staff from tourist enterprises (so they coordinate an appropriate response in case of an event) (Bird et al. 2010).

Regular and systematic documentation of both scientific, and communication activities is very important to ensure the generation of new knowledge, facilitate monitoring and reduce dependence on particular individuals, who may leave the process at any time. In this respect, no reports of risk communication campaigns by the SGC were made in the past because there was no awareness of its importance. The systematization of communicational experiences began only after 2004 as a result of the implementation in SGC of quality certification and corresponding ISO standard, which requires generating indicators of planned and executed activities on a regular basis. Today the SGC publishes on its website quarterly reports with results of the monitoring and volcanic hazard assessment, as well as the description of the developed communication activities.

3 Conclusions and Recommendations

In recent history, Colombia has experienced different volcanic crises with great social and economic impact, from fatal eruptions, to activity changes without reaching eruptive consequences. Volcanological institutions and scientific organizations in Colombia have consistently reiterated the need for communication activities on volcanic hazards and risks. However, on several occasions the local authorities have chosen not to support the communication. In the opportunities where there has been inter-agency support, such communication activities have generated positive results.

Technicians display great technical knowledge about eruption mechanism. Hence, they are important in the process of risk communication. The recommendation is to work together with social scientists, incorporating in the communication activities elements of the social science and techniques of public management to ensure that the message is understood and generates a consequent and timely reaction. A solid social science research program is necessary in order to get a better understanding of not only the impact of the communication of technical information in emergency management and risk reduction, but also to help SGC technicians and local authorities to understand the social processes that influence the decision-making of the public. Additionally, it is relevant to include an analysis of not only the volcanic hazard, but also the different components of vulnerability.

Multidisciplinary approaches that combine natural and social sciences could help address complex elements such as uncertainty and trust-building, as well as to understand what makes the authorities and the local community to react in volcanic crises. Constant feedback and monitoring are critical to improve any communicational processes. Therefore, it is essential to evaluate the cases when the community and local authorities have not supported the hazard and risk communication processes, and also to generate instruments to understand the impact and effectiveness of the communication activities. In addition, any communication and monitoring strategy should always respectfully incorporate the knowledge of communities and local and regional institutions. The previous steps favour trust building between different stakeholders, influencing key elements such as risk perception, effective communication and community reactions to government mandates. In the risk reduction process, it is also very important to include the media from the outset, as well as having non-linear communication protocols to ensure that the message reaches all potentially affected people, even if one link in the communication chain fails.

Regional cultural differences regarding land appropriation have an impact on the way people react. While North of Cauca ownership is low because the territory was occupied by people descended from other regions (as around Cerro Machín and NRV), ownership in the South is much higher due to traditional cultural attachment of several generations living in the area, as for example around NHVC. This favours or hinders the response to a volcanic alarm. Mining is an economic activity of high social impact in Colombia. In some regions of the country, local communities associate the volcanic research with mining activity, (which is translated into strong opposition and fieldwork obstruction, thus affecting the studies of hazard and risk).

Except for isolated conditions, the community tends to trust in the technical work of the scientific volcanic community of Colombia, particularly, in the scientific information provided by the SGC. However, an assessment of whether such information is understood and assimilated by the population has not been made, and therefore both positive and negative aspects on the likelihood of this information to promote effective response during volcanic crisis is unknown. Furthermore, assessment of the perceptions and knowledge of SGC technicians and main political authorities regarding the relevance of public involvement in risk management has not been carried out.

The Colombian Geological Survey—SGC, is a purely technical body, so its communicational work is limited to the transmission of technical knowledge about volcanic hazards. It would therefore be desirable to combine efforts with organisations or individuals in the social area to help assure responsiveness and understanding of scientific results by the communities and local authorities, in order to achieve the objectives of volcanic risk communication, which include both education and the promotion of risk reduction. A better understanding by geo-scientists and authorities of the social processes that influence the decision-making of the public is also necessary. The above can be achieved through feedback and simulations subsequent to the communication activities. These activities could be coordinated by the local committees of risk management with support of social scientists and should promote significant participation by the communities, as it was achieved in the successful case surrounding the eruption of the Nevado del Huila Volcano in 2008 described above.

Volcanic risk reduction is not responsibility of one single person or institution but of everyone, including volcanological observatories, technicians, social scientists, academics, politicians, planners, disaster relief agencies, communicators, and of course, the community (Duque 2005). This requires constant, clear, effective and organized communication.

References

Agudelo A, Narváez A, Ramírez Y (2012) Experiencias en la construcción de la gestión del riesgo en el cañón del río Páez. Red de Desastres Hidrometeorológicos y Climáticos (REDESClim). Felipe Muñoz (Compilador). Memorias II Congreso Regional SRA-LA. Capítulo Latinoamérica. Sociedad para el Análisis de Riesgos. Bogotá, Colombia, 9–12 de abril: 210–215

Bird DK, Gisladottir G, Dominey-Howes D (2010) Volcanic risk and tourism in southern iceland implications for hazard, risk and emergency response education and training. J Volcanol Geoth Res 189:33–48

Cepeda H, Mendez R, Murcia A, Vergara H (1986) Mapa preliminar de riesgos volcánicos potenciales del Nevado del Huila. Informe interno I-1981 INGEOMINAS. Gobernación del Huila e Instituto de Desarrollo del Huila – IDEHUILA. 48 pp

Cepeda H, Murcia LA, Monsalve ML, Mendez-Fajury R, Nuñez A (1996) Volcán Cerro Machín, Departamento del Tolima, Colombia: pasado, presente y futuro. INGEOMINAS Informe interno I-2305. Popayán

De Marchi B (2007) Not just a matter of knowledge. The Katrina debacle. Environ Hazards 7:141–149

Donovan A, Oppenheimer C (2014) Science, policy and place in volcanic disasters: insights from Montserrat. Environ Sci Policy 39:150–161. http://www.sciencedirect.com/science/article/pii/S1462901113001585

Duque Escobar G (2005) Las lecciones del Volcán Nevado del Ruiz a los 20 años del desastre de Armero. In: Conmemoración de los 20 años del desastre de Armero, 19 de noviembre de 2005, Auditorio de la Universidad Nacional de Colombia - sede Manizales. http://www.bdigital.unal.edu.co/2270/#sthash.vfBIrQ02.dpuf

Efe, Reuters (2008) Sigue emergencia por erupción del Huila. Dos pueblos están aislados. Caracol tv. http://www.caracoltv.com/articulo92278-sigue-emergencia-erupcion-del-huila-dos-pueblos-estan-aislados

El Tiempo (2008) 10 muertos deja la erupción del Nevado del Huila. http://www.nevados.org/index.php/es/satelite-88/53-colombia/volcanes-del-sur/241-10-muertos-deja-la-erupcion-del-nevado-del-huila.html

Garcia C, Fearnley C (2012) Evaluating critical links in early warning systems for natural hazards. Environ Hazards 12(2)

Hall M (1990) Chronology of the principal scientific and governmental actions leading up to the November 13, 1985 eruption of Nevado del Ruiz, Colombia. J Volcanol Geoth Res 42(1–2):101–115

Haynes K, Barclay J, Pidgeon N (2008) The issue of trust and its influence on risk communication during a volcanic crisis. Bull Volcanol 70:605–621

IFRC—International Federation of Red Cross and Red Crescent Societies (2009) World disasters report 2009—people-centred approach and the 'last mile'. IFRC, Geneva

INGEOMINAS (1996) Evaluación de amenaza y vigilancia volcánica del Complejo Volcánico Nevado del Huila. Convenio INGEOMINAS - Corporación NASA KIWE. 35 pp. http://www.sgc.gov.co/Popayan/Documentos/Informe_mapa_amenaza_VNH_PUBLICAR.aspx

INGEOMINAS (2006) 20 años del observatorio vulcanológico y sismológico de Manizales. 79 pp. http://www.sgc.gov.co/Manizales/Imagenes/OrigenesOVSM.aspx

INGEOMINAS (2012) Memoria explicativa del mapa de amenaza volcánica potencial del Nevado del Ruiz. 4 pp. http://www2.sgc.gov.co/Manizales/Imagenes/Mapas-de-Amenaza/VNR/Memorias_Nevado_del_Ruiz.aspx

Méndez R, Cortés GP, Cepeda H (2002) Evaluación de la amenaza volcánica potencial del Cerro Machín (Departamento del Tolima, Colombia). INGEOMINAS, Manizales, Colombia, 66 pp

Mileti D, Sorenson JH (1990) Communication of emergency public warnings: a social science perspective and state-of-the-art assessment. Oak ridge. Oak Ridge National Laboratory, U.S. Department of Energy, TN

Monsalve ML, Méndez R (1988) Memoria del Mapa Preliminar de Amenaza Volcánica Potencial del Complejo Volcánico del Cumbal. Convenio INGEOMINAS – Gobernación de Nariño, 18 pp

Monsalve ML, Méndez R (1995) Presentación de la metodología utilizada para la elaboración de mapas de amenaza volcánica en Colombia. INGEOMINAS, Manizales. http://aplicaciones1.ingeominas.gov.co/Bodega/i_raster/110/02/0000/18130/documento/pdf/0101181301101000.pdf

Ordoñez M (2011) El problema del Riesgo en los volcanes Galeras, Huila y Machín en Colombia. Instituto Colombiano de Geología y Minería - INGEOMINAS. PASI 2011: Open vent volcano hazards workshop. San Jose, Costa Rica

Parra E, Cepeda H (1990) Volcanic hazard maps of the Nevado del Ruiz volcano, Colombia. J Volcanol Geoth Res 42(1–2):117–127

Parra E, Cepeda H, Thouret JC (1986) Mapa de amenaza volcánica del Nevado del Ruiz. INGEOMINAS

Paton D, Smith L, Daly M, Johnston D (2008) Risk perception and volcanic hazard mitigation: individual and social perspectives. J Volcanol Geoth Res 172:179–188

Peralta H (2008) Planeación estratégica para la reducción integral de riesgos en el Municipio de Páez, Cauca, Colombia, por la reactivación del volcán Nevado del Huila en el marco del desarrollo local sostenible, desde la cosmovisión de las comunidades del pueblo Indígena Nasa. Revista EIRD Informa - Las Américas. Número 15. http://www.eird.org/esp/revista/No_15_2008/art13.html

Perry RW, Lindell MK (2008) Volcanic risk perception and adjustment in a multi-hazard environment. J Volcanol Geoth Res 172:170–178

Presidency of the Republic of Colombia (1986) El Volcán y la avalancha: 13 de noviembre de 1985. Secretaría de Información y Prensa de la Presidencia de la República. http://www.cridlac.org/digitalizacion/pdf/spa/doc126/doc126.htm

Punongbayan RS, Newhall CG (1998) Early Warning for the 1991 eruptions of Pinatubo volcano—a success story. Programme and abstracts, International IDNDR-conference on early warning systems for the reduction of natural disasters, EWC'98, Potsdam, Federal Republic of Germany, 7–11 Sept

Rodriguez H, Díaz W, Aguirre B (2004) Communicating risk and warnings integrated and interdisciplinary approach. University of Delaware Disaster Research Center. Preliminary Paper #337

Semana (2008) La pronta evacuación salvó miles de vidas de los vecinos del Nevado del Huila. http://www.semana.com/on-line/articulo/la-pronta-evacuacion-salvo-miles-vidas-vecinos-del-nevado-del-huila/84600-3

Villegas H (2003) Display of the Nevado del Ruiz volcanic hazard map using GIS. Geocarto Int 18(3). doi:10.1080/10106040308542276

Wisner B (2001) Socialism and storms. The Guardian, 14 Nov 2001

Challenges of Volcanic Crises on Small Islands States

Jean-Christophe Komorowski, Julie Morin,
Susanna Jenkins and Ilan Kelman

Abstract

Island communities frequently display specific risk-related characteristics that are attributable to the island locations of volcano-affected communities, in terms of exposure, vulnerability and living with the volcanic risk. This chapter examines volcanic crisis response and communication in island communities. We analyse lessons from volcanic crises in 1976 at La Soufrière (Guadeloupe, France), in 2005 and 2006 at Karthala (Grande Comore, Comoros), and in 1995 at Fogo (Cape Verde). Our analysis underscores the strong influence of deep-seated causes (historical, political, cultural, social, economic, and environmental) on the success and failure of volcanic risk communication, all of which are affected by island characteristics. The case studies demonstrate the intensity of politics that manifests in these instances—perhaps because of, rather than despite, the smallness and tightness of the communities, amongst other island characteristics. Consequently, improved information and less uncertainty would not straightforwardly lead to better communication or more harmonious acceptance of decision-making processes and of decisions.

J.-C. Komorowski (✉)
Institut de Physique Du Globe de Paris, Université Paris Diderot (Sorbonne Paris Cité), UMR CNRS 7154, Equipe "Systèmes Volcaniques" 1, Rue Jussieu, Paris 75238, Cedex 05, France
e-mail: komorow@ipgp.fr; jeanchristophe.komorowski@gmail.com

J. Morin
Laboratoire de Géographie Physique, UMR 8591, Université Paris 1 Panthéon-Sorbonne, Meudon, France
e-mail: julieapi@yahoo.fr

J. Morin
Institut de Physique Du Globe de Paris, UMR CNRS 7154, Equipe "Systèmes Volcaniques", Paris, France

S. Jenkins
School of Earth Sciences, University of Bristol, Bristol BS8 1RJ, UK
e-mail: Susanna.Jenkins@bristol.ac.uk

I. Kelman
Institute for Risk and Disaster Reduction and Institute for Global Health, University College London, Gower Street, London WC1E 6BT, UK
e-mail: ilan_kelman@hotmail.com

I. Kelman
Norwegian Institute of International Affairs (NUPI), Oslo, Norway

Advs in Volcanology (2018) 353–371
https://doi.org/10.1007/11157_2015_15
Published Online: 26 March 2017

One fundamental point is the need to engage with—not just consult—local populations regarding risk communication and decision-making, tailoring messages to the various audiences, and being clear regarding what is known and not known, plus what is feasible to do to fill in knowledge gaps to support decisions. Ultimately, it is necessary to foster synergies with communities to ensure that no party or knowledge dominates, but instead information is exchanged leading to decisions and decision-making processes that are better understood and accepted by all who are involved. Living with volcanic risk thus means working with communities on their terms.

Keywords

Small island · Volcanic eruptions · Volcanic risk · Crisis management · Guadeloupe · La Soufrière · Comoros · Karthala · Fogo · Cape Verde

1 Introduction

This chapter examines volcanic crisis response and communication in island communities. We analyse lessons from volcanic crises in 1976 at La Soufrière (Guadeloupe, France), in 2005 and 2006 at Karthala (Grande Comore, Comoros), and in 1995 at Fogo (Cape Verde). Our analysis underscores the strong influence of deep-seated causes (historical, political, cultural, social, economic, and environmental) on the success and failure of volcanic risk communication, all of which are affected by island characteristics.

Island communities frequently display specific risk-related characteristics that are attributable to the island locations of the volcano-affected communities, in terms of exposure, vulnerability and living with the volcanic risk (Lewis 1999, 2009; Méheux et al. 2007; Mercer and Kelman 2010; Morin 2012; Pelling and Uitto 2001). In fact, research and experience from island communities has been a significant foundation for understanding risk, disaster risk reduction, and disaster response, particularly in volcanic contexts (de Boer and Sanders 2002; Gaillard 2008;

Keesing 1952; Kelman et al. 2011; Pattullo 2000). Regarding specific vulnerabilities:

– small land size, despite large ocean territory, can mean that even moderately sized volcanic hazards threaten the entire territory and population, causing difficulties in planning for risk management, and often needing to consider the possibility of permanent evacuation of the entire island,
– relative inaccessibility and remoteness from centres can often cause problems for decision-making,
– small populations make it difficult to have local experts for all disciplines required, yet small populations do not necessarily increase the community's homogeneity meaning that diverse cultural responses could still be expected.

Regarding living with the risk:

– tight, kinship-based communities can build inherent trust, sometimes supporting rapid information dissemination and response,

- communities with no option but to live in close proximity to the volcano can build up local knowledge regarding volcano behaviour, warning signs, and responses,
- small, diversified economies—including offshore livelihoods such as migration and remittances—can sometimes permit more nimble adjustment to crisis.
- The volcano could have provided a baseline of robust livelihoods, such as through agriculture and tourism, meaning that the population might have resources available to support themselves for some time.

The legacy of the island case studies presented here, over the decades during which they ensued, has significant implications for crisis response plans and risk communication for potential future eruptions—or eruption threats. The lessons are outlined in terms of engaging with local populations regarding risk communication and decision-making, tailoring messages to the various audiences, and being clear regarding what is known and not known, plus what is feasible to do to fill in knowledge gaps to support decisions.

The bulk of this chapter is presented as the three case studies, with the lessons and island relevance woven into the descriptions. The final section synthesises some general points emerging and provides examples of ways forward.

2 Case Study 1: La Soufrière de Guadeloupe, Guadeloupe, Caribbean

Management of the volcanic crisis at La Soufrière of Guadeloupe in 1975-77 remains one of the most contentious in recent times (Fiske 1984; Komorowski et al. 2005; Beauducel 2006; Devès et al. 2015). The chronology of the phenomenology, and the actions taken by scientists and authorities to manage the crisis were summarized by Hincks et al. (2014). The mild but persistent seismic and fumarolic unrest that has slowly increased since 1992 at La Soufrière volcano (Komorowski et al. 2005; http://www.ipgp.fr/fr/ovsg/bulletins-mensuels-de-lovsg; Komorowski

et al. 2013b, 2014) has prompted renewed interest in geologic studies, monitoring, risk modelling, and crisis response planning (Komorowski et al. 2012, 2013a; https://sites.google.com/site/casavaanr/) in order to learn from the 1975–77 crisis and improve the management and response to any potential future crisis at this volcano.

At least 6 phreatic explosive eruptions (1690, 1797–98, 1812, 1836–38, 1956 and 1976–77) have occurred at La Soufrière de Guadeloupe during the historical period since AD 1635. In contrast to previous phreatic eruptions of La Soufrière and elsewhere in the Caribbean, a significant period of increasing volcanic seismicity was recorded and felt in Guadeloupe starting in July 1975, one year prior to the onset of the eruption. This unprecedented and rapidly escalating level of recorded and felt seismicity (Dorel and Feuillard 1980; Feuillard et al. 1983; Feuillard 2011), which in June 1976 reached levels that were 175 times the baseline monthly rate, was not accompanied by any modification of fumarolic activity. The eruption began with an unexpected explosion on 8 July 1976. The subsequent 9-month long period of explosive and ash-venting activity was interpreted as a still-born or failed magmatic event (Feuillard et al. 1983; Villemant et al. 2005; Boichu et al. 2011). Syn-eruptive degassing (H_2O, minor CO_2, H_2S, SO_2) with acid condensates (HCl, HF, Br) led to moderate environnmental impact with short-term public health implications.

A major controversy emerged among the scientific community (Fiske 1984) as to whether fresh juvenile magmatic components could be recognised in the ejecta, thus raising the probability of a transition from a purely phreatic, non-hazardous scenario to a highly hazardous magmatic 1902 Mount Pelée-style explosive scenario. The scientific disagreement was widely echoed in the media (Loubat and Pistolesi-Lafont 1977; Farugia 1977; De Vanssay 1979; Figs. 1 and 2). On August 15, the emergency plan was enacted by the authorities based on the systematic increase in seismicity and magnitude of explosions. About 70,000 people were evacuated from all of southern Basse-Terre for about six months until December 15, 1976 (Préfecture de

Guadeloupe 1977; De Vanssay 1979; Lepointe 1984, 1999). This evacuation remains controversial today; however, regardless of the interpretation of the eruption's evolution, evacuation of the population in the areas closest to and downwind from the erupting volcano was necessary due to the degassing and ash fallout.

The cost of the preventive evacuation was estimated at 60 % of the total annual per capita Gross Domestic Product (GDP) of Guadeloupe in 1976 (Lepointe 1984, 1999; Blérald 1986; Kokelaar 2002) excluding the losses of uninsured personal assets and open-grazing livestock. Hence, this eruption and the evacuation that ensued rank amongst the most costly for the 20th century (Annen and Wagner 2003) although the eruption itself did not cause any fatalities. A few years before the eruption, a public policy was implemented to move the banana export port facilities of the Basse-Terre harbour highly exposed to Caribbean swells to the more sheltered harbour and economic capital Pointe-à-Pitre. The widespread long-lasting evacuation reinforced this shift to Pointe-à-Pitre, contributing to the demise of the economy of the administrative capital, Basse-Terre, and to the feeling among the population of bitterness and of being forsaken.

Lack of a comprehensive and integrated monitoring network prior to and during the crisis, the then-limited knowledge of the eruptive history, and a tendency of caution exacerbated by the memory of past devastating Caribbean eruptions (e.g. the devastating explosive eruptions of Montagne Pelée in May 1902 and of Soufrière of St. Vincent in April 1902) all contributed to a high degree of scientific uncertainty alongside a publically-expressed lack of consensus and trust in available expertise. Hence, analysis, forecasting, and crisis response were highly challenging for scientists and authorities in the context of escalating and fluctuating activity as well as societal pressure in a small island community (Komorowski et al. 2005; Beauducel 2006). Given the uncertain evidence and the absence of scientific consensus on the likely outlook, and the lack prior to the crisis of a well-defined and accepted volcanic emergency response plan, the authorities felt impelled to adopt an approach involving zero risk of casualties from the volcano—without fully considering other social risks emerging due to the evacuation.

A binary manichean approach appeared in the scientific discourse through a major conflict between two leading, authoritative scientists, C. Allègre and H. Tazieff, that unravelled and escalated in the media and public (Farugia 1977; Loubat and Pistolesi-Lafont 1977; De Vanssay 1979). Within the context of the socio-cultural frustrations the islanders experienced, feeling to be an unimportant appendage to the mainland, the scientific disagreements engendered a distrust of mainland (colonial) science. The authorities expected and demanded that the scientists and volcanology would provide one clear answer, not a range of more likely or less likely outcomes. They operated in a binary framework and expected science to be precise, to provide the answers, and to render decisions easier.

In fact, this crisis epitomized the growing clash of scientific paradigms at the time: one of observational and intuitive empirical science, embodied by Tazieff, versus one of quantitative science relying on models and integrating a probabilistic framework and some degree of uncertainty, embodied by Allègre. In the end facing Hamlet's dilemma—will the eruption evolve into a paroxysmal explosive onset magmatic eruption or not (Fig. 1)—the manichean framework led to a very low acceptance of inherent uncertainty and the advocacy of the predominance and uniqueness of one or a few strands of evidence. The deep disagreement between opinionated scientists who often bypassed the responsible authority to speak directly to the population and the media—with the straightforwardness of attracting a substantial audience in a small, isolated community—thus vented the scientific debate directly to the public without critical and consensual appraisal, forcing authorities to seek alternative advice. Many "special" advisors provided their own cast on the uncertain processes developing at the volcano. This significantly undermined the integrity of the scientific judgement and the trust amongst

Fig. 1 a The Préfet Aurousseau and M. Feuillard reflect at the bedside of La Soufrière. After a drawing by C. Maillaud-Bourdan, published in France-Antilles on 27 September 1976; **b** H. Tazieff and R. Brousse quarrel over La Soufrière. After a drawing by C. Maillaud-Bourdan, published in Guadeloupe 2000 No. 46, October–November 1976 (Taken from De Vanssay 1979)

experts, decision-makers, and the exposed population.

In Guadeloupe, the societal perception and translation of these conflicting schools of scientific thinking saw the rise of the image of maverick, anti-conformist, and heroic scientists (e.g. Tazieff) against more precautionary, conservative scientists (e.g. Brousse and Allègre). A perceived contrast emerged of free, independent scientists fighting for a just cause (the former) versus scientists who became or were perceived to be aligned with a political agenda and to a system of governance (the latter). The public debate thus became political and polarized based on opposing "truths" from contrasting scientific expertise (Fig. 2) rather than on how science could help constrain epistemic and aleatory uncertainty in order to foster improved decision-making despite uncertainty. This situation acted as an ideal crucible for fuelling a media-hyped controversy on the crisis and its management (Loubat and Pistolesi-Lafont 1977; De Vanssay 1979; Farugia 1977) that was easily prone to wild conspiracy theories. The volcano hence became a catalyst of accumulated island-mainland disputes—a way to engage in a mutual catharsis of the remains of France's colonialist past in the region and the sequels in its post-colonialist policies.

The controversy and the conflation of science with long-standing political grievances had significant lasting effects on national and international volcanology. A portion of the population still collectively deems that the evacuation was unnecessary or that the need was largely exaggerated, with many believing that the politics of shifting towards Pointe-à-Pitre superseded science with the justification of applying at the maximum precautionary principle. This has fuelled conspiracy theories and perceptions that the hazard and risk assessment was imported from the mainland, that the scientists' mentality was disconnected from the localised island reality, and that it served a political agenda that had

Fig. 2 Photo taken from a television (TF1) debate on 11 November 1976 between C. Allègre, then Director of the Institut de Physique du Globe de Paris (*left*) and volcanologist H. Tazieff (*right*) with J. Besançon as the moderating journalist (Taken from Loubat and Pistolesi-Lafont 1977)

started prior to the eruption but was fortuitously reinforced by the eruption. In fact, there was a blatant underestimation by authorities, scientists, and media, that for the evacuated and affected population, the volcano being the source of their distress and livelihood loss rapidly faded behind the curtain of long-nurtured socio-political frustrations and issues of cultural identity and island empowerment.

The eruption marked a tipping point in the recent history of Guadeloupe and symbolized its unresolved dilemmas and injustices. The volcano became an excuse to publically express a profound dissatisfaction in public authorities and policy, in the centralized mainland-focused system of governance. All protagonists failed to recognize that politics became the real driver of risk perception, the local agenda, and the aspirations of civil society.

However, positive consequences included a major increase in national funding for volcano monitoring and research. Lessons learned significantly improved volcano crisis management worldwide and notably for the nearby 1979 St. Vincent eruption. In the context of increased population density on the volcano's flanks and ongoing major development plans for southern Basse-Terre, even a 1976–77 style eruption will likely pose major challenges to authorities and decision-makers despite the presence of a highly sophisticated monitoring network and the vast knowledge subsequently acquired about La Soufrière volcano. Thus, this infamous crisis exemplified the need for a structured and transparent approach to evidence-based decision-making in the presence of substantial scientific uncertainty (Aspinall 2010; Aspinall and Cooke 2013; Aspinall and Blong 2015).

3 Case Study 2: Karthala, Grande Comore, Comoros, Indian Ocean

Union of Comoros is a volcanic archipelago (Fig. 3) and one of the world's smallest and poorest countries (Taglioni 2003; UNDP 2014). Emigration is assumed by many to be the only solution to escape poverty, so 200,000

Comorians live in France (Da Cruz et al. 2004), which was Comoros' colonial power until 1975. When independence was proclaimed, France allegedly illegally kept Maore (one of the four Comorian Islands) under its political control, creating a legal battle with Comoros (Oraison 2004). Moreover, the governments of each island (Nzwani, Mwali, and Ngazidja) regularly contest their level of autonomy from the Union federal government. Consequently, it is difficult to define the exact prerogatives attributed to the islands' and Union's governments, including those linked to the volcanic crises management (Morin and Lavigne 2009).

Grande Comore Island (or Ngazidja; 1148 km^2) is dominated by Karthala, an active basaltic shield volcano which has erupted, on average, every seven years in recent decades. Volcanic hazards threaten the population, as effusive or explosive eruptions may impact upon the whole island (Bachèlery and de Coudray 1993), including coastal villages, Hahaya international airport, and the capital, Moroni (Morin et al. 2016). Electricity and telephone networks are easily and often saturated, disrupted, or inaccessible in some parts of the island that are out of range from the communications relay. Chouaybou (2010) underlines that after each major event on the island, diaspora members call their families, contributing to the network's saturation. As with other Comorian infrastructure, the monitoring network of the local volcano observatory (Observatoire Volcanologique du Karthala—OVK) is poorly developed and difficult to maintain, due to financial, technical and human resource constraints (Morin et al. 2009, 2016).

Additionally, within this context, Morin and Lavigne (2009), Morin (2012), Morin and Gaillard (2012) and Kelman et al. (2015) describe how some people consciously increase the volcanic risk level in their community to have an easier access to livelihoods. For example, some villagers repeatedly steal solar panels from the OVK monitoring network for easier access to electricity. Some other villagers vandalize the OVK network to contest foreign post-colonial "domination" (as the network was mainly funded

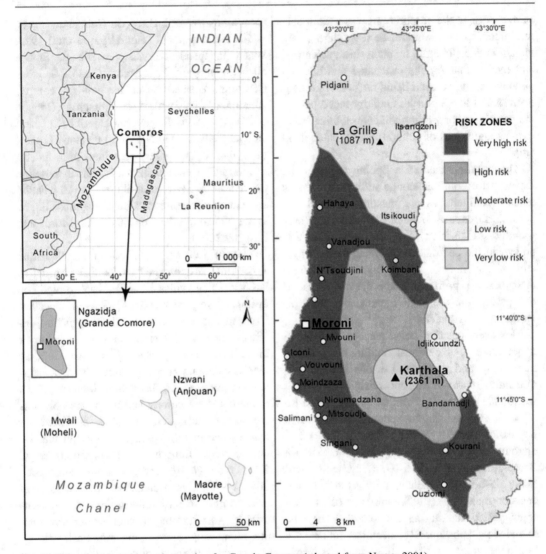

Fig. 3 Volcanic risk and hazards zonation for Grande Comore (adapted from Nassor 2001)

by France), or to claim that their deep links with the volcano and their customary chiefs' knowledge are enough to face the volcanic threat. Once the network is degraded, OVK has much reduced capacity to monitor and interpret seismic or volcanological activity. While authorities and OVK have targeted this issue in their last risk awareness campaigns, thefts and degradations continue.

The 2006 eruption demonstrated both the authorities' disorganization in facing crises and OVK's incapacity to deliver prompt, clear and updated information (Morin et al. 2009), making the communication process really poor. People

were informed of the eruption mainly by word-of-mouth or by their own observations; only a third of the population was warned by the media. The lack of warning systems, combined with an absence of any consistent prevention information, created anxiety amongst islanders and affected their responses, negatively impacting the crisis management. One consequence was that the OVK scientific team faced harder and more detailed communication tasks than expected (Fig. 4).

Several 'critical' periods for the OVK team were reported, the worst one being from 28 May 21:10 to 29 May 02:00, just after the eruption's

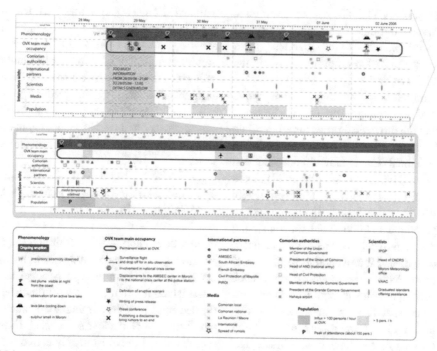

Fig. 4 Main components in the official schedule of the local volcano observatory team during the May 2006 eruption of Karthala (translated from Morin 2012). Due to the multiple and constant internal and external interactions of the six people involved in the May 2006 crisis management at OVK, it is estimated that 30–40 % of official meetings were not listed. Otherwise this register does not include exchanges with the officials who came to enquire personally about the situation, nor the constant calls received by the OVK team from inhabitants requesting or offering information. This 'informal' trade is a flow roughly equivalent to the 'formal' one

beginning. OVK faced 30 meetings and interviews with authorities of both Island and Union governments, and an agglomeration of hundreds of civilians at the OVK office, in addition to dozens of phone calls saturating OVK phone facilities, while no localization of the eruptive activity nor a scenario of its evolution was possible to provide at that time. People suspected that scientists did not communicate because of serious ongoing problems while the overwhelming number of journalists was difficult for the OVK team to manage, so journalists were sidelined for a few hours. The first information was given soon after 23:00, as soon as it had been decided to fly over the volcano at dawn in order to observe the nature and location of the eruption. Rumours circulated by a few local and international media (such as "lava is flowing down the slopes of Karthala, threatening

thousands of Comorians") immediately spread within villages and then to the international news. OVK team was forced to monitor the media three times a day (dawn, noon, and evening) to be able to quickly deny any wrong information. A protocol was established to improve communications efficiency, one of the rules being to limit the number of scientists communicating with the media: one in Shikomori (the local language) and another in French and English.

The second main problematic period was on 29 May from 09:00–12:30, just after the first flight over the volcano. Once more, rumours about the evolution and implications of the eruptive activity spread on the island. This indicates that the information given by the OVK team was interpreted incorrectly by some journalists. 47 interviews were given during this five-day-eruption, and OVK had to deny three rumours spread through the media. A disaster communication handbook for journalists was conceived after the 2006 eruption, based on the DFID handbook (2003).

The abnormally high number of interlocutors was difficult to manage for the OVK team, showing that the theoretical procedures described in the "*Karthala Plan*", 37-page appendix to PNPRU [National Emergency Preparation and Response Plan (Union des Comores 2004)], were not known. In fact, PNPRU has not been widely disseminated or updated since it was created in 2004. For example, in 2006, it did not contain the updated phone numbers of main risk management stakeholders. Moreover, due to Comoros' colonial past, PNPRU is modelled on French plans without taking into account island contexts. Its operational specification sheets are not necessary adapted to the traditional oral cultural context in Comoros for dealing with locals during a crisis (Morin et al. 2009). Warning dissemination and general information, for example, could rely more on the active local religious networks and non-profit organizations.

Paradoxically, because the monitoring network was still out of order, and despite its team already being overwhelmed, OVK had no other solution than to ask, through a media announcement, that the inhabitants deliver to OVK any testimony about seismicity or suspected volcanic activity. In response, inhabitants made dozens of phone calls to report felt phenomenon (e.g. smells and seismicity) and new "eruptive activity" (most of the time, bushfires). Calls peaked from May 30 to June 1 as earthquakes were felt in the Bahani area, a few kilometres north of Moroni.

The main difficulties in volcanic crisis communication on Grande Comore are mainly due to Comoros being a poor, small, remote island country. The small population implies difficulties in finding local experts in all fields, so that volcanological monitoring and understanding relies on foreign researchers from La Réunion and

mainland France. The whole territory is threatened by volcanic hazards while communication networks are very weak. OVK's inability to provide information, the inhabitants' and authorities' lack of preparedness, and the alarmist news disseminated by the media are amongst main weaknesses, leading to a failure of crisis management and crisis communication on Grande Comore during the last volcanic crises.

4   Case Study 3: Fogo Volcano, Cape Verde, Atlantic Ocean

Fogo volcano and island, within the Cape Verde archipelago of hotspot volcanism, lie approximately 600–800 km west of Senegal in Africa. Fogo Island is around 30 km in diameter and is formed entirely of the large stratovolcano of Fogo, which rises 2829 m above sea level and is the most prominent and only historically active volcano in Cape Verde. An approximately 9-km wide lateral collapse structure dominates the island, and is open to the east with very steep headwalls of up to 1 km (Fig. 4; Day et al. 1999). The central cone of Pico do Fogo rises more than 1 km above the partially infilled collapse scar and is home to 'Chã das Caldeiras' residents. One of the major hazards for Chã residents, aside from drought and a lack of basic services, is rockfalls from the steep headwalls; more direct volcanic eruption hazards (Fig. 5) have typically been of lesser concern.

Historical accounts indicate that Fogo volcano experienced regular eruptions during the early stages of Portuguese settlement (Fonseca et al. 2003) and that the summit cone of Pico do Fogo last erupted in 1680, producing large ash falls that rendered agricultural lands temporarily unusable and triggered mass emigration from the island (Ribeiro 1960). Recent activity has been more subdued with direct experience restricted to the effusive eruptions of 1951, 1995, and 2014–2015 that occurred from subsidiary flank vents and fissures near the base of Pico (Faria and Fonseca 2014).

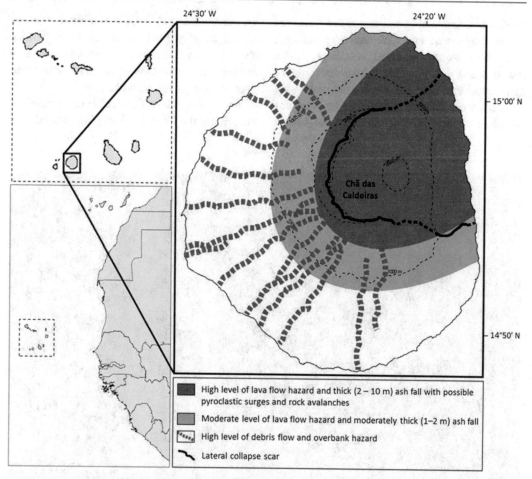

24°30′ W   24°20′ W

15°00′ N

Chã das
Caldeiras

14°50′ N

High level of lava flow hazard and thick (2 – 10 m) ash fall with possible
pyroclastic surges and rock avalanches

Moderate level of lava flow hazard and moderately thick (1-2 m) ash fall

High level of debris flow and overbank hazard

Lateral collapse scar

Fig. 5 Location of Fogo Island in the Cape Verde archipelago, with inset: provisional volcanic hazard map for a large explosive eruption similar to the 1680 event (Day 2009)

Official maps outlining potential future volcanic hazards are not publicly available for Fogo; however, Day and Faria (2009) identify three preliminary scenarios in an outline hazard analysis: (1) A continuation of post-1785 effusive activity on volcanic fissures within Chã das Caldeiras; (2) More intense and frequent fissure eruptions similar to those during the 17th and 18th centuries; and (3) A large explosive eruption from the summit of Pico do Fogo as in 1680 (Fig. 4). Additionally, the neighbouring island of Brava, 20 km to the west of Fogo shows geological evidence of Plinian volcanism and, in certain wind conditions, large eruptions from Brava may be expected to deposit significant ash falls on Fogo Island (Jenkins et al. 2014).

Communication of the volcanic hazard and risk on Fogo Island consists predominantly of top-down approaches supported by local monitoring institutions and the army-led National Civil Protection Service (SNPC), which was formed after the eruption in 1995. At the time of the 1995 eruption, Chã das Caldeiras was home to approximately 1,300 people in three villages; prior to the 2014 eruption, this number was estimated at nearly 1,500 (Global Volcanism Program 2014). A further 11,000 people in a number of villages on the steep eastern flanks of Fogo are exposed to overspilling flows.

Following intermittent, felt earthquakes through late March and early April 1995 at Fogo, a fissure eruption began on the night of 2–3 April

1995. Lava fountaining and associated flows cut off the main road into Chã and over the next week lava flows destroyed a village, water reservoir and a few square kilometres of fertile agricultural land (Fig. 6a). On 8 April, remaining residents were ordered to evacuate the Chã das Caldeiras or face arrest (Bulletin of the Global Volcanism Network 1995). The eruption ended around 28 May 1995 and in the years that followed communities and associated services such as health and education were permanently relocated outside of the collapse scar.

Within two years, Chã had been repopulated and prior to the 2014–2015 eruption was home to

Fig. 6 **a** A 2009 NASA Earth Observatory satellite image of Chã das Caldeiras showing the remaining villages and the extent of the 1995 lava flows; **b** 1995 permanent relocation sites within the red outline in the village of Achada Fuma (*image* Google Earth); **c** A building within the now partly abandoned relocation site (*photo* S. Jenkins)

around 800 people with most of the population living in Portela and Bangeira (Fig. 6a), as well as a burgeoning wine and tourism industry (Jenkins et al. 2014). The failure of people to permanently relocate following the 1995 eruption is attributed to the close social and economic ties that Chã residents have to their place of origin and because of unsuitable relocation sites that lacked sustainable sources for livelihood. Relocation sites, subsidised by the German government, were positioned outside of the collapse scar in areas of relatively low hazard to the south of the island; however, they provided very small living quarters and limited-to-no agricultural land (Fig. 6b, c). For the family-centred communities for whom farming was the main source of livelihoods for sustenance and trade, these sites offered an unviable alternative to Chã das Caldeiras (Fig. 7).

From November 2014 to February 2015, Fogo produced another effusive eruption similar to that of 1995. The eruption produced explosions and ash emissions, lava fountains and fast-moving lava flows moving up to 20 km/hour). All of the Chã residents were evacuated with lava flows

destroying more than 90 % of the main villages, Portela and Bangeira, a large area of agricultural land, communication poles and the only road into the area (Fig. 5). Intermittent ash explosions dispersed and deposited ash in the main city of São Filipe and across agricultural crops on the island; the airport and tourism industry were also impacted. As of writing, it is unclear if Chã residents will be able to return, but near-complete reconstruction would be required to enable repopulation of Chã.

One of the key issues facing risk communication on Fogo appears to be prioritising concerns about livelihoods for the affected communities. On Fogo, eruptions are not the major concern for at-risk communities; populations are more concerned with a lack of basic services, drought, and associated food shortages. The implementation of Fogo National Park in 2003, which aimed to protect the natural environment and regulate tourism through land-use and livestock restrictions, caused conflict between authorities and Chã residents who claimed that their livelihoods were adversely affected by such restrictions (Texier-Teixeira et al. 2013).

Fig. 7 Annotated photograph of December 2014 lava flows that buried Portela village. The view is looking approximately due South with Pico do Fogo to the left of the image and the steep collapse scar walls to the right. (*Photo* S. Jenkins)

More recent risk communication efforts by the authorities have included evacuation and communication exercises with a community-based focus. They have been carried out by SNPC and one of the local monitoring institutions, the National Institute of Meteorology and Geophysics, with partial support from the Italian Civil Protection and MIA-VITA (a European research project). Unfortunately, the exercises found that tour guides, women, and farmers, some of the most important groups in Chã das Caldeiras, were difficult to engage with the exercises and decision-making (Texier-Teixeira et al. 2013). Future volcanic risk management and communication therefore needs to first address socio-economic vulnerabilities and priorities in order to carry out fully successful risk reduction measures. The government of Cape Verde currently has plans to create a local Fogo civil protection agency to allow for more community-based activities, although this is not yet operational.

The 2014–2015 eruption and the destruction of villages within Chã may have changed the way volcanic hazard and risk at Fogo are communicated from that suggested by the pre-eruption studies. Future studies should investigate.

5 Risk and Crisis Communication for Island Communities

Island communities are frequently portrayed as being tight-knit with few people being strangers to each other, suggesting the potential that crisis response is more efficient because information travels quickly, people trust each other, and social structures are more nimble. The three case studies demonstrate that plenty of truth exists in such assumptions, but that does not necessarily ease the situation. Instead, as always with risk- and disaster-related communication and knowledge-policy interfaces, multi-faceted layers intertwine and information flow is not a linear, predictable process (Barclay et al. 2008; Gaillard and Mercer 2013; Weichselgartner 2003).

In all three cases presented here, it was not possible to reach a common denominator and

representation and perception of what a volcanic eruption is, what are the associated risks, what are the timescales of volcanic processes and the necessary timescales of risk prevention policies, and what exactly is happening and could happen in the particular situation under question. The media's role was particularly poignant, with difficulties often resulting for the scientists because the media and scientists did not have a trusting relationship. Particularly in small communities, it is important to avoid viewing the media as a separate or hostile entity, but instead to work them so that their needs are met but reports are accurate and informative. The relationship cannot be built during a crisis, but must start immediately so that all parties involved have already built an understanding of needs when the crisis emerges (see also Barclay et al. 2008; DFID 2003). The communication challenges can further be overcome through expanding the capacity and responsibility of the scientific community to explain its paradigms, its methodologies, different types of uncertainty (epistemic and aleatory), and the limitations to and for different audiences, namely the authorities, decision-makers, crisis responders, economic actors, the public, and the media.

An example from Guadeloupe aiming to do so is the widespread public dissemination of a monthly activity bulletin by the Observatoire Volcanologique et Sismologique de Guadeloupe (OVSG-IPGP) that was launched in 1999 and has continued ever since (http://www.ipgp.fr/fr/ovsg/bulletins-mensuels-de-lovsg; http://www.ipgp.fr/fr/ovsg/actualites-ovsg). A tri-monthly similar bulletin is issued by the Observatoire Volcanologique et Sismologique de Martinique (http://www.ipgp.fr/fr/ovsm/bilans-trimestriels-de-lovsm). It would be useful to enact studies to determine the reach of these bulletins and the understanding by and interest of those who read them. What information and styles could the bulletins include which would appeal to a wider readership and ensure that uptake of the information is increased? Such work will enhance participatory science, as with the media establishing a baseline and trust before an emergency so that crisis communication should be smoother and more widely accepted.

As well, as science evolves, new scientific approaches could assist in better integrating different sectors into volcanic crisis communication processes. Again from Guadeloupe, a recent retrospective Bayesian Belief Network analysis of the unrest of La Soufrière volcano in 1976 (Hincks et al. 2014; Komorowski et al. 2015) demonstrates that a formal evidence-based case could have been made to support the authorities' concerns about public safety and the decision to evacuate in 1976. Development of such novel probabilistic formalism for decision-making could help to reduce scientific uncertainty and better assist public officials in making urgent evacuation decisions or policy choices (Woo 2011) should the mild but sustained and slightly increasing seismic and fumarolic unrest that began in 1992, and further developed in 1998, evolve into renewed eruptive activity (OVSG-IPGP 2015; Villemant et al. 2014; Allard et al. 2014).

Yet the case studies demonstrate the intensity of politics that manifests in these instances—perhaps because of, rather than despite, the smallness and tightness of the communities, amongst other island characteristics. Consequently, improved information and less uncertainty do not straightforwardly lead to better communication or more harmonious acceptance of decision-making processes and of decisions. One fundamental point is the need to engage with —not just consult—local populations regarding risk communication and decision-making, tailoring messages to the various audiences, and being clear regarding what is known and not known, plus what is feasible to do to fill in knowledge gaps to support decisions. Activities to engage citizens in the science emerged to a large degree in the three case studies, suggesting that further possibilities be investigated, based on literature and past work in this area (e.g. Cadag and Gaillard 2012; DFID 2003; Kelman et al. 2015; Mercer and Kelman 2010; Texier-Teixeira et al. 2013). Examples of activities are collecting data, mapping exposure and vulnerability, assisting with communication, and organising town hall meetings between scientists and citizens—all of which should happen before a crisis so that a crisis simply continues this work and the

long-standing relationships, rather than starting anew during the most difficult time period. Methods exist for melding knowledge forms to ensure that local knowledge and non-local knowledge are treated equally, helping to garner respect and turn information conveyance into information exchange (Mercer et al. 2009)—a process implemented for a volcanic crisis in Papua New Guinea (Mercer et al. 2010).

Ultimately, this becomes a collaboration with communities to ensure that no party or knowledge dominates, but instead information is exchanged leading to decisions and decision-making processes which are better understood and accepted by all who are involved. That is, living with volcanic risk means working with communities on their terms.

Acknowledgements We would like to thank first and foremost the communities and authorities on the three case-study islands, who have supported studies of volcanic hazard and risk communication and perception on the islands. SFJ would like to thank João Fonseca, Bruno Faria and Simon Day and colleagues at the Laboratory for Civil Engineering of Cape Verde for providing information and field support at Fogo volcano. The work presented in this chapter was supported by the CASAVA Project (Agence Nationale pour la Recherche, ANR-09-RISK-02: JCK), the MIAVITA Project (EU FP7-ENV contract 211393: SFJ) and an AXA Research Fund fellowship (SFJ). JCK is grateful for stimulating and insightful discussions with M. Devès, F. Beauducel, T. Hincks, W. Aspinal, F. Leone, D. Grancher, M. Redon, M. Chenet and the participants of the CASAVA Consortium. JM is grateful to Patrick Bachèlery, Anthony Finizola, Nicolas Villeneuve, Magali Smietana and Hamidou Nassor from La Réunion Island University, Christopher Gomez from Christchurch University, Hamid Soule from OVK, and the AMISEC forces for their logistic support on Karthala volcano. We gratefully acknowledge the editors who invited us to contribute to this volume. This is IPGP contribution 3695.

References

Allard P, Aiuppa A, Beauducel F, Calabrese S, Di Napoli R, Crispi O, Gaudin D, Parello F (2014) Steam and gas emission rate from La Soufriere volcano, Guadeloupe (Lesser Antilles): implications for the magmatic supply during degassing unrest. Chem Geol 384:76–93

Annen C, Wagner J-J (2003) The impact of volcanic eruptions during the 1990s. Nat Hazards Rev 4(4): 169–175

Aspinall W (2010) A route to more tractable expert advice. Nature 463:294–295

Aspinall WP, Cooke R (2013) Expert Elicitation and Judgement. In: Hill L, Rougier JC, Sparks RSJ (eds) Risk and uncertainty assessment in natural Hazards. Cambridge University Press, Cambridge, pp 64–99

Aspinall WP, Blong R (2015) Chapter 70: volcanic risk assessment. In: H Sigurdsson, B Houghton, S McNutt, H Rymer, J Stix (eds) The Encyclopedia of volcanoes, 2nd edn. Elsevier, Academic Press, Amsterdam. doi:10.1016/B978-0-12-385938-9.01001-4

Bachèlery P, de Coudray J (1993) Carte volcano-tectonique (1/50000e) de la Grande Comore et notice explicative, République Fédérale Islamique des Comores, Centre national de documentation et de recherches scientifiques/République Française, Université de La Réunion, Département Sciences de la Terre, Saint-Denis de La Réunion, France

Barclay J, Haynes K, Mitchell T, Solana C, Teeuw R, Darnell A, Crosweller HS, Cole P, Pyle D, Lowe C, Fearnley C, Kelman I (2008) Framing volcanic risk communication within disaster risk reduction: finding ways for the social and physical sciences to work together. In: DGE. Liverman, CPG Pereira, B Marker (eds) Communicating environmental geoscience, Geological Society of London special publications, Geological Society of London, London, pp 163–177

Beauducel F (2006) A propos de la polémique de La Soufrière 1976. Available: http://www.ipgp.jussieu.fr/ ~beaudu/soufriere/forum76.html. Accessed 10 June 2015

Blérald A-P (1986) Histoire éruptive de la Guadeloupe et de la Martinique du XVIIème siècle à nos jours. Editions Karthala, Paris

Boichu M, Villemant B, Boudon G (2011) Degassing at La Soufrière de Guadeloupe volcano (Lesser Antilles) since the last eruptive crisis in 1976–77: result of a shallow magma intrusion. J Volcanol Geotherm Res 202:102–112

Bulletin of the Global Volcanism Network (1995) 03/1995 (BGVN 20:03) New eruption on 2 April generates lava flows within the caldera

Cadag JRD, Gaillard JC (2012) Integrating knowledge and actions in disaster risk reduction: the contribution of participatory mapping. Area 44(1):100–109

Chouaybou K (2010) Les risques volcaniques en Grande Comore, Master 2 Géographie, Université Paris 8. France, Paris

de Boer JZ, Sanders DT (2002) Volcanoes in human history. Princeton University Press, Princeton

Da Cruz V, Fengler W, Schwartzman A (2004) Remittances to comoros—volume, trends, impact and implications. Africa Region Working Paper Series, vol 75, pp 1–37

Day SF, Faria BVE (2009) A geological hazard map of the Island of Fogo showing broad distribution of volcanic and other hazards. MIA-VITA Workshop, São Filipe, Fogo

Day SJ, Heleno da Silva SIN, Fonseca JFBD (1999) A past giant lateral collapse and present-day flank instability of Fogo, Cape Verde Islands. Jour Volcanol Geotherm Res 94(1–4):191–218

De Vanssay B (1979) Les événements de 1976 en Guadeloupe: apparition d'une subculture du désastre. Centre Universitaire Antilles-Guyane (Pointe-à-Pitre, Guadeloupe) et Ecole des Hautes Etudes en Sciences Sociales, Université Paris 5, Thèse de Doctorat de 3ème cycle, pp 170. 7 Dec 1979

Devès MH, Ribémont T, Komorowski J-C (2015) Quand les sciences de la Terre rencontrent l'analyse de l'action publique: Retour sur les expériences des éruptions de la Soufrière de Guadeloupe et de Soufrière Hills (Montserrat). CoSPoF 2015, ST 18 "La science politique face aux objets complexes: pratiquent et défis de l'interdisciplinarité. 1–21 pages. Paper presented at the CoSPoF 2015, 6ème Congrès International des Associations Francophones de Science Politique, 5–7 February 2015, Institut d'études politiques et internationales, Université de Lausanne, Switzerland

DFID (2003) Communication during volcanic emergencies—an operations manual for the Caribbean. DFID Project R7406: Protecting small islands by improving forecasting and warning. Department for International Development, pp 37

Dorel J, Feuillard M (1980) Note sur la crise sismo-volcanique à la Soufriére de la Guadeloupe 1975–1977. Bull Volcanol 43(2):419–430

Faria B, Fonseca J (2014) Investigating volcanic hazard in Cape Verde Islands through geophysical monitoring: network description and first results. Nat Hazards Earth Syst Sci 14:485–499

Farugia L (1977) Soufrière 76. Edition Jeunes Antilles, pp 352

Feuillard M, Allègre C, Brandeis G, Gaulon R, Le Mouël J-L, Mercier J, Pozzi J, Semet M (1983) The 1975–1977 crisis of La Soufrière de Guadeloupe (FWI): A still-born magmatic eruption. our Volcanol Geotherm Res 16:317–334

Feuillard M (2011) La Soufrière de la Guadeloupe: un volcan et un peuple. Guadeloupe, Editions Jasor, Pointe-à-Pitre, p 246

Fiske RS (1984) Volcanologists, journalists, and the concerned local public: a tale of two crises in the eastern Caribbean. In: Geophysics Study Committee, Geophysics Research Forum, Commission on Physical Sciences, Mathematics, and Ressources, National Research Council, (eds) Explosive Volcanism. Inception, Evolution and hazards. Studies in Geophysics. National Academy Press, Washington, DC, p 170–176

Fonseca JFBD, Faria BVE, Lima NP, Heleno SIN, Lazaro C, d'Oreye NF, Ferreira AMG, Barros IJM, Santos P, Bandomo Z, Day SF, Osorio JP, Baio M, Matos JLG (2003) Multiparameter monitoring of Fogo Island, Cape Verde, for volcanic risk mitigation. J Volcanol Geotherm Res 2605:1–18

Gaillard JC (2008) Alternative paradigms of volcanic risk perception: the case of Mt Pinatubo in the Philippines. J Volcanol Geoth Res 172(3–4):315–328

Gaillard JC, Mercer J (2013) From knowledge to action Bridging gaps in disaster risk reduction. Prog Hum Geogr 37(1):93–114

Global Volcanism Program (2014) Weekly Reports: 3 to 9 December 2014, Weekly Reports of the Global Volcanism Network. Smithsonian Institution, Washington DC

Hincks TK, Komorowski J-C, Sparks RSJ, Aspinall W (2014) Retrospective analysis of uncertain eruption precursors at La Soufrière volcano, Guadeloupe, 1975–77: volcanic hazard assessment using a Bayesian Belief Network approach. J Appl Volcanol 3:1–26

https://sites.google.com/site/casavaanr/, Web-page of the CASAVA project, Understanding and assessing volcanic hazards, scenarios, and risks in the Lesser Antilles: implications for decision-making, crisis management, and pragmatic development, ANR-RISK-09-002; 2010-2014, Agence Nationale pour la Recherche, Paris, France, last accessed 10 June 2015

http://www.ipgp.fr/fr/ovsg/bulletins-mensuels-de-lovsg. Last accessed 10 June 2015

http://www.ipgp.fr/fr/ovsg/actualites-ovsg. Last accessed 10 June 2015

http://www.ipgp.fr/fr/ovsm/bilans-trimestriels-de-lovsm. Last accessed 10 June 2015

Jenkins SF, Spence RJS, Fonseca JFBD, Solidum RU, Wilson TM (2014) Volcanic risk assessment: Quantifying physical vulnerability in the built environment. J Volcanol Geotherm Res 276:105–120

Keesing FM (1952) The Papuan Orokaiva vs. Mt Lamington: Cultural Shock and its Aftermath, Human Organization 11(1):26–22

Kelman I, Lewis J, Gaillard JC, Mercer J (2011) Participatory action research for dealing with disasters on islands. Isl Stud J 6(1):59–86

Kelman I, Gaillard J-C, Mercer J, Crowley K, Marsh S, Morin J (2015) Culture's role in disaster risk reduction. Combining knowledge systems on Small Island Developing States (SIDS). Cultures and disasters, p 208–219 (in press)

Kokelaar BP (2002) Setting, chronology and consequences of the eruption of Soufrière Hills Volcano, Montserrat (1995–1999) In: Druitt TH, Kokelaar BP (eds) The eruption of Soufrière Hills Volcano, Montserrat, from 1995 to 1999. Mem Geol Soc Lond 21:1–44

Komorowski J-C, Boudon G, Semet M, Beauducel F, Anténor-Habazac C, Bazin S, Hammouya G (2005) Guadeloupe. In: Lindsay JM, Robertson REA, Shepherd JB, Ali S (eds) Volcanic Hazard Atlas of the Lesser Antilles. Seismic Research Unit, The University of the West Indies, Trinidad and Tobago, pp 65–102

Komorowski J-C, Beauducel F, Hincks T, Aspinall W, Sparks RSJ, Villemant B, Boudon G (2012) Lessons learned from the 1975–77 unrest-failed magmatic eruption at la Soufrière of Guadeloupe (West indies): implications for future unrest-eruption forecast planning and response, Invited Keynote speech, VUELCO workshop, "Volcanic unrest: uncertainties and scenario planning/forecasting» , Sunday 18 November, Cities on Volcanoes 7, 18-23 November 2012, IAVCEI, Colima, Mexico, abstract, oral

Komorowski J-C, Legendre Y, Barsotti S, Esposti Ongaro T, Jenkins S, Baxter P, Boudon G, Leone F, Neri A, Spence R, Aspinall W, Grancher D, Redon M, Chopineau C, de Chabalier J-B (2013a) Assessing long term hazards for la Soufriere of Guadeloupe volcano: insights from a new eruptive chronology, credible scenario definition, and integrated impact modelling. International Association of Volcanology and Chemistry of the Earth's Interior (IAVCEI), Scientific Assembly, Kagoshima, Japan, 19–24 Juy 2013, abstract, 4W_4D-P1, poster

Komorowski J-C, Beauducel F, Leone F. Redon M, Chopineau C, Bengoubou-Valerius M, and the CASAVA research consortium (C Antenor, W Aspinall, MD Baillard, S Barsotti, P Baxter, A Bitoun, G Boudon, M Burac, L Bruxelles, G Carazzo, J-B de Chaballier, M Chenet, A Chevallier, V Clouard, J-C Denain, C Dessert, T Esposti Ongaro, N Feuillet, S Fourmond, M Gherardi, D Grancher, J-R Gros-Desormeaux, T Hincks, S Jenkins, E Kaminski, A Le Friant, G Lalubie, F Lavigne, Y Legendre, T Lesales, M Mas, J-M Mompelat, J Morin, C Narteau, A. Neri, S Pelczar, R Robertson, S Sparks, R Spence, P Tinard, B Villemant, F Vinet, G Woo) (2013b) The scientific challenges of responding to potential eruptive scenarios at La Soufrière of Guadeloupe: Lessons learned from the 1975–77 unrest failed-magmatic eruption and insights from integrated interdisciplinary risk assessment in the framework of the CASAVA project, 42nd Workshop of the International School of Geophysics, Volcano Observatory Best Practices Workshop #2: Communicating Hazards, INGV and USGS, Erice, Sicily (IT), 02–06 Nov 2013, invited oral

Komorowski J-C, Beauducel F, Devès M, Dessert C, de Chabalier J-B, and the CASAVA research consortium (2014) Failed magmatic eruptions, uncertain precursors and false alarms: lessons learned from the 1976–77 La Soufrière of Guadeloupe volcano (French Antilles) crisis, Workshop of the Cost Action IS1304, "Expert Judgment Network: Bridging the Gap Between Scientific Uncertainty and Evidence-Based Decision Making". "Science, uncertainty and decision making in the mitigation of natural risks" Diparti-mento della Protezione Civile, Roma October 8-9-10, 2014, invited oral, http://www.expertsinuncertainty.net/

Komorowski J-C, Hincks T, Sparks RSJ, Aspinall W, and the CASAVA ANR project consortium (2015) CS5. Improving crisis decision-making at times of uncertain volcanic unrest (Guadeloupe, 1976). In: Brown SK, Loughlin SC, Sparks RSJ, Vye-Brown C et al (eds) Global volcanic hazards and risk, summary background paper for the UN-ISDR Global Assess-ment of Risks 2015 (GAR15) A report by the Global Volcano Model and the International Association of Volcanology and Chemistry of the Earth's Interior, pp 114–119. http://www.preventionweb.net/english/hyogo/gar/2015/en/bgdocs/GVM,%202014b.pdf

Lepointe E (1984) Essai sur la réponse sociale à une catastrope: La Soufrière de Guadeloupe en 1976, Thèse Doctorat d'Etat, Université Paris 10, Nanterre, vol 1 (pp 1–445) and vol 2 (pp 447–975)

Lepointe E (1999) Le réveil du volcan de la Soufriére en 1976: la population guadeloupéenne à l'épreuve du danger. In: Yacou A (ed) Les catastrophes naturelles aux Antilles—D'une Soufriére à l'autre. CERC Université Antilles et de la Guyane, Editions Karthala, Paris, pp 15–71

Lewis J (1999) Development in disaster-prone places: studies of vulnerability. Intermediate Technology Publications, London

Lewis J (2009) A island characteristic. Derivative vulnerabilities to indigenous and exogenous hazards. Shima: Int J Res Isl Cult 3(1):3–15

Loubat B, Pistolesi-Lafont A (1977) La Soufrière—à qui la faute ?. Presses de la Cité, Paris, p 220

Méheux K, Dominey-Howes D, Lloyd K (2007) Natural hazard impacts in small island developing states: a review of current knowledge and future research needs. Nat Hazards 40(2):429–446

Mercer J, Kelman I (2010) Living alongside a volcano in Baliau, Papua New Guinea". Disaster Prev Manag 9 (4):412–422

Mercer J, Kelman I, Suchet-Pearson S, Lloyd K (2009) Integrating indigenous and scientific knowledge bases for disaster risk reduction in Papua New Guinea. Geografiska Annaler: Series B, Hum Geogr 91(2): 157–183

Morin J (2012) Gestion institutionnelle et réponses des populations face aux crises volcaniques : études de cas à La Réunion et en Grande Comore. Thèse de doctorat, Université de La Réunion, Saint-Denis, La Réunion pp 368 (+annexes 88 pp)

Morin J, Gaillard J-C (2012) Lahar hazard and livelihood strategies on the foot slopes of Mt Karthala volcano, Comoros. In: Wisner B, Gaillard J-C, Kelman I

(eds) Handbook of hazards and disaster risk reduction. Routledge, London, pp 705–706

Morin J, Lavigne F (2009) Institutional and social responses to hazards related to Karthala volcano, Comoros—part II: the deep-seated root causes of Comorian vulnerabilities. SHIMA: Int J Res Isl Cult 3–1:54–71

Morin J, Lavigne F, Bachèlery P, Finizola A, Vil-leneuve N (2009) Institutional and social responses to hazards related to Karthala volcano, Comoros—part I: analysis of the May 2006 eruptive crisis. SHIMA: Int J Res Isl Cult 3–1:33–53

Morin J, Bachèlery P, Soule S, Nassor H (2016) Volcanic risk and crisis management on Grande Comore Island. In: Bachelery P, Lenat J-F, Di Muro A, Michon L (eds) Active Volcanoes of the Southwest Indian Ocean: Piton de la Fournaise and Karthala, Active Volcanoes of the World, Springer-Verlag, Berlin, Heidelberg, pp 208–221, doi: 10.1007/978-3-642-31395-0_24

Nassor H (2001) Contribution à l'étude du risque vol-canique sur les grands volcans boucliers basaltiques: le Karthala et le Piton de la Fournaise, unpublished PhD dissertation, University of La Réunion pp 218

Oraison A (2004) La mise en place des institutions de l ' «Union des Comores» prévues par la Constitution du 23 décembre 2001. L'avènement d'un régime de type présidentiel et fédéral dans un état francophone du canal de Mozambique Revue française de Droit Constitutionnel 4–60:771–795

OVSG-IPGP (2015) Le bilan annuel 2014 de l'activité volcanique de la Soufrière et de la sismicité régionale. http://www.ipgp.fr/fr/ovsg/actualites-ovsg. Last accessed 10 June 2015

Pattullo P (2000) Fire from the Mountain: the tragedy of montserrat and the betrayal of its people. Constable and Robinson, London

Pelling M, Uitto JI (2001) Small island developing states: natural disaster vulnerability and global change. Environ Hazards 3:49–62

Préfecture de Guadeloupe (1977) Volcan de la Soufrière en Guadeloupe: les événements de 1976. Service d'information de la Préfecture de Guadeloupe, janvier 1977, Basse-Terre, Guadeloupe pp 50

Ribeiro O (1960) A ilha do Fogo e as suas erupções. Junta de Investigaçôes do Ultramar

Taglioni F (2003) Recherches sur les petits espaces insulaires et sur leurs organisations régionales. Mém-oire d'HDR, Université Paris IV-Sorbonne, Paris 218

Texier-Teixeira P, Chouraqui F, Perrillat-Collomb A, Lavigne F, Cadag JR, Grancher D (2013) Reducing volcanic risk on Fogo Volcano, Cape-Verde, through a participatory approach: which out coming? Nat Hazards Earth Syst Sci 1:6559–6592

UNDP (2014) Human development reports. http://hdr.undp.org/fr/content/table-1-human-development-index-and-its-components

Union des Comores (2004) Plan National de Préparation et de Réponse à l'urgence. Union des Comores, Moroni, p 68

Villemant B, Hammouya G, Michel A, Semet M, Komorowski J-C, Boudon G, Cheminée J-L (2005) The memory of volcanic waters: shallow magma degassing revealed by halogen monitoring in thermal springs of La Soufrière volcano (Guadeloupe, Lesser Antilles), Earth Planet. Sci. Letters 237:710–728

Villemant B, Komorowski J-C, Dessert C, Michel A, Crispi O, Hammouya G, Beauducel F, de Chabalier JB (2014) Evidence for a new shallow magma intrusion at La Soufrière of Guadeloupe (Lesser Antilles). Insights from long-term geochemical monitoring of halogen-rich hydrothermal fluids. J Volcanol Geotherm Res 285:247–277

Weichselgartner J (2003) Toward a policy-relevant hazard geography: critical comments on geographic natural hazard research. Erde 134(2):181–193

Woo G (2011) Calculating catastrophe. Imperial College Press, London, p 355

Investigating the Management of Geological Hazards and Risks in the Mt Cameroon Area Using Focus Group Discussions

Mary-Ann del Marmol[ORCID], Karen Fontijn, Mary Atanga, Steve Njome, George Mafany, Aaron Tening, Mabel Nechia Wantim, Beatrice Fonge, Vivian Bih Che, Aka Festus, Gerald G. J. Ernst, Emmanuel Suh, Patric Jacobs and Matthieu Kervyn

Abstract

The scientific evaluation of hazards and risks remains a primary concern in poorly known volcanic regions. The use of such information to develop an effective risk management structure and risk reduction actions however also poses important challenges. We here present the results of a series of focus group discussions (FGDs) organised with city councillors from three municipalities around Mt Cameroon volcano, Cameroon. The Mt Cameroon area is a volcanically and tectonically active region regularly affected in the historical past by lava flows, landslides and earthquake swarms, and has a potential for crater lake outgassing. The lower flanks of the volcano are densely populated and the site of intense economic development. The FGDs were aimed at the elicitation of (1) the knowledge and perception of geological hazards, (2) the state of preparedness and the implementation of mitigation and prevention actions by the municipalities, (3) the evaluation of the effectiveness of the structure of communication channels established to respond to emergency situations, and (4) the recovery from an emergency. In all three municipalities stakeholders had good knowledge of the risks, except for

M.-A. del Marmol (✉) · K. Fontijn · G. G. J. Ernst · P. Jacobs
Department of Geology, Ghent University, Krijgslaan 281/S8, 9000 Ghent, Belgium
e-mail: jp.m.malingreau@gmail.com

K. Fontijn
Department of Earth Sciences, University of Oxford, South Parks Road, Oxford OX1 3AN, UK

S. Njome · G. Mafany · A. Tening · M. N. Wantim · B. Fonge · V. B. Che · A. Festus · E. Suh
Department of Geology and Environmental Science, University of Buea, P.O. Box 63, Buea, Cameroon

M. Atanga
Department of Nursing, University of Bamenda, P.O. Box 39, Bambili, NW Region, Bamenda, Cameroon

G. Mafany · A. Festus
Ministry of Scientific Research and Innovation, Institut de Recherches Géologiques et Minières, BP 4110 Yaounde, Cameroon

M. Kervyn
Department of Geography, Vrije Universiteit Brussel, 1050 Brussels, Belgium

Advs in Volcanology (2018) 373–394
https://doi.org/10.1007/11157_2017_3
Published Online: 12 April 2017

processes never experienced in the region. They generally grasped the causes of landslides or floods but were less familiar with volcano-tectonic processes. Stakeholders identified the lack of strategic planning to monitor hazards and mitigate their impacts as a major weakness, requesting additional education and scientific support. Response to natural hazards is mostly based on informal communication channels and is supported by a high level of trust between local scientists, decision makers and the population. Actions are taken to raise awareness and implement basic mitigation and prevention actions, based on the willingness of local political leaders. The strong centralisation of the risk management process at the national level and the lack of political and financial means at the local level are major limitations in the implementation of an effective risk management strategy adapted to local risk conditions. Our case study highlights the need for earth and social scientists to actively work together with national and local authorities to translate the findings of scientific hazard and risk assessment into improved risk management practices.

Keywords

Mount Cameroon · Focus group discussion · Natural hazard · Risk management · Landslide · Volcanic activity

1 Introduction

1.1 Mount Cameroon: General Setting and Types of Natural Hazards

Mount Cameroon in SW Cameroon is one of the largest (4095 m high) flow-dominated volcanoes on Earth, and one of the most frequently active volcanoes in Africa (Siebert et al. 2010). It has a NW-SE elongated shape of about 50 by 35 km and is part of the Cameroon Volcanic Line, a chain of Cenozoic volcanic structures extending about 2000 km from the Gulf of Guinea to the Adamawa Plateau in Tchad (e.g. Déruelle et al. 2007; Njome and de Wit 2014). Mount Cameroon erupted seven times in the 20th century, the last two confirmed eruptions taking place in 1999 and 2000 (Suh et al. 2003, 2008). The eruptive style generally comprises effusive and Strombolian-style activity, the latter mostly confined to the broad summit region. Basaltic lava flows occur predominantly along the NE and SW flanks of the volcano. These lava flows are relatively mobile, reaching lengths of up to 9 km (Bonne et al. 2008; Favalli et al. 2011; Njome et al. 2008; Wantim et al. 2013a, b), and thus pose a potential threat to communities at the base of the volcano. Favalli et al. (2011) and Wantim et al. (2013b) developed idealised lava flow models to be used as a base to alert and potentially evacuate communities at risk of advancing lava flows. Historical eruptions of Mount Cameroon were associated with destruction of plantations and farmland, critical infrastructure such as roads and bridges, as well as houses (e.g. in 1922 and 1999: Déruelle et al. 1987; Suh et al. 2003, 2011). Eruptions were typically also associated with health risks like respiratory problems and contaminated water supplies (e.g. in 1999: Atanga et al. 2009).

Other natural hazards occurring in the Mount Cameroon area include landslides (Che et al. 2011, 2012a, b), especially on the SE flank which records at least one landslide every year. These landslides occur on old hilly volcanic terrain with deeply weathered soils which are cultivated by a growing population (Che et al. 2012a, b).

However, because most of these landslides have not caused fatalities in the past, they are not always recorded or reported (Ayonghe et al. 2004; Diko 2012). Floods (Ndille and Belle 2014), crater lake outgassing and different types of earthquakes are other natural hazards that characterise the Cameroon Volcanic Line. Disastrous crater lake outgassing has occurred twice in the early 1980s at Lakes Monoun and Nyos (Freeth and Kay 1987; Issa et al. 2014). The local population strongly fears earthquakes because they have led to the destruction of houses and other infrastructure on numerous occasions in the past. Ateba et al. (2009) describe the seismic activity as "co-eruptive", noting that the 2000 eruption was characterised by sequences of earthquake swarms and volcanic tremor. The Global Network for Disaster Reduction reports in 2011 that about 25% of the natural hazards in Cameroon occur along the Cameroon Volcanic Line (GEADIRR 2011).

1.2 Project Motivation

The Mount Cameroon area is exposed to a variety of natural hazards and associated risks (Fig. 1). Donovan et al. (2014) suggest that Mount Cameroon is considered as a low risk potential, due to the predominantly effusive nature of its eruptions, but with an extremely high likelihood of an eruption in the next 30 years. A number of volcanic hazard and risk assessments have been performed (Bonne et al. 2008; Thierry et al. 2008; Favalli et al. 2011; Gehl et al. 2013) but these have largely been limited to scientific publications. Translation of relevant scientific information into understandable language for the local population is yet to be fully implemented in the area, and will facilitate the delivery of more efficient assistance in preparedness and response to natural hazards (e.g. Barclay et al. 2008). Previous risk awareness and perception studies have mostly been based on household surveys and found that (volcanic) risk is perceived differently among local scientists (at the University of Buea and the Cameroon Geological Survey) and the local population (Njome

et al. 2010; Pannaccione Apa et al. 2012). Atanga et al. (2009, 2010) assessed the health risks of Mount Cameroon volcanic ash, and studied mitigation approaches by community members and frontline workers. The ability or inability of the exposed population to cope with risk has not been assessed in detail, nor have the preparedness and mitigation efforts from the local authorities been evaluated.

Following the Mount Cameroon eruption crises in 1999 and 2000, a 5-year (2008–2013) bilateral capacity building project for geohazard research and management was established by the Flemish Interuniversity Council—University Development Cooperation (VLIR UOS, Belgium) between the University of Buea (Cameroon) and Ghent University (Belgium). The scientists from the University of Buea had limited training in geohazard crisis management as well as a shortage of laboratory facilities. The societal objectives were to raise the capacity and preparedness of the University of Buea, the local authorities and the population to improve the geohazard management strategy to the benefit of all relevant stakeholders. Scientific objectives included constraining the spatial distribution of lava flows (Wantim et al. 2011, 2013a, b) and landslide susceptibility (Che et al. 2012b, 2013). Another objective was to improve research and training capacity for monitoring volcanic and landslide hazards at the University of Buea.

Various activities were aimed at improving risk awareness and communication efficiency and were performed throughout the project (Fig. 2). All of these activities focused on volcanic hazards relevant to Mount Cameroon, as well as on landslide and crater lake hazards. Two stakeholder workshops were organised to raise awareness about geohazards and to discuss crisis management and early warning systems. The participants represented the Universities of Buea and Ghent, national research institutions, government services, municipal authorities, civil society (NGOs, CSOs, and farmer groups), and traditional authorities (e.g. village chiefs). The second workshop was concluded with the elaboration of various information billboards. Their content and design was informed by discussions

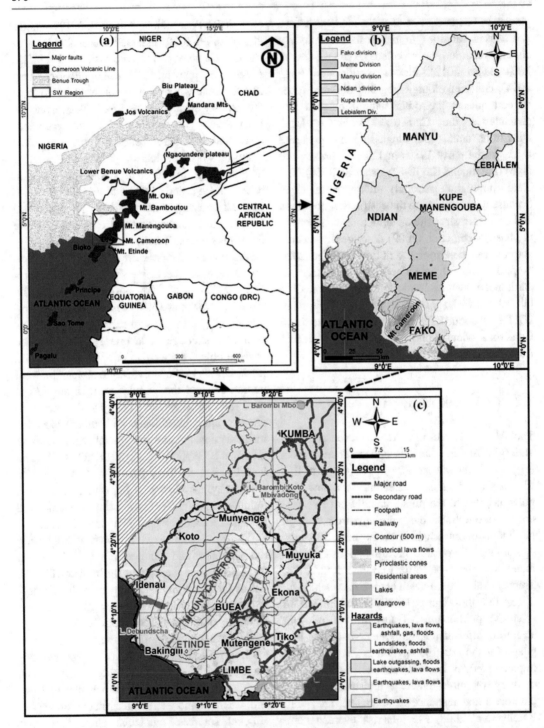

Fig. 1 **a** Location of the Cameroon Volcanic Line in West Africa; **b** Mount Cameroon with the surrounding—largely Anglophone—administrative divisions; **c** Mount Cameroon with main settlements at risk of a variety of natural hazards, including volcanic eruptions, landslides, floods, crater lake outgassing and earthquakes

Fig. 2 Photos illustrating awareness-raising actions undertaken as part of the 5-year VLIR-UOS project. **a** stakeholder workshop discussion session; **b** billboard in Bmbe illustrating volcanic hazards and risks; **c** billboard in

Limbe illustrating landslide hazards and risks; **d** billboard in Kumba illustrating crater lake outgassing hazards and risks

following introductions by the local and Belgian specialists of various geohazards and related (health) risks. For instance, it was decided that the morphology of the volcano drawn on the billboard (by a local artist from Buea) should represent the realistic view at the specific location where the billboard would be put in place. Further awareness-raising actions such as training of teachers were conducted in various schools and in front of the billboards. Additional sensitisation activities took place in villages and were led by local team members specialised in volcanic hazard and related health issues. A dedicated activity took place on International Women's Day (8th March) which is widely celebrated in Cameroon, and which provided the opportunity to reach a significant proportion of the population. Radio programs were found to be

a successful communication tool as they are very popular amongst the local population, and typically reach a much larger proportion of the population than the written press. Further training initiatives targeted at the press were undertaken to specifically improve communication efficiency and address risk awareness amongst the local population.

1.3 Focus Group Discussions

The merit of social studies on risks related to natural hazards lies in the fact that they are able to elicit information which can supplement results of purely scientific studies (e.g. hazard maps) and provide insights on how the population perceives risk (Atanga et al. 2010; Njome et al. 2010). The participation of communities has been found to be essential in identifying ways to mitigate risks (e.g. Gaillard and Dibben 2008). One way to interact with a community and collect data on the different aspects of the risk management cycle (preparedness, response, recovery, mitigation) is by means of focus group discussions (FGDs; Bohnsack 2004).

The FGD approach provides an open and interactive discussion forum, unlike a structured questionnaire, and is therefore particularly useful to obtain an in-depth understanding because discussions can range from simple to abstract ideas on a specific topic. In such circumstances all the relevant information can be discussed to the point of saturation. Moreover, it has been found that FGD participants tend to speak more freely and with more detail compared to a standard interview or structured questionnaire, which is believed to limit the extent to which people can express their own views (Mercer and Kelman 2010). The participants should be a cross-section of a community including both literate and illiterate people. FGDs provide a qualitative research tool for the scientists, but are particularly meaningful for the stakeholders as well, since they provide a platform for the participants to meet, interact, explore and cross-examine the matter at hand.

To understand how the local communities in the Mount Cameroon area are prepared for and cope with a variety of potential natural hazards, three FGDs were held at the city councils of three strategically selected towns. Each of these towns is prone to certain types of natural hazards (Fig. 1): (1) Buea (ca. 150,000 inhabitants) is located at the E foot of the volcano and has experienced ash fall from 20th century eruptions and significant earthquake damage in 1995; (2) Kumba (ca. 200,000 inhabitants) is located to the NE, too far to be directly impacted by an eruption but is downhill and downstream of a crater lake with the potential to outgas, and which provides the main water supply to the town; it is also prone to landslides and earthquakes; (3) Limbe (ca. 95,000 inhabitants; invited participants from Limbe 1 council) is located near the SE coast with a high risk of lava flow

invasion (Favalli et al. 2011) and has experienced multiple flooding and landslide events in recent decades (Ayonghe et al. 2004; Che et al. 2011). The risk imposed by floods and landslides is enhanced by unregulated building practices (Che et al. 2012b). The FGDs were organised in May 2010 as part of the risk assessment part of the VLIR UOS project. Between 4 and 6 participants were selected from each city council (Wilkinson 2004). Their level of responsibility within the council varied from councillor with or without technical education, to engineer, traditional ruler, local chief, and mayor.

In this study four cardinal points (preparedness, response, recovery and reconstruction) guided the FGDs (Fothergill 1996). The aim of the discussions was to collect the views of the councillors on each of the four points, for risks related to volcanic eruptions, landslides, floods and crater lakes. Each discussion began with a general presentation by a scientist on natural hazards which reminded the participants of some definitions such as hazard versus risk, using a simple every-day example (e.g. the occurrence of rain = hazard; getting wet = risk; carrying an umbrella = mitigating the risk), in order to guide the discussion process. It was established that, prior to and throughout the discussion, the concept of hazard versus risk was understood by all participants in the three cities. A local scientist further introduced the topic and moderated the conversation with the help of a predefined structure (Appendix). Volcanic, landslide or crater lake risks were not specifically mentioned or given as an example in the introduction, in order to avoid biased discussions. Notes were taken by another local scientist. The discussions were carried out in English, the local language mastered and preferred by all participants and researchers, and all discussions were audio- and video-recorded, and photographs were taken, all with consent from the participants. During the discussion it was ensured that the participants valued their role as representatives of their communities, and that they were able to speak on behalf of their community members. The discussion was free as all the participants were at the same level in discussing with researchers. Each discussion lasted between one to three hours, without predefined time limits.

2 Results

The councils of the three main cities around Mount Cameroon where the FGDs took place all have slightly different approaches to the preparation and response to hazards, and post-event reconstruction. Prominent and recurring themes are highlighted in Table 1, following the general structure (Appendix), and further discussed below.

2.1 Hazard and Risk Perception and Preparedness

The first section of the FGDs dealt with preparedness to natural hazards (Table 1). The different types of natural hazards constituting a threat in each area were well perceived except in Kumba. The outgassing of the crater lake was not mentioned as a potential hazard/risk. Floods and landslides were extensively discussed: this was not surprising given the high frequency of occurrence of these events.

In Buea, the participants expressed that they were "*at the foot of an active volcanic mountain and the mountain can erupt anytime*" and that

Table 1 Schematic overview of responses in each city council to each question of the FGD. Identities of respondents to specific questions have been omitted. The full details of each question can be found in Appendix. Some answers to questions appeared in discussions after subsequent questions, e.g. some specific hazard types (question A1a) were only mentioned in a discussion on associated risks (question A1c). Transcripts of the discussions are available upon request

Question		City		
Code	Theme	Buea	Kumba	Limbe 1
Aspect A: Preparedness				
A1—Natural hazards that could constitute risk to the municipality				
A1a	Types of hazard	– Volcanic eruption – Tremors – Landslides – Floods	– Landslides – Floods – Thunderstorms *Specific locations mentioned by name*	– Floods – Landslides – Volcanic eruption – Tsunami – Coastal subsidence/erosion
A1b	Causes of hazards	– Landslide: heavy rainfall – Floods: blocked gutters	– Human activity: construction on river embankments: narrowed water courses → Floods – Human activity: refuse disposal in rivers → Floods – Human activity: deforestation → Landslides – Human activity: reclaiming marshy areas – Heavy rainfall → Floods	– Topography: Limbe surrounded by hills – Topography: parts of Limbe below sea level – Human activity: deforestation, erosion → Landslides – Human activity: uncontrolled urbanisation, blocking natural waterways → Floods
A1c	Associated risks	– Building in mapped (landslide/flood) disaster zones – Damaged to houses due to vibration during an eruption	– Landslides: buildings on top of hills – Flooding (houses): transport of waste, risk of diseases	– Lost material property due to uncontrolled urbanisation – Coastal subsidence due to tremors during volcanic eruption

(continued)

Table 1 (continued)

Question		City		
Code	Theme	Buea	Kumba	Limbe 1
			– Clearing debris from the roads after landslide constitutes a risk in itself – *Lake Barumbi linked to Mount Cameroon: eruption affecting the lake water, which provides all the water supply for the town through the Mabungisse river at high risk of flooding*	– Coastal erosion: gradual inland movement of the coastline threatening the town – Heavy rainfall – water table rising above surface in some places
A1d	Warning signs	None known	– Landslides: cracks (lines) opening up in the soil – Flood-prone zones (near meanders): heavy rain, rising water levels, strange animal behaviour → People will move already once the water gets to a certain level	– Landslides, floods: several days of consecutive rain – Eruptions: only observed (fire, volcanic ash) when already taking place – Tremors – Vibrations sometimes before, during or after eruptions

A1e	Source of information	– Close collaboration with Ministry of Scientific Research
		– Scientists at University of Buea
		– Community knows about the volcano, but not about floods
			– Presence of town planners carrying out routine checks
			– Interaction with the people —who may base themselves on strange animal behaviour
			– Ministries:
			  • Environment
			  • Hygiene and Sanitation
			  • Mines, Water Resources and Energy
				– Scientists
				– Geological Research Centre
				– Surveillance system in place around the mountain

A2—Schemes in preparation for disasters				

A2	Scheme in preparation for disasters	– City council takes responsibility in case of an event, but no means are reserved specifically to prepare for disasters
		– Many councillors unaware of potential disasters
		– Spending scheme of council budget needs to be voted on – money more likely reserved for critical infrastructure than for disaster prevention, due to unawareness of councillors
			– Areas declared as risk zones (landslide-prone, regularly flooding): no building permits issued
			– Investigations in certain areas at regular risk as to what causes the problems
			– Regular (yearly) dredging of specific areas at high risk of flooding
			  • River course too narrow
			  • Surrounding vegetation too compact
			  • Refuse blocking the water flow
			– Some budget set aside for crisis management
			– Plans to terrace Ntoko Hill, at high risk of landslides
			– Most actions happen spontaneously, in case of an event. Preparation measures not systematically documented by the council. Documentation may exist
				None

(continued)

Table 1 (continued)

Question		City		
Code	Theme	Buea	Kumba	Limbe 1
			in the form of minutes of meetings. – Buildings in designated high-risk zones marked to be demolished	
A2a	Regulations towards reducing risk	– Laws/regulations exist, but not implemented due to lack of funding and more pressing basic needs of population	Administrative levels: – Government delegate or Mayor – will authorise budget allocation and action to take, based on technical reports written by responsible departments – Council to follow advise To approve building permits, the council departments in charge have to go and see what kind of area/environment/building the permit is being asked for.	– Building regulations, e.g. certain distance from stream banks – Certain areas declared as risk zones by the government These regulations are not respected by people
A2b	Unit in charge at municipality level	– Specific committee in charge of community protection – Close collaboration with the Mayor, Division Officer, Governor and Ministry in case of an event	Departments of Town Planning and Hygiene and Sanitation: both in charge of approving building permits	– Town Planning Services, in collaboration with related government services – Council due to take more responsibility and decisions related preparedness (decentralisation policy)
A2c	Facilities to manage risk	– Bulldozer to clear roads and gutters in case of floods – No other means	– Financial resources for major facilities (e.g. dredging works) put at disposal after authorisation by Government Delegate or Mayor based on technical reports – Basic facilities (equipment, transport, manpower) at disposal of Hygiene service for burials if needed	– First aid provided when anything happens – Nothing else—a crisis commission will be created in case of an event
A3—Warning systems				
A3a	Warning systems in place in case of an emergency	– None – (billboards explaining natural hazards in a more general way – not for emergencies)	– Local radio stations—paid on monthly basis to pass on council's announcements in an interactive programme where people can also call themselves and ask questions – Local churches – Letters to village chiefs, local quarter heads who will announce the information to their people – Meeting houses of traditional, tribal groups	– None – People tend to rely on themselves, helping each other – Chiefs and Mayor/Council in contact with each other—system filters down to village level

(continued)

Table 1 (continued)

Question		City		
Code	Theme	Buea	Kumba	Limbe 1
A3b	Population informed about these systems?	N/A	– Yes – Communication channels evaluated through population's response, revised where needed	N/A
A3c	(Other) possible tools to enhance warning efficacy	– Monitoring equipment on the volcano (was there before, but was taken away)	– Information boards with warning signals identifying high-risk zones, informing people they should not buy land or construct in those areas. These kinds of boards should be protected from demolition/removal	*– Aspect of spontaneous solidarity raised again* *– Mobile phones to communicate between chiefs, Mayor and councillors*
	Remarks	*Demand for information seminars at the council and community levels to create more awareness. At the council level, this could lead to more budgets being allocated to risk mitigation.*		

Aspect B: Response

B1—Risk communication to the population

B1	Means of communicating risk to population	– Media – Intervention unit going door to door to talk to people	– Radio – Churches – Quarter heads – Meeting houses All communication to and from the council passes through the Government Delegate	– Supervisory authority (administration) takes control – Going round villages, using e.g. whistle and bell to alert people
B1a	Anticipated response of population to warnings	– Some people are stubborn to evacuate (law enforcement used)	– People following warnings and moving away, especially for short-term hazards like floods – In other places, e.g. Ntoko Hill at high risk for landslides, people tend to be more adamant and stay. Usually less than 10% of the people. – About 10% of people have abandoned their houses permanently (usually forced by nature, e.g. permanent flooding) – Some people try to actively prevent floods by constructing dams to deviate the water (Need to be careful with forced action because of human rights action groups)	– They follow instructions on what has to be done – Spontaneous community response

(continued)

Table 1 (continued)

Question		City		
Code	Theme	Buea	Kumba	Limbe 1
B1b	Language used	– Pidgin english – Local dialect Depending on the area affected	– Official (written) communication all in English – Local delivery of the message usually in Pidgin English or local dialect, depending on tribal group – Signs (e.g. red paint marks), summon papers or abatement notices used to call on people and invite them to discuss building location	Pidgin english
B1c	Success of communication evaluated	– Go back and ensure people are safe – Council meeting discussing what has been done	Evaluated from the level of response	Message gets across
B2—Risk communication to higher authorities				
B2	Means of communicating risk to higher authorities	Council workers on site informing Mayor, who will then call the appropriate delegate at the Ministry or other services if more intervention is needed	– Mostly it is the higher authorities communicating risk to the council. The council depends purely on the population to report, but the government seems to have better mechanisms in place – Phone calls (for emergencies) and/or written letter to subdivision officer, not supported by documents (spontaneous action) – Final report made by the council after the events	– Phone calls – Administrative writings – Crisis meetings
B3—Risk communication to scientists				
B3	Liaison with researchers/scientists	First ones contacted as they are best placed to advise what to do	– One scientist once wrote to the council to warn about rising water levels at certain river. The governor set up a small commission to go out in the field and investigate – NGOs writing advisory reports on what can be done in terms of prevention, e.g. on Ntoko Hill – No set structure to interact with scientists and researchers – It is difficult to know who the scientists are, where they are based and what they are doing. The council expects the scientists to come to them and is then very willing to collaborate	Cordial relationship

(continued)

Table 1 (continued)

Question		City		
Code	Theme	Buea	Kumba	Limbe 1
B4—Immediate response				
B4	Immediate response in case of a crisis	– Observe on site – Look for means to evacuate people if needed, in collaboration with law enforcement officers – Contact media – Assistance with material or medical needs – Floods: open waterways and remove debris using bulldozer	– Go down to site and evaluate the degree of risk, the damage, etc. – Decide what can be done to support victims, e.g. assistance in burials – Provide help on the spot, e.g. trying to save people in danger – People themselves will also assist in relieving the situation – Ensure security to prevent looting – Action depends on the situation	– Go observe on site – People evacuated to safe place – Material and financial assistance – Population also assists
B5/B6—Last crisis				
B5	Last crisis experienced	Flood at the hospital 2 years ago (*2009*)	– Thunder strike about 1 month ago – Storm yesterday – Flood – Last eruption (Bakingli) affected the lake at the source of the Mabungisse river	– Frequent floods of different scales – 2001 floods most recent major crisis taking lives – 2009 landslide blocking the road
B6	Impacts of last crisis	– Damage to houses – Damage to hospital fence – Road to hospital blocked	– Yesterday's storm: · Roof collapses · Damage to crops – Floods: · people displaced · loss of property · loss of lives · reduced farmland (limited impact) · disease outbreaks (diarrhoea, malaria) – People uncomfortable living near place where decomposing bodies are buried (e.g. corpses only recovered from floods after a few days)	– Sketchy and unbalanced documentation of events – Loss of lives, property – People displaced, some people even moved to America – Depression, traumas – Financial burden – Thieves taking advantage
Aspect C: Recovery				
C1—Recovery strategy				
C1	Council strategy to help people overcome basic problems	– Look for temporary housing for evacuees – Delegation of Health to assess potential of disease outbreak – Extraordinary council session to discuss budget reallocation in case people need financial help to relocate or rebuild houses	– Council needs formal request for assistance before it can act, e.g. from health services requesting extra vaccinations, or from the population – People often afraid to come to the council if they were living in "no-go areas"	None

(continued)

Table 1 (continued)

Question		City		
Code	Theme	Buea	Kumba	Limbe 1
C2	Strategy used in last crisis?	– Yes	N/A	Other organisations (SONARA) came in and built houses to assist people, but in insufficient amounts. Those buildings are now used as a school—council was consulted at first but could not intervene

Aspect D: Mitigation

D1—Policy measures for long-lasting solutions to risk

D1	Policy	Areas indicated where construction is not allowed	– Contracts with dredger – No written policy documents guiding actions, all actions spontaneous – Budget is available from the government for councillors to attend seminars on risk management Proposals for the full dredging of a high-risk river, or terracing of a high-risk hill, but without financial means and resources, these proposals will not convert into policy	– Areas designated as risk zones – Relocation of people living in risk zones
D1a	Source of policy	Government	Government policy on risk management, none at the level of the council	Ministry of Town Planning and Housing

D2—Strategies to back up policy

		Keep waterways open	– Marking houses for demolition – Regular (small-scale) dredging	None

D3—Implementation of strategy

		N/A	– Regular dredging of river at high risk of flooding – Clearing areas – Marking houses of demolition	– Sensitisation and education – Encouraging people to stop farming but plant trees in risk areas – Stringent observance of building rules and regulations – People not actually relocated

D4—Other strategies and actions envisaged by the municipality

D4	Other strategies that could help	– Appropriate monitoring of seismic or volcanic activity – Sensitising communities in the form of seminars, to educate the population about the types of risks, and point out risky areas – Funding alone will not be sufficient	– Information seminars for the council	– Strategic planning – Relocation of people away from risk zones – Land now used for plantations freed for people to live; plantations moved to the forest – Decentralisation, more autonomy and power to the councils

"the community knows much about the volcanic mountain but does not know much about the flood". In Limbe, volcanic eruptions were also mentioned as a potential hazard, but in the other two towns, eruptions did not seem to be considered as a serious hazard. In Limbe it was further noted that coastal subsidence has occurred in the past due to tremors associated with an eruption. These tremors are felt only sometimes before an eruption and taken as a warning signal. Ateba et al. (2009) observed that the tremors are concomitant to an eruption. Although participants easily identified the cause and warning signs of landslides and floods, they did not mention any cause for earthquake and volcanic events nor did they know of warning signs except for tremors.

Heavy rainfall was considered the main natural cause of landslides (Buea and Limbe) and floods (Kumba and Limbe), and was generally also considered as a warning signal for floods. In Kumba it was noted that people will start to evacuate spontaneously once the water levels are rising above a critical point. Topography was also highlighted as a main cause of concern during the heavy rainy days. Nevertheless the causes of floods posing a risk to the population were largely ascribed to human activity: the waterways are narrowed and blocked as a result of "uncontrolled urbanisation" (Limbe and Kumba) with construction near river embankments. In marshy areas, the dumping of refuse in the rivers clogs up waterways. Deforestation is associated with a rise in erosion and landslide occurrence. An increased incidence of diseases as a result of waste transport in floods was highlighted as an associated population health risk in Kumba.

Damage to properties and loss of lives were mentioned as an impact of recent events. Construction in mapped disaster zones (Kumba) and on top of hills on old volcanic terrain (Limbe and Buea) was considered risky.

In Limbe, people have observed that three days of consecutive rainfall will cause people to worry about potential landslides. This timeframe fits well with that observed by Che et al. (2012b,) in their study of landslide occurrence and susceptibility in the Limbe area. In Kumba the population was aware that cracks opening up in the soil on a hill may also be a warning sign of an imminent landslide. Interestingly, the Kumba council acknowledged that removing landslide debris, e.g. clearing the roads, can in itself also pose a risk.

In Kumba, the Barombi crater lake flows into the frequently flooding Mabungise river. More importantly this river provides water and fish supplies to the city. Volcanic hazards were not mentioned in Kumba, which is understandable given the distance with respect to the volcano. In a previous workshop related to our VLIR-UOS project, the traditional chief of Kumba mentioned his concerns about the fish mortality and the risk of potential landslides of the crater walls and gas release from the lake. Previous occasions of fish mortality were related to the overturned stratified water after large rainfall leading to oxygen scarcity. Also as part of the same project, in 2010 information billboards on crater lake hazards were put up along the road towards the lake and at two entrances of the city (Fig. 2d). Nevertheless, gas release from the crater lake, i.e. similar to what happened at Lakes Nyos and Monoun in the 1980s, was not mentioned as a potential hazard by the FGD participants.

Tsunami was another type of hazard, mentioned in Limbe, a coastal town surrounded by hills with parts of the town lying below sea level. Coastal subsidence and erosion are active processes in Limbe, with a significant risk potential: sudden subsidence was observed in the late 1940s (cause unknown) as well as during past eruptive events of Mount Cameroon. In addition, gradual coastal erosion was observed through the rapid degradation of the coastal embankment that was built by the council.

The Buea and Limbe City Councils have well-established links with scientists at the University of Buea and/or the Geological Research Centre (also called Regional Research Centre) in Ekona from which they can obtain scientific information. The Buea councillors however admitted that the general level of hazard knowledge and awareness in the council was too low and dedicated information sessions would be most welcome. The Kumba Council is not connected to the Buea scientists but strongly welcomed researchers to approach the Council with information regarding hazards and risk. During the FGDs every council requested that the scientists actively provide information and possibly a training session about natural hazards and associated risk.

There is also collaboration with several Ministries in preparation for hazards, e.g. in the form of issuing building regulations and areas declared as high-risk zones for which building permits are in theory not issued. Nevertheless the councils all have limited financial resources, and limited to no facilities exist to manage hazardous events apart from providing basic first aid (Limbe).

A major and well-known problem enhancing the risk of floods and landslides is the disrespect of the building regulations and uncontrolled land use, with people still building houses in areas where they are not allowed to. In some cases those buildings are actively marked for demolition. However the implementation of the regulations, with actual evacuation of people from these zones and demolition of the buildings, does not typically seem to occur, except occasionally in Kumba. The local people are expected to go to the town planning service of the city council themselves to discuss their building plans and the regulations.

The town planning service, in charge of implementing building regulations, works with the related services at the national government level. The Mayor of Limbe indicated that the council is going through a process of increasing manpower to manage most of the regulations themselves, in light of new decentralisation policies. At the time of a disaster however, a special crisis commission is set up to manage it. No actual warning system exists and pre-event awareness-raising is limited to the existing billboards (specifically mentioned in Buea) and school education activities that were developed as part of our project. Commenting on the municipality regulations geared towards reducing risks, the participants agreed that, even if a scheme existed, decisions would still be largely taken "*spontaneously*" (Limbe).

In Kumba, an interactive communication system with a high penetration rate is in place, and is used to inform the population of the council's actions, but also to raise hazard awareness and issue active warnings in case of an event: "*We use the local radio stations, churches, tribal groups, letters and sub-chiefs (in charge of particular neighbourhoods) to inform the population. There are 3 radio stations present in Kumba city. We have already established partnerships with these radio stations and we pay them on a monthly basis to disseminate information.*" There are specific radio programmes with interviews of departmental heads of the council. During these programmes the population is encouraged to participate and make direct enquiries using their mobile phones. In case of an emergency the people know they have to listen to the radio for updates. The council measures the degree of awareness from the reaction of the people. If they find the message does not get across, authorities change their strategy. Limbe and Buea had not yet developed such a communication system when the FGDs took place.

With respect to the engineering means available in Buea, the following comments were made:

> It depends on the risk: for instance we have some means to manage something like the flood, we have the bulldozer if there is a flood; if there is a blockage we can rush and open up the gutters and so on but if it comes to natural disaster (e.g. the volcano) we do not have materials.

> We know that we have to clean all the gutters of the community. If we neglect it, stones will block it and this will cause floods."

> What is lacking is the knowledge that you people are giving us now. For instance, with the best of my knowledge we are 41 councillors but I don't *know if up to 20 are aware that we may have a disaster at any time*".

According to the participants from the three towns, little to no means are specifically reserved for hazard preparedness, mostly due to more pressing basic needs within the community. The councils will however take their responsibility in responding to an event and dealing with recovery as needed, and as is practically possible with the available resources.

2.2 Crisis Response Structure and Communication

Section B of the FGDs related to crisis management (Appendix, Table 1). In each town the participants explained the organisation of their respective council, although in Cameroon, the disaster risk management structure is centralised (Bang 2013, 2014).

In Buea, the councillors mentioned that, in case of an event, they *"work closely with the Mayor, the Delegate Officer, the Governor and every other Ministry that will be in charge of the disaster, such as the Ministry of Scientific Research"*.

On how the council communicates risk to higher authorities, in Buea, the councillors said that *"the committee in charge will go down to the field with the council workers in order to see what happened. They will alert the Mayor. He will be the one to call for any intervention from any Ministry... We have to report and then the Mayor takes it to higher quarters." "The mayor may now make a communiqué to the radio and then transmit it to the other services that are concerned"*. The power therefore seems to be relatively centralised in the hands of the Mayor in Buea.

When a significant event occurs, the Ministry of Scientific Research and Innovation and the University of Buea will provide information and work in close collaboration with the council and their monitoring units. When questioned about the existence of an early warning system however, the councillors did not really understand what this would encompass.

Kumba is not a city council sensu stricto but a municipality with a Government Delegate in charge of the urban councils. After receiving written technical reports from the various departments in case of an event, the Government Delegate decides and authorises the actions to solve the problems, since he holds the financial resources. In reality no written document is sent to the Government Delegate, as technically required, but communication occurs directly by telephone, especially in case of an emergency. During an emergency, basic equipment and facilities are made available to deal with immediate response on the ground. This includes equipment for dredging rivers where needed, e.g. in case of floods, but also assistance to the community with burying victims.

In Limbe the engineer reported that many disasters take the community by surprise, but that the council would organise an evacuation dependent on sufficient resources. This is the only time that evacuation is mentioned. Later in the discussion, the Mayor further described the communication to be easy these days: *"When a natural disaster occurs it goes around like wild fire"*. He referred to the fact that many people use mobile telephones for communicating via text messages (SMS). They may also make use of social media and informal communication channels within the neighbourhoods. The local authorities would also call the scientists. For instance, at the time of the 2010 landslide the Mayor immediately called the PhD researcher (author VBC) he had met at a previous workshop. He had at the time also requested immediate clearing of the road (main evacuation for the town along the ocean).

In every town, FGD participants requested education by the scientists on disaster preparedness. A better understanding of the causes of the hazards, e.g. supported by landslide susceptibility maps, would help in the evacuation of certain areas at critical moments before a disaster ensues, but also in implementing building regulations in the first place. Participants also wanted to know more about precursory signs of imminent events and the potential deployment of monitoring equipment.

2.3 Recovery and Reconstruction

None of the cities has a dedicated post-event recovery program of actions (Table 1). In Buea the participants mentioned that a special council session would be held to discuss housing and financial assistance needs after properties have been destroyed by an event. The councillors would then report to the Governor and the Minister in charge (e.g. Public Health, in case of a risk of cholera outbreak). The level of support provided by the regional and national authorities will depend on the scale of the disaster.

In Limbe, the participants complained about promises made by some private companies to rebuild houses after the last landslide crisis. These promises were not met, and the buildings were finally occupied by a school.

The Limbe council itself is performing sensitisation actions: the people should not farm or build houses in a landslide-prone area (e.g. also billboard shown in Fig. 2c). Instead, tree planting is encouraged. At the council level, they aim to implement stringent building rules and regulations. However they conceded to have a lack of strategic planning at the council level, and instead referred to the Department of Civil Protection at the Ministry of Territorial Administration and Decentralisation in Yaoundé.

In Kumba, funds for response and recovery are limited but directly available, e.g. for dredging rivers. The funds are provided by the Government Delegate and are part of the annually reviewed budget for risk management, and so rivers are regularly cleaned and dredged both before and after events. At the end of each FGD, participants agreed that a decentralisation of the risk management process, under the authority of the Ministry of Territorial Administration and Decentralisation, would be highly advisable, especially for the councils of Buea and Limbe. Discussions between the Ministries and Government Delegates and the local Mayors related to disaster response and recovery were reported to be limited.

3   Discussion

The FGDs revealed that the main hazards and risks, e.g. landslides, floods, seismic and volcanic hazards are generally well perceived. Precursory signals of floods and landslides are recognised as well as tremors as signs for volcanic activity. Because of their higher recurrence rates, landslides and floods are clearly the most pressing concerns to the local communities in terms of natural hazards. Volcanic hazards are of less concern, due to either the relatively frequent but small-scale eruptions typical for Mount Cameroon with limited impact on the urban infrastructure, or the infrequent nature of events such as crater lake outgassing, which do not occur in the living memory of most of the local population. This is reflected in the more limited awareness of these events relative to landslides and floods. We expect that the communication and response practices developed for frequent landslide/flood emergencies—with a background of limited financial resources—will be adopted at similar levels in case of a volcanic crisis. Most FGD participants however agreed that they need more technical information about the nature and causes of hazards that may affect their communities in the future, as well as about appropriate preventive measures.

The scientists realised from their side that they did not make enough pro-active effort on a regular basis to communicate the outcomes of their research. For example, the billboards installed in the framework of the VLIR-UOS project turned out to have limited impact to increase awareness. They were installed with consent of the councils, but councillors were not further informed about the message and purpose of these billboards. Such activities were instead limited to teachers and radio show hosts. Incorporating indigenous knowledge, e.g. passed on by traditional chiefs, into technical assessments and a range of awareness-raising actions may further help local inhabitants to better understand and appreciate natural events, and also behave adequately in case of an emergency.

Facilities to manage risk are practically non-existent except for assistance by the Civil Protection at the national level, intervening only in the case of large-scale events. This lack of adequate means and infrastructure in each of the cities remains a concern. Regional risk management is not yet effectively in charge of taking decisions to prevent disasters, to handle crises and to recover from or mitigate future events. Continuous occupation of risk zones by new buildings, and the absence of temporary evacuation or relocation plans at the city level are other bottlenecks in the effective reduction of risk impacts.

The general lack of prevention and preparedness actions of local councils is understandable in a country where risk management decisions are mostly taken by Ministries. The fact that the regulations regarding risk management are centralised makes it more complicated to apply actions and adapt communication strategies to the local context. It is clear that the communication chain to the population, e.g. with the use of frequent radio programmes, is best developed in Kumba, which is directly governed by a Government Delegate. The other local authorities are however poorly informed about any preparedness or prevention strategies that may exist at the national level. In addition to centralised regulations, illegal construction and lack of regulation enforcement on the ground are major bottlenecks for the efficiency of any prevention schemes.

The response scheme to hazardous events is largely characterised by spontaneous actions within the community and informal communication channels. Strong social networks typical for African communities strengthen the effectiveness of this informal communication. Although risk management commissions are officially defined, the communication and the decision process during events rely mostly on ad hoc communication between the affected population, local authorities and scientists. The Mayor and the City Council are the main actors for responding to impact, however with limited means. The leadership of the local Mayor and councillors, and their relationship to the national government, via the Government Delegate, control whether more resources and recovery support can be provided. As Kumba depends directly on the Government Delegate, more preventive and response actions are immediately undertaken thanks to rapid money allocation. In Buea and Limbe however, actions are delayed due to intermediate administrative levels between the Government Delegate and the City Council.

Effective collaboration between local and national authorities, as well as trust and frequent communication between local decision-makers, scientists and the population and their representatives are essential elements in the effectiveness of the management of hazardous events (Barclay et al. 2008). The awareness by the Mayors and councillors for the need of risk reduction actions, as well as the available means are other key elements. The personal participation of the Mayor of Limbe in the FGD demonstrated his interest and dedication. The FGD led to a fruitful exchange between the Mayor and several of the local chiefs, all of them feeling highly responsible for risk management in their local communities. The actions taken within the VLIR-UOS project, including the FGDs, have thus contributed to enhancing the relationships between the city councillors, mayors and scientists.

4 Conclusions

FGDs have the advantage of enabling interactions between all actors, with the opinion of all participants being considered at an equal level. Our FGDs contributed to increasing the awareness of risks among the councillors and to identifying the current state and limitations of the schemes aimed at preparing for, responding to and mitigating impacts of natural hazards in the vicinity of Mount Cameroon. The outcome of the FGDs is useful for the scientists and also contributes to raising participants' awareness about the need to address the different steps of the risk management cycle and the challenges faced in implementing them effectively (i.e. scientific knowledge, education and communication actions, preparedness and response plans, resource allocation and decision-making structure).

The scientists who are part of this project have realised they perhaps do not always make enough effort to actively communicate their relevant research results to the local communities and stakeholders. The relationship between the local authorities and the local scientists will hopefully continue to develop in the future, as requested by all FGD participants. Transfer of knowledge and leadership of the local authority are essential for implementing mitigation and preparedness actions and for effective coordination of the actions during and directly following a crisis.

Decentralisation of the decision process was the main wish expressed from all the councils that took part in our FGDs. Decentralisation of the governmental coordination of risk management related to local natural hazards would allow the development of a locally relevant plan of action to turn disaster prevention policies into practice. However decentralisation of the decision will lead to potential improvement only when associated with availability of sufficient funding earmarked to support implementation.

Acknowledgements This project was supported by the Flemish Interuniversity Council—University Development Cooperation (VLIR-UOS), granted to Ghent University and the University of Buea. Karen Fontijn is currently supported by the UK's Natural Environment Research Council grant NE/L013932/1 ("RiftVolc"). We are extremely grateful to all participants of the focus group discussions, as well as numerous people at the University of Buea, research institutes and city councils who have facilitated this work. Constructive reviews by Carina Fearnley and an anonymous reviewer have helped improve this manuscript.

Appendix: Focus Group Discussion Questions Guiding the Discussions Hosted by a Local Moderator

Code A — Preparedness

1. Do you perceive any natural hazard in your municipality that could constitute risk?
   - If yes, what are the hazards?
   - What are the causes?
   - How do these hazards constitute risk?
   - What are the warning signs of these events?
   - From where and whom do you expect to get reliable information on the possible risks associated to the hazard?

2. Is there any scheme in preparation for disaster?
   - Does your municipality have regulations geared towards reducing risk?
   - Is there anyone in the municipality in charge of implementing such a scheme?
   - Does the municipality possess facilities to manage risk?

3. Are there any warning systems in place to alert the population (incase of an emergency) of the risks associated with a natural hazard?
   - If yes, elaborate
   - Is the population properly informed about it?
   - Besides these, are you aware of any other tools that can be used to enhance the warning efficacy?

Code B — Response

1. How do you communicate risk to the population?
   - How do you anticipate the population's response to your warnings?
   - What language would you use to reach out to the threatened population?
   - How do you evaluate the success of your communication?

2. How do you communicate risk with higher authorities?

3. How do you liaise with researchers/scientists? About...
   - Effects to the physical environment
   - Psychological impacts
   - Economic distortions

4. What will be your immediate response when there is a crisis?

5. What is the last crisis that the municipality experienced?

6. What where the impacts of this crisis?
   - Effects to the physical environment
   - Psychological impacts
   - Physical health
   - Economic impacts
   - Social impacts.

Code C — Recovery

1. Does the council have a strategy for helping people to overcome the basic problems within a year of the crisis?
2. Was this strategy used during the last crisis?

Code D — Mitigation (Reconstruction)

1. Drawing from past experience and knowledge, are there any policy measures to guarantee long lasting solutions to managing risk resulting from natural hazards in your municipality?
   - Source of policy?
2. Are there strategies to back up this policy? (*probe for strategies and documentation*)
3. How are you implementing the strategies? (*probe for ongoing action*)
4. What other strategies and actions are envisaged by your municipality?

References

Atanga MBS, Van der Meerve A, Shemang EM, Suh CE, Kruger W, Njome MS, Asobo NE (2009) Volcanic ash from the 1999 eruption of Mount Cameroon volcano: characterization and implications to health hazards. J Cameroon Acad Sci 8:63–70

Atanga MBS, Van der Meerwe AS, Suh CE (2010) Health system preparedness for hazards associated with Mount Cameroon Eruptions: a community perspective from Bakingili Village, Cameroon. Int J Mass Emerg Disasters 28:298–325

Ateba B, Dorbath C, Dorbath L, Ntepe N, Frogneux M, Aka FT, Hell JV, Delmond JC, Manguelle D (2009) Eruptive and earthquake activities related to the 2000 eruption of Mount Cameroon volcano (West Africa). J Volcanol Geotherm Res 179:206–216. doi:10.1016/j.jvolgeores.2008.11.021

Ayonghe SN, Ntasin EB, Samalang P, Suh CE (2004) The June 27, 2001 landslide on volcanic cones in Limbe, Mount Cameroon, West Africa. J Afr Earth Sci 39:435–439. doi:10.1016/j.jafrearsci.2004.07.022

Bang HN (2013) Governance of disaster risk reduction in Cameroon: the need to empower local government. Jàmbá: J Disaster Risk Stud 5: Art. #77, 10pp. doi: 10.4102/jamba.v5i2.77X

Bang HN (2014) General overview of the disaster management framework in Cameroon. Disasters 38:562–586

Barclay J, Haynes K, Mitchell T, Solana C, Teeuw R, Darnell A, Crosweller HS, Cole P, Pyle D, Lowe C, Fearnley C, Kelman I (2008) Framing volcanic risk communication within disaster risk reduction: finding ways for the social and physical sciences to work together. Geol Soc London Spec Publ 305:163–177. doi:10.1144/SP305.14

Bohnsack R (2004) Group discussions and focus groups. In: Flick U, von Kardorff E, Steinke I (eds) A companion to qualitative research. Sage Publications, USA, pp 214–221

Bonne K, Kervyn M, Cascone L, Njome S, Van Ranst E, Suh E, Ayonghe S, Jacobs P, Ernst GGJ (2008) A new approach to assess long-term lava flow hazard and risk using GIS and low cost remote sensing: the case of Mount Cameroon, West Africa. Int J Remote Sens 29:6539–6564. doi:10.1080/01431160802167873

Che VB, Kervyn M, Ernst GGJ, Trefois P, Ayonghe S, Jacobs P, Van Ranst E, Suh CE (2011) Systematic documentation of landslide events in Limbe area (Cameroon): their geometry, sliding mechanism, controlling and triggering factors. Nat Hazards 59:47–74. doi:10.1007/s11069-11011-19738-11063

Che VB, Fontijn K, Ernst GGJ, Kervyn M, Elburg M, Van Ranst E, Suh CE (2012a) Evaluation of the degree of weathering in landslide-prone soils in the humid tropics: the case of Limbe (SW Cameroon). Geoderma 170:378–389. doi:10.1016/j.geoderma.2011.10.013

Che VB, Kervyn M, Suh CE, Fontijn K, Ernst GGJ, del Marmol M-A, Trefois P, Jacobs P (2012b) Landslide susceptibility assessment in Limbe region (SW Cameroon): a field calibrated seed cell and information value method. Catena 92:83–98. doi:10.1016/j.catena.2011.11.014

Che VB, Trefois P, Kervyn de Meerendre M, Ernst GC, Van Ranst E, Verbrugge J-C, Schroeder C, Jacobs P, Suh CE (2013) Geotechnical and mineralogical characterisation of soils from landslide scars and inferred sliding mechanism: case of Limbe, SW Cameroon. Glob Environ Change landslide Sci Pract 4: 43–49. http://link.springer.com/chapter/10.1007%2F978-3-642-31337-0_5

Déruelle B, N'ni J, Kambou R (1987) Mount Cameroon: an active volcano of the Cameroon Line. J Afr Earth Sci 6:197–214. doi:10.1016/0899-5362(87)90061-3

Déruelle B, Ngounouno I, Demaiffe D (2007) The 'Cameroon Hot Line' (CHL): a unique example of active alkaline intraplate structure in both oceanic and continental lithospheres. C R Geosci 339:589–600. doi:10.1016/j.crte.2007.07.007

Diko ML (2012) Community engagement in landslide risk assessment in Limbe, Southwest Cameroon. Sci Res Essays 7(32):2906–2912. doi 10.5897/SRE12.488 http://www.academicjournals.org/SRE

Donovan A, Eiser JR, Sparks RSJ (2014) Scientists' views about lay perceptions of volcanic hazard and risk. J Appl Volcanol 3:15. doi:10.1186/s13617-014-0015-5

Favalli M, Tarquini S, Papale P, Fornaciai A, Boschi E (2011) Lava flow hazard and risk at Mt. Cameroon volcano. Bull Volcanol 74:423–439. doi:10.1007/s00445-011-0540-6

Fothergill A (1996) Gender, risk, and disaster. Int J Mass Emerg Disasters 14(1):33–56

Freeth SJ, Kay RLF (1987) The lake Nyos gas disaster. Nature 325:104–105. doi:10.1038/325104a0

Gaillard J-C, Dibben CJL (2008) Volcanic risk perception and beyond. J Volcanol Geotherm Res 172:163–169

GEADIRR (2011) Geotechnology, environmental assessment and disaster risk reduction. Views from the Frontline (VFL) Cameroon report to the Global Network for Civil Society Organizations for Disaster Reduction (GNDR)

Gehl P, Quinet C, Le Cozannet GG, Kouokam E, Thierry P (2013) Potential and limitations of risk scenario tools in volcanic areas through an example at Mount Cameroon. Nat Hazards Earth Syst Sci 13:2409–2424. doi:10.5194/NHESS-13-2409-2013

Issa OT, Chako Tchamabé B, Padrón E, Hernández P, Eneke Takem EG, Barrancos J, Sighomnoun D, Ooki S, Nkamdjou Sigha, Kusakabe M, Yoshida Y, Dionis S (2014) Gas emission from diffuse degassing structures (DDS) of the Cameroon volcanic line (CVL): implications for the prevention of CO_2-related hazards. J Volcanol Geotherm Res 283:82–93. doi:10.1016/j.jvolgeores.2014.07.001

Mercer J, Kelman I (2010) Living alongside a volcano in Baliau, Papua New Guinea. Disaster Prev Manag 19:412–422

Ndille R, Belle JA (2014) Managing the Limbe floods, considerations for disaster risk reduction in Cameroon. Int J Disaster Risk Sci 5:147–156. doi:10.1007/s13753-014-0019-0

Njome MS, de Wit MJ (2014) The Cameroon line: analysis of an intraplate magmatic province transecting both oceanic and continental lithospheres: constraints, controversies and models. Earth Sci Rev 139:168–194. doi:10.1016/j.earscirev.2014.09.003

Njome MS, Suh CE, Sparks RSJ, Ayonghe SN, Fitton JG (2008) The Mount Cameroon 1959 compound lava flow field: morphology, petrography and geochemistry. Swiss J Geosci 101:85–98

Njome MS, Suh CE, Chuyong G, de Wit MJ (2010) Volcanic risk perception in rural communities along the slopes of Mount Cameroon, West-Central Africa. J Afr Earth Sci 58:608–622. doi:10.1016/j.jafrearsci.2010.08.007

Pannaccione Apa MI, Kouokam E, Akoko RM, Nana C, Buongiorno MF (2012) An ethical approach to socio-economic information sources in ongoing vulnerability and resilience studies: the Mount Cameroon case. Annal Geophys 55:3. doi:10.4401/ag-5569

Siebert L, Simkin T, Kimberly P (2010) Volcanoes of the world, 3rd edn. University of California Press, Berkeley

Suh CE, Sparks RSJ, Fitton JG, Ayonghe SN, Annen C, Nana R, Luckman A (2003) The 1999 and 2000 eruptions of Mount Cameroon: eruption behaviour and petrochemistry of lava. Bull Volcanol 65:267–281. doi:10.1007/s00445-002-0257-7

Suh CE, Luhr JF, Njome MS (2008) Olivine-hosted glass inclusions from scoriae erupted in 1954-2000 at Mount Cameroon volcano, West Africa. J Volcanol Geotherm Res 169:1–33. doi:10.1016/j.jvolgeores.2007.07.004

Suh CE, Stansfield SA, Sparks RSJ, Njome MS, Wantim MN, Ernst GGJ (2011) Morphology and structure of the 1999 lava flows at Mount Cameroon Volcano (West Africa) and their bearing on the emplacement dynamics of volume-limited flows. Geol Mag 148:22–34. doi:10.1017/S0016756810000312

Thierry P, Stieltjes L, Kouokam E, Ngueya P, Salley PM (2008) Multi-hazard risk mapping and assessment on an active volcano: the GRINP project at Mount Cameroon. Nat Hazards 45:429–456. doi:10.1007/s11069-007-9177-3

Wantim M, Suh CE, Ernst GC, Kervyn M, Jacobs P (2011) Characteristics of the 2000 fissure eruption and lava flow fields at Mount Cameroon volcano, West Africa: a combined field mapping and remote sensing approach. Geol J 46:344–363. doi:10.1002/gj.127

Wantim M, Kervyn M, Ernst GC, del Marmol M-A, Suh CE, Jacobs P (2013a) Morpho-structure of the 1982 lava flow field at Mount Cameroon volcano, West-Central Africa. Int J Geosci 4:564–583. doi:10.4236/ijg.2013.43052: http://www.scirrp.org/journal/ijg

Wantim MN, Kervyn M, Ernst GGJ, del Marmol M, Suh CE, Jacobs P (2013b) Numerical experiments on the dynamics of channelized lava flows at Mount Cameroon volcano with the FLOWGO thermo-rheological model. J Volcanol Geotherm Res 253:35–53. doi:10.1016/j.jvolgeores.2012.12.003

Wilkinson S (2004) Focus group research. Qualitative research: theory, method and practice. Sage Publications, London

Blaming Active Volcanoes or Active Volcanic Blame? Volcanic Crisis Communication and Blame Management in the Cameroon

Lee Miles, Richard Gordon and Henry Bang

Abstract

This chapter examines the key role of blame management and avoidance in crisis communication with particular reference to developing countries and areas that frequently experience volcanic episodes and disasters. In these contexts, the chapter explores a key paradox prevalent within crisis communication and blame management concepts that has been rarely tested in empirical terms (see De Vries 2004; Brändström 2016a). In particular, the chapter examines, what it calls, the 'paradox of frequency' where frequency of disasters leads to twin dispositions for crisis framed as either: (i) policy failure (active about volcanic blame on others), where issues of blame for internal incompetency takes centre stage, and blame management becomes a focus of disaster managers, and/or: (ii) as event failure (in this case, the blaming of lack of external capacity on active volcanoes and thereby the blame avoidance of disaster managers). Put simply, the authors investigate whether perceptions of frequency itself is a major determinant shaping the existence, operation, and even perceived success of crisis communication in developing regions, and countries experiencing regular disaster episodes. The authors argue frequency is important in shaping the behaviour of disaster managers and rather ironically as part of crisis communication can shape expectations of community resilience and (non)-compliance. In order to explore the implications of the 'paradox of frequency' further, the chapter examines the case of the Cameroon, where volcanic activity and events have been regular, paying particular attention to the major disasters in 1986 (Lake Nyos Disaster - LND) and 1999 (Mount Cameroon volcanic eruption - MCE).

L. Miles (✉) · R. Gordon · H. Bang
Bournemouth University, Dorset House, Poole,
Dorset, UK
e-mail: lmiles@bournemouth.ac.uk

Advs in Volcanology (2018) 395–409
https://doi.org/10.1007/11157_2017_2
© The Author(s) 2017
Published Online: 12 April 2017

Keywords
Blame games · Blame management · Paradox · Frequency of disasters ·
Cameroon · Volcano

1   Introduction

Many parts of the world today suffer from a
combination of high vulnerability to, and frequency of, natural hazards. In some instances,
this is because geological factors, such as the
existence of tectonic plate lines, result in repeated
occurrence of earthquakes, tremors, or volcanic
eruptions. In the Philippines, for example, the
complexity of geographical and geological factors prompts a 'diversity of hazards' from
earthquakes, volcanoes, tsunamis, to typhoons
and flash flooding. Yet, any frequency of recurring natural hazards does not automatically lead
to efficient and successful emergency planning or
disaster management. Even the experienced can
be caught out by the unexpected, leading to
public and media blame, and accusations of 'incompetency', amidst government claims of
insufficient capacity or 'incapacity' (Wooster
et al. 2005). Within this mix, developing effective crisis communication remains a constant
challenge, especially in the management of
expectations, and the avoidance of blame
(Brandström 2016a, b). Indeed, debates about
incompetency and/or incapacity are often heard
in economically developing countries, and thus
were resilience capacity is still evolving (Cutter
et al. 2008).

The chapter examines three aspects. First, it
explores the key relationship between crisis
communication and blame avoidance and management in the context of volcanoes within
developing countries and regions. Second, it
introduces the concept of a 'Paradox of Frequency'. Thirdly, it discusses the case study of
Cameroon where a major geological volcanic
line, characterized by active and comparatively
regular volcanic and gas activity has clearly
evidenced issues of crisis communication, blame
management and the 'Paradox of Frequency'.

2   Crisis Communication and Blame Management: Balancing Meaning Making, Framing and Blame Games

2.1   Meaning-Making

As Boin et al. (2005: 82) argues, *meaning making* in crisis and disasters 'is not just a matter of
following existing contingency plans or implementing strategic choices at the outset of a crisis.
It entails intuitive and improvised public communication by leaders who are suddenly cast into
the hectic pace of crisis reporting'. Disaster
managers and their political masters must
develop integrated 'framing' of crisis communication that successfully embodies, on the one
hand, response and recovery imperatives with a
strategy for the restoration and continuity of
economic activity, and on the other hand, business and national interests.

A frame is then a shared construction of
reality, and likewise, framing activity covers both
the use and the impact of frames. As Boin et al.
(2005: 88) highlights, framing represents 'the
production of facts, images and spectacles aimed
at manipulating the perception and reaction to a
crisis', and typically involves selective exploitation of data and arguments. In addition, framing
seeks to build (public) confidence in 'more or
less standardised sequences' and processes so
that participants feel part of disaster planning
working towards their own key interests (Boin
et al. 2005).

Meaning making can also, however, involve
'masking' where disaster managers conceal
and/or downplay aspects of a disaster (to reduce
the long term impacts) during and after crises.
Above all, the outcome of successful meaning
making should be avoidance, limitation, and
control of 'blame games' representing the

'struggle between protagonists inside and outside government about the allocation of responsibility for negative events' (Brändström et al. 2008: 114).

2.2 Blame-Gaming

Blame management is often seen as being about blame avoidance or at least managing and controlling blame-gaming. Blame games can be seen as situations where leaders (but could be extended to entities) protect their own self-interests by projecting negative aspects of the crisis onto other actors. The attribution of blame can be a major occupation of disaster managers and their policy leaders during a disaster, seeking to avoid or shape future accountability. Boin et al. (2010: 706) argues 'something or somebody must be blamed—for causing the crisis, failing to prevent it, or inadequately responding to it', pointing to the fact that the 'tragedy' of disaster as an 'Act of God' or 'beyond management' is no longer seen as a publically or socially acceptable explanation for a crisis. There has to be allocation of blame (Brändström and Kuipers 2003: 291).

For disaster managers, the incentive to inflate or diminish blame could be incentivized by a perceived threat of future demotion or dismissal or even by future progression and promotion (Boin et al. 2009: 99; Hood 2001: 8). This can be particularly true in developing countries, where disaster management frameworks may not be that well developed or resourced, and the pressure upon individuals may even be more intense.

According to Brändström (2016a: 34), blaming theory assumes that the blaming behavior of disaster managers and policy leaders is often determined by three factors—namely (i) the wider institutional and political conditions under which blame games occur, (ii) the blame management strategies that actors employ and, (iii) the skills used to apply these strategies in the public arena. In developing countries, the conditions affecting disaster management are quite challenging—with finite resources, immature institutional arrangements, unstable political conditions, and intense competition between governmental priorities. In other words, disposition towards blaming can be heightened principally because of the very institutional and political conditions that pertain in developing countries.

Moreover, the skills of disaster managers can also be influential, not least because a disaster provides opportunities to bring out the worst in people who seek to identify scapegoats in order to allocate blame (Ewart and McLean 2015: 169). It may also bring out the best in people and their entrepreneurial skills at times of crisis (see Miles and Petridou 2015; Miles 2016). Certainly, within the realms of crisis communication, disaster managers and policy leaders will be skillful at arming themselves with plentiful explanations to avoid blame (McGraw 1990: 119). Ewart and MacLean (2015: 168), for instance, catalogue six forms of identifiable explanations, from blaming lack of resources; the event itself; previous administration(s); the number of people and agencies involved; the delegated agency; or claiming ignorance to unforeseen consequences.

2.3 Framing

A critical feature within crisis communication is thus for disaster managers 'to position themselves in relation to what caused the event' (Brändström 2016a: 118) and distinguishing between framing causality as caused by internal (policy/political) or external (operational or other) factors. Often an external frame requires arguing 'credibly that the events may or may not have been foreseeable' and also challenge certainty that events were preventable or controllable once they had occurred. In particular, events affecting vulnerable disaster zones in developing countries can be portrayed as 'forces of nature'—where disaster managers argue that they 'cannot prevent them from happening and rarely are able to control them when they do' (Brändström 2016a: 118). Equally, opponents and critics will attempt to link operations to the internal workings of disaster planning, policy and politics.

While there are numerous models for understanding blaming behavior and impacts

(Brändström and Kuipers 2003; Brändström 2016a) certain aspects seem especially relevant for this chapter. First, there is political/policy failure (highlighting internal causes, and examples of individual or multiple mistakes or missed signals) which can be broadly equated with 'incompetency'. Second, there is systemic failure in relation to external factors—where the disaster management system cannot cope with the magnitude of the external event and thus there is 'incapacity' or 'incapability' to act effectively against such 'forces of nature' (adapted from Brändström 2016a). In this way, blame and accountability are either internalized or externalized. Frames indicating whether the events were foreseeable and controllable internalize accountability, meaning the blame is allocated to identifiable individuals and their policies. Whereas frames externalising accountability allow for disaster managers and policy makers to avoid blame and policies remain unchanged (Boin et al. 2009).

Balancing meaning making, framing, and blame gaming are therefore important for understanding crisis communication and dealing with the question of frequency of disasters also. In addition, managers framing the implications of the frequency of disasters in specific ways may lead to delegation to local actors in developing countries. Rather ironically, blame management may, directly or indirectly, facilitate narrative(s) of resilience in developing countries, where there is a bigger role for local communities and individuals doing more when confronted with ineffective or reluctant governmental action. In this way, the pressures of meaning making, framing and blame management facilitate official, and often unofficial, delegation of disaster management to others.

3 Paradox of Frequency: Policy Failure as 'Incompetency' and Event Failure as 'Incapacity'

The 'paradox of frequency' highlights how frequency can be framed within crisis communication, and in particular, as a paradoxical situation

where frequency of disasters facilitates twin dispositions for crisis framing.

First, there is the frame where the frequency of disaster leads to a stress and expectation on learning and competence building. Disaster managers, faced with a regularity of events, should be able to learn and hone their competencies in handling such events to a high level. They can internalise this competency within their emergency planning and policies, and even increase their own accountability in terms of blame management. Conversely, when things go wrong, the focus will be on *blame management of 'incompetency'*. Blame will often focus on not meeting expectations of competency and thus *policy failure* being strongly associated and framed as 'incompetency'. Thus, policy failure in volcanic crisis communication could be seen as *being active about volcanic blame on disasters managers* and others responsible for community resilience. Ironically, the existence of learning leads to assumptions that subsequent inadequacy to respond equates to incompetence (in not getting on top of problems) which build on rising, and at times unrealistic, expectations that successive disaster experiences internally and proportionately enhance resilience. In this case, the frequency of disaster events leads to ever more active volcanic blame management.

Second, there is the frame where the frequency of disasters leads to a stress on the magnitude, size and regularity of the event and on blame avoidance on the grounds of 'incapacity'. In this context, the ramifications of the disasters are so regular and/or so large that they require society and individuals to treat resilience towards disasters as part of their normal activities and as 'business as usual'. They need to do the best they can at these often frequent, but challenging times, resulting in limited expectations on government, and beliefs that recovery times may be long given the frequency of successive events. Expectations on government, agencies and disaster managers should therefore be constrained since there is only so much capacity (or incapacity) that can be provided in handling such frequent external occurrence. The dominant frame then is constructed around blaming the active volcanoes and not the disaster managers or policy leaders or even system. Hence, when things go wrong, it can be framed as simply 'system failure' where the frequency and size of disaster as an external event overwhelms the capacity of the disaster management system in place. In simple terms, *an event based failure where there is blame avoidance on the grounds of 'incapacity'*. Given this line of reasoning, the frequency of events means that disaster managers emphasise the importance of blame avoidance on them since, in operational terms, dispositions are firmly centred on blaming the frequent activity of the respective volcano.

This paradox is all the more important since in terms of crisis communication, disaster managers face a key challenge before, during and after disasters. There is, for example, a tendency for government agencies to want to both 'own the message' and 'be the messenger', especially as disasters, by definition, exceed the capacity of affected communities and thus the public will look to national or regional leadership for assistance. There is therefore a high propensity for such agencies—by adopting this approach—to be open to both blames on grounds of *incompetence* and *incapacity* simultaneously.

4 Communication Challenges in the Context of Volcanic Crisis Management

In theory, disaster managers should do their best work at crisis communicating in the period before a disaster, particularly when focusing on informing, educating and concentrating on themes such as risk reduction and disaster prevention. Campaigns often use printed leaflets (Bird et al. 2010), village elder gatherings (Cronin et al. 2004) and radio broadcasts and typically include schools' disaster awareness (Ronan et al. 2010), elderly or disabled social outreach, and early warning systems (Garcia and Fearnley 2012). Nevertheless, despite such public awareness campaigns, communities often remain reluctant to engage with government communication agendas. This may be a result of mistrust

from previous inadequate government actions or reactions to earlier disasters leading to an abiding cynicism and uncooperative attitude to subsequent public communication strategies (Haynes et al. 2008). As a result, future disaster related communications are perceived simply as government propaganda to protect reputation. It is arguable then whether governments should always be the sole owner of 'the message'. Crisis communication officers are thus faced with the twin challenges of not only having the right message but also employing the right messenger (McGuire et al. 2009); it may be necessary to think about incompetence and incapacity dynamics both in terms of *messaging* and in terms of the *messenger* when it comes to blame.

Crisis communication in volcanic crisis management is particularly challenging because volcanoes are highly complex scenarios scientifically, socially and politically with potential dire consequences to human, financial, social, physical and natural capital if not handled properly. This is critical for volcanic crises management in developing countries because action is required in uncertain circumstances where several gaps prevent efficient volcanic risk management. These include lack of adequate human resources and weak response structures; lack of understanding of the vulnerability of exposed elements; lack of assessment of vulnerability and community resilience or the capacity to recover after a catastrophe; lack of understanding of the vulnerability of exposed elements and generally weak disaster risk management frameworks (Bang 2014; MIAVITA 2012).

This complexity in communication (Fig. 1) is not made any easier by the fact that undertaking volcanic risk, hazard and vulnerability assessment ideally requires engagement of scientific agencies with diverse expertise (Brändström 2016a; Smale 2016; MIAVITA 2012; UN 1995), as well as the integration of information flows from stakeholders at the local, regional and national levels. Significant gaps remain in communication and information flows in volcanic crisis management in many parts of the world, including the Cameroon, which are prone to volcanic hazard risks. One commonly

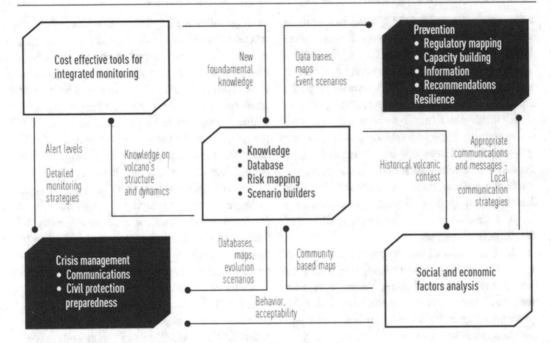

Fig. 1 Methodological framework with related information flows for managing volcanic events. *Source* MIAVITA (2012: 17)

identifiable problem remains the *transmission and translation of scientific early warning and monitoring*, whereby: (i) scientific monitoring is regular, providing not only constant information about the hazard to disaster managers or decision-makers, but: (ii) critical assessment that also feeds into key public communications, warnings and/or instructions on the level and kinds of actions to be taken (Volcano Observatory 2016; Smale 2016). Arguably, the former is well developed for a few hundred of the world's active volcanoes (Simkin and Siebert 1994; UN 1995), while the latter is often poorly developed or absent, inadequate and/or ineffective (Sanderson 1998; Clay et al. 1999; Kokelar 2002).

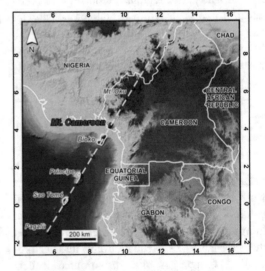

Fig. 2 The Cameroon Volcanic Line (*dashed yellow line*). *Source* Favalli et al. (2012: 424)

5   Volcanic Hazards in the Cameroon

Cameroon is prone to natural hazards mainly due to a geologic/topographic feature in the country known as the Cameroon Volcanic Line (CVL) (Fig. 2). For this chapter, the 1600 km long CVL also fits the criteria of frequency, with

regular occurrence of landslides, floods, earth tremors, toxic gas emissions (as happened in 1984 and 1986 in Lake Monoum and Lake Nyos respectively) and frequent volcanic eruptions. Located on the CVL, is Mt. Cameroon/Fako, the largest, most active volcano in West and Central

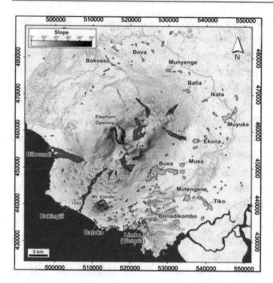

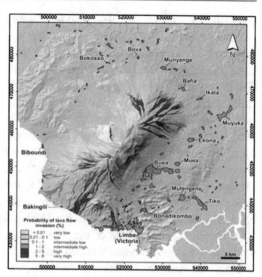

Fig. 3 Map showing Lava flows produced during eruptions of Mt. Cameroon in the 20th century (*red areas*), and towns and villages in the region (*crosshatched areas*). *Source* Favalli et al. (2012: 424)

Fig. 4 Lava flow hazard map around Mt. Cameroon volcano. *Source* Favalli et al. (2012: 432)

Africa (Kling et al. 1987; Duruelle et al. 1987; Fitton 1987; Freeth and Kay 1987; Bang 2012, 2013), having erupted eight times in the 20th century and most recently in 2012 (Global Volcanism Program 2012). The historical record of Mt. Cameroon shows an average period of 17 years (Njome et al. 2010) between successive eruptions. Moreover, Mt. Cameroon is also notable due to its size as an elliptical volcano, straddling the continental margin at the bottom of the Gulf of Guinea in the South West Region (SWR) of Cameroon (Fig. 3), and with a height of 4095 m above sea level and an area of about 3000 km^2.

Generally, Mt. Cameroon is characterised by three types of eruptions: explosive types, moderately explosive types (that have built the cinder cones) and effusive types, which are responsible for lava flows (Tchoua 1971; Tsafack et al. 2009). Voluminous lava flow, rather than pyroclastic materials is the greatest threat from Mt. Cameroon, often from summit and flank eruptions (Pyle 1999). Recent studies, mapping the risk of lava flow inundation (Fig. 4) and other hazards around Mt. Cameroon (Bonne et al. 2008; Thierry et al. 2008; Favalli et al. 2012; Wantim et al. 2013),

highlight the notable vulnerability of the two biggest towns, and largest population centres in the region. Buea town (90,000 people)—capital of the SWR and administrative headquarters, and Limbe town (85,000 people)—the main tourist town situated along the Atlantic coast, and other villages closer to the Volcano are susceptible to inundation by lava flow. Crisis communication is very important given that there is significant risk to strategic and critical buildings essential to disaster management located in Buea, the capital and regional headquarters of the SWR, and located at the foot of Mt. Cameroon. It should also be noted that such vulnerability also applies to earthquakes and landslides since both felt and instrumentally recorded earthquakes have been documented with the vast majority along or close to the CVL, and largely concentrated in the Mt. Cameroon region (Ateba and Tabod 2009). Volcanic eruptions are usually preceded/or accompanied by volcanic and tectonic earthquakes, indicating that earthquake monitoring remains very important for predicting MCEs.

In terms of crisis communication in the Cameroon, several aspects are also important. First, although the area was seismically active prior to the 1999 eruption, there was no extensive pre-warning or early warning system in place to warn the population of the threat—severely reducing crisis communication. Secondly, the frequency of volcanic eruptions and the importance of earthquake monitoring have led to some developments in the Cameroon. In the Mt. Cameroon region, for example, the first network of six permanent seismic stations was setup in 1984. Thirdly, the extensiveness of the monitoring has varied at differing points in time. After the Lake Nyos Disaster (LND) in 1986, seismic monitoring of Mt. Cameroon was extended to the region on the Oku Volcanic field where Lake Nyos is situated. However, prior to the 1999 and 2000 eruptions, all but one of the sensors was working, due to lack of maintenance (Ateba et al. 2009). The number of seismic stations was increased after the 1999, and 2000 eruptions, culminating in 32 broadband stations being installed (2005–07) over the CVL and the Congo Craton. All the stations, however, were dismantled in 2007 because they were not operational, except for two, one in Ekona at the foot of Mt. Cameroon and the other in Yaounde (Ateba and Tabod 2009). The volcano is now monitored using a network of six telemeter seismic broadband stations that detect the magnitude and location of earthquakes, and data is processed at a monitoring centre, located at Ekona not far from Buea (Lenhardt and Oppenheimer 2014).

Fourthly, the crisis management framework of the Cameroon is relatively new, with most of the significant developments after the 1999/2000 volcanic eruptions. Clear institutional structures, including communications have been attempted, with volcanic crisis management falling under the remit of civil protection. The government retains the primary responsibility, and has instituted a national policy on crises management that recognises a multi-agency, interdisciplinary and inter-cooperation. The nodal coordinating agency for civil protection is the Department/Directorate of Civil Protection (DCP) in the powerful Ministry of Territorial Administration and Decentralisation (MTAD). Most notably, a multi-agency and multi-disciplinary approach to natural and other hazards has been operational only since 2005. This is supported by governmental legislation setting out a general national

strategy for risk reduction and disaster management that includes a National Risk Observatory, and emphases three phases of pre-crisis, disaster response and recovery/rehabilitation—all of which stress the importance of information flows and crisis communication (Bang 2012, 2014). There is also a National Disaster Prevention and Management Programme, which ideally, should liaise with the DCP in coordinating all the local, regional, national and international stakeholders in disaster management, and envisages a decentralised structure where authority lies with chief government administrators in these administrative divisions, who double as the main crisis/disaster managers (Bang 2014). Yet as Bang (2014) notes, these policies are only as good as they appear since most have not been implemented in volcanic crisis situations in recent decades.

6 Crisis Communication: The Case of the 1999 Eruption of Mt. Cameroon

The 28 March–22 April 1999 eruption of Mt. Cameroon is well documented. Although the Cameroon's scientists and authorities were aware that the volcano was indeed active, the actual eruption took everyone by surprise. In terms of crisis communication when the eruption started, the local community was informed through the official government run regional state radio in Buea (CRTV Buea). The eruption, however, was not forecast in advance and the population was not pre-warned of any impending eruption, highlighting the lack of an early warning system and in tandem, effective crisis communication strategies.

This was very surprising since reports of seismic activities leading up to the eruption were available, and painted a picture of an impending eruption. The eruption, which started on 28 March, was explosive, emitting gases and pyroclastic lava flow. On 30 March, a second vent opened, releasing huge quantities of lava that flowed for about 14 km south-southwest towards the village of Bakingili (Suh et al. 2003). In

2000, a brief fissure eruption at the summit produced a lava flow that spread mostly south eastwards and stopped 4 km from the outskirts of Buea (Favalli et al. 2012). Fortunately, the lava flow rate was slow, providing sufficient time for any 'at risk' population on its flow path to be evacuated.

The effects of the 1999 MCE were reasonably profound. Fortunately, there were no human casualties and the damage was mainly restricted to infrastructure and economic activity. The magnitude 4 earthquake damaged houses in Buea, leaving some people homeless. Most notably the lava flow that moved towards the coast, particularly the village of Bakingili, affected over 1000 people including 600 inhabitants of Bakingili village who were subsequently evacuated; the first time ever in the history of MCEs (Atanga et al. 2010). Luckily, the 10–12 m thick lava flow narrowly missed Bakingili village, whose population had been evacuated just a few days earlier, but severely damaged infrastructure, and affected the local economy and tourism resorts of the West Coast. Volcanic ash affected the coastal villages of Batoke, Debundsha, Bakingili and Idenau, causing eye and respiratory problems (Afane et al. 2001). Gas and ash emissions also polluted drinking water for about 2600 residents in the area (Atanga et al. 2010). Although there was no human casualty, the eruption caused a total economic loss estimated at about US$790,000 (Lenhardt and Oppenheimer 2014).

6.1 Flaws in Crisis Communication

Three observations can be readily made in terms of the 1999 MCE. Firstly, crisis communication and early warning was found lacking. In taking everybody by surprise, the 1999 MCE exposed flaws in Cameroon's disaster management system, including both scientific and governmental lack of preparedness, despite frequency of eruptions on Mt. Cameroon. Although various scientific studies had been carried out prior to 1999,

the Cameroonian authorities had no clear idea of the level of risks associated with the volcano (Thierry et al. 2008). In addition, there was no warning system in place to alert the population. Although, and following the 1986 LND, a carbon dioxide detection system was adopted to alert the population (Bang 2012), a more extensive early warning system had not been introduced in other hazard prone regions of the country. Indeed, this was only put in place on Mt. Cameroon after the 1999 MCE (Thierry et al. 2008). Hence, the culture of disaster management was reactive, and did not place sufficient emphasis on preparedness for natural hazards that the government, scientists and the public recognised were frequent in the Cameroon. In short, there were ready-made grounds for claims of incompetency in terms of blame management. Second, even when emergency management effectively began as part of the response phase (Fig. 5), the practical reality was that disaster response was widely dispersed providing multiple points of confusion and tension on crisis communication.

From the perspective of crisis communication, the structure centred on the scientific committee, which provided feedback of its monitoring activities to the governor of the SWR during daily meetings as the eruption continued. The meetings were open, attended by members of the public and the press, who received updates from the scientists/government and gave interviews to heads of the committees/chairpersons who consequently updated the public. Although this was an opportunity to eliminate false rumours or wrong information (Ateba and Tabod 2009), the management of the eruption revealed a plethora of problems. In practice, and as shown in Fig. 5, a complex array of actors participated in the disaster management contributing to multiple information flows and communication. Reports also suggested a striking lack of coordination in terms of sharing results and information even inside the committee, resulting in significant confusion (Ateba and Tabod 2009).

Third, the 1999 experience highlighted that there were major deficiencies in how crisis

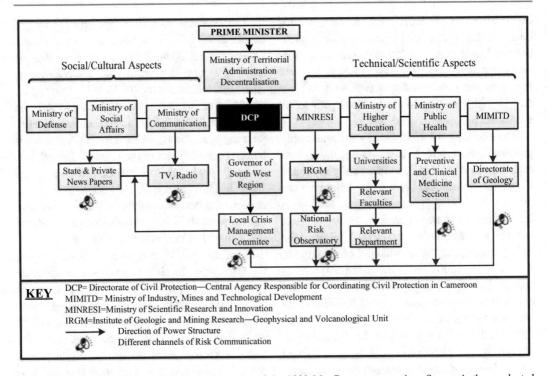

Fig. 5 Organisational structure for the management of the 1999 Mt. Cameroon eruption. *Source* Authors, adapted from Ateba and Tabod (2009: 45)

communication fed into and shaped key decision-making. A good example of this relates to the specific case of the evacuation of 600 people from Bakingili Village, which occurred comparatively late, and was marred by poor preparations, and revealed many communication problems. Although the lava flow had been approaching the village for weeks, evacuation was not considered a priority, regardless of the fact that scientists monitoring the eruption had warned of the threat to the local population in the area. If the lava flow speed was faster, it would have reached Bakingili village before the evacuation. The authorities were divided about whether to inform people living on the SW flank of the Mountain that lava flow was a threat to their settlements because of fear of panic, which neither the national nor regional authorities were sufficiently prepared to handle. Simultaneously, there were radio, TV and media announcements, reassuring the villagers that any dangers were minimal and urging the population to 'stay vigilant', directly contradicting messages from

scientists monitoring the eruption who had identified that the village was along the flow path of the lava, and there was no sign of the eruption stopping soon. Moreover, field scientists close to Bakingili were warning the villagers of the danger, resulting in an overall picture where the local residents were getting different messages and mixed signals from various media. When the decision was finally taken to evacuate the residents of Bakingili village due to fear of a possible inundation by the lava flow, the confusion and delay meant that the temporary camps hastily built in Tiko, some 40 km away, lacked basic provision and/or facilities for emergency relief operations. This prompted anger amongst the relocated villagers, who subsequently blamed the government for lack of readiness, inadequate resourcing, and/or an unwillingness to adequately cater for their needs in the camps. The consequence was mistrust and miscommunication between the communities, local scientists and emergency managers (Atanga et al. 2010; Njome et al. 2010).

6.2 Blame Gaming and Blame Avoidance

From the blame perspective, several aspects were evident. First, blame in terms of 'incompetency' was directed at the authorities and identified the lack of preparedness including any emergency plan to evacuate people. This was largely framed as apportioning blame for internal policy failure and incompetence given the threat posed by lava flow in the region was already known from previous eruptions. The blame associated with such internal incompetency and policy failure far outstripped the ability of disaster managers to use blame avoidance strategies based on incapacity and system failure due to the magnitude of external events. In addition, communication on the rate of the advancing flows was not regarded as essential by the authorities, who stressed that prior eruptions had not threatened a settlement in the region before. Although the lava flow rate towards the SW coast was being regularly communicated by scientists to the authorities, nothing was forecast or reported for the ash fall, which caused health hazards affecting many communities. Above all, even as the eruption unfolded, the population in the region was not warned at any stage that this health hazard existed. Hence, the framing and blaming of incompetency was also associated with very poor communication management by the authorities with the local indigenes (Atanga et al. 2010). Compounding this was the setting up of temporary evacuation camps to host the displaced population that lacked basic facilities. Finally, the delay in the final decision to evacuate 'at risk' populations, and its then slow transmission to the local population is indicative of incompetence. Here, the authorities were blamed for inept decision-making, incompetent policy implementation and poor communication framed as policy rather than event failure, particularly in relation to the delay in the evacuation of Bakingili residents.

Framing and blaming of 'incompetency' compounded, what Atanga et al. (2010) has identified as a culture of limited or non-compliance among the Cameroonian

villages and settlements. Since the villagers had never been informed of scientific studies about risks on Mt. Cameroon, there was a reluctance to accept evacuation orders from the government notwithstanding the strong community support for enhanced crisis communication. Villagers strongly viewed effective crisis communication as the best way to enhance further and optimal co-operation with the government, including executing strategies, which would strengthen community resilience. Above all, the 1999 eruption also highlighted the need to integrate emergency planning with respect for local customs in order to avoid conflicts, which had bearing on attitudes towards compliance. Here crisis communication was also deemed to play a key role; principally in providing feedback on governmental planning and implementation. Following this line of reasoning, modern information dissemination methods need to be accompanied by and integrated with local/traditional methods to facilitate crisis communication; for example, use of the village's traditional announcement system—'the gong',[1] to facilitate information flow in the event of an emergency.

7 Discussion and Conclusions: Cameroon and the Paradox of Frequency

The experience of the 1999 Mt. Cameroon eruption leaves us with several important observations regarding the 'paradox of frequency'. First, the lack of contingency emergency plan for a frequent event like an eruption reinforces blames framing centred on incompetency rather than incapacity. Second, the 1999 eruption highlights the importance of factoring in the cultural perception of MCEs. The local tribes around the mountain, for instance, believe eruptions are caused by the mountain God (Epasa Moto). Consequently, when he is angry, their

---

[1]The gong is s traditional instrument, which when played or sounded; everyone is alerted and comes out to get a message.

tribal chief offers sacrifices to appease 'Epasa Moto' to prevent the destruction of their villages (Atanga et al. 2010; Njome et al. 2010). Volcanic eruptions on Mt. Cameroon thus forms part of the cultural fabric of the communities, affecting the perception of volcanic hazard among indigenous populations and thus the role of government in mitigating, preparing and responding to it. It is also appropriate to factor such narratives into crisis communication in order to contextualise messaging, facilitate compliance and develop community resilience and orderly volcanic hazard response among communities around the mountain.

Third, the experience of the 1999/2000 volcanic eruptions showed how little has been learnt and embedded from the previous experience of the LND of the 1980s. Functions were duplicated and financial and material destined for the disaster survivors were embezzled (Bang 2012). There were problems with inadequate needs assessments of the disaster survivors. Similarly, in 1999, no staff from the Ministry of Health sat on the crisis committee to provide advice on health risks from the eruption thereby demonstrating deficiencies in learning about the role of scientific advice in shaping crisis communication.

According to Bang (2012), part of the reason for this lack of learning—in spite of the frequency of disasters—was that there was a notable lack of follow-up projects on the social aspects of the LND. Disaster managers failed to learn from internal policy failure and thereby how *to turn incompetency into practical competency* via *lessons learned or at least counter accusation of incompetency* via *more sophisticated communication strategies*. In fact, government officials used framing strategies emphasising the external nature and magnitude of the LND and the later MCE events in a quick, but largely futile, attempt to shift criticisms about policy failures in the management of the crisis from themselves. They highlighted a 'systemic

failure' narrative, where they attempted blame avoidance by highlighting the poor financial state of the country, combined with the magnitude of external natural events.

Other factors can be explained by concepts of blame management. Field observations by Bang reveal also that the political context is important in the Cameroon case. Government authorities and officials appointed to manage the various administrative units of the country—who also double as disaster managers—usually behaved to protect/defend their jobs/positions rather than accepting mistakes. Equally, they were resistant to delegating to others more knowledgeable in crisis management to take control, because they might be sympathisers of opposition parties and may take credit for any good job done (Bang 2012).

As a conclusion, there is empirical evidence that demonstrates that frequency of events is not a guarantee of effective learning and enhanced preparedness for the future. Later blame and blame gaming will continue especially since issues of competency remains at the fore even today in the Cameroon. Issues, such as, lack of political resolve or will, and inadequacy of human and financial resources (Bang 2014) remain as relevant today—suggesting that frequency of occurrence is not necessarily the main factor determining levels of preparedness and resilience. At the very least there needs to be continuous commitment and political will as part of the 'bouncing forwards' that embodies a quest for change, improvement and innovation (Miles 2016). Learning and review of experience must therefore accompany frequency of events. Simply experiencing frequent events will not automatically lead to effectiveness in handling those events. Only by incorporating more sophisticated ideas of crisis communication can any resistance to learning from previous disasters be addressed within Cameroon's disaster management system. There is no time to waste. One thing the paradox of frequency tells us—is that time will not wait until the next disaster is upon the Cameroon.

References

Afane E, Sende N, Biowole J, Akoh-Arrey M, Muna F (2001) Irritation respiratoire Cameroun en Mars 1999. A quarterly publication of the Faculty of Medicine and Biomedical Sciences, University of Yaounde 1, Cameroon, vol 3, pp 62–63

Atanga M, Merve A, Njome M, Kruger W, Suh E (2010) Health system preparedness for hazards associated with Mount Cameroon eruptions: a case study of Bakingili village. Int J Mass Emerg Disasters 28 (3):298–325

Ateba B, Tabod C (2009) Monitoring seismic and volcanic activity in Cameroon: advanced workshop on evaluating, monitoring and communicating volcanic and seismic hazards in East Africa, 17–28 Aug 2009. ICTP, Trieste. http://indico.ictp.it/event/a08176/session/19/contribution/15/material/0/0.pdf

Ateba B, Dorbath C, Dorbath L, Ntepe N, Frogneux M, Aka F, Hell J, Delmond J, Manguelle D (2009) Eruptive and earthquake activities related to the 2000 eruption of Mount Cameroon volcano (West Africa). J Volcanol Geoth Res 179:206–216

Bang H (2012) Disaster management in Cameroon: the Lake Nyos disaster experience. Disaster Prev Manag 21(4):489–506

Bang H (2013) Governance of disaster risk reduction in Cameroon: the need to empower local government. Jàmbá J Disaster Risk Stud. http://dx.doi.org/10.4102/jamba.v5i2.77

Bang H (2014) General overview of the disaster management framework in Cameroon. Disasters 38:562–586

Bird DK, Gísladóttir G, Dominey-Howes D (2010) Volcanic risk and tourism in southern Iceland: implications for hazard, risk and emergency response education and training. J Volcanol Geoth Res 189:33–48

Boin A, 't Hart P, Stern E, Sundelius B (2005) The politics of crisis management: public leadership under pressure. Cambridge University Press, Cambridge

Boin A, 't Hart P, McConnell A (2009) Crisis exploitation: political and policy impacts of framing contests. J Eur Publ Policy 16(1):81–106

Boin A, 't Hart P, McConnell A, Preston T (2010) Leadership style, crisis response and blame management: the case of Hurricane Katrina. Publ Adm 88 (3):706–723

Bonne K, Kervyn M, Cascone L, Njome S, Van Ranst E, Suh E, Ayonghe S, Jacobs P, Ernst G (2008) A new approach to assess long-term lava flow hazard and risk using GIS and low-cost remote sensing: the case of Mount Cameroon, West Africa. Int J Remote Sens 29:6539–6564

Brandström A (2016a) Crisis, accountability and blame management. CRISMART, Stockholm

Brändström A (2016b) Crisis, accountability and blame management. Strategies and survival of political office-holders. http://fhs.diva-portal.org/smash/get/diva2:896367/FULLTEXT01.pdf

Brändström A, Kuipers S (2003) From 'Normal Incidents' to political crises: understanding the selective politicization of policy failures. Gov Opposition 38(3):279–305

Brändström A, Kuipers S, Daleus P (2008) The politics of tsunami responses: comparing patterns of blame management in Scandinavia. In: Boin A, McConnell A, 't Hart P (eds) Governing after crisis: the politics of investigation, accountability and learning. Cambridge University Press, Cambridge, pp 114–147

Clay E, Barrow C, Benson C, Dempster J, Kokelaar P, Pillai N, Seaman J (1999) An evaluation of HMG's response to the Montserrat volcanic emergency. https://www.gov.uk/government/uploads/system/uploads/attachment_data/file/67966/ev635.pdf

Cronin SJ, Gaylord DR, Charley D, Alloway BV, Wallez S, Esau JW (2004) Participatory methods of incorporating scientific with traditional knowledge for volcanic hazard management on Ambae Island, Vanuatu. Bull Volcanol 66:652–668

Cutter SL, Barnes L, Berry M, Burton C, Evans E, Tate E, Webb J (2008) A place-based model for understanding community resilience to natural disasters. Glob Environ Change 18(4):598–606

De Vries M (2004) Framing crises: response patterns to explosions from firework factories. Administration and Society 36:594–614

Duruelle B, N'nI J, Kambou R (1987) Mount Cameroon: an active volcano of the Cameroon Line. J Afr Earth Sci 6:197–214

Ewart J, McLean H (2015) Ducking for cover in the 'blame game': news framing of the findings of two reports into the 2010–11 Queensland floods. Disasters 39(1):166–184

Favalli M, Tarquini S, Papale P, Fornaciai A, Boschi E (2012) Lava flow hazard and risk at Mt. Cameroon volcano. Bull Volcanol 74:423–439

Fitton J (1987) The Cameroon Line, West Africa: a comparison between oceanic and continental alkaline volcanism. In: Fitton J, Upton B (eds) Alkaline igneous rocks. Geological Society of London Special Publication, pp 273–291

Freeth S, Kay R (1987) The Lake Nyos disaster. Nature 325:104–105

Garcia C, Fearnley C (2012) Evaluating critical links in early warning systems for natural hazards. Environ Hazards 11:123–137

Global Volcanism Program (2012) Report on Cameroon. In: Sennert S (ed) Weekly volcanic activity report, 1–7 Feb 2012. Smithsonian Institution and US Geological Survey. http://volcano.si.edu/showreport.cfm?doi=GVP.WVAR20120201-224010

Haynes K, Barclay J, Pidgeon N (2008) The issue of trust and its influence on risk communication during a volcanic crisis. Bull Volcanol 70:605–621

Hood C (2001) The blame game: spin, bureaucracy and self-preservation. Princeton University Press, Woodstock

Kling G, Clark M, Compton R, Devine D, Evans W, Humphrey A, Tuttle M (1987) The 1986 Lake Nyos gas disaster, Cameroon, West Africa. Science 236 (4798):169–175

Kokelar B (2002) Setting, chronology and consequences of the 1995–1999 eruption of Soufriere Hills Volcano, Montserrat. In: Druitt T, Kokelaar B (eds) The eruption of Soufriere Hills Volcano, Montserrat, from 1995 to 1999. Geological Society of London Memoirs, pp 1–43

Lenhardt N, Oppenheimer C (2014) Volcanism in Africa: geological perspectives, hazards, and societal implications. In: Ismail-Zadeh A, Urrutia-Fucugauchi J, Kijko A, Takeuchi K, Zaliapin I (eds) Extreme natural hazards, disaster risks and societal implications. Cambridge University Press IUGG Special Publication Series, Cambridge, pp 169–199

McGuire W, Solana M, Kilburn C, Sanderson D (2009) Improving communication during volcanic crises on small vulnerable islands. J Volcanol Geoth Res 183:63–75

McGraw K (1990) Avoiding blame: an experimental investigation of political excuses and justifications. Br J Polit Sci 20(1):119–131

MIAVITA (2012) Handbook of volcanic risk management: prevention, crisis management and resilience. http://miavita.brgm.fr/Documents/Handbook-VolcRiskMgt-lr.pdf

Miles L (2016) Entrepreneurial resilience. Crisis Response J 11(4)

Miles L, Petridou E (2015) Entrepreneurial resilience: the role of policy entrepreneurship in the political perspective of crisis management. In: Bhamra R (ed) Organisational resilience. CRC Press, London, pp 67–81

Njome M, Suh E, Chuyong G, de Wit M (2010) Volcanic risk perception in rural communities along the slopes of Mount Cameroon, West-Central Africa. J Afr Earth Sci 58:608–622

Pyle D (1999) Widely dispersed quaternary tephra in Africa. Global Planet Change 21:1–15

Ronan K, Crellin K, Johnston D (2010) Correlates of hazards education for youth: a replication study. Nat Hazards 53:503–526

Sanderson D (1998) Volcanic warning dissemination in Montserrat. In: Lee B, Davis I (eds) Forecasts and warnings. National Coordination Committee for the International Decade for Natural Disaster Reduction, London, Thomas Telford, pp 534–541

Simkin T, Siebert L (1994) Volcanoes of the world: a regional directory, gazetteer, and chronology of volcanism during the last 10,000 years, 2nd edn. Geoscience Press, Tucson, AZ, 349 pp

Smale L (2016) Preventing volcanic disasters: the critical nature of communication. http://london-nerc-dtp.org/2016/05/09/preventing-volcanic-disasters-the-critical-nature-of-communication/

Suh E, Sparks J, Fitton J, Ayonghe S, Annen C, Nana R, Luckman A (2003) The 1999 and 2000 eruptions of Mount Cameroon: eruption behaviour and petrochemistry of lava. Bull Volcanol 65:267–281

Tchoua F (1971) Le volcanism estrombolien de la plaine de Tombel (Cameroun). Annales de la Faculté des sciences du Cameroun, pp 53–78

Thierry P, Stieltjes L, Kouokam E, Ngueya P, Salley P (2008) Multi-hazard risk mapping and assessment on an active volcano: the GRINP project at Mount Cameroon. Nat Hazards 45:429–456

Tsafack J, Wandji P, Bardintzeff J, Bellon H, Guillou H (2009) The Mount Cameroon stratovolcano (Cameroon Volcanic Line, Central Africa): petrology, geochemistry, isotope and age data, geochemistry. Min Petrol 47:65–78

UN (United Nations) (1995) Early-warning capacities of the United Nations system with regard to natural disasters. Secretary General Report A/50/526, 28 pp

Volcano Observatory (2016) Cameroon volcano. https://www.volcanodiscovery.com/mt-cameroon.html

Wantim M, Kervyn M, del Ernst G, Marmol M, Suh E, Jacobs P (2013) Numerical experiments on the dynamics of channelized lava flows at Mount Cameroon volcano with the FLOWGO thermos rheological model. J Volcanol Geoth Res 253:35–55

Wooster M, Demeritt D, Dill K, Webley P (2005) Enhancing volcanic hazard avoidance capacity in central America through local remote sensing and improved risk communication. DFID, London

Part Two Summary: Observing Volcanic Crises

Gill Jolly and Carina J. Fearnley

Numerous volcanic crises have been observed, from small and local in scale, to large regional events. Each crisis has its own contingencies, both spatially and temporally. Differing cultures, politics, economics, population sizes, scales of events, types of volcanic hazards, and geographical constraints shape the dynamics of a crisis. It is potentially only with hindsight can the variables be recognised that may have played a significant role in the success or failure of a crisis. Each event can provide valuable insights of issues to be weary of in future crisis, but equally numerous examples can help build a picture of good practices or procedures that can help foster strong links during a crisis, particularly in relation to communication between the various stakeholders.

The chapters that form Part Two of this volume illustrate some of the key issues that emerge in crises, with examples from 6 continents to illustrate the diverse and wide range of issues. These chapters demonstrate that sharing knowledge and experience is vital as long as this is done in a transparent sensitive manner, preferably prior to a crisis, and with some humility.

Effective crisis communication is particularly challenging when dealing with caldera eruptions, especially during large phases of unrest when it is unclear as to whether an eruption is imminent. Given the infrequency of caldera eruptions, experiences remain limited, yet from recent caldera eruptions such as at Rabaul in Papua New Guinea, unrest can result in eruptive activity within very short time frames, in this case <27 h. This results in very little time to respond to a state of unrest, particularly when an eruption could be particularly powerful. The challenges that Hill et al., Chapter "Volcanic Unrest and Hazard Communication in Long Valley Caldera, California" experienced, during the 1980s unrest at Long Valley Caldera, highlight the need for effective messages that can easily be understood by the public at large. This includes open transparency, and being careful with media and public relations. The importance of good sound science is critical to building effective forecasts and also to building effective and credible working relations. Maintaining long-term relationships becomes increasingly challenging in an ever changing, globalised job market where continuity is hard to find. Hill et al. outline the 'narrow path' involved in such crises, where there is a small space between overreactions and false alarms, and conservative responses. An important lesson is that although mistakes can be costly, poor relations can be rebuilt over time.

Mt Pinatubo was an extraordinary event that balanced this 'narrow path' exceptionally well. Newhall and Solidum, Chapter "Volcanic Hazard Communication at Pinatubo from 1991 to 2015" speak of the 'pressure-cooker' that

C. J. Fearnley (✉)
Department of Science and Technology,
University College London, Gower Street,
London WC1E 6BT, UK
e-mail: c.fearnley@ucl.ac.uk

G. Jolly
GNS Science, 1 Fairway Drive, Avalon,
Lower Hutt 5010, New Zealand
e-mail: G.Jolly@gns.cri.nz

Advs in Volcanology (2018) 411–415
https://doi.org/10.1007/11157_2017_25
© The Author(s) 2017
Published Online: 11 November 2017

scientist's face, the battle between the desires to continue to obtain scientific data to reduce the uncertainties, versus the pressures to communicate immediately. Indeed the immediacy of Pinatubo was key to its success; starting educational programmes early when there were still significant levels of uncertainty, employing rigorous checklists, and building in flexibility to response plans. Some of the challenges of Pinatubo were that this caldera eruption occurred following a period of long dormancy, generating scepticism. Embracing potential false alarms, reviewing the various possible scenarios, and striving for scientific coordination are clearly vital to the success of these large scale events. Indeed the recognition of trust and personal connections between the various stakeholders are sometimes more important than the warning itself. The challenges of calderas often only begin with the eruptions, but can continue with various hazards for decades after.

The 1994 eruption of Rabaul in Papua New Guinea is a positive example of an informed responsive community living around an active caldera. McKee et al., Chapter "The 1994 eruptions at Rabaul Volcano, Papua New Guinea: the roles of instrumental volcano surveillance and community awareness in preparing for the outbreak of the eruptions" outline the experiences of those living at Rabaul, many of whom have had to deal with several periods of unrest, most with only a few hours indication of an imminent eruption, across numerous vents. Whilst Papua New Guinea would benefit from the use of automated and continuous data to aid the monitoring process, with raised community awareness surrounding the volcano, crisis communication can become somewhat surplus, particularly when populations decide to self-evacuate. One of the challenges faced in Rabaul was the looting that occurred when the mass evacuation began, which also affected observatory staff. It easy to forget that observatory and civil defence personnel are also affected by looting, and civil unrest can lead to significant logistical issues for those who are trying to do their job, whilst keeping family safe.

Prolonged, low-level volcanic crises produce very different challenges to those of calderas. Popocatepetl volcano in Mexico, surrounded by over 20 million people, and persistently active over the last 20 years presents some of the most challenging conditions to effectively communicate. De la Cruz-Reyna et al., Chapter "Challenges in Responding to a Sustained, Continuing Volcanic Crisis: The Case of Popocatépetl Volcano, Mexico, 1994-Present" outlines potential solutions to maintaining a responsive community despite indifference and weariness at times among the vulnerable populations: use of social media, increasing use of web cameras and the use of their Volcano Traffic Light Alert Systems (VTLAS). The VTLAS translate the level of volcanic hazard into three alert levels for the populations, leaving no room for uncertainty about what needs to be done in response to the warning. Yet, the lessons learnt in Mexico illustrate the value of consensual approaches and deliberations between the scientists and civil protection organisations to manage the uncertainties of forecasting potential activity.

Personal connections are a recurring theme in Part 2, where at Mt Ruapehu in New Zealand, years of preparedness, communication, and building understanding through planning and simulation exercises have resulted in effective lahar management around Mt Ruapehu. Becker et al., Chapter "Organisational Response to the 2007 Ruapehu Crater Lake Dam-Break Lahar in New Zealand: Use of Communication in Creating an Effective Response" highlight the increase in communication during the crisis relative to the simulation. Despite the successful preparation, increased communication presented new challenges to the communication flows between the various stakeholders that was averted by using a diversity of communication channels. This highlights the need to have numerous channels in places, not only as a back up, but to enable multi-actor network communication and collaboration.

The Eyjafjallajökull 2010 eruption was a relatively small infrequent event that caused a significant crisis globally. The management of the crisis demonstrated that pre-eruption risk communication is critical to building trust between emergency management officials and the general public. Bird et al., in "Crisis Coordination and Communication During the 2010 Eyjafjallajökull Eruption" outline the value of speaking with 'one voice', a challenge when many institutions work together, but in this case it was successful with clearly defined roles and responsibilities, not just within Iceland, but also across Europe. This case study highlights the need for dedicated media centers, in order to meet the enormous demands for information, particularly for such an event that causes international chaos and travel disruption.

For both frequently erupting and long-dormant volcanoes, Solana et al., Chapter "Supporting the Development of Procedures for Communications During Volcanic Emergencies: Lessons Learnt from the Canary Islands (Spain) and Etna and Stromboli (Italy)" highlight the challenges of managing these uncertain events. The inexperience of all involved (from the scientists to the public) can create significant tensions and misunderstandings that can result in poor communication, and lead to poor decision making processes. To overcome these challenges, the key is to clearly define roles and responsibilities between responding groups, along with detailing the communication protocols prior to the event. In particular the liability of scientific groups needs to be carefully outlined. To facilitate this, coordinating bodies that sit outside the scientific community are vital to generate consensus and recognise the different cultures, needs and limitations of the responding groups. Solana et al. also highlight the advantages of working with external researcher, sharing resources, and engaging the public, and outline an Italian model that encourages collaboration and camaraderie.

With over 127 active volcanoes, from the supervolcanic to the effusive style, Chapter "Integrating The Social And Physical Perspective In Mitigation Policy And Practice" by Andreastuti et al. outlines that Indonesia has much to teach us about the importance of communicating hazard information. To achieve mutual understandings across different actors and cultures informal approaches are often seen as the most effective, this is done via 'socialization' using participatory knowledge dissemination. This enables local populations to engage with, and understand the hazards and risks they face, empowering them to conduct their own assessments and make their own decisions. The strong focus on public participation helps maintain strong communication networks and lasting relationships. Whilst each volcano is different, each community around a volcano also has different features that result in different responses, that should influence the way scientists and government agencies communicate and manage volcanic crises. Where the social and physical perspectives of the volcano are integrated, meaningful communication that adapts over time is developed.

The experiences of Volcano Colima in Mexico as discussed by Cuevas-Muniz and Gavilanes Ruiz, "Social Representation of Human Resettlement Associated with Risk from Volcán de Colima, Mexico" describe the importance of the social representation of volcanic risk and how it is redefined by daily life. A key lesson learnt from these experiences is that rather than communication processes that omit the knowledge, values, and desires formed by the vulnerable populations, there is a need to find a way to improve social representation. This needs to be established within both the policy and practice of crisis communication and management, particularly when linked to the resettlement of local populations, who need a voice in the decision-making processes. The study of La Yerbauena highlights the consequences of mistrust and resentment that can result from a breakdown in this vital communication.

"If I Understand, I am Understood: Experiences of Volcanic Risk Communication in Colombia" is a poignant phrase from the work of Garcia and Mendez-Fajury that explores risk communication experiences during volcanic emergencies in Colombia. This reflective highlights the importance of a solid social science research program, to not only increase the impact of crisis communication, but also help the authorities and scientists improve their decision-making processes by understanding the social processes involved. There has been significant focus on various education programmes to foster better understandings around the science of volcanic eruptions. Whilst the community tends to trust in the technical work of the scientific volcanic community of Colombia, further assessment of the relevance and perception of technical scientific reports by the public is required to establish how much of this information is understood and assimilated.

For island communities living in the shadows of volcanoes, there is a particularly strong influence of historical, political, cultural, social, economic, and environmental factors influencing the success and failure of volcanic risk communication. Komorowski et al., Chapter "Challenges of Volcanic Crises on Small Islands States" outline the particular intensity of politics that manifests in island communities, 'perhaps because of, rather than despite, the smallness and tightness of the communities, amongst other island characteristics'. Given the small scale and often independent communication, it is vital these local communities are engaged rather than 'consulted' so to be able to work with communities on their terms. This provides the opportunities for collaboration with communities rather than being shaped or dominated by one political party or group of actors. It is vital to be clear what is known and unknown and what can reasonably be done to fill the gaps.

Mt Cameroon volcano, Cameroon, like many volcanic regions around the world is still poorly understood. Marmol et al., Chapter "Investigating the management of geological hazards and risks in the Mt Cameroon area using Focus Group Discussions" provide valuable guidance on how to develop an effective risk management structure and generate risk reduction actions when political, social and economic conditions are challenging. Problems at Mt Cameroon range from the physical hazard aspects, to a lack of resources or capacity to monitor and mitigate against hazards and little willingness of political leaders to raise awareness and implement effective policy. To move to overcome these obstacles, Marmol et al. describe how they conducted a series of focus group discussions (FGDs) with city councillors from three municipalities around Mt Cameroon volcano, Cameroon. The authors highlight a strong need in such cases for scientists to work together with national and local authorities to translate the findings of scientific hazard and risk assessment into improved risk management practices. This chapter demonstrates that FGDs can provide an excellent framework in which to implement these aspirations through enabling interactions between all actors, with the opinion of all participants being considered at an equal level.

Whilst still focusing on Mt Cameroon volcanoe, Bang et al., "Blaming Active Volcanoes or Active Volcanic Blame? Volcanic Crisis Communication and Blame Management in the Cameroon" outlines the challenges of a 'blame' culture; they explore how blame can be managed and avoided, particularly in developing countries with frequent volcanic activity. Here, 'a paradox of frequency' can occur whereby a crisis can be blamed on either policy failure, or event failure. Blame and conflict between competing scientific groups can be highly destructive for all stakeholders and often arises due to a lack of clear and inclusive protocols. Frequency of crises can be important in shaping the behaviours of disaster managers and also of the local populations. Often Cameroon government authorities act to protect and defend their jobs rather than accepting mistakes and applying learnings. The chapter concludes by noting that it may not be the case that the frequency of events guarantees effective learning and enhanced preparedness, but that continued lack of political will and financial and human resources may hinder learning those vital lessons.

Part Three
Communicating into the Future
Deanne K. Bird and Katharine Haynes

Communicating Information on Eruptions and Their Impacts from the Earliest Times Until the Late Twentieth Century

David Chester, Angus Duncan, Rui Coutinho, Nicolau Wallenstein and Stefano Branca

Abstract

Volcanoes hold a fascination for human beings and, before they were recorded by literate observers, eruptions were portrayed in art, were recalled in legend and became incorporated into religious practices: being viewed as agents of punishment, bounty or intimidation depending upon their state of activity and the culture involved. In the Middle East the earliest record dates from the third millennium BCE and knowledge of volcanoes increased progressively over time. In the first century CE written records noted nine volcanoes in the Mediterranean region plus Mount Cameroon in West Africa, yet by 1380 AD the record only totalled 48, with volcanoes in Japan, Indonesia and Iceland being added. After this the list of continued to increase, but important regions such as New Zealand and Hawaii were only added during the last 200 years. Only from 1900 did the rate of growth decline significantly, but it is sobering to recall that in the twentieth century major eruptions have occurred from volcanoes that were considered inactive or extinct, examples including: Mount Lamington—Papua New Guinea, 1951; Mount Arenal—Costa Rica, 1968 and Nyos—Cameroon, 1986. Although there were instances where the human impact of historical eruptions were studied in detail, with

D. Chester (✉)
Department of Geography and Environmental Science, Liverpool Hope University, Liverpool L16 9JD, UK
e-mail: jg54@liv.ac.uk

D. Chester
Department and Planning, University of Liverpool, Liverpool L69 3BX, UK

A. Duncan
Department of Geography and Planning, University of Liverpool, Liverpool, UK
e-mail: A.M.Duncan@liverpool.ac.uk

R. Coutinho · N. Wallenstein
Centro de Vulcanologia e Avaliação de Riscos Geológicos, Universidade dos Açores, Rua da Mãe de Deus, 9501-801 Ponta Delgada, Portugal
e-mail: Rui.MS.Coutinho@azores.gov.pt

N. Wallenstein
e-mail: Nicolau.MB.Wallenstein@azores.gov.pt

S. Branca
Istituto Nazionale di Geofisica e Vulcanologia, 95125 Catania, Sicily, Italy
e-mail: branca@ct.ingv.it

Advs in Volcanology (2018) 419–443
https://doi.org/10.1007/11157_2016_30
© The Author(s) 2017
Published Online: 26 March 2017

examples including the 1883 eruption of Krakatau and 1943–1952 eruption of Parícutin, these were exceptions and before 1980 there was a significant knowledge gap about both the short and long-term effects of major eruptions on societies. Following a global review, this chapter provides a discussion of the ways in which information has been collected, compiled and disseminated from the earliest times until the 1980s in two case study areas: the Azores Islands (Portugal) and southern Italy. In Italy information on eruptions stretches back to prehistoric times and has become progressively better known over more than 2,000 years, yet even here there remain significant gaps in the record even for events that took place between 1900 and 1990. In contrast, located in the middle of the Atlantic, the Azores have been isolated for much of their history and illustrate the difficulties involved in using indigenous sources to compile, not only assessments of impact, but also at a more basic level a complete list of historical events with accurate dates.

Keywords

History of eruptions · Increase in global knowledge of eruptions · Volcanoes of Southern Italy and the Azores

1 Introduction: The Global Picture

The purpose of this chapter is two fold. First, to review the communication of information on eruptions and its dissemination from the earliest times until the onset of the 'modern' era of volcanology, which for the purpose of this volume is taken to be the latter part of the 20th century. In the last three decades of the twentieth century volcanology experienced a major change in its scientific status, with events such as the eruption of Mount St. Helens in 1980 and space missions to the terrestrial planets highlighting the important role of volcanism as a planetary process. This in turn focused interest on volcanology and stimulated research funding which, *inter alia*, placed the communication of information on eruptions and their impacts on a more secure footing. The authors are aware that, given the present state of published research, especially on early historic eruptions, this account cannot be truly 'global' and will of necessity be strongly biased towards the acquisition and dissemination of information within countries that are part of the 'western' intellectual tradition, though where possible we have endeavoured to spread the net more widely.

Secondly, the issues surrounding communication will be explored by means of two case studies: the Azores (Portugal) and Mt. Etna in Sicily. These case studies provide good—in some respects contrasting—examples of the ways in which responses to eruptions and their impacts have evolved, from the earliest recorded eruptions to the situation obtaining towards the close of the twentieth century. Although both the Azores and Mt. Etna provide documentary accounts of activity, in each area the distinctive written source and varying eruptive styles and impacts, allow differing insights to emerge. Etna has one of the most extensive records of volcanic activity which stretches over some 2000 years of recorded history. Early accounts from the classical era sought mythical explanations of activity, although some later Greek and Roman writers suggested more rational explanations, though even these were not usually based on detailed observation (Chester et al. 2000). Sicily is located at the centre of the Mediterranean Region and historically had excellent communications with the wider European world. The Renaissance led to the early development of empirical science and a letter, for instance, read to the *Royal Society of London* shows insightful observations of the 1669 eruption of Etna (Anon 1669). In the 19th century and following the eighteenth century Enlightenment, Etna became an important field area for many European scientists, including Carlo Gemmellaro (1787–1866), Charles Lyell (1794–1875), George Poulett Scrope (1797–1876) and Wolfgang Sartorius von Waltershausen (1809–1876), who went on to make major contributions to the understanding of geology and its development as a discipline. More recently, Etna was selected: as one of *decade volcanoes* for detailed interdisciplinary study as part of the United Nations designated *International Decade of Natural Disaster Reduction* (1990–2000); and as a *Laboratory Volcano* jointly sponsored by the *European Economic Communities*[1] and the *European Science Foundation*. In contrast, the Azores have been historically isolated from mainland Europe. Eruptions before the twentieth century did not attract international attention or, indeed, much interest from mainland Portugal. Records were mostly archived locally and generally did not contribute to the development of scientific thinking more widely until the second half of the twentieth century.

2 Communicating Scientific Information: Prehistoric and Historical Perspectives

In the early 1980s it was well established that, even before there were written records of volcanoes and their activity, eruptions were depicted in art, remembered orally and formed part of often elaborate religious rituals, volcanoes being viewed as 'agents of benevolence, fear or vengeance depending on their state of activity and the society involved' (Chester 2005, p. 404;

Blong 1982, 1984). In the Middle East, it is often claimed by volcanologists and others that the earliest record of an eruption is a wall painting from the Neolithic town of Çatal Hüyük in Anatolia, showing an eruption with the ejection of blocks and bombs (Mellaart 1967, pp. 59–60, 176–177; Chester 2005). More recently the archaeological team working at the site have concluded that the putative 'volcano' is actually a leopard skin with spots (Hodder 2015, personal communication). The earliest definite records of volcanic activity date, however, from Mesopotamia in the third millennium BCE (Foster 1996; Polinger-Foster and Ritner 1996)[2] and were soon followed by accounts of volcanic activity by Greek, Roman and Islamic writers (Sigurdsson 1999, pp. 14–17).[3] The eruption of Aso volcano in 553 CE was the earliest recorded eruption in Japan (Simkin et al. 1981, p. 66). It occurred a year after Buddhism had been introduced into the country (Keys 1999, pp. 323–324) and may reflect the importance of religious functionaries in providing written accounts about important events.

Most of the data currently available on historic volcanic activity has been collected by the Smithsonian Institution through the *Global Volcanism Program* (GVP), which collates information on current and past activity over the past 10,000 years. Accounts of current activity are provided by the Smithsonian—USGS *Weekly Activity Report*, and comprehensive summaries of past activity are available in the *Bulletin of the*

---

[1]Now the European Union.

[2]Accounts include 'Starry Mountain' in the Khabur Region. This is probably Kawbab volcano in Mesopotamia (Polinger-Foster and Ritner 1996).

[3]Special attention was paid to Santorini (ancient Thera), Vesuvius and Etna. One of the most significant episodes in Islamic intellectual history was the translation of classical texts into Arabic, some of which included accounts of extreme natural events. This took place in Baghdad and other centres of scholarship between the beginning of the eighth and the close of the tenth centuries CE. In some cases this ensured preservation of important information on eruptions. Especially in Spain, the role of Islamic scholars needs to be acknowledged. Authors, who including the philosopher Ibn Rushd (Latin—Averroes), not only engaged with these texts, but also added their own observations on natural phenomena (Stone 2003; Akasoy 2007).

Global Volcanism Network. These data are periodically archived in successive editions of *Volcanoes of the World* (Simkin et al. 1981; Simkin and Siebert 1994; Seibert et al. 2010). Progress up to the 1980s in both locating volcanoes and cataloguing their eruptions is summarised succinctly in a quotation from the first (1981) edition of *Volcanoes of the World*:

> if a list of... volcanoes had been continually kept, it would, at the time of Christ, have contained only the names of 9 Mediterranean volcanoes and West Africa's Mount Cameroon.[4] In the next 10 centuries the list would have grown by only 17 names, 14 of them Japanese. The first historic eruptions of Indonesia were in 1000 and 1006, and newly settled Iceland soon added 9 volcanoes to help swell the list to 48 by 1380 AD... The list has continued to grow, with several important volcanic regions such as Hawaii and New Zealand being completely unrepresented until the last 200 years. Only in the present century has the *rate* of growth declined significantly (Simkin et al. 1981, p. 23).

Later editions of *Volcanoes of the World* show that advances in knowledge have not slackened and that, whereas in 1981 there were 627 volcanoes with recorded eruptions, by 1994 this figure had risen to 719 and reached 858 in 2010 (Simkin and Siebert 1994; Seibert et al. 2010), largely due to: better monitoring of present day eruptions—especially of those occurring in isolated areas through the use of satellite-based remote sensing; and improved knowledge of events which occurred in antiquity. With regards to the latter, there have been few years in the past three decades that have not seen major publications dealing with pre-historic eruptions, although reviewing these works is beyond the scope of this chapter (see: Firth and McGuire 1999; Sigurdsson 1999; Harris 2000; McCoy and Heiken 2000; McGuire et al. 2000; Balmuth et al. 2005; Grattan 2006; Grattan and Torrence 2007; Oppenheimer 2011).

In terms of better recording, not only has the *Smithsonian Institution* continued its invaluable work since 1981 in collecting eruption information and disseminating it to the volcanological research community, but data especially on

human impact has also become more widely available. This was not just from academic authors (e.g. Tanguy et al. 1998; Witham 2005; Cashman and Giordano 2008), but from organizations which have included: the Brussels-based *Centre for Research on the Epidemiology of Disasters* (or CRED); and re-insurance companies, in particular Munich Re (Auker et al. 2013). Advances continue to be made in the second decade of the twenty-first century as witnessed by the new data-base of large magnitude explosive volcanic eruptions (LaMEVE), which forms part of the larger *Volcanic Global Risk Identification and Analysis Project* (VOGRIPA) (Crosweller et al. 2012), and the improved catalogue of fatalities caused by volcanic activity from 1600 to 2010 (Auker et al. 2013).

When examining the written record between 1400 CE when around 50 volcanoes were identified and 1980, when the figure reached 627, there are several points which require more detailed discussion. First, it is clear from the quotation from *Volcanoes of the World* (1981) which is cited above, that progress did not occur at a constant rate during the 580 years which elapsed from 1400 to 1980. There were two periods of marked growth in the list of known active volcanoes: a steady increase from the beginning of the 16th century to the mid-18th century and a second episode of more rapid growth from the mid-18th century to the mid-20th century (Fig. 1). The first period coincides with the *Renaissance* and *Age of Exploration*, especially Spanish and Portuguese penetration into the New World and the invention of the printing press. The second phase reflects a number of developments which include: the easier dissemination of information because of the more widespread use and distribution of printed material particularly newspapers and magazines; the great increase in industrialization, scientific understanding, technology and global trade and the more open intellectual climate associated with the *Enlightenment*.[5] This was facilitated by major advances

---

[4]Mount Cameroon was first observed in eruption by Hannon, a Carthaginian navigator in the fifth century BCE (Anon 2015).

[5]The *Enlightenment* was an intellectual movement which began in the British Isles in seventeenth century and

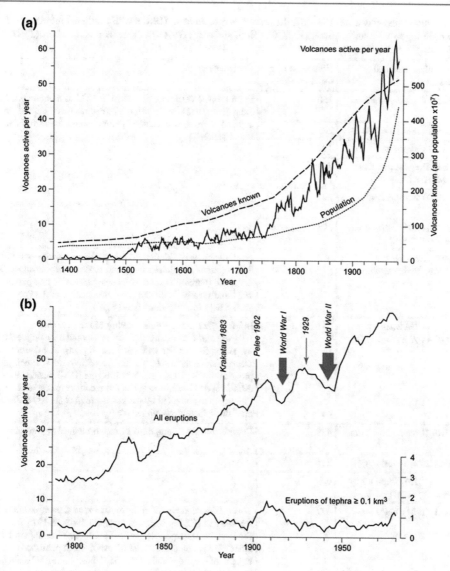

Fig. 1 The reporting of global volcanic activity up to 1980 CE. *Top* known and active volcanoes, and world population 1400–1980. Volcanoes active per year are plotted as a 10 year running mean. 'Volcanoes known' represents the total number to have had historic eruptions. *Bottom* the number of eruptions (*right* hand scale) producing ≥0.1 km³ of tephra, plotted as a 10-year running mean. Eruptions ≥0.1 km³ of tephra equates to a *VEI* (i.e. *Volcanic Explosivity Index*) of 4—Newhall and Self (1982). From Siebert et al. (2010, Fig. 9, p. 32 and Fig. 10, p. 33) and reproduced by permission of the University of California Press

(Footnote 5 continued)

developed in other European and North American countries in the eighteenth century. Its leading doctrines were *inter-alia*: a commitment to reason; the notion of progress, based on education, science and the arts; the rejection of the authority of tradition including religious tradition and a stress on nature which can be studied using empirical methods (Inwood 1995).

in communication; first daily newspapers and from the mid-19th century the electric telegraph. Growth continued up to 1980 as these technologies together with others which came into use—including telex and rapid reliable air-transport—were further developed. By 2010 a flattening in the trend line of the number of volcanoes recorded as active per year, implies

Table 1 Eruption catalogues from 1825 to 1981 (based on Simkin et al. (1981), with additional information from the references cited in the table and: Simkin (1993, 1994), Sigurdsson (1999, 2000), Simkin and Siebert (2000), Simkin et al. (2001 and the references cited in the table)

Catalogue (author, date and abbreviated title)[a]	Volcanoes with dated eruptions	Comments
Scrope (1825) *Considerations on volcanoes*	150	From scope's catalogue published in 1825 to that compiled by Schneider (1911), most of the information was derived from archival sources, with relatively little being added from field investigations
von Humboldt (1858) *Kosmas...*	225	
Scrope (1862) *Volcanoes...*	191	
Fuchs (1865) Die *vulkanischen ersheinungun der Erde*	270	
Mercalli (1907) *Vulcani attivi della terra*	231	
Schneider (1911) *Die vulkanischen erscheinungen der Erde*	298	
Sapper (1917) *Katalog der geschichtlichen Vulkanausbruche*	430	From the close of the first decade of the twentieth century, field investigations recognised more and more active volcanoes. The dedicated journal, *Bulletin Volcanologique*, was first published in 1924 and greatly facilitated the dissemination of data collected during field investigations (see below)
Catalogue of active volcanoes of the world (CAVW 1951–1975)	441	In May 1922 at the Rome meeting of the *International Union of Geodesy and Geophysics* (IUGG)[b], proposed a catalogue authored by geologists familiar with particular regions. The *Great Depression* of the 1930s and the Second World War delayed this project until the 1950s. Later in 1960 the IUGG decided to publish a *Bulletin of Volcanic Eruptions* to record eruptions each year. Reports were compiled by the *Volcanological Society of Japan* and publishing in *Bulletin Volcanologique*
Lamb (1970) *Volcanic dust in the atmosphere...*	435	Compiled to study eruptions of meteorological significance
Macdonald (1972) *Volcanoes*	516	
Gushchenko (1979) *Eruptions of the volcanoes of the world...*	609	
Simkin et al. (1981) *Volcanoes of the world*	627	This catalogue grew out of the *Smithsonian Center for Short-Lived Phenomena* (CSLP) which as set up in 1968. In 1975 the Smithsonian incorporated the CSLP into the Washington-based *Scientific Event Alert Network* (SEAN), which published a monthly bulletin of eruptions and summarised these in the journal *Geotimes*. The full SEAN was available through subscription

[a]There is also the catalogue of von Hoff and Berghaus (1840/41). This is not included because the coverage of not global, but is strongly biased towards the 'old world'
[b]Later this became the *International Association of Volcanology and Chemistry of the Earth's Interior* (IAVCEI)

that virtually all volcanic eruptions on land have been identified (Seibert et al. 2010).

With respect to the scientific community, communication was greatly facilitated by the publication of summary catalogues. In the 17th century, the pioneering work of Bernhardus Varenius (1650)[6] was noteworthy, but from the

beginning of the 19th century the number of catalogues burgeoned (Table 1), being formalized in the mid-20th century with the publication of the *Catalog of Active Volcanoes of the World* (1951–1975). As Table 1 shows, in the early

---

[6]A German geographer, who was also known as Bernhardus Varen (1622–1650), worked mostly in

(Footnote 6 continued)
Amsterdam. He was aware of the discoveries of many contemporary Dutch navigators. He listed 21 volcanoes with dated eruptions (Simkin et al. 1981: 1).

1980s communication was still dominated by paper-based media with circulation being restricted to academic institutions and government research organisations who were both willing and had the financial means to subscribe to learned journals and reports.

Several authors the majority of whom are or have been associated with the *Smithsonian Institution* (Simkin 1993, 1994; Simkin and Siebert 2000; Simkin et al. 2001), have commented further on the historic trends plotted in Fig. 1. They note, *inter alia*:

a. That the increase in the number of active volcanoes[7] over time is related to growth in the world's population and better communications, and not to any increase in the frequency of volcanic activity.

b. When the period from ca. 1800 is examined in more detail (Fig. 1) peaks and troughs become apparent. Peaks can be seen to follow newsworthy eruptions such as the three large magnitude events of 1902 (i.e. Mont Peleé, Martinique; La Soufriére, St. Vincent and Santa Maria, Guatemala) and Krakatau, Indonesia in 1883. Simkin and Siebert (1994) argue that such peaks are due to increased post-eruption reporting when there was a heightened awareness of activity. Not only did these eruptions produce voluminous newspaper reports in many countries, the London *Times* for example publishing 14 reports on Krakatau alone some of them very detailed,[8] but also major scientific studies (e.g. Lacroix 1904; Verbeek 1884). Troughs, in contrast, coincide with the disruption of global science brought about by the First and Second World Wars, and the Great Depression of the 1930s which followed the Wall Street stock market crash of 1929.

c. One measure of the incomplete character of the eruption archive before the last few centuries, is that even the record of large eruptions decays rapidly from an average of more than 5 per decade in recent centuries to 0.7 per decade before the 15th century (Simkin and Siebert 1994). Other studies of under-reporting have been published by Deligne et al. (2010) and Furlan (2010).

d. Fatal eruptions are far more likely to be preserved in the historical record than non-fatal ones (Simkin et al. 2001). For eruptions occurring since 1500 CE, detailed information on human impacts is generally more sparse than data on eruptive processes and their effects, although this inbalance was beginning to be redressed albeit in an inchoate fashion in the years leading up to 1980 (e.g. Furneaux 1965; Nolan 1979; Simkin and Fiske 1983; Blong 1984).

e. In many part of the world eruption records are short—often less than one hundred years—whereas repose periods of many volcanoes are much longer (Tazieff 1983). This means that even in the decades immediately before 1980, several large eruptions occurred from volcanoes which were thought by local populations to have been inactive. Examples included: Mount Lamington, in Papua New Guinea which killed ca. 5000 people in 1951; Mount Arenal (Costa Rica) in 1968 and Heimaey (Iceland) in 1973 (Chester 2005). This raises the important question of how scientists can effectively communicate risk to communities who do not consider a given volcano to be active. Long repose is often associated, moreover, with the silicic volcanism of subduction zones, regions in which there are dense clusters of population.

Discussion of the historic increase of information about volcanoes and their eruptions has been focused so far at the global scale, but in the sections that follow progress will be reviewed with reference to the two case studies: the Azores in Portugal and Mount Etna, Sicily (southern

---

[7]An active volcano is frequently defined as one showing historic activity. This definition introduces a lack of consistency, because the span of historic records varies across the world from thousands to less than 200 years. We follow *Volcanoes of the World* in including volcanoes with dated eruptions that have occurred during the Holocene i.e. the last 10,000 years (Simkin and Siebert 1994, p. 12).

[8]These are available electronically from *The Times Digital Archive*—last accessed 9/5/14.

Italy). Although the general trends already identified are present and, while both regions are culturally southern European, here the similarity ends and their histories are contrasting in terms of both the recording and communication of data on eruptions. The Azores Islands for most of their history have been characterised by isolation; both geographic and intellectual. Located in the middle of the Atlantic Ocean, the islands were only settled in the 15th century and for much of their subsequent history have never held a central or even a secondary place within the scientific mainstream. The Azores exemplify the difficulties of using indigenous data sources to compile, not just assessments of impact but, also at a more basic level, a complete list of events with accurate dates. Many of the accounts which are available were collected by a limited number of literate observers, who were often priests, and require much careful interpretation to extract usable information on eruption processes and their effects.

In contrast, information on Mount Etna and its eruptions stretches back to prehistoric times and has become progressively better known over more than 2000 years of written history. In Greek and Roman times it was at the centre of attempts by classical authors to make sense of the natural world with interpretation often involving aspects of mythology (Chester et al. 2000), in the *Renaissance* major studies of Etna and its eruptions were published, the *Enlightenment* and the nineteenth century saw the volcano being studied by both indigenous and distinguished foreign scientists of the calibre of Sir William Hamilton, Charles Lyell and Sartorius von Waltershausen and this continued through to he 1980s. Etna and Vesuvius were part of the European Grand Tour, Sicily was strategically important during the Napoleonic Wars, later in the nineteenth century visits to the volcano were eased by the spread of railways and stream ships and, following the introduction of the electric telegraph, educated appetites in the USA and Western Europe could be satisfied by detailed reports in newspapers of record which often appearing only hours after the events being described (Chester et al. 1985, 2012). Yet even on Etna significant gaps remained in the record—not least on human impacts—even for events that took place between 1900 and 1980.

---

3 The Azores: Communicating Eruption Information from an Isolated Region

The settlement of the Azores was part of the voyages of discovery undertaken by Portuguese navigators from the 15th century and, although there is debate over whether or not there were earlier visitors to the islands (Ashe 1813; Admiralty 1945; De Meneses 2012), it is generally accepted that in the autumn of 1431 an expedition led by Gonçalo Velho Cabral established a settlement on Santa Maria and that by 1457 all nine islands were known (Fig. 2). Indeed the arrival of the first settlers on São Miguel ca. 1439–1443, probably coincided with the dome-forming final stage of a sub-plinian phreatomagmatic eruption of Furnas volcano (Queiroz et al. 1995; Guest et al. 1999). Between settlement of the archipelago and 1980 a further 26 eruptions took place (Fig. 2 and Table 2), with one subsequent submarine occurring between 1998 and 2001 (Gaspar et al. 2015).

The communication of information about eruptions was not marked by steady progress and even in the second half of the 20th century there were still major gaps in the record. For instance the Azores volume of the *Catalogue of Active Volcanoes of the World* was published in the 1960s (Neumann van Padang et al. 1967), but when compared with what is known today (Gaspar et al. 2015) shows a lack of detail about the volcanological character of historic events, human impact is ignored and some eruptions—most notably that of ca. 1439–1443—are absent. Another standard reference from the period (Weston 1964), also ignores the ca. 1439–1443 event, lists an eruption of Sete Cidades (São Miguel Island) in 1439 that did not occur and reports a lava flowing in the direction of Rabo de Peixe (São Miguel Island) in 1652 which is incorrect. Maps and memoirs produced by the

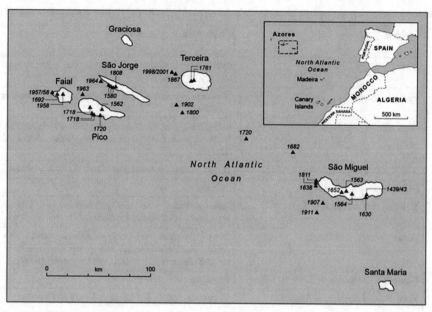

Fig. 2 The Azores archipelago: general location and the position and dates of historic eruptions (data from Gaspar et al. 2015)

Serviços Geológicos de Portugal[9] (e.g. Zbys-zewski et al. 1958, 1959; Zbyszewski 1961) were published as part of the 1:50,000 scale mapping of Portugal and the Atlantic Islands and are likewise partial in their treatment of historical eruptions and their impacts.

Between 1950 and 1980 a few studies of past eruptions were published in international scientific journals, but these were either focused on: the most recent 1957/8 phreatomagmatic eruption at Capelinhos on Faial Island, but again strongly themed on volcanological aspects rather than human impacts (e.g. Machado et al. 1962); or concerned large plinian events that took place long before the islands were settled and are, therefore, outside the scope of this review (e.g. Walker and Croasdale 1971; Booth et al. 1978). Establishing a detailed record such as that provided by Gaspar et al. (2015), requires detailed analysis of all

accessible archival sources and its cross-checking with field evidence so that historical accounts may be either verified or eliminated.

Although the Azores lie some 1360 km to the west of the coast of Portugal, lack of transmission of information cannot be blamed solely on physical isolation. Admittedly before the era of mass air transport visiting the islands was difficult and time consuming, but in comparison with other mid-oceanic islands seaborne communication was well developed, with the ports of Ponta Delgada (São Miguel), Horta (Faial) and Angra do Heroísmo (Terceira) in particular being important staging posts in Atlantic trade. In the era before steam power the agricultural economy of the archipelago was highly specialized and was concerned not only with subsistence, but also with provisioning the many ships that visited the islands (Admiralty 1945; Callender and Henshall 1968). The volcanoes of the Azores tend to give rise to short-lived subaerial activity with activity lasting no more than few days or weeks, and in the 17th and 19th centuries there is little evidence that the eruptions of: Pico 1718 and 1720; Terceira 1761 and São Jorge 1808,

---

[9]Later part of the *Instituto Geológico Mineiro* (Institute for Geology and Mining), from 2004 to 2007 the *Instituto da Engenharia Tecnologia e Inovação* (Institute of Engineering Technology and Innovation) and, thereafter, the *Laboratório Nacional de Energie e Geologia* (National Laboratory of Geology and Energy).

Table 2 Major eruptions on the islands of the Azores: 1439/1443 to 1980 CE (Based on Gaspar et al. 2015)

Eruption date	Location	Characteristics of eruption
1439/1443	Furnas volcano, S. Miguel	Sub-plinian, phreatomagmatic associated with trachytic pumice and a lava dome
1562–64	Fissural System, Pico	Hawaiian and strombolian activity, producing lava flows and pyroclasts
1563	Fogo volcano, S. Miguel	Sub-plinian, phreatomagmatic eruption. Products include trachytic pumice lapilli/ash and surges. Two deaths due to gas inhalation
1563	NW Flank of Fogo volcano, S. Miguel	Hawaiian activity, with basaltic lava flows, spatter and pyroclasts
1564	Fogo volcano, S. Miguel	Phreatic
1580	Manadas Fissural System, S. Jorge	Hawaiian/strombolian activity, producing lava flows and a pyroclastic flow. About 15 deaths
1630	Furnas volcano, S. Miguel	Sub-plinian/phreatomagmatic eruption, associated with lava domes, pumice, ash/lapilli, lava domes and pyroclastic density currents. At least 195 deaths from surges and collapsed buildings
1638	Candelária submarine volcano, S. Miguel	Surtseyan
1652	Picos Fissural System, S. Miguel	Vulcanian—ashes, blocks, lava domes and flows
1672–1673	Capelo Fissural System, Faial	Basaltic lava flows and pyroclasts. A least 3 deaths and around 1200 persons displaced
1682	Crista João Valadão Submarine Volcanic System, off the west coast of S. Miguel	Few details
1718	Pico Volcano	Hawaiian and strombolian eruption, producing submarine and sub-aerial pyroclasts and lava flows. 2 deaths
1720	Fissural Volcanic System, Pico	Basaltic pyroclastic and lava flows
1720	D. João de Castro Submarine Volcano, between S. Miguel and the Central Islands	Surtseyan
1761	Santa Bárbara Volcano, Terceira	Vulcanian (?) Trachytic ashes, blocks and lava domes
1761	Terceira Fissural Volcanic System	Hawaiian and strombolian activity, producing basaltic bombs, lapilli, ash and lava flows
1800	Submarine SSW of Terceira	?
1808	Manadas Fissural Volcanic System, S. Jorge	Hawaiian/strombolian and phreatomagmatic. Basaltic lapilli/ash, lava and pyroclastic flows. More than 30 deaths
1811	Sabrina Submarine Volcanic System, S. Miguel	Basaltic submarine eruption, ash and blocks
1867	Crista da Serreta Submarine Volcanic System, west of Terceira	Basaltic submarine eruption
1902	Submarine, SW of Terceira Island	?
1907	Submarine, SW of S. Miguel Island	?
1911	Submarine, SW of S. Miguel	?
1957/8	Capelo Fissural Volcanic System, Faial	Surtseyan, hawaiian and strombolian activity. Submarine and sub-aerial basaltic pyroclasts, surges and lava flows
1958	Caldera Volcano, Faial	Phreatic and phreatomagmatic. Ash produced
1963	Cachorro Submarine Volcano, offshore, north of Pico	Submarine activity
1964	Velas Submarine Volcano, S. Jorge	Submarine activity
1998/2001	Crista da Serreta Submarine Volcanic System, west of Terceira	Submarine lava balloons, submarine ashes and volcanic gases

attracted interest from outside the islands. Volcanoes on the Azores did not show phases of persistent activity as was the case with volcanoes like Etna or Vesuvius and this, together with isolation, may explain the lack of foreign visitors. It is interesting to note that a detailed description of a submarine eruption in June 1811, which occurred off the shore of São Miguel, was undertaken by the captain of a Royal Navy frigate and communicated to the *Royal Society of London* (Tillard 1812). This report was fortuitous because the ship was in the area by pure chance and the eruption was named *Sabrina* after the frigate. In the 1890s the islands (and especially Faial) became major nodes on the worldwide telegraph network, with personnel from Germany, the UK and USA being billeted on the islands particularly in Horta the principal settlement of Faial. Despite ships and later telegrams arriving in Europe and North America on a daily basis searches of newspapers of record—such as the London *Times* (1785–2008) and *New York Times* (1851–2007) and daily publications of a more popularist character contained no eruption reports until well into the 20th century, although it must be admitted that between 1808 and 1957/58 there were no eruptions on land, only a number of submarine eruptions. The 1957/58 eruption of Capelinhos, Faial (Coutinho et al. 2010) was the first to be covered in any detail.[10]

Isolation may take several forms and in the Azores was expressed intellectually, scientifically and politically. In Portugal both the institutional and professional development of science and more particularly the geological sciences, lagged behind that in other European countries

and greatly inhibited the contemporary collection and transmission of eruption data. In a devoutly Catholic country it might be thought that the development of geology would have been inhibited by religious considerations, especially following the publication of Lyell's *Principles of Geology* and in 1830 Darwin's *Origin of Species* in 1859, but as Carneiro et al. (2013) have argued this was not the case. Universities had been outside religious jurisdiction since 1772 and there was a long tradition of independence between science and religion among educated elites across Portugal. In the second decade of the twentieth century, 'despite the close relationship between the newly established regime[11] and the Roman Catholic Church, António Salazar (1889–1970) maintained the separation of Church and State; in addition, the State held no official position regarding scientific matters' (Carneiro et al. 2013, p. 333).

As Mota and Carneiro (2013, p. 24) have noted, 'despite the creation of the national geological survey being coeval to other European countries, the teaching of geology and geological research became effective only as late as the mid-twentieth century and the Portuguese Geological Society was only founded in 1940.' In addition until 1911 there was only one university in Portugal at Coimbra, after this Oporto and Lisbon were added and the University of the Azores only dates from two years after the Portuguese Revolution of 1974. Until well into the twentieth century, people—almost invariably men—who wanted careers in science only had three choices: 'medicine; the clergy; and, in the nineteenth century, military engineering' (Carneiro et al. 2013, p. 332).[12]

---

[10]The *London Times, Nineteenth Century British Library Newspapers* and *Nineteenth Century US Newspapers* are available from Gale *GENGAGE*—see Footnote 4. The New York Times is available from ProQuest http://www.proquest.com/. Accessed 16 May 2014. It is interesting that, in contrast to eruptions, Azorean earthquakes were extensively reported in the international press. Earthquakes frequently caused death and destruction to places in the Azores visited by European and North American seafarers and, in the nineteenth century alone, earthquakes of intensity VIII or greater affected Terceira in 1800, 1801 and 1841 and São Miguel in 1810, 1811 and 1852.

[11]This is the authoritarian *Estado Novo* regime (1928–1974) which was led by António Salazar until 1968.

[12]It is notable that neither of the two leading figures in Azorean volcanology in the first six decades of the twentieth century was a geologist by training. José Agostinho (1888–1978), who published on a variety of volcanological topics (e.g. Agostinho 1932) was an army officer and meteorologist. Frederico Machado (1918–2000) was a civil engineer, who made notable contributions to both recording and managing the 1957/8 eruption and earthquake on Faial Island (Machado et al. 1962).

A lack of trained personnel and of focused science policy can be clearly seen as late as the 1950s. By the time of the Capelinhos eruption (Faial—1957/58), the current authors have noted that leading roles were played by engineers and other professionals, rather than geologists and that the regime was open to and encouraged scientific enquiry, which was financially supported and involved both Portuguese and foreign scholars. There were no 'in-depth studies of the impacts of and government responses to the emergency. Evaluative and potentially critical studies of policy were not welcomed by the government' (Coutinho et al. 2010, p. 266).

These features of Portuguese intellectual history meant that in the Azores the intellectual elite, who included State officials, was diligent in recording eruptions in locally published books, monographs, academic journals and freely discussed scientific ideas—including geological advances—in the local press. They did not, however, disseminate information more widely. Many early eruptions and their effects were recorded by the priest historian, Gaspar Frutuoso (ca. 1522–1591—Fig. 3), in Book IV (São Miguel) and Book VI (Terceira, Faial, Pico, Flores, Graciosa and São Jorge) of his multi-volume work: Saudades da Terra (English translation—A Nostalgic Longing for the Land—Frutuoso 2005). This work remained in manuscript form until the late nineteenth century, was published from 1873 onwards (Luz 1996) and did not become widely available outside Portugal until much later. For instance, in the United States the Library of Congress Catalogue only lists Book IV (1876) before the complete work appears in 1978, while in the UK the position is only slightly better with only three locations being listed for the 1873 edition.[13]

Later in the nineteenth century the historian and politician Ernesto do Canto[14] (1831–1900—

Fig. 3), collected many accounts of historical eruptions and published these in a journal, Arquivo dos Açores, which he both launched and printed. This publication had limited dissemination outside the Azores and in fact has only been widely accessible since it has become available in digital form from the University of the Azores (http://www.sdoc.uac.pt/pt/publicacoes).

The role of the press in reporting science in general and particularly natural events has recently been reviewed by Simões et al. (2012). Although rates of illiteracy in Portugal were ca. 79% in 1900, in 1894 there were 23 newspapers and periodicals published in Ponta Delgada (São Miguel), the capital of the Azores, which were read by the elite and made available for public readings in taverns, cafes and shops. In reviewing one newspaper, the progressive Diário dos Açores—founded in 1870, Simões et al. (2012, p. 314) show how science was extensively reported, with the motivation being 'an attempt by the scientific and political communities to gain the support of the general public'. Although hygiene and public health were the dominant themes, there was also reporting of geological topics, with the Messina earthquake of 1908 being afforded extensive coverage. Indeed following the 1909 Benavente earthquake near to Lisbon (Degg and Doornkamp 1994), there was lobbying for a seismic network to be established on the islands and, indeed, two seismological stations had been set up in the Azores in 1902; one at Ponta Delgada and the other at Horta. Although the Diário dos Açores contains coverage of volcanology, progressive lobbying through the paper was less successful than was the case with earthquakes and did not lead to any significant new research being undertaken on contemporary activity, or the impacts of historical eruptions becoming more widely known (see Simões et al. 2012, pp. 318–325). It is perhaps not coincidental that eruptions in the second half of the 19th and the early 20th centuries were small-scale submarine events and the loss of 200 houses and landslides in Terceira in 1866 was caused by volcano-related seismic activity, rather than by volcanic activity per se (Gaspar et al. 2015).

---

[13]Oxford and Cambridge University libraries and the library of Kings College London.

[14]Also known as Ernesto do Canto e Castro.

Fig. 3 Two pioneers of Azorean volcanology. **a** Statue (*left*) of Gaspar Frutuosa at Ribeira Grande (São Miguel) his place of birth ca. 1522. He was the son of a local landowner and trained for the priesthood at the University of Salamanca (Rodrigues 1991). **b** Photograph of Ernesto do Canto in 1884 (*right*). Canto was a notable Azorean intellectual and was born into one of the most influential families on the islands. As well as founding the *Arquivo dos Açores*, Canto was member of the Portuguese *Academy of Sciences*, the *Lisbon Geographical Society* and made valuable donations to the public library in Ponta Delgada. For further details of his life and work see Dias (1931). Photographs Nicolau Wallenstein

Fig. 3 (continued) **(b)**

4   Mount Etna, Sicily: The Accumulation of Knowledge Over 2000 Years

The volcanoes of Italy and especially Etna and Vesuvius were amongst the first to be known to literate European observers. Etna (Fig. 4) a huge volcano in comparison with Vesuvius, covering an area of 1178 km^2 and standing 3328 m in height (Branca et al. 2011), was virtually continually in eruption during classical times, with activity being typically effusive and strombolian in character, though there were more explosive events such as the plinian eruption of 122 BCE which produced considerable tephra fall and caused severe damage in Catania (Coltelli et al. 1998). Although most activity occurred away from inhabited areas, at

Fig. 4 Mount Etna: location map

times larger flank eruptions caused widespread destruction. Etna and its eruptions feature in the literature of the classical age through reflections of a predominantly mythological and legendary character and, indeed, a list of authors includes some of the era's greatest writers (see Chester et al. 2000; Duncan et al. 2005; Johnston 2005; Smolenaars 2005). Records of eruptions can also be extracted from Greek and Roman sources, particularly the works of Pindar, Diodorus Siculus, Thucydides, Virgil, Pliny the Elder, Suetonius, and Lucretius, but as Branca and Del Carlo (2004, pp. 2–3) have noted: only major eruptions were recorded; information has often to be 'translated from poetic language into hard fact'; there is an issue in linking eruption reports to deposits found in the field and, although some eruptions are associated with important historical events—most notably the death of Julius Caesar in 44 BCE—before the 17th century the dating of the lava flows is problematic. The publication of the new 1:50,000 scale map of Etna in 2011 acted as a stimulus for ancient texts to be examined afresh. New dates for eruptions occurring in the classical and medieval eras have been proposed, that are supported by both archaeomagnetic and radiometric dating of volcanic products (Branca et al. 2011; Tanguy et al. 2012) and which allows a comprehensive eruption history to be defined for the last 2500 years.

What Branca and Del Carlo (2004: 2) have termed the *ancient period* continued until the 11th century and the end of the era of Islamic[15] domination of Sicily. Thereafter until the close of the 12th century, there was Norman control of Sicily that in the 13th century gave way to Spanish domination which broadly coincides with the *Renaissance*. Two decades ago it was thought that the record of flank eruptions was fairly robust from the 14th century (Romano and Sturiale 1982), though there was questioning of the status of some 16th and early 17th century eruptions by Chester et al. (1985, Table 3.3), but on the basis of information collected during the construction of the 2011 geological map this conclusion can no longer be sustained. As Branca and Del Carlo (2004, p. 5), note 'we can only be certain about eruptions that occurred below 1800 m …on the southeastern flank, because information on volcanic activity was only available in the cultivated and inhabited zones on the outskirts of Catania' (Fig. 4). In spite of this lack of comprehensiveness, knowledge of Etna's eruptions improved greatly especially from the start of the 16th century and mirrored the pattern for the *Renaissance* in general, with the more widespread dissemination of printed volumes written by distinguished polymaths such as: Fazzello (1558),[16] Filoteo (1590),[17] and Carrera (1636).[18] From 1600, a growing interest in the natural sciences meant that almost all flank eruptions—though not all summit activity—was recorded. In fact geological and archaeomagnetic data have revealed that some flank eruptions in the 17th century, and in the late 18th and early 19th centuries some events on the upper northwest flank, were not recorded in contemporary sources (Tanguy et al. 2012). As noted in the introduction large eruptions stimulate research, reporting and communication and the

1669 eruption was no exception. Etna's most voluminous historical event, the 1669 eruption, devastated the south-eastern flank of the volcano, destroyed the western part of Catania, the large village of Nicolosi and many smaller settlements (Fig. 4). It was reported by Sicilian and mainland Italian authors (Monaco 1669; Squillaci 1669; Tedeschi Paternò 1669; Borelli 1670), together with foreign writers (e.g. Anon 1669; Winchilsea 1669).[19] There were reports at the time in international scientific publications (e.g. *Gazette de France* and the *Philosophical Transactions of the Royal Society of London*). These sources provide descriptive material on aspects of the eruption and some of its immediate impacts. Recently, Branca et al. (2015) have undertaken a detailed study of this material which shows that the impact of the 1669 eruption was complicated by the catastrophic earthquake that devastated eastern Sicily in 1693.

Following 1669 and reflecting the strong empirical interest in nature which characterized the *Enlightenment* (see Footnote 4), the manner of describing eruptions changed and became more detailed, complete and for the first time summit activity began to be recorded, though it was not until the establishment of the Etnean Volcanology Institute by Gaetano Ponte in 1926, that this was undertaken in a systematic manner (Branca and Del Carlo 2004, p. 3). Within this tradition two major studies were published by Recupero (1815) and Ferrara (1818). They included observations made during the late 18th century and earlier and their publications were widely distributed across western Europe, appearing in library catalogues in France and Great Britain. An increasing number of foreign visitors were arriving in Sicily. At the start these were young aristocrats and other equally wealthy tourists who were experiencing the *Grand Tour* of classical sites and historic cities. Some took an active interest in and published widely on Etna (Vaccari 2008), most notably the Scot Brydone (1773) and later

---

[15]It is often termed the Arabic period, but the Islamic settlers also included Berbers and Spanish Muslims.

[16]Tommaso Fazzello (or Fazello) 1498–1570—Priest, historian and orator and known as the father of Sicilian history (Anon 2014a).

[17]Antonio Filoteo (unknown—1573)—Jurist and historian. Also known as Antonio Filoteo Degli Omadei. (Anon 2014b).

[18]Pietro Carrera (1573–1647)—Chess player, priest, historian and Italian author (Anon 2014c).

[19]The English Earl of Winchilsea (1669) was sailing off the coast of Etna at the time of the eruption. An *English Merchants' Report* to the *Royal Society of London* (Anon 1669) was sent by business men resident in Sicily.

during the brief British occupation of Sicily (1806–1815) an officer visiting the volcano sent home a report of the 1809 flank eruption which was subsequently published in a Scottish newspaper (Anon 1809). By the time of the 1865 flank eruption visits by foreigners were well established, were facilitated by the advent of railways and are described as 'pouring into the region'(Anon 1865). In fact this event was the first eruption to be photographed by the Frenchman, Paul M Berthier, who accompanied the volcanologist Orazio Silvestri in a visit to the vent (Abate et al. 2013). By the early years of the 20th century, 'dark tourism' was causing problems for the authorities in managing the visitor influx (Anon 1923).[20] Tourism stimulated a thirst for news first in Europe and later in the USA, and during the course of the 19th century there was marked increase in eruption reporting in both newspapers of record and in the more popular press (Chester et al. 2012).

During the 19th century rational investigation of Etna made a significant contribution to understanding within the earth sciences. This is an apparent paradox, because in a devoutly Catholic region in which every eruption was and is still associated with elaborate rituals supposedly to assuage divine wrath (Chester et al. 2008), *Enlightenment* science prospered so well. Giuseppe Recupero (1720–1778), Francesco Ferrara (1767–1850) and later Giuseppe Alessi (1774–1837) were priests yet, with one exception, encountered no opposition from the ecclesiastical hierarchy. The exception was Canon Recupero (Fig. 5), famous for his published work and for introducing the Naples-based diplomat and volcanologist Sir William Hamilton to Etna. Many years before the publication of Lyell's *Principles of Geology* (1830), Recupero speculated about some lavas being much older than the generally accepted biblical chronology for the age of the earth. Brydone (1773, p. 132) writes:

Recupero tells me he is exceedingly embarrassed, by these discoveries…. Moses[21] hangs like a dead weight upon him, and blunts his zeal for inquiry; for that really he has not the conscience to make the mountain so young, as that prophet makes the world. What do you think of these sentiments from a Roman Catholic divine? The bishop, who is famously orthodox …. has already warned him to be upon his guard; and not to pretend to be a better natural historian than Moses; not to presume to urge anything that may in the smallest degree be deemed contradictory to his sacred authority.

Clearly observation and study of the natural world—God's created order—was acceptable, but speculation about its origin and when it came into existence was not. Following the publication of Brydone's book Recupero was censured and lost his canonical preferment (Rodwell 1878).

Giuseppe Alessi was also a priest/scientist and published extensively on Etna and its eruptions (Alessi 1829–1835), but in the first half of the 19th century the baton passed to secular scholars, in particular to the brothers Gemmellaro (Mario 1773–1838, Carlo 1787–1866 and Giuseppe 1788–1876). Between them the brothers observed and described Etna's activity and speculated about its volcanic phenomena (Guest et al. 2003, p. 180), Giuseppe accompanying Lyell on his visits and Carlo producing a major academic study (Gemmellaro 1858). Lyell's *Principles of Geology*, his theory of uniformitarianism and the methodology espoused by him and other leading contemporary geologists of the time, found ready acceptance in Italy (Vaccari 1998) and by the close of the century, not only was the eruptive history of Etna known in considerable detail from ca. 1600, but flows had been mapped with accuracy most notably by the German geologist Wolfgang Sartorius von Waltershausen. The 1:50,000 map produced by Waltershausen, was published as the *Atlas des Aetna* (von Waltershausen 1844–1859) and was the world's first geological map of a large active volcano. A scholarly society, the *Accademia Gioenia*,[22]

---

[20]Dark tourism involves visits to sites associated with death, war and other tragedies. Not all the tourism was 'dark', however, and major studies were published by some visitors not least Rodwell (1878).

[21]It was generally accepted at the time that Moses was the author of the Pentateuch, the first five books of the Old Testament or Hebrew Bible, beginning with Genesis.

[22]The academy was named after the Sicilian scholar, Giuseppe Gioeni d' Angio (1743–1822).

(a)

JOSEPH CANONICUS RECUPERO

Etnæ prodigium flammæ simul igne et unde
Scripsit, quid Valles, quid juga Montis habent

Fig. 5 Two pioneers of Etnean volcanology. **a** Canon Giuseppe Recupero (1720–1778). **b** Alfred Rittmann (1893–1980). *Photographs* Stefano Branca and the *Istituto Nazionale di Geofisica e Vulcanologia* (INGV), Catania

was established in Catania in 1824, many leading geologists were members and its journal quickly found a place in the libraries of learned societies and major universities in Italy and across Europe.

The years from ca. 1850 to 1928, the year in which the well studied eruption destroyed the village of Mascali, were ones of consolidation. Major flank eruptions occurred in: 1852/3; 1865; 1879; 1883; 1886; 1892; 1910; 1911 and 1923, and there were progressively more detailed accounts in newspapers of record particularly in the United States and United Kingdom. Sicily was first linked to the Italian mainland by the electric telegraph in the 1850s, with the first reliable trans-Atlantic links dating from a few years later (Chester et al. 2012, pp. 75–76). By the twentieth century, newsreels

Fig. 5 (continued)

were also depicting Etnean eruptions for an international as well as a national audience. Detailed studies of eruptions were published by a number of scientists among whom were Orazio Silvestri, Annibale Riccò, Ottorino De Fiore and Gaetano Ponte. In 1926 Gaetano Ponte founded the *Etnean Volcanological Institute* within the University of Catania, and under his leadership a complete inventory of eruptive activity was maintained throughout the economically difficult 1930s and during the Second World War (see Branca and Del Carlo 2004, pp. 3–4), although until the 1950s international contacts were much reduced and there was little communication of information outside Italy other than through the pages of *Atti Accademia Gioenia di Scienze Naturali*.

The appointment in 1958 of the eminent Swiss volcanologist, Alfred Rittmann (1893–1980—Fig. 5), to the *Etnean Volcanological Institute* re-launched the international profile of volcanic studies. In 1967 with the financial support of the *Consiglio Nazionale delle Richerche* (CNR—National Research Council) and with the patronage of UNESCO, the institute became the *International Institute of Volcanology* and Rittmann encouraged collaboration between Italian, French and British scientists, that culminated in both: the production of a 1:50,000 scale geological map and memoir (Romano 1982) and the publication of a research volume (Chester et al. 1985), which grew out of this collaboration.

5 Concluding Remarks

Italy and in particularly Etna has unrivalled historical records of volcanic activity extending back well over 2500 years to classical times. There are large gaps in the record, however, with accounts dating from Greeks and Romans times being typically mythological and/or speculative and containing relatively little usable observational detail. For both Etna and the Azores it was priests who from the 16th century provided written descriptive accounts of eruptions and to a more limited extent their impact. They showed an ability not only to describe, but also to utilize their observations to develop an understanding of volcanological processes, which in the case of Canon Recupero brought him into conflict with his ecclesiastical superiors. Sicily is situated at the cross-roads of Mediterranean trade routes and the 1669 eruption of Etna attracted international interest (Azzaro and Castelli 2013). Observers showed a scientific approach in their description, observing and measuring lava flows and making comments on their impacts on local communities. Results were published first in letters and reports of limited circulation, later in research volumes and, beginning with the *English Merchants' Report* to the *Royal Society of London* in 1669, in international scientific journals (Anon 1669).

With the exception of the submarine eruption off the coast of São Miguel which was fortuitously observed by Captain Tillard, the Azores were more remote and eruptions were neither witnessed nor described by international scholars to the same extent. This remoteness, combined with the distinctive intellectual history of Portugal and the Azores, meant that, although there were major eruptions in the 16–18th centuries, documentation of activity was largely restricted to the islands, sometimes not published until much later and not communicated to the wider world to anything like the same extent as occurred on Etna. Indeed until the eruption of Capelinhos in 1957/8, there is virtually no reference to eruptions in the Azores in volcanological texts, despite better communications —steam ships and the electric telegraph—being important elements in the islands' economy from the mid-19th century.

At the global scale the ways in which information on eruptions has been communicated between 1980 and the middle of the 2010s is the subject of another chapter in this volume, but for reasons of completeness the situation in the

Azores and Etna can be brought up-to-date. For the Azores the designation of Furnas as a *Laboratory Volcano* acted as a powerful stimulus and brought together a number of scholars from a variety of European countries and from a variety of disciplines, who researched topics which ranging from the 'traditional' (e.g. volcanic geology), to the innovative (e.g. hydrogeology, health hazards and human vulnerability). These were subsequently published in a special 'issue' of *Journal of Volcanology and Geothermal Research* (Duncan et al. 1999). Today volcanology in the Azores is studied both by the *Departamento de Geociências, Universidade dos Açores* and the university hosted: *Centro de Vulcanologia e Avaliação de Riscos Geológicos* (*CVARG* Centre of Volcanology and Evaluation of Geological Risk); and *CIVISA* (*Centro de Informação de Vigilância Sismovulcânica dos Açores* or Centre for Information and Seismovolcanic Surveillance of the Azores) (Gaspar et al. 2011). These bodies have continued to communicate information both, locally through liaison with the Civil Defence authorities, and internationally by means of publication in peer-reviewed academic volumes/journals and having a strong on-line presence (http://www.cvarg.azores.gov.pt/Paginas/home-cvarg.aspx).

The situation in Sicily is broadly similar to that in the Azores, with attention being focused, *inter alia*, on: observing and communicating information on contemporary eruptions; liaising with the Civil Defence authorities; reconstructing the impact of historic eruptions (e.g. Branca et al. 2013); geophysical monitoring of Etna and increasing public awareness. From 1999 the *International Institute of Volcanology* became a section of the *Instituto Nazionale di Geofisica e Vulcanologia*, its website is available in Italian and English (http://www.ct.ingv.it/en/) and, as in the Azores, there is a strong emphasis on international publishing and attendance at academic conferences.

Even in well studied areas, such as Japan, Hawaii, the Cascades and Italy, in 1980 the data available to the international research community were almost exclusively focused on the physical characteristics of eruptions, the spatial extent of their products and petrological aspects of lava flows and pyroclastic deposits. This deficiency was, however, becoming recognized, for example by Gordon Macdonald in his seminal book (Macdonald 1972, pp. 427) notes, 'it is time for volcanologists to put less emphasis on purely scientific aspects of their science, such as the generation and modification of magmas, and to give more attention to humanistic aspects—prediction and control of volcanic eruptions and the utilization of volcanic energy'. There were exceptions, such as Sheets and Grayson's (1979) edited volume *Volcanic Activity and Human Ecology* and Murton and Shimabakuro's (1974) paper on human adjustments to volcanic hazard in Hawaii—both published in the decade before 1980, but overall volcanology was dominated by the concerns of the geologist and igneous petrologist. Therefore, before the designation of the 1990s as the *International Decade for Natural Disaster Reduction* (*IDNDR*) and the subsequent *International Strategy for Disaster Reduction Communication* (*ISDR*) from 2000, the information communicated by means of the international peer reviewed literature was different in kind as well as being more restricted in volume. For instance in the eight years between 1982 and 1990, a survey of research output carried out in the context of the European *Laboratory Volcanoes* initiative,[23] showed that academic publication was dominated by the concerns of pure researchers, with papers in academic journals on more applied topics (i.e. prediction, social impact, policy implication and civil defence), constituting but a very small proportion of the total (Chester et al. 2002, pp. 419–420). Similar comments apply to other volcanic regions, with much of the information on social impact and civil protection being restricted to official reports and conference

papers. This 'grey literature' was neither widely known outside its country of origin nor was it peer—reviewed by members of the international research community. Today in the wake of the *IDNDR* and the current *ISDR*, not only has volcanology become more focused on hazard reduction, but the personnel has also widened to include social scientists, health professionals and experts in civil defence (Chester 2005).

---

[23]This was an initiative of the European Union/European Science Foundation during the 1990s. The volcanoes chosen for detailed study were: Mount Etna; Furnas (São Miguel, Azores); Piton da la Fournaise (Réunion, Indian Ocean); Teide (Tenerife) and Santorini (Greece). Later Krafla in Iceland was added. This initiative paralleled the 'Decade Volcanoes' research programme sponsored by IAVCEI.

References

Abate T, Branca S, Monaco C (2013) Le eruzione dell'Etna nell'opera di Orazio Silvestri (1835–1890): Il disegno come strumento per l'osservazione scientifica. Caracol, Palermo

Admiralty (1945) Spain and Portugal, The Atlantic Islands (Br 502, vol 5). Naval Intelligence Division, London

Agostinho J (1932) Vulcanismo dos Açores. Vista geral. A Terra 4:32–36

Akasoy A (2007) Islamic attitudes to disaster in the middle ages: a comparison of earthquakes and plagues. Medieval Hist J 10(1–2):387–410

Alessi G (1829–1835) Storia delle eruzione dell'Etna. *Atti Accademia Gioenia di Scienze Naturali Catania* Ser.1 (3): 17–75; (4): 23–74; (5): 43–72; (6): 85–116; (7): 21–65; (8): 99–148 and (9): 127–206

Anon (1669) An answer to some inquiries concerning the eruption of Mount Aetna, An. 1669. Communicated by some inquisitive English merchants, now residing in Sicily. Phil Trans R Soc Lond 4:1028–1034

Anon (1809) Letter from an officer of he 27th regiment dated 12th April 1809. Caledonian Mercury, June 19 (Issue 13653)

Anon (1865) The eruption of Etna. Boston daily advertiser March 7 (Issue 56), column F (no page numbers)

Anon (1923) Etna eruption ends. June 26 crater still active. Trainloads of sightseers. The Times (London), June 26, p 12, column c

Anon (2014a) Tommaso Fazello. L Enciclopedia Italiana. Treccani.it. http://www.treccani.it/enciclopedia/tag/tommaso-fazello/. Accessed 18 July 2014

Anon (2014b) Antonio Filoteo (or Antonio Filoteo degli Omodei) L Enciclopedia Italiana. Treccani.it. http://www.treccani.it/enciclopedia/antonio-filoteo-degli-omodei. Accessed 28 May 2014

Anon (2014c) Pietro Carrera. http://en.wikipedia.org/wiki/Pietro_Carrera. Accessed 28 May 2014

Anon (2015) Mount Cameroon. University of Oregon. http://volcano.oregonstate.edu/oldroot/volcanoes/cameroon/. Accessed 7 Apr 2015

Ashe T (1813) The history of the Azores or Western Islands. Sherwood, Neely and Jones, London

Auker MR, Sparks RSJ, Siebert L et al (2013) A statistical analysis of the global historical volcanic fatalities record. J Appl Volcanol 2(2): 24 p. http://www.appliedvolc.com/content/2/12

Azzaro R, Castelli V (2013) L' eruzione etnea del 1669 nelle relazioni giornalistiche contemporanee. INGV, Le Nove Muse Editrice, Catania

Balmuth MS, Chester DK, Johnston PA (2005) Cultural responses to the volcanic landscape: the mediterranean and beyond. Archaeological Institute of America, Boston

Blong RA (1982) The time of darkness, local legends and volcanic reality in Papua New Guinea. University of Washington Press, Seattle and Australian National University Press, Canberra

Blong RA (1984) Volcanic hazards. Academic Press, Sydney

Booth B, Croasdale R, Walker G (1978) A quantitative study of five thousand years of volcanism on S. Miguel, Azores. Phil T R Soc Lond 228:271–319

Borelli GA 1670 (reprinted 2001) Storia e meteorologia dell'eruzione dell'Etna dell 1669. Giunti, Firenze

Branca S, Azzaro R, De Beni E, Chester D, Duncan A (2015) Impacts of the 1669 and the 1693 earthquakes of the Etna Region Eastern Sicily, Italy: an example of recovery and response of a small area to extreme events. J Volcanol Geoth Res 303:25–40

Branca S, Del Carlo P (2004) Eruptions of Mt. Etna during the past 3,200 years: a revised compilation integrating the historical and stratigraphic records. In: Bonaccorso A, Calvari S, Coltelli et al (eds) Mt. Etna Volcano Laboratory, vol 143. Geophysical Monograph Series, American Geophysical Union, Washington DC, pp 1–27

Branca S, Coltelli M, Groppelli G et al (2011) Geological map of Etna volcano, 1:50,000 scale. Ital J Geosci 130:265–291

Branca S, De Beni E, Proietti C (2013) The large and destructive 1669 AD eruption at Etna volcano: reconstruction of the lava flow field evolution and effusion rate trend. Bull Volcanol 75:694. doi:10.1007/s00445-013-0694-5

Brydone P (1773) A tour through Sicily and Malta. W. Strahan and T. Cadell, London

Callender JM, Henshall JD (1968) The land use of Faial in the Azores. In: Cook AN (ed) Four island studies. The world land use survey, Monograph 5. Geographical Publications Ltd., Bude, Chapter II, p 23

Carneiro A, Simões A, Diogo MP, Mota TS (2013) Geology and religion in Portugal. Notes Rec R Soc 67:331–354

Carrera P (1636) Il Mongibello descritto in tre libri. G. Rossi, Catania

Cashman KV, Giordano G (2008) Volcanoes and human history. J Volcanol Geotherm Res 176:325–329

CAVW (1951–1975) Catalogue of active volcanoes of the world. International association of volcanology and chemistry of the earth's interior, Napoli, vol 1–22

Chester DK (2005) Volcanoes, society and culture. In: Marti J, Ernst GJ (eds) Volcanoes and the environment. Cambridge University Press, Cambridge, pp 404–439

Chester DK, Duncan AM, Guest JE, Kilburn CRJ (1985) Mount Etna: the anatomy of a volcano. Chapman and Hall, London

Chester DK, Duncan AM, Guest JE, Johnston PA, Smolenaars JJL (2000) Human responses to Etna volcano during the classical period. In: McGuire WG, Griffiths DR, Hancock PL, Stewart IS (eds) The archaeology of geological catastrophes. Geological Society of London, Special Publication 171, pp 179–188

Chester DK, Dibben CJL, Duncan AM (2002) Volcanic hazard assessment in Western Europe. J Volcanol Geotherm Res 115:411–435

Chester DK, Duncan AM, Dibben CRJ (2008) The importance of religion in shaping volcanic risk perceptions in Italy, with special reference to Vesuvius and Etna. J Volcanol Geotherm Res 172:216–228

Chester DK, Duncan AM, Sangster H (2012) Human responses to eruptions of Etna (Sicily) during the late-Pre-Industrial Era and their implications for present-day disaster planning. J Volcanol Geotherm Res 225–226:65–80

Coltelli M, Del Carlo P, Vezzoli L (1998) The discovery of a Plinian basaltic eruption of Roman age at Etna volcano, Italy. Geology 26:1095–1098

Coutinho R, Chester DK, Wallenstein N et al (2010) Responses to, and the short and long-term impacts of the 1957/1958 Capelinhos volcanic eruption and associated earthquake activity on Faial, Azores. J Volcanol Geotherm Res 196:265–280

Crosweller HS, Arora B, Brown SK et al (2012) Global database on large magnitude explosive volcanic eruptions (LaMEVE). J Appl Volcanol 1:4. http://www.appliedvolc.com/content/1/1/4

De Meneses A de F (2012) The discovery. In: Simas RM (trans) Cultural itinerary of the Azores. Regional Directorate for Culture, Governo dos Açores, Ponta Delgada

Degg M, Doornkamp J (1994) Earthquake hazard atlas: 7. Iberia (Portugal and Spain). London Insurance and Reinsurance Market Association, London

Deligne NI, Coles SG, Sparks RSJ (2010) Recurrence rates of large explosive volcanic eruptions. J Geophys Res 115:B06203. doi:10.1029/2009JB006554

Dias U de M (1931) Literatos dos Açores—Estudo histórico sobre os escritores açorianos. Emp. Tipografico Vila-Franca do Campo, Ponta Delgada

Duncan AM, Gaspar JL, Guest JE et al (1999) Furnas Volcano, São Miguel, Azores. J Volcanol Geotherm Res 92 (1–2), special volume, p 214

Duncan AM, Chester DK, Guest JE (2005) Eruptive activity of Etna before A.D. 1600, with particular reference to the classical period. In: Balmuth MS, Chester D, Johnston PA (eds) Cultural responses to the volcanic landscape. Archaeological Institute of America Colloquia and conference papers 8, Boston, pp 57–69

Fazzello T (1558) De Rebus Sicilis decades duae, nunc primum in lucum editae, his accessit totius operis index locupletissimus. Folio Panormi, Catania

Ferrara F (1818) Descrizione dell' Etna, con la storia delle eruzione e il catalogo dei prodotti. Lorenzo Dias, Palermo

Filoteo A (1590) Aetna topografia atque ejus incendirum istoria. Folio, Perugia

Firth CR, McGuire WJ (1999) Volcanoes in the quaternary. Geological Society Special Publication 161, Geological Society, London

Foster BR (1996) Volcanic phenomena in Mesopotamian sources. Appendix B. In: Polinger-Foster K, Ritner RK (eds) Texts, storms and the Thera eruption. J Near East Stud 55(1):1–14

Frutuoso G (2005) Saudades da Terra. Instituto, Cultural de Ponta Delgada, Pont Delgada, Açores

Fuchs CW (1865) Die vulkanischen Erscheinungen der Erde. Verlagsort, Leipzig

Furlan C (2010) Extreme value methods for modelling historical series of large volcanic magnitudes. Stat Modell 10(2):113–132

Furneaux R (1965) Krakatoa. Secker and Warburg, London

Gaspar JL, Queiroz G, Ferreira T, Amaral P, Viveiros F, Marques R, Silva C, Wallenstein N (2011) Geological hazards and monitoring at the Azores (Portugal). In: Earthzine—fostering earth observation and global awareness. http://www.earthzine.org/2011/04/12/geological-hazards-and-monitoring-at-the-azores-portugal/. Accessed 1 June 2014

Gaspar JL, Queiroz G, Ferreira T, Medeiros AR, Goulart C, Medeiros J (2015) Earthquakes and volcanic eruptions in the Azores region: geodynamic implications from major historical events and instrumental seismicity. In: Gaspar JL, Duncan AM, Chester DK (eds) Volcanic Geology of São Miguel Island (Azores Archipelago). Geological Society, London

Gemmellaro C (1858) La vulcanolgia dell'Etna. Tipografia dell'Accademia Gioenia, Catania

Grattan J (2006) Aspects of Armageddon: an exploration of the role of volcanic eruptions in human history and civilization. Quat Int 151:10–18

Grattan J, Torrence R (2007) Living under the shadow. The cultural impacts of volcanic eruptions. Left Coast Press, Walnut Creek, California

Guest JE, Gaspar JL, Cole PD et al (1999) Volcanic geology of Furnas Volcano, São Miguel, Azores. J Volcanol Geotherm Res 92:1–29

Guest J, Cole P, Duncan A, Chester D (2003) Volcanoes of Southern Italy. Geological Society, London

Gushchenko II (1979) Eruptions of volcanoes of the world: a catalog. Academy of USSR Far East Science Center. Nauka Publishing, Moscow

Harris SL (2000) Archaeology and volcanism. In: Sigurdsson H, Houghton B, Mc Nutt SR, Rymer H, Stix J (eds) Encyclopedia of volcanoes. Academic Press, San Diego, pp 1301–1314

Hodder I (2015) Personal communication Professor Ian Hodder Stanford University, California, dated 28 Mar 2015

Inwood MJ (1995) Enlightenment. In: Honderich T (ed) The Oxford companion to philosophy. Oxford University Press, Oxford, pp 236–237

Johnston PA (2005) Volcanoes in classical mythology. In: Balmuth MS, Chester DK, Johnston PA (eds) Cultural responses to the volcanic landscape. Archaeological Institute of America Colloquia and conference papers 8, Boston, pp 297–310

Keys D (1999) Catastrophe: an investigation into the origins of the modern world. Century, London

Lacroix A (1904) La Montagne Peleé et ses éruption. Masson et Cie, Paris

Lamb HH (1970) Volcanic dust in the atmosphere; with a chronology and assessment of its meteorological significance. Phil T R Soc A 266:425–533

Luz JB (1996) O homem e a história em Gaspar Frutuoso. Homenagem ao Prof. Doutor Lúcio Craveiro da Silva. Rev Port Filosofia 52:475–486

Lyell C (1830) Principles of geology. John Murray, London

Macdonald GA (1972) Volcanoes. Prentice Hall, Englewood Cliffs, New York

Machado F, Richards AF, Mulford JW et al (1962) Capelinhos eruption of Fayal volcano Azores. J Geophys Res 67(9):3519–3529

McCoy F, Heiken G (2000) Introduction. In: Mc Coy F, Heiken G (eds) Volcanic hazards and disasters in human antiquity. Geological Society of America special paper 345, Boulder CO, pp v–vi

McGuire WJ, Griffiths DR, Hancock PL, Stewart IS (2000) The archaeology of geological catastrophes. Geological Society Special Publication 171, Geological Society, London

Mellaart J (1967) Çatal Hüyük: a Neolithic town in Anatolia. Thames and Hudson, London

Mercalli G (1907) I vulcani attivi della terra: morfologia, dinamismo, prodotti, distribuzione geografica. Hoepli, Milano

Monaco F (1669) Cataclysmus aetnaeus sive inundation ignea Aetnae montis anni. Hertz, Venezia

Mota TS, Carneiro A (2013) 'A time for engineers and a time for geologists': scientific lives and different pathways in the history of Portuguese geology. Earth Sci Hist 32(1):23–38

Murton BJ, Shimabakuro S (1974) Human adjustment to volcanic hazard in the Puna district, Hawaii. In: White GF (ed) Natural hazards: local, national, global. Oxford University Press, New York, pp 151–161

Neumann van Padang M, Richards AF, Machado F, Bravo T, Baker E, Le Maitre W (1967) Atlantic ocean: catalogue of the active volcanoes of the world. International Association of Volcanology, Rome, Italy, Part XXI

Newhall CG, Self S (1982) The Volcanic Explosivity Index (VEI): an estimate of the explosive magnitude for historical volcanism. J Geophys Res 87(C):1231–1238

Nolan ML (1979) Impact of Parícutin on five communities. In: Sheets PD, Grayson DK (eds) Volcanic activity and human ecology. Academic Press, New York, pp 293–335

Oppenheimer C (2011) Eruptions that shook the world. Cambridge University Press, Cambridge

Polinger-Foster K, Ritner RK (1996) Texts, storms and the Thera eruption. J Near East Stud 55(1):1–14

Queiroz G, Gaspar JL, Cole PD, Guest JE et al (1995) Erupções vulcânicas no valle das Furnas (Ilha de S. Miguel, Açores), na primeira metade do seculo XV. Açoreana 8(1):159–165

Recupero G (1815) Storia naturale e generale dell'Etna. Ed. Dafni, Tringale Editore (Catania 1970)

Rodrigues R (1991) Notícia Biográfica do Dr. Gaspar Frutuoso. Cultural Institute of Ponta Delgada, Ponta Delgada

Rodwell GF (1878) Etna: a history of the Mountain and its eruptions. Kegan Paul, London

Romano R (ed) (1982) Mount Etna Volcano. Memorie Della Società Geologica Italiana XXIII

Romano R, Sturiale C (1982) The historical eruption of Mt. Etna (volcanological data). In: Romano E (ed) Mount Etna Volcano. Estratto da Memorie Della Società Geologica Italiana XXIII, pp 75–97

Sapper KT (1917) Katalog der geschichtlichen Vulkanausbrüche. Karl J. Trubner, Strasburg

Schneider K (1911) Die vulkanischen Erscheinungen der Erde. Gebrüder Borntraeger, Berlin

Scrope GP (1825) Considerations of volcanoes: the probable causes of their phenomena, the laws which determine their march, the disposition of their products, and their connexion with the present state and past history of the globe; leading to the establishment of a new theory of the Earth. W. Phillips, London

Scrope GP (1862) Volcanoes: the character of their phenomena, their share in the structure and composition of the surface of the globe, and their relation to its internal forces. Longman, Green, Longmans and Robert, London

Seibert L, Simkin T, Kimberly P (2010) Volcanoes of the world, 3rd edn. Smithsonian Institution, Washington DC

Sheets PD, Grayson DK (1979) Volcanic activity and human ecology. Academic Press, New York

Sigurdsson H (1999) Melting the Earth: the history of ideas on volcanic eruptions. Oxford University Press, New York and Oxford

Sigurdsson H (2000) The history of volcanology. In: Sigurdsson H, Houghton BF, McNutt SR, Rymer H, Stix J (eds) Encyclopedia of volcanoes. Academic Press, San Diego, pp 15–37

Simkin T (1993) Terrestrial volcanism in space and time. Annu Rev Earth Planet Sci 21:427–452

Simkin T (1994) Distant effects of volcanism—how big and how often. Science 264:913–914

Simkin T, Fiske R (1983) Krakatau 1883: the volcanic eruption and its effects. Smithsonian Institution, Washington DC

Simkin T, Siebert L (1994) Volcanoes of the world, 2nd edn. Smithsonian Institution, Washington DC

Simkin T, Siebert L (2000) Earth's volcanoes and eruptions: an overview. In: Sigurdsson H, Houghton BF, McNutt SR, Rymer H, Stix J (eds) Encyclopedia of volcanoes. Academic Press, San Diego, pp 249–261

Simkin T, Siebert L, McClelland L, Bridge D, Newhall C, Latter JH (1981) Volcanoes of the world. Smithsonian Institution, Washington DC

Simkin T, Siebert L, Blong R (2001) Volcano fatalities—lessons from the historical record. Science 291:255

Simões A, Carneiro A, Diogo MP (2012) Riding the wave to reach the masses: natural events in the early twentieth century Portuguese daily press. Sci Educ 21:311–333

Smolenaars JJL (2005) Earthquakes and volcanic eruptions in Latin literature: reflections and emotional responses. In: Balmuth MS, Chester DK, Johnston PA (eds) Cultural responses to the volcanic landscape. Archaeological Institute of America Colloquia and conference papers 8, Boston, pp 311–329

Squillaci P (1669) Relatione del fuoco di Mongibello, e di quel che segui. Stamperia del Dragondelli, Roma

Stone C (2003) Doctor, philosophy and renaissance man, Saudi Aramco World, pp 8–15

Tanguy J-C, Ribière Ch, Scarth A et al (1998) Victims of volcanic eruptions: a revised database. B Volcanol 60:137–144

Tanguy J-C, Condomines M, Branca S et al (2012) New archeomagnetic and ^{226}Ra-^{230}Th dating of recent lavas for the geological map of Etna volcano. Ital J Geosci 131:241–257

Tazieff H (1983) Some general points about volcanism. In: Tazieff H, Sabroux JC (eds) Forecasting volcanic events. Developments in volcanology, vol 1. Elsevier, Amsterdam, pp 9–25

Tedeschi Paternò T (1669) Breve Raguaglio degl'Incendi di Mongibello avvenuti in quest'anno 1669. Egidio Longo, Napoli

Tillard S (1812) A narrative of the eruption of a volcano in the sea off the Island of St. Michael. Phil T R Soc 102:152–158

Vaccari E (1998) Lyell's reception on the continent of Europe: a contribution to an open historiographical problem. Geological Society of London Special Publication 143, pp 39–52

Vaccari E (2008) "Volcanic travels" and the development of volcanology in 18th century Europe. Proc California Acad Sci 59 (suppl. 1:3):37–50

Varen PR (1650) Geographia generalis, in qua affections generales telluris explicantur. L Elzevirium, Amstelodami

Verbeek RDM (1884) Kort verslag over de uitbarsting van Krakatau op 26, 27 en 28 Augustus 1883. Landsdrukkerij, Batavia

von Hoff KEA, Berghaus HKW (1840/1841) Chronik der Erdbeben und Vulcan-Ausbrüche mit vorausgehender Abhandlung über die Natur dieser Ercheinungen. Bei J Perthes, Gotha, 2. Vols

von Humboldt A (1858) Kosmos: Entwurf einer physis-
chen Weltbeschreibung. Mit einer boigraphischen
Einleitung, vol 5. J.G. Cotta, Stuttgart

von Waltershausen WS (1844–1859) Atlas des Aetna.
Vandenhoeck, Göttingen and Berlin

Walker GPL, Croasdale R (1971) Two Plinian-type
eruptions in the Azores. Q J Geol Soc Lond 127:17–
55

Weston FS (1964) A list of recorded volcanic eruptions in
the Azores with brief reports. Boletim do Museu e
Laboratório Mineralógico e Geológico da Faculdade
de Ciêcias, Universidade de Lisboa 10:3–18

Winchilsea (Earle of) (1669) A true and exact relation of
the late prodigious earthquake and eruption of Mount
Etna or Montegibello as it came in a letter written to
his majesty from Naples. Together with a more
particular narrative of the same, as it is collected out

of several relations sent from Catania. Newcomb,
London

Witham CS (2005) Volcanic disasters and incidents: a
new database. J Volcanol Geotherm Res 148:191–233

Zbyszewski G (1961) Étude geologique de l'ile de S.
Miguel (Açores). Comunicações Serviços Geológicos
de Portugal 45:5–79

Zbyszewski G, Moitinho de Almeida F, Veiga Ferreira O,
Torre de Assunção C (1958) Notícia explicativa da
Folha "B", da ilha S. Miguel (Açores) da Carta
Geológica de Portugal na escala 1:50000. Serviços
Geológicos de Portugal, Lisboa

Zbyszewski G, Ferreira OV, Assunção CT (1959) Notícia
explicativa da Folha "A", da ilha S. Miguel (Açores)
da Carta Geológica de Portugal na escala 1:50000.
Publ. Serviços. Geológicos de Portugal, Lisboa

What Can We Learn from Records of Past Eruptions to Better Prepare for the Future?

David M. Pyle[iD]

Abstract

There is a long and rich record of historical observations of volcanic activity that has the potential to enhance current understanding of volcanic eruptions and their impacts, and to inform planning of responses to future events. However, apart from a small number of well documented examples, much of this broader material remains unread and little used. In this chapter, we explore examples of contemporary observations and accounts of volcanic eruptions at Santorini (Greece) and the Soufrière, St. Vincent, in the 18th, 19th and early 20th centuries. We show how these sorts of data could be used to inform and advance our understanding of, and approach to, volcanic crises; and to better understand the roles that communication—of hazards, of past events, or during an emerging crisis—may play in helping to prepare for the future.

1 Introduction

Despite considerable advances in scientific methodologies, monitoring techniques and modelling capacity, many aspects of the science of volcanology remain empirical; in particular the anticipation of 'what may happen next'. This presents a critical challenge to those charged with the management of emerging volcanic events, and those responsible for communications and crisis management as events unfold during an eruptive sequence. At volcanoes with a long legacy of monitored activity, recognising and developing an empirical understanding of the way the volcano behaves, and being able to read the signs of what might be unfolding, is one of the core skills of the staff at the local volcano observatory. However, instrumental records of volcano observation are short compared to the typical intervals between large eruptions; and what about volcanoes not known to have had unrest or activity in the recent past? It is in these contexts that accounts of prior activity, perhaps deep in the past, have particular value, both in providing a qualitative picture of what may have happened in the past; and in forming a narrative

D. M. Pyle (✉)
Department of Earth Sciences, University of Oxford, South Parks Road, Oxford OX1 3AN, UK
e-mail: david.pyle@earth.ox.ac.uk

Advs in Volcanology (2018) 445–462
https://doi.org/10.1007/11157_2017_5
© The Author(s) 2017
Published Online: 18 May 2017

with which to engage communities who may be affected by volcanic activity.

In the context of a volcano which is starting to show unrest, the conventional approach to the understanding of the emerging volcanic hazard still relies to a great extent on a combination of (i) the mapping of the likely hazard, based on the physical deposits of prior eruptions, (ii) assessment of the timing of past eruptions and (iii) effective monitoring of the volcano, whether from ground-based sensors, or by satellite remote sensing, combined with an analysis of instrumental records of past eruptions (e.g. Scarpa and Tilling 1996; Haynes et al. 2007; Pyle et al. 2013; Hicks et al. 2014). The next step is often based around the analysis of likely eruptive scenarios, which will in turn be based on an understanding either of the past eruptive history of that particular volcano; or, in the case of a volcano showing unrest for the first time, may be based on scientists' judgements about the sorts of eruptive scenarios that might be typical of the volcano. In recent years, this process has been developed and applied to great effect. For example at Montserrat, West Indies, an iterative and repeated process of expert elicitation has been used to form a consensus view of the state of the volcano (Aspinall et al. 2002, 2003; Aspinall 2006; Wadge and Aspinall 2014). The development of 'Bayesian Belief Network' approaches represents a valuable tool for using the learning from past events to inform decision making about future events (Hincks et al. 2014).

As other work on Montserrat and elsewhere has also shown, a poorly understood aspect of the response to emerging volcanic events is the role played by communication: not only in terms of the formal pronouncements on hazards, risks and 'alerting' processes (e.g. Barclay et al. 2008, 2015; Wadge and Aspinall 2014; Donovan et al. 2012, 2014), but also in terms of the communications with and within the diverse communities affected by the emerging crisis; and the impacts on those affected by the activity (e.g. Hicks and Few 2015). This is a significant gap which would certainly be worthy of future investigation.

In this chapter, I argue that volcanologists could add considerably to the evidence base relating to past eruptions, and consequently improve our capacity to manage impending or future crises, by seeking out the wider historical records of past eruptions. In particular, there is much to be learned from records (for example contemporary accounts, written by the people who experienced the event) that document not only how the past physical events unfolded, but also the social, economic and political consequences of the event. These same resources may also record the roles played by communication during these past crises: from the immediate response (when, why and how did people affected by events respond?), to the diffusion of news of the events, and the response of external actors and agencies.

2 Volcanic Eruptions and Their Consequences

The rich sources of contextual information around volcanoes and the impacts of their historical eruptions have only rarely been exploited. Three better known examples, where diverse source materials have been brought together for analysis sometime after the primary event, include studies of the eruption of Tambora, in 1815 (e.g. Stommel and Stommel 1983); of Krakatoa in 1883 (Simkin and Fiske 1983); and, on a smaller scale, of Parícutin, Mexico in 1943 (Luhr and Simkin 1993). These examples are briefly described in the following sections.

2.1 Tambora, 1815

The Tambora eruption of April 1815 was one of the largest eruptions of the past 500 years, but was understudied until relatively recently (Self et al. 1984; Self and Gertisser 2015). Eyewitness accounts of the eruption were gathered at the time by the regional Governor, Stamford Raffles, who dispatched a team to deliver emergency relief to affected communities, and collect information on what had happened (see Oppenheimer 2011; Pyle 2017). It was only many decades later that others attempted to assess the wider impacts of the eruption. Heinrich Zollinger, a Swiss botanist, climbed Tambora in 1847, and later documented the scale and severity of the eruption, including the casualties (Zollinger 1855). The extent of the global consequences of the eruption became clearer once the stories of the 'Year without a Summer' of 1816 had been gathered, and linked to the eruption of Tambora (e.g. Milham 1924; Stommel and Stommel 1983; Stothers 1984), while the narratives of links to global health crises and economic collapse are still emerging (e.g. Post 1977; D'Arcy Wood 2014; Oppenheimer 2015). The Tambora event is a case study of a globally-disruptive event; and one that would benefit from further analysis, particularly in the context of communication and crisis management.

2.2 Krakatoa, 1883

Krakatoa was the first volcanic eruption with a global impact where news of the event travelled faster than the spread of the ash cloud. The newly completed international network of submarine telegraph cables ensured that the opening phases of the Krakatau eruption in summer 1883 were reported in *The Times* newspaper in London within 36 h of the event (Simkin and Fiske 1983; Winchester 2003). The aftermath of the climatic phase of the Krakatoa eruption in August 1883 was an early example of the use of crowd-sourcing to gather information about the far-flung impacts of an eruption. George Symons, a British meteorologist, chaired the Royal Society's Krakatoa Committee. The remit of the committee was to collect information on the scientific phenomena attending the eruption, and Symons placed calls in early 1884 for 'the communication of authenticated facts respecting the fall of pumice and dust… unusual disturbances of barometric pressure and sea-level (etc.)' (Symons 1888, p. iv). The final report contained only a very brief outline of the relief efforts that followed the eruption (Symons 1888, p. 2), alongside more extensive eyewitness reports from those both at sea, and on land. To mark the centenary of the eruption, Simkin and

Fiske (1983) collected together these and many other accounts, and refined the timeline for the unfolding events. The existence of a telegraph network meant that some of the otherwise transient records of events were recorded for Krakatoa in ways that had never before happened for an eruption of this scale. However, while analysis of this material could provide valuable insights into the nature of emergency communication and response during a large-scale volcanic emergency, the full potential of these records has not yet been realised.

2.3 Parícutin, 1943

The eruption of Parícutin, Mexico, was on a smaller scale, but had dramatic consequences for farming communities of the high Mexican plains. In February 1943, an eruption began without warning in the corner of a corn field. Over the next nine years, the new volcanic cone of Parícutin grew, eventually covering 25 km^2 of land with lava. To mark the 50th anniversary of the eruption, Luhr and Simkin (1993) collected and edited a volume of papers and reports—some contemporary with the eruption; others offering a retrospective analysis of the wider impacts of the events on those affected. Accounts document the first-hand experiences of farmers who first had to cope with the fallout of ash and cinders; then a 'rain of mud', when the rains arrived, and finally resettlement in new locations that had no connection to their original homes, and required them to adapt their farming practices to lower elevations. The value of this volume in bringing together a diverse range of papers on the physical, environmental and social impacts of a one-off eruption was considerable; but, as with the examples of Tambora and Krakatoa, lacked any formal analysis of crisis communications.

2.4 Other Examples

While there are volcanological studies of many past eruptions, there are rather fewer that extend to an analysis of the human impacts and responses. Recent work on historical eruptions of Vesuvius in 1906 and 1944 that explore human impacts, crisis management and emergency response (e.g. Chester et al. 2007, 2015) reveal some important lessons for dealing with future volcanic crises, at Vesuvius and elsewhere.

Deeper in the past, volcanic events and their impacts may only have been recorded in official and other records if the event was of sufficient scale to have required a response (evacuation, aid and assistance, rebuilding or relocation); and if the local governance systems collected and recorded such information. There are extensive records of natural hazard events from the Spanish colonial period of Latin America (e.g. Petit-Breulh Sepúlveda 2004, 2006; Hutchison et al. 2016), but scant records from, for example, Ethiopia (e.g. Gouin 1979; Wiart and Oppenheimer 2000): differences that in part reflect the spatial locations and numbers of chroniclers, and the security of the archives. Beyond this, large-scale natural disasters may be captured in oral histories that require patience, luck and persistence to piece together (e.g. Blong 1982; Johnson 2013).

3 Why the Instrumental Records of Past Eruptions Is Deficient

Two major challenges face scientists in their efforts to extend and develop an evidence-based approach to anticipate the behaviour of less-well-known volcanic systems that are starting to show their first known episodes of unrest. Even for systems where the recent volcanic record is well preserved, challenges include:

- The preserved geological record, and much of the physical observational record, for most volcanoes is almost exclusively a record of past *eruptive activity*. In itself, this is a record biased towards activity of a scale or style, or deposited into an environment that can be preserved (e.g. Pyle 2016). There is often a lack of evidence from which to make inferences about the nature of either non-eruptive

unrest, or of long-term precursors to subsequent activity (Moran et al. 2011). Non-eruptive unrest leaves no accessible trace in the geological record, and as result most of our current knowledge about unrest dates from the modern, instrumental, era.

- The catalogued records of past activity are almost exclusively records of eruptions, rather than of non-eruptive unrest (e.g. Siebert et al. 2010); and these records rarely, if ever, document the wider social, political and economic consequences of these past eruptions. While catalogues of volcanic activity have evolved an effective set of tools for recording the totality of an eruption—start date, size, end date—there are as yet no agreed standards for recording, preserving and making accessible the long-termtime-series of measurements (and associated metadata) and observations of volcanic events, and their consequences, beyond the daily, weekly or monthly 'status' bulletins; despite valiant efforts by individuals (e.g. Perret 1924), international projects including WOVOdat (Venezky and Newhall 2007), the Global Volcanism Programme (Siebert et al. 2010) and the Global Volcano Model (Loughlin et al. 2015); and a handful of prominent case studies (e.g. Mount St. Helens—Lipman and Mullineaux 1981; Pinatubo—Newhall and Punongbayan 1996; and Montserrat—Druitt and Kokelaar 2002; Wadge et al. 2014). As a result, fine-grained details about the evolution of eruptive activity during a crisis are prone to being lost, and will be hard, if not impossible, to recover after the event. Thus, our understanding of even the 'volcanic' part of past volcanic crises (leave alone their social imprint) is far from complete (e.g. Hicks and Few 2015); and our understanding of the run-up to eruptions is even more fragmentary.

Volcanology is now entering a period of time when it is possible to monitor volcanoes globally using satellite remote-sensing; and where near-real-time automated remote detection of changes in behaviour (whether thermal, geodetic, seismic or gas) is becoming a reality (Hooper et al. 2012; Pyle et al. 2013). The challenge will be to match this step change in our *capacity* to monitor changes in behaviour, with our ability to interpret these changes in behaviour, in a way that is useful to those charged with managing volcanic risk.

An indication of the scale of this problem can be seen from an analysis of global patterns of volcano deformation (Biggs et al. 2014). Biggs et al. analysed the published satellite geodetic (InSAR) studies of 198 systematically observed volcanoes in the 18 year period up to 2013, distinguishing between volcanoes that erupted in that same period; and those that did not. Critically, they found that only 44% of detected 'deforming volcanoes' actually erupted during this period; meaning that the majority of deforming volcanoes are not poised to erupt; while 6% of 'non-deforming' volcanoes erupted, meaning that some restless volcanoes may show little sign of being restless, before erupting. This result shows that the successful recognition of pre-eruptive unrest will continue to require multiple sources of observation; while the interpretation of the signals of unrest will continue to require an understanding of the specific attributes of the system that is showing unrest.

4 Retrospective Analysis of Volcanic Crises

Here, we use records from historical eruptions at two typical subduction-zone volcanoes (Santorini, Greece; Soufrière, St. Vincent) to illustrate how retrospective analysis of the wider archives and records of past eruptions might help in the management of ongoing and future crises.

4.1 The Kameni Islands, Santorini, Greece

First we consider post-calderadome-forming eruptions on the Greek island of Santorini; a restless caldera in the Mediterranean. The volcano is well known to the local residents (e.g. Dominey-Howes and Minos-Minopoulos2004), and intimately linked to the deep archaeological history of both Santorini, and the Aegean Bronze Age 'Minoan' culture (e.g. Marinatos 1939). It has a significant transient summer population of tourists; some of whom will have come to see the volcano and its hot springs.

Over the past 3000 years, a series of dacite lava and tephra eruptions have progressively built the Kameni islands; a 4 km^3 edifice which now emerges above sea level within the caldera (Druitt et al. 1999; Nomikou et al. 2014; Table 1; Fig. 1). The first eruptions to form the nascent Kameni islands would have been exclusively submarine, but there have been at least 8 subaerial eruptions since 46 AD which have progressively enlarged the Kameni islands (Fytikas et al. 1990; Pyle and Elliott 2006; Nomikou et al. 2014). Of these eruptions, there are written contemporary accounts of six, and detailed accounts of all known eruptions since 1707.

An eruption from 1707–11 was the first such event on Santorini to be documented in a journal (Box 1, Gorée 1710); and is the earliest eruption of Santorini for which early maps or sketches exist (Fig. 2). 150 years later, from 1866 to 1870, a major dome-forming eruption became the focus of a significant amount of contemporary observation and writing. This eruption led to the first treatise on the medical effects of volcanic eruptions and their gas emissions (da Corogna 1867), and was documented in considerable detail by Ferdinand Fouqué, leading to the modern ideas on the origin and evolution of calderas (Fouqué1879). Contemporary observations of eruptions in 1866–70 and 1925–1928 detail the eruption progress, including rates of dome growth, of explosions, and lava extrusion. These datasets provide the essential quantitative information that underpin later forecasts for the style of future activity (e.g. Watt et al. 2007; Jenkins et al. 2015). The last of the eruptions of Santorini was in 1950, three decades before the installation of the first instrumental monitoring networks on the islands.

Table 1 Historical activity of the Kameni islands, Santorini, Greece

Eruption date	Location	Primary sources
10 Jan–2 Feb 1950	Nea Kameni	Georgalas (1953)
20 Aug 1939–July 1941	Nea Kameni	Georgalas and Papastamatiou (1951)
23 Jan–17 Mar 1928	Nea Kameni	Reck (1936)
11 Aug 1925–Jan 1926	Nea Kameni	Reck (1936)
26 Jan 1866–15 Oct 1870	Nea Kameni	Fouqué (1879)
23 May 1707–14 Sep 1711	Nea Kameni	Gorée (1710)
1570 or 1573	Mikra Kameni	No primary records
1457	Palea Kameni	No primary records
726	Palea Kameni	No primary records
46–47	Palea Kameni (Thia)	No primary records
199–197 BC	Hiera (or Lera)	No primary records

Notes Full sources listed in Fouqué (1879); Stothers and Rampino (1983); Pyle and Elliott (2006); Nomikou et al. (2014)

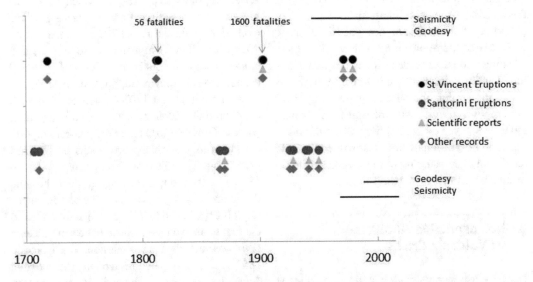

Fig. 1 Summary timeline of eruptions of the Kameni islands, Santorini, and the Soufriere, St. Vincent, since 1700 AD, showing those events for which contemporary records (including newspaper reports, or sketches, or diaries, or official reports) and scientific reports exist. In contrast to the ca. 300 year-long record of contemporary observations, the scientific instrumental monitoring record (e.g. seismicity, or ground deformation/geodesy) is much shorter. There are no systematic instrumental monitoring records of the most significant eruptions of either system (1902–3 eruption of St. Vincent; 1866–1870 eruptions of Santorini)

Box 1: The 1707 eruption of Nea Kameni, Santorini
Excerpt from Gorée Fr (1710) 'A relation of a new island, which was raised up from the bottom of the sea on the 23rd of May 1707, in the Bay of Santorin, in the Archipelago'.

'Five days before it appeared, viz on the 18th of May between one and two of the clock in the afternoon, there was at Santorini an earthquake, which was not violent and continued but a moment; and in the night between the 22nd and 23rd there was also another, which was yet less sensible

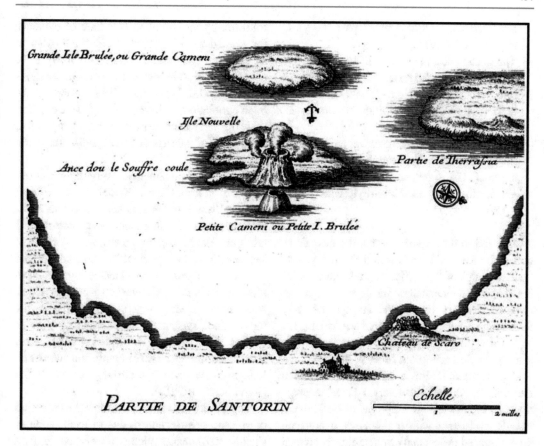

Fig. 2 Map of the Santorini caldera, showing the Kameni islands during the eruption of 1707–1711. The Ile Nouvelle (new island) is the lava dome and flow formed during the 1707 eruption. This now forms a part of the island of Nea Kameni. From Tarillon (1715)

than the former [...] Add to this, that a long time before the earthquakes the fishermen perceived an ill smell every time they passed by that place [...] Notwithstanding it is very certain that there have not been any other earthquakes at Santorini than those which, 14 or 15 years ago, continued for several days and were very violent. However it was, some seamen discover'd this island early in the morning, but not being to distinguish what it was they imagined it to be some sort of vessel that that had suffered shipwrack.

The smoak appeared first upon the 16th of July: at which time there rose up a ridge of black stones and which was afterwards not only the centre of the while island, but also of the fire, and smoke and great noise

that was heard some time after. The smoak which issued out of this ridge of stones [...] was very thick and white.'

4.2 Eruption Progress

Work on the Kameni islands since the mid-1800s documenting both the nature of eruptions and the petrology of the erupted products meant that by the early 21st century, volcanologists were able to develop a simple but detailed conceptual model of what a future eruption might look like. The key factors that fed into the model were:

– The observations that the compositions of the erupted products, and the style of eruption have varied very little over the past 500 years (e.g. Nicholls 1971);

- The observation that many of these eruptions may have been 'triggered' by the influx of hot, volatile-rich magma some short time before (e.g. Higgins 1996; Martin et al. 2006);
- The observation that there is an apparent linear relationship between the period elapsed since the last eruption, and the volume erupted in the next eruption (Pyle and Elliott 2006).

This led to the hypothesis that the next eruption of the Kameni islands would be preceded by 'general uplift of the edifice and discoloration of the sea, and anticipated some days to weeks in advance. The eruption would involve the early formation of lava domes, which would later act as a focus for vigorous, intermittent explosive activity' (Pyle and Elliott 2006, p. 266).

The recognition of a relationship between eruption length and the interval between eruptions for the last 4 large eruptions of the Kameni islands, consistent with a model of a constant time-averaged deep supply of magma, also meant that it was possible to forecast the duration of a future eruption. While this analysis is entirely empirical—based on collation of observations from eruptions over the past 300 years—this approach is typical of the way in which much modern volcanology still operates; this analysis would not have been possible without the depth and breadth of contemporary descriptive materials accompanying the historical eruptions.

4.3 Precursors

Although the eruption record of the Kameni islands is well known, the record of anything that may have happened in between eruptions is almost non-existent. We know of precursors, or pre-eruption changes, for several eruptions (see Box 1), but apart from a mention of an earthquake swarm a few years prior to the 1707 eruption, there are no known reports or records of any sort of unrest that wasn't subsequently followed by an eruption. This lack of evidence for 'precursory behaviour that didn't culminate in an eruption' (non-eruptive unrest) is likely to be a common feature of many volcanic systems (Moran et al. 2011)—and likely to explain the poor correlation between 'run up' and eruption scale that is evident in the most recent analysis of these sorts of datasets (e.g. Passarelli and Brodsky 2012).

The lack of documented non-eruptive unrest and the volcanological model for 'the next eruption' was brought into sharp relief in 2011, with the start of the first modern volcano-seismic crisis on Santorini since the eruption of 1950. In January 2011, the first small earthquakes located within the caldera were detected by the local seismic network. This was clearly anomalous, since most of the detected seismicity in the region since systematic measurements began in the late 1980s has been associated with structures *outside* the caldera, most notably the submarine Kolombos volcano (Dimitriadis et al. 2005; Nomikou et al. 2012).

The event was readily identifiable as a period of volcano-seismic unrest, due to localisation of shallow earthquakes along a well-known fault system thought to have been responsible for the delivery of magma to the surface during previous eruptions; and from the patterns of uplift and ground deformation detected from the network of continuous GPS instruments and analysis of satellite radar interferograms (e.g. Newman et al. 2012; Parks et al. 2012). Later, field evidence also showed changes in the nature of the diffuse degassing around the summit craters. In early 2012, the authorities convened a Special Scientific Committee for the Monitoring of Santorini Volcano, and oversaw the deployment of a host of new instruments across and around Santorini, but fairly soon the unrest came to an end and nothing further happened (Aspinall and Woo 2014). No formal notice of the unrest was declared by the authorities until the event was effectively over (see Vougioukalakis et al. 2016), and although there are now a dozen or more scientific papers describing these events, there was no eruption and there is no record of the event in the Smithsonian Institution's Global

Volcano Programme dataset or associated reports.

Best estimates of the scale of the magmatic anomaly associated with the 2011–2012 unrest suggest that the shallow magmatic pressure source increased in volume by ca. 14–23 million cubic metres (Parks et al. 2015); equivalent to a couple of decade's worth of 'steady state magma accumulation'. This presents a challenge to the previous consensus model for the volcano: here was evidence for a large, shallow intrusion which did not lead to eruption. Whether this is the typical behaviour of the system, or not, is something that cannot yet be determined—because although the rich documentary record of past events furnishes us with the evidence for how the next eruption *might* proceed, it provides us with no information, one way or the other, about the frequency and style of episodes of unrest. However, the episode of unrest has stimulated new work on conceptual physical models for repeated eruptions driven by pressurisation and failure of the shallow magma reservoir by intrusion (e.g. Browning et al. 2015; Degruyter et al. 2016), and stimulated retrospective analysis of bathymetric maps, that chart the ups and downs of the volcano since the 1850s (Watts et al. 2015). The unrest has also opened up a public discussion of how the scientific community, civil defence and local authorities should plan for a future crisis; including how to manage hazards, risk and communications (Vougioukalakis et al. 2016).

4.4 The Soufrière, St. Vincent

The Soufrière, St. Vincent, is another example of a lava-dome forming volcano in a subduction-zone setting. In contrast to the Kameni islands, recent and historical eruptions of St. Vincent have been of a more mafic magma (basaltic andesite, rather than dacite), and eruptions have tended to be significantly more explosive than those of the Kameni islands, with much more serious consequences (Fig. 1; Table 2). Below, we briefly introduce the chronology of the St. Vincent eruptions, and their broader consequences, before discussing the challenges for the future.

4.5 Eruptive History and Impacts

St. Vincent lies in the eastern Caribbean, and is one of the southern islands of the Lesser Antilles Arc. The eruptive history of the presently active volcano—the Soufrière, a complex of craters at the top of the Morne Garou—is not well known prior to about 1700. Since then, St. Vincent has

Table 2 Historical activity of the Soufrière, St. Vincent

Date	Notes	Phenomena
	Summit crater is dry, with an exposed lava dome	
13 April–October 1979	Explosive eruptions, and dome extrusion	P, E, L
17 May 1971–1972	Minor effusive eruption	L
	Summit crater is water filled	
6 May 1902–1903	Major explosive eruptions	P, E
1880	Crater lake level increases, temperatures rise	
	Summit crater is water filled	
1814	Possible minor eruption	E?
27 April–6 May 1812	Major explosive eruption	P, E
ca. 1784?	Possible minor eruption	L?
26 March 1718	Major explosive eruption	P, E

P precursor seismicity; *E* explosive eruption; *L* lava dome Compiled from Shepherd (1831), Anderson and Flett (1903), Robertson (1995) and Richardson (1997)

experienced four major explosive eruptions, and at least one minor dome-forming eruption (Fig. 1; Table 2; Robertson 1995). Eruptions in 1718, 1812 and 1902–3 each had major consequences for the northern sectors of the island—with widespread tephra fallout, explosive ejection of ballistic blocks, column-collapse pyroclastic density-currents, and lahars deposited across or coursing down the flanks of the volcano. Each of these explosive eruptions led to significant ash fallout across other islands of the Caribbean; notably Barbados.

The eruptions of 1812 (at least 56 fatalities; Robertson 1995) and 1902–3 (1600 fatalities) were severe, with much damage to properties, many associated with the sugar plantations in the northern parts of the island; and deaths in the communities which provided labour for those plantations (Anderson and Flett 1903; Smith 2011; Pyle et al. 2017). The details of the social and economic consequences of these eruptions can be reconstructed in an extraordinarily fine-grained way, principally because of the preservation of official correspondence and reports from the time in colonial archives (Gullick 1985; Smith 2011; Richardson 1997; Pyle et al. 2017).

There are no known reports, however, of the consequences of the explosive eruption of 1718 for the inhabitants of St. Vincent. The only written record of the eruption was gathered from the descriptions of European mariners, and published anonymously in a pamphlet that declared 'the entire desolation of the island of St. Vincent', and that 'the island was no more' (Box 2; Defoe 1718). This account was not mentioned by Shepherd (1831) in his history of St. Vincent, even though by that time the 1718 eruption would have been well known to natural historians from the works of de Humboldt and Bonpland (1825).

Box 2: The 1718 eruption of St. Vincent

Daniel Defoe 'The destruction of St. Vincent' from Mist's Journal, July 5, 1718.

'On the 27th [March] in the morning the air was darkened in a dreadful manner; which darkness by all accounts seems to have extended over all the colonies and the islands which were within 100 miles of the place [...] The sum of [reports from ships] 'they saw in the night that terrible flash of fire and after that innumerable clashes of thunder [...] a thousand times as loud a thunder or cannon. As the day came on, still the darkness increased.

In the afternoon they were surprised with the falling upon them as thick as smoke, but fine as dust and yet solid as sand; this fell thicker and faster as they were nearer or farther off—some ships had it nine inchers, others a foot thick, upon their decks. The island of Martinico is covered with it at about seven to nine inches thick; at Barbadoes it is frightful, even St. Christopher's is exceeded four inches'.'

The eruption of April–May 1812 and its consequences were recorded in detail in a range of primary contemporary sources (letters and diaries) and published reports and accounts (see Smith 2011, for a detailed analysis). The diaries and reports of a British barrister and plantation owner, Hugh Perry Keane (Box 3) are thought to also have provided the materials for *The Times* leader story on the eruption, published six weeks later in late June (Smith 2011). Keane's sketches of the eruption were later used by JMW Turner to inform his 1815 painting of the volcano in eruption. One notable feature of this eruption is the contrast between the published accounts that stress the 'dreadful' scale of the eruption, and the apparent light loss of life (Smith 2011). Estate owner Alexander Cruikshank wrote that 'there has [not] been so violent an Eruption recorded since the destruction of Herculaneum and Pompeii' (Blue Book 1813, p. 4), and a committee of landowners petitioned the Crown for compensation, on the basis that these 'distressed Memorialists [had suffered a] severe visitation of Divine

Providence... unexampled in any of His Majesty's dominions for much more than a century' (Blue Book 1813, p. 9). These accounts were presumably motivated by a desire for financial recompense, and stress the physical impacts of the eruption on their crops, estate lands and property. In contrast, the same reports are very thin on the consequences of the eruption either for their enslaved workers (one sugar estate owner appended the brief statement 'but most providentially not many lives were lost'; Blue Book 1813, p. 4), or for the island's native residents. Reports suggest that many people from the Carib communities who lived around the flanks of the volcano evacuated spontaneously within a couple of days of the eruption starting, but before the eruption reached a climax (Blue Book 1813; Smith 2011). It is not known whether this reflected a response based on oral histories of prior eruptions; or that the effects of the tremor and tephra fallout had become intolerable.

Box 3: The 1812 eruption of St. Vincent Excerpts from the diary of Hugh Perry Keane [Virginia Historical Society mss1 k197 a23], transcribed in Smith (2011), Hamilton (2012) and Pyle (2017).

Weds 29 [April] Then to see the Souffrier, involved in dark clouds and vomiting black sands. Landed at Wallilabou. Spent the evening in contemplation of the volcano, and slept there.

Thurs 30th in the afternoon the rousing of the Mountain increased and at 7 o'clock the flames burst forth and the dreadful eruption began. All night watching it between 2 and 5 o'clock in the morning showers of stones and earthquakes threatened our immediate destruction.

May 1 The day did not give light till nearly 9—the whole island involved in gloom. The mountain was quiet all night.

May 2 Rose at 7, Drawing up my narrative for the register.

Sun 3rd Rose at 7 and after gathering some Bfast... Proceeded to Wallibu—strange and dismal sight, the river dried up

---

and the land covered with cinders and sulphur. Morne Ronde Hid in smoke and ashes—the track covered with trees and a new formation given to it—burnt carcasses of cattle lying everywhere.

Mon: 4 rose at 6 and took a cup of coffee... and returned to town. Kingston in great confusion.

Wed 6 the volcano again blazed away from 7 till 1/2 past 8.

The catastrophic eruptions of 1902–3, in which about 1600 people were killed, left dense records of official and other communications during both the immediate crisis, and the relief and recovery efforts (Blue Book 1903, 1904; Anderson and Flett 1903). Eyewitness accounts of the build up to the eruption again suggest that significant numbers of people from communities living on the flanks of the volcano evacuated spontaneously during the 24 h before it reached a climax (Pyle et al. 2017). Indeed, by the time the scale of the eruption became apparent to colonial officials in Kingstown, telephone lines to the north of the island had already been interrupted and submarine telegraph lines from St. Vincent to the neighbouring island of St. Lucia had been severed (Report of Edward Cameron, Administrator; Blue Book 1903). A combination of the infrastructure of the Colonial Government; the relative ease of international communication (telegrams) and the coincidence of the devastating eruption of Mont Pelée, Martinique, just one day after the eruption of the Soufrière of St. Vincent meant that these two volcanic disasters in the Caribbean attracted world-wide interest at the time, and a rapid international relief effort (Pyle et al. 2017). Scientists and journalists arrived with some of the relief boats, and dramatic reports of the aftermath of the eruption were soon widely available (e.g. Russell 1902; Morris 1903; Anderson 1903; Hovey 1903a, b). Since that time, though, little of this material has been re-considered by volcanologists; and a full analysis of both the entire eruption sequence, and the recovery process has yet to be completed.

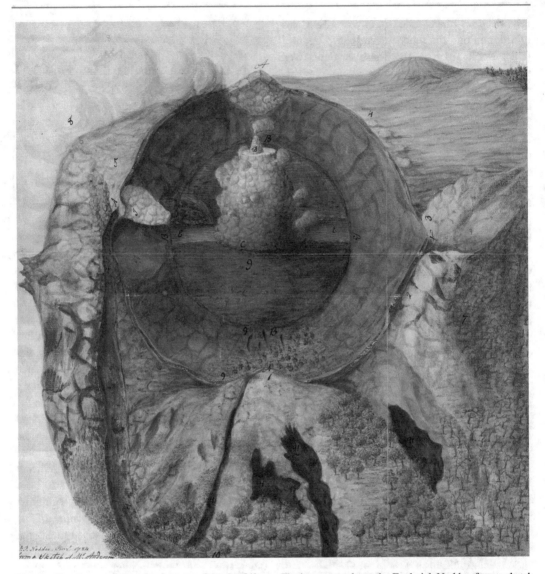

Fig. 3 Sketch map of the summit crater of the Soufrière of St. Vincent in 1784, showing a steaming lava dome (*top, centre*) within a steep-sided summit crater, flanked by two small crater lakes (Anderson and Yonge 1785). The image was drawn by Frederick Nodder, from a sketch by J. Anderson. Image RS 9780, Map of Morne Garou, reproduced by permission of the Royal Society

In between these eruptions, each of which were preceded by many months of felt seismicity in areas close to the volcano, there may well have been other small to moderate eruptions which were non-explosive, and which produced few noticeable consequences—whether in terms of detectable seismicity, or other manifestations. There has been speculation that an eruption in 1784 formed a lava dome which was subsequently described after the first known ascent by a European in late 1784. Anderson, who was then keeper of the newly established Botanic Station on St. Vincent, describes the nature of the crater in detail, along with a sketch map of the summit area (Fig. 3). This can readily be interpreted as a lava dome with actively degassing fumaroles—not greatly different from the current status of the dome, which extruded in 1979.

During the 'modern era' there were two eruptions of St. Vincent, both of which were

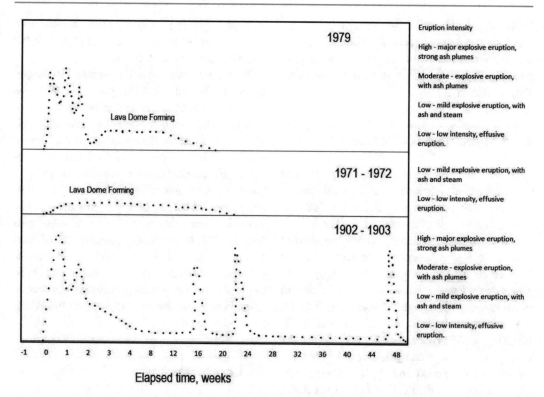

Fig. 4 Schematic timelines of the eruptive sequences at St. Vincent for each of the last 3 eruptions, based on contemporary reports (Blue Book 1902, 1903; Aspinall et al. 1973; Shepherd and Aspinall 1982; Shepherd and Sigurdsson 1982)

closely observed and documented. In 1971–1972 a 'quiet' effusive eruption led to the emplacement of a lava dome within the summit crater-lake. This was followed, just 8 years later, by a short but violent series of explosive eruptions (from April 13–26, 1979), followed by six months of the emplacement of a lava dome (Aspinall et al. 1973; Shepherd et al. 1979; Shepherd and Sigurdsson 1982; Huppert et al. 1982).

4.6 Challenges for the Future

In the context of anticipating future events, the detailed records of these past eruptions and their timelines (Fig. 4), would certainly inform the development of future eruption scenarios. But our understanding of the nature of any precursors is, as with the case of Santorini, rather limited, with just the two most recent eruptions having any instrumental monitoring record—the first of

which made barely a trace in terms of detectable precursory seismicity (Aspinall et al. 1973).

Each of the last four explosive eruptions were preceded by felt earthquakes in the north of the island; each of the eruptions for which the crater was flooded also showed short-timescale changes in the nature of the lake waters. There are no quantitative data on the nature of any pre-eruption gas emissions, since the most recent eruptions in 1979 occurred only just in time for the satellite monitoring era (Carn et al. 2016), and before the widespread adoption of volcanic gas measurement technologies. Nonetheless, there have been local crises on St. Vincent triggered by the appearance of sulfurous odors and haze. The most recent of these events occurred in mid-February 2005, and was detected both on the island of St. Vincent and on the Grenadines (50–75 km S). In the event, Seismic Research Unit scientists and the Soufrière Monitoring Unit determined that the cause of the event was purely meteorological in origin: the typical winds in the area take the fumarole emissions out to sea, and this event was ascribed to an unusual wind field (Weekly Report, Smithsonian Global Volcanism Project, March 2005).

5 A 'Typology' for Volcanoes?

Until we have accumulated a few more decades of instrumental monitoring and global survey data for both restless and erupting volcanoes, volcanologists will of necessity have to fall back on the empirical approach of trying to understand how individual volcanoes might behave in the lead up to a future eruption. With larger and longer-term datasets, we should soon be able to make progress developing quantitative 'typologies' of volcano behaviour (c.f. Hone et al. 2007; Grosse et al. 2014), and also to test whether the concept of an 'eruption cycle' for volcanoes can usefully be developed and applied more generally (e.g. Luhr and Carmichael 1980; Luhr 2002; Pyle et al. 2013).

Until that time, statistical approaches to managing both long-term activity and poorly-constrained future activity might best be developed by building on the 'evidence-based learning' approaches to managing volcanic unrest developed by Aspinall and others, (Aspinall et al. 2003; Aspinall 2012; Hicks et al. 2014). This approach has been widely used on Montserrat as an element of the regular external scrutineering of the state of the volcano. Advances in machine learning and the use of the 'crowd' to harness the potential of citizen science volunteer communities for collecting, analysing and interpreting digital data of all sorts (both structured, and unstructured; quantitative and qualitative) have already begun to make an impact in the area of rapid disaster response (e.g. Ramchurn et al. 2015). The new field of 'human agent interaction' offers significant potential for developing new ways of converting contextual information on past volcanic events (including that currently buried in archives) into a form that can be used to develop and test conceptual and quantitative models of hazard and risk; and as a way of beginning to analyse the networks of people, agencies and organisations affected by or responding to volcanic events.

Both approaches have the potential to deepen understanding of what happens to volcanoes when they enter into a new phase of unrest or eruption, and to aid the detection and diagnosis of these changes in time-series of observations. They also have potential to feed into the processes of communication of hazards and risk, by providing a framework within which to understand the parallels between volcanoes. For example, connecting with the narratives of past events at a re-activating volcano, or of past hazard events at a volcano with shared characteristics, will be important elements in helping to frame discussions and decision-making processes about how to prepare for and mitigate the effects of a future event.

6 Conclusions

To make the best use of evidence-based approaches to the management of emerging volcanic unrest, volcanologists would benefit from making fuller use of the wider contextual 'data' that may exist that documents the consequences of prior volcanic activity, or unrest, at that volcano. Retrospective analysis both of the formal scientific literature, and reading of a wider range of contemporary sources that document the broader personal, social, economic and political impacts of prior events will enrich and add to our capacity to anticipate, prepare for and mitigate the consequences of future events; and to advance the ways in which these learnings may be communicated.

Acknowledgements This work has been supported by the NERC-ESRC funded STREVA project, 'strengthening resilience in volcanic areas', and has benefitted enormously from discussions with Jenni Barclay, Teresa Armijos Burneo, Paul Cole, Anna Hicks, Chris Kilburn, Sue Loughlin, Tamsin Mather, Richie Robertson, Mel Rodgers, Simon Smith, Jon Stone, Caroline Williams, Emily Wilkinson and many others. I thank St. Anne's College and the University of Oxford for permitting a period of sabbatical leave; Cambridge University Library, the Bodleian Libraries, Oxford, and the British Library for access to their archives; the Bodleian Libraries

Exhibitions Team for providing a creative outlet and focus for some of this work; and Carina Fearnley and the editors for their patience.

References

Anderson A, Yonge G (1785) An account of Morne Garou, a Mountain in the Island of St. Vincent, with a description of the volcano on its summit. Phil Trans R Soc London 75:16–31

Anderson T (1903) Volcanic studies in many lands. John Murray, London, p 202

Anderson T, Flett J (1903) Report on the eruptions of the Soufriere, in St. Vincent, in 1902, and on a visit to Montagne Pelée, in Martinique. Part I. Phil Trans R Soc London A 200:353–553

Aspinall WP (2006) Structured elicitation of expert judgement for probabilistic hazard and risk assessment in volcanic eruptions. In: Mader HM et al (eds), Statistics in volcanology, IAVCEI Proc Volcanol 1:15–30

Aspinall WP (2012) Comment on "Social studies of volcanology: knowledge generation and expert advice on active volcanoes" by Amy Donovan, Clive Oppenheimer and Michael Bravo. Bull Volcanol 74:1569–1570

Aspinall WP, Woo G (2014) Santorini unrest 2011–2012: an immediate Bayesian belief network analysis of eruption scenario probabilities for urgent decision support under uncertainty. J Appl Volcanol 3:12

Aspinall WP, Sigurdsson H, Shepherd JB (1973) Eruption of Soufriere volcano on St. Vincent Island, 1971–1972. Science 181:117–181

Aspinall WP, Loughlin SC, Michael FV et al (2002) The Montserrat volcano observatory: its evolution, organization, role and activities. In: Druitt TH, Kokelaar BP (eds) The eruption of Soufrière Hills Volcano, Montserrat, from 1995 to 1999, Geol Soc London Mem 21:71–92

Aspinall WP, Woo G, Voight B, Baxter PJ (2003) Evidence-based volcanology: application to eruption crises. J Volcanol Geotherm Res 128:273–285

Barclay J, Haynes K, Mitchell T, Solana C, Teeuw R, Darnell A, Crosweller HS, Cole P, Pyle D, Lowe C, Fearnley C, Kelman I (2008) Framing volcanic risk communication within disaster risk reduction: finding ways for the social and physical sciences to work together. In: Liverman DGE, Pereira CPG, Marker B (eds) Communicating environmental geoscience, Geol Soc London Spec Pub 305:163–177

Barclay J, Haynes K, Houghton B, Johnston D (2015) Social processes and volcanic risk reduction. In: Sigurdsson H et al (eds) The encyclopedia of volcanoes, 2nd ed, pp 1203–1214

Biggs J, Ebmeier SK, Aspinall WP et al (2014) Global link between deformation and volcanic eruption quantified by satellite imagery. Nat Comms 5:3471

Blong RJ (1982) The time of darkness. University of Washington Press, USA, 242 pp

Blong RJ (1982) The time of darkness: local legends and volcanic reality in Papua New Guinea. University of Washington Press, Seattle

Blue Book (1813) Report from committee on petition of persons interested in estates in the Island of Saint Vincent. Parliamentary paper by command 182, house of commons, London

Blue Book (1903) Correspondence relating to the volcanic eruptions in St. Vincent and Martinique in May 1902, with map and appendix. parliamentary paper by command 1201, HMSO, Darling and Son, London

Blue Book (1904) Further correspondence relating to the volcanic eruptions in St. Vincent and Martinique in 1902 and 1903. Parliamentary paper by command 1783, HMSO, Darling and Son, London

Browning J, Drymoni K, Gudmundsson A (2015) Forecasting magma-chamber rupture at Santorini volcano, Greece. Sci Rep 5:15785

Carn SA, Clarisse L, Prata AJ (2016) Multi-decadal satellite measurements of global volcanic degassing. J Volcanol Geoth Res 311:99–134

Chester DK, Duncan AM, Wetton P, Wetton R (2007) Responses of the Anglo-American military authorities to the eruption of Vesuvius, March 1944. J Hist Geogr 33:168–196

Chester DK, Duncan A, Kilburn C, Sangster H, Solana C (2015) Human responses to the 1906 eruption of Vesuvius, southern Italy. J Volcanol Geoth Res 296:1–18

da Corogna L (1867) De l'influence des emanations volcaniques sur les etres organises, particulierement etudiée a Santorin pendant l'eruption de 1866. Adrien Delahaye, Paris (In French)

D'Arcy Wood G (2014) Tambora: the eruption that changed the world. Princeton University Press, USA, 312 pp

Defoe D (1718) An account of the island of St. Vincent in the West Indies and of its entire destruction on the 26th March last, with some rational suggestions concerning the causes and manner of it. Mist's Wkly J, Issue 82, July 5. [Reprinted in Aitken, ed., 1895, Romances and Narratives vol. 15]

Degruyter W, Huber C, Bachmann O, Cooper KM, Kent AJR (2016) Magma reservoir response to transient recharge events: the case of Santorini Volcano (Greece). Geology 44:23–26

Dimitriadis IN, Panagiotopoulos DG, Papazachos CB et al (2005) Recent seismic activity (1994–2002) of the Santorini volcano using data from local seismological network. In: Fytikas M, Vougioukalakis GE (eds) South Aegean Volcanic Arc, Dev Volcanol 7:185–203

Dominey-Howes D, Minos-Minopoulos D (2004) Perceptions of hazard and risk on Santorini. J Volcanol Geotherm Res 137:285–310

Donovan AR, Oppenheimer C, Bravo M (2012) Social studies of volcanology: knowledge-generation and expert advice on active volcanoes. Bull Volcanol 74:677–689

Donovan A, Oppenheimer C, Bravo M (2014) Reflexive volcanology: 15 years of communicating risk and uncertainty in scientific advice on Montserrat. In: Wadge G, Robertson REA, Voight B (Eds), Eruption of Soufrière hills volcano, Montserrat from 2000 to 2010, Geol Soc London Mem 39:457–470

Druitt TH, Edwards L, Mellors RM et al (1999) Santorini Volcano. Geol Soc London Mem 19:1–178

Druitt TH, Kokelaar BP (eds) (2002) Eruption of Soufrière hills volcano, Montserrat from 2000 to 2010. Geol Soc London Mem 39:457–470

Fouqué F (1879) Santorin et ses éruptions. Masson et compagnie, Paris (In French)

Fytikas M, Kolios N, Vougioukalakis G (1990) Post-Minoan activity of the Santorini volcano: volcanic hazard and risk, forecasting possibilities. In: Hardy DA, Keller J, Galanopoulos VP et al (eds) Thera and the Aegean world III, vol 2. The Thera Foundation, London, pp 183–198

Georgalas GC (1953) L'éruption du volcan de Santorin en 1950. Bull Volcanol 13:39–55

Georgalas GC, Papastamatiou J (1951) Uber den Ausbruch des Santorinvulkanes von 1939–1941. Der Kténas Ausbruch. Bull Volcanol 11:1–37

Gorée F (1710) A relation of a new island, which was raised up from the bottom of the sea on the 23rd of May 1707, in the Bay of Santorin, in the Archipelago. Phil Trans Roy Soc London 27:354–375

Gouin P (1979) Earthquake history of Ethiopia and the Horn of Africa. IDRC, Ottawa, vol 118, 259 pp

Grosse P, Euillades PA, Euillades LD, van Wyk de Vries B (2014) A global database of composite volcano morphometry. Bull Volcanol 76:784

Gullick CJMR (1985) Myths of a minority: the changing traditions of the Vincentian Caribs. Studies of Developing Countries, vol 30, van Gorcum, Assen, Netherlands

Hamilton J (2012) Volcano: nature and culture. Reaktion books, London, p 208

Haynes K, Barclay J, Pidgeon N (2007) Volcanic hazard communication using maps: an evaluation of their effectiveness. Bull Volcanol 70:123–238

Hicks A, Barclay J, Simmons P, Loughlin S (2014) An interdisciplinary approach to volcanic risk reduction under conditions of uncertainty: a case study of Tristan da Cunha. Nat Hazards Earth Syst Sci 14:1871–1887

Hicks A, Few R (2015) Trajectories of social vulnerability during the Soufrière Hills volcanic crisis. J Applied Volcanol 4:10

Higgins MD (1996) Magma dynamics beneath Kameni volcano, Thera, Greece, as revealed by crystal size and shape measurement. J Volcanol Geotherm Res 70:37–48

Hincks TK, Komorowski J-C, Sparks RSJ, Aspinall WP (2014) Retrospective analysis of uncertain eruption precursors at La Soufrière volcano, Guadeloupe, 1975–77: volcanic hazard assessment using a Bayesian Belief Network approach. J Appl Volcanol 3:3

Hone DWE, Mahoney SH, Sparks RSJ, Martin KT (2007) Cladistic analysis applied to the classification of volcanoes. Bull Volcanol 70:203–220

Hooper A, Prata F, Sigmundsson F (2012) Remote sensing of volcanic hazards and their precursors. Proc IEEE 100:2908–2930

Hovey EO (1903a) An American report upon the West Indian eruptions. Nature 67:256–259

Hovey EO (1903b) Bibliography of literature of the West Indian eruptions published in the United States. Bull Geol Soc America 15:562–566

de Humboldt A, Bonpland A (1825) Personal narratives of travels to the equinoctial regions of the new continent during the years 1799–1804, 2nd ed, Translated by Williams HM, vol 4. Longman et al, London, p 573

Huppert HE, Shepherd JB, Sigurdsson H, Sparks RSJ (1982) On lava dome growth, with application to the 1979 lava extrusion of the Soufrière of St. Vincent. J Volcanol Geotherm Res 13:119–130

Hutchison AA, Cashman KV, Williams CA, Rust AC (2016) The 1717 eruption of Volcán de Fuego, Guatemala: cascading hazards and societal response. Quatern Int 394:69–78

Jenkins SF, Barsotti S, Hincks TK, Neri A, Phillips JC, Sparks RSJ, Sheldrake T, Vougioukalakis G (2015) Rapid emergency assessment of ash and gas hazard for future eruptions at Santorini Volcano, Greece. J Appl Volcanol 4:16

Johnson RW (2013) Fire mountains of the islands—a history of volcanic eruptions and disaster management in Papua New Guinea and the Solomon Islands. Australian National University ePress, Australia

Lipman PW, Mullineaux DR (eds) (1981) The 1980 eruptions of Mount St. Helens, Washington, US geological survey professional paper 1250, 844 pp

Loughlin SC, Sparks RSJ, Brown SK, Jenkins SF, Vye-Brown C (2015) Global Volcanic hazards and risk. Cambridge University Press, UK, 410 pp. doi:10.1017/CBO9781316276273

Luhr JF, Carmichael ISE (1980) The colima volcanic complex, México: part I. Post-caldera andesites from Volcán Colima. Contrib Mineral Petrol 71:343–372

Luhr JF, Simkin T (1993) Paricutín: the volcano born in a Mexican cornfield. Geoscience Press, Phoenix, Arizona, p 427

Luhr JF (2002) Petrology and geochemistry of the 1991 and 1998–1999 lava flows from Volcán de Colima, México: implications for the end of the current eruptive cycle. J Volcanol Geotherm Res 117:169–194

Marinatos S (1939) The volcanic destruction of minoan crete. Antiquity 13:425–439

Martin VM, Holness MB, Pyle DM (2006) Textural analysis of magmatic enclaves from the Kameni islands, Santorini, Greece. J Volcanol Geotherm Res 154:89–102

Milham WI (1924) The year 1816—the causes of abnormalities. Mon Weather Rev 52:563–570

Moran SC, Newhall C, Roman DC (2011) Failed magmatic eruptions: late-stage cessation of magma ascent. Bull Volcanol 73:115–122

Morris C (1903) The volcano's deadly work: from the fall of Pompeii to the destruction of St. Pierre, WE Scull, 448 pp

Newhall CG, Punongbayan RS (Eds) (1996) Fire and mud: eruptions and lahars of Mount Pinatubo, Philippines, University of Washington Press, USA, 1126 pp

Newman AV, Stiros S, Feng L et al (2012) Recent geodetic unrest at Santorini Caldera, Greece. Geophys Res Lett 39:L06309. doi:10.1029/2012GL051286

Nicholls IA (1971) Petrology of Santorini volcanic rocks. J Petrol 12:67–119

Nomikou P, Carey S, Papanikolaou D et al (2012) Submarine volcanoes of the Kolumbo volcanic zone NE of Santorini Caldera, Greece. Global Planet Change 90–91:135–151

Nomikou P, Parks MM, Papanikolaou D et al (2014) The emergence and growth of a submarine volcano: the Kameni islands, Santorini, Greece. GeoResJ 1–2:8–18

Oppenheimer C (2011) Eruptions that shook the world. Cambridge University Press, USA, 392 pp

Oppenheimer C (2015) Eruption politics. Nat Geosci 8:244–245

Parks MM, Biggs J, England P et al (2012) Evolution of Santorini Volcano dominated by episodic and rapid fluxes of melt from depth. Nat Geosci 5:749–754

Parks MM, Moore J, Papanikolaou X, Biggs J, Mather TA, Pyle DM, Raptakis C, Paradissis D, Hooper A, Parsons B, Nomikou P (2015) From quiescence to unrest—20 years of satellite geodetic measurements at Santorini volcano, Greece. J Geophys Res 120:1309–1328

Passarelli L, Brodsky EE (2012) The correlation between run-up and repose times of volcanic eruptions. Geophys J Int 188:1025–1045

Perret FA (1924) The Vesuvius eruption of 1906: study of a volcanic cycle, Carnegie Institution of Washington, USA, 151 pp

Petit-Breuhl Sepúlveda ME (2004) La historia eruptiva de los volcanes hispanoamericanos (siglos XVI al XX): el modelo Chileno. Casa de las Volcanes 8, Cabildo de Lanzarote

Petit-Breuhl Sepúlveda ME (2006) Naturaleza y desastres en Hispanoamérica. La visión de los Indíginas, Silex Ediciones, Madrid, p 160

Post J (1977) The last great subsistence crisis. The Johns Hopkins University Press, Baltimore, p 256

Pyle DM, Elliott J (2006) Quantitative morphology, recent evolution, and future activity of the Kameni Islands volcano, Santorini, Greece. Geosphere 2:253–268

Pyle DM, Mather TA, Biggs J (2013) Remote sensing of volcanoes and volcanic processes: integrating observation and modelling—introduction. In: Pyle DM, Mather TA, Biggs J (eds) Remote sensing of volcanoes and volcanic processes: integrating observation and modelling, Geol Soc London Spec Pub 380:1–13

Pyle DM (2016) Field observations of tephra fallout deposits. In: Cashman K, Ricketts H, Rust A, Watson M, Mackie S (eds) Volcanic Ash: Hazard Observation. Elsevier, Netherlands, pp 25–37

Pyle DM (2017) Volcanoes: encounters through the ages. Bodleian Publishing, England, 224 pp

Pyle DM, Barclay J, Armijos MT (2017) The 1902–3 eruptions of the Soufrière, St. Vincent: impacts, relief and response. J Hist Geogr (in review)

Ramchurn SD, Wu F, Jiang W, Fischer J, Reece S, Roberts S, Rodden T, Greenhaigh C, Jennings NR (2015) Human-agent collaboration for disaster response. J Auton Agents Multi-Agent Syst 30:82–111

Reck H (ed) (1936) Santorin. Das Werdegang eines Inselvulkans und sein Ausbruch, 1925–1928. Ergebnisse einer deutsch-griechischen Arbeitgemeinschaft. Berlin. 3 vols. (In German)

Richardson BC (1997) Economy and environment in the Caribbean: Barbados and the Windwards in the late 1800s. Univ West Indies Press, West Indies

Robertson REA (1995) An assessment of the risk from future eruptions of the Soufrière volcano of St. Vincent, West Indies. Nat Hazards 11:163–191

Russell IC (1902) The recent volcanic eruptions in the West Indies. Nat Geogr Mag 13:267–285

Scarpa R, Tilling RI (1996) Monitoring and mitigation of volcano hazards. Springer, Berlin, p 841

Self S, Rampino MR, Newton MS, Wolff JA (1984) Volcanological study of the great Tambora eruption of 1815. Geology 12:659–663

Self S, Gertisser R (2015) Tying down eruption risk. Nat Geosci 8:248–250

Shepherd C (1831) An historical account of the island of Saint Vincent. W Nicol, London

Shepherd JB, Aspinall WP (1982) Seismological studies of the Soufrière of St. Vincent, 1953–1979: implications for volcanic surveillance in the Lesser Antilles. J Volcanol Geotherm Res 12:37–55

Shepherd JB, Aspinall WP, Rowley KC et al (1979) The eruption of Soufrière volcano, St. Vincent, April–June 1979. Nature 282:24–28

Shepherd JB, Sigurdsson H (1982) Mechanism of the 1979 explosive eruption of Soufrière volcano, St. Vincent. J Volcanol Geotherm Res 13:119–130

Siebert L, Simkin T, Kimberly P (2010) Volcanoes of the world, 3rd edn. University of California Press, Berkeley

Simkin T, Fiske RS (1983) Krakatau 1883: the volcanic eruption and its effects. Smithsonian Institution Press, Washington, DC, p 464

Smith SD (2011) Volcanic hazard in a slave society: the 1812 eruption of Mount Soufrière in St. Vincent. J Hist Geogr 37:55–67

Stommel H, Stommel E (1983) Volcano weather: the story of 1816, the year without a summer. Seven Seas Press, Newport RI, 177 pp

Stothers RB, Rampino MR (1983) Volcanic eruptions in the Mediterranean before A.D. 630 from written and archaeological sources. J Geophys Res 88:6357–6371

Stothers RB (1984) The great Tambora eruption in 1815 and its aftermath. Science 224:1191–1198

Symons GJ (ed) (1888) The eruption of Krakatoa and subsequent phenomena. Trübner, London

Tarillon (1715) Relation en forme de journal de la nouvelle isle sortie de la mer dans le Golfe du Santorin, in, Fleurian D'Armenonville, Th. C., editor, Nouveaux memoires des missions de la compagnie de Jesus, dans le Levant, Paris: Nicolas Le Clerc, pp 126–161. (In French)

Venezky DY, Newhall CG (2007) WOVOdat design document; the schema, table descriptions, and create table statements for the database of worldwide volcanic unrest (WOVOdat Version 1.0): US geological survey open file report 2007–1117, 184 pp

Vougioukalakis G, Sparks RSJ, Pyle D, Druitt T, Barberi F, Papzachos C, Fytikas M (2016) Volcanic hazard assessment at Santorini volcano: a review and a synthesis in the light of the 2011–2012 Santorini unrest. In: Proceedings of the 14th international conference of the geological society of Greece, Bull Geol Soc Greece L: 274–283

Wadge G, Aspinall WP (2014) A review of volcanic hazard and risk-assessment praxis at the Soufriere Hills Volcano, Montserrat from 1997 to 2011. In: Wadge G, Robertson REA, Voight B (eds) Eruption of Soufrière Hills volcano, Montserrat from 2000 to 2010. Geol Soc London Mem 39:439–456

Wadge G, Robertson REA, Voight B (eds) (2014) Eruption of Soufrière Hills volcano, Montserrat from 2000 to 2010, Geol Soc London Mem 39:501 pp

Watt SFL, Mather TA, Pyle DM (2007) Vulcanian explosion cycles: patterns and predictability. Geology 35:839–842

Watts AB, Nomikou P, Moore JDP, Parks MM, Alexandri M (2015) Historical bathymetric charts and the evolution of Santorini submarine volcano, Greece. Geochem Geophys Geosystems 16:847–869

Wiart P, Oppenheimer C (2000) Eruptive history of Dubbi volcano, northeast Afar (Eritrea), revealed by optical and SAR image interpretation. Int J Remote Sensing 21:911–936

Winchester S (2003) Krakatoa: the day the world exploded, 27 August 1883. Viking, London, p 432

Zollinger H (1855) Besteigung des Vulkanes Tambora auf der Insel Sumbawa, und schilderung der Erupzion desselben im Jahr 1815. Winterthur, Switzerland, p 22

Reflections from an Indigenous Community on Volcanic Event Management, Communications and Resilience

H. Gabrielsen, J. Procter, H. Rainforth, T. Black,
G. Harmsworth and N. Pardo

Abstract

Ngāti Rangi, an indigenous tribe of Aotearoa New Zealand, live on the southern flanks of their ancestral mountain, Ruapehu, an active volcano. Ruapehu has erupted and caused lahars within living memory, and nearby Tongariro erupted as recently as 2012. Ngāti Rangi and other tribes affiliated to these mountains are intimately connected to and familiar with the moods, signs, and language of the mountains and have valuable knowledge to contribute to decision-making and warning systems during volcanic events. To date this knowledge or *mātauranga* Māori has been somewhat under-utilised, and Ngāti Rangi have not always been included in decision-making processes during volcanic events. But communication is improving, and Ngāti Rangi have begun a journey of building their own monitoring, information collection, and communication systems. Past and present monitoring, warning systems, communications and tribal civil defence resources are examined to determine how Ngāti Rangi and their tribal knowledge can be better recognised, communications with governmental volcanic hazard management agencies improved to ultimately work together to improve outcomes for the tribe and local community.

H. Gabrielsen · H. Rainforth
Te Kahui O Paerangi—Ngāti Rangi Trust,
P.O. Box 195, Ohakune, New Zealand

H. Gabrielsen · J. Procter (✉) · N. Pardo
IAE, Massey University, Private Bag 11 222,
Palmerston North 4442, New Zealand
e-mail: J.N.Procter@massey.ac.nz

T. Black
Te Whare Wānanga O Awanuiārangi,
13 Domain Road, Whakatāne, New Zealand

G. Harmsworth
Landcare Research-Manaaki Whenua, Private Bag
11052, Palmerston North 4442, New Zealand

1 Introduction

Koro Ruapehu is constantly changing. Sometimes he's sleeping, sometimes he's active – sometimes he erupts

(pers. comm. Ngāti Rangi Trust 2014)

Despite a plethora of initiatives internationally, regionally and locally to reduce risk or increase resilience to natural hazards (e.g. the United Nations International Strategy for

Advs in Volcanology (2018) 463–479
https://doi.org/10.1007/11157_2016_44
© The Author(s) 2017
Published Online: 05 April 2017

Disaster Reduction), indigenous communities and peoples are not well provided for.

Like many indigenous cultures around the world the indigenous people of Aotearoa New Zealand, have observed and monitored and then responded to and recovered from numerous hazardous volcanic events. The indigenous knowledge gained from these experiences is rarely considered when scientifically identifying volcanic hazards or developing emergency management plans, yet mātauranga Māori (Māori knowledge) does contain a unique, valid epistemology and data source. The mātauranga has driven Māori decision making to endure and adapt to the natural hazards they face (Durie 2005). This has not been fully recognised by current hazard and emergency management regimes in New Zealand and has resulted in a disconnect in communication between the indigenous populations and Government agencies (Jolly et al. 2014). This disconnect has become more evident over time and particularly in relation to the 1995–6 sequence of eruptions of Ruapehu, the 2007 lahar and eruptions of Ruapehu and the 2012 eruption of Te Maari, Mt. Tongariro. Is it then feasible to communicate hazard and risk in today's world to Māori living in these areas within a knowledge framework that is spatially and temporally consistent with their past understandings? Simply, is there a means to desegregate methods to create an understanding of risk unique to our volcanic areas that is universally acceptable by all?

The case study chosen for this research is Ngāti Rangi, a central North Island iwi (tribe) who have held unbroken occupation over the area for over 1000 years. They have an intense and living relationship with their ancestral maunga (mountain), which they refer to as Matua te Mana ("*prestige of the father*") and is located within the Tongariro National Park, one of New Zealand's United Nations Educational, Scientific and Cultural Organization (UNESCO) World Heritage sites. The Tongariro National Park is recognised not only for the natural values of the landscape, but also for the cultural values associated with these maunga (mountain) (UNESCO 2014). This recognition demonstrates the cultural

importance placed on this area by local iwi and the acknowledgement of this by UNESCO, who govern the World Heritage List and locations worldwide, in the awarding of dual status in 1993 (recognition of cultural and environmental values) (Keys and Green 2008). The Tongariro National Park is managed by the Department of Conservation, a Crown entity (an organisation that forms part of New Zealand's state sector). Despite the formal recognition of the cultural significance to local iwi of the Tongariro National Park, iwi have little involvement in its management (Gabrielsen 2014).

To understand and develop resilience within Māori communities requires an examination of the role of traditional knowledge within volcanological hazards, risk communication, and emergency management. This research combines several disciplines and therefore requires a distinct method to acquire and analyse data. A mixture of qualitative and semi quantitative research data collection techniques were applied that is based upon recognised kaupapa Māori techniques (Smith 1997). The research undertaken for this study is based on an analysis of the current emergency management framework, analysis of marae survey data, marae assessments, conversations with Ngāti Rangi leaders and elders and an assessment of volcanic based data.

An important aspect of this research is that all researchers are Māori with knowledge of Māori language culture and customs. Secondly, the case study proposed was based on iwi and their interactions with their environment. This meant that iwi determined the progression of the research throughout which was paramount. This process also allows the iwi to be the decision makers, to provide what information they want and to decide how it is used. Consequently, special processes that protect the iwi and their knowledge base were crucial to this research. This research involved the interviewing of iwi leaders, iwi environmental management staff and elders within the Ngati Rangi Tribe. Over the course of a year (2014) approximately 10 individuals were interviewed and 3 wananga (traditional workshops of 3–5 people) were held on

Ngati Rangi Marae (traditional meeting house) in rural areas located within the Ruapehu Volcano ring plain in recently active volcanic areas. In conjunction with interviews, oral traditions, waiata (traditional songs), purakau (ancestral accounts) were also examined. Another unique source of information was written records or minutes of traditional church meetings from the 1800s. Due to traditional practice iwi members are reluctant to have their information publically quoted.

The qualitative aspect of this research was used largely to seek a human perspective. The aim was to identify perspectives from the ground, from people that lived in the area, and from people with a relationship to the land, and to the volcano. Iwi in general have large repositories of knowledge coded in local waiata (songs) and karakia (prayer) and held by those in particular deemed worthy of holding on to such knowledge. Historical knowledge of volcanic episodes should be confined within iwi history and korero (speech). This assumption was made purely based on the fact that Ngāti Rangi have long lived within the lands of their ancestors, for over a thousand years, and therefore will have experienced and recorded in some way, volcanic events.

Gaining a better understanding of historical occurrences and responses is beneficial for current research on volcanic hazards and for emergency management. A qualitative approach was able to unearth to some degree the current gap in the knowledge base regarding iwi and volcanic hazards, and understanding what aspects contribute to iwi resilience to natural hazards. This approach was also required as a means for some freedom of movement in the type of method utilised to better support the dynamic nature of iwi and the preference with which iwi choose to be consulted. The ultimate outcome is elucidating some potential indicators of what resilience means to Māori and can it be strengthened within current emergency management frameworks.

1.1 Resilience and Indigenous Communities

Various aspects of communication with, and the resilience of, indigenous communities are contained within natural hazards research, but are described in the following ways: disaster prevention (Alcántara-Ayala 2002), disaster risk reduction (Mercer et al. 2010), and assessing the vulnerability of communities to natural hazards (Cutter et al. 2010). Work with indigenous communities within the Pacific region has provided examples of capacity building within indigenous communities and incorporating their own cultural knowledge into scientific methodology to adequately prepare for, and deal with, natural hazards (Petterson et al. 2003). The community of Savo Island in the Solomon Islands is exposed to a high level of volcanic activity with a history of large fatalities (Petterson et al. 2003). Outside expertise and assistance was sought to initiate the development of strategies to address the risks from volcanic activity on the island (Petterson et al. 2003).

The development of these strategies to address the risks from volcanic activity on Savo encompassed in-depth work with the local community. This work included workshops and identifying and using local knowledge of hazards in conjunction with science to develop a disaster management plan. This process identified in some respects how crucial the political, economic and infrastructural climate is in supporting the resilience of these island nations; political drivers secured the expertise thus enabling the development and implementation of the disaster management plan. Despite the best intentions in aiding indigenous communities in developing strategies to deal with natural hazards, there can be a multitude of barriers in undertaking this work. One of the challenges identified in work undertaken on Ambae Island, Vanuatu, was initially the lack of acceptance by the local population of scientific knowledge (Cronin et al. 2004b).

Breaking through the barrier of the dominance of science is essential for indigenous cultures, as there are a significant number of deep-seated issues surrounding research, intellectual property and exploitation. These issues have led to indigenous communities distrusting researchers, their methods and their desired outcomes. As a means to alleviate such issues researchers have used the principles of the Participatory Rural Appraisal to alter the attitudes and approach of the specialists and to promote community input and knowledge (Bird and Gísladóttir 2012; Cronin et al. 2004a, b). These principles perhaps parallel Māori research initiatives (e.g. kaupapa Māori; Smith 1997), which emphasises elements central to the Treaty of Waitangi, such as participation, partnership, and protection (Robertson 1999). In Vanuatu, Cronin et al. (2004b) observed that strong cultural customs prevented the indigenous peoples from accommodating standard scientific methods, but also that these methods were inconsistent with those customs and the knowledge and beliefs of the people. The researchers envisaged that PRA would act as the instrument to incorporate traditional knowledge into the development of a hazard management plan without the risk of jeopardising the indigenous communities' local belief structures (Cronin et al. 2004b).

Despite work by scientists and disaster management researchers in their aim to understand and improve the resilience of many indigenous cultures, Campbell (2009) indicates that many Pacific Island nations were once inherently resilient to natural hazards. Traditional disaster reduction measures describe the ways through which indigenous communities succeeded in living with natural hazards. Colonisation introduced changes to these societies (Zimmet et al. 1990) that removed the importance of their traditional and highly social practices and left communities unprepared and ill-equipped to deal with natural hazards. Globalisation and other external pressures may be processes that are out of these communities' control, but still have far reaching impacts on their internal processes and traditions (Mercer et al. 2010; Pelling and Uitto 2001). Among these pressures, Paulinson (1993) found the market forces to be at fault. These aspects may inhibit indigenous communities from being resilient. Despite this, the traditional disaster reduction measures that are representative of inherent qualities central to communities living with natural hazards promote resilience.

2 Ruapehu Volcano

The convergence of the Australian and Pacific tectonic plates is the driver of volcanism within New Zealand. Situated within the Tongariro Volcanic Centre (TVC) (Acoella et al. 2003) (Fig. 1), Ruapehu is the largest and most active volcano in the North Island standing at 2797 m high (Lecointre et al. 2004; Neall et al. 1999). Upon Ruapehu are three summit craters that have all been active over the last 10,000 years (Donoghue and Neall 2001). This includes the current Crater Lake, Te Wai-a-moe, which is situated over the active southern Crater. Ruapehu's periodic activity causes a number of hazardous events with evidence of these volcanic hazards recorded in the landscape and represented by the surrounding volcaniclastic ring plain, (Fig. 1) which is made up of fragmented rocks deposited by historical lahars, debris avalanches and some fluvial and glacial deposits (Donoghue and Neall 2001).

The Department of Conservation management of this area is guided by the Tongariro National Park Management Plan 2006–2016 (Department of Conservation 2006a, b). This document outlines the roles and responsibilities of the Department and the policies that guide the use of this area. The responsibility of managing natural hazards lies with the District and Regional Councils, while monitoring volcanic activity is undertaken in conjunction with research providers such as GNS Science and Massey University. The Department of Conservation describes the risks from natural hazards as taking '*two main forms*' (Department of Conservation

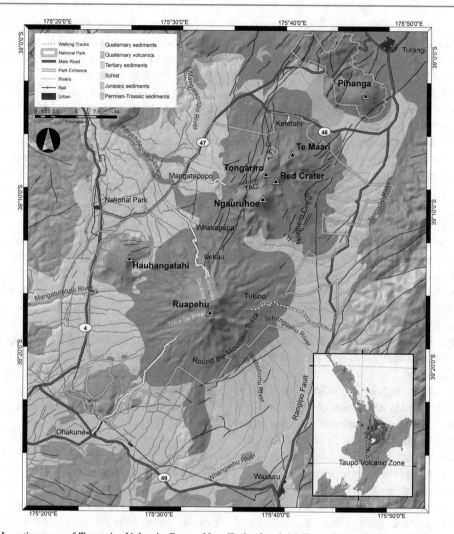

Fig. 1 Location map of Tongariro Volcanic Centre, New Zealand and the Taupo Volcanic Zone (*inset*)

2006b): (1) flows such as lahars, pyroclastic flows, lateral blasts, landslides/floods and lava flows and (2) air borne materials such as rocks, tephra and toxic gases. Ruapehu is unique in the sense that the Crater Lake, Te Wai-a-moe, is located over the current active vent of the volcano. Keys and Green (2008) mention that only one other crater lake (i.e. Kelut in Java, Indonesia) similar to that of Ruapehu exists within the world where there has been an intense focus on physical lahar mitigation. Both of these crater lakes and the research undertaken in reference to lahar hazards have provided important scientific data for hazards research.

3 Māori/Iwi in New Zealand and Ngāti Rangi

Indigenous cultures around the world have unique world views that inform their cultural values, belief systems and link them to the natural world. Royal (2005) argues this worldview sets Māori and other indigenous cultures apart from the mainstream populations. The common denominator among indigenous cultures is that the natural world is perceived as a living being (Royal 2005; Hart 2010) and this connection with the environment ties indigenous peoples around the world with one another, epistemologically.

Although world views varies from iwi to iwi, a strong thread uniting iwi Māori viewpoints is the genealogical connection to the physical and spiritual world. In the Māori world view, human beings are the last creation of the god of the forest, Tāne Māhuta. Rather than this furnishing humans with primacy over creation, as is the common interpretation within the European tradition, for Māori this position renders them as subordinates of all other creatures in the natural world. This world view has two implications. Firstly, Māori are connected to all things within nature. This includes physical features, such as mountains, rivers, rocks, and land, as well as what is traditionally thought of as living things. For Māori everything has a life force, an essence or a mauri. Māori are bound to be continually respectful to those who have precedence, i.e. the rocks, the mountains, etc. that precede them in their genealogy.

Mātauranga Māori is defined as 'Māori knowledge', and is a term that places importance on Māori histories, knowledge, and language, and refers to Māori ways of thinking, doing, and acting (e.g. Smith 1990). It is a multi-faceted and complex concept that is connected to a multitude of sources of language, culture, land, customary and intellectual knowledge sources. Furthermore, mātauranga Māori bridges both traditional and contemporary Māori knowledge and philosophies through which Māori history and knowledge are uncompromisingly told. The platform of mātauranga Māori advocates for a system of Māori knowledge that recognises cultural identity and cultural affirmation as important foundations that are connected to Māori world views. Mātauranga Māori is a crucial element in chronicling perspectives, experiences and knowledge of specific landscape and volcanic events within iwi history in Aotearoa New Zealand. Information surrounding volcanic hazards exists within the mātauranga-ā-iwi (specific tribal knowledge), for Ngati Rangi principally through waiata (song) and karakia (prayer). As a repository of cultural knowledge and information

over the generations of Ngāti Rangi existence, these examples provide the opportunity to review past events and provide a basis of knowledge to recover from future events.

At the time of the arrival of European explorers and traders, Māori may have numbered about 100,000 (citation Pool 1991). Māori numbers plummeted to around 42,000 in 1896, due to war and disease, before recovering in the early 20th century (Pool 1991). A large proportion of Māori land was alienated, often through land confiscations by the Crown, suspect land purchases, or as a result of debt accrued, in association with the Crown programme of converting customary titles to fit with the British land title system; this loss removed the economic base of the people and severed connections to traditional lands. Māori culture was severely affected through the imposition of European belief systems, practices, and laws. Despite that, iwi across New Zealand maintained cultural knowledge and connections to place. For Ngāti Rangi, this means chiefly or in part a connection to Ruapehu, their ancestral mountain and active volcano.

3.1 Māori and Hazards

There is limited literature on Māori and natural hazards within New Zealand. Few unpublished documents have provided further insight into the relationship that exists between iwi and the natural hazards present within their rohe (tribal area). Proctor (2010) explores how the principles of tikanga (traditional practices) can be applied to the management of natural hazards, particularly flooding in Pawarenga in Northland, New Zealand. Proctor (2010) found that tikanga was a valued resource used by locals and concluded that 'tikanga Māori is an inherent part of … resiliency' (Proctor 2010).

King et al. (2007) and Lowe et al. (2002) are the few who have explored the relationships between iwi and natural hazards. They found that iwi and hapū (sub-tribe) hold a store of information throughout oral narratives such as 'mōteatea (laments), pēpeha (quotations), wha-katauki (proverbs) and waiata (songs)' (King et al. 2007). These repositories of information not only tell stories but can contribute information on historical events and natural hazard occurrences to natural hazard management. King et al. (2007) outlines three specific ways that Māori environmental knowledge can be applied to natural hazard management: (1) as described previously, stories, songs and place names hold a wealth of knowledge based on experiences and recollections of events; (2) the information extracted from these avenues can thus be mapped in relation to natural hazards; and (3) it can also provide for Māori involvement in planning for hazards.

Most of the current volcanic based knowledge that exists in New Zealand is largely derived from the European context. Lowe et al. (2002) suggest that the lack of information is partly due to the late settlement of New Zealand by humans. Consequently, the recorded history of interactions between people and volcanism is short. There is a paucity of information that has been published on the actual experiences of early Māori prior to colonization; however, Lowe et al. (2002) assume that Māori must have experienced numerous volcanic events from many of New Zealand's volcanic centers (Table 1).

3.2 Ngāti Rangi and "Their" Volcano

In terms of volcanoes, the Māori world view results in an approach where humans are connected through genealogy to mountains, and in particular for Ngāti Rangi they are descendants of Mount Ruapehu. To Ngāti Rangi Mt. Ruape-huis referred to as the grandfather. For Ngāti Rangi, specific connections to the maunga come through Te Rau-hā-moa and Paerangi. When Te Rau-hā-moa brought Paerangi to Aotearoa (New Zealand), the alighting of the bird ignited the fires waiting in Ruapehu, waking up the volcanic life of the mountain. Paerangi himself made the mountain his home, giving rise to one of the names of the mountain—Paerangi i te Whare Toka, or Paerangi of the House of Stone. Today, Ngāti Rangi numbers are at an estimated 8000 people (citation). Fifteen percent of those live in the tribal area (rohe) (Fig. 2), while the others live elsewhere in New Zealand and overseas. The iwi is supported by an iwi authority, Ngāti Rangi Trust, and guided by a tribal council representing all subtribes, Te Kāhui o Paerangi. The Trust is responsible for supporting the day to day work of the tribe, from social support programmes, to tribal events, to upholding the environmental responsibilities and cultural knowledge of the tribe.

Table 1 Volcanic hazards probably experienced or witnessed by prehistoric Māori

Hazard type	Volcano or centre associated with event
Pyroclastic fall	Taranaki, Tongariro, Whakaari, Auckland, Okataina
Pyroclastic flows	Tarakaki, Tongariro, Okataina
Pyroclastic surges	Okataina
Lava flows	Tongariro, Auckland, Okataina
Lava dome building	Tarakani, Tongariro, Okataina, Tuhua
Lahars	Taranaki, Tongariro
Post-eruptive flooding	Taranaki, Tongariro, Okataina
Debris avalanches	Taranaki, Tongariro, Whakaari
Volcanogenic earthquakes	Taranaki, Tongariro, Auckland, Okataina
Lightning, forest fires	Taranaki, Tongariro, Okataina
Hydrothermal eruptions	High-temp. geothermal systems in the Taupo volcanic zone (e.g. Ketetahi Springs)
Acidic rain/volcanic gases	Ruapehu, Tongariro

Adapted from Lowe et al. (2002)

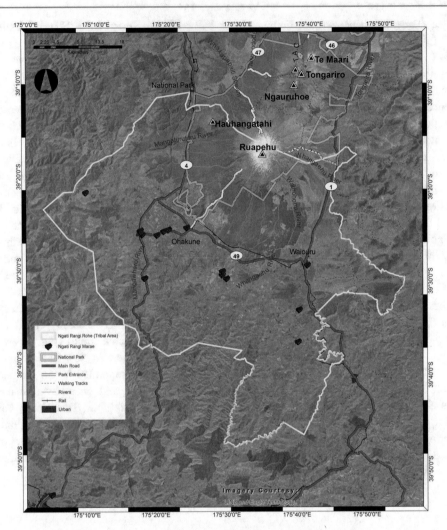

Fig. 2 Ngāti Rangi tribal area or rohe and the Tongariro Volcanic Centre including the location of Ngati Rangi marae or traditional meeting house

For countless generations Ngāti Rangi have inhabited the southern flanks of Ruapehu (Fig. 2). They have born witness to his volcanic activity since human settlement in New Zealand. There are early written accounts of lahars within the Whangaehu River, one such by Reverend Richard Taylor (*as cited in* Hodgson 1993) who in 1861 reported uncharacteristic flooding within the Whangaehu River. Ngāti Rangi has within their oral narratives records of events, records by way of mōteatea (poetry, accounts of ancestors, ngeri (chants), karakia (prayer), and other forms of waiata (songs). For example, the waiata below is from Ngāti Rangi and was written with reference to the eruption of 1945:

Moimoi Tahuārangi te pikinga i Tuhirangi
Ka whakamau te hiwi ki Murimotu ee
Kei tuahiwi taku rori haerenga ki roto Ōhāpopo
Takoto whāroa ngā mānia ki Karioi ee
Kia tū wātea taku titiro te puke ki Ruapehu
Te whakaingo mai he tau pakipaki
Papaki rawa i taku uma
He puke nohoanga nō te keukeu roa
He roa te tāringa kia whakaaria mai ngā tohu tukutuku
Tukutahi te puehu turaki whakatua
Ka whakahoki mai hei tāpora mō te nohoanga ia koutou mā eei

The guardians cry as they ascend Tuhirangi
And then continue on to Murimotu
Over yonder is the path to Ōhāpopo
Where the plains of Karioi open up
So that I can clearly view Ruapehu
Oh the majesty as I wait for a clear period of
weather
And his majesty also reminds
That he originated from the great surges of the
ocean
I stand waiting for activity
Behold! An eruption of ash.
Do not fear, this ash will cloak and replenish the
land and help us live as one.

The waiata is another example of traditional knowledge; waiata is a medium where large tracks of information are stored and repeated throughout the generations. Through this translation provided by Ngāti Rangi, the waiata (songs) relays a number of factors describing the iwi and their relationship with Ruapehu, and their reaction to volcanic activity. This has the potential to reveal the nature of the relationship between Ngāti Rangi and their ancestral maunga and the associated volcanic processes or hazards.

4 Ngāti Rangi Experiences

The communication of Ngāti Rangi perceptions of volcanic 'hazards' or mātauranga Māori (Maori knowledge) or mātauranga-ā-iwi (iwi knowledge) is articulated in their internal iwi korero (oral tradition; language), but also officially stated within their Ngāti Rangi Taiao Management Plan 2014, a management plan developed to address environmental issues within the Ngāti Rangi rohe or region (Gabrielsen 2014). Ngāti Rangi rejects the use of the term hazard when describing the consequence of volcanic activity (Rainforth et al. 2012; Gabrielsen 2014). The perspective is that it is a natural event that should not be restrained, diverted or withheld. This position is communicated throughout the generations and therefore is widely accepted throughout the iwi. This viewpoint is based on the acceptance of Matua te Mana (in general terms, the power and prestige associated with the volcano as an ancestor) and in a wider context, Rūaumoko, as natural entities

and processes involved in volcanic activity. A Ngāti Rangi pao, a very concise song usually sung for entertainment, describes Ruapehu in his eruptive state.

O rongo Ruapehu
Turaki auahi
Puahiri Whakarunga
Ki whai tua ee

If you ever hear Ruapehu
Erupting with ash
You can be comforted knowing
The prevailing wind takes it elsewhere.

Key words and phrases within this pao such as 'puahiri whakarunga' is a descriptive word for an eruption, 'turaki auahi' could also signify the visual experience of witnessing a plume of ash—'auahi' meaning smoke and 'turaki' to throw down. 'Rongo' could also represent reputation; the reputation or fame of Ruapehu, perhaps an indication of the perceptions and understanding of Ruapehu as a volcanic entity.

The Maori perceptions of the volcano as being an active, living entity provides a focus and a need for engagement with that entity on a practical level through the designation of tapu (sacred) areas in relation to the mountain. Practical measures such as exclusion areas or explicit prescriptions of behavior (i.e. not to stop in lahar channels when walking through) were put in place to signify the dangers and risks of places around the maunga, such as Te One Tapu and the kaitiaki (guardian) of the mountain that exist in this area known as Te Ririo and Takakā. Tapu is a belief, a notion that educates one to respect the natural world as 'Māori things involve the whole of nature' (Pewhairangi 1992).

In the past, Ngāti Rangi has taken advantage of the active nature of Ruapehu as a means to deliver specific kōrero (speech) and historical knowledge concerning volcanic activity. Despite the lull between periods of volcanic activity, there is a continuation of internal iwi knowledge sharing which extends throughout these periods of quiescence. There are a variety of techniques Ngāti Rangi use to share knowledge and historical experiences regarding the maunga, which is not always specific to volcanic activity. These are through wānanga (workshops) and rā wairua (religious services or more specifically the Rā māramatanga—an annual event dedicated to the spiritual experiences of Ngāti Rangi). Knowledge and experiences are also transmitted via tikanga, iwi stories, karakia and waiata. The exchange of knowledge is also crucial to the understanding and sharing of cultural and scientific language on a bi-cultural level.

4.1 1945 Eruption

The eruption of 1945 had substantial impacts on the daily lives of those living at the foot of the mountain, largely revolving around ash and its resulting impacts: skin, eye and throat issues, crop failure, issues with stock feed, shearing blades dulling during shearing season, impacts on driving visibility and corrosive impacts on vehicles and machinery (Johnston 1997). A prominent Ngāti Rangi kuia recollected the eruption clearly 'we were covered in ash' (pers comm. 2014). Johnston (1997) sourced 13 separate references of ash fall within Ohākune in 1945 from July through to September with the final date of ash fall occurring over a three-day period. The tribal account of this event is also captured within a waiata 'Moimoi Tahuarangi' which pays homage to the prestige of the mountain and the celebration of eruptive events. During this time, the relationship between the tribe and their mountain was not really recognized by authorities, and cultural use of the mountain was nil due to his state of tapu (sacredness) or exclusion areas were defined.

4.2 Tangiwai Disaster 1953

The Tangiwai railway disaster of 1953 (where a lahar removed a rail bridge across the channel moments before a train crossed killing 151 people) is long held within the memory of Ngāti Rangi and is potentially the most memorable lahar, due to the present elder generation being alive during this time. Many whānau (families) hold stories about this night, as many were scheduled to

travel on the train, and others had a local dance interrupted by news of the event. Many families also contributed to the cleanup of the awa post-disaster and for some time Ngāti Rangi did not visit the Whangaehu River for their cultural and spiritual purposes. A cultural a rāhui (period of prohibition) over the area was implemented by the iwi until a time when it was deemed culturally appropriate. The communication of this event was predominantly by word of mouth. As a result of this event, the New Zealand Railways Department installed a lahar warning device upstream of the Tangiwai Bridge in order to detect any future rise in river level and acidity in existing stream flow due to the introduction of Crater Lake water which could signal an alarm to halt railway traffic until inspection (Neall 1976).

4.3 1995/1996 Series of Eruptions

The 1995/1996 volcanic activity initiated internal discussions within Ngāti Rangi around historical iwi kōrero relating to volcanic activity. This period was integral for iwi revisiting local knowledge and perhaps gaining a more scientific understanding of volcanic activity. This again would have fortified the iwi and their relationship with their ancestral mountain. Notwithstanding this, activity still stimulated practical questions regarding the safety of the iwi marae, homes and infrastructure from volcanic flows (leading up to the 2007 lahar). Many prominent key Ngāti Rangi leaders had belief in their ancestral maunga that he would look after the iwi, which is resounded throughout the iwi. This also contributes to the trust in the experience of Ngāti Rangi tupuna (tribal ancestors) in the placement of marae in the rohe.

4.4 Consultation and Involvement Pre-2007 Lahar

The flow of communication between the Crown and Ngāti Rangi became an integral part of consultation following the 1995/1996 eruptions due to the risk of an eminent Crater Lake dam break, which occurred on 18 March 2007. The consultation between Ngāti Rangi and the government (by way of the Department of Conservation, Minister of Conservation Sandra Lee and the Ministry of Civil Defence and Emergency Management) was viewed as successful. The success was due to the sharing of knowledge, communication and decision-making by Ngāti Rangi. Ngāti Rangi took a stance on the engineering solutions that were proposed and deemed them unacceptable on a cultural and spiritual level, as well as practically.

The position Ngāti Rangi held and still hold was that no intervention would be undertaken on the maunga when discussions turned to a proposed engineering solution at the Crater Lake. An alternative was chosen with the creation of a bund on an apex of the laharic fan or Te One Tapu in consultation with neighboring iwi and the Crown. Emergency management preparation was central to the lead up to the 2007 lahar. The local Karioi Forestry and the timber mill of Winstone Pulp International (WPI) participated in planning for the event as the Whangaehu River meanders through the forest and both of their industrial sites (Karioi and the Timbermill) are located within a 1–2 km vicinity to the Whangaehu River and the Tangiwai Bridge.

The community at Karioi was involved as part of preparatory measures to ensure the safety of the community as the lahar made its way past the Tirorangi Marae bridge. Two community meetings were held for the Karioi residents largely to discuss traffic safety measures and the movement of stock. Pagers were the main means of communicating instructions, timeframes and—monitoring the lahar flow. The main concern for the residents was the safety of the bridge, as the potential impact from its ruin would have resounding economic and social impacts. The local community was empowered through the process of consultation and heavy involvement in the response to the lahar.

4.5 Current Communication

The Ngāti Rangi Trust website provides users with direct links via the internet to a variety of current volcanic surveillance and monitoring tools of Ruapehu, which are:

- Links to GeoNet for current volcanic alert levels;
- Link to Horizon Regional Council's maunga camera;
- and The Ngāti Rangi installed Te Wai-ā-moe (Crater Lake, Ruapehu) camera, which feeds directly to the Trust.

High-level communication occurs at the top level, among research and monitoring institutes such as GNS Science and Massey University, local authorities such as Horizons Regional Council and Waikato Regional Council and Crown entities such as the Department of Conservation and the Ministry of Civil Defence and Emergency Management and the iwi authority, Ngāti Rangi Trust. Through these interactions active involvement in current monitoring of culturally significant sites has been initiated by the iwi and supported by these organisations. The iwi initiated monitoring of Te Wai-ā-moe, Whangaehu River and potentially Lake Rotokura in the near future. The ongoing dissemination of information through these high-level personnel is crucial.

5 Discussion

Tobin and Montz (1997) describe natural hazards as the possibility of interaction between natural events and humans. Therefore, based on this description, a natural hazard is described based on its potential to impact people and property. It is clear that the use of the term 'hazards' to describe a volcanic eruption and its resulting impact on surrounding populations by scientists and emergency managers in New Zealand and internationally is a subject of contention for Ngāti Rangi. Their relationship with Ruapehu expresses their acceptance and understanding of him as a powerful being of nature and awareness of their place within nature and te ao Māori (the Maori world). This recognition and understanding of Ruapehu ultimately means that Ngāti Rangi people accept him as an active volcanic entity and celebrate and welcome his volcanic processes. Ruapehu is Matua te Mana, the guardian of mana (prestige/authority) who uses volcanic activity to share part of his mana with the people, and to replenish and revitalise the land and Ngāti Rangi.

Traditional practices and cultural traits have allowed iwi to endure within New Zealand. Maintaining aspects of their cultural and spiritual traditions and oral narratives have also aided the ability for iwi to adapt to external changes such as colonialism. Ulluwishewa et al. (2008) and Harmsworth and Awatere (2013) maintain that iwi hold a distinct worldview that guides their daily lives. This worldview can be described as an outlook that is heavily embedded in the past but merges with the present; a combination of traditional and modern concepts and beliefs. This worldview is often used as a basis for future decision-making and involves looking to the past to better plan for the future. The teachings of ancestors are prominent in this worldview. There remains a close connection with the local environment, which is representative of a link to the wider holistic aspect of whakapapa (simply defined as genealogy) and whānau (family). Iwi also take this focus on and reverence of their history and fuse it with modern lifestyles. Thus this worldview demonstrates a mixture of both modern and traditional aspects that represent iwi and the Māori culture.

Both Durie (2005) and Walker (2004) describe Māori resilience to natural hazards, more specifically as endurance. They both discuss the struggles of Māori throughout the colonisation phase and the later stages of growth within Aotearoa/New Zealand. Their work highlights the endurance of Māori to survive and adapt specific cultural traits and practices to

flourish and be present in this day and age within Aotearoa/New Zealand.

5.1 Traditional Knowledge as a Tool for Building Resilience

To adequately prepare for hazards specifically with indigenous communities, traditional knowledge should be used to provide a unique insight into information on historical events, as well as previous response methods. It has long been acknowledged that communities residing in hazard prone areas over a number of generations understand hazard processes, and potentially some previous methods of response towards hazards (Campbell 2009; Cashman and Cronin 2008; Cashman and Giordano 2008). Traditional knowledge and oral traditions, which derive from oral narratives (Cashman and Cronin 2008), are valuable tools that represent an awareness and understanding of the locality. They provide an account of historical methods used to avoid, mitigate or reduce the impacts associated with natural hazards. In Iceland, historical accounts of ash fall indicated the level of severity and the resulting impact on visibility (Bird and Gísladóttir 2012), therefore demonstrating for example, what work needs to be done prior to the lack of visibility setting in. These historical accounts can provide local communities with moral support. In the study by Bird and Gísladóttir (2012) one participant said: '*I just thought about the past, the stories. How good it was that we had heard the stories, I knew that it had happened again, I knew that it wouldn't last forever*' (Bird and Gísladóttir 2012, p. 1271).

Recounting those stories from Iceland outlined a natural hazard event, its impacts on the local communities, mitigation measures to undertake and perhaps some indication of its duration. Place names also hold some merit in indicating further insight into a location and its history (King et al. 2007), as representative of an event that left an imprint on the landscape and the people. In contrast, there is still a lack of understanding on the nature of the hazards in volcanic zones, as well as a real understanding of all possible hazard types. In Java, Indonesia, Lavigne et al. (2008) identified that there was little to no actual understanding of volcanic processes; therefore the local population was not aware of the entire volcanic hazard types, their associated risks and more importantly the areas they impacted. This research highlights peoples' perceptions of risks and the importance of bringing into account the human dimension with regards to natural hazard management.

It has been heavily emphasised of late how imperative the human dimension is to natural hazard management (Bird et al. 2011). Understanding the interaction and relationship local communities have with the land may describe the continued existence of people in the vicinity of volcanoes. Lavigne et al. (2008) supports this view by noting the rise in research relating to the human dimension of natural hazard management and more specifically, the behaviour of people in the face of natural hazards. They outlined three significant areas to further understanding the human dimension of hazards and the reactions to natural hazard events: (1) the perception of risk, (2) cultural beliefs and (3) socio-economic constraints.

Individual and community perception of risk is based on a number of differing factors, such as the nature of the hazard, its frequency, duration, past experiences and exposure to the hazard. These factors do contribute to risk perception; however, the lack of understanding of volcanic processes, their low frequency and duration combined with limited or no exposure to past events despite living in an active volcanic zone all contribute to lower perception of risk regarding volcanic hazards. Hazard knowledge and risk perception of Katla by locals in Iceland demonstrated the results of hazard knowledge inherited from their forebears (Bird et al. 2009). This study indicated that exposure to and experience of volcanic hazards was discussed and recognised by the younger generation and contributed to their level of risk perception. Gregg et al. (2004) note that, in some cases, hazard awareness is not an indication of hazard knowledge and does not carry over to preparedness or responsiveness. Limited knowledge of the threat

that Kona in Hawai'i has on the local population is mirrored with the low level of preparedness at the individual level (Gregg et al. 2004).

The cultural belief system of the human dimension as described by Lavigne et al. (2008) focuses on the ties that individuals and communities have with their local environment. These cultural beliefs can often dictate the decisions of the individuals regarding their residence in these hazard zones, and can also act as an avenue to describe the events.

6 Conclusions

Ngāti Rangi have resided at the southern side of Ruapehu from time immemorial. Their exposure to volcanic activity throughout the generations has meant that they have adjusted their livelihoods and developed strategies and practices to live safely with a volcano. Subsequently, their traditional kainga and pā (traditional homes and fortified villages) are located in areas considered safe by the people. Ngāti Rangi are the human voice of their maunga, and they speak for him when needed but will watch and listen for the tohu (indicators) that will indicate to them their level of safety. Any precautionary measures against the potential impacts of volcanic activity and eruptions that involve alterations have the ability to distort their connection to their maunga. Therefore, Ngāti Rangi is adamant in their stance to protect Mt. Ruapehu against any human alteration. It goes against Iwi cultural constructs (or termed their tikanga) to demean the mana of Matua te Mana by altering his geomorphological nature. People chose to settle in areas along flood plains and along lahar channels, perhaps without prior knowledge and understanding of volcanic processes. People chose to position themselves on the landscape and therefore make the maunga and its natural processes hazardous. However, from the perspective of Ngāti Rangi, moving a mountain to protect their home is unacceptable; people should move their homes to protect the mountain. Finally, their belief is Ruapehu is their ancestor (tūpuna), their koroheke (most senior elder) and as such, he is the key to their cultural identity, and history and the link to the realm of ngā atua through whakapapa. Matua te Mana provides strength to the iwi, who are strong in the belief that Ruapehu is a maunga and has the right to be able to erupt, shake, and express his emotions without restraint from humans.

The gap between western-scientific based hazard management/monitoring frameworks and mātauranga Māori adaption strategies can be bridged by determining mātauranga Māori-based cultural descriptors or indicators that are traditionally used to monitor volcanic hazards and plan for risk. Combined with the documentation of strategies of recovery/adaptation by Māori communities in volcanic landscapes utilising Māori researchers, Māori language/Te Reo and applying participatory research techniques, Ngāti Rangi has developed its own mitigation, response and recovery strategies to be applied in the future, alongside and in tandem with western-science based hazard management plans. The recognition and application of traditional knowledge and practices to act as a basis for future planning and decision making in volcanic emergencies will ultimately increase participation of the indigenous communities and contribute to increasing resilience of the community.

Acknowledgements We would like to thank the support of Ngati Rangi Iwi, Kaumatua and Kuia that contributed to this work and provided a valuable contribution to this work. We would also like to thank the various Ngati Rangi Marae that provided hospitality and a culturally appropriate venue to share knowledge. This would not have been possible without the support of the New Zealand Natural Hazards Research Platform or MBIE through the Vision Matauranga Capability Fund.

Glossary

Māori Term Description

Ahi tipua Fire demons who bought volcanic activity to Aotearoa 'Te Hoata' and 'Te Pupū' according to Ngāti Tūwharetoa tradition

Ahi-kā-roa Burning fires of occupation

Aotearoa New Zealand

Atua Godsdeities

Hapū Sub-tribe

Haungaroa Sister to Ngātoroirangi (Ngātoroirangi—tohunga of the Te Arawa waka)

Hawaiiki The ancestral homeland of Māori

Iwi Tribe

Karakia Incantationprayer

Kawa Protocols

Kōrero Speakspeech, narrative

Koroua Eldergrandfather

Kuiwai Sister to Ngātoroirangi

Mana Prestigeauthority

Marae Term used to describe the traditional meeting house and entire complex of Māori

Mauri Life force

Mouri See 'Mauri'

Ngā atua 'The gods' but refers to the gods in Te Ao Māori (the Maori World)

Ngāti Turumakina Hapū of Ngāti Tūwharetoa

Ngāti Tūwharetoa Iwi in the central North IslandNew Zealand, of the Te Arawa Waka

Ngauruhoe Mountain of the central North Island

Paerangi-i-te-Whare-Toka Paerangi's house of stone (Paerangi—the eponymous ancestor of Ngāti Rangi)

Papatūānuku Earth mother

Pareitetaitonga Shelter from the southern winds. Peak on Ruapehu

Pēpeha Proverb

Ranginui Sky father

Ritenga Customs

Ruapehu Mountain of the central North Island

Rūaumoko God of earthquakes

Tane Mahuta Son of Rangi and PapaGod of the Forest

Tangata Whenua People of the land

Tangihanga Funeral

Tapu Sacred

Tawhirimatea God of the Wind

Te Ao Marama The world of light

Te Arawa Waka and iwi of the Bay of Plenty

Te Ika ā Māui The big fish of Māui (also known as the North Island of New Zealand)

Te kore Nothingness energy, the void

Te po The night

Te reo Māori The Māori language

Te Wai-ā-moe Ruapehu's Crater Lake

Tikanga Customs

Tohu Signs

Tohunga Spiritual leader

Tongariro Mountain of the central North Island

Tupuna Ancestor (Tūpuna = Ancestors)

Tūrangawaewae Place to stand

Uri Descendants

Waiata Songs

Waka Canoeor rather, the large double hulled waka that journeyed from Hawaiiki to Aotearoa

Wānanga Specific gathering devoted to learning and sharing knowledge

Whakapapa Geneology (simple translation)

Whānau Family

Wharepuni Meeting house of a marae

Whenua Land

References

Acoella V, Spinks K, Cole J, Nicol A (2003) Oblique back arc rifting of Taupo Volcanic Zone, New Zealand. Techtonics, 22(4):19-1, 19-18

Alcántara-Ayala I (2002) Geomorphology, natural hazards, vulnerability and prevention of natural disasters in developing countries. Geomorphology 47:107–124

Bird DK, Gísladóttir G (2012) Residents' attitudes and behavior before and after the 2010 Eyjafjallajökull eruptions—a case study from southern Iceland. Bull Volc 74:1263–1279

Bird DK, Gísladóttir G, Dominey-Howes D (2009) Resident perpection of volcanic hazards and evacuation procedures. Nat Hazards Earth Syst Sci 9:251–266

Bird DK, Gísladóttir G, Dominey-Howes D (2011) Different communities, different perspectives: issues affecting residents' response to a volcanic eruption in southern Iceland. Bull Volc 73(9):1209–1227

Campbell J (2009) Islandness: vulnerability and resilience in Oceania. Shima Int J Res I Cult 3:85–97

Cashman KV, Cronin SJ (2008) Welcoming a monster into the world: myths, oral traditions, and modern societal response to volcanic disasters. J Volcanol Geoth Res 176:407–418

Cashman KV, Giordano G (2008) Volcanoes and human history: an introduction. J Volcanol Geoth Res 176:325–329

Cronin SJ, Petterson MG, Taylor PW, Biliki R (2004a) Maximising multi-stakeholder participation in government and community volcanic hazard management programs; a case study from Savo, Solomon Islands. Nat Hazards 33:105–136

Cronin SJ, Gaylord DR, Charley D, Alloway BV, Wallez S, Esau JW (2004b) Participatory methods of incorporating scientific with traditional knowledge for volcanic hazard management on Ambae Island, Vanuatu. Bull Volcanol 66:652–668

Cutter SL, Burton CG, Emrich CT (2010) Disaster resilience indicators for benchmarking baseline conditions. J Homel Secur Emerg Manage 7(1), Article 51

Department of Conservation (2006a) Lahars from Mt. Ruapehu: Tongariro/Taupo. Retrieved from http://www.doc.govt.nz/documents/about-doc/concessions-and-permits/conservation-revealed/lahars-from-mt-ruapehu-lowres.pdf

Department of Conservation (2006b) Tongariro National Park Management Park: Te Kaupapa Whakahaere mo Te Papa Rēhia o Tongariro. Retrieved from http://www.doc.govt.nz/Documents/about-doc/role/policies-and-plans/national-park-management-plans/tongariro-national-park/tongariro-national-park-management-plan.pdf

Donoghue SL, Neall VE (2001) Late quaternary constructional history of the southeastern Ruapehu ring plain, New Zealand. N Z J Geol Geophys 44(3):439–466

Durie M (2005) Ngā tai matatū: Tides of Māori endurance. Oxford University Press, Melbourne, Australia

Gabrielsen H (2014) Ngāti Rangi Taiao Management Plan. Ngāti Rangi Trust, Ohākune, New Zealand

Gregg CE, Houghton BF, Johnston DM, Paton D, Swanson DA (2004) The perception of volcanic risk in Kona communities from Manua Loa and Hualālai volcanoes, Hawai`i. J Volcanol Geoth Res 130:179–196

Harmsworth GR, Awatere S (2013) Indigenous māori knowledge and perspectives of ecosystems. In: Dymond JR (ed) Ecosystem services in New Zealand—conditions and trends. Manaaki Whenua Press, Lincoln, New Zealand

Hart MA (2010) Indigenous worldviews, knowledge, and research: the development of an indigenous research paradigm. J Indigenous Voices Soc Work 1(1):1–16

Hodgson KA (1993) Late quaternary lahars from Mount Ruapehu in the Whangaehu River Valley, North

Island, New Zealand. Unpublished doctoral dissertation, Massey University, Palmerston North, New Zealand

Johnston DM (1997) Physical and social impacts of past and future volcanic eruptions in New Zealand. Unpublished doctoral dissertation, Massey University, Palmerston North, New Zealand

Jolly GE, Keys HJR, Procter JN, Deligne NI (2014) Overview of the co-ordinated risk-based approach to science and management response and recovery for the 2012 eruptions of Tongariro volcano, New Zealand. J Volcanol Geoth Res 286:184–207

Keys HJ, Green PM (2008) Ruapehu lahar New Zealand 18 March 2007: lessons for hazard assessment and risk mitigation 1995–2007. J Disaster Res 3(4):284–296

King DNT, Goff J, Skipper A (2007) Māori environmental knowledge and natural hazards in Aotearoa—New Zealand. J R Soc N Z 37(2):59–73

Lavigne F, De Coster B, Juvin N, Flohic F, Gaillard JC, Morin J, Sartohadi J (2008) People's behaviour in the face of volcanic hazards: perspectives from Javanese communities, Indonesia. J Volcanol Geoth Res 172(3–4):273–287

Lecointre J, Hodgson K, Neall V, Cronin S (2004) Lahar-triggering mechanisms and hazard at Ruapehu volcano, New Zealand. Nat Hazards 31:85–109

Lowe DJ, Newnham RM, McCraw JD (2002) Volcanism and early Māori society in New Zealand. In: Grattan J, Torrence R (eds) Natural disasters and cultural change. Routledge, London, England, pp. 126–134

Mercer J, Kelman I, Taranis L, Suchet-Pearson S (2010) Framework for integrating indigenous and scientific knowledge for disaster reduction. Disasters 34 (1):214–239

Neall VE (1976) Lahars as major geological hazards. Bull Int Assoc Eng Geol 14:233–240

Neall VE, Houghton BF, Cronin SJ, Donoghue SL, Hodgson KA, Johnston DM, Lecointre JA, Mitchell AR (1999) Volcanic hazards at Ruapehu Volcano. Wellington: Ministry of Civil Defence. Volcanic hazards information series 8. 30 p

Paulinson DD (1993) Hurricane hazard in Western Samoa. Geogr Rev 83:45–53

Pelling M, Uitto JI (2001) Small island developing states: natural disaster vulnerability and global change. Glob Environ Change Part B Environ Hazards 3(2):49–62

Petterson MG, Cronin SJ, Taylor PW, Toila D, Papabatu A, Toba T, Qopoto C (2003) The eruptive history and volcanic hazards of Savo, Solomon Islands. Bull Volc 65:165–181

Pewhairangi N (1992) Learning and Tapu. In: King M (ed) Te Ao Hurihuri: aspects of Māoritanga. Octopus Publishing Group (NZ) Ltd, Auckland, New Zealand, pp 9–14

Pool I (1991) Te iwi Maori: a New Zealand population, past, present & projected. Auckland University Press, Auckland

Proctor EM (2010) Toi tu te whenua, toi tu te tangata: a holistic Māori approach to flood management in Pawarenga. Unpublished master's thesis, Waikato University, Hamilton, New Zealand

Rainforth H, Procter J, Black T, Harmsworth G, Pardo N (2012) Exploring indigenous knowledge for assessing volcanic hazards and improving monitoring approaches. In: 7th cities on volcanoes conference, 19–23 Nov 2012, Colima, Mexico

Robertson N (1999) Māori and psychology: research and practice. In: The proceedings of a symposium sponsored by the Māori and psychology research unit. Māori & Psychology Research Unit, Hamilton, New Zealand

Royal TAC (2005) An organic arising: an interpretation of tikanga based upon the Māori creation traditions. Paper presented at the Tikanga Rangahau Mātauranga Tuku Iho: traditional knowledge and research ethics conference, 10–12 June 2004, Te Papa Tongarewa, Wellington, New Zealand

Smith GH (1990) Research issues related to Māori education. Paper presented to NZARE Special Interest Conference, Massey University, reprinted in 1992, The Issue of Research and Māori, Research Unit for Māori Education, The University of Auckland

Smith GH (1997) The development of kaupapa Māori: theory and praxis. Doctoral dissertation, Research Space, Auckland

Tobin GA, Montz BE (1997) Natural hazards: explanation and integration. The Guildford Press, New York City, NY

Ulluwishewa R, Roskruge N, Harmsworth G, Antaran B (2008) Indigenous knowledge for natural resource management: a comparative study of Māori in New Zealand and Dusun in Brunei Darussalam. GeoJournal 73(4):271–284

UNESCO (2014) World heritage list. Retrieved from http://whc.unesco.org/?cid=31&mode=table

Walker R (2004) Ka whawhai tonu matou: struggle without end. Penguin Books, Auckland, New Zealand

Zimmet P, Dowse G, Finch C, Serjeantson S, King H (1990) The epidemiology and natural history of niddm-lessons from the South Pacific. Diab Metab Rev 6(2):91–124

Fostering Participation of Local Actors in Volcanic Disaster Risk Reduction

Jake Rom Cadag, Carolyn Driedger, Carolina Garcia, Melanie Duncan, J.C. Gaillard, Jan Lindsay and Katharine Haynes

Abstract

Studies of recent volcanic crises have revealed that official evacuation and contingency plans are often not followed by communities at risk. This is primarily attributable to a lack of long-term coordination and planning among concerned stakeholders, and in particular, a lack of participation of local populations in disaster risk reduction (DRR). A lack of participation suggests the prevalence of top-down approaches, wherein local people are disengaged or even excluded in the development of DRR plans. It is not surprising, therefore, that existing plans are often non-operational, nor acceptable to the people for whom they are intended. Through an investigation of case studies at Mount Rainier (USA) and Bulusan (Philippines), and references to volcanoes elsewhere, this chapter aims to determine the key principles and important considerations to ensure peoples' participation in volcanic DRR. The chapter discusses key factors that encourage local empowerment, engagement, influence, and control in development of plans and actions. It adds information to the existing literature about how participatory approaches can encourage contributions by both local and outside actors, the latter providing knowledge, resources and skills when unavailable at local levels. Such approaches promote dialogue and co-production of knowledge between the community and outside actors. Contributions from multiple and diverse stakeholders

J.R. Cadag (✉)
University of the Philippines Diliman, Quezon City, Philippines
e-mail: jdcadag@up.edu.ph

C. Driedger
U.S. Geological Survey, Reston, USA

C. Garcia
IU Colegio Mayor de Antioquia, Medellin, Colombia

M. Duncan
UCL/CAFOD, London, UK

J.C. Gaillard · J. Lindsay
The University of Auckland, Auckland, New Zealand

K. Haynes
Risk Frontiers, Macquarie Park, Australia

Advs in Volcanology (2018) 481–497
https://doi.org/10.1007/11157_2016_39
© The Author(s) 2017
Published Online: 06 April 2017

further enable all groups to address the underlying social, economic, political and cultural issues that contribute to the vulnerabilities of local people. Consequently, DRR becomes more sustainable because local actors are not fully dependent upon outside actors and resources, relying instead on local capacities.

1 Introduction

Volcanoes, especially active ones, are generally perceived as sources of hazards by outside actors of disaster risk reduction (DRR), particularly scientists and government authorities (Cashman and Giordano 2008). For local inhabitants, however, volcanoes often hold deeper meaning because they become emblems of a homeland, often with long lived and deep cultural significance; and are a source of livelihoods and spiritual strength (Donovan 2010). This duality also explains, to some extent, the different perceptions and understanding of DRR between local actors (in particular local authorities and people), and outside actors (e.g., scientists, government agencies, and non-government organizations).

In development and disaster studies, local actors often refers to individuals and groups of people occupying or attached to a specific community and/or territory which include but are not limited to inhabitants, officials, local organizations, and different social groups including the most marginalized (e.g., Gujit and Shah 1998; Heijmans 2009). Here, we refer to local actors as a "collection of people in a geographical area" who "share a particular social structure", "have a sense of belonging", and whose "daily activities take place within the geographical area" (Abercrombie et al. 2006, p. 71). It is important to note, however, that a community can also be "relational" referring to a "quality characteristic of human relationship, without reference to location" (Gusfield 1975, p. xvi). This also suggests that 'local actors' cannot be used to refer to a set of homogenous groups in different contexts.

Outside actors, unlike local actors, do not exemplify a sense of community and/or identify themselves as part of a set of relationships within a specific geographical area. Sometimes, outside actors unknowingly insist on implementing actions based on plans and policies that are at times contradictory to local actors' views and needs. During times of crisis, these plans may fail. Programmes initiated and maintained solely by outside actors can result in an ineffective DRR process, loss of local knowledge and deepening mistrust between the different actors of DRR (Haynes et al. 2008; Mercer and Kelman 2010). There are, however, stories of success of communication between local and outside actors despite the great complexity of a volcanic crisis. During the reawakening of Mount Pinatubo in 1991, local people, unaware that they lived on the slopes of a volcano noted steaming and ground cracks. This information was relayed to authorities, who found it necessary to initiate a rapid top-down education campaign with the eventual valuable inclusion of local actors (Punongbayan et al. 1996).

Reducing disaster risk requires the participation of local actors in many aspects and stages of volcanic DRR (Wisner et al. 2012). During a crisis, volcanic activity contingency plans created only by outside actors can become non-operational due to being unacceptable or unfamiliar to local actors. Some well-recognized examples of disasters that resulted in great human casualties due to lack of collaboration between outside and local actors include the eruptions of Nevado del Ruiz volcano in 1985 (Voight 1990) and Merapi Volcano in 2010 (Kusumayudha 2012; Mei et al. 2013). To foster the participation of local actors in volcanic

DRR, community-based and participatory approaches have been employed by scientists, government agencies, and NGOs.

This chapter provides a rationale for the inclusion of local actors in reducing disaster risk and reaffirms the importance of integrating bottom-up and top-down actions in the entire DRR process. Some questions that this chapter aims to address are how local actors can be integrated in components of volcanic DRR. What are the key principles and important considerations for policy and practice to ensure peoples' participation in volcanic DRR?

2 Participatory Approaches to Volcanic DRR

Fostering local actors' participation through bottom-up and community-based initiatives is an alternative to isolated technocratic, top-down, command-and-control approaches to DRR. Participation refers to "a voluntary process by which people (…) influence or control the decisions that affect them" (Saxena 1998, p. 111). It is often defined along a continuum, ranging from total lack of control to self-mobilising initiatives where local actors own and control decision making (Arnstein 1969; Chambers 2005). Participation therefore refers to a process, rather than an outcome, and includes sharing and redistribution of power among stakeholders of DRR.

Since the 1970s, Civil Society Organisations (CSOs: non-state actors such as non-governmental organisations (NGOs), non-profit organizations (NPOs), social and religious organizations, among others; Kaldor 2003), have been promoting a shift in power relations to the benefit of local actors who face volcanic risk. In a few countries, such as New Zealand, the USA, and recently, Colombia, national and regional governments promote and even mandate participatory engagement, but the majority of volcanic regions around the world are still subject to top-down mitigation approaches. The practice of community-based and participatory DRR was widely promoted in the 1980s as Community-Based Disaster Risk Reduction (CBDRR), through the creation of national and international networks

of CSOs involved in grassroots activities (Heijmans 2009; Delica-Willison and Gaillard 2012). Proponents of CBDRR advocate that local actors are better placed than a central government to implement DRR actions as, in addition to considerable local knowledge and cultural understanding, their lives and livelihoods are at stake, therefore providing greater incentive to plan and take action. Both the scientific and practitioner literature acknowledge the capacities of local actors in responding to volcanic hazards on their own, as long as they are empowered with adequate organizational resources (e.g., Quarantelli and Dynes 1972; Delica-Willison and Willison 2004; Bowman and White 2012).

CBDRR consists of self-developed, culturally and socially acceptable, economically and politically feasible ways of coping with and avoiding disasters (e.g., endogenous resources, skills and local knowledge) (Maskrey 1984). This does not necessarily exclude external support, but provides access to external knowledge about hazards and risk, and educational and preparedness resources where needed, without perpetuating a cycle of dependency. CBDRR thus requires the participation of outside actors. In CBDRR, participatory approaches are frequently adopted for hazard, vulnerability and capacity analysis and the subsequent development of strategies and actions, for example to assess risk, raise hazard awareness and develop community-based warning systems. In some localities CBDRR evolves, with the occasional guidance of outside actors. Lessons drawn from practice are always considered to improve CBDRR, thereby ensuring it is flexible and adaptive to adjust to changing physical and social environments. It ultimately aims to empower people, which requires "transformation of existing social, political and economic structures and relations in ways that empower the previously excluded or exploited" (Hickey and Mohan 2005, p. 238).

Fostering people's participation in CBDRR requires innovative and flexible methodologies such as those featured in the Participatory Learning and Action (PLA) approach. PLA is "a growing family of approaches, methods, attitudes and behaviours to enable and empower people to share, analyse and enhance their knowledge of life and conditions and to plan, act, monitor, evaluate and reflect" (Chambers 2002, p. 2). Outside actors do not dominate the process but provide support to initiatives of local actors who know local issues best. CBDRR is thus a means to flip power relationships and encourage more meaningful participation through downward accountability towards local actors (Chamber 1983; Cornwall et al. 2000; Breett 2003).

Unfortunately, as Cornwall (2008, p. 269) states, "participation' can be used to evoke—and to signify—almost anything that involves people. As such, it can easily be reframed to meet almost any demand made of it". In many instances, participation is in fact seen as an outcome, rather than a process (the 'tyranny of participation', Cooke and Kothari (2001)). In cases where CBDRR is driven exclusively by outsider interests (White 1996), and that marginalized groups and "disadvantaged individuals" remain "excluded from participatory decision-making" (Pelling 1998, p. 484). Projects and activities are pre-designed by outsiders who make sure that enough local actors take part to report alleged "participation" upward to funding agencies (see Bowman and White 2012). This skewed approach to participation is evident in the many assessments of vulnerability and capacity that provide statistics based on standardized frameworks (demographics, gender characteristics, incomes, resources, health, etc.), from which plans are made and imposed upon local actors (Twigg 1998; Heijmans 2004). In many instances, although potentially useful on a governmental level for rapid prioritization of resources, these alien frameworks do not make much sense to local people in the context of the reality of their everyday life (Bhatt 1998; Delica-Willison and Willison 2004), and thus discourage participation, especially when concerns for survival take highest precedence (see the case of volcano Cerro Machin, Colombia Chap. 16).

Participatory approaches can rely heavily on the skill of one or more facilitators, who play a key role in the process (Duncan 2014). Challenges for the facilitator include ensuring the inclusion of the most marginalized people, managing the community's expectations of the process and balancing their role as a facilitator and as educator (e.g., Cronin et al. 2004). CBDRR has also been criticised for reinforcing the interests of the already powerful within communities, as it often proves difficult to reach the less powerful, more marginalised people that it is meant to empower (Cooke and Kothari 2001). This concept is balanced with the recognition that there is value in working with local actors who possess leadership qualities and who are opinion leaders within the community.

The following sections address the issues mentioned above in the context of volcanic environments in two different regions of the world: Mount Rainier, USA and Bulusan Volcano, Philippines. These cases were selected because they respectively provide accounts of long-term and short-term participatory approaches to CBDRR in disaster preparedness and crisis management based on first hand in-depth research from some of the chapter authors. The two case studies are not meant to be compared. They are considered examples of good practices of participatory volcanic DRR in two different contexts— disaster preparedness and crisis management. They also serve as a means of exploring the strengths and limitations of CBDRR in fostering disaster preparedness and crisis management.

3 Disaster Preparedness at Mount Rainier, USA

Mount Rainier is a 3392 m high volcano in the Cascade Range and the highest mountain in Washington State, USA (Fig. 1). It is recognized as one of the nation's most hazardous volcanoes (Ewert et al. 2005), with 78,000 people residing in the lahar-prone Puyallup River Valley (Wood and Soulard 2009). In some localities, the next lahar could reach communities with only about one-half hour of warning.

During the 1990s, a series of new publications (Scott et al. 1995; Scott and Vallance 1995; Hoblitt et al. 1998) highlighted Mount Rainier's hazards, especially its severe lahar hazard, and it motivated scientists to inform local officials and the public

Fig. 1 Mount Rainier dominates the landscape over the Puyallup River valley and the city of Orting (foreground). Around A.D. 1500 a landslide-driven lahar flowed down the west flank of Mount Rainier and inundated the valley floor. *Photograph by* E. Ruttledge, USGS, January 2014

through multiple presentations. The scientists' aim was to advise local people about the risks of living in lahar-hazard zones so that they could visualize undesirable outcomes and assume responsibility for CBDRR. They recognized that participatory methods might sustain "a long-term conversation" (Mileti 1999), and that people at risk might progress from initial hazard awareness to understanding of the risk, and belief in their ability to take effective mitigative action. Such progression comes from personalizing and then, confirming the risk with others, developing intentions for action, and making mitigative actions, as categorized variously in social models (e.g., Sorensen 1982; Sorensen and Mileti 1987; Paton 2003). At Mount Rainier, the resulting effort is driven by three groups, each contributing to the larger effort according to their organizational mission, resources, and needs.

(1) The first group of local actors consists of local emergency managers, at county, city, and fire district level, with professional responsibilities for the safety of the local community. During the mid-1990s, this group, in conjunction with state officials, called into being the Mount Rainier Volcano Work Group, which led first to development of an emergency response plan (Pierce County 1998, updated in 2008). They worked with the US Geological Survey (USGS) to install a lahar-detection system, followed by county and state efforts to build a public notification system

consisting of emergency broadcasts, personal electronics notifications, and sirens (Pierce County 2014). This effort is augmented by a series of volcano-evacuation route signs, which point to high ground and safety during a lahar. Community-based emergency educators added volcano hazards to neighbourhood multi-hazard emergency preparedness training, including interactions with marginalized populations. In conjunction with local actors, the county developed lahar-evacuation routes that are displayed on a new inter-agency website (Pierce County 2014). In the words of one local safety official within the lahar-hazard zone, "We will not be victims of the next lahar. Our agency will aid the community in the best way possible because we have taken the time now to plan and prepare".

(2) A second important group of local actors consists of enthusiastic community members and school safety officers who have developed a series of resident-driven efforts to mitigate problems associated with potential lahars. After scientists in the mid-1990s informed them of the lahar risks, local residents initiated a long-term sequence of lahar evacuation drills for thousands of students (Fig. 2) in the towns of Puyallup and Sumner (Caffazo 2014), and in Orting (Orting School District 2015). In the community of Orting, local residents raised funds for initial design of a system of efficient but costly walkways and pedestrian bridges across a highway

Fig. 2 Students practice evacuation during a 2002 lahar drill in downtown Orting. During an actual lahar, the lahar detection and notification system would provide residents of the city with approximately 30 min of warning, which is marginally sufficient for evacuation from some lahars. *Photograph by* C. Driedger, USGS, October 2002

Fig. 3 Public meetings concerning the threat of lahars prompted local teachers to outline the scope, messaging, and content of Mount Rainier volcano teaching materials such as this poster with an activity guide on the back side. *Photograph by* C. Driedger, USGS, 2003

and river that, if built, would shorten evacuation routes by enabling rapid egress to high ground (Bridge4kids 2014; Plog 2014). Local teachers proposed and participated in development of teaching materials (Driedger et al. 1998, 2005, 2014) (Fig. 3). School students developed a lighting system to improve night time visibility of some volcano evacuation route signs which are installed in the Puyallup, Carbon and Nisqually River valleys (Fig. 4).

(3) As outside actors, state and federal emergency managers, scientists, and park staff provide as-needed technical, organizational, and occasional financial assistance as needed (Pierson et al. 2014). Since the mid-1990s, this has required almost half-time involvement by one outreach specialist at USGS, who attends local meetings, answer inquiries, and aids in product development. Staff at USGS and Mount Rainier National Park (MRNP) sponsor an annual teacher training. USGS scientists train park staff, and aid with development of geohazard-oriented displays (Driedger et al. 2002). Federal funds supported development of a "web portal" that indicates hazards of individual property parcels (Washington Department of Natural Resources 2014). USGS produced an assessment of risk (Wood and Soulard 2009), and provides volcano trainings for officials and the public. Washington

Emergency Management Division and USGS assembled a media guidebook (Driedger and Scott 2010). Local and outside actors developed an outdoor interpretive sign about Mount Rainier hazards (Schelling et al. 2014). Product development methodology is based upon the premise that no single agency can know the needs of residents unless representative users are involved in determination of need, design, development, review, and implementation (Perry et al. 2016). Several important observations emerged. Multi-level participation in CBDRR allows each entity to make contributions that strengthen the entire effort, and promote long-term continuity. Enthusiasm and creative ideas from local actors whose lives and livelihoods are at stake provide long-term motivation for continual mitigation plan upgrades. CBDRR efforts are stronger because of the long-term commitment of scientific, organizational, and occasional financial support from outside actors. A motivated hazards-aware citizenry can initiate mitigation efforts that meet community needs, yet are

Fig. 4 In the Puyallup, Carbon and Nisqually River valleys, volcano evacuation route signs direct drivers towards high ground, and they serve the additional educational purpose of reminding local residents of the hazards and/or of the need for protective action. Students in the Orting School District developed the idea of enhancing some of the signs with placing flashing orange lights powered by solar panel to improve use during darkness. They developed a proposal and submitted it to authorities who funded the project. *Photograph by* C. Driedger, USGS, October 7, 2014

beyond the financial means of local governments. Community officials are considering a variety of funding sources and multiple options for rapid lahar evacuation. In the words of one resident activist, *"The people have led and many leaders have truly heard, taken to heart, and acted upon the concerns and solutions proposed by its citizenry."*

4 Locally-Led Crisis and Evacuation Management at Bulusan Volcano, Philippines

Bulusan Volcano is a 1559 m high stratovolcano formed inside a caldera. It is one of the most active volcanoes in the Philippines having erupted at least 16 times since late 1800s. Recent eruptions, such as in November 2010 to November 2011, were characterized mainly by ash ejection and volcanic earthquake swarms and resulted in recurrent mass evacuations of nearby towns (PHIVOLCS 2014). At least six municipalities and hundreds of *barangays* (villages) under political jurisdiction of the province of Sorsogon

are situated at the foot of the volcano. Barangay Cogon—the nearest village to the summit of the volcano—has a total population of 1020 people in 211 households and is within the probable danger zone of the volcano, defined as 4–10 km from the summit (Municipality of Irosin 2012). However, agricultural areas especially coconut plantations, the backbone of the village economy, are within the 4 km permanent danger zone.

Between 18 and 20 February 2011, a CBDRR was implemented in Cogon involving officials and representatives from multiple sectors of the community. The activity was initiated by Integrated Rural Development Foundation of the Philippines (IRDF), a local NGO advocating for the participation of local actors in DRR. The objectives of the activity were twofold: risk assessment through Participatory 3-Dimensional Mapping (P3DM) and development of a volcanic activity contingency plan for the village. The 3D map provides local actors with a bird's eye view of their territory, giving them a clear picture of important community information in order to determine their vulnerabilities, capacities and exposure to volcanic hazards (Fig. 5) (Cadag and

Fig. 5 Large-scale participatory 3-dimensional map (1:1250) of Cogon, Irosin, Philippines showing hazard-prone areas (shaded with *grey paint*), vulnerable assets and people and local resources (both depicted with push pins), February 2011 (adapted from Cadag et al. 2012: 84)

Gaillard 2012). Local participants traced information on the 3D map that are useful for risk assessment. The information was then subsequently confirmed and further improved by outside actors (e.g., municipal DRR officer, NGO personnel, and local scientists) to ensure precision (in terms of location) and compatibility with their plans. Likewise, the contingency plan developed during the CBDRR details the roles of local actors (e.g., village chief and councillors, health workers, village police and representatives from different sectors of the community) in the entire evacuation process, particularly in the management of the evacuation area. The approach and tool (i.e. P3DM) was highly appreciated by the participants because of its effectiveness in engaging actors from the different sectors and in combining their plans for DRR. According to the representative of the Disaster Risk Reduction and Management Office of the municipality, "CBDRR is not new to us... But it is the first time that we assess risk and plan actions (for DRR) using a single tool (3D map) that we all understand."

On 21 February, 2011 at 9:12 a.m., only a day after the CBDRR activity, the volcano suddenly erupted and ejected volcanic ash for several minutes (Fig. 6). There were no warnings from the Philippine Institute of Volcanology and Seismology (PHIVOLCS) nor municipal officials. It took only about 15 min for ash fall to reach the village of Cogon, reducing visibility to zero and rendering lamps and flashlights useless. Evacuation vehicles from the municipal center were unable to reach the village. Community members, particularly the local leaders, were thus the first to facilitate the evacuation. Three hours later, with the help of municipal rescue units and other volunteers, most of the residents of the village were evacuated to a school at the municipal center, situated 10 km away from Cogon.

At the onset of the crisis, the evacuation center was managed by the municipal officials and school coordinators. Yet, despite their best efforts, observations of participants and informal interviews revealed that the evacuation center was chaotic and under-prepared but only for the first half day. For example, only a few rooms and toilets were available; food distribution was delayed; and trashcans were full. This made evacuees uneasy; they did not have any idea of the government's efforts, nor of what was going

Fig. 6 Eruption of Mt. Bulusan, Philippines on February 21, 2011 at 9:12 a.m. *Photograph by* J. Cadag, 21 Feb 2011

to happen to them. Lack of coordination and communication among the affected populations and the authorities was quite evident (Cadag et al. 2012). In the late afternoon of the same day, village officials of Cogon decided to implement their newly conceived contingency plan which was a part of the recent CBDRR activity in the village (Fig. 7). Firstly, local officials coordinated with the school coordinators and municipal officials and helped to arrange and organise the rooms for the evacuees. They reassigned the rooms so that families from the same hamlets were reunited, making it easier for the village police and health workers (assigned to particular hamlets) to monitor the evacuees, for curfews and cleanliness, respectively. Mothers helped the government authorities in food preparations, which then became easier, faster and more efficient. Pregnant and nursing women, and older and sick people were allocated rooms. The village chief and councillors gave regular updates to the evacuees on the situation in the evacuated village, particularly on the damages incurred.

The successful management by local actors of the evacuation center was attributed to the recent CBDRR program in the village. Aware of the new village contingency plan, school coordinators and municipal officials decided to entrust the management of the evacuation center to the evacuees. The contingency plan thus underpinned the local officials' and evacuees' actions during the evacuation. According to an elected official from the village, "When we made the (contingency) plan, we thought it was for

compliance purposes only … But now we know we can use it to make our situation better in the evacuation center and to justify our actions."

Moreover, the 3-dimensional map assisted the initial assessment of volcanic impacts immediately after the eruption. It aided local officials in locating the areas most affected by ash fall, and in assessing the damage to shelters and farms. Damage and needs assessment by the local people and authorities and delivery of reports to concerned higher government authorities then became faster and more efficient. Altogether, the CBDRR program contributed to the success of the management of the evacuation center and post-disaster damage assessment through the leadership of the local actors and with the support of the outside actors.

5 Participation, Inclusion, and Empowerment of Local Actors in Volcanic DRR

This chapter has emphasized that participation of local actors offers numerous potential means to improve many aspects of DRR. Dialogue during participatory activities plays a vital role in the integration of knowledge across the different actors of DRR. Eventually, this integration leads to combination of top-down and bottom-up actions and is likely to be more efficient, context-appropriate, and sustainable (Wisner et al. 2012). If properly facilitated, it may result in local empowerment that allows local actors to

Fig. 7 Local leaders from the village of Cogon, Irosin, Philippines discussing with a municipal health officer the implementation of the village's evacuation plan (*left*). A village health worker conducting the registration of evacuees for easy health monitoring and distribution of relief goods (*right*). *Photograph by* J. Cadag, 22 Feb 2011

assess disaster risk, enhance their capacities and reduce underlying vulnerabilities (Pelling 2007; Maskrey 2011; Cadag and Gaillard 2012).

The case study of Mount Rainier reaffirms four important aspects of participatory approaches that are relevant in volcanic DRR. Firstly, collaboration among local and outside actors of DRR is possible when participation is sought. Local actors have been successfully integrated in DRR through long-term engagement and dialogue using a variety of participatory approaches. While enthusiasm and resources for various projects within the CBDRR have waxed and waned over the years, it is the long-term commitment by local and outside actors that has sustained the CBDRR effort. Secondly, participatory approaches should not be evaluated solely on the basis of immediate results but also on long-term positive outcomes. It is therefore important to reemphasize that process is equally or, in the long term, even more important than the original desired short-term outcomes. Thirdly, participatory approaches can empower local actors and encourage them to become key actors of DRR. The Rainier case study involved local actors who are self-motivated and resourceful, who are aware of the hazard, and who can develop plans and take actions. This example illustrates that participatory approaches, accomplished with sufficient intention, vigor, resources, and commitment by local and outside actors can produce positive outcomes.

Fourthly, sometimes CBDRR requires outside actors to make shifts in power relation in unconventional ways. As an example, at Mount Rainier, local officials and media requested successfully that scientists modify their usage of traditional scientific terminology to reduce miscommunication in education and during crises. At their request, the term 'debris flow' is applied only to small seasonal events that can not directly impact communities, while the term 'lahar', similar in structure but vastly larger in scale, refers to events that could create serious impacts, principally during eruptions and debris avalanches. Similarly, the term 'active' is applied consistently to Mount Rainier to reflect the internal volcanic processes present, even during quiescence. More recently, these specific and process-oriented applications of terminology are applied broadly by officials in other volcano-hazard work groups within the Cascade Range. In this manner, the expeditious nature of top-down decision-making is traded for authenticity and intentional efficiency within the larger mitigation effort. In the case of Bulusan Volcano, local leaders and residents were involved in risk assessment and contingency planning prior to the eruption, and then took on the role as managers during the crisis. The successful management of the evacuation center by the local actors (in cooperation with outside stakeholders) so early in the crisis is commendable. This positive outcome highlights the importance of participation by local actors including the marginalized sectors (i.e. homeless, people with disabilities, the economically disadvantaged, women, children, people of a variety of sexual orientations, etc.) in all aspects and stages of volcanic DRR. Moreover, volcanic crisis and evacuation management plans were localized yet consistent with the plans of outside actors, putting emphasis on the role of local actors as first responders in times of crisis, whilst reinforcing the importance of outside actors in fulfilling the lack of resources at the local level, particularly in dealing with large scale crises (Delica-Willison and Willison 2004; Cadag and Gaillard 2012).

Some of the important insights emerging from these two case studies relate specifically to trust, dialogue, participatory methods, and empowerment. This study is consistent with the findings of previous work on trust and risk communication (e.g., Haynes et al. 2008). Generally, trust means confidence in the reliability of someone or something. In risk communication, trust is determined by several factors such as general trustworthiness (e.g., competence, care, fairness, and openness) and scepticism (e.g., credibility, reliability, and integrity) (Poortinga and Pidgeon 2003, p. 607). The case studies reinforced that participation builds trust among local and outside actors of DRR, which eventually results in more fruitful collaboration and better DRR. Programs that involve community participation empower local actors and eventually encourage them to trust government authorities and their information (Paton et al. 2008), and vice versa—outside actors gain respect for local knowledge.

Achieving trust, however, is difficult. Trust is only possible when actors of DRR are fully engaged in a process of dialogue, i.e. the continuous exchange of knowledge, ideas, and opinions. Dialogue is a means for DRR actors (particularly marginalized sectors) to be heard on equal footing with other actors (Heijmans 2009) so as to: "respect the diversity of opinions" (Abarquez and Murshed 2004, p. 81). Dialogue, therefore, promotes integration of knowledge and action in DRR (Wisner et al. 2012; Gaillard and Mercer 2013).

To sustain trust and dialogue among actors, participatory methods and tools must also be sustainable, i.e. maintained and adapted at the local level by local actors who recognise (without being dependent upon) the contributions of outside actors. This is best achieved when local actors have active roles in the conceptualization, conduct, and maintenance of participatory approaches (Cadag and Gaillard 2012). This is the case of CBDRR in Bulusan Volcano where local actors have indicated greater interest in improving their 3D map and local plans to further strengthen their disaster preparedness.

In spite of their successes, in these case studies, challenges extend to both local and outside actors. As noted in the Mount Rainier example, during long-lasting efforts, it is inevitable that politics play a strong role; volunteers reach fatigue; agency personnel change; and competing priorities threaten the main objectives of DRR efforts. Outside actors are often prohibited by the institutionalized top-down and command-and-control paradigm of DRR, which logically contradict the idea of local participation and empowerment. Meaningful efforts to promote empowerment and participation require that outsiders recognise the need for change within their institutions, which is emphasised by Chambers (1995, p. 197):

> Participation "by them" ['local actors particularly the marginalised sectors] will not be sustainable or strong unless we ['outside actors'] too are participatory. "Ownership" by them means non-ownership by us. Empowerment for them means disempowerment for us. In consequence, management cultures, styles of personal interaction and procedures all have to change.

6 Considerations for Policy and Practice of DRR in Volcanic Environments

This chapter highlights a number of lessons for CBDRR policy and practice in volcanic environments. It reaffirms that:

1. Participatory approaches in CBDRR initiate the personal and community progression from personalizing and then, confirming the risk with others, developing intentions for action, that are required to take action during a crisis, as noted variously by several authors (i.e. Sorensen 1982; Sorensen and Mileti 1987; and Paton 2003).

2. CBDRR can be well suited for volcanically hazardous areas as some appropriate emergency responses such as residents identifying locally understood and recognizable hazardous phenomena, developing locally-based neighborhood notification methods, self evacuation, and sheltering often requires the individual or community to be self reliant.

3. Dialogue developed through volcanic CBDRR can promote trust among all actors, which in turn sustains CBDRR efforts.

4. Co-development of hazard and response messages early on, and consistent use of them by local and outside actors can facilitate educational processes, and lay a foundation for sustainable CBDRR.

5. Participatory approaches can invite inquiry, such as the search for information that reinforces people's recognition of the hazard, and discussion about DRR. As with any educational activity, participation is the best teacher because it provides local actors with the knowledge to educate and empower others.

6. CBDRR can operate over short and long durations. The key components are that the process is collaborative and emphasises local actors in the processes of organization and planning for future necessary actions. While the resources of outside actors can strengthen CBDRR, planning should be accomplished, communicated, and practiced principally by local actors.

7. The value may be limited by vested interests within the community, fatigue of all actors, changing personnel, as well as other cultural and political or hazard issues that compete for focus. The actors may not understand each other's culture and resources, leading to unrealistic goals and expectations. Local actors can become insular and exclude new ideas and/or the needs of marginalized people, thus becoming, for all practical purposes, another group of 'outside actors.'

8. While governments and organizations may profess support for CBDRR, there is always a threat that, on a more personal level, outside actors will withdraw from CBDRR precepts for the sake of immediate efficiency, such that they damage outside/local actor relationships and the CBDRR that they seek to engender.

9. In addition to participatory approaches to CBDDR, in volcanic environments it is important to consider the value of legislative instruments and the legal responsibilities of government to protect the life of their citizens, thus directly affecting governmental decision-making. However, in some countries, legislation to manage volcanic DRR focuses mainly on crisis management often detached from larger risk reduction efforts such as land-use policy, protection of infrastructure and overcoming unequal power relations within society. Furthermore, crisis management planning is often restricted to government authorities with little or no community consultation and participation.

10. To achieve effective DRR in volcanic environments, an intense process of facilitation and negotiation is required of scientists, policymakers and the public. This is especially necessary during long term crises characterized by shifting political, cultural and scientific landscapes (Donovan and Oppenheimer 2014). Scientific knowledge can be enhanced by the participation of local actors as citizen scientists (see Irwin 1995) and observers of volcanic activity (e.g., Stone et al. 2014).

11. In order to be effective, legislative instruments for DRR should be created and applied well in advance of a crisis and should involve all stakeholders, including local people and others exposed to the prevailing hazards.

7 Conclusion

Although limited to two case studies, this study provides examples and discussions which support the supposition that volcanic DRR is more effective when local actors participate, regardless of volcanic environments or contexts. CBDRR exists in varying forms across volcanic regions around the world, as exemplified by the long CBDRR ongoing at Mount Rainier and the short-term initiative at Bulusan volcano. Given that each volcano and surrounding communities have their own specific context, it is difficult and against best practice to determine a rigid and standardized procedure for conducting CBDRR. Rather than seeing this as an obstacle, it should be embraced as an opportunity to develop customized means to fulfil community needs.

Whilst it is not possible to standardise CBDRR, a number of guiding principles for fostering the participation of local actors in DRR have been identified. Firstly, participation is a process, not an outcome, and it should empower local people and build dialogues. This reduces dependence upon outside actors and resources, and encourages reliance on local capacities. Participation must be flexible and is only as good as the knowledge, intentions and resources available to local actors. At their best, participatory approaches encourage collaborative contributions by both local and outside actors, the latter providing knowledge, resources and skills to complement the strengths of the former. Such approaches promote dialogue and co-production of knowledge between the community and outside actors. CBDRR should work in tandem with top-down, legislative processes, with local and outsider actors holding each other accountable. The personal and professional relationships developed with local actors can spur outside actors to continue to support efforts, whilst the attention from outside actors can motivate local actors to maintain CBDRR. The process should, therefore, be mutually beneficial. Finally, whilst it can be challenging to engage and maintain the participation of both local and outside actors, CBDRR has been demonstrated as an essential component in sustaining ownership and communication between key actors in volcanic settings. Indeed, the true value of CBDRR is not only measured in products or documents, but also in creating a conducive environment for collaboration where the hearts, minds and trust of the people are devoted. It is an environment where local actors are empowered to implement DRR plans and actions and where policies that institutionalize peoples' participation and multi-actor collaboration are in place.

References

Abarquez I, Murshed Z (2004) Community-based disaster risk management: field practitioners' handbook. Asian Disaster Preparedness Center (ADPC), Klong Luang

Abercrombie N, Hill S, Turner BS (2006) Community. The Penguin dictionary of sociology. Penguin, London, pp 71–72

Arnstein SR (1969) A Ladder of citizen participation. J Am Inst Plan 35(4):216–224

Bhatt MR (1998) Can vulnerability be understood? In: Twigg J, Bhatt MR (eds) Understanding vulnerability: South Asian perspectives. Intermediate Technology Publications, Colombo, pp 68–77

Breett EA (2003) Participation and accountability in development management. J Dev Stud 40(2):1–29

Bridge4Kids (2014) http://www.bridge4kids.org. Accessed 14 Oct 2014

Bowman L, White P (2012) Community' perceptions of a disaster risk reduction intervention at Santa Ana

(Ilamatepec) Volcano, El Salvador. Environ Hazards 11:138–154

Cadag JRD, Gaillard JC (2012) Integrating knowledge and actions in disaster risk reduction: the contribution of participatory mapping. Area 44(1):100–109

Cadag JRD, Gaillard JC, Francisco A, Glipo A (2012) Reducing the risk of disaster through participatory mapping in Irosin, Philippines. In: Garcia-Acosta V, Briones F (eds) Social strategies of prevention and adaptation. Ciesas, Mexico City, pp 81–87

Caffazo D (2014) Puyallup students earn praise during massive school lahar drill. The News Tribune September 30 2014. http://www.thenewstribune.com/2014/09/30/3408050_puyallup-students-earn-praise.html?rh=1. Accessed 11 Nov 2014

Cashman KV, Giordano G (2008) Volcanoes and human history. J Volcanol Geotherm Res 176(3):325–329

Chamber R (1983) Rural development: putting the last first. Longman Scientific and Technical, Harlow

Chambers R (1995) Poverty and livelihoods: whose reality counts? Environ Urban 7(1):173–204

Chamber R (2002) Relaxed and participatory appraisal: notes on practical approaches and methods for participants in PRA/PLA related familiarisation workshops. Institute of Development Studies, Brighton

Chambers R (2005) Ideas for development. Earthscan, London

Chambers R (2007) Who counts? The quiet revolution of participation and numbers. Institute of Development Studies at the University of Sussex, Brighton

Cooke B, Kothari U (2001) Participation: the new tyranny?. Zed Books, London

Cornwall A, Lucas H, Pasteur K (2000) Accountability through participation: developing workable partnership models in the health sector. IDS Bull 31(1):1–13

Cornwall A (2008) Unpacking 'participation': models, meanings and practices. Community Dev J 43(3):269–283

Cronin S, Gaylord D, Charley D, Alloway B, Wallez S, Esau J (2004) Participatory methods of incorporating scientific with traditional knowledge for volcanic hazard management on Ambae Island, Vanuatu. Bull Volcanol 66:652–668

Delica-Willison Z, Willison R (2004) Vulnerability reduction: a task for the vulnerable people themselves. In: Bankoff G, Frerks G, Hilhorst D (eds) Mapping vulnerability: disasters, development and people. Earthscan, London, pp 145–158

Delica-Willison Z, Gaillard JC (2012) Community action and disaster. In: Wisner B, Gaillard JC, Kelman I (eds) Handbook of hazards and disaster risk reduction. Routledge, London, pp 711–722

Donovan K (2010) Doing social volcanology: exploring volcanic culture in Indonesia. Area 42(1):117–126

Donovan A, Oppenheimer C (2014) Science, policy and place in volcanic disasters: insights from Montserrat. Environ Sci Policy 39:150–161

Driedger CL, Faust L, Lane L, Smith M, Smith R (1998) Mount Rainier—the volcano in your back yard, poster and activity guide for educators. U.S. Geological Survey Miscellaneous Publication, p 2

Driedger CL, Stout T, Hawk J (2002) The Mountain is a Volcano!—addressing geohazards at Mount Rainier. J Assoc Natl Park Rangers 18:14–15

Driedger CL, Scott WE (2010) Volcano hazards. In: Schelling J, Nelson D (eds) Media guidebook for natural hazards in Washington—addressing the threats of tsunamis and volcanoes. Military Department Emergency Management Division, Washington, 44 p. http://www.emd.wa.gov/about/documents/haz_Volcano_MediaGuide_2013.pdf. Accessed 11 Nov 2014

Driedger CL, Doherty A, Dixon C, Faust L (2005) Living with a volcano in your backyard—an educator's guide with emphasis on Mount Rainier. U.S. Geological Survey and National Park Service, General Information Product. p 19

Duncan M (2014) Multi-hazard assessments for disaster risk reduction: lessons from the Philippines and applications for non-governmental organisations. Dissertation, University College, London

Ewert JW, Guffanti M, Murray TL (2005) An assessment of volcanic threat and monitoring capabilities in the United States: framework for a national volcano early warning system. U.S. Geological Survey Open-File Report 2005–1164. 62 p

Gaillard JC, Mercer J (2013) From knowledge to action: bridging gaps in disaster risk education. Prog Hum Geogr 37(1):93–114

Gaillard JC, Dibben CJL (2008) Volcanic risk perception and beyond. J Volcanol Geotherm Res 172(3–4):163–169

García C, Mendez-Fajuri R (Chapter 16 in this edition) Experiences about volcanic risk socialization in Colombia. In: Fearnley C, Bird C, Haynes K, Jolly G, McGuire B (eds) Volcanic crisis communication: observing the volcano world

Gujit I, Shah MK (1998) The myth of community: gender issues in participatory development. ITDG Publishing, London

Gusfield JR (1975) The community: a critical response. Harper Colophon, New York

Haynes K, Barclay J, Pidgeon N (2008) The issue of trust and its influence on risk communication during a volcanic crisis. Bull Volcanol 70(5):605–621

Heijmans A (2004) From vulnerability to empowerment. In: Bankoff G, Frerks G, Hilhorst D (eds) Mapping vulnerability: disasters, development and people. Earthscan, London, pp 115–127

Heijmans A (2009) The social life of community-based disaster risk management: origins, politics and framing policies. Working paper No. 20, Aon Benfield UCL Hazard Research Centre, London

Hickey S, Mohan G (2005) Relocating participation within a radical politics of development. Dev Change 36(2):237–262

Hoblitt RP, Walder JS, Driedger CL, Scott KM, Pringle PT, Vallance JW (1998) Volcano hazards from Mount Rainier, Washington, revised 1998. U.S. Geological Survey Open-File Report 98-428

Irwin A (1995) Citizen science: a study of people, expertise and sustainable development. Routledge, London

Johnston D, Becker J, Coomer M, Ronan K, Davis M, Gregg C (2006) Children's risk perceptions and preparedness: Mt. Rainier 2006 hazard education assessment tabulated results. GNS science report 2006/16, p 30

Kaldor M (2003) Civil society and accountability. J Hum Dev 4(1):5–27. doi:10.1080/1464988032000051469

Kusumayudha SB (2012) Analysis on the capacity building for mitigating volcanic hazards versus the 2010 eruption of Mount Merapi, Central Java, Indonesia. In: Proceedings of the 34th international geological congress, 5–10 Aug 2012, pp 1–11

Maskrey A (1984) Community based hazard mitigation. In: Proceedings of the international conference on disaster mitigation program implementation. Ocho Rios, Jamaica, 12–16 Nov 1984, pp 25–39

Maskrey A (2011) Revisiting community-based disaster risk management. Environ Hazards 10(1):42–52

Mei ETW, Lavigne F, Picquout A, de Bélizal E, Brunstein D, Grancher, D, Sartohadi J, Cholik N, Vidal C (2013) Lessons learned from the 2010 evacuations at Merapi volcano. J Volcanol Geotherm Res 261:348–365

Mercer J, Kelman I, Lloyd K, Suchet-Pearson S (2008) Reflections on use of participatory research for disaster risk reduction. Area 40(2):172–183

Mercer J, Kelman I (2010) Living alongside a volcano in Baliau, Papua New Guinea. Disaster Prev Manag 19 (4):412–422

Mileti DS (1999) Disasters by design: a reassessment of natural hazards in the United States. Joseph Henry Press, Washington, DC

Municipality of Irosin (2012) 2010 Population and demography. Municipal profile of Irosin. Sorsogon, Philippines

Orting school district emergency response plan, OSD Disaster Awareness Response Team (DART). http://www.ortingschools.org/Page/101, p 17. Accessed 10 Jan 2015

Paton D (2003) Disaster preparedness: a social-cognitive perspective. Disaster Prev Manag 12:210–216

Paton D, Smith L, Daly M, Johnston D (2008) Risk perception and volcanic hazard mitigation: individual and social perspectives. J Volcanol Geotherm Res 172 (3–4):179–188

Pelling M (2007) Learning from others: the scope and challenges for participatory disaster risk assessment. Disasters 31(4):373–385

Pelling M (1998) Participation, social capital and vulnerability to urban flooding in Guyana. J Int Dev 10 (4):469–486

Perry SC, Blanpied ML, Burkett ER, Campbell NM, Carlson A, Cox DA, Driedger CL, Eisenman DP, Fox-Glassman KT, Hoffman S, Hoffman SM, Jaiswal KS, Jones LM, Luco N, Marx SM,

McGowan SM, Mileti DS, Moschetti MP, Ozman D, Pastor E, Petersen MD, Porter KA, Ramsey DW, Ritchie LA, Fitzpatrick JK, Rukstales KS, Sellnow TS, Vaughon WL, Wald DJ, Wald LA, Wein A, Zarcadoolas C (2016) Get your science used—six guidelines to improve your products: U.S. Geological Survey Circular 1419, p 37. http://dx.doi.org/10.3133/cir1419

PHIVOLCS (Philippine Institute of Volcanology and Seismology) (2014) Bulusan volcano. http://www.phivolcs.dost.gov.ph/html/update_VMEPD/Volcano/VolcanoList/bulusan.htm. Accessed 12 Oct 2014

Pierce County (2008) Mount Rainier volcanic hazards plan DEM. Pierce County (Washington). Department of Emergency Management (working draft). http://www.co.pierce.wa.us/documentcenter/view/3499. Accessed 13 Oct 2014

Pierce County (2014) Mount Rainier active volcano. Pierce County (Washington). Department of Emergency Management. http://www.piercecountywa.org/activevolcano. Accessed 13 Oct 2014

Pierson, TC, Wood NJ, Driedger CL. (2014). Reducing risk from lahar hazards: concepts, case studies, and roles for scientists. J Appl Volcanol 3:16. doi:10.1186/s13617-014-0016-4

Plog K (2014) Politicians, scientists, local leaders revisit Orting's Bridge for kids at summit. The News Tribune, October 25, 2014. http://www.thenewstribune.com/2014/10/25/3450135_politicians-scientists-local-leaders.html?sp=/99/296/331/&rh=1. Accessed 11 Nov 2014

Poortinga W, Pidgeon N (2003) Exploring the dimensionality of trust in risk regulation. Risk Anal 23:961–972

Punongbayan R, Newhall C, Bautista L, Garcia D, Harlow D, Hoblitt R, Sabit J, Solidum R (1996) Eruption hazard assessments and warnings in Fire and Mud. University of Washington Press, Seattle, pp 67–85

Quarantelli EL, Dynes RR (1972) When disaster strikes: it isn't much like what you've heard and read about. Psychol Today 5(9):66–70

Saxena NC (1998) What is meant by people's participation? J Rural Dev 17(1):111–113

Schelling J, Prado L, Driedger CL, Faust L, Lovellford P, Norman D, Schroedel R, Walsh T, Westby L (2014)

Mount Rainier is an active volcano—are you ready for an eruption? Washington Emergency Management Division (exhibit)

Scott KM, Vallance JW (1995) Debris flow, debris avalanche and flood hazards at and downstream from Mount Rainier, Washington. U.S. Geological Survey Hydrologic Atlas: 729, 2 maps, pamphlet 9 p

Scott KM, Vallance JW, Pringle PT (1995) Sedimentology, behavior, and hazards of debris flows at Mount Rainier, Washington. U.S. Geological Survey Professional Paper 1547, p 56

Sorensen JH (1982) Evaluation of emergency warning systems at Ft. Oak Ridge National Laboratory, St. Vrain Nuclear Power Plant, Oak Ridge TN

Sorensen JH, Mileti DS (1987) Decision-making uncertainties in emergency warning system organizations. Int J Mass Emerg Disasters 5:33–61

Stone J, Barclay J, Simmons P, Cole PD, Loughlin SC, Ramón P, Mothes P (2014) Risk reduction through community-based monitoring: the vigías of Tungurahua, Ecuador. J Appl Volcanol 2014(3):11

Twigg J (1998) Understanding vulnerability: an introduction. In: Twigg J, Bhatt MR (eds) Understanding vulnerability: South Asian perspectives. Intermediate Technology Publications, Colombo, pp 1–11

Voight B (1990) The 1985 Nevado del Ruiz volcano catastrophe: anatomy and retrospection. J Volcanol Geotherm Res 44:349–386

Washington Department of Natural Resources (2014) Natural hazards web portal. http://www.dnr.wa.gov/ResearchScience/Topics/GeosciencesData/Pages/geology_portal.aspx. Accessed 11 Nov 2014

White SC (1996) Depoliticising development: the uses and abuses of participation. Dev Pract 6(1):6–15

Wisner B, Gaillard JC, Kelman I (2012) Framing disaster: theories and stories seeking to understand hazards, vulnerability and risk. In: Wisner B, Gaillard JC, Kelman I (eds) Handbook of hazards and disaster risk reduction. Routledge, London, pp 18–33

Wood NJ, Soulard CE (2009) Community exposure to lahar hazards from Mount Rainier, Washington. U.S. Geological Survey Scientific Investigations Report, 2009–5211

"There's no Plastic in Our Volcano": A Story About Losing and Finding a Path to Participatory Volcanic Risk Management in Colombia

Jacqui Wilmshurst

Abstract

This chapter tells the story of a group of stakeholders who came together to collaborate on developing a more effective risk management strategy at Galeras volcano; an active and potentially extremely dangerous volcano in southern Colombia. It tells of how they came together, lost their way and then finally found the path to a truly participatory process. Woven into the story is a history of the main phases of risk communication in its widest sense, including some of the lessons learned throughout recent decades. It also extends an invitation to those involved in volcanic risk management to explore aspects of their own psychology, as an extension to the growing body of work that seeks to understand the psychology of those living with the risks. Relevant factors suggested for such an exploration include assumptions, biases, perceptions and worldviews and how these might influence, for better or for worse, the contributions being made to the field. Alongside the valuable lessons drawn from the story itself, they include the pitfalls of unexamined assumptions, the importance and value of collaborative and participatory approaches, and the essential task of ensuring that everyone is truly speaking the same language. Further applicable insights are offered from a range of other fields beyond Disaster Risk Reduction including leadership development, psychotherapy and behavioural safety.

Keywords

Risk communication · Psychology · Transactional analysis · Behavioural safety · Participatory approaches · Collaboration

J. Wilmshurst (✉)
83 Morley Street, Sheffield S6 2PL, UK
e-mail: jakwilmshurst@gmail.com

Advs in Volcanology (2018) 499–514
https://doi.org/10.1007/11157_2017_16
© The Author(s) 2017
Published Online: 15 July 2017

Introduction

On the fourth day of a week-long workshop designed to promote dialogue and co-operation between stakeholders at the base of Galeras, an active volcano just outside the city of Pasto in Nariño province, southern Colombia (see Fig. 1), an indigenous elder took the stage in an agitated state. He began to outline all the reasons that it was ridiculous to suggest that their volcano, on whose flanks they had lived and worked for several generations, could possibly contain plastic. The many stakeholders attending the University-led workshop included scientists, local and national government agencies, emergency services and at-risk communities. The indigenous elder stated that he knew there to be rocks, ash and all sorts of other materials for sure, but categorically not plastic. He spoke so fast and with such force that the official translators in the box at the rear of the auditorium gave up after the first few minutes. At this point, most of the foreign scientists left the auditorium for a tea break and the attending community members stated that they no longer wished to remain for the final days of the workshop. On discovering this, the scientists, both local and international, congregated outside to discuss how such a well-intentioned and carefully planned opportunity to create new collaboration and what was believed to be a participatory path to managing risk could have broken down so completely.

At the same time, a Colombian scientist with a sound understanding of (and concern for) the social issues surrounding risk management at Galeras, asked me whether I would be prepared to spend some time with the community members to listen to their concerns. As a psychologist, whose reason for being at the workshop in the first place was to understand better the psychological issues relating to risk management and communication, I readily agreed and a meeting was set up with key community representatives at my hotel that evening. During this meeting, I spent over two hours simply listening to the attendees of the meeting express their views, needs and frustrations; something that they had been expecting to have the opportunity to do in front of a far larger audience during the workshop itself. This airing

Fig. 1 Map of major volcanoes in Colombia (*Source* USGS)

of opinion and opportunity to be heard was to become the basis for the renewed process of dialogue communication and the rebuilding of trust that is central to this chapter.

There is an ever-expanding body of literature on the psychology of risk communication and management in the field of Disaster Risk Reduction (DRR), the primary focus of which has traditionally been on the psychology of those 'at risk'. This is of, course, an essential field of study. As a psychologist working in DRR I argue that for us to understand better a scenario such as the one presented above, there is a balance to be redressed in understanding the psychology of those whose role is to communicate and manage the risks. How those inhabiting these roles think and behave also influences the outcomes of their vital work; for better or for worse.

If you are involved in the management and communication of risks in this context, or indeed any other, then this chapter is designed both to offer you an interesting and valuable story about a journey in risk communication and management, and to invite you to take the next step towards a

Fig. 2 Galeras volcano viewed from the eastern side of the city of Pasto (*Photo* Wilmshurst 2009)

better understanding of your own thoughts, perceptions and behaviours and how they may influence the work that you do. I use this story to illustrate and introduce a range of contributions to effective practice from a variety of contexts that translate usefully to work in DRR. Learning is drawn from other applied fields such as health psychology, leadership development, psychotherapy and industrial safety, all of which also seek to understand human communication and behaviour in their respective contexts.

This story is told from my own perspective and based on my personal interpretation of events, with all the accompanying filters, biases and as yet undiscovered (and therefore unexamined) assumptions. There is absolutely no intention to criticise any of those involved, although I do aim to challenge constructively at times. The intention is to generate reflection, discussion and perhaps even changes in practice in a vitally important field of study and application.

Risk Perception Around Galeras

We will pick up the story again shortly, but first it is useful to offer some more information about the wider context in which the story unfolds.

This enables a broader understanding as to how and why initial attempts to engage with communities to create more collaborative approaches can be fraught with unexpected challenges. Especially for those who dare to try first!

The field of DRR is generally dominated in the academic realm by the applied physical sciences and is itself a relatively young field. It has taken significant time to recognise the importance of psychology and the social sciences and, as a result, the role of these disciplines is still unfolding. Early work concentrated on understanding the psychology of communities living at risk and relied heavily on work carried out in risk perception from the 1960s onwards, particularly in relation to nuclear power during those post war decades in which public consciousness of the potential dangers were at an understandable high (e.g. see Douglas 1992; Slovic 2000; Pidgeon et al. 2003).

Understanding how those living with risks make sense of those risks, and then make decisions accordingly, is certainly a good place to start. I would argue, however, that understanding the psychology of all stakeholders is not only desirable, but essential for the creation of truly effective risk management strategies that fully respect the needs and perspectives of all of those stakeholders. The British Psychological Society (BPS) Crisis, Disaster and Trauma section state that "recognising that the role of psychology is not only to assist in managing the psychological impact of disasters but also to play a key part in understanding how people behave (or do not behave) in the events leading up to a disaster; and engaging in planning at all stages" (ICSU 2008, p. 38). Here, 'people' can be interpreted to be all those involved in risk communication and management, not just those who may be impacted upon by the risks. There are clearly several groups of 'stakeholders' who fall into the remit of communicating and managing the risks and therefore they are no more of a homogenous entity than 'people living at risk' or 'the public'. With this acknowledgement, the intended lessons learnt from my experience is both for scientists, who primarily study the risks and seek to communicate them onwards, and for those who seek to use this scientific information to manage those risks and whose roles are often 'sandwiched' in between the scientists and those living with the risks.

Those of us who have tried it know that turning attention to an examination of oneself is generally a lot more uncomfortable than seeking to analyse others (if done honestly). The rewards are potentially enormous however; it is arguably a moral duty of anyone who seeks to involve themselves in the lives of others in ways that can have huge influence over their welfare. In other words, there is as much to be gained by scientists and risk managers taking an honest look at their (often unconscious until examined) beliefs, assumptions and biases as well as those of the people they seek to help. Otherwise they can end up, for example, openly despairing at the 'irrationality' of people who choose to live on an active volcano despite the warnings, whilst flying all over the world to warn these people of their erroneous ways at the same time as absolutely believing in human-induced climate change and yet racking up the air miles as they go! We are all contradictory by virtue of being human, and our unexamined assumptions (with consequent behaviours) can lead to the appearance of an attitude often referenced (with an intentional injection of irony) in the world of personal development consulting as; 'take my advice, I'm not using it…'.

Volcanic risk management in Colombia in the years leading up to the point in time at which the workshop took place is a complex story. The Colombian government, and indeed the population of the whole country, was dealt a huge blow when another volcano, Nevado del Ruiz (Fig. 1), erupted in 1985 killing over 23,000 people in just a few hours. Scientists had for some time been warning of an eruption and the potential for catastrophic consequences, but the country and the government had other priorities weighing heavily on them at the time, not least the civil conflict that had been raging for several decades with huge loss of life and drastic economic, social and political consequences. A retrospective examination of the events leading up to the disaster led researchers to conclude that it had been caused by "…cumulative human error—by misjudgement, indecision and bureaucratic shortsightedness." (Voight 1990, p. 1; see also Hall 1990). Only three years later, whilst this devastating disaster was still exceptionally raw and very much at the forefront of the nation's attention and psyche, Galeras volcano (Fig. 2) became active again after 10 years of dormancy. Scientists realised quickly that an eruption could occur of a great enough magnitude to cause another catastrophic disaster. It is no surprise, therefore, that the government wasted no time in giving Galeras emergency status and mobilised a great deal of resource towards averting a repeat of such a devastating event.

A feeling of panic was, of course, a fully understandable and very human response to the Nevado del Ruiz disaster. Such an emotive backdrop to the reactivation of Galeras would make it difficult for anyone (with their humanity intact) to respond with any kind dispassionate and methodical examination of the new risks posed. It is easy to see how some of the decisions made in this context would end up creating a legacy for those whose job it later became to manage the emerging risks at Galeras. One such decision made unilaterally and with lasting consequences, was to put in place a plan to relocate all communities who lived in an area deemed to be 'high risk' by the scientific community. The plan was designed to ensure minimal loss of life in the event of the type of eruption that scientists' research had suggested was possible, although incredibly uncertain.

There was an assumption that minimising loss of life absolutely had to be the top priority of any risk management strategy, therefore it came as a huge surprise to those who had devised the plan that there would be resistance from the communities whose lives were being so carefully protected. It turned out that there were other considerations alongside the protection of life that were featuring just as strongly, for a number of key reasons, and these were discovered later once dialogue became more open. And so it was that for a number of years an increasing 'stalemate' developed between those who felt that their role was to protect and those whose lives were being protected. It was in this context that finally, in 2009, it was decided that an opportunity must be created to bring together scientists, risk managers and communities in order to facilitate new dialogue and co-operation in relation to the potentially devastating risks posed by Galeras. This brings us back to the workshop, with the angry and distressed elder on the stage and the future of the workshop was found to be hanging in the balance. So where did it go from here?

"There Is no Plastic…"

For three days, geologists from the UK, USA and Colombia had presented their latest scientific knowledge and understanding, relating to the unique behaviour of Galeras volcano, to the communities living on its flanks. The intention of the organisers was to give the scientists an opportunity to educate the communities as to the nature and magnitude of the risks and, as a result, to convince them of the importance of heeding the advice of the government and risk managers and, ultimately, to accept the need to be permanently relocated out of the areas deemed to be at greatest risk.

Here a wider moral question as to what we really believe the true purpose of risk communication is

raised. In my experience this can be diverse and often based on unconscious assumptions on the part of those responsible for informing and shaping it. I have heard the overall purpose of risk communication and management in DRR explained as all of the following at various times; (a) to use expert information to make decisions on behalf of people living at risk should they be deemed (by those experts) not to be making optimal decisions for themselves; (b) to attempt to persuade people to act in whatever way risk communicators deem best for them; (c) to offer timely, accurate and appropriately formulated risk information so that people can choose their own actions in a fully informed way, based on integrating this information with what else that they know; or (d) to work together as equals to understand the risks from all perspectives and to design collaborative and inclusive ways in which to manage them together.

Each of these aspirations is best met using very different approaches, informed by a diverse range of work on human behaviour and relationships. For example, where better to turn to if one's goal is to persuade than the world of commercial advertising? Yet if the desire is to collaborate with those at risk such that they can make empowered decisions for themselves, consciously rather than through clever manipulation by the concerned 'experts', then the principles are entirely different. Thankfully there are ever fewer who subscribe to the intentions and accompanying beliefs contained in option (a) above. It is worth taking a moment here to consider what your approach has been so far and whether it might change once considered more carefully and consciously.

This brings us back to the business of how to examine and build a better understanding of our own psychology. When working as a consultant I use a model known as the 'conscious competence model' (Gordon International Training). The basic structure of this model is as follows; when one has not yet considered learning something new, one is necessarily then in a state of 'unconscious incompetence' (simply never having tried whatever it is and never having considered doing so—i.e. 'we don't know what we don't know'). Think of learning to drive a car. In order to begin learning, and before reaching the states of first 'conscious competence' (managing to drive, but still concentrating on finding the 'biting point' with each gear change) and then eventually 'unconscious competence' (arriving back home in your car via your usual route having actually meant on this occasion to go to the supermarket), it is first necessary to spend some time in a state of 'conscious incompetence' (think of early attempts at hill starts with frequent subsequent stalling!). This stage of 'conscious incompetence', however long it lasts, is a very uncomfortable one and creates a feeling known as 'cognitive dissonance'. In other words, we now know what we want to do but either cannot yet do it or are not yet doing it, for whatever reason. In health psychology and risk psychology, understanding cognitive dissonance has been instrumental in making sense of how, when we have a choice to change our behaviour or our belief (e.g. stop smoking, or convince ourselves that Auntie Ethel lived to 101 years old despite smoking heavily so it cannot be that dangerous after all). See, for example, Feather (1962), McMaster and Lee (1991), Conner and Norman (2001).

It is the same both when we decide we want to examine and better understand ourselves and when we decide we want to approach something in a different way. Like, for example, taking risk communication from an 'information deficit model' of one way communication, from 'experts' to target 'subjects', forward into a democratic, empowering and collaborative journey of discovery (The 'information deficit model' will be explained in more detail shortly.). Doing something like this requires a lot of hard work and it is very easy to slip back into what we previously knew and felt comfortable doing. This is summarised rather well by a well-loved British author of humorous science fiction Douglas Adams: "it can be very dangerous to see things from someone else's point of view without the proper training" (1995).

Another useful model for assisting a move towards better understanding the impact of our own unconscious behaviours and assumptions on those with whom we are communicating belongs to the field of Transactional Analysis (TA). Originally developed in the 1950s by a psychiatrist named Eric Berne for use in psychotherapeutic practice, TA has been taken successfully (and enduringly) into fields such as leadership and management development (e.g. Wagner 1996) and industrial safety (e.g. Marsh 2014). In fact, the latter offers a good description of a facet of the model known as 'ego states', which is a way of explaining styles of communication than can influence significantly the effectiveness of our relationships (Fig. 3). The following explanation comes from the field of behavioural safety, a field which is also essentially about risk communication and management, and therefore from which we can draw much valuable learning: "the basic model is like a snowman with three bubbles on top of each other. The lower bubble represents passive, sulking behaviour. The top bubble represents aggressive, authoritarian behaviour. The middle bubble, however, is where you want to be—firm, fair, analytical and reasonable" (Marsh 2014, p. 97). Even more importantly for us, "the theory also talks of the 'nurturing parent'". This is still top bubble but without the aggression. The trouble is the side effects are that your paternal attitude may well be seen as patronising (because you are talking down to people). This mindset "will get in the way of your listening and communication" (Marsh 2014, p. 98). In understanding which bubble we are in when attempting to communicate something to another, we can gain a much better understanding of how people are likely to respond. If we are

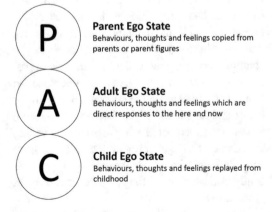

Parent Ego State
Behaviours, thoughts and feelings copied from parents or parent figures

Adult Ego State
Behaviours, thoughts and feelings which are direct responses to the here and now

Child Ego State
Behaviours, thoughts and feelings replayed from childhood

Fig. 3 Transactional Analysis Ego States Model (*Source* Davidson and Mountain (2017))

Supervisor   Direct Report

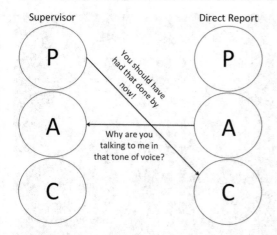

Fig. 4 Transactional Analysis Ego States: Crossed Transactions (*Source* Bush 2015)

able to communicate from the middle bubble, we are most likely to elicit a middle bubble response from whomever we are addressing. When we communicate (consciously and intentionally or not; more often the latter), we risk what is known as a 'crossed transaction' (Fig. 4). When this happens, we are in a situation where, for example, an inadvertently parental style may elicit a rather child-like response—for example defiance and rebellion in the face of feeling patronised and told what to do. Similarly, a crossed transaction can go the other way; a defiant or rebellious stance can elicit a parental response. Suffice to say that crossed-transactions are far messier and difficult to manage that straightforward ones! Once one understands how crossed-transactions occur, it is far easier to adjust one's own position and therefore encourage the other to move into the adult bubble also.

The plan to give the scientists an opportunity to educate the communities as to the nature and magnitude of the risks falls in line with the previously mentioned 'information deficit model'. The assumption of such a model in a risk context is that if those at risk are not taking action to avoid the identified risks, there must be a deficit either in their understanding and perception of those risks. Without examining the assumptions underlying such a deficit model, it can appear to be the most rational and obvious approach to risk communication and it is for this reason that it

was the most dominant model for many years. Much has now been learned, however, about the limitations of such an approach due to the realities of human behaviour, emotion and cognition, especially in such complex risk contexts. "Information-deficit models of risk communication are simply inadequate to deal with the accumulation of data that illustrates the significance of emotional state and affective imagery" (Breakwell 2007, p. 172).

When information held by an 'expert' is deemed essential for the welfare of another it is communicated from an assumption (conscious or otherwise) of 'information deficit' and transmitted in a one-way communication designed to address this deficit, it could be said in terms of the model that one is in the 'nurturing parent' bubble. Whilst the intention is benign and caring, communicating from this bubble is more likely to trigger a 'crossed transaction' and often this is in the form of a rebellious and resistant 'child' response.

TA can help us to understand why the indigenous elder perhaps took to the stage in such an emotional state to declare that there was no plastic in their volcano. He may have been triggered into a defiant state by feeling that he was being told with great authority, by people from a faraway place, about his beloved volcano. In this state, it would have been much easier for him to hear the very technical, scientific and unfamiliar term 'pyroclastic flow' as 'plastic'.

The workshop had been set up explicitly as an opportunity for dialogue and co-operation on risk management at the Galeras volcano, with the invitation extended to all communities living in the high risk zones. Yet for the first three days these community members were given no opportunity to contribute, only to listen to a long list of scientific presentations using technical language. Great care had been taken to ensure effective translation from English to Spanish and vice versa, but the issue of translating scientific language into that which could be understood by non-scientists had been completely overlooked. This combination of oversights and miscommunications resulted in community members who felt, beyond having been invited in the first place, utterly disrespected and unheard (Fig. 5).

Fig. 5 Elder from the
indigenous community of
Jenoy, featured in this story,
on the stage at the University
of Nariño talking about
'plastic' in their volcano
(*Photo* Wilmshurst 2009)

Over time, being communicated to by someone in that parental bubble, which can sometimes appear patronising and controlling, can lead to what is known as 'learned helplessness' (frequently seen in depression) where individuals simply perceive that they have no control over what is happening to them in a certain context and will stop trying to engage or change what is happening. It is worth noting that it is not especially relevant whether their perception of control continues to be accurate or not, what matters is that they stop attempting to have agency based on their perception that it is outside of their control.

The intention of all of this is not, I reiterate, to detract from the obvious importance of understanding and explaining the nature of the risks and of the way in which those risks are understood by those living with them. The challenge is how to integrate scientific information into an approach that stands the most chance of being 'effective', whatever that is decided to mean. Much has been learned about how to do this, and more of a consensus reached on the aim of risk communication in many contexts, over the past few decades of research and practice. Fischoff (1998, p. 134) offers a summary of the history of the evolution of risk communication and presents these as the first five phases:

1. All we have to do is get the numbers right.
2. All we have to do is tell them the numbers.
3. All we have to do is explain what we mean by the numbers.
4. All we have to do is show them that they've accepted similar risks in the past.
5. All we have to do is show them that it is a good deal for them.

So, if the workshop was designed and approached from the perspective of these phases, then it offers another possible part of the explanation as to why those running it encountered the same pitfalls that were encountered during this period of evolution, which ultimately led to the current understanding of what more is required to achieve better outcomes for all.

Working with Different 'Ways of Knowing'

When communicating and managing risk, trust can be undermined and therefore affect the likelihood of scientific information being well received and acted upon in a positive way. It has been found that losing trust is generally a great deal easier than gaining it, so there is much value in understanding how best to preserve it. This is known as 'the asymmetry principle' (Poortinga and Pidgeon 2004). Trust is a complex and multi-dimensional thing and absolutely key in effective risk communication (see Poortinga and Pidgeon 2003; Wilmshurst 2010; Crosweller and Wilmshurst 2013 for further discussion of trust in a DRR context). See also Paton's (2010) related work on people's relationships with information sources.

A more in depth understanding of biases in our decision-making, conscious or unconscious, is available through the work of Daniel Kahneman and his late research partner Amos Tversky (see Kahneman 2011). Perhaps the most common of the biases they propose and discuss is 'confirmation bias', in which we inadvertently filter out information that goes against what we already believe while seeking out and keeping only that which agrees with it.

An understanding of the biases brought into this field can offer much to the creation of a joint understanding of the perspectives and assumptions we all bring. In the case of science, an example of bias (to some degree including confirmation bias) can be in the form of what is sometimes known as 'scientism'. Broadly "scientism is a matter of putting too high a value on natural science in comparison with other branches of learning or culture" (Sorell 1994). An over-emphasis on the scientific method and its findings, leading to it being viewed as the only meaningful way of knowing about the world, can be extremely damaging when it leads to the dismissal of other forms of knowledge, including experiential. Participatory approaches in research allow for other forms of knowledge to be integrated alongside science and lead to much more collaborative and inclusive solutions, not to mention the ability to draw on a much more broader range of resources when addressing a given problem. They intentionally blur the lines between researcher and subject, creating conditions for all parties to become involved in constructing the questions, seeking methods to answer these questions and bringing different forms of knowledge to shape the answers. For more on participatory approaches, see Reason and Bradbury 2006.

One example of scientism in the context of volcanic risk, once noticed and appreciated, led to a public and very important apology. In Hawaii, myths shared and transmitted across generations about the behaviour of the volcanoes were dismissed for a long time by geologists working on the islands. Eventually, one of these geologists was forced to realise the mistake and was willing to acknowledge it openly and to apologise: "the cultural memory was right and our scientific surveys were wrong." "We were very clearly wrong and we only realised this recently. It's pretty embarrassing that geologists failed to take the Pele-Hi'iaka chants into account because we hadn't believed that the chants had any real meaning." This realisation and apology did not end there, but in fact led to a commitment to work with the locals and their myths in order to seek more insight from their own 'ways of knowing': "Swanson believes that many more scientific treasures lie in the Hawaiian chants, ready for scientists to decipher" (Swanson 2008; Palmer 2015).

I was offered anecdotal information about how the value of local knowledge at Galeras was also discovered. Stories that told of clear skies over the volcano preceding an eruption were initially dismissed by scientists, because their understanding at the time was that gas emissions (and therefore steam and cloud) would increase up to the point of eruption. It later turned out, I was told, that at Galeras there are indeed often clear skies up until the point of eruption. This is because a feature of the volcano is that as the pressure builds, the build-up of magma results in a 'plugging' effect thereby stopping gas emissions (and therefore preventing steam and therefore the resulting cloud sitting about the crater) until the pressure builds sufficiently to cause an eruption. I am not a geologist, so I cannot verify this story in terms of scientific data on the subject, but the story illustrates the importance to local people of their myths and cultural knowledge being respected alongside scientific knowledge.

As it happens, there have only been nine deaths caused by eruptions since Galeras became active again in 1989. Seven of those killed were scientists and two were tourists who went up to the crater with them (their story is told in Williams and Montaigne 2001). What is relevant here is the fact that the only fatalities so far have been scientists (and those accompanying them), rather than any of those living on the volcano, and this has created an unfortunate legacy in relation to local perceptions of, and trust in, scientists studying Galeras. This was only ever expressed to me alongside deep regret about the lives lost, I must add.

For more on the evolving understanding of the role of myth and other non-science based sources of knowledge see, for example, Chester (2005). There has also been relevant and interesting work on cultural theory in relation to risk perception. Douglas (1992) pointed out that the Japanese did not have a word for risk (or didn't at the time of her writing) because they did not approach danger from the viewpoint of establishing probability. This has significant and obvious implications for the subject matter of this chapter. See also Smith (1999) on the importance of uncovering and examining cultural bias when conducting research with those from other cultures who may hold world views very different than our predominantly scientific paradigm in the so-called 'West'.

Rebuilding the Trust

There has also been widespread belief among scientists that the public are unable to conceptualise uncertainty in relation to risk assessment. Many scientists have thought, as a result of this belief, that communicating uncertainty would therefore increase distrust in science and cause confusion (Frewer et al. 2003). This further points to an ongoing adherence, in some cases, to information deficit assumptions that may be taking time to shift. In contrast, research to understand effective leadership and management has repeatedly shown the value and importance of transparency and fallibility by those who seek to influence others. See also Gigerenzer (2002) for more on how different groups of people, including health professionals responsible for

communicating risk, understand risk and uncertainty themselves. This links back also to the work mentioned earlier on risk as a cultural concept that is not necessarily held by all those with whom DRR professionals seek to engage (Douglas 1992).

There are a number of additional phases in Fischoff's observed evolution of risk communication, based on more recent lessons learned across the whole of the risk communication field, to which we will return later. As we will see, the workshop described above appears to have entered into the 'stakeholder engagement' phase of risk management from an (inadvertent) place of unequal status between the scientists and risk managers on the one hand, and those they were seeking to inform and protect on the other. This approach to risk communication also generally demonstrates an intention towards persuasion. In her book on the psychology of risk, Breakwell notes that "…risk communication research has focused on what happens not simply when information is being transmitted but when that information is part of a message that is designed to persuade. The persuasion is aimed at making the recipients see the hazard in a different way and as a consequence change their attitude or behaviour towards it" (2007, p. 131). There is a huge and growing body of work on how to persuade people to engage in 'behaviour change' in various contexts (see Marsh 2014 to learn about 'nudge' theory and the very effective behavioural change intervention using a plastic fly in a urinal to improve hygiene in public toilets). For the purposes of this chapter and because of the nature of the DRR context, however, we are going to move on to those approaches that seek to be fully inclusive and participatory.

As a psychologist, I had been invited along to the workshop to assess whether it would be a viable case study for a research project to explore psychological risk factors in volcanic risk management. Clearly I had my answer at this point! The intervention by one of the local scientists that I mentioned earlier, who was really concerned that the hard work and good intentions were potentially about to be lost, was a major turning point. Once I had offered to spend some time learning more about the perspectives, concerns and experiences that the community members had come along hoping to share, the energy changed and community participants agreed to re-engage in the workshop. Following this necessary and important clearing of emotional energy the way was finally cleared to get a range of stakeholders in a room to begin new dialogue.

Another meeting was organised for the next day and more emotional expression was needed and allowed before things could move on. This was especially important as this next meeting happened to fall on the anniversary of the Nevado del Ruiz disaster and there were people in the room who had lived through the devastation and loss first hand. Eventually, the subject matter turned to the future and the desire to find a more effective may forward was universally shared. From this new place, it was possible for the group to produce three lists, together, outlining the issues that all stakeholders present agreed on. These lists were; 'Points of Agreement', 'Points of Conflict' (they were surprised to find that they actually agreed on what the points of disagreement were and this provided a significant breakthrough in trust and understanding between the group members at this point) and 'Conclusions of the Workshop' (which included the need to review current decrees, laws, policies etc. and most notably the proposed plans for enforced relocation). The group included representation from national and local risk management bodies, communities living in the high risk zones and local scientists. International scientists were not represented as they were in a parallel session to discuss the communication of scientific uncertainty.

From the production of these lists, and agreement that all now felt that they had been given an opportunity to be heard more fully, came an agreement to embark on a new process of dialogue and collaboration between the stakeholders. The process would be facilitated and explore what could be done to take them forward, using their lists of agreement and of agreed areas of continued disagreement and conflict as a guide. Thus the step was taken from

an approach based on 'information deficit' and the need for persuasion and therefore from the one-way presentation of scientific data of Fischoff's step five, into the final three phases of his journey of risk communication:

6. All we have to do is treat them nice.
7. All we have to do is make them partners.
8. All of the above.

Thankfully most risk management strategies have now started to move away from an approach that suggests that it's all about risk perception, a deficit of knowledge (and the accompanying need to educate people about the risks until they understand enough to do as the scientists are telling them) towards approaches that are collaborative and participatory (e.g. John Twigg 2009). There is still a legacy of the old approaches apparent in some areas, however, and therefore more work to be done.

Participatory methods have been gaining ground rapidly in DRR in recent years, not least due to the evidence of how effective they are in creating sustainable strategies that draw on the wide range of knowledge and resource made available by such collaborative methods. One cautionary note worth making here is that they do, however, need to be truly participatory rather than approaches that appear to engage all stakeholders but that ultimately consist of officials seeking endorsement of decisions already made (confirmation bias at work) and mistaking this for collaboration and participation. Breakwell (2007, p. 172) notes that:

> Risk communication is a mammoth topic…in moving from an examination of the classic literature on persuasion through to the discussion of consultative and participation methods, it echoes the journey made by risk communication and philosophy over the last half-century. There has been a move from seeing the public as targets for influence to recognising them as partners of in the process of risk management. Of course, not all institutions have made this move and not all risks are particularly amenable to it.

Her last point is vital, and there will no doubt continue to be spirited debate about which hazards and risks are, and are not, amenable to participatory approaches in the field of DRR and beyond. What should be avoided, however, is the use of non-participatory approaches simply because they are perceived to be less time and resource intensive and more comfortable and familiar, rather than because they best suit the context and needs of the people affected by them.

So, as the workshop concluded in the beautiful surroundings of La Cocha Lake in the hills to the east of Pasto, the participants of the group that had come together as a result of the breakdown part way through had made a commitment to meet again and to build an ongoing partnership. Here began a three month facilitated process of dialogue and collaboration back at the university in Pasto, to build a more participatory approach to risk management at Galeras volcano. Members of the group represented the local science community, the university, local and national disaster management groups and organisations and three of the communities living in the high risk zone on the flanks of the volcano. The process was named by the group, very aptly, 'Speaking the Same Language'. Whilst I guided and facilitated the process, all content was created and decisions made by the members of the group. It was agreed that the group would follow, albeit loosely to allow for flexibility, an 'action research' approach

(Reason and Bradbury 2006). As part of this, the members drew up and reached consensus on a group objective and a set of group working 'rules' for the facilitation period (Figs. 6 and 7).

At the start of the process, solutions suggested by the officials responsible for managing volcanic risk centred on two main areas: (a) permanent relocation in the medium to long term for those living in the designated high risk areas of 'red zone' and (b) in the more immediate term and for those living outside the 'red zone' (high risk zone), having a well-practised evacuation plan in the event of a significant eruption being forecast. Many of the community members wanted to start from a place of open discussion about the suitability of the proposed (and in some areas partially implemented) plans. This included reaching agreement between all stakeholders on the nature and extent of the risks posed. Also, arriving at solutions that would satisfy 'officials' that the people whose interests they sought to serve were making sound and informed decisions about their safety and welfare. In a wider DRR context, the moral question remains as to who should have the final say in what constitutes sound and informed decisions and optimal behaviours, and with whom the responsibility to ensure behaviours are carried out ultimately lies.

Fig. 6 Members of the group working together on a more inclusive risk management strategy (*Photo* Wilmshurst 2009)

This is a question that permeates much of the risk literature in relation to health and safety as a whole, but certainly there are many more ambiguities in a situation such as this. There is, however, plenty of evidence to show that enforcing compliance in a 'top down' manner without genuine engagement with all parties is generally far less successful than approaches that allow all to have a genuine part to play in how a culture of good risk management is developed (Marsh 2014). For this reason, despite some obvious and important differences, there is much of value to DRR to be found in the world of industrial safety management.

The process ran from September until December 2009. At end of this first facilitated period, the group felt no further need for me as the facilitator (as was hoped and intended at the start, as the process was designed to work towards empowering the group to continue in a self-regulating manner). They had by this point identified and brought together a much wider group of important stakeholders, including representatives from further disaster management bodies and various government departments including, importantly given the nature of the risks, the department of health.

Things did not always go smoothly, of course, and we lost important representatives from one of the communities about one month into the process. This was due to an impasse relating to the presence of officials deemed by them to be essential for the process. These community members, from the indigenous community of Jenoy, were also at this time working hard to gain recognition from the government of their indigenous heritage and membership of the Quillasinga people. I am pleased to report that they had made significant progress on this when I visited them the following year. Following both their departure and the end of the facilitated period, it was also decided that 'communities' could no longer be a homogenous part of the group membership, as each community had different environments (literally in terms of living at different altitudes on the flanks of the volcano and therefore producing different crops and keeping different livestock). They also had different cultural, needs and priorities. It was therefore decided that representatives from each community would engage with risk managers independently until such time as they felt able to come back together as one group. As well as differences between them, there also emerged

Fig. 7 Galeras volcano overseeing our work: the view from our meeting room within the University of Nariño (*Photo* Wilmshurst 2009)

interesting and important differences in perspective amongst the young and the elders within the respective communities. A further very important consideration that emerged from the process was the fact that whilst the scientists and risk managers were mostly viewing the volcano as a potential hazard, the communities living on its flanks were seeing it as both a potential hazard and a huge source of opportunity. The opportunities include very fertile soil for growing crops, hot springs for bathing and their many myths and stories that being them close to their 'Taita Galeras' as a spiritual connection to their culture and their roots. For more on the more recent acknowledgement of the importance of volcanoes as opportunities as well as hazards, and how these perceived benefits can offset the risks, see Kelman and Mather 2008.

It emerged too that the way in which the community members were considering the risks was from a far broader perspective, and were taking into account other threats and relative considerations, than those of the risk managers. This is not surprising given that the focus of risk managers and communicators is usually from the point of view of a particular hazard, risk or set of priorities. Again, an examination of the assumptions inherent in that is hugely valuable and can avert a number of problems. In this case, it emerged from the conversations that despite the obvious fear of a large volcanic eruption following the Nevado del Ruiz disaster, there was a recognition that the risks of an earthquake are also very significant given that Pasto sits very close to a major fault line and has been affected by a number of large earthquakes in the past. This has important potential implications when considering relocating these communities from the flanks of the volcano into a city living at risk of powerful earthquakes. Community members resisting relocation to the city also pointed out that, to them, the loss of their culture and practices and their community cohesion was potentially a bigger threat to them than losing their lives to an eruption. This again surfaced the assumption that preventing loss of life should necessarily be the sole focus of risk management strategies. The request was instead that quality of life be held in equal regard by those managing the risks.

I left the process at the end of the agreed phase of facilitated work, at which point there was a fantastic level of motivation and commitment to continue to work in a collaborative and participatory way, in spite of the many hurdles and frustrations inherent in such an approach. The last time I had contact with members of the group they were continuing to progress, learn and further develop strategies together, although not without significant challenges. There was however thankfully still no desire to return to earlier approaches after all that had been gained and learned through such an inclusive approach.

Conclusions: Speaking the Same Language

This experience has taken us on a journey through Fischoff's stages of the evolution of risk communication and collected additional learning from other fields along the way. The intention was two-fold: firstly, to tell you a story about how a group of people came together from a place of mistrust and miscommunication to build a participatory approach to managing risks at Galeras volcano. Secondly, to use this story to bring to life a range of lessons and contributions from DRR and beyond as an invitation to you to think about your own role in DRR and the range of psychological factors you bring to what you do, including a whole collection of assumptions, biases, perceptions and beliefs.

The lessons learned through getting lost and finding the way again are many, but they can be centred around a few key areas. The journey of risk communication has been one of building layers, rather than of moving through and beyond distinct stages. This is important because it means that we need to hold on to the learning and contribution of each stage and keep learning about who is best placed to achieve each and how. This allows for all stakeholders to play their part in risk management processes that are truly inclusive and integrated, and which allow us to benefit from the immensely rich learning available from, for example, cultural memory, lived experience and lay-persons' observations alongside scientific inquiry. In bringing together stakeholders to work together, mistakes are inevitably made and many more lessons are learned, usually more quickly if at times a little more painfully! When the conditions are created for open, honest and respectful communication, assumptions are surfaced and can be worked with and through consciously. Through this, inadvertent biases can be removed, intentional ones can be challenged, and relationships will be both more effective and usually more enjoyable.

It is important, alongside continuing to develop the interdisciplinary nature of DRR by bringing in relevant academic disciplines, to continue to seek applicable learning from other fields involved in understanding human behaviour, relationships and communication. This not only opens the door to many more sources of rich learning, but helps us to minimise how many painful lessons we must learn for ourselves when others have already been there first.

When we make this into a conscious journey and maintain an open mind and a collaborative spirit, we stand the best chance of bringing about solutions in which we truly all are speaking the same language.

Acknowledgements Sincere thanks to all of those with whom I have worked at Galeras volcano in Colombia. May your perseverance, humility and ever-evolving insight continue to inform and inspire.

References

Adams D (1995) The hitch hiker's guide to the galaxy: a trilogy in five parts. William Heinemann

Breakwell GM (2007) The psychology of risk. Cambridge University Press

Bush D (2015) 'When Supervision Becomes Parenting' https://www.linkedin.com/pulse/when-supervision-becomes-parenting-douglas-bush-m-a- (Accessed 7 May)

Chester DK (2005) Theology and disaster studies: the need for dialogue. J Volcanol Geoth Res 172 (3–4):319–328

Conner M, Norman P (eds) (2001) Predicting health behaviour. Open University Press

Crosweller H, Wilmshurst J (2013) Natural hazards and risk: the human perspective. In: Sparks S, Hill L (eds) Risk and uncertainty assessment for natural hazards, Cambridge University Press

Davidson C, Mountain A (2017) "Transactional analysis". http://www.businessballs.com/transactionalanalysis.htm (Accessed 15 Mar 2017)

Douglas M (1992) Risk and blame: essays in cultural theory. Routledge

Feather NT (1962) Cigarette smoking and lung cancer: a study of cognitive dissonance. Aust J Psychol 14(1):55–64

Fischoff B (1998) Risk Communication. In: Lofstedt R, Frewer L (eds) Risk and modern society. Earthscan, London

Frewer LJ, Scholderer J, Bredahl L (2003) Communicating about the risks and benefits of genetically modified foods: the mediating role of trust. Risk Anal 23:1117–1133

Gigerenzer G (2002) Reckoning with risk: learning to live with uncertainty. Penguin Books

Gordon Training International. 'The Conscious Competence Model': www.mindtools.com/pages/article/newISS_96.htm

Hall M (1990) Chronology of the principle scientific and governmental actions leading up to the November 13, 1985 eruption of Nevado del Ruiz, Colombia. J Volcanol Geoth Res 42(1–2):101–115

ICSU (International Council for Science) (2008) A science plan for integrated research on disaster risk: addressing the challenge of natural and human-induced environmental hazards. ICSU, Paris www.icsu.org/publications/reports-and-reviews/IRDR-science-plan

Kahneman D (2011) Thinking fast and slow. Farrar, Straus and Giroux

Kelmanm I, Mather TA (2008) Living with volcanoes: the sustainable livelihoods approach for volcano-related opportunities. J Volcanol Geoth Res 172(3–4):189–198

Marsh T (2014) Talking safety: a user's guide to world class safety conversation. Gower

McMaster C, Lee C (1991) Cognitive dissonance in tobacco smokers. Addict Behav 16(5):349–353

Palmer J (2015) Why ancient myths about volcanoes are often true. http://www.bbc.co.uk/earth/story/20150318-why-volcano-myths-are-true (Accessed 6 May 2017)

Pidgeon N, Kasperson RE, Slovic P (eds) (2003) The social amplification of risk. Cambridge University Press

Poortinga W, Pidgeon NF (2003) Exploring the dimensionality of trust in risk regulation. Risk Anal 23(5):961–972

Poortinga W, Pidgeon N (2004) Trust, the asymmetry principle, and the role of prior beliefs. Risk Anal 24 (6):1475–1486

Reason P, Bradbury H (eds) (2006) The handbook of action research. Sage

Slovic P (2000) The perception of risk. Earthscan, London

Smith LT (1999) Decolonising methodologies. Zed Books

Sorell T (1994) Scientism: philosophy and the infatuation with science. Routledge

Swanson D (2008) Hawaiian oral tradition describes 400 years of volcanic activity at Kīlauea. J Volcanol Geoth Res 176:427–431

The Information Deficit Model. https://en.wikipedia.org/wiki/Information_deficit_model. (Accessed 30 Apr 2017)

Twigg J (2009) Characteristics of a disaster-resilient community: a guidance note. University College London, www.abuhrc.org/research/dsm/Pages/project_view.aspx?project=13 (Accessed 15 Mar 2017)

Voight B (1990) The 1985 Nevado del Ruiz volcano catastrophe: anatomy and retrospection. J Volcanol Geoth Res 42(1–2):151–188

Wagner A (1996) The transactional manager. The Industrial Society

Williams S, Montaigne F (2001) Surviving galeras. Houghton Mifflin

Wilmshurst J (2010) Living with extreme weather events: an exploratory study of psychological factors in at-risk communities in the UK and Belize. Unpublished Ph. D. thesis

Cultural Differences and the Importance of Trust Between Volcanologists and Partners in Volcanic Risk Mitigation

Chris Newhall ⓘ

Abstract

A challenge in volcanic hazards communication is to bridge the cultural and language gaps between volcanologists and those who use volcanological information. We might be nominally from a single culture, e.g., Japanese, American, Italian, etc., but the cultural gaps between volcanologists and those who use volcano information can be as wide or wider than those from one country to the next. We have different goals or agendas, different approaches to solving problems, different terminologies, different definitions of success, and different reward systems. The first step toward bridging gaps is to recognize and accept the differences—valuing each other's goals and agreeing to work as a team to satisfy both. This acceptance plus involving information users in the information gathering helps to build trust. Without such trust, players are unlikely to accept each other's advice. Mainly from personal experience, I note commonly encountered cultural differences. Then, given the cultural differences, I note the critical importance of bridging those differences with trust. Finally, I give three short case histories—from Mount St. Helens, Pinatubo, and Usu—in which trust was built and differences were successfully overcome.

1 Introduction

At Mount St. Helens in 1980, Cowlitz County Sheriff Les Nelson once lamented: *"Trying to get a straight answer from a geologist is like trying to corner a rat in a round house."*

Geologists were not trying to hide anything, but they knew that Mount St. Helens could show a variety of behavior so they couched their answers in the kind of caveats that all scientists are taught to use. *"There are several possible scenarios..." "Uncertainties are high..."* and similar. Sheriff Nelson, and others, wanted simpler answers: yes, no, or, at least, most likely, probably yes, probably not, or similar.

C. Newhall (✉)
Mirisbiris Garden and Nature Center, Santo
Domingo, Philippines
e-mail: cgnewhall@gmail.com

Advs in Volcanology (2018) 515–527
https://doi.org/10.1007/11157_2016_40

It is easy to imagine additional words that might be said about volcanologists, to their face or behind their backs. Here are some that I can imagine:

> Volcanologists are a strange lot … Some of them hike up and down volcanoes, digging in the dirt or whacking off pieces of rock and telling us that they know how this volcano works. Others go around planting sensors in the ground, or sniffing the gases, and tell us they know how this volcano works. Sometimes they agree, often they don't. We're told there are still others who spend all their time heating and squeezing rocks in their laboratories, or writing computer programs they say simulate real life volcanoes. They publish papers in scientific journals that only they can read, and consider their work done. If we ask them "Is it dangerous?" their answers are so ambiguous that we might do just as well with a pair of dice. Do they think we care about what will happen in the next thousand years? And if we ask them "What should we do?" they answer "Ask someone else. That's not our mandate or expertise!"

Just as easily, I can imagine words of volcanologists about officials and other non-scientists:

> Officials and the public are a strange lot. They think that if we are good scientists, we should know exactly when and how each volcano will erupt. If we explain how complicated volcanoes are, and how science focuses on what isn't known, they roll their eyes in disbelief. Why, some of them even want us to give them 24-hour warnings! We aren't like seers who can divine the future! If we talk about the different ways that a volcano can kill them, they say "We don't care HOW we might get killed – just if or when. For us, a bomb falling on our head is the same as being toasted front or back." If we offer probabilities of various scenarios, they don't understand. And they keep resisting us… saying they want to stay in their homes, or on their jobs, or protect their cows. How can they expect us to keep them safe if they themselves won't take precautions?

Everyone who has travelled or worked in another culture knows that there are significant differences in values and customs from one culture to the next. This paper is a short cross-cultural look at differences within single cultures, just between scientists and those who use scientific information.

2 The Literature of Scientific Communication

Many books have been written about scientific communication, and how to bridge between the worlds of science and everyday life. Many excellent tips are given by Hayes and Grossman (2006), Manning (2006), Dean (2009), Olson (2009), Kennedy and Overholser (2010), Bultitude (2011), Fischhoff (2011), Graveline (2013), and other references in The Earth Institute (2014). Broad concepts of credibility and trust in risk communication are discussed by Renn and Levine (1991). Ways by which scientific uncertainty can be communicated in ways that enhance scientists' credibility are discussed by Morgan and Henrion (1990), Morgan (1998), Moss (2011), Mastrandrea et al. (2010), Pidgeon and Fischhoff (2011), and Socolow (2011). Ways by which scientists can communicate well across disciplines are discussed by Harris and Lyon (2013), among others. The present paper does not pretend to review the available literature, much less be a scholarly treatise. Rather, it presents personal experiences, which can be put by others into more general lessons of how best to communicate across professions.

3 Cultural Differences Among Players at Volcanoes?

Here are six cultural differences between scientists and non-scientists, as noticed during volcanic crises and as seen through my eyes as a volcanologist. By the generic term "scientist" I refer to both physical and social scientists, though my personal focus is naturally in physical science, specifically in volcanology. Some of you will correctly note stereotyping and sweeping generalizations, made in the interest of brevity. But every user group is a body of individuals. Generalizations will not apply to all, but in my experience they do apply to many!

First, we are concerned with different problems. *Scientists* ask "What, when, how, and especially, WHY?" *Civil defence officials and land managers* are concerned with "What, when, how serious, and what can we do to keep people safe?" *Engineers* ask "What is the problem and what can we build to fix it?" *Politicians* ask "How can we balance many competing interests (and be re-elected)?" *News media* ask "How can we translate and convey this in interesting ways?" And *citizens* ask, "What should my family do?" None of these are easy questions or tasks!

Second, scientists and non-scientists have different goals and reward systems. Everyone shares the common goals of public safety and well-being, doing one's job well, and advancing one's career and pay scale. In addition, differing goals include, *for research scientists*, to satisfy intense curiosity and to have fun in doing so. Much of the reward for scientists is the simple satisfaction of making new discoveries. However, these days, scientists must also compete with peers for professional recognition, including metrics of academic achievement. Those at universities and research-oriented volcano observatories get rewarded primarily for their research publications or contributions to such publications, how often their publications are cited by other scientists, and their success in garnering research grants. Scientists at public-service oriented organizations may be rewarded primarily for smooth operations and providing high quality advice to those who seek it.

Career public servants such as civil defence officials or land managers typically get rewarded for protecting the lives, infrastructure, property, economies, natural resources, and well-being of communities. Local officers will be rewarded for answering directly to people of the community while officers at higher levels of government may be rewarded for aiding the development of policy and/or funding prospects for the organization. *Politicians* seek a variety of rewards. They seek re-election, yes, and funds for re-election, but also the satisfaction of successfully balancing between competing interests. A common requirement in times of volcanic crises is to

successfully guard the safety of their constituents yet, at the same time, help them continue their lives as normally as possible, and to minimize disruption of business. *Engineers* are typically tasked with designing and implementing structural measures to reduce risk, and are rewarded with contracts, positive evaluations, and promotions if the project is successful. Most engineers take great pride in what they build or fix, so some of their reward is also internal. The *owners of news media* are driven variously by commitment to inform and serve the public and by the profit motive. *News reporters* are driven by similar commitment to public service, but are rewarded for column inches or minutes of airtime and editorial, peer, and public recognition.

Citizens, the most diverse group of all, need to balance keeping their families safe from the volcano and safe from other threats, including loss of crops, jobs, income, schooling, friends and social support networks, and other pillars of daily life.

Third, we speak different languages. Every field has its own specialized jargon. The jargon of *scientists* is a shorthand that is generally understood ONLY by scientists. Volcanological terms for major hazards like pyroclastic flows, tephra fall, and lahars all need clear definition, ideally in videos; more technical terms like magma compositions, extrusion rates, earthquake types, or monitoring technologies are best reserved for audiences who will appreciate them. Similarly, social science has its own jargon.

Engineering also has a specialized vocabulary, though more widely used and understood than that of scientists. *Civil defence officials* have their own jargon and acronyms, mostly non-technical but equally baffling to scientists. *News media and the public* use the language of everyday life.

The ways we view and draw the world are also different. Historically, *geoscientists* used maps and cross-sections to visualize the world in three dimensions, though increasingly those can be combined into fancy 3D graphics. Traditional maps (in plan view) and cross-sections are fine for *geoscientists and engineers*, but too abstract for *most others* (Haynes et al. 2007). Modern 3D visualizations, including oblique aerial views, are much better. GIS technology and the ready availability of 3D visualization tools through local-host GIS, Google Earth, Bing Maps, and other services are wonderful tools to help geoscientists communicate with non-geoscientists.

The meanings we attach to adjectives and other descriptors may be very different. *Geologists* have a very long view of time, far longer than of interest to most who seek advice. "Soon" to a *layman* might mean tomorrow while "soon" to a geologist might mean 100 years from now! Fast or slow are in the same category. The terms high and low, and possible, probable, likely and unlikely are notoriously ambiguous. Social scientists and others have documented wide ranges of numeric probabilities that different people attach to the same terms (Morgan 1998; Mastrandrea et al. 2010). Doyle and Potter (2015) recently suggested a table that translates from adjectives to probabilities of geologic hazards. In my own experience, the best ways to avoid the ambiguity of adjectives is to define them quantitatively, as in Mastrandrea et al. (2010) and Doyle and Potter (2015), or avoid them entirely by using a ladder of comparable risks as discussed later in this paper.

Fourth, we have different approaches to solving problems. In the scientific method, *scientists* identify a problem, propose hypotheses, gather data, and test the hypotheses (including forecasts of the future). Typically, the data gathering and interpretation involves a strong emphasis on observational skills, measurement, and quantitative assessment. Indeed, scientists like to quantify everything, including hazard and risk. *Non-scientists* (except engineers and a few others) are often wary of numbers, either because they don't understand them or they don't trust them. In an exception, those making policy for climate adaptation reportedly prefer numbers to qualitative descriptors (Moss 2011).

Scientists also strive for very low levels of uncertainty, such as might be acceptable for publication in peer-reviewed journals. Notwith-standing pressures to publish several papers per year or meet deadlines for project funding, scientists pride themselves in taking as much time as is needed before they offer advice, often months or years. Often, scientists resist calls to provide quick advice; at the same time, they should recognize that some decisions simply must be made quickly.

In contrast, *Civil Defence leaders and land managers* identify the problem, consider alternative solutions, prepare a decision matrix (e.g., cost benefit, etc.), and then make the optimal decision, sometimes within just hours or days. These decision-makers typically have higher tolerance than scientists for uncertainty, though in very high-stakes decisions such as siting of a nuclear facility, they too will pay great attention to uncertainty. *Politicians* follow a similar approach, though with more attention to public opinion. Accordingly, the metrics and weightings may be different. Scientific facts and advice may be just a small part of a political decision. For example, on matters of hazards and risk, scientists can evaluate hazard (and sometimes risk), but it is inevitably a political matter to evaluate how much risk the public (and the politician) is willing to accept. Decisions about acceptable risk, in turn, depend on the trade-offs between the benefits of taking the risk versus the potential losses if one takes the risk and loses. Scientific probabilities play a role, but only alongside many other factors. The role of *citizens* is to tell politicians and other decision makers how much risk they are willing to tolerate, and/or to 'vote with their feet' by self-evacuating if they so decide.

Engineers define the problem as best as they can in the time available, then design and implement a solution. The process may include consideration of several different designs and eventually choosing one. Sometimes, engineers express frustration with scientists if the starting or input parameters for what they are supposed to design keep changing. At some point, an engineer must lock in a design, whereas scientists keep on gathering data and, in some cases, call for a different or more flexible design. Scientists expect Nature and people to change; engineers may also expect change, but for their design they need a snapshot in time. Scientists emphasize uncertainty; engineers typically include a factor of safety to account for uncertainty and are ready to move on. *News media* generally limit themselves to reporting on how others solve problems, though some opinion or editorial pieces and some committed local reporters will actively facilitate communication between scientists, engineers, decision makers and the public. Members of the media bring special expertise in interviewing, listening, and comparing various points of view or approaches to problem solving.

Fifth, we differ in how we know what we know (epistemology). Nearly all who want to learn about a topic—scientists and non-scientists alike—start with published knowledge. The primary knowledge might be in academic journals, or in more accessible forms like books, popular magazines, Wikipedia and other sources on the internet, documentary videos, public lectures, and the like. While most of what is published is correct, all of us who publish know that there is misleading information too. The "library"—sensu latu—is a great place to start, but must be read with a critical mind.

Most *physical scientists* believe that the world operates according to well-known physical laws and that most everything can be explained in terms of physics and chemistry. I am among them. We have strong faith in the scientific method, observing, testing multiple hypotheses, and throwing them out in order until we accept the surviving hypothesis (or few). Scientists are always studying the world, learning, and discovering. We accept that the process should involve high levels of self-critique and peer review. We should be glad to disprove our own hypotheses or have our hypotheses be disproven, as that invariably leads to formulation of better hypotheses and brings us closer to the truth.

Jasanoff (1996) and Bäckstrand (2003) describe civic science: participatory, democratic, and addressing often-controversial societal problems without simple right or wrong answers, and with meaning derived from both absolute knowledge and human context. Civic science stands in contrast to conventional academic science that seeks to prove or disprove hypotheses, creating "absolute" knowledge within the scientific community that may or may not be used by decision makers, and that almost never involves citizens. On controversial matters involving big business, civic science also stands in contrast to science which promotes corporate interests. Most scientists are trained for conventional academic science, and can easily enter the proprietary world of corporate science. Those who will join in civic science must expect greater democracy and relativism than in standard university training.

What does civic science have to do with natural hazards? After all, aren't natural hazards wholly apolitical? No, decisions must still be made, sometimes even controversial decisions, and the more transparent and democratic the scientific process, the more trust will be established. One excellent example of civic volcanology is the network of *vigías* around Tungurahua Volcano in Ecuador, where an early overestimation of hazard created distrust of volcanologists that had to eventually be reversed, and an important part of that reversal was inclusion of local residents as scientific observers and alerters (Stone et al. 2014).

Where *decision makers and engineers* do rely on scientists for information on hazards and risk, there comes a serious responsibility for those scientists to resolve normal differences of interpretation. Scientific teams must try to resolve scientific debates and then present a consensus view to decision makers. If some issues cannot yet be resolved, it is fine to present them as competing hypotheses and explain how we will try to test them [from the field of climate change, see advice by Socolow (2011)]. Scientific credibility will still be intact. But if several different scientists offer competing advice, decision makers must decide who to believe. Personalities and trust, as much or more than evidence, become deciding factors. If scientific debate still rages in public, officials will lose faith in all scientists.

Many citizens "learn" by trusting a charismatic public figure—be it a politician, a cleric, a media figure, or anyone else. Sometimes, a scientist with unusually good communication skills and a knack for simplifying and popularizing scientific concepts can be the charismatic figure, but there is a risk that such scientists will fall into the trap of dogmatism unless they listen carefully to all scientific views before making public pronouncements. Charisma should be paired with humility, as I have seen cases in which good scientists became overly confident of their own expertise and Nature has proven them wrong.

Other citizens learn by their own observations, and by oral traditions. Traditional knowledge includes non-scientific explanations of the natural world that may or may not have a basis in physics or chemistry. For example, traditional knowledge around some volcanoes that water wells dry up before eruptions has a good physical explanation. But other "knowledge" (belief) has no physical explanation and is, instead, based on religious faith. This group of *citizens* will be equally or more convinced by traditional explanations as by those from scientists. A good case in point involved the late Mbah Marijan, spiritual gatekeeper of Merapi, who was trusted by a group of followers to know from conversations with the spirits of Merapi whether their place would be in danger or not (Schlehe 2010; Donovan et al. 2012) and, more broadly, how that augured for future national events (Dove 2010). Once differences in jargon are overcome, social scientists can help physical scientists to understand that the latter's physical explanations may or may not trump traditional knowledge.

Sixth, we have different resources and tools at our disposal. Typically, *scientists* have moderate to good resources for literature review, gathering of new data, computing, and sharing results through scientific meetings and publications. We may also have extensive experience with other volcanic crises—something that is rare for those with whom we work.

Engineers have resources for design and implementation of structural measures—often much more, in monetary terms, than resources of scientists. This is a natural consequence of the cost of such structural measures, and in most cases is accepted as necessary. However, this discrepancy between funding for science and engineering intervention can become a sore subject for scientists if engineers are not utilizing the best available scientific information, and end up wasting large amounts of money on structures that scientists anticipate will not work. At Pinatubo, much money was spent on building woefully inadequate sediment control structures before engineering measures eventually grew large enough to handle the threat (Janda et al. 1996). At Merapi, in retrospect, sediment control structures actually caused pyroclastic flows to jump out of stream channels and thereby increased the death toll (Lube et al. 2011; Baxter et al. in press).

Civil defence officials, land managers, and politicians typically have the authority and resources to control public access to areas near a volcano, to support an evacuation if needed and, sometimes, to fund engineering intervention. In a few countries, civil defence agencies provide substantial funding to scientists and, in return, can expect projects that address their very practical concerns. The *news media* have, by the nature of their work, great communication resources. They have the ear and the eye of the public and, if they choose to do so, they can be wonderfully effective in translating scientific information into terms that others can use and facilitating two-way communication with scientists.

Citizens have only their own eyes and ears, but often have the advantage of living on the volcano and being able to spot changes that escape modern instruments. A case in point is that local farmers reported the start of the flank eruption of Eyjafjallajökull in 2010 to police and scientists, not vice versa (Bird and Gísladóttir 2012). A number of eruptions in remote areas like the Aleutians are reported first by airline pilots who are constantly scanning their horizons for any in-flight hazards.

4 TRUST Between Scientists and Non-scientists Is Critical for Successful Risk Mitigation

Given the significant cultural differences between players in a volcanic crisis, there will inevitably be scepticism and a period of adjustment before each group is comfortable with the others. Perhaps the biggest challenge is to develop trust between the various players. More specifically, trust between scientists and those who use scientific information is *essential* if that information is to be accepted and used (Paton 2007; Haynes et al. 2008a, b; Donovan et al. 2014; Stone et al. 2014). Trying to understand and accept the cultural differences among the various groups, and involving users in the scientific process whenever feasible, are the best ways I know to develop this trust. If the scientific team with an official mandate for advice doesn't reach out to develop such trust, decision-makers may very well look elsewhere for advice, including to scientists perhaps less qualified but more communicative, or even to pseudoscientists who seem to speak with authority.

Here are three examples where trust was critical for volcanic risk mitigation.

4.1 Mount St. Helens 1980

Capsule timeline: March 20, 1980, earthquakes start. March 27, first phreatic eruption. More phreatic eruptions and strong bulging of North flank. Sunday, May 18, massive sector collapse and laterally directed blast. Most forest workers were off work on Sunday; most would-be tourists had been kept out of the area but there was access for some tourists along a myriad of small logging roads. A lateral blast far larger than expected led to 57 deaths.

Years of friendly interaction between geologists of the US Geological Survey (USGS) and officials of the US Forest Service (USFS) had already establish good trust and credibility even before the volcano became restless in 1980. An excellent long-range hazard assessment had already been published in 1978, albeit not promoted or read as widely as it might have been.

Trust had also been developed over nearly a decade of interaction between seismologists of the University of Washington (UW) and a few USGS seismologists who worked at the University, and that served as a good introduction between the UW seismologists and other USGS scientists who came quickly to Vancouver, Washington.

As the crisis evolved, USGS and UW scientists and USFS/State officials grew to understand and appreciate each other's roles. Most of these roles were never in doubt: scientists would try to anticipate what the volcano might do; USFS and State officials would decide how to manage the risk. Aside from a few early hiccups in which officials asked the USGS what they should do and the USGS declined as a matter of agency policy, these complementary roles were well understood and accepted.

The USFS quickly established an Emergency Coordination Center (ECC), where they provided desks and phones for all of the other key parties (State of Washington Emergency Services, County Sheriffs, representatives of hydroelectric and nuclear power utilities, major timber companies). The Forest Service also organized daily briefings for these representatives and for the press. The ECC was great for building trust, as it also afforded opportunities for 1:1 consultations with the scientists about specific places, e.g., a specific bridge, road intersection, etc., free from the glare of TV cameras. Even before the giant landslide and blast on May 18, 1980, trust between scientists and USFS was strong.

The public, the news media, and the timber companies clamoured for unrestricted access. Fortunately for most, the USFS resisted and declared a red zone off limits to all and a blue zone with only limited access. Unfortunately, the Governor of the State of Washington bowed to political pressure and kept areas under her jurisdiction officially open. The Washington State Patrol tried to block access anyway, but couldn't legally do so, and most fatalities in the eruption occurred on land under State control (Saarinen and Sell 1985).

After the landslide and blast, the shock of events practically glued everyone together. The USGS stationed two scientists at the ECC— one as liaison between the scientific team and the risk managers, and the other to provide consistent information to the news media. I came to Mount St. Helens not long after the landslide and blast, as the liaison to risk managers, and worked side-by-side for several years with USFS and State personnel. Because we worked in the same office, shared the same coffee pot, and heard the same conversations, we developed a mutual understanding of each other's needs and competencies, even idiosyncrasies. We learned to "read" each other, and to learn from each other.

As mentioned above, USGS policy strictly forbade scientists from suggesting to public officials how they should manage risk. Any recommendation regarding evacuations or other mitigation measures contains an implicit assessment of acceptable risk, including personal, economic, and political matters well beyond the expertise or mandate of scientists. In the US, the public is staunchly, vehemently protective about personal freedoms, including the right to make most decisions about their own personal safety. While private citizens might take recommendations or orders from land-managers and law-enforcement, they certainly would not wish for their freedom to be "managed" by scientists. Recommendations about evacuations and other mitigation are political matters, all agreed. Nevertheless, there were times when Forest Service or other risk managers struggled to understand the hazard, and to decide on their response. In those cases, because we trusted each other, we held "off-the-record" conversations in which scientists actually went beyond their official limits to guide risk managers. We did not suggest what officials should do, but used personal risk tolerance as bridge. The Forest Service staff would ask us, "Would you personally stay in this place, or would you let your own family stay in this place?" Those were questions we could answer without going beyond our brief, yet our answers gave them the information they needed to make their own decisions.

As another example, we struggled together to reach a shared understanding of the magnitude of remaining risk, using probabilities. The Forest Service, timber companies, and loggers pushed us to simplify the presentation and to make the bottom line something that they could easily understand. The result was a chart ("risk ladder") on which volcanic risk, once calculated, could be compared at a glance to familiar risks (Fig. 1). The chart includes occupational risks (soldier in war, helicopter pilot, logger, office worker, etc.), lifestyle risks (lung cancer from smoking), and accident risks (traffic risks, risks from floods, lightning strikes, etc.). Probabilities per se meant almost nothing; locating one's risk on this chart made it instantly simple to decide whether a risk was acceptable or not. None of the parties—the timber company, loggers' union, or Forest Service—was concerned about high uncertainties or about caveats in comparing voluntary and involuntary risk.

Although the post-May 18 period had far less threat and drama than that during and before May 18, trust and personal friendship that developed in the course of working together has continued to serve all parties well right up to the present, including the 2004–2008 eruptive period.

4.2 Pinatubo 1991

Capsule timeline: July 16, 1990: M 7.8 strike-slip earthquake with epicenter 100 km NE of Pinatubo. April 2, 1991, phreatic explosions along a fissure across north-northeast flank of Pinatubo. Fluctuating unrest until escalation in early June. Extrusion of lava dome June 7–12. Strong VEI 3 scale eruptions June 12–14. Climactic VEI 6 scale eruption and caldera formation on June 15. Most of those at risk from pyroclastic flows had been evacuated by June 14, just barely in time.

Because Pinatubo had no historic eruptions, officials of Central Luzon had no established relations or trust with Philippine scientists. There was, however, a well-established relationship and

Annual Risk	Age	Occupation	Disease	Accident
10^0				
	90			
10^{-1}				
	80	Soldier in war		
10^{-2}	60			
	50	Helicopter pilots	Heart disease	
		Logging	Cancer	
10^{-3}	20		AIDS, Sub-Sahara	
		Mining		All accidents
		Agriculture		Car accidents
10^{-4}		Transport, construction	AIDS, industrial'zd	
		All workers (avg.)		
		Manufact'g, retail, gov't		
10^{-5}				Drowning
...				Hurricanes

Fig. 1 Risk ladder showing comparative annual risks to life, in the US, circa 2000. Simplified and updated from that originally used in Newhall (1982) to help loggers and timber company managers understand volcanic risk they faced in salvaging blown-down timber. Assuming good volcano monitoring, communications, and willingness to stop work during elevated volcanic unrest, the added annual risk to each logger's life was ∼0.0012/y (± one order of magnitude), or approximately the same as a logger's normal occupational risk though with higher uncertainty

trust between the National Disaster Coordinating Council (civil defence, NDCC) and the Philippine Institute of Volcanology and Seismology (PHIVOLCS). Starting from the top, this relationship and trust was brought down to the provincial and municipal levels. Initially, local scepticism was extremely high, so trust at the local level took a long time to develop, but a combination of many briefings and increasingly visible signs from the volcano itself eventually turned the tide and allowed evacuation of the riskiest areas before the eruption (Newhall and Punongbayan 1996; Punongbayan et al. 1996; Newhall and Solidum, this volume).

The slopes of Pinatubo were home to an indigenous people, the Pinatubo Aytas. Centuries of discrimination and distrust led most Aytas to shun interaction with lowland Filipinos. Few of the Aytas understood anything of modern science, and most believed that Mount Pinatubo was the home and domain of their god Apo Namalyari, so anything that happened at Pinatubo would have to be explained as an action of Apo Namalyari. A few of the responding PHIVOLCS scientists spoke Kapampangan or Ilocano, languages familiar to all of the Ayta, but none of the scientists spoke any of the Ayta dialects, nor were there any longstanding relationships. Fortunately, there were several trusted religious missions on Pinatubo—one of the Franciscan Sisters of Mary northwest of the summit, and two of evangelical Protestant missionaries north and south of the summit. The communication gap between scientists and indigenous Aytas was bridged by these trusted missionaries.

Longstanding trust between USGS and PHIVOLCS scientists also helped greatly in this crisis, and I suspect that it helped in the larger task of developing trust and credibility with local officials. Had we scientists ourselves not been completely unified, our task to dissolve official and public scepticism and promote preparedness could easily have foundered.

The matter of *trust* with commanders of the US military bases was particularly complicated. The Pinatubo crisis arose in the midst of a tense renegotiation to extend the lease for US bases in the Philippines, and there was little trust between officials of US military bases and those of the

Philippine government. Initial warnings from Philippine scientists were dismissed, and US commanders only began to pay attention when USGS scientists were invited in by PHIVOLCS. Even then, scepticism remained high, as the US scientists had to be conscious of their first duty to the Philippine government and populace. Gradually, though, daily interactions and a growing body of worrisome facts about the volcano dissolved scepticism one officer at a time. The joint PHIVOLCS-USGS team received critical logistical support from the US military (especially, the US Air Force), and the US military in return received notices at the same time as Philippine civil defence officials, and had access to the scientific team for additional discussions.

Two seemingly minor incidents at our makeshift observatory on Clark Air Base greatly improved trust and understanding of our message. The first was the anniversary of the big eruption of Mount St. Helens, on May 18. After working day and night since early April, the team took a day off on May 18 to relax, and invited military officers over for a BBQ and beer. There, we could talk about our families, fun, and matters other than the volcano, and we noticed a palpable improvement in relations with the military officers. They saw, for the first time, that we scientists were human and not much different from themselves! We even had a sense of humour. Amazing ☺

The other incident was soon thereafter, when scientists started looking for a fall-back position on Clark Air Base, as far from the volcano as possible. The very fact that we were concerned for our own safety in the middle of Clark Air Base made the officers take notice. Suddenly, things we had discussed many times before, including the possibility of pyroclastic flows reaching Clark Air Base, took on new meaning. On June 10, shortly after evacuation of most personnel from Clark Air Base to Subic Naval Station, the scientific team moved to the fallback position at the far edge of Clark Air Base, and some of the remaining military officers moved with the scientists.

4.3 Usu 2000

Capsule timeline: March 27, strong earthquake swarm and pronounced ground fracturing begins. March 31, explosive eruptions begin and continue for ~3 weeks while a cryptodome is simultaneously being emplaced. The main hazards were ballistic fragments, formation of new craters, widespread ground fracturing, and hot lahars. Timely evacuations kept everyone safe.

Usu Volcano, in Hokkaido, Japan, has a history of explosive eruptions and cryptodome growth over many centuries, most recently in 1910, 1943–44, and 1977–1982. In those same crises, there was unusually good trust between scientists, police officials, and mayors. In the case of 1910, the police chief Mr. Iida had been a student of the leading volcano and earthquake scientist of the time, Prof. Omori at Univ. of Tokyo, so the contact and respect was already established. The crisis of 1943–44 came during WW II, and scientific response was led by another senior professor from Univ. of Tokyo, Prof. T. Minakami. The postmaster, Mr. S. Mimatsu, was an amateur scientist and worked closely with Prof. Minakami. (Mr. Mimatsu's son still maintains a volcano museum near Usu).

After the 1943–44 eruption, tourist development and population increased significantly. Risk at Usu arises because towns lie extraordinarily close to the volcano, just 2–3 km from the summit and even less from flank vents. For fear of scaring away tourists, there was strong resistance to discussion of volcanic hazards. However, Prof. Hiromu Okada had learned from study of previous crises that the key to risk mitigation had been trust and respect between early scientists and police officials. So, Prof. Okada took on the personal challenge of frequent interaction and trust-building with local officials, especially local mayors. He managed to change the conversation from "Don't mention volcanic hazards!" to "OK, if we listen to the scientists, educate the public, and back off briefly when needed, we can safely co-exist with these volcanic hazards!" Mayor Okamura of Abuta town was an early convert. Prof. Okada's personality is excellent for building trust and, in 2000, when another magma intrusion (cryptodome) was rising, people were safely evacuated ahead of phreatic explosions that damaged many buildings. It was an outstandingly successful case of volcanic risk mitigation in Japan.

5 Concluding Tips

When a volcano awakens, volcanologists and other scientists will need to work with civil defence leaders, politicians, business leaders, engineers, news media and, sometimes, citizens. Most scientists are not trained for such interaction, and the interaction can be challenging. There will be an initial period in which each group is simply trying to get to know the other, and to assess each other's motives, technical competence, and judgment. Scientists will be under a spotlight, since scientific information will become a major factor in mitigation decisions.

Wide cultural differences exist between volcanologists and those who use volcanological information, and these differences must be understood and respected if there is to be trust between both groups. How can this understanding be achieved? The following tips are especially for volcanologists, but may apply as well to other players.

- Work in close proximity as much as you can, and try to understand each other's culture and needs. Volunteer to work side-by-side in space organized by and for those needing volcano information.
- Expect scepticism, and do not take it personally. Things that may signal obvious danger to a scientist, e.g., a town built on a pyroclastic flow deposit or unusual squiggles on a seismogram, may not be at all obvious to others, and the onus is on those who see danger to convince others that it is real. Try to relate it to what the audience knows and cares about.
- Involve those who need volcano information in the gathering and dissemination of that

information. Users become partners in the scientific process.
- Be professional, patient, and show that you want to help those at risk and those who must make mitigation decisions. Without pre-empting the decision maker's responsibility, share insights into your own personal risk tolerance.
- Remember, we scientists may seem to be from another planet. Bring humility and a sense of humour to the table. Share personal aspirations and worries, notes about families, and other common interests, and food and drink as well. Invite your counterparts out for a beer, or a picnic, or birthday party, or make any other simple, personal gestures to move your interaction from strictly formal to space in which you see the more human sides of each other.
- Personal trust that grows out of such interaction is a critical prerequisite before officials will make the necessary, hard decisions for mitigation.
- Once trust has been built, be careful to maintain it. Obviously, correct forecasts will build and maintain trust, but so, too, can humility and honest statements of uncertainty when we don't know what the volcano will do next.

In every interaction, professionalism, cross-cultural sensitivity, and personal touches make a powerful combination. Good luck!

Acknowledgements Thanks to Kevin Krajick, Earth Institute of Columbia University, for helpful pre-review comments, and to anonymous reviewers who pointed the author to relevant social science literature, the importance of distinguishing opinions from evidence, and necessary clarifications.

References

Bäckstrand K (2003) Civic science for sustainability: Reframing the role of experts, policy-makers, and citizens in environmental governance. Global Environ Politics 3–4:24–41

Baxter P, Jenkins S, Rosadi S, Komorowski J-C, Dunn K, Purser D, Voight B, Shelley I (in press) Human survival in volcanic eruptions: thermal injuries in pyroclastic surges, their causes, prognosis and emergency management. Burns Accessed on 21 Feb 2017

Bird DK, Gísladóttir G (2012) Residents' attitudes and behaviour before and after the 2010 Eyjafjallajökull eruptions—a case study from southern Iceland. Bull Volc 74:1263–1279

Bultitude K (2011) The why and how of science communication. In: Rosulek P (ed) Science Communication. European Commission, Pilsen, 18 p. https://www.ucl.ac.uk/sts/staff/bultitude/KB_TB/Karen_Bultitude_-_Science_Communication_Why_and_How.pdf. Accessed 10 Nov 2014

Dean C (2009) Am I making myself clear? A scientist's guide to talking to the public. Harvard University Press

Donovan K, Suryanto A, Utami P (2012) Mapping cultural vulnerability in volcanic regions: the practical application of social volcanology at Mt. Merapi, Indonesia. Env Hazards 11:303–323

Donovan A, Eiser JR, Sparks RSJ (2014) Scientists' views about lay perceptions of volcanic hazard and risk. J Appl Volcanol 3:15, 14 p

Dove MR (2010) The panoptic gaze in a non-western setting: self-surveillance on Merapi volcano, Central Java. Religion 40:121–127

Doyle EEH, Potter SH (2015) Methodology for the development of a probability translation table for GeoNet. GNS Science Report 2015/67, 18 p

Fischhoff B (2011) Applying the science of communication to the communication of science. Clim Change 108(4):701–705. doi:10.1007/s10584-011-0183-9

Graveline D (2013) 7 ineffective habits of scientists who communicate with public audiences. http://www.dontgetcaught.com, 19 Feb 2013. Accessed 10 Nov 2014

Harris F, Lyon F (2013) Transdisciplinary environmental research: building trust across professional cultures. Environ Sci Policy 31:109–119. doi:10.1016/j.envsci.2013.02.006

Hayes R, Grossman D (2006) A scientist's guide to talking with the media: practical advice from the Union of Concerned Scientists. Rutgers University Press. Excerpts at http://www.ucsusa.org/sites/default/files/legacy/assets/documents/global_warming/UCS_Desk_Reference_Scientists_Guide.pdf

Haynes K, Barclay J, Pidgeon N (2007) Volcanic hazard communication using maps: an evaluation of their effectiveness. Bull Volc 70:123–138

Haynes K, Barclay J, Pidgeon N (2008a) The issue of trust and its influence on risk communication during a volcanic crisis. Bull Volc 70(5):605–621

Haynes K, Barclay J, Pidgeon N (2008b) Whose reality counts? Factors affecting the perception of volcanic risk. J Volcanol Geoth Res 172:259–272

Janda RJ, Daag AS, Delos Reyes PJ, Newhall CG, Pierson TC, Punongbayan RS, Rodolfo KS, Umbal JV (1996) Assessment and response to lahar hazard at Mount Pinatubo. In: Newhall CG, Punongbayan RS (eds) Fire and mud, eruptions and lahars of Mount Pinatubo, Philippines. Quezon City, PHIVOLCS and Seattle, University of Washington Press, pp 107–139

Jasanoff S (1996) Beyond epistemology: relativism and engagement in the politics of science. Soc Stud Sci 26:393–418

Kennedy D, Overholser G (eds) (2010) Science and the media. American Academy of Arts and Sciences, Cambridge, MA, 95 p. www.amacad.org/pdfs/sciencemedia.pdf

Lube G, Cronin SJ, Thouret J-C, Surono (2011) Kinematic characteristics of pyroclastic density currents at Merapi and controls on their avulsion from natural and engineered channels. Geol Soc Am Bull 123:1127–1140. doi:10.1130/B30244.1

Manning P (2006) Communicating effectively with politicians. Physics in Canada. June-July 2006, pp 163–168. http://www.interactions.org/pdf/SLAC_pavan_manning.pdf

Mastrandrea MD, Field CB, Stocker TF, Edenhofer O, Ebi KL, Frame DJ, Held H, Kriegler E, Mach KJ, Matschoss PR, Plattner G-K, Yohe GW, Zwiers FW (2010) Guidance note for lead authors of the IPCC fifth assessment report on consistent treatment of uncertainties. Intergovernmental panel on climate change (IPCC). Available at: http://www.ipcc-wg2.gov/meetings/CGCs/Uncertainties-GN_IPCCbrochure_lo.pdf

Morgan MG (1998) Uncertainty analysis in risk assessment. Human Ecol Risk Assess 4(1):25–39

Morgan MG, Henrion M (1990) Uncertainty: a guide to dealing with uncertainty in quantitative risk and policy analysis. Cambridge University Press, 348 p

Moss RH (2011) Reducing doubt about uncertainty: guidance for IPCC's third assessment. Clim Change 108:641–658. doi:10.1007/s10584-011-0182-x

Newhall CG (1982) A method for estimating intermediate- and long-term risks from volcanic activity, with an example from Mount St. Helens, Washington. US Geological Survey Open-File Report 82–396, 59 p

Newhall CG, Punongbayan RS (1996) The narrow margin of successful volcanic-risk mitigation. In: Scarpa R, Tilling RI (eds) Monitoring and mitigation of volcanic hazards. Springer, Berlin, pp 807–838

Newhall CG, Solidum RU (this volume) Volcanic hazard communication at Pinatubo from 1991 to present. In: Fearnley CJ, McGuire B, Jolly G, Bird D, Haynes K (eds) Observing the volcano world: volcano crisis communication. Springer

Olson R (2009) Don't be such a scientist. Island Press, Washington 206 p

Paton D (2007) Preparing for natural hazards: the role of community trust. Disaster Prev Manag 16(3):370–379

Pidgeon N, Fischhoff B (2011) The role of social and decision sciences in communicating uncertain climate risks. Nat Clim Change 1:35–41

Punongbayan RS, Bautista MLP, Harlow DH, Newhall CG, Hoblitt RP (1996) Pre-eruption hazard assessments and warnings. In: Newhall CG, Punongbayan RS (eds) Fire and mud: eruptions and lahars of Mount Pinatubo, Quezon City, Philippine Institute of

Volcanology and Seismology, and Seattle, University of Washington Press, Philippines, pp 67–85

Renn O, Levine D (1991) Credibility and trust in risk communication (Chapter 9). In: Kasperson RE, Stallen PJM (eds) Communicating Risks to the Public. Kluwer Academic Public, Netherlands, pp 175–218

Saarinen TF, Sell JL (1985) Warning and response to the Mount St. Helens eruption. Albany, State University of New York Press, 240 p

Schlehe J (2010) Anthropology of religion: disasters and the representations of tradition and modernity. Religion 40:112–120

Socolow RH (2011) High-consequence outcomes and internal disagreements: tell us more please. Clim Change 108(4):775–790. doi:10.1007/s10584-011-0183-9

Stone J, Barclay J, Simmons P, Cole PD, Loughlin SC, Ramón P, Mothes P (2014) Risk reduction through community-based monitoring: the *vigías* of Tungurahua, Ecuador. J Appl Volcanol 3:14 p. doi:10.1186/s13617-014-0011-9

The Earth Institute, Columbia University (2014) Communicating science. http://www.earth.columbia.edu/articles/view/2636. Accessed 10 Nov 2014

International Coordination in Managing Airborne Ash Hazards: Lessons from the Northern Pacific

Yohko Igarashi, Olga Girina, Jeffrey Osiensky
and Donald Moore

Abstract

Airborne volcanic ash is one of the most common, far-travelled, direct hazards associated with explosive volcanic eruptions worldwide. Management of volcanic ash cloud hazards often requires coordinated efforts of meteorological, volcanological, and aviation authorities from multiple countries. These international collaborations during eruptions pose particular challenges due to variable crisis response protocols, uneven agency responsibilities and technical capacities, language differences, and the expense of travel to establish and maintain relationships over the long term. This report introduces some of the recent efforts in enhancing international cooperation and collaboration in the Northern Pacific region.

1 Introduction

Airborne volcanic ash is one of the most common, far-travelled, direct hazards associated with explosive volcanic eruptions worldwide. Management of volcanic ash cloud hazards often require coordinated efforts of meteorological, volcanological, and civil aviation authorities from multiple countries. These international collaborations during eruptions pose particular challenges due to variable crisis response protocols, uneven agency responsibilities and technical capacities, language differences, and the expense of travel to establish and maintain relationships over the long term. The steady rise in global aviation, particularly on the remote routes between North America and Asia that overfly more than 100 potentially active volcanoes in the

Y. Igarashi (✉)
Seismology and Volcanology Department, Japan
Meteorological Agency (JMA), Tokyo, Japan
e-mail: y_igarashi@met.kishou.go.jp

O. Girina
Kamchatka Volcanic Eruption Response Team
(KVERT), Petropavlovsk-Kamchatsky, Russia

J. Osiensky
Environmental and Scientific Services Division,
National Oceanic and Atmospheric Administration,
National Weather Service (NOAA/NWS) Alaska
Region, Silver Spring, USA

D. Moore
Volcanic Ash Advisory Centre (VAAC) Anchorage,
National Oceanic and Atmospheric Administration,
National Weather Service (NOAA/NWS) Alaska
Region, Silver Spring, USA

Advs in Volcanology (2018) 529–547
https://doi.org/10.1007/11157_2016_45
Published Online: 29 March 2017

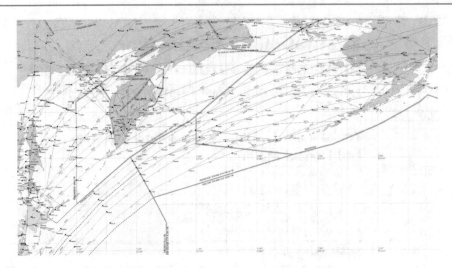

Fig. 1 Route map in the Northern Pacific region

United States (Alaska and Aleutian Islands) and the Russian Federation (Kamchatka and Kurile Islands), means that more and more aircraft are at risk from the impacts of airborne volcanic ash.

The Northern Pacific (NOPAC) air routes connecting Alaska to the far east (Fig. 1) carry 10,000 people per day and up to 50,000 aircraft per year with some routes passing over the Kamchatka Peninsula with around 30 volcanoes (Gordeev and Girina 2014; VAAC Anchorage 2015a). Commercial aircraft in this region are required to operate on a fixed route and flight level approved by Air Navigation Service Providers (ANSPs). They need an approval before or during a flight when they change their route and/or flight level; re-routing to avoid a volcanic eruption is no exception. Over the past two decades, more than 60 strong explosive eruptions in the Russian Far East (Girina et al. 2007, 2009, 2014a, b; Gordeev and Girina 2014; McGimsey and Neal 1996; McGimsey and Wallace 1999; McGimsey et al. 2003, 2004a, b, 2005, 2008, 2011, 2014; Neal and McGimsey 1997; Neal et al. 2004, 2005a, b, 2009a, b, 2011, 2014) have tested coordination among relevant agencies and institutions in Japan, the Russian Federation, the United States, and Canada, prompting ongoing testing of existing systems with a number of lessons learned. Critical to meeting this challenge of a rapid, international response to volcanic ash

cloud hazard is development of written, updated, and practiced response plans or agreements detailing roles and responsibilities.

Frequent exercises that test the readiness and procedures, involving representatives of international air carriers, are important tools to continually refine the response process. A mechanism to engage air carriers and critically evaluate individual eruption responses to events is also necessary to focus these improvements. Each of the three main components of ash cloud response: meteorology, volcanology, and air traffic management, have different challenges in obtaining a seamless coordinated response. The establishment of a worldwide system of Volcanic Ash Advisory Centres (VAACs) in the mid-1990s assisted greatly in development of a consistent meteorological response and warning product suite. Similarly, the mature system of international conventions in air traffic management contributes significantly to coordinated handling of air traffic during eruptions that may disrupt the air routes. However, there remains strong variability in the adequacy of volcano surveillance and alerting by appropriate regional volcanological authorities, a challenge increasingly met by the growing use of remote and satellite based monitoring and eruption detection techniques. Recent eruptions of Sarychev-Peak Volcano in 2009 (McGimsey et al. 2014) and Kliuchevskoi

Volcano in 2013 (Girina et al. 2014a, b) illustrate aspects of both the successes and ongoing challenges of international eruption response in the Northern Pacific, as well as worldwide.

2 VAACs and Volcano Observatories Related to Volcanic Ash Clouds in the Northern Pacific Region

To avoid aircraft-related disasters caused by volcanic ash clouds, a framework for the International Airways Volcano Watch (IAVW) was established in 1993 by the International Civil Aviation Organization (ICAO). Under this framework, nine VAACs were designated as centres to monitor volcanic eruptions and to provide information on the locations and movement of volcanic ash clouds as well as an outlook for their regions of responsibility (Fig. 2). In the Northern Pacific region, there are four VAACs: Anchorage, Montreal, Tokyo and Washington. Among them, VAAC Anchorage has the area of responsibility covering the entire Anchorage Flight Information Region (FIR) as well as an area bounded on the west by 150°E Longitude

and on the south by 60°N Latitude, which includes all the volcanoes within the State of Alaska. VAAC Anchorage's area of responsibility is adjacent to volcanoes located in Kamchatka Peninsula and the Northern Kurile Islands, which are in the area of responsibility of VAAC Tokyo that covers the East Asia and Northwest Pacific regions. Several volcanoes in the region are quite active and ash clouds often move across the boundary of the area of responsibility of these two VAACs where the VAACs hand over the responsibility of information issuance through close coordination and communication. This chapter mainly highlights the activities of these two VAACs and related organizations in the region.

VAAC Anchorage was established by the United States Department of Commerce (DOC) National Oceanic and Atmospheric Administration's (NOAA) National Weather Service (NWS) at the request of the Federal Aviation Administration (FAA). It has been providing information on volcanic ash clouds in the form of Volcanic Ash Advisories (VAAs) around the clock, supporting the Anchorage Meteorological Watch Offices (MWO) and the Anchorage Area Control Center (ACC).

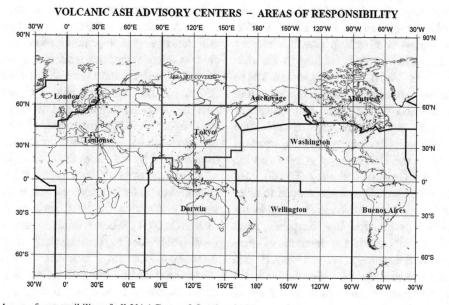

VOLCANIC ASH ADVISORY CENTERS – AREAS OF RESPONSIBILITY

Fig. 2 Areas of responsibility of all VAACs as of October 2016

VAAC Tokyo has been monitoring volcanoes around the clock and issuing VAAs since 1997. It is a part of the Japan Meteorological Agency (JMA). VAAC Tokyo was originally established in the Tokyo Aviation Weather Service Center, the branch office of JMA located at Haneda Airport, and was transferred to JMA headquarters in Tokyo in 2006. Through the experience of several significant eruptions such as that of Sarychev Peak in 2009, VAAC Tokyo strengthened the operations by allocating five forecasters specific for volcanic ash in 2011, while not all VAACs have their own forecasters (in such VAACs, forecasters share other aviation duties, too). VAAC Tokyo supports MWO Tokyo, also a part of JMA, as well as the ACCs in the area (JMA/VAAC Tokyo 2015).

As for VAAC operations, early detection of volcanic eruptions is crucial. To enable timely VAA provision, a VAAC monitors satellite imagery for volcanic ash clouds around the clock. Whenever a new eruption is identified, it announces the possibility of eruption through VAA issuance and continues VAA provisions until the volcanic ash cloud is dissipated. However, as satellite imagery is not continuous data, the initial detection can be delayed. For example, the new satellite of HIMAWARI-8, that was launched on October 7, 2014 and has been operational since July 7, 2015, principally provides imagery every ten minutes, and MTSAT-2, still in operation as a back-up, provides imagery twice per hour. Therefore, adding to the data from satellites, VAACs Anchorage and Tokyo also receive information about eruptions from relevant volcano organizations and sometimes from aircraft in operation. For example, the Alaska Volcano Observatory (AVO), operated conjointly by the University of Alaska and the United States Geological Survey (USGS), provides daily and weekly volcano reports to adjacent VAACs. Volcano observatories such as the Kamchatka Volcanic Eruption Response Team (KVERT) on behalf of the Institute of Volcanology and Seismology (IVS) Far East Branch of Russian Academy of Sciences and the Sakhalin Volcanic Eruption Response Team of the Institute of Marine Geology and Geophysics,

provide not only daily and weekly volcano reports but also timely eruption information about volcanoes in Kamchatka Peninsula and Kurile Islands.

Volcano activities in Japan are monitored by four JMA Volcanic Observations and Warning Centers (VOWCs) located in Sapporo, Sendai, Tokyo and Fukuoka. VAAC Tokyo receives volcanic activity reports from these VOWCs immediately after eruptions. As the area of responsibility of VAAC Tokyo extends to the Philippines, the centre also receives timely information from the Philippine Institute of Volcanology and Seismology (PHIVOLCS).

Information provided by most of these volcano observatories is called "Volcano Observatory Notice for Aviation" (VONA). While periodical information is useful to grasp the latest volcanic conditions, VONAs are indispensable for timely VAA provision by VAACs, especially with regard to the initial issuances. Given volcanoes in Kamchatka Peninsula are remarkably active, VAACs Tokyo and Anchorage often issue VAAs regarding volcanic ash clouds from those volcanoes based on VONAs from KVERT.

KVERT was established in 1993 aiming at improving safety for aviation during explosive eruptions. It has a similar goal to VAACs to reduce the risk of aircraft encountering volcanic ash clouds in the Northern Pacific region through timely detection of volcanic unrest, tracking of ash clouds, and prompt notification of airlines, civil aviation authorities, and others about the hazards (Gordeev and Girina 2014; Neal et al. 2009a, b). The complex analysis of published data on volcanic activity and the data from 22 years of KVERT's continuous monitoring of volcanoes allows a quantitative evaluation of the hazard posed by volcanoes to aviation. The level of hazard to aviation from each of the Kamchatkan volcanoes is communicated by KVERT using the Aviation Colour Code recommended by ICAO (2004). When KVERT issues a VONA, it is automatically disseminated to VAACs Anchorage, Darwin, Montreal, Tokyo and Washington, and all international and local users of the Northern Pacific region such as ICAO, FAA, NOAA, AVO, USGS, the Yelizovo

Airport Meteorological Center (Yelizovo AMC), the Kamchatka Hydro-Meteorological Center (KHMC), the Kamchatka Branch of the Ministry for Emergency Situations (KB MES), and mass media. It is also automatically uploaded on the KVERT website: http://www.kscnet.ru/ivs/kvert/van/ (Girina and Romanova 2015).

Adding to the volcanoes in Kamchatka Peninsula, several volcanoes in Japan are also very active; therefore, VAAC Tokyo frequently issues VAAs in this regard based on VONAs from JMA's VOWCs.

Based on VONAs from volcano observatories adding to volcanic ash clouds detection in satellite imagery, VAACs issue VAAs designed to assist MWOs in preparing international standard Significant Meteorological Information (SIGMET) on volcanic ash clouds. VAAs, describing the latest extent and forecast trajectories of volcanic ash clouds, are updated every six hours so long as ash clouds are identified by satellite imagery. The VAAs are issued within the six hours if unforeseen changes occur in observations. The roles of relevant organizations and regulations of operations are given by ICAO (2007).

3 Case Study of Impacts of a Volcano Eruption onto Air Traffic

The impacts of volcanic eruptions are the strongest triggers for improvements in volcanic ash responses by relevant organizations because they give a true account of tasks to overcome as well as successful operations. This section shows two case study examples of major volcano eruptions and subsequent actions taken by relevant organizations.

3.1 Case Study #1

Eruption of Sarychev-Peak Volcano in 2009

An eruption of Sarychev-Peak Volcano in the Kurile Islands was detected at 01:59 UTC on June 12, 2009. VAAC Tokyo identified the eruption from satellite imagery and issued the first VAA at 06:49 UTC with an observed volcanic ash cloud at 34,000 ft extending to the east.

On the day of the eruption, only five aircraft requested re-routing; however, the volcanic ash cloud in the VAAC Tokyo's area of responsibility reached 54,000 ft the next day according to the VAA issued by VAAC Tokyo, and the volcanic ash cloud continued to be observed at that height for one and a half days. As a result of the VAAs, most flights avoided the NOPAC route and flew through Russian airspace instead.

As it was a continuous eruption with ash emission, the volcanic ash cloud extended more and more widely. It migrated into VAAC Washington's area of responsibility, also covering a small part of VAAC Anchorage's region, while volcanic ash clouds due to subsequent emissions covered VAAC Tokyo's area of responsibility. VAACs Tokyo and Washington (and later VAAC Anchorage) issued VAAs and advisories in graphic format (Volcanic Ash Graphic: VAGs) (Fig. 3). As the volcanic ash cloud covered a wide area across the NOPAC region, the Air Traffic Management Center in the Japan Civil Aviation Bureau (JCAB/ATMC) set the Pacific Organized Track System (PACOTS) avoiding the NOPAC route based on the VAAC Tokyo's advisories. Oakland Air Route Traffic Control Center (ARTCC) set westward PACOTS based on the advisories provided by VAAC Washington and requested JCAB/ATMC also to set eastward PACOTS in the same way.

The volcanic ash cloud remained relatively high even after it lowered from the maximum height. Following a request from an airline, VAAC Tokyo therefore began providing VAA/VAG every three hours instead of the regular six-hour interval from around 09:00 UTC on June 16. Under this situation, some irregular incidents occurred and JCAB/ATMC as well as the Oakland ARTCC responded to each case. For example, a particular aircraft headed into the volcanic ash cloud area without knowing the situation and JCAB/ATMC was obliged to advise re-routing. Another example was that one particular airline, flying on a regional route in

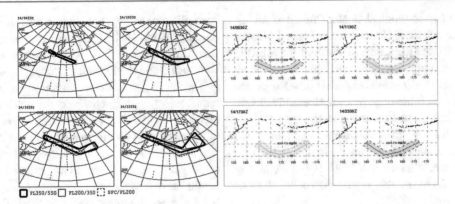

□ FL350/550 □ FL200/350 ⌈⌉ SFC/FL200

Fig. 3 VAGs by VAACs Tokyo (*left*) and Washington (*right*) issued at 06:22 UTC and 06:24 UTC on 14 June 2009, respectively, for a volcanic ash cloud covering a wide area from Sarychev-Peak Volcano in Kamchatka peninsula in VAAC Tokyo's region to VAAC Washington's area of responsibility. VAG is composed of four maps with volcanic ash cloud areas at present (*left top*), six hours ahead (*right top*), twelve hours ahead (*left bottom*) and eighteen hours ahead (*right bottom*), and explanatory in text. Here, only maps are extracted from the original VAGs (he texted information is not shown)

Southeast Asia, wished to request rerouting, but did not know where to make a request, so JCAB/ATMC in Japan and the Oakland ARTCC coordinated with JCAB/ATMC taking the role to respond to the request. Considering the situation, JCAB/ATMC issued a NOTAM at 12:53 UTC on June 22 describing the need to collect information about the volcanic ash cloud caused by the Sarychev-Peak Volcano eruption.

VAAC Washington ended advisory provision in its area of responsibility at 05:00 UTC on June 19. VAAC Tokyo announced the volcanic ash cloud dissipation at 02:52 UTC on June 23 and the VAAC Anchorage at 07:30 UTC on June 25. As the volcanic ash cloud remained for more than ten days, the impact on aviation operations was significant.

During this event, close communication between JCAB/ATMC and the Oakland ARTCC assisted successful collaborative operations. The flexibility of VAAC Tokyo in providing VAAs every three hours instead of the regular six-hour interval was also user-friendly because airlines were able to re-route with minimum detours based on the frequently-updated advisories. However, at the same time, this event highlighted a necessity for thorough information distribution so that no aircraft operates toward the volcanic ash cloud area (McGimsey et al. 2014; JMA/VAAC Tokyo

2010; NOAA/VAAC Anchorage 2015b; NOAA/VAAC Washington 2015).

3.2 Case Study #2

Eruption of Kliuchevskoi Volcano in 2013

A strong explosive and effusive eruption of Kliuchevskoi Volcano in the middle of Kamchatka Peninsula started on August 15, 2013 and lasted until December 20, 2013, with repeated eruptions and ash dissipation. VONA and VAA/VAGs on November 19 reported that the ash reached above 40,000 ft; however, the paroxysmal phase of eruption on October 15–20, in which explosions sent volcanic ash up to around 30,000–33,000 ft each time, was probably the most significant phase of the activity, because the volcanic ash cloud extended across the boundary of VAACs Tokyo and Anchorage's areas of responsibility and both VAACs issued advisories for their respective areas of responsibility (Girina et al. 2014a, b; JMA/VAAC Tokyo 2014; KVERT/VONA 2013; NOAA/VAAC Anchorage 2015b).

The volcanic ash cloud moved to the southeast and partially migrated into VAAC Anchorage's area of responsibility at around 18:00 UTC on October 18. VAAC Anchorage then consulted with VAAC Tokyo on future plans of VAA/VAG issuances. VAAC Tokyo decided to continue VAA/VAG issuances but from 00:54 UTC on October 19, VAAC Tokyo handed over the responsibility for some part of the volcanic ash cloud, which had migrated into VAAC Anchorage's area of responsibility. The volcanic eruption continued with volcanic ash clouds continuously produced. Once VAAC Tokyo handed over the responsibility for some part of a volcanic ash cloud that had migrated into VAAC Anchorage's region, another volcanic ash cloud extended across the boundary of the VAACs areas of responsibility a few hours later, instigating another handover. In this way, the VAACs provided VAA/VAGs for their respective airspaces, that is, the ash area was divided into two following their areas of responsibility (the entire volcanic ash cloud area could not be obtained in either single VAA/VAG). In addition, each VAAC uses its own diffusion model for forecasting volcanic ash cloud areas, so the results differ slightly. For this event, the results of the two VAACs were inconsistent, each having issued VAA/VAGs for their own region without coordination with the other (Fig. 4). An airline made an inquiry as to the difference of volcanic ash cloud extent in the VAA/VAGs issued by VAACs Tokyo and Anchorage, which highlighted the difficulty for airline users when two or more VAACs issue advisories for volcanic ash clouds caused by the same eruption individually without adequate coordination. In such cases, users need to obtain and monitor two advisories for one event. If the advisories have a gap that cannot be ignored, users will have difficulty in understanding the situation.

As the volcanic ash cloud moved to the southeast and approached VAACs Washington and Montreal, those VAACs also issued transfer VAAs. When an ash cloud is approaching within 300 nautical miles from the boundary of areas of responsibility, VAACs to which the ash is approaching are required to transfer VAAs from a VAAC with volcanic ash in its airspace (ICAO 2004). In this way, as the volcanic ash cloud covered a wide area over the boundaries of

multiple VAACs' areas of responsibility, not only the volcanic ash cloud itself but also the situation that multiple VAACs issued VAAs and/or VAGs for its respective area of responsibility had an impact on users (JMA/VAAC Tokyo 2014; NOAA/VAAC Anchorage 2015b; NOAA/VAAC Washington 2015).

4 Particular Challenges of International Coordination in Volcanic Ash and Visions of the Future

As volcanic ash clouds flow regardless of borders, international cooperation/coordination is indispensable. Smooth communication between VAACs as well as among all the related organizations is essential in order to ensure safety. It is essential for volcano observatories, MWOs and VAACs to provide information on volcanic eruptions and sequential volcanic ash cloud diffusions to users such as airlines, civil aviation authorities, and relevant organizations in a way that users can grasp the situation easily. Users need to prepare effective risk management procedures and protocols for such cases. Close coordination/communication between information providers and users is required for smooth response against volcanic ash cloud emissions. Both information providers and users need to be prepared for various types of cases considering volcano locations, eruption duration, volcanic ash cloud propagation and coverage, as preferable responses may differ for each type. Adding to such preparation, a language skill is also required. As English is the standard international aviation language, some organizations in non-native English speaking countries encounter a language barrier that makes it difficult to coordinate/communicate smoothly, speedily and in detail.

Therefore, both information providers and users have been undertaking various efforts regarding the requirements, including efforts of eliminating language barriers. This section introduces some of the particular challenges being addressed by them.

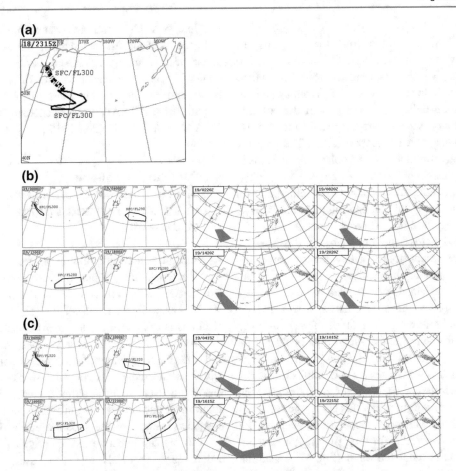

Fig. 4 **a** Observed (present) volcanic ash cloud extent in a *left top* of a VAG from VAAC Tokyo before handover which was issued at 23:59 UTC on 18 October 2013. The volcanic ash cloud area surrounded by a *dotted line* was in VAAC Tokyo's region and that surrounded by a *solid line* was in VAAC Anchorage's region. **b** VAGs from VAACs Tokyo (*left*) and Anchorage (*right*) after the handover for the same volcanic ash cloud issued at 00:54 UTC and 02:20 UTC on 19 October 2013, respectively. From these issuances, the VAACs started to provide VAA/VAGs for the volcanic ash cloud in their respective areas of responsibility individually. The volcanic ash extent in the VAGs had a big gap especially for forecasts at 12 and 18 h ahead. **c** VAGs from the VAACs at 04:45 UTC and 04:15 UTC on 19 October 2013, respectively. The gap in the VAGs became smaller but still inconsistency remained

Challenges being taken by information providers

When a volcanic ash cloud flows from the area of responsibility of a certain VAAC to another, the responsibility to issue VAAs is to be handed over. This situation frequently occurs between VAACs Tokyo and Anchorage: when a volcano in Kamchatka or Kuril Islands erupts, the volcanic ash cloud often migrates into the area of responsibility of VAAC Anchorage. VAAC Tokyo then hands over its responsibility to VAAC Anchorage. One important aspect to note here is, aircraft need to continue their flights across the Northern Pacific region under a consistent risk management approach, regardless of which VAAC is responsible. Therefore, the forecast extent of volcanic ash clouds in VAA/VAGs from VAAC Tokyo before a handover and from VAAC Anchorage after the handover should not have inconsistencies. Considering the frequent occurrence of a handover as well as the necessity of providing consistent advisories between the two VAACs before and after the handover, there are particular challenges in coordination.

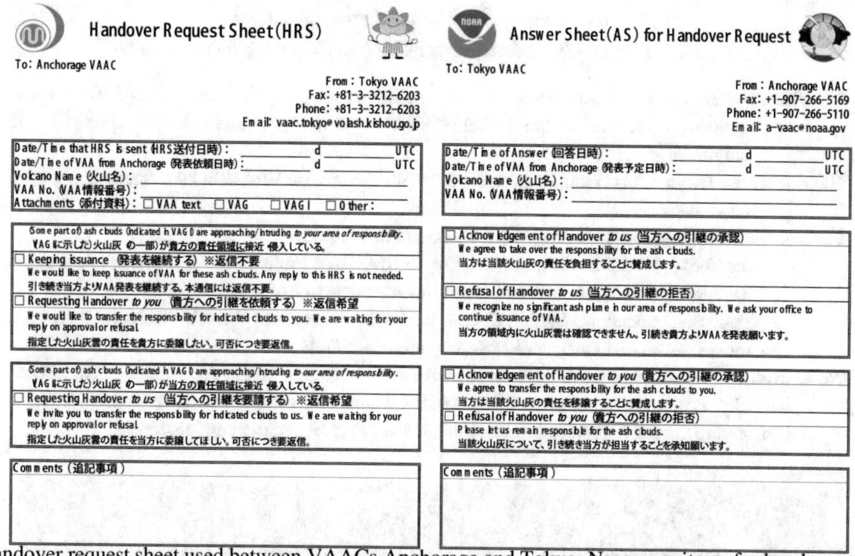
Handover Request Sheet (HRS)

To: Anchorage VAAC

From: Tokyo VAAC
Fax: +81-3-3212-6203
Phone: +81-3-3212-6203
Email: vaac.tokyo@volash.kishou.go.jp

Date/Time that HRS is sent (HRS送付日時): ______ d ______ UTC
Date/Time of VAA from Anchorage (発表依頼日時): ______ d ______ UTC
Volcano Name (火山名): ______
VAA No. (VAA情報番号): ______
Attachments (添付資料): ☐ VAA text ☐ VAG ☐ VAGI ☐ Other:

(Some part of) ash clouds (indicated in VAGI) are approaching/intruding *to your area of responsibility*.
(VAGIに示した)火山灰(の一部)が貴方の責任領域に接近/侵入している。
☐ Keeping issuance (発表を継続する) ※返信不要
We would like to keep issuance of VAA for these ash clouds. Any reply to this HRS is not needed.
引き続き当方よりVAA発表を継続する。本通信には返信不要。
☐ Requesting Handover *to you* (貴方への引継を依頼する) ※返信希望
We would like to transfer the responsibility for indicated clouds to you. We are waiting for your reply on approval or refusal.
指定した火山灰雲の責任を貴方に委譲したい。可否につき要返信。

(Some part of) ash clouds (indicated in VAGI) are approaching/intruding *to our area of responsibility*.
(VAGIに示した)火山灰(の一部)が当方の責任領域に接近/侵入している。
☐ Requesting Handover *to us* (当方への引継を要請する) ※返信希望
We invite you to transfer the responsibility for indicated clouds to us. We are waiting for your reply on approval or refusal.
指定した火山灰雲の責任を当方に委譲してほしい。可否につき要返信。

Comments (追記事項)

Answer Sheet (AS) for Handover Request

To: Tokyo VAAC

From: Anchorage VAAC
Fax: +1-907-266-5169
Phone: +1-907-266-5110
Email: a-vaac@noaa.gov

Date/Time of Answer (回答日時): ______ d ______ UTC
Date/Time of VAA from Anchorage (発表予定日時): ______ d ______ UTC
Volcano Name (火山名): ______
VAA No. (VAA情報番号): ______

☐ Acknowledgement of Handover *to us* (当方への引継の承認)
We agree to take over the responsibility for the ash clouds.
当方は当該火山灰の責任を負担することに賛成します。
☐ Refusal of Handover *to us* (当方への引継の拒否)
We recognize no significant ash plume in our area of responsibility. We ask your office to continue issuance of VAA.
当方の領域内に火山灰雲は確認できません。引続き貴方よりVAAを発表願います。

☐ Acknowledgement of Handover *to you* (貴方への引継の承認)
We agree to transfer the responsibility for the ash clouds to you.
当方は当該火山灰の責任を移譲することに賛成します。
☐ Refusal of Handover *to you* (貴方への引継の拒否)
Please let us remain responsible for the ash clouds.
当該火山灰について、引き続き当方が担当することを承知願います。

Comments (追記事項)

Fig. 5 Handover request sheet used between VAACs Anchorage and Tokyo. Necessary items for handover procedures are already in the sheet both in English and Japanese

HANDOVER PROCEDURES
BETWEEN
THE VOLCANIC ASH ADVISORY CENTRES
ANCHORAGE AND TOKYO

May 06, 2015

Fig. 6 Guideline of handover procedures. Criteria to conduct a handover is documented for eruption types such as single (short duration) eruptions, intermittent eruptions and continuous eruptions with a volcanic ash cloud moving to the east, south or west

(1) Guideline of handover procedures

VAACs Anchorage and Tokyo have prepared a specific form called "Handover Request Sheet (HRS)" in which necessary items are already included both in English and Japanese (Fig. 5). When a case that requires a handover occurs, the VAACs complete necessary parts on the sheet

Handover procedures from Tokyo to Anchorage
- Type 3: Continuous (long duration) eruption (case 2: propagating eastward) -

Condition: some part of an ash cloud due to a continuous eruption diffuses eastward, crosses the boundary of the AoRs from the Tokyo VAAC's side and approaches the meridian of 180 E longitude.

Handover: the Tokyo VAAC requests a hand-over to the Anchorage VAAC and the Anchorage VAAC sends back an AS.

Further action in case the Anchorage VAAC accepts the request: the Tokyo VAAC issues a VAA stating in RMK that some part of the volcanic ash cloud has moved out of its AoR and it notifies the necessity of checking VAAs both from the Tokyo and Anchorage VAAC. Once a handover is done, both VAACs will issue VAAs only for the ash cloud in their own AoR, regardless of 180 E.

Note: even after some parts of the ash cloud crosses the boundary of the AORs, the Tokyo VAAC basically continues to issue a VAA for the whole ash cloud until it approaches 180 E considering the convenience for users, though there may be some exceptional cases.

Scenario of Eruption:

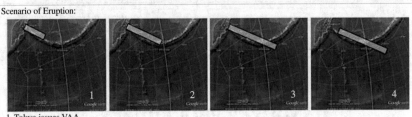

1. Tokyo issues VAA
2. Tokyo keeps issuing VAA and hands over to Anchorage once the cloud reaches 180E
3. Both Anchorage and Tokyo issue VAA for their own AoRs
4. When the eruption ends and the cloud becomes obviously apart from the volcano, Anchorage issues VAA for the whole cloud

Example of a VAA

```
FVFE01 RJTD DDhhmm
VA ADVISORY
DTG: YYYYMMDD/hhmmZ
VAAC: TOKYO
VOLCANO: SHEVELUCH 300270
PSN: N5639E16122
AREA: RUSSIA
SUMMIT ELEV: 3283M
ADVISORY NR: YYYY/nnn
INFO SOURCE: MTSAT-2
AVIATION COLOUR CODE: NIL
ERUPTION DETAILS: VA EMISSIONS CONTINUING. (or other comments depending on the situation.)
OBS VA DTG: DD/hhmmZ
OBS VA CLD: (observed VA area)
FCST VA CLD +6 HR: (forecast VA area)
FCST VA CLD +12 HR: (forecast VA area)
FCST VA CLD +18 HR: (forecast VA area)
RMK: THE RESPONSIBILITY FOR SOME PART OF ASH IS BEING TRANSFERRED TO ANCHORAGE. PLS SEE FVAK21
PAWU ISSUED BY ANCHORAGE WHICH DESCRIBES CONDITION OVER OR NEAR THE TOKYO AREA. WE KEEP
ISSUING VAA FOR THE VA CLD IN OUR AREA.
NXT ADVISORY:   YYYYMMDD/hhmm=
```

Fig. 6 continued

and exchange it in order to simplify and speed up the procedures. Additionally, the two VAACs, in advance, shared information on decision-making criteria on how and when to conduct handover procedures. This is because the timing to handover may well be different between VAACs depending on the situation, especially when a volcanic ash cloud extends across the areas of responsibility of both VAACs. The criteria have been coordinated and documented as a guideline (Fig. 6) in order that both VAACs can expect beforehand how the other centre will act with

Handover procedures from Tokyo to Anchorage
- Type 3: Continuous (long duration) eruption (case 3: propagating southward) -

Condition: some part of an ash cloud due to a continuous eruption diffuses southward, crosses the boundary of the AoRs from the Tokyo VAAC's side and is expected to enter the AoR of the Washington VAAC.

Handover: the Tokyo VAAC requests a hand-over to the Anchorage VAAC and the Anchorage VAAC sends back an AS. The Tokyo VAAC informs the Washington VAAC of the situation.

Further action in case the Anchorage VAAC accepts the request:
the Tokyo VAAC issues a VAA stating in RMK that some part of the volcanic ash cloud has moved out of its AoR and it notifies the necessity of checking VAAs from the Tokyo, Anchorage and Washington VAAC.

Note: even after some parts of the ash cloud crosses the boundary of the AORs, the Tokyo VAAC basically continues to issue a VAA for the whole ash cloud. Discussion on heights and areal coverage of the plume will be held whenever necessary in order to reach a reasonable agreement between Anchorage and Tokyo.

Scenario of Eruption:

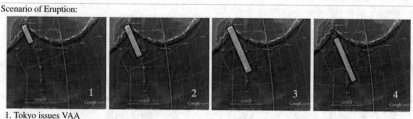

1. Tokyo issues VAA
2. 3. Tokyo keeps issuing VAA for the whole ash cloud
4. When the eruption ends and the cloud becomes obviously apart from the volcano, Anchorage issues VAA for the whole cloud

Example of a VAA

```
FVFE01 RJTD DDhhmm
VA ADVISORY
DTG: YYYYMMDD/hhmmZ
VAAC: TOKYO
VOLCANO: SHEVELUCH 300270
PSN: N5639E16122
AREA: RUSSIA
SUMMIT ELEV: 3283M
ADVISORY NR: YYYY/nnn
INFO SOURCE: MTSAT-2
AVIATION COLOUR CODE: NIL
ERUPTION DETAILS: VA EMISSIONS CONTINUING. (or other comments depending on the situation.)
OBS VA DTG: DD/hhmmZ
OBS VA CLD: (observed VA area)
FCST VA CLD +6 HR: (forecast VA area)
FCST VA CLD +12 HR: (forecast VA area)
FCST VA CLD +18 HR: (forecast VA area)
RMK: THE RESPONSIBILITY FOR SOME PART OF ASH IS BEING TRANSFERRED TO ANCHORAGE. PLS SEE FVAK21
PAWU ISSUED BY ANCHORAGE AS WELL AS FVXX21 KNES BY WASHINGTON WHICH DESCRIBE CONDITION FOR
THE VA CLD IN EACH AREA. WE KEEP ISSUING VAA FOR THE VA CLD IN OUR AREA.
NXT ADVISORY:   YYYYMMDD/hhmm=
```

Fig. 6 continued

volcanic ash clouds moving towards/across the border of their areas of responsibility.

As described in the case study of the eruption of Kliuchevskoi Volcano in 2013, it is not user-friendly if two VAACs provide VAA/VAGs with a volcanic ash cloud area for their own individual areas of responsibility and/or if VAA/VAGs from two VAACs are inconsistent. Therefore, for continuous eruptions, VAACs Anchorage and Tokyo agreed to issue VAA/VAGs from one VAAC as much as possible even after the volcanic ash cloud area

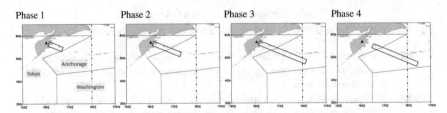

Fig. 7 Handover procedures for a continuous eruption at a volcano in VAAC Tokyo's area of responsibility with a volcanic ash cloud extending to the east. *Triangle* and *rectangle* indicate a volcano and a volcanic ash cloud area, respectively. The boundary of VAACs is drawn with a *solid line* and 180°E is drown with a *dotted line*. Phase 1: volcanic ash cloud is in VAAC Tokyo's area of responsibility and VAAC Tokyo issues VAA/VAG. Phase 2: volcanic ash cloud migrated into VAAC Anchorage's area of responsibility extending from the volcano in VAAC Tokyo's region, but it remains west of 180°E, so VAAC Tokyo issues VAA/VAG for the entire volcanic ash cloud. Phase 3: volcanic ash cloud crossed 180°E so both VAACs Anchorage and Tokyo issue VAA/VAGs for their own areas of responsibility. Phase 4: eruption ended with volcanic ash cloud obviously apart from the volcano, so VAAC Tokyo conducts a handover and VAAC Anchorage issues VAA/VAGs for the entire volcanic ash cloud

extends across the boundary of their areas of responsibility.

For example, when a volcanic ash cloud, due to a continuous eruption at a certain volcano in VAAC Tokyo's airspace, extends to the east crossing the boundary of the areas of responsibility and covers a large area from the volcano to VAAC Anchorage's region, VAAC Tokyo continues to issue advisories for the entire volcanic ash cloud until it reaches 180°E so that airlines can grasp the current and future extent of volcanic ash cloud from VAA/VAGs provided by one VAAC (Tokyo). When it crosses 180°, VAAC Tokyo hands over the responsibility to VAAC Anchorage for a part of the volcanic ash cloud which has migrated into VAAC Anchorage's region. The VAACs cannot avoid providing VAA/VAGs from the two VAACs for a while, but once the eruption ends and the volcanic ash cloud separates from the volcano, VAAC Tokyo immediately conducts a handover for the entire volcanic ash cloud to VAAC Anchorage (Fig. 7).

When a volcanic ash cloud from a continuous eruption at a certain volcano in VAAC Tokyo's airspace extends to the south crossing the boundary of VAACs Tokyo and Anchorage's areas of responsibility and/or VAACs Anchorage and Washington's regions, VAAC Tokyo continues to issue advisories for the entire volcanic

ash cloud with necessary coordination with the other two VAACs about the height and extent of the volcanic ash cloud. The timing of handover varies, depending on the situation in this case, but when the eruption ends and the volcanic ash cloud moves to the south separated from the volcano, it is agreed that VAAC Tokyo immediately hands over the responsibility for the entire volcanic ash cloud to VAAC Anchorage, and VAAC Anchorage sequentially conducts a handover to VAAC Washington if the volcanic ash cloud still exists and is moving to the south (Fig. 8).

When a continuous eruption at a certain volcano in VAAC Anchorage's airspace produces a volcanic ash cloud to the west migrating into VAAC Tokyo's area of responsibility, VAAC Anchorage will continue issuing VAA/VAGs until it reaches 160°E, though this situation seldom occurs. Then, if the volcanic ash cloud continues moving to the west across 160°E and migrates into VAAC Tokyo's region, VAAC Anchorage hands over the responsibility for a part of the volcanic ash cloud that crossed 160°E to VAAC Tokyo. In the same way as mentioned previously, once the eruption ends and the volcanic ash cloud moves to the west separated from the volcano, VAAC Anchorage immediately conducts a handover for the entire volcanic ash cloud to VAAC Tokyo (Fig. 9).

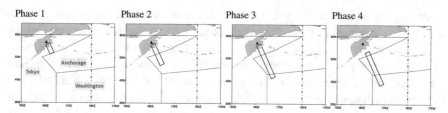

Fig. 8 Handover procedures for a continuous eruption at a volcano in VAAC Tokyo's area of responsibility with a volcanic ash cloud extending to the south. *Triangle* and *rectangle* indicate a volcano and a volcanic ash cloud area, respectively. The boundary of VAACs is drawn with a *solid line* and 180°E is drown with a *dotted line*. Phase 1: volcanic ash cloud is in VAAC Tokyo's area of responsibility and VAAC Tokyo issues VAA/VAG. Phase 2–3: volcanic ash cloud migrates in neighbouring VAACs' areas of responsibility but is still extending from the volcano in VAAC Tokyo's region, so VAAC Tokyo issues VAA/VAGs for the entire volcanic ash cloud with necessary coordination among the relevant VAACs for its height and extent. Phase 4: eruption ended with volcanic ash cloud obviously apart from the volcano, so VAAC Tokyo conducts a handover and VAAC Anchorage issues VAA/VAGs for the entire volcanic ash cloud

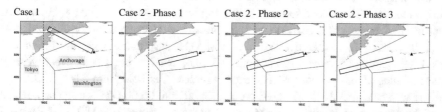

Fig. 9 Handover procedures for a continuous eruption at a volcano in VAAC Anchorage's area of responsibility with a volcanic ash cloud extending to the west. *Triangle* and *rectangle* indicate a volcano and a volcanic ash cloud area, respectively. The boundary of VAACs is drawn with a *solid line* and 160°E is drawn with a *dotted line*. Case 1: volcanic ash cloud migrates into VAAC Tokyo's area of responsibility extending from the volcano in VAAC Anchorage's region, but it remains east of 160°E, so VAAC Anchorage continues issuing VAA/VAGs for the entire volcanic ash cloud. Case 2: (Phase 1) volcanic ash cloud is in VAAC Anchorage's area of responsibility and VAAC Anchorage issues VAA/VAG. (Phase 2) volcanic ash cloud crosses 160°E and VAAC Anchorage conducts a handover for a part of the volcanic ash cloud that has migrated in VAAC Tokyo's region. Both VAACs Anchorage and Tokyo issue VAA/VAGs for their own areas of responsibility. (Phase 3) eruption ended with volcanic ash cloud obviously apart from the volcano, so VAAC Anchorage conducts a handover and VAAC Tokyo issues VAA/VAGs for the entire volcanic ash cloud

Information sharing on decision-making criteria is also done for single (short duration) eruptions and intermittent eruptions. The procedures for single (short duration) eruptions are more straight-forward. VAACs Anchorage and Tokyo agreed to hand over the entire volcanic ash cloud when more than half of it has migrated into the neighbouring VAAC's area of responsibility. The procedures for intermittent eruptions are also relatively straight-forward, because intermittent eruptions are, as it were, repeated single eruptions. VAACs agreed to repeat the procedures for single eruptions applying to newer ash clouds generated by intermittent eruptions.

(2) Challenge to collaborative decision analysis and forecast via chat system

In order to provide consistent advisories before and after the handover, it is better to share forecasters' thoughts before volcanic ash clouds actually cross the border of the areas of responsibility, especially for a complicated or exceptional situation. Therefore, VAACs Anchorage and Tokyo have started testing a chat system for closer and more flexible communication. NOAA has provided its proprietary chat system and created an account for this challenge (Osiensky et al. 2014). As part of the test, the VAACs are aiming at finding necessary specific patterns of phenomena as well as phrases of questions and answers corresponding to them, and creating a template like a frequently-asked questions-sheet so that the communication will be smooth between members including non-native English speakers.

The first test was held in July 2014 based on a scenario of the past eruption at Kliuchevskoi in October 2013, in which a volcanic ash cloud moved far southeast. As the VAA/VAGs were not user-friendly in those days as described earlier, the VAACs prepared a scenario following the current guidelines on handover procedures introduced previously that had been established between them in spring 2014, as shown in Fig. 6, and conducted the test. The second test was held in December 2014 based on a scenario of the eruption at Sheveluch in September 2014, in which a volcanic ash cloud moved to the north and where the timing of dissipation was not clear. The third test was held in July 2015 based on a scenario of the eruption at Sheveluch in March 2015, in which a volcanic ash cloud moved to the south and migrated into both VAACs Anchorage and Washington's areas of responsibility. Not only VAACs Anchorage and Tokyo but also VAAC Washington took part in the third test to check if communication/coordination among the three VAACs would work well via a chat system. After that, operational use of the chat system was utilized on a trial basis instead of through scheduled tests that required coordination ahead of time. If this trial proves to be successful and becomes fully operational in these VAACs, it could be used as a model case and applied to coordination/communication, not only between VAAC Tokyo and other organizations, but between other VAACs and volcano observatories particularly in the area where English is not the native language.

Challenges being taken by information users

The cooperation/coordination introduced before is undertaken by VAACs essentially as information providers. Considering the importance of international cooperation and

coordination mentioned earlier, Volcanic Ash Exercises are conducted in some regions under the framework of ICAO. The first exercise was established in ICAO European and North Atlantic (EUR/NAT) region called the VOLCEX and has been conducted since 2008. Realizing the effectiveness of the VOLCEX, a similar exercise in the EUR (EAST) Region including Kamchatka Peninsula started in 2013 recognising that this region experiences frequent volcanic eruptions that often affect aviation operations especially around the NOPAC routes. Therefore, an exercise in this region, named VOLKAM, has been conducted and is making good progress in coordination procedures between all participating parties (air navigation service providers, air traffic management centres, aeronautical information services, volcano observatories, VAACs, MWOs and users such as airlines). So far, VOLKAM has been held every year: the first exercise was held from 21:00 UTC on January 15, 2013 to 06:00 UTC on January 16, 2013, the second one from 21:00 UTC on March 4, 2014 to 04:00 UTC on March 5, 2014 and the third one from 22:00 UTC on April 15, 2015 to 04:00 UTC on April 16, 2015. The exercises have a different focus each time and participants test new challenges during the exercises (ICAO 2014a, b, 2015a, b; JCAB 2015).

In addition to the volcanic ash exercises, when airlines make a detour at an actual eruption, they need to coordinate with relevant organizations for re-routing. The Cross Polar Trans East Air Traffic Management Providers Working Group (CPWG) is dealing with the topic of international coordination for re-routing and JCAB is one of the members of CPWG.

The following are examples of the challenges being met by participants of the exercises including airlines, and/or members of CPWG.

(1) Determination of a re-route according to a scenario and a matrix on a response for a re-routing request

Once a notification of an eruption is received by a dispatcher, the potential impact to flights that are already en-route is evaluated and if the impact is expected, re-routing procedures will be taken. Re-routing should be conducted immediately because an encounter with a volcanic ash cloud may cause a fatal accident; even a small amount of volcanic ash can cause enormous financial costs with respect to repairing engines and other parts. As all flights in the volcanic ash-affected region undertake re-routing procedures, it should be well organized to accommodate all of them in a limited number of routes, considering the issue of remaining fuel. Additionally, there are regulations and/or restrictions in each State, such that re-routing options are not always accepted. Hence, it is quite effective to prepare a possible contingency route based on an assumed eruption in advance, even if it is a paper-plan and only used during an exercise. As this route has cleared the political issue and other conditions (like a fuel amount), it could be a realistic alternative route in case of a sudden eruption, surely saving time in coordination and implementation.

However, procedures for responding to a request for re-routing are not currently standardized as described in the case study for the eruption of Sarychev-Peak Volcano in 2009; they differ depending on the ANSP. It may be better if a standardized procedure among all ANSPs is prepared, but this is difficult because of various restrictions in each country, and it will take time to achieve. For example, airlines expect re-routing procedures for aircraft in flight to be conducted using the Air Traffic Service Communication (ATSC) via air traffic control centres, while the ground system of air traffic control centres in some countries cannot process re-routing messages from aircraft in flight because transaction between pilots and air traffic controllers are prioritized. Another example is that some countries apply a license system and requires aircraft to obtain permission from an authorized organization when they fly over particular airspaces. If an aircraft requests re-routing over such countries in order to avoid volcanic ash cloud, it needs to obtain permission that will take time. Therefore, before pursuing this ideal to prepare a standardized procedure among all ANSPs, it has been set as a primary goal to create a matrix on each ANSP's status when it receives

a request for re-routing so that airlines can easily grasp the present situation. This work originated from the volcanic ash exercise VOLKAM. Currently the task has been dealt with in the framework of CPWG, so all the members of CPWG including FAA and JCAB can work on this issue. In addition, it is also regarded as an important aspect to consider how to enable organizations related to the matrix to obtain the volcanic eruption information; this is an on-going task as well. The matrix may be tested in VOLKAM sometime in the future once a draft version is prepared (ICAO 2014a, b).

(2) Use of VOLKAM sheet

Similar to the collaborative decision analysis and forecast via a chat system being conducted between VAACs Anchorage, Tokyo and Washington, a spreadsheet named VOLKAM Sheet, prepared by JCAB/ATMC, was workshopped by participants of the volcanic ash exercise in 2015, in an effort to organize relevant information in one sheet chronologically and reduce the issue of language barriers. The VOLKAM Sheet contains chronological information on a present situation for the eruption phase, volcanic ash cloud area, influence in traffic flow and aircraft operations based on the volcanic ash cloud conditions, as well as the information about the expected coordination and actions among the relevant organizations such as a flow control of aircraft, resetting PACOTS and the timing of the next VAA/VAG, and other information issuances. A remarks column is prepared in the sheet in case there are any special notes to share (Fig. 10).

During the exercise, each organization sent this VOLKAM Sheet to all participants via e-mail, with an organization name and a version number so that everybody understood which spreadsheet was the latest one to add new information about the present situation and/or planned actions. The usability of this sheet to improve situation awareness among the relevant organizations was tested during the exercise in 2015. The participants understood the idea that it would be better to prepare a communication method rather than a phone call, considering that

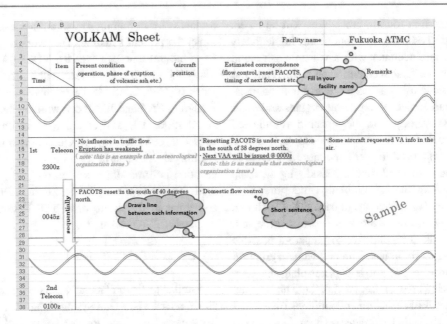

Fig. 10 VOLKAM Sheet used in the volcanic ash exercise in 2015: it was prepared by JCAB/ATMC with columns for the information on a present situation, expected actions and remarks. Participants of the exercise put some information that they have and shared the sheet during the exercise. Information was added in order of time

they respectively have three languages as a native tongue, English, Russian and Japanese. When a spreadsheet with necessary information is shared, the merits are that it can at least avoid mishearing and misunderstanding, and the participants can read what was discussed again later.

The exercise in 2015, which tested the usability of written information on the VOLKAM sheet, highlighted areas for improvement. One prospect for improvement is to share the information via a website rather than a spreadsheet. The website would have access limited only to relevant organizations, where participants would have the ability to directly edit and update the website ensuring it remains current with the latest information.

5 Summary

Volcanic ash cloud can seriously affect aircraft and air services by causing engine failure, poor visibility due to ash-related scouring of aircraft windshields and take-off/landing delays due to ash accumulation at airports. As volcanic ash may cause a fatal accident and also as it crosses borders, it is essential to provide coherent and consistent volcanic ash-related information to airlines, civil aviation authorities, MWOs and other relevant organizations to avoid aviation disasters. Hence, international cooperation and coordination with the efforts of meteorological, volcanological and civil aviation authorities from multiple countries is indispensable.

Along with the requirement for proper information issuances and smooth coordination among relevant organizations, various efforts have been taken in some regions. As for the Northern Pacific region, volcanoes in Kamchatka Peninsula are remarkably active: this brings particular challenges for related organizations because a volcanic ash cloud frequently moves across the boundary of VAACs Tokyo and Anchorage's areas of responsibility. They need to conduct handover procedures whenever it occurs, and especially when the volcanic ash cloud covers the NOPAC routes, ATMCs, ACCs and airlines coordinate for re-routing.

Among recent eruptions, those of Sarychev-Peak Volcano in 2009 and Kliuchevskoi Volcano in 2013 are introduced in this chapter as examples to illustrate successes and ongoing challenges of international eruption response coordination.

In this respect, VAACs Anchorage and Tokyo are working on a smoother handover and consistent information issuances using English. JCAB/ATMC and other aviation-related organizations are making similar efforts in a volcanic ash exercise in 2015 to grasp all the relevant information in one sheet in chronological order. The aim is to overcome language barriers, using a VOLKAM Sheet mainly for discussion on re-routing by sharing the present situation of the eruption phase, volcanic ash area, influence onto the traffic flow and aviation operations, as well as the expected coordination and actions by relevant organizations. The planning process itself for re-routing in the exercise is also meaningful: that will contribute to shorten the time required for coordination in a real case. In addition, an effort to share the present situation of each ANSP, responding to a re-routing request, by creating a matrix on such information has begun; members of CPWG are working on this task aiming at establishing a standardized procedure among all the ANSPs, as a long-term ideal outcome.

In this way, the volcanic ash- and/or aviation-related organizations will continue their work seeking for better coordination and operations, respectively.

Acknowledgements Information as for the on-going efforts in JCAB given by Mr. Takayuki Harada, the deputy chief air traffic controller in JCAB, has been a great help to flesh out the article as well as making it well-balanced in highlighting efforts taken by both information providers and users.Some of the figures were created using Generic Mapping Tools open-source software (GMT; Wessel, P. and W.H.F. Smith, New, improved version of Generic Mapping Tools released, EOS Trans. Amer. Geophys. U., Vol. 79, No. 47, p. 579, 1998).

References

Anchorage AK (2015a) Volcanic Ash Advisory Center, the National Oceanic and Atmospheric Administration (NOAA/VAAC Anchorage) AAWU—Volcanic Ash

Advisory Center. Retrieved 7 Aug 2015, from http://vaac.arh.noaa.gov/vaac_info.php

Anchorage AK (2015b) Volcanic Ash Advisory Center, the National Oceanic and Atmospheric Administration (NOAA/VAAC Anchorage) VAAs/VAGs. Retrieved 7 Aug 2015, from http://vaac.arh.noaa.gov/list_vaas.php

Girina OA, Romanova IM (2015) Activity of Kamchatkan and Northern Kuriles volcanoes database of Kamchatkan Volcanic Eruption Response Team. In: 26th IUGG general assembly, 22 June–02 July 2015. IUGG/IAVCEI, Prague, P. VS10p-456

Girina OA, Manevich AG, Malik NA, Melnikov DV, Ushakov SV, Demyanchuk YuV, Kotenko LV (2007) Active volcanoes of Kamchatka and Northern Kurils in 2005. J Volcanol Seismolog 1(4):237–247

Girina OA, Ushakov SV, Malik NA, Manevich AG, Melnikov DV, Nuzhdaev AA, Demyanchuk YuV, Kotenko LV (2009) The active volcanoes of Kamchatka and Paramushir Island, North Kurils in 2007. J Volcanol Seismolog 3(1):1–17

Girina OA, Manevich AG, Melnikov DV, Demyanchuk YuV, Petrova E (2014a) Explosive eruptions of Kamchatkan volcanoes in 2013 and danger to aviation, EGU2014. Austria, Vienna

Girina OA, Manevich A, Melnikov D, Nuzhdaev A, Demyanchuk Y (2014b) Activity of Kamchatkan volcanoes in 2012-2013 and danger to aviation. In: International workshop "JKASP-8", Sapporo, Japan. 22–26 Sept 2014

Gordeev EI, Girina OA (2014) Volcanoes and their hazard to aviation. Herald Russ Acad Sci 84(2):134–142

ICAO (2004) Handbook of the international airways volcano watch (IAVW), 2nd edn, from http://www.icao.int/publications/Documents/IAVW%20Handbook%20Doc%209766_en.pdf

ICAO (2007) ICAO annex 3 meteorological service for international air navigation, 16th edn, from http://www.wmo.int/pages/prog/www/ISS/Meetings/CT-MTDCF-ET-DRC_Geneva2008/Annex3_16ed.pdf

ICAO (2014a) Summary of discussions of the eighteenth meeting of the cross polar trans east air traffic management providers working group (draft version) from https://www.faa.gov/about/office_org/headquarters_offices/ato/service_units/systemops/ato_intl/documents/cross_polar/CPWG19/CPWG19_WP02_CPWG18_Summary_of_Discussions.pdf

ICAO (2014b) Summary of discussions of the VOLKAM14 Debrief & EUR (EAST) VOLCEX/SG/4 Meetings

ICAO (2015a) Exercise directive for volcanic ash exercise in Kamchatka in 2015 (VOLKAM15)

ICAO (2015b) Outcomes from other ICAO volcanic ash exercises from http://www.icao.int/APAC/Meetings/2015%20VOLCEX_SG1/IP02_ICAO%20AI.3%20-%20Outcomes%20from%20other%20ICAO%20VA%20exercises.pdf

JCAB (2015) VOLKAM15 outcomes at JCAB from http://www.icao.int/APAC/Meetings/2015%20VOLCEX_

SG1/SP03_Japan%20-%20VOLKAM15-Outcomes.
pdf

KVERT/Volcano Observatory Notification to Aviation
(KVERT/VONA) (2013) Retrieved 7 Aug 2015, from
http://www.kscnet.ru/ivs/kvert/van/index.php

McGimsey RG, Neal CA (1996) 1995 volcanic activity in
Alaska and Kamchatka: summary of events and
response of the Alaska Volcano Observatory, U.S.
Geological Survey Open-File Report OF 96-0738, 22 p

McGimsey RG, Wallace KL (1999) 1997 volcanic activity
in Alaska and Kamchatka: summary of events and
response of the Alaska Volcano Observatory, U.S.
Geological Survey Open-File Report OF 99-0448, 42 p

McGimsey RG, Neal CA, Girina OA (2003) 1998
volcanic activity in Alaska and Kamchatka: summary
of events and response of the Alaska Volcano
observatory, open-file report 2004-1033. Department
of the Interior, U.S. Geological Survey, U.S, p 35

McGimsey RG, Neal CA, Girina OA (2004a) 1999
volcanic activity in Alaska and Kamchatka: summary
of events and response of the Alaska Volcano
observatory, open-file report 2004-1033. Department
of the Interior, U.S. Geological Survey, U.S, p 45

McGimsey RG, Neal CA, Girina OA (2004b) 2001
volcanic activity in Alaska and Kamchatka: summary
of events and response of the Alaska Volcano
Observatory, open-file report 2004-1453. Department
of the Interior, U.S. Geological Survey, U.S, p 53

McGimsey RG, Neal CA, Girina OA (2005) 2003
volcanic activity in Alaska and Kamchatka: summary
of events and response of the Alaska Volcano
Observatory, open-file report 2005-1310. Department
of the Interior, U.S. Geological Survey, U.S, p 58

McGimsey RG, Neal CA, Dixon JP, Ushakov SV (2008)
2005 volcanic activity in Alaska, Kamchatka, and the
Kurile Islands: summary of events and response of the
Alaska Volcano observatory, U.S. Geological Survey
Scientific Investigations Report 2007-5269, 94 p

McGimsey RG, Neal CA, Dixon JP, Malik NV,
Chibisova MV (2011) 2007 volcanic activity in
Alaska, Kamchatka and the Kurile Islands: summary
of events and response of the Alaska Volcano
observatory, U.S. Geological Survey Scientific Investigations Report 2010-5242, 110 p

McGimsey RG, Neal CA, Girina OA, Chivisova MV,
Rybin AV (2014) 2009 volcanic activity in Alaska,
Kamchatka, and the Kurile Islands: summary of events
and response of the Alaska Volcano observatory, U.S.
Geological Survey Scientific Investigations Report
2013-5213, 125 p

Neal CA, McGimsey RG (1997) 1996 volcanic activity in
Alaska and Kamchatka: summary of events and

response of the Alaska Volcano Observatory, U.S.
Geological Survey Open-File Report OF 97-0433, 34 p

Neal CA, McGimsey RG, Chubarova OS (2004) 2000
volcanic activity in Alaska and Kamchatka: summary
of events and response of the Alaska Volcano
observatory, U.S. Geological Survey Open-File
Report OF 2004-1034, 37 p

Neal CA, McGimsey RG, Dixon JP, Melnikov DV
(2005a) 2004 volcanic activity in Alaska and Kamchatka: summary of events and response of the Alaska
Volcano Observatory, U.S. Geological Survey
Open-File Report 2005-1308, 71 p

Neal CA, McGimsey RG, Girina OA (2005b) 2002
volcanic activity in Alaska and Kamchatka: summary
of events and response of the Alaska Volcano Observatory, Open-File Report 2004-1058. Department of
the Interior, U.S. Geological Survey, U.S, p 55

Neal CA, Girina OA, Senyukov SL, Rybin AV, Osiensky J, Izbekov P, Ferguson G (2009a) Russian
eruption warning systems for aviation. Nat Hazards
51(2):245–262

Neal CA, McGimsey RG, Dixon JP, Manevich AG,
Rybin AV (2009b) 2006 volcanic activity in Alaska,
Kamchatka, and the Kurile Islands: summary of events
and response of the Alaska volcano observatory, U.S.
Geological Survey Scientific Investigations Report
2008-5214, 102 p

Neal CA, McGimsey RG, Dixon JP, Cameron CE,
Nuzhdaev AA, Chibisova MV (2011) 2008 volcanic
activity in Alaska, Kamchatka, and the Kurile Islands:
summary of events and response of the Alaska
Volcano Observatory, U.S. Geological Survey Scientific Investigations Report 2010-5243, 94 p

Neal CA, Herrick J, Girina OA, Chivisova MV,
Rybin AV, McGimsey RG, Dixon JP (2014) 2010
volcanic activity in Alaska, Kamchatka, and the Kurile
Islands: summary of events and response of the Alaska
Volcano Observatory, U.S. Geological Survey Scien-
tific Investigations Report 2014-5034, 76 p

Osiensky J, Moore D, Igarashi Y (2014) Moving toward a
globally harmonized volcanic ash forecast system:
anchorage and Tokyo VAAC best practices on
collaboration, AGU fall meeting from https://agu.
confex.com/agu/fm14/meetingapp.cgi#Paper/27123

Tokyo Volcanic Ash Advisory Center, The Japan Meteorological Agency (JMA/VAAC Tokyo) (2010) Volcanic ash advisories (2009), from http://ds.data.jma.
go.jp/svd/vaac/data/Archives/2009_vaac_list.html

Tokyo Volcanic Ash Advisory Center, The Japan Meteorological Agency (JMA/VAAC Tokyo) (2014) Volcanic ash advisories (2013), from http://ds.data.jma.
go.jp/svd/vaac/data/Archives/2013_vaac_list.html

Tokyo Volcanic Ash Advisory Center, The Japan Meteorological Agency (JMA/VAAC Tokyo) (2015) The roles of the Tokyo VAAC. Retrieved 7 Aug 2015, from http://ds.data.jma.go.jp/svd/vaac/data/Inquiry/vaac_operation.html

Washington Volcanic Ash Advisory Center, The National Oceanic and Atmospheric Administration (NOAA/VAAC Washington) (2015) Washington VAAC 2009 volcano ash advisory archive, from http://www.ssd.noaa.gov/VAAC/ARCH09/archive.html

Decision-Making: Preventing Miscommunication and Creating Shared Meaning Between Stakeholders

Emma E.H. Doyle and Douglas Paton

Abstract

The effective management and response to either volcanic eruptions or (often prolonged) periods of heightened unrest, is fundamentally dependent upon effective relationships and communication between science advisors, emergency managers and key decision makers. To optimise the effectiveness of the scientific contribution to effective prediction and management decision making, it is important for science advisors or scientific advisory bodies to be cognisant of the many different perspectives, needs and goals of the diverse organisations involved in the response. Challenges arise for scientists as they may need to be embedded members of the wider response multi-agency team, rather than independent contributors of essential information. Thus they must add to their competencies an understanding of the different roles, responsibilities, and needs of each member organisation, such that they can start to provide information implicitly rather than in response to explicit requests. To build this shared understanding, the team situational awareness (understanding of the situation in time and space), and the wider team mental model (a representation of the team functions and responsibilities), requires participating in a response environment together. Facilitating the availability of this capability has training and organizational development implications for scientific agencies and introduces a need for developing new inter-agency relationships and liaison mechanisms well before a

E.E.H. Doyle (✉)
Joint Centre for Disaster Research, Massey University, PO Box 756, Wellington 6140, New Zealand
e-mail: e.e.hudson-doyle@massey.ac.nz

D. Paton
Faculty of Engineering, Health, Science and the Environment, School of Psychological and Clinical Sciences, Charles Darwin University, Darwin, NT 0909, Australia
e-mail: Douglas.Paton@cdu.edu.au

Advs in Volcanology (2018) 549–570
https://doi.org/10.1007/11157_2016_31
Published Online: 26 March 2017

volcanic crisis occurs. In this chapter, we review individual and team decision making, and the role of situational awareness and mental models in creating "shared meaning" between agencies. The aim is to improve communication and information sharing, as well as furthering the understanding of the impact that uncertainty has upon communication and ways to manage this. We then review personal and organisational factors that can impact response and conclude with a brief review of methods available to improve future response capability, and the importance of protocols and guidelines to assist this in a national or international context.

1 Introduction

Whether it involves a period of unrest (e.g., Long Valley, CA in 1982), an ongoing eruption (e.g., Soufrière Hills Volcano, WI), or responding to a blue sky eruption (e.g., Ruapehu, NZ, 2007; Mt Ontake, Japan, 2014), the response to complex volcanic crises requires the coordinated and complementary contributions of numerous organizations and agencies. The degree to which this can effectively be achieved depends on whether the quality and degree of relationship and network building conducted before, during, and after, a crisis can facilitate the shared understanding required for communicated information to enhance effective decision making.

The challenge in this task is twofold. The first relates to the need to bring representatives from diverse sources together (Paton et al. 1998; Doyle et al. 2015), including technical advisors (such as geologists, geophysicists, engineers, and social scientists), emergency management (civil defence, fire service, police, army, national and local government), lifeline organisations (lifelines companies, transport, water), as well as community organisations and special interest groups (e.g. neighbourhood support and volunteer groups, Rotary, Lions club, etc.). A major challenge to developing effective crisis management arises because these representatives bring with them different objectives, priorities and interpretive and operational beliefs (Paton et al. 1998; Doyle et al. 2015). The second task is thus how to facilitate the ability of these representatives to collaborate and share knowledge in order to effectively respond to a crisis.

Recognition of the diverse consequences volcanic crises create can result in organisations appreciating why they need to be part of a multi-agency group response. However, this appreciation does not automatically translate into acceptance of either the need to develop new roles and responsibilities, or that crisis response goals may need to be reconciled with the political or economic pressures that each representative brings with them to the crisis response environment. A further challenge to the scientific community arises from the need for some of them to be embedded members of the wider response multi-agency team, rather than independent contributors of essential information. For example, to enhance interagency communications during recent hazard events and exercises in NZ, members of the GNS Science team were situated as liaison officers within the Emergency Operations Centre (EOC) and responded within the emergency management team itself. In addition, the crisis response context can introduce a need to deal with demands that would rarely, if ever, be encountered in routine work contexts and that can elevate levels of stress and interfere with decision making.

The atypical demands that can impair response during a disaster were evident in evaluation of the multi-agency response to the Ruapehu 1995–1996 eruptions. These demands included:

Table 1 Potential stressors that negatively impact on response capability and personal and team performance when responding to or managing crisis events and disasters (after Paton 1996)

- Degree of warning or change in conditions (low warning times or rapid change increases physical and psychological demands)
- Degree of uncertainty from event and organizational sources
- Time of day (stress greater at night and when having to respond at the end of a working day)
- Presence of traumatic stimuli (such as sensationalised news coverage)
- Lack of opportunity for effective action (attributions about perceived response failure can be internalised rather than more accurately attributed to environmental factors outside of their control)
- Knowing victims or families
- Intense media interest or public scrutiny directed at event management and those responsible
- Higher than usual or expected responsibility
- Higher than usual physical, time and emotional demands (including cumulative stress over time)
- Contact with those affected
- Resource availability and adequacy (and how these change over time)
- Co-ordination problems
- Conflict between agencies
- Inadequate and changing role definition
- Inappropriate leadership practices
- Single versus multiple threats

intense media interest or public scrutiny, resource availability and adequacy, co-ordination problems, a lack of defined responsibility for co-ordination of response, inadequate communication with other organisations, conflict between agencies, and inadequate and changing role definition (Paton 1996; Paton et al. 1998, 1999) (Table 1). The IAVCEI Subcommittee for Crisis Protocols (IAVCEI 1999) provides additional examples of problems commonly experienced in volcanological response (see Table 2), each of which correspond to the disaster stressors identified by Paton (1996; see Table 1).

While some potential event-related stressors reflect the dynamics of hazard impacts (e.g., volcanic ash affecting communication infrastructure), others reflect inadequacies in crisis communication systems and the expertise available to use them (Paton et al. 1998, 1999; Johnston et al. 1999). In the absence of the development of appropriate crisis management procedures and training in crisis management, which is the norm, the associated negative reactions can detrimentally affect performance and decision making (e.g., physiological and psychological symptoms of anxiety and fear, "tunnel vision", failure to prioritise, "freezing" and loss of concentration; Flin 1996; Flin et al. 1997; Klein 1997; Paton et al. 1999). It thus becomes important to identify the management systems and procedures, and personal and team capabilities, required to facilitate effective multi-agency response and use this to inform the training needs and training strategies adopted in all response agencies. Evaluation of previous volcanic crisis management experiences can provide a good starting point for this process.

Mitigating these issues prior to a crisis, particularly at the science:decision-maker interface is important, as effective multi-organisational management needs to be built on "consensus about task goals and priorities; co-operation and team framework; a sense of group identity; a strong sense of community within the organization; and the breakdown of bureaucracy and formalities" (Paton et al. 1999, p. 17). Addressing these issues requires an appreciation of how a shared understanding of response needs can be achieved prior to a crisis, such that this multi-organisational, multi-level, multi-team

Table 2 Common problems of professional interaction of volcanologists during crises, as identified in the IAVCEI subcommittee for crisis protocols (IAVCEI 1999)

Problem	Detail
Poor communication and teamwork among scientists	• Failure to value diverse scientific expertise, approach, and experience • Overselling of new methods • Failure to honour prior work on a volcano, and, in the reverse direction, failure to share study opportunities • Failure to share information and scarce logistical resources • Failure to work as a single scientific team, and thus loss of potential synergism, i.e., loss of a cooperative result that is greater than the sum of individual results • Failure of scientists to use a single voice for public statements • Failure of science-funding agencies, job supervisors, and promotion panels to give full credit for self-sacrifice and teamwork during volcanic crises
Leadership problems	• Leaders without leadership skills • Failure of leaders to recognize the limits of their own technical expertise • Confusion about team roles, policies, and procedures • Failure to encourage those who can and wish to help • Failure to develop (a) respect for scientific differences within a team, (b) a method for developing consensus, and (c) a means for acknowledging differences that cannot be resolved • Failure to balance risk and rewards of dangerous field work • Failure to recognize and minimize fatigue
Issues for visiting scientists, invited and uninvited	• Scientists who arrive at a crisis without invitation • Invitations from other than the primary scientific team, e.g., from a competing or peripheral local group • Unilateral foreign funding decisions • Cultural differences regarding scientific discussion and decision making • Public statements by visiting scientists • Pre-emption of research and publication opportunities by visitors, while local scientists are still busy managing the crisis
Unwise and unwelcome warnings	• Warnings from pseudo-scientists • Warnings or forecasts from scientists from other fields • Warnings or forecasts by volcanologists working in isolation, either on-site or far from the volcano in question • Exaggerated statements of risk, or, conversely, overly reassuring statements about safety of an area when significant risk exists • Outdated warnings or forecasts in need of change

Poor communication between scientists and public officials	• Unfamiliarity with each other's needs and expectations, methods, expertise, and limits
	• A conscious decision to withhold or delay some hazards information
	• Official scepticism of scientific advice
	• Procedural failures in communication with public officials:
	  – Failure to put warnings in writing, for clarity and later accountability
	  – Failure to distribute warnings to all key parties. Failure to establish a clear "chain of communication" between scientists, public officials, and external agencies such as civil defines
	  – Failure to confirm that officials truly understand our warnings

Ineffective relations with news media	• Inadequate interaction with the news media • Premature or excessive interaction with the news media

response is managed effectively, and how communication impacts the quality of decision making. In lieu of real events, these capabilities can be developed through shared exercises, scenario planning, and other relationship building activities, which have training and organizational development implications for scientific agencies.

In this chapter we review the fundamentals of decision making at the individual, team, multi-team, organisational and agency levels (Sect. 2), by drawing on psychological and critical incident management research. We review the concepts of mental models, which are an individual's representation of a situation such as a response environment including needs, responsibilities, and interdependencies; or a representation of a system such as volcanic unrest which incorporates their internal, personalised, experiential, and contextual understanding of how the volcanic system operates. We discuss how these mental models contribute to a "shared meaning" between agencies (Sect. 3), and how that relates to communication and information sharing; as well as the impact that uncertainty has upon communication and ways to manage this. We then review a number of personal and organisational factors that can impact response (Sect. 4), and conclude (in Sect. 5) with a brief review of the methods available to enhance response, and the importance of protocols and guidelines to assist this in a national or international context. Throughout this chapter we focus on the response phase of a crisis. There are however many other complementary approaches to enhance risk communication with communities living with volcanic risk, which we do not consider here, including community-based disaster risk management and other participatory techniques (see review in Barclay et al. 2008; and also Williams and Dunn 2003; Cronin et al. 2004; Gaillard 2006; Cadag and Gaillard 2014).

2 Introducing Decision Making

During a volcanic crisis, several decision making styles and processes are required. Decision making itself has been studied extensively across a range of fields. Here we focus on those

supported through research in crisis and risk management contexts (Lipshitz et al. 2001; Doyle and Johnston 2011).

2.1 Individual Processing Systems

Considering first the processes occurring at an individual level, the field of psychology offers us understanding of the theory of two "parallel processing systems" (Epstein 1994; Sloman 1996; Chaiken and Trope 1999; Slovic et al. 2004). The first, known as either Type 1 or the *affective* processing system, involves rapid, unconscious, action-oriented processing, and results in people interpreting risk as an emotional state or feeling (e.g., fear, dread, anxiety; Epstein 1994; Loewenstein et al. 2001; Slovic et al. 2004; Doyle et al. 2014b), and can thus reduce or increase risk perceptions. These are assumed to be the default response "unless intervened by distinctive higher order" Type 2 processes (Evans and Stanovich 2013a), or *analytical* processing systems (Epstein 1994), which heavily load working memory, and utilise hypothetical thinking, more deliberate computational cognitive processes (and thus longer decision times). These are learnt processes that apply rules and procedures (algorithms, normative rules and logic) to the analyses of data and to justify actions (i.e., to respond to demands rather than reacting to them).

As of Doyle et al. (2014b, p. 78) we consider that "the adoption of the affective and analytical processing systems [is] not an either-or situation, but rather a more complex balancing act influenced by the degree of uncertainty or threat in the decision context, and ... relative experiences" (Keren and Schul 2009; Kruglanski and Gigerenzer 2011; Evans and Stanovich 2013a, b; Keren 2013; Osman 2013; Thompson 2013). Thus, if time permits, scientists tend to adopt the analytical process due to their formal training in data analysis and decision making. Meanwhile, non-scientists adopt a more affective process dependent upon prior experience, time pressures and operating procedures. However, when scientists are called upon to respond to atypical demands (particularly in a multi-agency context), if there are no formal or procedural rules to abide by, the affective system usually prevails (Loewenstein et al. 2001) and decision making effectiveness is compromised as a result (Weber 2006). Mitigating this problem calls for all those interacting in decision making to receive training which develops competency in different decision making styles, and, importantly, practice using them in simulated and actual crisis events.

2.2 Incident Management and Naturalistic Decision Making

The analytical (or Type 2) decision making has been identified as having four steps (Flin 1996, p. 141–142): (1) identifying the problem; (2) generating a set of options; (3) evaluating these options; (4) implementing the preferred option (Saaty 2008). However, this assumes a 'perfect' environment. In reality, most decisions are made in uncertain 'naturalistic settings' defined by: ill-structured problems; uncertain dynamic environments; shifting, ill-defined, or competing goals; action/feedback loops; time stress; high stakes; multiple players; and influences from organizational goals and norms (Zsambok 1997; Crichton and Flin 2001; Klein 2008; Doyle and Johnston 2011). Research into incident management has identified four distinct 'naturalistic decision making' processes seen in these conditions (Crego and Spinks 1997; Pascual and Henderson 1997; Crichton and Flin 2002): (1) recognition primed and intuition led action; (2) action based on written or memorized procedures; (3) analytical comparison of different options; and (4) creative designing of a novel course of action; ordered in terms of decreasing pressure and time commitments.

Within a crisis, an individual decision maker (whether scientist or emergency manager) may move along this spectrum of decision processes depending upon the evolving conditions, and will not be limited to just one decision making style (see Martin et al. 1997, p. 283; Doyle and Johnston 2011). Those operating at a strategic level should use the analytic style to accommodate the broader perspective required under these circumstances (Paton et al. 1998, 1999; Paton and Flin 1999). For those working at a tactical/coordinating level, an analytical approach should be adopted in (relatively) high time, low risk circumstances (such as when planning courses of action between eruption episodes and identifying future eruptive scenarios). However, in an eruption phase, rapid decisions need to be taken in minutes, making adoption of naturalistic decision making styles essential. For example, in Exercise Ruaumoko (which simulated the response to the lead up to an eruption in Auckland; MCDEM 2008; McDowell 2008), on site science-liaison officers often found themselves having to give almost instantaneous responses to officials during the peak of the crisis, which would have encouraged more recognition primed decision making.

Fundamental to all of these decision making processes, and an effective decision resulting from those processes, is individual and team situational awareness (SA, Endsley 1997; Martin et al. 1997), which is the understanding of the situation and needs in both time and space. This encompasses a capacity to use key environmental cues to comprehend the current situation (in relation to goals) and to project future status. Training in this essential competence enhances the ability of decision makers to anticipate and make proactive decisions that deal more effectively with emergent issues. During the initial and on-going situation assessment, individuals and team members play crucial roles in this process (Sarna 2002). As stated by Doyle and Johnston (2011, p. 75), "a decision maker may make the correct decision based on his or her perception of the situation, but if his or her situation assessment is incorrect, this may negatively influence his or her decision (Crichton and Flin 2002)". Because the inputs into decision making can come from different professionals and/or from team members that may be geographically dispersed (e.g., in an EOC, and at the volcano), decision making training must include the development of distributed decision-making skills (where the decision-making responsibility does not lie with a single entity, but rather is distributed throughout the responding organisations; Flin 1996; Paton et al. 1998; Kapucu and Garayev 2011). For distributed decision making to work effectively, decision makers must have some shared meaning (mental model) about the event and their respective roles in defining and resolving response problems.

3 Shared Meaning in Multi-agency Response: Mental Models

An individual's mental model of a hazard is defined by Bostrom et al. (2008, p. 308) as "how people understand and think about the hazard, and their causal beliefs". For incident management, this represents a mental "map" of the operating environment. This must encompass event characteristics and hazard consequences and each other's differing needs, responsibilities, roles, and demands, as well as the interdependencies that contribute to effective problem solving and decision making (Rogalski and Samurçay 1993; Flin 1996; Paton and Jackson 2002). A shared mental model allows a distributed team to share understanding of the task at hand, anticipate and proactively respond to information needs (Lipshitz et al. 2001; Pollock et al. 2003), and make shared decisions (Orasanu 1994; Salas et al. 1994).

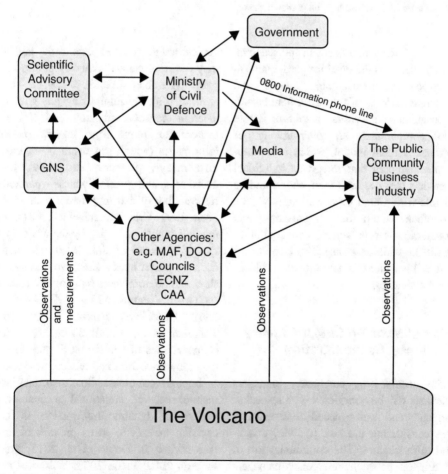

Fig. 1 Example of the information flow, communication network and many agencies involved in an eruption, as observed from the response to the 1995 Ruapehu eruption by Paton et al. (1999). *CAA* Civil Aviation Authority; *DOC* Department of Conservation; *ECNZ* Electricity Corporation of New Zealand; *GNS* Institute of Geological and Nuclear Sciences (now GNS Science); *MAF* Ministry of Agriculture and Fisheries

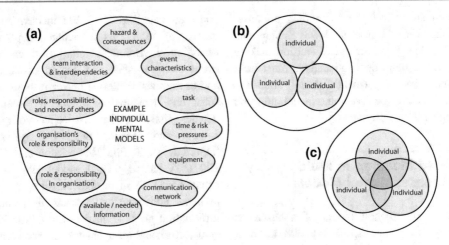

Fig. 2 **a** Examples of the various mental models within an individual's over-arching or super-ordinate mental model during a volcanic crisis. **b** A poor shared mental model between individuals. **c** A good shared mental model between individuals

From an incident management perspective, the major challenge is the need for participants to be able to continue to use their "routine" and "expert" mental models of the constituent hazard consequences and individual response. These need to complement an over-arching or super-ordinate mental model, which is a cooperative mental model that integrates individual mental models that describe their understanding of: their role within their organisation, how they relate to others within their organisation, and their organisation's role within, and communication with, the wider response team (see Figs. 1 and 2). It is this super-ordinate model that facilitates communication.

3.1 Shared Meaning in Multi-agency Response: Communication

Information received by emergency managers from scientists will be considered with respect to other demands and issues placed on emergency managers considering the risk to lives, economies and infrastructure. The communication of risk information between emergency managers and the public, or scientists and emergency managers, is also subject to the mental models gulf (Morgan et al. 2002). This is where there is a

gap between "what experts know and the plan they develop, versus what key public know and prefer" (Heath et al. 2009, p. 129; see also Doyle et al. 2014a). By minimising this gulf, communication between advisors and key decision makers can move from *explicit* requests for information (which can result in an increase in time delays, pressures and stress, impacting decision making effectiveness—particularly if reformatting of that information is also required; Klein 1997; Crichton and Flin 2002) through to *implicit* supply of advice by the advisors as they recognise ahead of time what information the decision maker needs. Effective teams have been shown to be dominated by communication styles such as this (Paton and Flin 1999; Lipshitz et al. 2001; Paton and Jackson 2002; Kowalski-Trakofler et al. 2003). Table 3 describes the characteristics of teams that display these effective communication and advice provision styles.

Implicit communication also facilitates the maintenance of situational awareness during periods of dynamic information as it allows decision makers to focus on task management (see review in Doyle et al. 2015; Paton and Jackson 2002; Paton 2003; Wilson et al. 2007; Owen et al. 2013). For this to occur, these scientists and experts must recognise and understand the needs of the decision makers, as well as

Table 3 Characteristics of teams that display effective communication and advice provision styles during a multi-agency crisis response

Provision of science advice should involve (Paton and Jackson 2002; Paton 2003)	Effective team communication involves (Wilson et al. 2007; Owen et al. 2013)	Effective team and interagency science advice communication (Doyle et al. 2015)
Anticipation and definition of information needs	Accurate and timely information exchange, correct phraseology and 'closed-loop' communication techniques	Should consider the ability of diverse stakeholders to interpret data communicated, and apply to resolve response issues while operating in a collaborative environment
Organized networks between information providers and recipients	Coordinated behaviour based on shared knowledge, performance monitoring, back up and adaptability	Needs the development of a super-ordinate team identity and the ability to switch between agency and shared mental models as required
Established capability to "provide, access, collate, interpret and disseminate information compatible with decision needs and systems"	Co-operative team orientation, efficacy, trust and cohesion	May involve stakeholders who rarely interact with one another outside the context of managing a volcanic crisis, and so depends on emergent team dynamics and concepts such as "swift-trust" (Meyerson et al. 1996)

their timelines and thresholds, supplying information that is *useful, useable, and used* (Rovins et al. 2014).

During the Ruapehu 1995–1996 eruptions, comparison of pre-existing networks with information providers revealed both incomplete networks and inconsistencies with respect to information sources, in particular with the main information provider GNS Science (Paton et al. 1998, 1999). This resulted in agencies seeking information in an ad hoc basis (i.e. through explicit requests) which would have contributed to communication difficulties, as evidenced by the 37% of organisations who reported inadequate communication during the response (ibid). One way of mitigating this issue involves advisors developing a capacity to, where possible, anticipate others information needs (Doyle and Johnston 2011; Doyle et al. 2015).

Advisors must recognise the very specific information and advice needs of decision makers prior to an event and have procedures in place in advance of an event to provide that information in a timely manner directly where it is needed within the organisational structure. For this to effectively occur scientists (either as on-site science advisors, or off-site expert panels) must not just be external experts to a multi-agency response team, but must be considered part of

the extended and distributed team handling the emergency management response (Doyle and Johnston 2011). Integrating Science Advisory Groups (SAGs) into a wider response team offers the opportunity for technical and scientific experts to directly inform effective planning, intelligence gathering, and decision making of the emergency personnel and government officials.

The Auckland Volcanic Science Advisory Group (AVSAG), an example of such a SAG, was tested during Exercise Ruaumoko (MCDEM 2008; McDowell 2008). From this it was identified that the AVSAG process facilitated the provision of valuable advice in a clear, timely manner. A clear advantage was the presence of a science advisor in both the National Crisis Management Centre and the Auckland Civil Defence and Emergency Management (CDEM) Group EOC, maintaining shared situational awareness between the SAG and the emergency management team. However, having two on-site liaison officers at different EOCs did result in a divergence of advice in these two locations at the peak of the crisis (Cronin 2008), as shared situational awareness could not be maintained between them.

These past experiences, through both response and exercises, has identified that this SAG approach is beneficial as it provides "one trusted source" for science information during a crisis (MCDEM 2008; Smith 2009), facilitates an integration of a wide range of expert opinions required to manage uncertainty during decision making (as recommended by Lipshitz et al. 2001) and can help combat issues that may arise due to conflict between scientists (Barclay et al. 2008). It also enabled the volcanologists to speak with "a single voice" to reduce confusion, as advocated by the IAVCEI protocols (1999).

3.2 Shared Meaning: Uncertainty

If the volcanic crisis environment was perfectly predictable, the development of the relationships and competencies discussed above would be a relatively straightforward task. However, volcanic crises present evolving, emergent demands, and a highly uncertain response environment. The IAVCEI protocols (1999) highlight that uncertainty should always be acknowledged. This raises two issues. Firstly, how uncertainty should be communicated. Secondly, how uncertainty influences the quality of the relationships between individuals.

Regarding the first, reviews by Doyle et al. (2014a, b), identified that there is much discourse as to whether revealing the uncertainties associated with a risk assessment will strengthen or decrease trust in a risk assessor and their message, and how it impacts decision-making behaviour (Miles and Frewer 2003; Wiedemann et al. 2008). On the one hand, the communication of uncertainty has been suggested to enhance credibility and trustworthiness. On the other, however, studies have suggested that it can decrease people's trust and the credibility of the provider, as it can allow people to justify inaction or their own agenda, or to perceive the risk as being higher or lower than it is depending on their personal attitudes (Johnson and Slovic 1995, 1998; Smithson 1999; Miles and Frewer 2003; Johnson 2003; Wiedemann et al. 2008; Doyle et al. 2014b). The role of ethics in whether or not to communicate uncertainty has also become a focus of recent discussion across

disciplines, including whether communicating this uncertainty actually enhances or diminishes the autonomy of the receiver of the message, and whether it produces an overall benefit or can actually cause harm (Han 2013; Austin et al. 2015; Grasso and Markowitz 2015). Keohane et al. (2014) suggest that scientists 'should understand their own ethical choices in using scientific information to communicate to audiences' (p. 343), and identify five principles for scientific communication under uncertain conditions: honesty, precision, audience relevance, process transparency, and specification of uncertainty about conclusions.

To address how to manage and communicate uncertainty, many disciplines including volcanology, climate change, and meteorology (IAVCEI 1999; Moss and Schneider 2000; Gill et al. 2008; Mastrandrea et al. 2010; see also Moss and Schneider 2000; Patt and Dessai 2005; Budescu et al. 2009; Doyle and Potter 2016), have established guidelines that advocate the clear and transparent communication of uncertainty, a documentation of all processes related to uncertainty, and the use of formalised probabilistic terms and frameworks for assessment and communication (see Table 4). In volcanology, it has become increasingly popular to use probability statements in communications (Doyle et al. 2014a), which involve knowledge of both the dynamical phenomena and the uncertainties involved (Sparks 2003). Further, the use of probabilistic cost benefit analysis and Bayesian Event Trees has been driven by a desire to make objective and traceable decisions via quantitative volcanic risk metrics (Aspinall and Cooke 1998; Marzocchi and Woo 2007; Woo 2008; Lindsay et al. 2009).

However, probabilistic statements, whether in numeric or linguistic formats, can commonly be misinterpreted because their framing, directionality and probabilistic format can bias people's understanding, thereby affecting people's action choices (see Fig. 3; e.g., Teigen and Brun 1999; Karelitz and Budescu 2004; Honda and Yamagishi 2006; Joslyn et al. 2009; Budescu et al. 2009; Lipkus 2010). This has been identified as a particular issue in volcanic crisis

Table 4 A summary of the existing guidelines for communicating uncertainty from the volcanological, weather and climate change communities

Budescu et al. (2009)	• "Make every possible effort to differentiate between the ambiguity of a target event and its underlying uncertainty" • "Specify the various sources of uncertainty underlying key events and outline their nature and magnitude, to the degree that this is possible" • "Use both verbal terms and numerical values to communicate uncertainty" • "Adjust the width of the numerical ranges to match the uncertainty of the target events"
World meteorological office (Gill et al. 2008)	Uncertainties should be communicated: • "For improved decision making"—especially when they have many options available to them, to weigh up contingencies • "Helps manage user expectations"—a more open, honest and effective relationship • "Promotes user confidence"—If users understand forecasts have a degree of uncertainty... they can tune their decision-making to manage this uncertainty ... • "As it reflects the state of science"

Moss and Schneider (2000, p. 37) (as used by the IPCC; see also Patt and Dessai 2005; Mastrandrea et al. 2010)	• Identify the most important factors and uncertainties that are likely to affect the conclusions
	• Document ranges and distributions in the literature [... the key causes of uncertainty ...]
	• Make an initial determination of the appropriate level of precision [... after considering the state of the science and the nature of the uncertainties...]
	• Characterize the distribution of values that a parameter, variable, or outcome may take
	• Using standard terms, rate and describe the state of scientific information on which the conclusions and/or estimates are based
	• Prepare a "traceable account" of how the estimates were constructed
	• Use formal probabilistic frameworks for assessing expert judgment
	• Consider target audience and develop a pluralistic approach (Patt and Dessai 2005):
	  – Sophisticated part in numeric format
	  – General chapters using verbal and narrative phrase
	  – Formalise translation between numerical and verbal probabilities

IAVCEI (1999) subcommittee for crisis protocols	• Uncertainty should be acknowledged • Forecasts, warnings, and other important public statements are best when written first • Date-stamped, team-approved hazard maps, together with their assumptions, should also be entered into the formal record of warnings. Competing or uncoordinated, multiple hazards maps are confusing to the public and should be avoided • Scientific caution in the face of uncertainty is good, but it needs to be balanced against the legitimate information needs of decision makers and the public at risk. If the data do not allow a definitive forecast,

(continued)

Table 4 (continued)

	factual statements about what is known are an important step. Warnings of serious events that are known to be possible, issued before such events can be forecast as probable, may hasten precautions and save lives • Use probabilities to calibrate qualitative assessments of risk. Avoid commonly used adjectives such as "soon" or "high-" or "low-(risk)," because they mean different things to different people. Probabilities and comparisons to familiar non-volcanic risks help to avert misunderstanding that risk is higher or lower than it actually is • Under no circumstances should hazard be intentionally overstated or understated. Any decision to "err on the safe side" should be a conscious, openly discussed decision. Never disregard what seems like a low-probability, "worst case" event, because such events can and do occur (e.g., Mount St. Helens and Pinatubo). Instead, estimate the probabilities of worst-case and lesser scenarios, as above, to put the "worst-case scenario" in proper perspective
Doyle et al. (2014a, b) see also the operational guidelines in Doyle and Potter (2016)	• There is a need to adopt formal numerical and verbal probability translation tables that are specific to volcanology • If communicating time window forecasts, be consistent in the use of either "within" or "in" throughout out all statements, bulletins and reports; particularly for long window forecast statements where "within the next X days" has a statistically significant different interpretation to "in the next X days" • Scientists should make all possible efforts to communicate forecasts, likelihoods, and probabilities over a range of relevant time windows, including a probability forecast for a shorter immediate time window in particular (such as the first 24 h) • Any formalised communication strategy should be accompanied by exercises, simulations, and education programs with both the decision-makers and the public to help facilitate a greater understanding of the complexities inherent in these uncertain forecasts

communications (IAVCEI Subcomittee for Crisis Protocols 1999; Cronin 2008; Haynes et al. 2008; Solana et al. 2008; McGuire et al. 2009; Doyle et al. 2014a; Doyle and Potter 2016). Thus scientists communicating in a multi-agency volcanic response must consider the best practice guidelines that have been established across a range of disciplines to address this issue (see Table 4).

The second issue regarding uncertainty in a crisis is its influence on the quality of the relationships between responding individuals, and thus the quality of mental models and performance, discussed next.

4 People and Organizations

A common thread running through the previous sections concerned the crucial role shared mental models played in cooperative action. It is important to appreciate how organizational and work-family relationships influence the ease with which respective agency representatives can

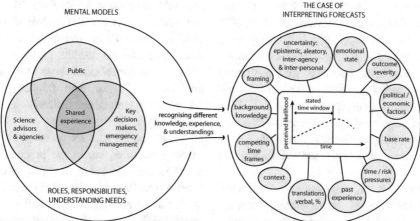

Fig. 3 The factors that affect the interpretation of forecasts, and the influences on resultant decision making (Doyle et al. 2014b, after the graphical abstract of Doyle et al. 2014a)

"participate" in a super-ordinate mental model and the degree to which they cope with and adapt to crisis demands (i.e., how organizational factors influence susceptibility to stress and thus performance).

4.1 Organizational Characteristics

Organizational characteristics influence the mental models that agency representatives bring to the crisis management environment. The organizational socialization (the norms, customs, and ideologies of that organisation) and the organizational cultural change that occurs when organisations interact, will influence the thinking and behaviour of people within an organization, as well as their mental models of communication between organizations. These processes then spill over to affect how volcanic crisis events and their consequences are responded to and interpreted (Paton et al. 2009; Paton and Norris 2014). The knowledge and interpretive processes that representatives bring to the crisis response role as a result of their organizational cultural provides the foundation (capabilities represented by pre-existing mental models) for their individual contribution. However, as introduced above, the need for diverse organizational representatives to integrate their respective contributions in complementary ways requires the development of a super-ordinate mental model defined by the collective and multifaceted demands of the crisis event. The ease with which this can be achieved is complicated by the fact that this is done in a climate of uncertainty. Interaction under uncertainty can result in pre-existing beliefs dominating which may or may not be amenable to alteration. This then prevents the development of interagency trust and undermines collective performance (Paton et al. 2009). When dealing with atypical, challenging crisis events, emergency management representatives, by virtue of the need for them to play complementary roles in defining and managing complex events, become more reliant on others for information and guidance about how to respond.

4.2 Trust

Faced with uncertainty, when decision makers are reliant on one another for information and decision making, trust plays a pivotal role in facilitating sustainable collaboration in multi-agency crisis response contexts (Siegrist and Cvetkovich 2000; Mayer et al. 1995). Trust influences organizational intention to collaborate and share information between stakeholders (Mohr and Spekman 1994; Kapucu 2006).

Trust was identified as a key issue for relationships between scientists and officials during the eruptions of Soufriere Hills Volcano, Montserrat, WI, where Haynes et al. (2007) identified that this trust was influenced by a number of factors including: competence; integrity; value similarity; openness; and conflicting messages of safety and danger. Trust in the scientists was based upon high perceived reliability, competence, openness and integrity; while trust in government authorities was based upon high perceived levels of competence, reliability, and fairness.

The diversity of, for example, agencies, organizational cultures and operating practices brought together into a crisis management environment, and the need for diverse volcanic hazard consequences to be managed by agency representatives that differ in levels of familiarity with each other, can threaten the degree to which crisis management activities are characterized by trust (Dietz et al. 2010). In part, this reflects the lack of familiarity between interacting agencies. It is also influenced by how organizational cultural characteristics ingrained in routine work, such as hierarchical reporting practices and levels of bureaucracy (including command and control expectations), influence the relationships that emerge in an inter-agency crisis management context (Dietz et al. 2010; Dirks and Ferrin 2001).

Scientists tend to bring experience of working in organizational cultures that emerge in flatter, more organic organizational cultures in which information flow is common. This makes it easier for them to engage in practices that focus on sharing information and building trust. However, representatives from government departments and emergency services agencies, whose routine culture is characterized by generally higher levels of hierarchical relationships and reporting, tend to be predisposed towards maintaining their own agency-based independence. This fosters an emergent culture of rivalry among organizations in ways that work against information sharing between agencies during a crisis (Waugh and Streib 2006; Iannella and Henricksen 2007; Marcus et al. 2006; Marincioni 2007). Furthermore, these predisposing cultural features can result in relationships that are characterized by

in- and out-group differentiation. These in- and out- groups then affect the quality of information flow and increases the likelihood of information being restricted to members within their own organization (or includes those with whom they are familiar), rather than sharing information across all stakeholders (Militello et al. 2007; Owen 2013).

These factors interact to not only constrain information flow, but in the process, introduce significant challenges to the development of the level of trust required for effective collaboration and decision making under conditions of uncertainty (McKnight et al. 1988; Banai and Reisel 1999; Siegrist and Cvetkovich 2000). Thus the agency representatives brought together for response needs may not have the mutual interaction experiences needed to forge trust in each other, with aspects such as cultural diversity adding to this challenge. Thus, there is an important need to develop trust in situ while responding in a high demand environment.

If trust is absent, those working in an EOC setting are more likely to focus on task demands in ways that reflect their core expertise and normal operating practices, rather than functional collaboration in ways that ensure they work in complementary ways to resolve multi-faceted response problems (Pollock et al. 2003). This reduces their capacity to contribute effectively to the emerging needs of the response (Pollock et al. 2003). The dynamic, prolonged nature of volcanic crisis response thus requires different ways of ensuring the development of trust. The concept of *swift trust* represents an approach to trust building in situations where people must collaborate on complex, evolving volcanic crisis tasks under high risk, low time constraints that preclude the development of trust through normal means (Goodman and Goodman 1976; Meyerson et al. 1996; Hyllengren et al. 2011; Faraj and Xiao 2006; Robert et al. 2009; Lester and Vogelgesang 2012; Crisp and Jarvenpaa 2013).

For swift trust to develop, "team" members must be assigned specific roles (that align with key response issues and needs) in the temporary work group (Meyerson et al. 1996). This can be facilitated by establishing a super-ordinate mental model that makes the key contributions of all agencies to effective whole-of-incident management clear. It can be developed pre-event via use of techniques such as cross training in organizational crisis management training (Blickensderfer et al. 1998). Second, swift trust emerges if members are informed that there is a high likelihood of future collaboration (in incident reviews, simulations) with those with whom they are collaborating (Goodman and Goodman 1976; Meyerson et al. 1996). Finally, swift trust develops by ensuring that all participants identify that success relates to the super-ordinate management as much as it does to how they contribute their personal expertise (Curnin et al. 2015). This developmental task focuses on how the input of different representatives is necessary to develop a holistic response to multifaceted demands that exceed the expertise of any one agency. Doing so facilitates role clarification, and increases the capacity of team members to understand that each stakeholder brings to the collective task, their specialist skills and knowledge as required to fulfil one of several specialist roles in the multi-agency team (Kramer 1999). This, in turn, enables the development and performance of collaborative working practices and supports emergent multi-agency coordination.

4.3 Work-Family

Another element that has a significant bearing on people's stress management and performance in high risk settings and that has not been considered in volcanic crisis response contexts concerns how "work-family" relationships affect how well people cope with working in high risk, high stress contexts (Paton and Norris 2014). Organizations that take steps to facilitate family involvement in the employment experience, for their personnel who work in high risk settings, record better communication and trust between personnel and management, and thus more effective stress management in high risk work settings. Family involvement includes, for example, providing support services for partners

and children when personnel are deployed, providing regular updates for partners and other family members when personnel are deployed, providing roles (e.g., administrative roles, media liaison roles) and setting up peer support programs for partners (Paton and Kelso 1991; Paton and Norris 2014). In contrast, in organizations where these kinds of family engagements are not offered, factors such as lack of information about what is happening in the field, and the resultant increased anxiety amongst family members, will increase perceived psychological stress in partners and children and lessens their availability and effectiveness as support resources for personnel deployed to deal with volcanic crises (Paton et al. 2009). Awareness of these issues is imperative for volcanic crisis response, where responding scientists may be geographically dispersed and away from family during periods of high risk and high stress work. Under such circumstances, ensuring partners are involved as much as possible to facilitate opportunities for personnel to remain connected with family, and access social support from this quarter, can provide cost-effective stress management resources for those deployed (Paton and Kelso 1991). This aspect of a comprehensive crisis management strategy plays a key role in supporting performance and well-being, particularly when personnel can be deployed and working in high demand, high stress contexts over prolonged periods of time.

---

5 Concluding Remarks: Developing Future Response Capacity

We have highlighted how good shared mental models of the response situation between individuals within and across organisations, characterised by good situational awareness, strong inter-organisational networks, and high trust between responding organisations and individuals, have been shown to enhance communication and thus decision making. However, developing and maintaining such shared mental models is itself an important task. Research has shown that shared experience, through training, can help improve the quality of such mental models (Cannon-Bowers and Bell 1997; Crego and Spinks 1997; Paton et al. 2000; Pliske et al. 2001; Borodzicz and van Haperen 2002).

Ideally, multi-organisational and multi-disciplinary planning activities, collaborative exercises and simulations should be undertaken with all team members and advisors to help in the development of similar mental models of the task (see review in Paton and Jackson 2002; Doyle and Johnston 2011; Doyle et al. 2015). A comprehensive suite of training and relationship building activities prior to an event, and a detailed analysis of event and exercise response, can help enhance this future response capability and identify areas for improvement. This is particularly important given the rarity of volcanic and other hazard events, and thus a lack of opportunity for real world experience. This training and exercising needs to develop both individual and team situational awareness and explore how and when each is appropriate for response, within evolving, dynamic response environments (Doyle et al. 2015). Team situational awareness can be developed in post-event and post-exercise reviews that include identifying inter-agency relationship issues as opportunities for development (and not as problems). Through the analysis of past events, lessons for successful communication, advice provisions and distributed decision-making can also be learnt.

The above processes describe group learning from crises, exercises and training, which is identified by Borodzicz and van Haperen (2002) to occur along three dimensions: personal, interpersonal, and institutional. Several training methods have been identified that can enhance naturalistic decision-making (Cannon-Bowers and Bell 1997), enhance decision skills (Pliske et al. 2001), train effective teams (Salas et al. 1997), and develop effective critical incident and team based simulations (Flin 1996; Crego and Spinks 1997) all of which are relevant for volcanologists and scientific advisory groups. These include cross training, positional rotation, scenario planning, collaborative exercises and simulations, shared exercise writing tasks including co-writing, swapped writing and 'train the

trainer' type tasks, in addition to workshops, seminars and specific knowledge sharing activities (Doyle and Johnston 2011; Doyle et al. 2015).

Adopting such an evaluative approach has greatly enhanced the response environment in New Zealand, resulting in a formation of a number of scientific advisory groups with formalised Terms of Reference, protocols for communication and networking with emergency management and key response organisations (e.g., CPVAG 2009). These, accompanied by regular workshops and meetings to facilitate relationship building and shared understanding, will help improve the communication and information flow and thus the shared situational awareness in a crisis. Being part of an exercise schedule (i.e. Exercise Ruaumoko; MCDEM 2008), also provided Auckland CDEM and the associated science agencies a focus to develop the Auckland Volcanic Science Advisory Group structure, including arranging formal contract agreements for the participating scientists (McDowell 2008; Cronin 2008).

The development of formalised protocols, Terms of Reference and the use of established guidelines for response and communication (such as those issued by IAVCEI, WMO, and IPCC; IAVCEI 1999; Gill et al. 2008; Mastrandrea et al. 2010) can greatly enhance response processes by reducing ambiguity about 'what to do', 'what to communicate', 'how to communicate', and 'who to communicate to', particularly in high stress, high pressure, high consequence events. As stated by IAVCEI (1999), responding scientists must also identify a team plan for crisis response (Table 5), and we suggest that such plans and procedures should be tested prior to an event. Simulations should aim to reproduce reality as closely as possible, reflecting the realities of advisory processes in turbulent conditions (Rosenthal and 't Hart 1989; Borodzicz and van Haperen 2002). However, evaluation of exercises and events must minimise the risk of creating an optimistic bias that overestimates future response preparedness and capability (Paton et al. 1998).

In conclusion, it is also important for future work to consider the role that international

Table 5 Steps identified by IAVCEI (1999, p. 332) to form a team plan for volcanologists responding to a crisis

1. Clear identification of scientific, warning, and other tasks (including communications with civil defence, news media, and others)

2. Clear identification of responsibility (group or individual) for each task, including that of team leader

3. Clear identification of a mechanism for selecting team leader(s)

4. Procedures and policies on likely issues of scientific interaction, including:
    (a) Rights and responsibilities for data and sample sharing
    (b) Resolution of differences in scientific approach and/or interpretation
    (c) Preparation and release of forecasts, warnings, and other public statements
    (d) Restrictions of access to hazardous areas (and application/approval procedures for access permits)
    (e) Requirements and roles of visiting scientists
    (f) Communication, within and outside the scientific team; and
    (g) Publication of scientific results, and distribution of authorship

frameworks and initiatives will have on any protocols and procedures developed for volcanic science advisors at a local or national level. For example, recently the UN Office for Disaster Risk Reduction's Hyogo Framework for Action (HFA) 2005–2015 (United Nations International Strategy for Disaster Reduction (UNISDR) 2007) has been reviewed. Changes include a reconsideration of the role of science and technology, the role of local science and local knowledge, and the role of international science advice mechanisms as ratified in the Sendai Framework for Disaster Risk Reduction 2015–2030 (SFDRR: UNISDR 2015). Volcanologists must be aware of the impact of any changes to international frameworks such as the HFA and the SFDRR, as they will affect regional and local frameworks and support, including both funding, resources, and the legitimacy of any formalised local approaches that may be developed to maintain "shared meaning" with stakeholders during a volcanic crisis.

Acknowledgements EEHD was supported by a Foundation for Research Science and Technology NZ S&T Postdoctoral Fellowship MAUX0910 2010–2014, and funding from EQC and GNSScience 2014–2016.

References

Aspinall W, Cooke R (1998) Expert judgement and the Montserrat Volcano eruption. In: Mosleh A, Bari RA (eds) Proceedings of the 4th international conference on probabilistic safety assessment and management PSAM4, September 13–18. New York, USA, pp 2113–2118

Austin J, Gray G, Hilbert J, Poulson D (2015) The ethics of communicating scientific uncertainty. Environ Law Report 45(2):10105

Banai M, Reisel W (1999) Would you trust your foreign manager? An empirical investigation. Int J Hum Res Manag 10(3):477–487

Barclay J, Haynes K, Mitchell T, Solana C, Teeuw R, Darnell A, Crosweller HS, Cole P, Pyle DM, Lowe C, Fearnley C, Kelman I (2008) Framing volcanic risk communication within disaster risk reduction: finding ways for the social and physical sciences to work together. Geol Soc London, Spec Publ 305:163–177. doi:10.1144/SP305.14

Blickensderfer E, Cannon-Bowers JA, Salas E (1998) Cross-training and team performance. In: Cannon-Bowers JA, Salas E (eds) Making decisions under stress: implications for individual and team training. American Psychological Association, Washington, D.C., USA, pp 299–311

Borodzicz E, van Haperen K (2002) Individual and group learning in crisis simulations. J Contingencies Cris Manag 10:139–147. doi:10.1111/1468-5973.00190

Bostrom A, French S, Gottlieb S (2008) Risk assessment, modeling and decision support: strategic directions. Springer, Berlin

Budescu DV, Broomell S, Por H-H (2009) Improving communication of uncertainty in the reports of the intergovernmental panel on climate change. Psychol Sci 20(3):299–308

Cadag JR, Gaillard JC (2014) Integrating people's capacities in disaster risk reduction through participatory mapping. In: Lopez-Carresi A, Fordham M, Wisner B, Kelman I, Gaillard JC (eds) Disaster management: international lessons in risk reduction, response and recovery. Routledge, New York, pp 269–286

Cannon-Bowers JA, Bell HE (1997) Training decision makers for complex environments: implications of the naturalistic decision making perspective. In: Zsambok CE, Klein G (eds) Naturalistic decision making. Lawrence Erlbaum Associates, Mahwah, NJ, pp 99–110

Chaiken S, Trope Y (1999) Dual process theories in social psychology. Guilford Press, New York

CPVAG (2009) Central Plateau Volcanic Advisory Group Strategy, October 2009. Report No 2010/EXT/1117. In: Morris B (ed) Horizons Regional Council, Palmerston North, NZ

Crego J, Spinks T (1997) Critical incident management simulation. In: Flin R, Salas E, Strub M, Martin L (eds) Decision making under stress emerging themes and applications. Ashgate Publishing Limited, Aldershot, England, pp 85–94

Crichton M, Flin R (2001) Training for emergency management: tactical decision games. J Hazard Mater 88(2–3):255–266

Crichton M, Flin R (2002) Command decision making. In: Flin R, Arbuthnot K (eds) Incident command: tales from the hot seat. Ashgate Publishing Limited, Aldershot, England, pp 201–238

Crisp CB, Jarvenpaa SL (2013) Swift trust in global virtual teams. J Pers Psychol 12(1):45–56

Cronin SJ (2008) The Auckland Volcano Scientific Advisory Group during Exercise Ruaumoko: observations and recommendations. In: Civil defence emergency management: exercise Ruaumoko. Auckland Regional Council, Auckland

Cronin SJ, Gaylord DR, Charley D, Alloway BV, Wallez S, Esau JW (2004) Participatory methods of incorporating scientific with traditional knowledge for volcanic hazard management on Ambae Island, Vanuatu. Bull Volcanol 66:652–668

Curnin S, Owen C, Brooks B, Paton D (2015) A theoretical framework for negotiating the path of emergency management multi-agency coordination. Appl Ergon 47:300–307

Dietz G, Gillespie N, Chao G (2010) Unravelling the complexities of trust and culture. In: Saunders M, Skinner D, Dietz G, Gillespie N, Lewicki R (eds) Organizational trust: a cultural perspective Cambridge. Cambridge University Press, UK, pp 3–41

Dirks K, Ferrin D (2001) The role of trust in organizational settings. Organ Sci 12:450–467

Doyle EE, Johnston DM (2011) Science advice for critical decision-making. In: Paton D, Violanti J (eds) Working in high risk environments: developing sustained resilience. Charles C Thomas, Springfield, pp 69–92

Doyle EEH, McClure J, Johnston DM, Paton D (2014a) Communicating likelihoods and probabilities in forecasts of volcanic eruptions. J Volcanol Geotherm Res 272:1–15. doi:10.1016/j.jvolgeores.2013.12.006

Doyle EEH, McClure J, Paton D, Johnston DM (2014b) Uncertainty and decision making: volcanic crisis scenarios. Int J Disaster Risk Reduct 10:75–101

Doyle EEH, Paton D, Johnston DM (2015) Effective management of volcanic crises: evidence-based approaches to enhance scientific response. J Appl Volcanol 4:1

Doyle EEH, Potter HS (2016) Methodology for the development of a probability translation table for GeoNet. GNS Science Report 2015/67. GNS Science, Lower Hutt, NZ

Endsley MR (1997) The role of situation awareness in naturalistic decision making. In: Zsambok CE,

Klein G (eds) Naturalistic decision making. Lawrence Erlbaum Associates, Mahwah, pp 269–284

Epstein S (1994) Integration of the cognitive and the psychodynamic unconscious. Am Psychol 49(8):709–724

Evans JSBT, Stanovich KE (2013a) Dual process theories of cognition: advancing the debate. Perspect Psychol Sci 8:223–241

Evans JSBT, Stanovich KE (2013b) Theory and metatheory in the study of dual processing: reply to comments. Perspect Psychol Sci 8:263–271

Faraj S, Xiao Y (2006) Coordination in fast-response organizations. Manage Sci 52(8):1155–1169

Flin R (1996) Sitting in the hot seat: leaders and teams for critical incident management. Wiley, Chichester

Flin R, Salas E, Strub M, Martin L (eds) (1997) Decision making under stress: emerging themes and applications. Ashgate Publishing Limited, Aldershot

Gaillard J-C (2006) Traditional communities in the face of natural hazards: the 1991 Mount Pinatubo eruption and the Aetas of the Philippines. Int J Mass Emerg Disasters 24:5–43

Gill J, Rubiera H, Martin C, Cacic I, Mylne K, Dehui C, et al (2008) World meteorological organization guidelines on communicating forecast uncertainty. World Meteorological Organization, WMO/TD No. 4122

Goodman R, Goodman L (1976) Some management issues in temporary systems: a study of professional development and manpower-the theater case. Adm Sci Q 21(3):494–501

Grasso M, Markowitz EM (2015) The moral complexity of climate change and the need for a multidisciplinary perspective on climate ethics. Clim Change 130:327–334

Han P (2013) Conceptual, methodological, and ethical problems in communicating uncertainty in clinical evidence. Med Care Res Rev 70(1):14S–36S

Haynes K, Barclay J, Pidgeon N (2007) The issue of trust and its influence on risk communication during a volcanic crisis. Bull Volcanol 70(5):605–621

Haynes K, Barclay J, Pidgeon N (2008) Whose reality counts? Factors affecting the perception of volcanic risk. J Volcanol Geotherm Res 172(3–4):259–272

Heath R, Lee J, Ni L (2009) Crisis and risk approaches to emergency management planning and communication: the role of similarity and sensitivity. J Public Relat Res 21(2):123–141

Honda H, Yamagishi K (2006) Directional verbal probabilities. Exp Psychol 53(3):161–170

Hyllengren P, Larsson G, Fors M, Sjöberg M, Eid J, Olsen OK (2011) Swift trust in leaders in temporary military groups. Team Perform Manage 17(7/8):354–368

Iannella R, Henricksen K (2007) Managing information in the disaster coordination centre : lessons and opportunities. In: Van de Walle B, Burghardt P, Nieuwenhuis C (eds) Proceedings of the 4th international ISCRAM conference, May 2007, pp 1–11

International Association for Volcanology and Chemistry of the Earth's Interior (1999) IAVCEI subcommittee for crisis protocols, professional conduct of scientists during volcanic crises. Bull Volcanol 60:323–334

Johnson BB (2003) Further notes on public response to uncertainty in risks and science. Risk Anal 23(4):781–789

Johnson BB, Slovic P (1995) Presenting uncertainty in health risk assessment: initial studies of its effects on risk perception and trust. Risk Anal 15(4):485–494

Johnson BB, Slovic P (1998) Lay views on uncertainty in environmental health risk assessment. J Risk Res 1(4):261–279

Johnston DM, Paton D, Houghton BF (1999) Volcanic hazard management: promoting integration and communication. In: Ingleton J (ed) Natural disaster management. United Nations (IDNDR), Coventry, pp 243–245

Joslyn SL, Nadav-Greenberg L, Taing MU, Nichols RM (2009) The effects of wording on the understanding and use of uncertainty information in a threshold forecasting decision. Appl Cognitive Psychol 23(1):55–72

Kapucu N (2006) Interagency communication networks during emergencies: boundary spanners in multiagency coordination. Am Rev Publ Admin 36(2):207–225

Kapucu N, Garayev V (2011) Collaborative decision-making in emergency and crisis management. Int J Publ Admin 34(6):366–375

Karelitz TM, Budescu DV (2004) You say "probable" and I say "likely": improving interpersonal communication with verbal probability phrases. J Exp Psychol: Appl 10(1):25–41

Keohane R, Lane M, Oppenheimer M (2014) The ethics of scientific communication under uncertainty. Politics Philos Econ 1394:343–367

Keren G (2013) A tale of two systems: a scientific advance or a theoretical stone soup? Commentary on Evans & Stanovich (2013). Perspect Psychol Sci 8:257–262

Keren G, Schul Y (2009) Two is not always better than one: a critical evaluation of two-system theories. Perspect Psychol Sci 4:533–550

Klein G (1997) The current status of the naturalistic decision making framework. In: Flin R et al (eds) Decision making under stress: emerging themes and applications. Ashgate Publishing Limited, Aldershot, pp 11–28

Klein G (2008) Naturalistic decision making. Hum Factors J Hum Factors Ergon Soc 50:456–460. doi:10.1518/001872008X288385

Kowalski-Trakofler KM, Vaught C, Scharf T (2003) Judgment and decision making under stress: an overview for emergency managers. Int J Emerg Manag 1:278–289. doi:10.1504/IJEM.2003.003297

Kramer R (1999) Trust and distrust in organizations: emerging perspectives, enduring questions. Annu Rev Psychol 50(1):569–598

Kruglanski AW, Gigerenzer G (2011) Intuitive and deliberative judgements are based on common principles. Psychol Rev 118:97–109

Lester P, Vogelgesang G (2012) Swift trust in ad hoc military organizations. In: Laurence J, Michael M (eds) The Oxford handbook of military psychology. Oxford University Press, New York, pp 176–186

Lindsay J, Marzocchi W, Jolly G, Constantinescu R, Selva J, Sandri L (2009) Towards real-time eruption forecasting in the Auckland Volcanic Field: application of BET_EF during the New Zealand National Disaster Exercise "Ruaumoko". Bull Volcanol 72:185–204. doi:10.1007/s00445-009-0311-9

Lipkus IM (2010) Numeric, verbal, and visual formats of conveying health risks: suggested best practices and future recommendations. Med Decis Mak 27(5):696–713

Lipshitz R, Klein G, Orasanu J, Salas E (2001) Focus Article: taking stock of naturalistic decision making. J Behav Decis Mak 14:331–352. doi:10.1002/bdm.381

Loewenstein GF, Weber EU, Hsee CK, Welch N (2001) Risk as feelings. Psychol Bull 127(2):267–286

Marcus LJ, Dorn BC, Henderson JM (2006) Meta-leadership and national emergency preparedness: a model to build government connectivity. Biosecurity Bioterrorism: Biodefense Strategy Pract Sci 4(2):128–134

Marincioni F (2007) Information technologies and the sharing of disaster knowledge: the critical role of professional culture. Disasters 31(4):459–476

Martin L, Flin R, Skriver J (1997) Emergency decision making—a wider decision framework? In: Flin R, Salas E, Strub M, Martin L (eds) Decision making under stress emerging themes and applications. Ashgate Publishing Limited, Aldershot, pp 280–290

Marzocchi W, Woo G (2007) Probabilistic eruption forecasting and the call for an evacuation. Geophys Res Lett 34:1–4. doi:10.1029/2007GL031922

Mastrandrea MD, Field CB, Stocker TF, Edenhofer O, Ebi KL, Frame DJ et al (2010) Guidance note for lead authors of the IPCC fifth assessment report on consisten treatment of uncertainties. IPCC cross-working group meeting on consistent treatment of uncertainties. Jasper Ridge, CA USA http://www.ipcc.ch/pdf/supporting-material/uncertainty-guidance-note.pdf. Last accessed 22 Oct 2014

Mayer R, Davis J, Schhorman F (1995) An integrative model of organizational trust. Acad Manag Rev 20(3):709–734

MCDEM (2008) Exercise Ruaumoko ' 08 final exercise report. Ministry of Civil Defence and Emergency Management, Wellington 79 pp

McDowell S (2008) Exercise Ruaumoko: evaluation report. Auckland Civil Defence Emergency Management Group, Auckland 24 pp

McGuire WJ, Solana MC, Kilburn CRJ, Sanderson D (2009) Improving communication during volcanic crises on small, vulnerable islands. J Volcanol Geotherm Res 183(1–2):63–75

McKnight D, Cummings L, Chervany N (1988) Initial trust formation in new organizational relationships. Acad Manag Rev 23(3):473–490

Meyerson D, Weick KE, Kramer RM (1996) Swift trust and temporary groups. In: Kramer RE, Tyler TR (eds) Trust in organisations: frontiers of theory and research. Sage Publications Inc, Thousand Oaks, pp 166–195

Miles S, Frewer LJ (2003) Public perception of scientific uncertainty in relation to food hazards. J Risk Res 6 (3):267–283

Militello L, Patterson ES, Bowman L, Wears RL (2007) Information flow during crisis management: challenges to coordination in the emergency operations center. Cognit Technol Work 9(1):25–31

Mohr J, Spekman R (1994) Characteristics of partnership success: partnership attributes, communication behavior, and conflict resolution techniques. Strateg Manag J 15(2):135–152

Morgan MG, Fischhoff B, Bostrom A, Atman CJ (2002) Risk communication: a mental models approach. Cambridge University Press, Cambridge

Moss RH, Schneider SH (2000) Uncertainties in the IPCC TAR: recommendations to lead authors for more consistent assessment and reporting. In: Pachauri R, Taniguchi T, Tanaka K (eds) IPCC supporting material, guidance papers on the cross cutting issues of the third assessment report of the IPCC. pp 33–51

Orasanu J (1994) Shared problem models and flight crew performance. In: Johnston N, McDonald N, Fuller R (eds) Aviation psychology in practice. Aldershot, England, pp 255–285

Osman M (2013) A case study: dual-process theories of higher cognition—commentary on Evans & Stanovich (2013). Perspect Psychol Sci 8:248–252

Owen C (2013) Gendered communication and public safety: women, men and incident management. Aust J Emerg Manag 28(2):3–10

Owen C, Campus SB, Brooks B, Chapman J, Paton D, Hossain L (2013) Developing a research framework for complex multi-team coordination in emergency management. Int J Emerg Manag 9:1–17

Pascual R, Henderson S (1997) Evidence of naturalistic decision making in military command and control. In: Zsambok CE, Klein G (eds) Naturalistic decision making. Lawrence Erlbaum Associates, Mawhah, pp 217–226

Paton D (1996) Training disaster workers: promoting wellbeing and operational effectiveness. Disaster Prev Manag 5(5):11–18

Paton D (2003) Stress in disaster response: a risk management approach. Disaster Prev Manag 12:203–209. doi:10.1108/09653560310480677

Paton D, Flin R (1999) Disaster stress: an emergency management perspective. Disaster Prev Manag 8 (4):261–267

Paton D, Jackson D (2002) Developing disaster management capability: an assessment centre approach. Disaster Prev Manag 11:115–122. doi:10.1108/09653560210426795

Paton D, Kelso BA (1991) Disaster stress: the impact on the wives and family. Counselling Psychol Quart 4:221–227

Paton D, Norris K (2014) Vulnerability to work-related posttraumatic stress: family and organizational influences. In: Violanti JM (ed) Dying for the Job: police work exposure and health. Charles C. Thomas, Springfield, pp 126–141

Paton D, Johnston DM, Houghton BF (1998) Organisational response to a volcanic eruption. Disaster Prev Manag 7:5–13. doi:10.1108/09653569810206226

Paton D, Johnston DM, Houghton BF, Flin R, Ronan K, Scott B (1999) Managing natural hazard consequences: planning for information management and decision making. J Am Soc Prof Emerg Plan 6:37–47

Paton D, Ronan KR, Johnston DM, Houghton BF, Pezzullo L (2000) La Riduzione del Rischio Vulcanico: Integrare le prospettive psicologiche e geologiche. Psychomedia, Mental Health and Communication, 5, (May), [Online Serial], URL http://www.psychomedia. it/pm/grpind/social/paton-it.htm. Accessed 14 Oct 2014

Paton D, Violanti JM, Burke K, Gherke A (2009) Traumatic stress in police officers: a career length assessment from recruitment to retirement. Charles C Thomas, Springfield

Patt A, Dessai S (2005) Communicating uncertainty: lessons learned and suggestions for climate change assessment. Comptes Rendus Geosci 337(4):425–441

Pliske RM, McCloskey MJ, Klein G (2001) Decision skills training: facilitating learning from experience. In: Salas E, Klein G (eds) Linking expertise and naturalistic decision making. Lawrence Erlbaum Associates, Mahwah, pp 37–53

Pollock C, Paton D, Smith D, Violanti J (2003) Team resilience. In: Paton D, Violanti J, Smith L (eds) Promoting capabilities to manage posttraumatic stress: perspectives on resilience. Charles C. Thomas, Springfield, pp 74–88

Robert LP, Denis AR, Hung Y-TC (2009) Individual swift trust and knowledge-based trust in face-to-face and virtual team members. J Manage Inf Syst 26 (2):241–279

Rogalski J, Samurçay R (1993) A method for tactical reasoning (MTR) in emergency managment: analysis of individual acquisition and collective implementation. In: Rasmussen B, Brehmer B, Leplat J (eds) Distributed decision making: cognitive models for co-operative work. Wiley, New York, pp 287–298

Rosenthal U, 't Hart P (1989) Managing terrorism: the south Moluccan hostage takings. In: Rosenthal U, Charles MT, 't Hart P (eds) Coping with crises: the management of disasters, riots and terrorism. Charles C Thomas, Springfield, pp 340–366

Rovins JE, Doyle EEH, Huggins TJ (2014) 2nd integrated research on disaster risk conference—integrated disaster risk science: a tool for sustainability. In: Planet@Risk, vol 2, Issue 5, Special Issue for the Post-2015 Framework for DRR, Global Risk Forum GRF Davos, Davos. pp 332–336

Saaty TL (2008) Decision making with the analytic hierarchy process. Int J Serv Sci 1:83–98

Salas E, Stout RJ, Cannon-Bowers JA (1994) The role of shared mental models in developing shared situational awareness. In: Gilson RD, Garland DJ, Koonce JM (eds) Situational awareness in complex systems: proceedings of a Cahfa conference. Embry-Riddle Aeronautical University Press, Daytona Beach, FL, pp 298–304

Salas E, Cannon-Bowers JA, Johnston JH (1997) How can you turn a team of experts into an expert team? Emerging training strategies. In: Zsambok CE, Klein G (eds) Naturalistic decision making. Lawrence Erlbaum Associates, Mahwah, pp 359–370

Sarna P (2002) Managing the spike: the command perspective in critical incidents. In: Flin R, Arbuthnot K (eds) Incident command tales from the hot seat. Ashgate Publishing Limited, Aldershot, pp 32–57

Siegrist M, Cvetkovich G (2000) Perception of hazards: the role of social trust and knowledge. Risk Anal 20:713–719

Sloman SA (1996) The empirical case for two systems of reasoning. Psychol Bull 119(1):3–22

Slovic P, Finucane M, Peters E, MacGregor DG (2004) Risk as analysis and risk as feelings: some thoughts about affect, reason, risk, and rationality. Risk Anal 24(2):311–322

Smith R (2009) Research, science and emergency management: partnering for resilience. Tephra, community resilience: research, planning and civil defence emergency management. Ministry of Civil Defence & Emergency Management, Wellington, New Zealand, pp 71–78

Smithson M (1999) Conflict aversion: preference for ambiguity vs conflict in sources and evidence. Organ Behav Hum Dec Processes 79(3):179–198

Solana MC, Kilburn CRJ, Rolandi G (2008) Communicating eruption and hazard forecasts on Vesuvius, Southern Italy. J Volcanol Geotherm Res 172(3–4):308–314

Sparks RSJ (2003) Forecasting volcanic eruptions. Earth Planet Sci Lett 210(1–2):1–15

Teigen KH, Brun W (1999) The directionality of verbal probability expressions: effects on decisions, predictions, and probabilistic reasoning. Organ Behav Hum Dec Processes 80(2):155–190

Thompson VA (2013) Why it matters: the implications of autonomous processes for dual-process theories—commentary on Evans & Stanovich (2013). Perspect Psychol Sci 8:253–256

United Nations International Strategy fo Disaster Reduction (2007) Hyogo framework for action 2005–2015: Building the resilience of nations and communities to disasters. Extract from the final report of the world conference on disaster reduction (A/CONF:206/6). UNISDR Secretariat, Geneva. http://www.unisdr.org/files/1037_hyogoframeworkforactionenglish.pdf Last accessed 22 Oct 2014

United Nations International Strategy for Disaster Reduction (2015) Sendai framework for disaster risk reduction 2015–2030. UNISDR Secretariat, Geneva. http://www.preventionweb.net/files/43291_sendaiframeworkfordrren.pdf Last accessed 21 Dec 2015

Waugh WLJ, Streib G (2006) Collaboration and leadership for effective emergency management. Public Adm Rev 66(s1):131–140

Weber EU (2006) Experience-based and description-based perceptions of long-term risk: why global warming does not scare us (yet). Clim Change 77(1–2):103–120

Wiedemann P, Borner F, Schultz H (2008) Lessons learned: recommendations for communicating conflicting evidence for risk characterization. In: Wiedemann PM, Schultz H (eds) The role fo evidence in risk characterisation: making sense of conflicting data. Wilery-VCH Verlag GmbH & Co. KGaA, Weinheim, pp 205–213

Williams C, Dunn CE (2003) GIS in participatory research: assessing the impact of landmines on communities in North-west Cambodia. Trans Geo Inf Syst 7:393–410

Wilson KA, Salas E, Priest HA, Andrews D (2007) Errors in the heat of battle: taking a closer look at shared cognition breakdowns through teamwork. Hum Factors 49:243–256

Woo G (2008) Probabilistic criteria for volcano evacuation decisions. Nat Hazards 45(1):87–97

Zsambok C (1997) Naturalistic decision making: where are we now? In: Zsambok CE, Klein G (eds) Naturalistic decision making. Lawrence Erlbaum Associates, Mahwah, pp 3–16

Using Statistics to Quantify and Communicate Uncertainty During Volcanic Crises

Rosa Sobradelo and Joan Martí

Abstract

For decades, and especially in recent years, there has been an increasing amount of research using statistical modelling to produce volcanic forecasts, so that people could make better decisions. This research aims to add confidence by arming users with quantitative summaries of the chaos and uncertainty of extreme situations, in the form of probabilities—that is to say the measure of the likeliness that an event will occur.

Introduction

Probabilistic terms and associated jargon are often part of the working environment of volcanologists. Research activities about volcanic hazard and the quantification of volcanic risk even led to officially defining volcanic hazard in terms of probability (Blong 2000). The last decade has produced a comprehensive framework of studies, surveys and computer-assisted procedures for transforming field data into probabilities of occurrence of a particular scenario (Newhall and Hoblitt 2002; Marzocchi et al. 2004, 2008, 2010; Aspinall 2006; Martí et al. 2008; Neri et al. 2008; Sobradelo and Martí 2010, 2015; Sobradelo et al. 2013). Following the successful development of probabilistic tools, came the challenge of communicating their results. Research and operational strategies started to incorporate the enhancement of the communication of these probabilistic forecasts to decision makers and the public (Marzocchi and Woo 2007; Marzocchi et al. 2012; Sobradelo et al. 2014). At the same time, extensive work has been done in the psychological and sociological aspects on the perception and interpretation of uncertainty, for both volcanology and across other hazards. Despite this extensive use, sometimes there is confusion surrounding the statistical interpretation of probabilities, partly due to unclear statistical concepts like: What is a probability? What is statistical science? How much can I rely on a probability estimate? What are they used for?

R. Sobradelo (✉)
Willis Research Network (Willis Towers Watson), London, UK
e-mail: sobradelo@gmail.com

R. Sobradelo
UCL Hazard Centre Honorary Fellow Member, University College London, London, UK

J. Martí
Institute of Earth Sciences Jaume Almera, CSIC, Barcelona, Spain

Advs in Volcanology (2018) 571–583
https://doi.org/10.1007/11157_2017_15
Published Online: 15 July 2017

What is uncertainty? How does uncertainty and probability relate to each other? Why are statistics and probabilities sometimes misunderstood? Why is it that scientists and/or users (officials) don't fully appreciate the uncertainty surrounding a probability estimate?

In this chapter we try to address the above questions by focusing on the statistical meaning of probability estimates and their role in the quantification and communication of uncertainty. We hope to provide some insights into best practices for the use and communication of statistics during volcanic crises.

Quantifying and Communicating Uncertainty in Volcanology

Volcanology is by nature an inexact science. Deciphering the nature of unrest signals (volcanic reactivation), and determining whether or not an unrest episode may be an indication of a new eruption, requires knowledge on the volcano's past, current and future behaviour. In order to achieve such a complex objective experts in field studies, volcano monitoring, experimental and probabilistic modelling, amongst other, work together under pressure and tight time constrains. It is important that these stakeholders communicate on a level that caters for the needs and expectations of all disciplines; in other words, it is important to agree on a common technical language. This is particularly relevant when volcano monitoring is carried out on a systematic survey basis without continuous scientific scrutiny of monitoring protocols or interpretation of data.

By definition, uncertainty is the state of being uncertain. It is used to refer to something that is doubtful or unknown. It means lack of confidence about something. Hence, it is directly related to the amount of knowledge we have about a process. A forecast, in the form of a probability estimate, is an attempt to quantify this uncertainty and support decision-making. Forecasting potential outcomes of volcanic reactivation (unrest) usually implies high levels of scientific uncertainty. Anticipating whether a particular volcanic unrest will end with an eruption and where

(temporal and spatial uncertainty) requires scientific knowledge of how the volcano has behaved in the past, and scientific interpretation of precursory signals. Whilst this may be less challenging for volcanoes that erupt often, it is far more difficult for volcanoes with long eruptive recurrence and less data available, and even more so for those without historical records.

The main goal of volcano (eruption) forecasting is to be able to respond to questions of *how*, *where*, and *when* an eruption will happen (Sparks 2003). To address those questions we often use probabilities in an attempt to quantify the intrinsic variability due to the complexity of the process. The communication of those probabilities will have to adapt to the recipient of that information. Making predictions on the future behaviour of a volcano follows similar reasoning as in other natural phenomena (storms, landslides, earthquakes, tsunamis, etc.). Each volcano has its own characteristics depending on magma composition, physics, rock rheology, stress field, geodynamic environment, local geology, etc., which makes its behaviour unique. What is indicative in one volcano may not be relevant in another. All this makes the task of volcano forecasting challenging and difficult, especially when it comes to communicating uncertainty to population and decision-makers.

During a volcanic emergency, relevant questions are first how to quantify the uncertainty that accompanies any scientific forecast, and second, how to communicate it to policy-makers, the media and the public. Scientific communication during volcanic crises is incredibly challenging, with no standardized procedures on how this should be done among the stakeholders involved (scientists, governmental agencies, media and local populations). Of particular importance is the communication link between scientists and decision-makers (often Civil Protection agents). It is necessary to translate the scientific understanding of volcanic activity into a series of scenarios that are clear to decision-making authorities. Direct interaction between volcanologists and the general public is also important both during times of quiescence and activity. Information that comes directly from the scientific community has a special impact on risk perception and on the trust that people place on scientific information. Therefore, the effective management of a volcanic crisis requires the identification of practical actions, to improve communication strategies at different stages and across different stakeholders: scientists-to-scientists, scientists-to-technicians, scientists-to-Civil Protection, scientists-to-decision makers, and scientists-to-the general public.

The Role of Statistics and Probabilities in the Quantification of Uncertainty

Concepts, Definitions and Misconceptions

Formally speaking, *Statistics* is a body of principles and methods for extracting useful information from data, assessing the reliability of that information, measuring and managing risk, and supporting decision-making in the face of uncertainty. Rather than drowning in a flood of numbers, statistics helps to make better management decisions and gives a competitive advantage over intuition, experience and hunches alone.

Probability shows the likelihood, or chances, for each of the various future outcomes, based on a set of assumptions about how the world works. It allows handling randomness (uncertainty) in a consistent, rational manner and forms the foundation for statistical inference (drawing conclusions from data), sampling, linear regression, forecasting, and risk management.

With statistics, we go from observed data to generalizations about how the world works. For example, if we observe that the seven hottest years on record occurred in the most recent decade, we may conclude (perhaps without justification) that there is global warming. With probability, we start from an assumption about how the world works, and then figure out what type of data we are likely to see under that assumption. In the above example, we could assume the null hypothesis, H_0: *There is no global warming*, and then test how likely is it to observe the seven hottest years within the last decade if H_0 was true. We then use the observed data to look for significant statistical evidence to reject H_0 in favour of the alternative, H_1: *Some phenomena related to global warming may be ongoing*. To some extent, we could say that probability provides the justification for statistics.

However, there is no precise definition for probability. All attempts to define it must ultimately rely on circular reasoning. According to the Oxford Dictionary, probability is "the state of being probable; the extent to which something is likely to happen or be the case". Roughly speaking, the probability of a random event is the "chance" or "likelihood" that the event will occur. To each random event A we attach a number $P(A)$, called the probability of A, which represents the likelihood that A will occur. The three most useful approaches to obtaining a definition of probability are: the classical, the relative frequency, and the subjective (Jaynes 2003; Colyvan 2008), discussed further below.

The number of volcanic eruptions of magnitude greater than 1 in the next t years in a particular area is an example of a random variable, Y. When we try to quantify the value of Y we are implying that a *true value* exists, and we want to anticipate to it, so that we can make advanced decisions. That is, we want to estimate a range of values that we think will contain the *true value* of the random variable Y. The most common way of showing this range of values is by presenting a *best estimate* ± *confidence margin*. Here, we could distinguish between two types of uncertainty, the one surrounding the *best estimate*, type A, and the one that accounts for the level of confidence that we have in that best estimate, type B. It is not enough to provide a *best guess* (point estimate) for a parameter, we also need to say something about how far from the true parameter value such an estimator is likely to be. The confidence interval is one way of conveying our uncertainty about a parameter. With that, we report a range of numbers, in which we hope the true parameter will lie.

Measures of Uncertainty

Probability can be used as a measure of uncertainty, both type *A* and *B*. The way we understand probabilities depends on the degree of numeracy we have. It is common in our daily lives to make choices with some level of uncertainty, for instance, whether or not to order the fish of the day in a new restaurant, or whether to buy one or two bags of fruit in a new shop. To make those simple decisions, we unconsciously go through previous knowledge on similar experiences to work out some kind of *odds* of making the right choice. Suppose now that we are being rushed to make up our mind at the restaurant, we will have to rush our decision. The main difference between this and the decision of whether to evacuate a populated area threatened by a destructive volcanic event is the *penalty* or *loss* for making the wrong decision. In the first case, the loss is negligible to our daily lives, but a wrongly timed evacuation decision could have serious consequences. For this reason, the interpretation of *probability* must be in the context of *how much we are willing to lose if we make the wrong decision*. The difference between probability, the extent to which something is likely to happen; and risk, a situation involving exposure to danger; means that the relevance of a probability estimate for the occurrence of an event will depend on the associated risk, this is, on how much exposure to danger is in the occurrence of the event. Suppose the odds are one to ninety nine (1:99) that our car breaks down in the middle of a trip. We would most likely still take our family on that trip. Instead, suppose we are given the same odds for an airplane crash. We would most likely not want to take our loved ones on that plane. In both cases the probability is the same, but the risk is different. This illustrates how probability estimates must be interpreted in the context of their associated risk.

Clearly emotions, values, beliefs, culture and interpersonal dynamics play a significant role in decision-making processes. Extensive work in the field of psychology and sociology has examined perceptions and interpretation of uncertainty for both volcanology and across other hazards (weather, tsunami, operational earthquake forecasting, climate change) (Fischhoff 1994; Cosmides and Tooby 1996; Kuhberger 1998; Windschitl and Weber 1999; Bruine De Bruin et al. 2000; Gigerenzer et al. 2005; Patt and Dessai 2005; Risbey and Kandlikar 2007; Morss et al. 2008; Budescu et al. 2009; McClure et al. 2009; Joslyn et al. 2009; Mastrandrea et al. 2010; Jordan et al. 2011; Eiser et al. 2012; Doyle et al. 2014a, b). However, that is not the scope of this chapter. For the purpose of our argument, we focus on the 'rational side' of decision-making. That is, the quantification of uncertainty using statistical theory.

What makes statistics so unique is its ability to quantify uncertainty, so that statisticians can make a categorical statement about their level of uncertainty, with complete assurance. But the statements have to be made taking into account all possible factors (sources of uncertainty) and making sure the data are correctly selected to eliminate all sources of bias. These could have a significant impact and involve matters of life and death. So far we assumed that the probability estimates have been calculated using the right methods. For the restaurant or supermarket examples this could be a simple arithmetic mean. Forecasting the occurrence of a volcanic event will require more elaborated mathematical modelling. The accuracy in the probability estimate will depend largely on the model selection.

Disciplines and Schools of Thought

To quantify uncertainty using statistics there are three main disciplines statisticians rely on: (i) data analysis, (ii) probability, and (iii) statistical inference (Cooke 1991; Pollack 2003; Kirkup and Frenkel 2006). The first step is always the data analysis, that is, the gathering, display and summary of the data. In the case of volcanoes, we look at past and monitoring data, and we make the necessary adjustments for any inconsistencies (e.g.: Sobradelo and Martí 2015). The second step is the formal study of the laws of chance, also called the laws of probability, whose birthplace is in the 17th century for no other reason than to be used in gambling (Cooke 1991). Probabilities are the result of applying probability models to describe the world, and this is done using the concept of random variables, that is, the numerical outcome of a random experiment or a random process we are trying to understand, so that we can forecast its future outcome (height, weight, income, eruptive events in the last 500 years, number of seismic events in one day, etc.). Finally, we use the above so that we can make inferences in the real world with a certain degree of confidence (Rice 2006).

Approaches to developing probability models, associated with different schools of thought, are: (1) the classical, based on gambling ideas, which assumes that the game is fair and all elementary outcomes have the same probability; (2) the relative (objective) frequency approach which believes that if an experiment can be repeated, then the probability estimate that an event will occur is equivalent to the proportion of times the event occurs in the long run; and (3) the personal (subjective) probability approach which believes that most of the events in life are not repeatable (Cooke 1991; Jaynes 2003). They base the probability on their personal belief of the likelihood of an outcome, and then update that probability as they receive new evidence (Cosmides and Tooby 1996). An objectivist uses either the classical or the frequency definition of probability. Subjectivists, also called Bayesians, apply formal laws of chance to their own personal probabilities. What makes the Bayesian approach subjective is the choice of models and a priori beliefs to define the prior probabilities, even if the rules and observed data to update and compute the posterior probabilities are quite "objective". The Bayesian approach claims that any state of uncertainty can be described with a probability distribution, making it suitable for the study of volcanic areas where very little or no data exists, other than theoretical models or expert scientific beliefs. These initial probabilities get updated each time new information arrives, making the approach quite dynamic and easy to apply.

For many years there has been controversy over the "frequentist" versus "Bayesian" methods.

However, neither the Bayesian nor the frequentist approaches are universally applicable (Jaynes 2003). For each situation, some approaches and models are more suitable than others to produce probability estimates as accurately as possible with high confidence. It is the task of the statistician to decide and justify the model selection to ensure reliability of the results. But a brilliant analysis is worthless unless the results are successfully communicated, including its degree of statistical uncertainty.

Often presented as an alternative to the probabilistic approach, is the deterministic approach. Events are completely determined by cause-effect chains (causality), without any room for random variation. Here, a given input will always produce the same output, as opposed to probabilistic models that use ranges of values for variables in the form of probability distributions. This approach is sometimes used in fields with a lot of data, like in weather forecasting, or where the underlying process can be explained with physics-based models, such as in seismology. In any case, the reliability of probabilistic versus deterministic forecasts is sometimes a cause of debate, and is often a mixed of both, a deterministic and a probabilistic approach, the preferred option.

How Reliable Is a Forecast: Data and Methodology

By giving an expected value for a forecast we are already quantifying a measure of uncertainty. This value will have an interpretation based on the degree of confidence which the estimate is made with, which will depend on the type, amount, quality and consistency of the evidence upon which the estimate is made, usually past data or theoretical models.

The degree of confidence, or certainty, is quantified and expressed via the variance or standard deviation (squared root of the variance). Suppose we have three measurements of a random process (e.g. inter-event time in years) of 2, 3, and 4 years, and want to draw some conclusion about the inter-event time based on these values.

We use 3 years, a simple arithmetic mean, as the estimate of the inter-event time. The three measurements are equally distant and symmetrical around the mean. The variance, which measures the dispersion of the values around the mean, is 1, and the median, which is the value in the middle, is 3 as well as the mean. Suppose we do the same exercise with measurements 1, 3, and 5, we still get a mean of 3, but now we can see the values 1, and 5 are two units away from the mean, and so the variance, as a measure of dispersion around the mean, is now 4, instead of 1. Note, however, that the values are symmetrically distributed around the mean, and that the mean and median are still the same as before, 3. The only thing that has changed is the variance, now larger. The lower the variability around the point estimate, the more reliable is our estimate. Let's take a sample with 10 measurements: 1, 1.4, 2, 2.1, 2.2, 2.3, 3, 4, 5, 7. The estimated inter event time, based on a simple arithmetic mean, is still 3, but we based this estimate on 10 rather than 3 observations. The more data we have to compute our estimates, the more confident we are in these results (Rice 2006).

Apart from the reliability of the data to produce an estimate, a crucial aspect of a forecast is the correct choice of methodology to model this. Most of the time we do not know the underlying distribution of a random process (e.g. number of volcanic eruptions in a time interval and particular area, assumed to be random), and so we make assumptions to help us find a function within a family of known distributions (Normal or Gaussian, Exponential, Binomial, Beta, Poisson, Chi-Squared, Log-normal, etc.) that would be suitable to model this unknown process (see Rice 2006; Gonick and Smith 2008; McKillup and Dyar 2010 for details on these distributions). This facilitates making inferences and forecasts based on the conveniently known properties of these functions. The choice of the distribution family depends on the characteristics of the sample data (how many observations are there, whether it is a symmetrical or a skewed distribution, what type of measurement was used, etc.). To select the most appropriate distribution, it is important that the data is an unbiased and representative sample of the population. Therefore, the data gathering process and a preliminary and exhaustive analysis of the dataset are crucial to reduce uncertainty and increase confidence in the final results. Needless to say, the choice of distribution and assumptions about the sample data add uncertainty to the results, and must be taken into account when presenting the final outcome.

Arithmetical means are pure descriptive measures used to sum up the information from the data sample. In practice, we would not use a simple arithmetical mean to estimate probabilities and make inferences about complex processes. There are a large number of statistical modelling techniques (not the scope of this chapter) based on the type of data we have, its distribution, quality and quantity and the type of question we want to answer. In the end, the reliability of the probability estimate (whereas an inter event time of 3 years or not) will depend on the accuracy, reliability and amount of data used to reach that conclusion, together with the statistical model and approach. That is why a probability estimate should always be presented with some measure of its variability (estimated error, usually given by the variance or standard deviation) and it should be made clear that it is an estimate based on the available data, and that we have assumed that a future behaviour of the random event will follow the same pattern we have observed in this dataset. This might in fact not be the case, and that is why sometimes we hear about time series data being "stationary or not", meaning that depending on what time interval the data comes from, the pattern observed may be different. In short, there are many assumptions and sources of uncertainty around a probability estimate that have to be taken into consideration when interpreting probability.

Taking a bigger picture view, ultimately all we are doing is drawing some general conclusions about an unknown process (the inter-event time) from some samples of observations. We do not have access to all the possible observations of this process, but still want to anticipate the future value of this event, so we can be better prepared should the event strike. This is the reason why we use statistical approaches to model random events, unless we can see into the future, a probability estimate can never be either 0 or 100%.

Using Probabilities to Communicate Uncertainty

Since the late 1990s there has been significant focus on improving communications during volcanic crises (IAVCEI 1999; McGuire et al. 2009; Aspinall 2010; Donovan et al. 2012a, b; Marzocchi et al. 2012; Sobradelo et al. 2014). A common factor that emerges is the value of probabilities as a way to communicate scientific forecasts and their associated uncertainties, for natural hazards in general (Cooke 1991; Colyvan 2008; Stein and Stein 2013), or more specific for volcano forecasting (Aspinall and Cook 1998; Marzocchi et al. 2004; Aspinall 2006; Sobradelo and Martí 2010; Marzocchi and Bebbington 2012; Donovan et al. 2012c). However, it also requires the need to communicate the uncertainty that accompanies any forecast on the future behaviour of a natural system.

Making predictions on the future behaviour of a volcano involves analysis of past data, monitoring of the current situation and identification of possible scenarios. Quite often, these predictions are challenging to quantify and communicate due to lack of data and past experience. An added source of complexity is when the probability estimates are very small, <1%. Most lay people are not familiar with decimals or small fractions. A layman will easily understand a probability of 0.2 or 20%, but not so well one of 0.0002 or 0.02%, even when both are associated to the same level of risk. Scientists responsible for the communication of volcanic forecasts have the difficult task of selecting the scientific language to deliver a clear message to a non-scientific audience.

The uncertainty that accompanies the identification and interpretation of eruption precursors derives from the unpredictably of the volcano as a natural system (aleatory or deep uncertainties) and from our lack of knowledge on the behaviour of the system (epistemic or shallow uncertainties) (Cox 2012; Stein and Stein 2013). These

uncertainties will depend on how well we know the volcanic system. Active volcanoes with high eruption frequencies can be more easily predicted (i.e. they are reasonably well known and so past events are good predictors of future ones, shallow uncertainties). In contrast, deep uncertainties are associated to probability estimates based on poorly known parameters or poor understanding of the system, this is usually the case for volcanoes characterised by low eruption frequencies.

In everyday life we are often quite unaware that we use probabilities (commonly known as "common sense") to evaluate the degree of uncertainty we face. The question is whether we prefer or understand better the mathematical expression of probability (e.g.: 20% chance of an event occurring) or more verbal statements such as likely, improbable, certainly, to make our decisions. Greater precision does not necessarily imply greater understanding of what the message really is, as it will be perceived differently (Slovic 2016).

Some countries, like USA, prefer to use probabilities to express uncertainties with weather forecasts, while some European countries prefer to use verbal expressions. In both cases, people react according to the forecast. There are different ways in which probabilities (and uncertainties) can be described. These include words, numbers, or graphics. The use of words to explain probabilities tend to use language that appeals to people's intuition and emotions (Lipkus 2007). However, it usually lacks precision as it tends to introduce significant ambiguity by the use of non-precise words such us "probable", "likely", "doubtful", etc. A probability is the "measure" of the likeliness that an event will occur, so it makes sense to expect a numerical value (e.g. percentages) associated to that measure. However, in volcanology most of the time there is insufficient observational data to present probabilistic forecasts with enough level of confidence. Using only numerical expressions may fail when the audience has a low level of numeracy. The interpretation of probabilistic terms can vary greatly depending on the educational level of the receptor and whether verbal or numerical expressions are used (Budescu et al. 2009; Spiegelhalter et al. 2011; Doyle et al. 2014a; Gigerenzer 2014). To minimise this problem, a combination of verbal uncertainty terms (e.g.: very likely) with quantitative specifications (e.g.: <90% probability) has been recommended, for example, to better understand results from Intergovernmental Panel on Climate Change (IPCC) (Budescu et al. 2009, 2012). Climate scientists working within the IPCC have adopted a lexicon to communicate uncertainty through verbal probability expressions ranging from "very likely", "likely", "about as likely as not unlikely", "very unlikely" and "exceptionally unlikely" to refer to probabilities (e.g. IPCC 2005, 2007). The terms are assigned specific numerical meanings but are typically presented in verbal format only, so that a probability of occurrence of 1% will be interpreted as "very unlikely" for that particular event, and a probability of 66% will be seen as "likely" for the event to happen. Similarly, anything in the range of 33–66% would be perceived as "about as likely as not unlikely".

Since 2011 it has been increasingly common to use graphics to represent probabilities in natural hazards (Kunz et al. 2011; Spiegelhalter et al. 2011; Stein and Geller 2012). The advantage of communicating uncertainties (or probabilities) visually is that people are everyday better prepared and trained to use and understand infographics, as an immediate consequence of the globalised use of internet and informatics, and a graphic can be adapted to stress the importance of the content of the communication and can be adapted to the needs and capabilities of the audience (Spiegelhalter et al. 2011).

In addition to considering the way probabilities (and uncertainties) are communicated, there is a need to consider the local context of the particular society in which the volcanic crisis is occurring. "Odds" is an expression of relative probability that is well understood by many communities (e.g. gambling, games of chance) and can be effective also to communicate volcano forecasting if it is correctly adapted for the purpose. Regulations (i.e. legal and commonly accepted norms) frequently determine the articulation of uncertainty and risk used to manage environmental and natural hazards. Finally, culture is of key importance in communication (Oliver-Smith and Hoffmann 1999; Eiser et al. 2012). The way in which risk is

perceived may change depending on cultural beliefs of each society, and in the same way the cultural diversity of societies facing a volcanic threat may imply that communication methods that work in one country or culture may not work in another. Therefore, it is important to investigate and gain in-depth understanding of the particular cultural aspects of each society in order to define the best communication procedures and languages in each case. There are numerous studies that demonstrate the importance of public education, pre-crisis education programmes, and risk perception to better understand scientific communication during crisis (e.g. Bird et al. 2009; Budescu et al. 2012; Dohaney et al. 2015). Most of them agree that better educated populations on natural hazards understand better risk communication and behave in a more orderly way for managing a crisis. There are additional sociological and qualitative aspects to consider when communicating probabilities beyond the scope of this chapter, but address issues around risk perception, trust, decision-making, and managing disasters e.g. Kilburn 1978; Fiske 1984; Tazieff 1977; Paton et al. 1999; Chester et al. 2002; Sparks 2003; Haynes et al. 2007, 2008; Baxter et al. 2008; Solana et al. 2008; Fearnley 2013; Doyle et al. 2015.

What Should Be Communicated?

The key questions focus around what can be forecasted. Should the forecasting of the outcome of a volcano be determining whether it will erupt of not? How big or explosive will it be? When? Where? What is the dimension of the problem? These are basic questions that civil protection asks to the scientist once an alert has been declared, and the process of managing a volcanic crisis has started (IAVCEI 1999; McGuire et al. 2009; Aspinall 2010; Donovan et al. 2012a, b; Marzocchi et al. 2012; Sobradelo et al. 2014). Usually, scientists can answer these questions with approximations (probabilities) based on knowledge of previous cases from the same volcano, or from other volcanoes with similar characteristics, knowledge of the past eruptive history of the volcano, warning signals (geophysical and geo-chemical monitoring), and knowledge about the significance of these warning signs. Whilst giving probabilities as an outcome of a volcano forecast may be relatively easy for the scientist (depending on the degree of information available), it may not be fully understood by the decision-maker or any other recipient of such information. It is necessary to find a clear and precise way to communicate this information between scientists and key decision-makers, to avoid misunderstandings and misinterpretations that could lead to an incorrect management of the volcanic emergency and, consequently, to a disaster.

In recent years, a way used to improve the communication of statistics, as well as decision-maker needs, is through the development of exercises where a volcanic crisis is simulated and all key players involved in risk management, such as scientists, civil protection, decision-makers, population and media are invited to participate, as in a real case. Exercises have been carried out at different volcanoes such as Vesuvius (MESIMEX, Barberi and Zuccaro 2004), or Campi Flegrei, Cotopaxi and Dominica (VUELCO Project, www.vuelco.com), New Zealand (DEVORA), among others. These simulations facilitate interaction and cooperation between the stakeholders, and the sharing and exchanging of procedures, method-ologies and technologies among them, including scientific communication. They present an opportunity for learning the exact role and responsibilities that each key player has in the management of a volcanic crisis, as well as exchanging concerns and feedback on specific matters.

Whilst volcanic forecasts centre on scientific data and probabilities as much as possible, sci-entists may also recommend safe behaviour directly to the public, providing advice that saves people's lives (e.g. going up a hill if a lahar threatens). Often this is beyond the legal requirements of the scientists, who are required to comment on the volcanic science only, but they could feel a moral duty to assist (Fearnley 2013). However, this should not imply or be confused with making decisions on how to manage a volcanic emergency (e.g. evacuation),

as this frequently falls under the remit of civil protection (or other such government organisa-tions), although in some countries such as Indonesia the scientists and the civil protection organisations work together rather than having distinct roles; it is dependent on the governance structures of the country.

When Should a Volcano Forecast Be Communicated?

Ideally, forecasts should be communicated as early as possible, and then with increasing fre-quency if, or when, an eruption nears. This means there should be a permanent flow of information between scientists, the vulnerable populations, and policy-makers on the eruptive characteristics of the volcano, its current state of activity, and its associated hazards, even when volcanoes do not show signs for alarm. This is to aid preparation for when an emergency starts and things need to move much faster. However, in many cases sci-entific communication in hazard assessment and volcano forecasting is just restricted to volcanic emergencies. When volcanic unrest starts and escalates, the origin of this unrest needs to be investigated to assess the level of hazard expected. Good detection and interpretation of precursors will help predict what will happen with a con-siderable degree of confidence. This means that scientific communication during a volcanic crisis needs to be constant and permanently updated with the arrival of each new piece of data. The longer it takes to make a decision, the greater the potential losses are likely to be as vulnerability increases. This constitutes the main challenge in communicating forecasts and probabilities during a volcanic crisis. In essence, the relationship between the decrease of uncertainty in the inter-pretation of the warning signs of pre-eruptive processes to acceptable (reliable) levels, and the time required to make a correct decision, is a function of the degree of the scientific knowledge of the volcanic process and of the effectiveness of scientific communication. Therefore, scientific communication during a volcanic crisis needs to be effective from the start.

Conclusion

In order to improve scientific communication during a volcanic crisis it is recommended that the communication protocols and procedures used by the different volcano observatories and scientific advisory committees are compared for each level of communication: scientist-scientist, scientist-technician, scientist-Civil Protection, scientist-general public. Experience from other natural hazards helps, as do clear and effective ways to show probabilities and associated uncertainties. Although each cultural and socio-economic situation will have different communication requirements, comparing different experiences will help improve each particular communication approach, thus reducing uncertainty in communicating volcano forecasts.

Finally it is worth mentioning that a crucial aspect in facilitating risk communication is education. This, however, is a long-term task that requires to be conducted permanently in societies threatened by natural hazards. Risk perception depends on cultural beliefs but also on whether or not a society has been educated on its natural environment and potential hazards. In the same way scientific communication is better perceived and understood when the population have previous knowledge on the existence and potential impacts of natural hazards. There are numerous studies that demonstrate the importance of public education, pre-crisis education programmes, and risk perception to better understand scientific communication during crisis (e.g. Bird et al. 2009; Budescu et al. 2012; Dohaney et al. 2015). Most of them agree that better educated populations on natural hazards understand better risk communication and behave in a more orderly way for managing a crisis. Therefore, best practices on communication should also consider improving education of population on natural hazards, their potential impacts and the ways to minimise the associated risks, as well as on how to behave during the implementation of emergency plans in a crisis.

References

Mastrandrea MD et al (2010) Guidance note for lead authors of the IPCC fifth assessment report on consistent treatment of uncertainties. IPCC cross-working group meeting on consistent treatment of uncertainties. Jasper Ridge, CA, USA, 6–7 July 2010. Accessed from http://www.ipcc.ch/pdf/supporting-material/uncertainty-guidance-note.pdf

Jordan TH et al (2011) Operational earthquake forecasting: state of knowledge and guidelines for utilization. A report by the international commission on earthquake forecasting for civil protection. Ann Geophys 54(4). doi:10.4401/ag-5350

Aspinall WP (2006) Structured elicitation of expert judgment for probabilistic hazard and risk assessment in volcanic eruptions. In: Mader HM et al.(eds) Statistics in volcanology. Special Publication of IAVCEI # 1, Geological Society of London, pp 15–30

Aspinall WP (2010) A route to more tractable expert advice. Nature 463:294–295

Aspinall W, Cook R (1998) Expert judgement and the Montserrat volcano eruption. In: Mosleh A et al (eds) Proceedings of the 4th international conference on probabilistic safety assessment and management PSAM4, 13–14 Sept. Springer, New York, pp 2113–2118

Barberi F, Zuccaro G (2004) Soma Vesuvius MESIMEX. Final technical implementation report, 2004/393427 (ec.europa.eu/echo/files/policies/prevention:preparedness/mesimex.pdf)

Baxter PJ, Aspinall WP, Neri A, Zuccaro G, Spence RJS, Cioni R, Woo G (2008) Emergency planning and mitigation at Vesuvius: a new evidence-based approach. J Volcanol Geoth Res 178:454–473

Bird DK, Gisladottir G, Dminey-Howes D (2009) Resident perception of volcanic hazards and evacuation procedures. Nat Hazards Earth Syst Sci 9:251–266. www.nat-hazards-earth-syst-sci.net/9/251/2009

Blong R (2000) Volcanic hazards and risk management. In: Sigurdsson H et al (eds) Encyclopedia of Volcanoes. Academic, San Diego, pp 1215–1227

Bruine De Bruin W et al (2000) Verbal and numerical expressions of probability: "It's a fifty-fifty chance'. Organ Behav Hum Decis Process 81(1):115–131

Budescu DV, Broomell S, Por HH (2009) Improving communication of uncertainty in the reports of the intergovernmental panel on climate change. Psychol Sci 20(3):299–308

Budescu DV, Por HH, Broomell S (2012) Effective communication of uncertainty in the IPCC reports. Clim Change 113:181–200

Chester DK, Dibben CJL, Duncan AM (2002) Volcanic hazard assessment in Western Europe. J Volcanol Geoth Res 115:411–435

Colyvan M (2008) Is probability the only coherent approach to uncertainty? Risk Anal 28(3):645–652

Cooke RM (1991) Experts in uncertainty: opinion and subjective probability in science. Oxford University Press, Oxford

Cosmides L, Tooby J (1996) Are humans good intuitive statisticians after all? Rethinking some conclusions from the literature on judgment under uncertainty. Cognition 58(1):1–73

Cox lA Jr (2012) Confronting deep uncertainties in risk analysis. Risk Anal 32:1607–1629

Dohaney J, Brogt E, Kennedy B, Wilson TM, Linsay JM (2015) Training in crisis communication and volcanic eruption forecasting: design and evaluation of an authentic role-play simulation. J Appl Volcanol 4:12. doi:10.1186/s13617-015-0030-1

Donovan AR, Oppenheimer C, Bravo M (2012a) Science at the policy interface: volcano monitoring technologies and volcanic hazard management. Bull Volcanol 74:1–18

Donovan AR, Oppenheimer C, Bravo M (2012b) Social studies of volcanology: knowledge generation and expert advice on active volcanoes. Bull Volcanol 74:677–689

Donovan AR, Oppenheimer C, Bravo M (2012c) The use of belief-based probabilistic methods in volcanology: scientists' views and implications for risk assessments. J Volcanol Geotherm Res 247–248:168–180

Doyle EEH, McClure J, Johnston DM, Paton D (2014a) Communicating likelihoods and probabilities in forecasts of volcanic eruptions. J Volcanol Geoth Res 272 (2014):1–15

Doyle EEH, McClure J, Paton D, Johnston DM (2014b) Uncertainty and decision-making: volcanic crisis scenarios. Int J Disaster Risk Reduct 10:75–101

Doyle EEH, Paton D, Johnston D (2015) Enhancing scientific response in a crisis: evidence-based approaches from emergency management in New Zealand. J Appl Volcanol 4:1

Eiser JR, Bostrom A, Burton I, Johnston DM, McClure J, Paton D, van der Pligt J, White MP (2012) Risk interpretation and action: a conceptual framework for responses to natural hazards. Int J Disaster Risk Reduct 1:5–16

Fearnley CJ (2013) Assigning a volcano alert level: negotiating uncertainty, risk, and complexity in decision-making processes. Environ Plan A 45(8): 1891–1911

Fischhoff B (1994) What forecasts (seem to) mean. Int J Forecast 10:387–403

Fiske RS (1984) Volcanologists, journalists, and the concerned local public: a tale of two crises in the eastern Caribbean: in studies in geophysics explosive volcanism: inception, evolution and hazards. National Academy Press, Washington, pp 170–176

Gigerenzer G (2014) Risky savvy: how to make good decisions. Penguin, 312 pp

Gigerenzer G, Hertwig R, van den Broek E, Fasolo B, Katsikopoulos KV (2005) "A 30% chance of rain tomorrow": how does the public understand probabilistic weather forecasts? Risk Anal 25(3)

Gonick L, Smith W (2008) The cartoon guide to statistics. Paw Prints, 240 pp

Haynes K, Barclay J, Pidgeon N (2007) The issue of trust and its influence on risk communication during a volcanic crisis. Bull Volc 70(5):605–621

Haynes K, Barclay J, Pidgeon N (2008) Whose reality counts? Factors affecting the perception of volcanic risk. J Volcanol Geoth Res 172(3–4):259–272

IAVCEI Subcommittee for Crisis Protocols (1999) Professional conduct of scientists during volcanic crises. Bull Volcanol 60:323–334, comment and reply in Bull Volcanol 62:62–64

Intergovernmental Panel on Climate Change (2005) Guidance notes for lead authors of the IPCC fourth assessment report on addressing uncertainties. Retrieved from http://www.ipcc.ch/pdf/supporting-material/uncertainty-guidance-note.pdf

Intergovernmental Panel on Climate Change (2007) A report of working group I of the intergovernmental panel on climate change: summary for policymakers. Retrieved from http://www.ipcc.ch/pdf/assessment-report/ar4/wg1/ar4-wg1-spm.pdf

Jaynes ET (2003) Probability theory: the logic of science. Cambridge University Press, Cambridge, 753 pp

Joslyn SL, Nadav-Greenberg L, Taing MU, Nichols RM (2009) The effects of wording on the understanding and use of uncertainty information in a threshold forecasting decision. Appl Cognitive Psychol 23(1): 55–72

Kilburn CRJ (1978) Volcanoes and the fate of forecasting. New Scientist 80:511–513

Kirkup L, Frenkel RB (2006) An introduction to uncertainty in measurement. Cambridge University Press, Cambridge, 248 pp

Kuhberger A (1998) The influence of framing on risky decisions: a meta-analysis. Organ Behav Hum Dec 75 (1):23–55

Kunz M, Gret-Regamey A, Hurni L (2011) Visualization of uncertainty in natural hazards assessments using an interactive cartographic information system. Nat Hazards 59:1735–1751

Lipkus M (2007) Numeric, verbal, and visual formats of conveying health risks: suggested best practices and future recommendations. Med Decis Making 27(5): 696–713

Martí J, Aspinall WP, Sobradelo R, Felpeto A, Geyer A, Ortiz R, Baxter P, Cole P, Pacheco JM, Blanco MJ, Lopez C (2008) A long-term volcanic hazard event tree for Teide-Pico Viejo stratovolcanoes (Tenerife, Canary Islands). J Volcanol Geotherm Res 178:543–552

Marzocchi W, Bebbington M (2012) Probabilistic eruption forecasting at short and long time scales. Bull Volc 74:1777–1805

Marzocchi W, Woo G (2007) Probabilistic eruption forecasting and the call for an evacuation. Geophys Res Lett 34

Marzocchi W, Sandri L, Gasparini P, Newhall C, Boschi E (2004) Quantifying probabilities of volcanic events: the example of volcanic hazard at Mount Vesuvius. J Geophys Res 109. doi:10.1029/2004JB003155

Marzocchi W, Sandri L, Selva J (2008) BET EF: a probabilistic tool for long- and short-term eruption forecasting. Bull Volcanol 70(5):623–632

Marzocchi W, Sandri L, Selva J (2010) BET VH: a probabilistic tool for long-term volcanic hazard assessment. Bull Volcanol 72:705–716

Marzocchi W, Newhall C, Woo G (2012) The scientific management of volcanic crises. J Volcanol Geotherm Res 247–248:181–189

McClure J, White J, Sibley CG (2009) Framing effects on preparation intentions: distinguishing actions and outcomes. Disaster Prev Manage 18:187–199

McGuire WJ, Solana MC, Kilburn CRJ, Sanderson D (2009) Improving communication during volcanic crises on small, vulnerable islands. J Volcanol Geotherm Res 183(1–2):63–75

McKillup S, Dyar MD (2010) Geostatistics explained: an introductory guide for earth scientists. Cambridge University Press, Cambridge, 414 pp

Morss RE, Demuth JL, Lazo JK (2008) Communicating uncertainty in weather forecasts: a survey of the U.S. Public. Weather Forecast 23(5):974

Neri A, Aspinall WP, Cioni R, Bertagnini A, Baxter PJ, Zuccaro G, Andronico D, Barsotti S, Cole PD, Esposti Ongaro T, Hincks TK, Macedonio G, Papale P, Rosi M, Santacroce R, Woo G (2008) Developing an event tree for probabilistic hazard and risk assessment at vesuvius. J Volcanol Geoth Res 178:397–415

Newhall C, Hoblitt R (2002) Constructing event trees for volcanic crises. Bull Volc 64(1):3–20

Oliver-Smith A, Hoffmann SM (eds) (1999) The angry earth: disaster in anthropological perspective. Routledge, New York

Paton D, Johnston D, Houghton B, Flin R, Ronan K, Scott B (1999) Managing natural hazard consequences: information management and decision making. J Am Soc Prof Emerg Managers 6:37–48

Patt A, Dessai S (2005) Communicating uncertainty: lessons learned and suggestions for climate change assessment. Comptes Rendus Geosci 337(4):425–441

Pollack HN (2003) Uncertain science... uncertain world, UK. Cambridge University Press, Cambridge, p 256

Rice JA (2006) Mathematical statistics and data analysis. Duxbury Press, Belmont

Risbey JS, Kandlikar M (2007) Expressions of likelihood and confidence in the IPCC uncertainty assessment process. Clim Change 85(1–2):19–31

Slovic P (2016) The perception of risk. Routledge

Sobradelo R, Martí J (2010) Bayesian event tree for long-term volcanic hazard assessment: application to Teide-Pico Viejo stratovolcanoes, Tenerife, Canary islands. J Geophys Res 115. doi:10.1029/2009JB006566

Sobradelo R, Martí J (2015) Short-term volcanic hazard assessment through Bayesian inference: retrospective application to the Pinatubo 1991 volcanic crisis. J Volcanol Geoth Res 290:1–11

Sobradelo R, Bartolini S, Martí J (2013) HASSET: a probability event tree tool to evaluate future volcanic scenarios using Bayesian inference. Bull Volcanol 76(1):1–15

Sobradelo R, Martí J, Kilburn CRJ, López C (2014) Probabilistic approach to decision-making under uncertainty during volcanic crises: retrospective application to the El Hierro (Spain) 2011 volcanic crisis. Nat Hazards. doi:10.1007/s11069-014-1530-8

Solana MC, Kilburn CRJ, Rolandi G (2008) Communicating eruption and hazard forecasts on Vesuvius, Southern Italy. J Volcanol Geoth Res 172:308–314

Sparks RSJ (2003) Forecasting volcanic eruptions. Earth Planet Sci Lett 210(1–2):1–15

Spiegelhalter D, Pearson M, Short I (2011) Visualizing uncertainty about the future. Science 333(6048):1393–1400

Stein S, Geller RJ (2012) Communicating Uncertainties in natural hazard forecasts. Eos 93(38):361–362

Stein S, Stein JL (2013) How good do natural hazard assessments need to be? GSA Today 23(4/5):60–61

Tazieff H (1977) La Soufrière, volcanology and forecasting. Nature 269:96–97

Windschitl PD, Weber EU (1999) The interpretation of "likely" depends on the context, but "70%" is 70%—right? The influence of associative processes on perceived certainty. J Exp Psychol Learn 25(6):1514–1533

Insurance and a Volcanic Crisis—A Tale of One (Big) Eruption, Two Insurers, and Innumerable Insureds

Russell Blong, Catherine Tillyard and George Attard

Abstract

Although probabilistic insurance loss models, particularly for ash fall, are currently being developed volcanic risk has been widely ignored by insurers and policy holders alike. Volcanic eruption cover is often grouped in insurance and reinsurance policies with earthquake and tsunami cover. Many volcanic eruptions include several perils occurring in different spaces around the volcano, with widely varying intensities and consequences, sometimes all at once, sometimes sequentially, and sometimes repeatedly. Given the possibly large differences in hazard characteristics, event durations and potential losses the policy alignment with earthquake and tsunami covers can be unfortunate. Does 'volcanic activity' have the same meaning as 'volcanic eruption'? Do the terms 'ash fall' and 'pyroclastic fall' have identical meanings to an insurer—or to a volcanologist? Some policies cover all volcanic perils while others include only named volcanic perils such as pyroclastic flows, ash falls, and/or lava flows. Often the intent of the coverage is not clear—were some volcanic perils missing from a list excluded by accident or design? Does a policy that covers damage occasioned by a fall of volcanic ash also cover the cost of clean-up, removal, transport and appropriate storage of the ash—even if the fall of 5–10 mm of ash causes almost no property damage? Clear communication between the insurance sector and policy holders (and the media) is dependent upon informed understanding of the nature of volcanic perils and volcanic eruptions, insurance wordings, and the potential losses to property and business interruption covers. This chapter explores these issues using examples of policy wordings, evidence from past eruptions, insurance case law, and potential losses in future eruptions.

R. Blong (✉)
Aon Benfield, Sydney, Australia
e-mail: russell.blong@aonbenfield.com

C. Tillyard
London, UK

G. Attard
Singapore, Singapore

Advs in Volcanology (2018) 585–599
https://doi.org/10.1007/11157_2016_42
© The Author(s) 2017
Published Online: 15 March 2017

1 Introduction

Our aim is to highlight the challenges presented by volcanic eruptions with respect to insurance and communication. In particular, we wish to provide stakeholders with a range of perspectives on (i) the important issues as the insurance industry might see them; and (ii) the (possible) views of some other stakeholders including the policy holder, re/insurer and regulator. We begin by focussing on a single realistic—yet hypothetical—disaster scenario for a large (VEI 6) eruption, the sort of eruption that might occur somewhere in the world on average once in 50 years or so (Deligne et al. 2010). Clearly, this is just one eruption scenario out of the thousands that are possible—even more likely—but it emphasises that there are numerous insurance-related issues to consider. Our main concern is with the possible approaches of two insurance companies at opposite ends of the corporate resilience spectrum—one that has an embedded strategy to quickly adapt to disruptions while maintaining continuous business operations and safeguarding people, assets and overall brand equity, and another that has considered risk resilience in less detail and may have difficulty continuing to function (or remaining solvent) in the aftermath of a large event where thousands of policyholders want to make a claim.

We then delve into three insurance-related issues:

- modelling and communicating risk from volcanic activity
- the Contract Wordings used in insurance policies and what they might mean in relation to an eruption and its aftermath; and
- examples of insurance case law from around the world and what these court determinations might mean for insurers, policy holders and others experiencing the consequences of a large eruption.

While the eruption scenario and its insurance consequences are set in the (near) future, the details are firmly grounded in (recent) experience.

For the purposes of this chapter, we have not considered governments' role in bridging the insurance protection gap or disaster risk financing, nor have we delved deeply into the ability of insurance as a signalling device to drive risk mitigation and risk management practices. In addition, we have not considered broader macroeconomic impacts—natural hazards have unexpected consequences beyond direct economic and insured loss including broad disruption from operational challenges, effects on counterparties, trading relationships and financial markets.

2 Modelling and Communicating Volcanic Risk

Assessment and communication of natural catastrophe risk within the insurance industry is often through the use of various catastrophe models. Traditionally the perils covered by these models are: earthquake (and now correlating tsunami); hurricane; windstorms; winterstorms; tornado; and flood. Increasingly there is a move towards modelling terrorism, pandemic and cyber risks.

Models vary in their levels of sophistication; from fully probabilistic models to simpler deterministic, scenario type models. The aim is to give a re/insurer a view on the probability of loss against their portfolio (of say residential houses in the UK) from one or more of these perils. For example, once every two hundred years an insurer might incur a loss of £100 m or more from flood. Companies use these results to understand the risk from natural catastrophes that they are exposed to, and to help them determine how much reinsurance (insurance for insurers) they need to buy.

There are currently few commercially available probabilistic catastrophe models for volcanic risk. Those that are available tend to focus on one hazard such as ash or pyroclastic density currents (PDCs). Development of new models is driven by industry demand; to date volcanoes have not caused sufficient insured losses to create a demand.

Fig. 1 Example of a volcano
RDS for Mexico City and
Popocatepetl volcano. The
exposure of two separate
insurance portfolios has been
overlain in blue and green on
scenario hazard layers for
(i) ash and pumice fall
>10 cm, and (ii) lahars

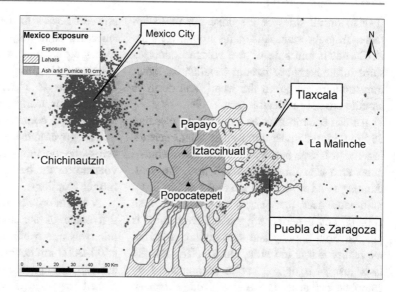

Developing deterministic scenario models is
much more straightforward. Realistic Disaster
Scenarios (RDSs) are commonly used within
re/insurance to give companies a view of loss
from a single event (Fig. 1). However, these do
not always include a view on how frequently
such an event might be expected.

Two of the challenges with respect to communication of volcanic risk within a re/insurance
organisation are:

- Volcanic risk often goes unmodelled. Therefore, do re/insurers have an appreciation of
  their exposure to volcanic eruptions and the
  potential for loss? Non-modelled (and poorly
  modelled) perils have gained more focus in
  recent years and there is a growing pressure
  on the industry to form a view (e.g. ABI
  2014)
- If an insurer has a view on volcanic risk, how
  does it compare that to the natural catastrophe
  risks that companies and boards are used to
  dealing with (e.g. earthquakes and hurricanes)? The return periods of volcanic eruptions can be very long (thousands of years)
  and may seem insignificant when compared
  to more frequently occurring perils such as
  earthquake or hurricane. However, volcanic
  eruptions have the potential to cause very
  large losses.

If a company does manage all of the above in
terms of modelling and view of volcanic risk,
there are many other challenges that still remain.

3 An Eruption Scenario

This scenario is based on a VEI 6 eruption. No
location is specified, but it could be in almost any
volcanic part of the world. We limit discussion to
just one hypothetical country, but the reality may
be that more than a single country is affected by
ash fall, thus giving rise to even more insurance
issues as policy conditions, interpretations and
government responses vary significantly around
the world.

A couple of middle-sized earthquakes stir the
volcano to life producing a few throat-clearing
but minor eruptions over a period of a few years.
Minor ground shaking (volcanic tremor) continues intermittently throughout this time. Minor
damage is limited to areas on the volcano's
slopes. Mandatory evacuation areas extend only
five kilometres from the vent. Local 'experts',
including volcanologists, speculate privately and
argue publically about the future course of the
eruption.

Then, one bright early spring morning the
volcano stirs to life with a massive Plinian
eruption, producing pyroclastic density currents

(PDCs) on all sides of the cone out to 15 km radius from the vent. A tsunami produced by the PDCs down one side of the volcano damages ships in the harbour, harbour installations, other infrastructure, cargo on the wharf, and destroys an elite coastal residential area.

An ash column rises to an altitude exceeding 25 km with the upper atmosphere winds spreading the ash fall nearly 1000 km downwind—more than 1 m thick a few km from the vent, thinning to a few cm 500 km away and to just a millimetre near the margin of the more than 100,000 km^2 experiencing ash fall. Evacuation is ordered from the city and surrounding areas but the reality is it is too little, too late. The ash fall lasts just 24 h or so, but is accompanied by complete darkness. Property and other damage are widespread across a sizeable city's commercial centre and most but not all suburbs and surrounding areas. In areas near the volcano, PDCs destroy property. Where ash falls are more than 300–400 mm thick severe damage to buildings and other insured property is widespread; with falls of 100 mm or less only weaker buildings are damaged significantly but even these falls are enough to hinder transport and create substantial issues for communications, electricity, water distribution and sewage networks and other associated impacts from volcanic ash (Wilson et al. 2015). Across the very large areas where ash fall are only a few mm thick, some agricultural crops are severely damaged and livestock are distressed.

At the height of the eruption the volcanic cone collapses leaving a caldera 5 km across. Over the next few days rainfall compacts the ash to about half its freshly-fallen thickness and increases its density significantly; a hardened surface crust forms, strengthened by sulphur and other volatiles. Rainfall erodes and redistributes some of the ash, further clogging roads and drainage systems.

Late advice, clogged roads, a reluctance to evacuate, and a desire to remain to protect property meant hundreds of citizens didn't evacuate. In more severely affected areas the remaining citizens are presumed dead but the threat of further PDCs and the heat retained in the deposits prevents access to the city or emergency assistance for some days, longer in some districts. Elsewhere emergency services are overloaded but the cleanup begins.

Airspace is closed across most of three countries for nearly a week. There is widespread disruption to commerce and the tourist industry.

The government enforces an exclusion zone across the remaining suburbs near the foot of the volcano for the best part of two months, fearing further eruptions.

A month or two after the eruption it is evident that some of the remaining metal roofs, even relatively new roofs, are corroding quickly. From 1500–2000 km downwind, beyond the region of ash fall, insurance claims are emerging for tarnished metal and silver, clothes and other exposed fabrics destroyed by acidic vapours.

Five months later, although the volcano is now doing nothing more than quietly steaming, the rainy season begins. Secondary lahars (volcanic mudflows), resulting from intense rainstorms on the volcano's slopes, carve deep channels, threaten life, limb, property, and infrastructure (especially bridges) even at distances of 20–30 km from the caldera, with burial (or erosion) of some remaining insured property.

4 A Range of Insurance Responses

It is unrealistic to regard all insurers as similar. Table 1 shows a selection of eruption-related issues that are likely to arise, characterising an insurer at the 'well-prepared' end of the insurance company response spectrum.

At the other end of the insurance industry spectrum are the insurers who have not thought much about volcanic risk even though their policy wordings imply that they will compensate policy holders for volcanic eruption-related damage. As the volcano moves "inexorably" towards an eruption some companies will fail to think strategically or operationally about the impending damage, the myriad claims, communication with their clients and reinsurers or to recognise the potential consequences for the company itself.

Table 1 The prepared insurer

Timeframe	The prepared insurer
Before the eruption—internal	1. Aware the volcano is restive; engaged with local volcano and emergency management experts; developed reasonable understanding of potential eruption styles, magnitudes and associated types of damage; reviewed and understood implications of eruption-related policy wordings; modelled potential losses associated with each scenario; established realistic estimates of the number of claims anticipated and assessments of the time required to close 50, 90% of claims 2. Discussed requirements, potential issues and possible problems with claims adjusters and repairers; communicated a range of scenarios and their implications to the Board and across senior management; informed all staff of crisis roles and taken staff feedback into account 3. Revisited operational, tactical and strategic aspects of crisis plan—test off-site data backup facilities, work from home/off site arrangements, denial of access to office buildings, absence of key personnel, alternative communications, support for staff experiencing losses/damage/stress 4. Engage with media organisations. Provision of information to insureds using email and social media. Assess transport and infrastructure needs/priorities
Before the eruption—engagement with insureds	1. Generated footprints for several key scenarios; identified at-risk policy holders using GIS technologies 2. Communicated information regarding policy coverage (including FAQ on making temporary repairs, deductibles, limits on alternative accommodation, looting, underinsurance and the limitations of cover), social media information sites, preparedness, evacuation and damage mitigation advice, how to contact your insurer, and how to make a claim should damage occur 3. Drew policyholders attention to valuable sources of information such as http://volcanoes.usgs.gov/hazards/index.php and http://www.gns.cri.nz/Home/Learning/Science-Topics/Volcanoes/Volcanic-Hazards and http://link.springer.com/article/10.1186%2Fs13617-014-0010-x#page-1 4. Announces refusal to extend cover under existing policies or to offer new policies in areas within 200 km of the volcano
After the eruption	Mobilised staff support procedures. Engaged with media; put social media plan into operation. Small team communicated frequently to deal with any unforseen/emerging issues and to discuss issues arising from possible ongoing eruptions which could last for weeks, months, years
Summary	The prepared insurer has considered reputational, credit, market, liquidity, operational and environmental risks long before the rare catastrophe occurs. The staff has been engaged in corporate resilience decision making. This insurer can be fairly confident that the policyholder who insured with them this year will be back next year

Nearly all insurance companies will lie somewhere on the spectrum between the extremely well-prepared and proactive company described in the table and the rare 'response' just summarised.

5 Insurance Policy Wordings— Some Examples and Issues

Insurance policies contain a number of clauses or definitions that may not have been tested or thought about in detail until after an event occurs. While policies differ from country to country and insurer to insurer, all contain a standard clause defining a 'loss occurrence'; this usually refers to all individual losses arising out of and directly occasioned by one catastrophe. Commonly, the duration and extent of any 'loss occurrence' is limited to '*72 consecutive hours as regards earthquake, seaquake, tidal wave and/or volcanic eruption*' (LP098A; see further discussion regarding IUA01-033 at http://www.iuaclauses.com/site/cms/contentDocumentView.asp?chapter=9).

Around the world many wordings are similar: 'volcanic eruption' is often grouped together with earthquake and tsunami, even with 'seaquake' and 'other convulsions of nature'. Some policies/contracts refer to 'volcanic activity', 'volcanic eruption' (seismic events) or even 'losses caused by volcano'. In plain English, these different wordings do not have the same meaning—that may or may not have been the intent.

Some New Zealand examples help to highlight potential interpretation and communication issues with wordings:

i. On Christmas Eve 1953, water in volcano Ruapehu's crater lake breached the crater wall. The resulting mudflow/lahar destroyed a rail bridge pier and an overnight express train (Fig. 2) plunged into the Whangaehu River, taking 151 lives. The insured loss was minimal but could be considerably more today.

The Smithsonian Institution Global Volcanism Program in Washington D.C. is the world repository for volcanic information. The GVP database holds information on every known eruption in the world in the last 10,000 years. Volcanologists, historians and others contribute, check and update information while GVP staff scour the relevant literature to ensure the database is as accurate as possible.

The Smithsonian database shows that Ruapehu was <u>not</u> in eruption between 1952 and 1956 (or to be more precise between July 1952 and the 18th November 1956) (http://volcano.si.edu/volcano.cfm?vn=241100). Certainly the Ruapehu crater wall was breached on December 24th 1953 and a lahar rushed down the Whangaehu Valley ultimately taking 151 lives, but the breach was not the result of an eruption.

Would an insurance contract defining a loss occurrence using the expression 'volcanic eruption' provide cover? How would a policy relying on the seemingly broader expression 'volcanic activity' respond?

ii. In 2005 several houses in Rotorua, New Zealand suffered damage caused by hydrothermal activity (Fig. 3)—steam, gas and hot water heated by igneous activity—at several locations within the city. There were no insurance issues as the New Zealand Earthquake Commission (EQC) policy, which provides additional cover to all privately insured residential property in the

Fig. 2 The locomotive destroyed in the Ruapehu, New Zealand, lahar 1953 (AAVK W3493 D1952 Archives New Zealand The Department of Internal Affairs Te Tari Taiwhenua)

Fig. 3 Suburban hydrothermal activity, Rotorua, New Zealand (*Photo* B Scott, GNS Sciences, with permission)

country, states: "If your house is damaged by earthquake, natural landslip, tsunami, volcanic eruption or hydrothermal activity (as defined in the Earthquake Commission Act 1993 and any amendments) we will pay."

However, if the words 'hydrothermal activity' had been omitted from the EQC policy so that any claim relied on 'volcanic eruption' (or on 'volcanic activity') would the damage have been covered?

iii. Another policy wording, probably used only once, stated: *"volcanic activity and resulting earthquake, tsunami, lahar, lava flow, ash fall and/or fire following any of these perils"* Notwithstanding it is agreed that no loss occurrence shall last more than 672 h [28 days].

This clause apparently aimed to be all-inclusive, but perhaps having the implication that if a hazard associated with 'volcanic activity' was not named, then it was not covered under this loss occurrence clause. If this implication is correct, it is unfortunate that the list does not include 'pyroclastic density current' or similar phenomena as the insurance claim for a building impacted by a PDC is likely to be around 110%

of the sum insured: 100% damage + 10% for debris removal.

We might also note that the above definition may raise other issues. Does 'resulting earthquakes' mean that only damage produced by earthquakes that occur <u>after</u> volcanic activity has begun is covered? Would 'ash fall' (Fig. 4) include cover for damage produced by 'volcanic bombs'? A volcanologist or sedimentologist could readily argue (with the support of scientific literature and practice) that 'volcanic ash' refers only to particles less than 2 mm in diameter (less than 4 mm in diameter in some definitions). The Greek word 'tephra' (used by Aristotle) which covers all airborne volcanic particles from fine dust and ash to blocks the size of houses would be more all-encompassing than 'ash fall'. We needn't worry about adding a Greek word to the definition; after all 'lava' is Greek, 'tsunami' is Japanese, and 'lahar' is Indonesian!

The Text Box below gives further examples from real-life insurance policies with a few questions highlighted in red. All of these examples emphasise that insurance Wordings need to keep the coverage simple and re/insurers need to think carefully about what their policies (intentionally or otherwise) include or exclude. Policy holders also need to think carefully about the Wordings in policies and, if necessary, to engage in dialogue with their insurers.

Fig. 4 A Rabaul (Papua New Guinea) suburban house severely damaged (total loss) by 600–800 mm ash fall from Tavurvur volcano in 1994. The timber-framed roof has collapsed inwards and the walls spread outwards. The sub-floor area has been buried by air fall ash and minor lahars (*Photo* R Blong)

- **Definition of Volcanic Eruption:** *Volcanic Eruption* is a form of a volcanic activity ejecting volcanic materials i.e. such as lava flows, pyroclastics and or volcanic gasses onto the earth's surface either from a central vent or from fissures of a volcano. What about Lahars? Ground deformation? Was the intent to exclude damage produced by these and other volcanic hazards?

- **Definition of Tsunami:** *Tsunami* is a great sea wave produced by submarine earth movement such as subduction of crustal plates or by submarine volcanic eruption. The Krakatau eruption of 1883 was not a submarine eruption but the tsunami produced on August 26 (either by massive pyroclastic flows striking the sea surface or by collapse of the volcano) killed more than 34,000 people and produced extensive damage on the nearby coasts of Sumatra and Java. Was the underwriting intent to include all volcanically-generated tsunami?

- "However, earthquake or volcanic eruption shall happen by natural cause, but excluding earthquake or volcanic eruption caused by any material from space (Earthquake or Volcanic Eruption or Tidal Wave or Tsunami Endorsement)." Very likely, something was lost in translation (possibly three languages were involved in this translation).

- **Definition of Loss:** A key question for insurers on Montserrat during the 1990s was the definitions in the original contracts with the insureds on the island, and *physical damage* versus *loss of use*. A number of the properties on the island had suffered no physical damage after eruptions in 1997; however they were included in the exclusion zone. Is this covered under the insurance contract? The insurer in this example paid claims to the household owners on the basis of denial of access/loss of use.

6 The Hours Clause

As implied in the foregoing examples 'loss occurrence' clauses in insurance policies (particularly reinsurance policies) usually contain an Hours Clause; commonly this refers to a loss occurrence being limited to a period such as 72 consecutive hours (3 days), 168 h (1 week), or occasionally 672 h (28 days). Usually, the insurer can specify to the reinsurer when the Loss Occurrence begins, but if the damaging event lasts longer than the specified Hours Clause then a new Loss Occurrence event begins. Generally, for each Loss Occurrence the insurer will pay the retention (which might be tens of millions of dollars) with the reinsurers footing the bill for the rest of the damage. Clearly, with the large sums potentially at stake the number of Loss Occurrences agreed is of major concern to both insurers and reinsurers. This will also be of interest to individual insurance policyholders when, say, a fall of tephra produces damage, followed two months later by another damaging tephra fall. In most cases the policyholder will have to pay the deductible twice.

The eruption of Soufriere Hills on Montserrat is a good example of how even defining a loss occurrence can be complicated when considering volcanic eruptions.

In 1995 Soufrière Hills, the volcano on Montserrat in the Caribbean began erupting. The eruption continues today. In 1997 parts of the capital, Plymouth, were destroyed by PDCs. Most of the surrounding areas in the southern part of the island were relatively undamaged. In 2014 most of the houses in areas near Plymouth remain only lightly damaged by ash fall (Fig. 5) and the occasional hurricane but stand empty as the southern end of the island is located in a government-ordained exclusion zone.

During June to August 1997 a number of explosive eruptions led to the expansion of the exclusion zone on the island. Properties within this zone, however, had not necessarily suffered any physical damage from the eruptions.

A large part of the discussion between one insurer on the island and its reinsurers was how to define 'loss occurrence'. The loss in this example was 'loss of use'. The question was whether the June 25th 1997 eruption was the 'occurrence' that led to those properties being in the exclusion zone, and effectively a total loss.

Explosive stages of volcanic eruptions can last from a few minutes to more than a year with a median value of less than 10 h (Jenkins et al. 2007). In many cases it is too dangerous for loss assessors to determine damage for days or weeks after an eruption where explosive phases are irregularly interspersed between less violent stages of activity. PDCs only a few tens of cm thick are probably too hot and dangerous to walk on for days or weeks. After an eruption has ceased heavy rains can redistribute ash as mudflows down streets and through buildings. Volcanic eruptions can produce long periods where loss assessment is not possible. This can become even more complicated when a policy is renewed with a different insurer between eruptions spaced a few days or weeks apart.

When large eruptions occur huge volumes of unconsolidated volcanic ash are deposited on surrounding slopes. As ash fall or PDCs have usually destroyed all vegetation there is little to hold the volcanic sediments in place so erosion is usually rapid—as a rough rule of thumb as much as half the sediment will be eroded within a year or two of the eruption. This means that huge volumes of sediment move into rivers and overflow burying lower-lying land. This process can continue for years as illustrated below. Only the upper half of the Bacolor (Philippines) church remains above ground (Fig. 6), buried over a period of several years by lahars from the cataclysmic eruption of Pinatubo in 1991 (one of the two largest eruptions in the world in the 20th century).

Both the Montserrat example and the burial of the Bacolor church, and numerous other buildings, villages and infrastructure, more than 30 km east of Pinatubo, illustrates that the direct

Fig. 5 A lightly-damaged but abandoned shoe shop in the government-ordained exclusion zone in Plymouth, Montserrat in 2004 (*Photo* C Tillyard)

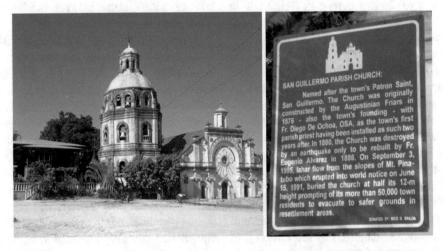

Fig. 6 The San Guillermo parish church, Bacolor (Philippines), buried by lahars years after the 1991 Pinatubo eruption (*Photos* R Blong)

consequences of a volcanic eruption can continue for years even after the volcano has ceased erupting. Such scenes also illustrate that insurers, policy writers and insureds need to understand the consequences and the timeframes of volcanic eruptions to ensure that policies reflect not only underwriting intent but also volcanic reality.

7 Clean-up Costs

Clean up costs resulting from ash fall over an urban area remains an important issue for policyholders, insurers and reinsurers to consider.

We can imagine an ash fall about 10 mm thick across a city of about 200,000 people that covers an area of roughly 100 km^2. The ash has to be removed before it blows and washes into the storm water system which probably has (illegal) links to the sewage system and/or blows around aggravating respiratory problems, creating other health issues, and exacerbating wear and tear on machinery, vegetation including commercial crops, and nerves (Hayes et al. 2015).

A 10 mm fall of volcanic ash will likely cause little damage to well-constructed buildings, but the ash needs to be removed from building roofs and gutters, prevented from damaging sensitive equipment including most electronics, the electrical generation and distribution network, communication networks, airport runways, and roads (where a few mm obscures road markings and makes the surface slippery). The cleanup process may need to be repeated several times to remove most of the ash, or because ash continues to fall.

For our scenario city of 200,000 people the 10 mm ash fall has a volume of about 1 million cubic meters (say 10,000 truckloads); it is not simply a matter of trucking the ash to another location where it can continue to blow around. The clean-up will require planning. Suitable dump sites might be 20 km outside the city. Dump sites require maintenance so that ash doesn't continue to blow around.

Cleanup can be expensive. A repeat of the 1707 eruption of Fuji in Japan, for example, would spread ash across the Tokyo and Yokohama urban areas (and elsewhere). The cost of cleanup and removal of ash from the urban areas has been estimated to cost more than USD10 billion (Christina Magill, Risk Frontiers, pers. comm., September 2015). The damage bill could be quite limited but the cleanup costs will be substantial.

Not all of this cost would fall to insurers but have re/insurers considered the potential costs? Would policies cover the cost of debris removal and cleanup when there is no or little material damage? After the 1980 Mount St. Helens (USA) eruption, around 90% of insurers in eastern Washington paid policyholders an hourly rate to remove ash from roofs and building surrounds. Would this practice continue?

8 Other Stakeholders

Here we briefly summarise some of the possible issues and considerations of other insurance industry stakeholders:

- Loss adjusters are likely to have minimal (if any) experience of forensically examining the myriad claims' issues arising in the aftermath of an eruption.
- Like insurers, reinsurers may not have considered the implications of the contract wording, impacts of volcanic eruption in their pricing, or which aspects of eruption-related damage might be covered under the catastrophe bonds they issued.
- Regulators will be working through the impacts of the event on the industry and specifically policyholders including the solvency positions of insurers.
- The media are focused on a lot of local examples of tragedy, isolated insurance issues and massive generalisations, failing to appreciate the complexity of numerous issues from insurers' points of view or the diversity in insurance response.
- Government involvement is significant. Local governments have clean-up issues and desperately need support from state government officials to contribute funds to reinstate infrastructure, assist the clean-up process, provide handouts, and deal with those who are uninsured and/or underinsured. Nobody much has thought about where all the cleaned up ash is going to be dumped. Political motivation may also play a part, for example the assumption that re/insurance companies have deep pockets with politicians publically urging the industry to be generous, even outside policy conditions/limitations.
- It is really difficult to contemplate the legal responses to the insurance issues generated by our eruption scenario, but the following

examples from insurance case law provide insights for both insurers and insureds into the arcane worlds of policy wordings and the law.

9 Insurance Case Law

There is a rather limited amount of case law relating to insurance and volcanic eruptions available. Contract wordings (and their intent) are usually of paramount importance. The examples below allow few conclusions to be reached. Nonetheless, these brief summaries provide valuable examples of some of the issues that arise in interpreting insurance contracts.

9.1 Philippines, 1991

The eruption of Pinatubo in early June, 1991, covered a wide area in ash fall including two insured properties in Angeles City. Six days later Typhoon 'Diding' (known internationally as Typhoon Yunya) swept across central Luzon bringing significant rainfall to Angeles City. Subsequently the rooves of the insured properties collapsed. The insureds claimed that the proximate cause of the damage was the typhoon whereas the insurer claimed the damage resulted from the eruption of Pinatubo, that volcanic eruption was an excluded peril and that the losses were not covered under the policy.

Clause 6 of the policy read (in part): 'This insurance does not cover any loss or damage occasioned by or through or in consequence, directly or indirectly, of any of the following occurrences, namely:

(a) Earthquake, volcanic eruption or other convulsion of nature'.

The Appeal Court found: 'An examination of the records reveals that no damage was sustained by the insured properties at the height of the volcanic upheaval. True, there may have been volcanic ashes deposited on the roofs but these

did not result in any untoward incident until the typhoon came on June 15, 1991 which bought more significant amount of ash fall in the affected area, caused the same to be soaked with rainwater thereby making it heavy which lead to the damage of the insured properties. Consequently, it can be deduced that the proximate cause of the roofs caving in and the subsequent entry of the water inside the insured premises was the typhoon and not the volcanic debris' (Court of Appeals, Manila 1993).

While this was bad news for the insurer, the Appeal Court also found that the lower court's awarding of attorney's fees to the defendant-appellant was uncalled for (Court of Appeals, Manila 1993).

9.2 New Zealand, 1995

Eruptions of Ruapehu in 1995 and 1996 during the ski seasons caused Ruapehu Alpine Lifts (RAL), a ski lift operator, to cease operating because of ash fall on the snow. The ash was corrosive, the ski field was closed and RAL lost business. RAL had a material damage policy and the material damage claim was met, but claims under Business Interruption (BI) were declined even though the cover provided for 'Earthquake, geothermal activity or volcanic eruption'. The quantum of loss of NZD4.669 million plus GST + interest + any licence fees properly payable was not in dispute. The standard BI policy wording did not exclude snow from being 'property' whether natural or artificially created (possibly an important point as RAL had groomed and shaped the snow).

The issue revolved around the sense and meaning of the terms in the policy and the intention of the parties to the policy. Did ash fall onto snow constitute damage to 'building and other property' so as to fall within the cover of the consequential loss policy?; the argument was over whether snow is 'property', whether 'building' in 'building and other property' limits the nature of 'other property' and whether the cover was also limited by 'for the purposes of the Business'. 'Other property' it appears is not confined to fixed assets but includes plant, machinery, and stock; in its wider sense 'property' can include debts, goodwill, rights, interests and claims.

RAL won the case with costs in June 1998 (High Court of New Zealand 1998). State Insurance appealed to the New Zealand High Court but lost in May 1999.

9.3 United States, 1980

The appeal arose from a dispute between two insurance companies and their insureds following the May 18 1980 eruption of Mount St. Helens in which pyroclastic density currents melted snow and ice on the volcano. These PDCs combined with sediment eroded from the large debris avalanche deposit and river channels and with torrential rain from the eruption cloud, groundwater and the waters of Spirit Lake to produce mudflows down the Toutle Valley. At a distance of 30–40 km from the volcano and about 10 h after the eruption began, the appellants homes were destroyed by mudflows/lahars or by mudflows/lahars preceded by water damage from flooding.

All three policies stated the following:

Section 1—Exclusions

We do not cover loss resulting directly or indirectly from:

...

2. Earth Movement. Direct loss by fire, explosion, theft, or breakage of glass or safety glazing materials resulting from earth movement is covered.

3. Water damage, meaning:

(a) flood ....

The term 'Earth Movement' was not specifically defined (evidently it had been defined in earlier years but had been omitted from this policy in order to simplify the language).

The insurers rejected the insured's claims on the basis that the damage was excluded as 'earth movement'. The trial court assumed the movement of Mount St. Helens was an 'explosion' within the terms of the insurance policies, and noted that the true meaning of 'explosion' was to be determined by jurors. It further determined that the mudflows which destroyed the appellants' homes would not have occurred without the eruption of Mount St. Helens; that is, the eruption was a proximate cause of damage to the appellants' homes. However, this was not a unanimous decision by the bench (Supreme Court of Washington 1983).

9.4 Iceland, 2010

The minor eruption of Eyjafjallojökull in 2010 produced an ash cloud which grounded more than 100,000 flights across Europe, disrupted the travel plans of around 10 million airline passengers and produced substantial economic losses to passengers, airlines and third parties (Alexander 2013).

Two insurance-related issues are of interest.

i. Costs of the disruption to insurers were relatively small as most BI policies require physical damage to an insured item to trigger the policy. As there was no damage to airplanes or to airports outside Iceland insurers were generally not liable for the losses sustained. However, policies do tend to cover 'damage caused to third parties by negligent air traffic control guidance' and it remains a question as to whether airspace was closed for longer than was strictly necessary— although the hazards to aviation (and jet engines and modern aircraft in particular) presented by volcanic ash have been well-known for at least 30 years, and a global network of VAAC (Volcanic Ash Advisory Centres) has been established for about two decades, surprisingly little effort was made to determine the 'safe' concentrations of volcanic ash through which aircraft could fly before the Eyjafjallajökull eruption.

More recently, non-damage Business Interruption (NDBI) policies have been offered whereby components of a BI policy without preceding property damage have been available as extensions to existing policies. NDBI policies indemnify an airline for any cancelled flight arising from airspace or airport closure by a third-party authority or airport operator caused by non-manmade events. However, there has been surprisingly limited uptake of such policies, perhaps emphasising (a) the extensions are too restrictive or too expensive; (b) a disconnect between the ways insurers see policy wording and the ways risk managers in industry view wordings; (c) short-term thinking on long-term issues; and/or (d) the need for all parties to view risk within a holistic corporate resilience framework.

ii. Under European Union regulations air passengers are not entitled to compensation when 'force majeure' is involved. However, not all travel policies are the same (Australian travel policies, for example, tended to provide cover for the disruption resulting from the Eyjafjallajökull ash cloud). Moreover, some European insurers made refunds on abandoned travel and some insurers paid goodwill gestures.

More significantly, both a UK lower court and the UK Financial Ombudsman Service regarded the ash cloud as falling under 'poor weather conditions' which were covered by at least some travel policies. The FOS Ombudsman concluded that 'it would be fair and reasonable for the insurer to treat the wind-borne ash cloud as poor weather conditions under Ms B's travel policy; it would not be fair and reasonable for the insurer to decline Ms B's claim; and Ms B's claim should succeed and the insurer should pay the benefit available under her policy plus interest (at 8% per year simple)' (UK FOS Final Decision, March 2011; see also Brannigan 2010).

More recently there have been discussions among various parties regarding travel insurance cover for events such as the Eyjafjallajökull ash cloud but it will still be important to read the fine print.

Most interestingly for insurers and insureds alike the UK Financial Ombudsman Services (2011) made an important point in providing an opinion on Ms B's claim: 'It is a general principle of English courts that an ambiguous contractual term must be given the interpretation that is less favourable to the party who supplied the wording, which was the insurer in this case. So although I consider the "poor weather" encompasses ash on the wind, if there is any ambiguity about it, this principle will apply'.

10 Conclusions

Within the re/insurance industry, risk from volcanic hazards is often unmodelled and poorly understood. We need to understand more about the nature of volcanic hazards and to recognize that there are quite a few things happening before, during, and after volcanic eruptions that produce consequences including damage to a wide range of insured assets.

Not all insurance policies are the same. Insurers, reinsurers, loss adjusters, policy holders, and other players in the insurance space, need to ensure that policy wordings reflect both volcanic reality and underwriting intent.

Not all insurance companies are the same. Many are well-organised, experienced, pro-active and resilient, whilst others may have difficulty responding or even surviving an uncommon event that produces a large number of claims well beyond their experience. In that sense, insurers and the other insurance-related players in a volcanic crisis, are no different to other commercial organisations (or governments, or non-government organisations, or families, or individuals for that matter) experiencing an unusual event—some insurers have limited resilience; some have what we might call 'planned resilience'; others will adapt quickly to changed circumstances. Some insurers (and some policyholders) will survive, some will thrive.

References

ABI, Association of British Insurers 2014 Non-modelled risks—a guide to more complete catastrophe risk assessment for (re)insurers. https://www.abi.org.uk/~/media/Files/Documents/Publications/Public/2014/prudential%20regulation/Nonmodelled%20risks%20a%20guide%20to%20more%20complete%20catastrophe%20risk%20assessment%20for%20reinsurers.pdf

Alexander D (2013) Volcanic ash in the atmosphere and risks for civil aviation: a study in european crisis management. Int J Disaster Risk Sci 4(1):9–19

Brannigan VM (2010) Alice's adventures in volcano land: the use and abuse of expert knowledge in safety regulation. Eur J Risk Regul 107–113

Court of Appeals, Manila (1993) Republic of the Philippines Court of Appeals, Manila, Sixteenth Division, Leonor Infante and CA G.R> CV No 43449 Swagman Hotel and Travel, Inc., Decision, 7 p

Deligne NI, Coles SG, Sparks RSJ (2010) Recurrence rates of large explosive volcanic eruptions. J Geophys Res 115:B06203. doi:10.1029/2009JB006554

Hayes JL, Wilson TM, Magill C (2015) Tephra fall clean-up in urban environments. J Volcanol Geoth Res 304:359–377

High Court of New Zealand (1998) 61-404 Ruapehu Alpine Lifts Limited v State Insurance Limited, June 1998, CCH Australia Limited, 61-404, 74,432–74,443

Jenkins SF, Magill CR, McAneney KJ (2007) Multi-stage volcanic events: a statistical investigation. J Volcanol Geoth Res 161(4):275–288

Supreme Court of Washington (1983), 98 Wn.2d 533; 656 P.2d 1077; 1983 Wash. LEXIS 1340

UK Financial Ombudsman Service (2011) Final Decision, Complaint by Ms B, March 2011, 10 p

Wilson TM, Jenkins S, Stewart C (2015) Impacts from volcanic ash fall. In: Papale P (ed) Volcanic hazards, risks, and disasters. Elsevier, Amsterdam, pp 47–86

Challenges and Benefits of Standardising Early Warning Systems: A Case Study of New Zealand's Volcanic Alert Level System

Sally H. Potter, Bradley J. Scott, Carina J. Fearnley⊙, Graham S. Leonard and Christopher E. Gregg

Abstract

Volcano early warning systems are used globally to communicate volcano-related information to diverse stakeholders ranging from specific user groups to the general public, or both. Within the framework of a volcano early warning system, Volcano Alert Level (VAL) systems are commonly used as a simple communication tool to inform society about the status of activity at a specific volcano. Establishing a VAL system that is effective for multiple volcanoes can be challenging, given that each volcano has specific behavioural characteristics. New Zealand has a wide range of volcano types and geological settings, including rhyolitic calderas capable of very large eruptions (>500 km^3) and frequent unrest episodes, explosive andesitic stratovolcanoes, and effusive basaltic eruptions at both caldera and volcanic field settings. There is also a range in eruption frequency, requiring the VAL system to be used for both frequently active 'open-vent' volcanoes, and reawakening 'closed-vent' volcanoes. Furthermore, New Zealand's volcanoes are situated in a variety of risk settings ranging from the Auckland Volcanic Field, which lies beneath a city of 1.4 million people; to Mt. Ruapehu, the location of popular ski fields that are occasionally impacted by ballistics and lahars, and produces tephra that falls in distant cities. These wide-ranging characteristics and their impact on society provide opportunities to learn from New Zealand's

S.H. Potter (✉) · B.J. Scott · C.J. Fearnley · G.S. Leonard · C.E. Gregg
Department of Science and Technology, University College London, Gower Street, London WC1E 6BT, UK
e-mail: S.Potter@gns.cri.nz

C.J. Fearnley
e-mail: c.fearnley@ucl.ac.uk

S.H. Potter · G.S. Leonard
GNS Science, 1 Fairway Drive, Avalon, Lower Hutt 5010, New Zealand
e-mail: G.Leonard@gns.cri.nz

B.J. Scott
GNS Science, 114 Karetoto Road, RD4, Taupo 3384, New Zealand
e-mail: B.Scott@gns.cri.nz

C.E. Gregg
Department of Geosciences, East Tennessee State University, Box 70357, Johnson, TN 37614, New Zealand
e-mail: GREGG@mail.etsu.edu

Advs in Volcanology (2018) 601–620
https://doi.org/10.1007/11157_2017_18
© The Author(s) 2017
Published Online: 30 August 2017

experience with VAL systems, and the adoption of a standardised single VAL system for all of New Zealand's volcanoes following a review in 2014. This chapter outlines the results of qualitative research conducted in 2010–2014 with key stakeholders and scientists, including from the volcano observatory at GNS Science, to ensure that the resulting standardised VAL system is an effective communication tool. A number of difficulties were faced in revising the VAL system so that it remains effective for all of the volcanic settings that exist in New Zealand. If warning products are standardised too much, end-user decision making and action can be limited when unusual situations occur, e.g., there may be loss of specific relevance in the alert message. Specific decision-making should be based on more specific parameters than the VAL alone, however wider VAL system standardisation can increase credibility, a known requirement for effective warning, by ensuring that warning sources are clear, trusted and widely understood. With a credible source, user groups are less likely to look for alternatives or confirmation, leading to faster action. Here we consider volcanic warnings within the wider concept of end-to-end multi-hazard early warning systems including detection, evaluation, notification, decision-making and action elements (based on Carsell et al. 2004).

Keywords

Volcanic Alert Level · Early warning system · Standardisation · New Zealand

1 Early Warning Systems and Standardisation

An Early Warning System (EWS) can be defined as a system designed to provide "hazard monitoring, forecasting and prediction, disaster risk assessment, communication and preparedness activities, systems and processes that enables individuals, communities, governments, businesses and others to take timely action to reduce disaster risks in advance of hazardous events" (UN 2003). They are recognized as "a means of getting information about an impending emergency, communicating that information to those that need it, and facilitating good decisions and timely response by people in danger" (Mileti and Sorensen 1990, p. 2–1). Essentially EWSs facilitate the provision of timely warnings to minimize loss of life and to reduce economic and

social impact on vulnerable populations (Garcia and Fearnley 2012).

The operation of an EWS presents numerous challenges due to variations in: scale (global, national, regional, local); temporality (rapid onset, slow onset, frequent, infrequent); function (safety, property, environment); and hazard (e.g., weather, climate, geohazards). A Volcano Alert Level (VAL) system is a communication tool within a volcano EWS, which simplifies the communication of volcanologists' interpretation of data (Newhall 2000). The VAL system is disseminated with supporting information that provides more specific details and local context to enable responding agencies, the public, and other stakeholders to make informed decisions (Fearnley 2011; Potter et al. 2014). The levels can be labelled using words, numbers, colours, and/or symbols, and summarise information from 'background' activity (no unrest), through to the highest level of activity (usually a large eruption) (Newhall 2000; Fearnley 2011).

Volcano observatories play a key role in the management and assignment of alert levels for volcanoes. However, with over 80 volcano observatories around the world, it is understandable that VAL systems operate in very different ways. Some provide only scientific advice on what a volcano is doing, others forecast activity, and others provide guidance on what vulnerable populations should do. There is a growing discussion globally around the role of VAL systems and whether they should be standardised, either nationally and/or internationally.

In 1989, the member states of the United Nations declared the period from 1990 to the year 2000 to be the International Decade for Natural Disaster Reduction (IDNDR) to focus attention on reducing loss of life, and social and economic disruption caused by natural disasters. Of fundamental importance was the recognition in 1991 of early warnings as a key objective of disaster reduction practices (Maskrey 1997). The United Nations has called for more effective procedures via standardisation and the application of new technologies and enhanced scientific understanding (United Nations 2006) but few hazards have an EWS operating beyond a national or regional scale. The most successful example is the Pacific Tsunami Warning Centre (PTWC), established in Hawaii in 1949. Until recently, PTWC provided Warning and Watch alert level services for numerous Pacific nations and beyond, but currently they provide nations with threat level information (e.g., wave amplitude forecasts), which the countries then use in-house to develop and disseminate specific tsunami alert levels (i.e., Warning, Watch).

The growing pressure on volcano observatories to standardise warnings is felt especially in relation to provision of advice and alert levels to the aviation sector. In 2006, the International Civil Aviation Organization (ICAO) globally standardised a considerable portion of products, including the alert levels (via the Aviation Colour Code (ACC); Table 1), messages to airlines that observatories are expected to provide (via

the Volcano Observatory Notice for Aviation; VONA), and the framework for monitoring and alerting related to ash clouds (via the Volcanic Ash Advisory Centres; VAACs). While VAACs provide the aviation community with information regarding where ash currently is in the air, the role of the ACC is more about warning (Gardner and Guffanti 2006). The ACC allows a recognition of the level of volcano activity for the purpose of attention by the aviation industry, and to inform their decisions, such as regarding re-routing or extra fuel (Gardner and Guffanti 2006). VONAs are standardised plain-English messages aimed at dispatchers, pilots, and air-traffic controllers 'produced by Volcano Observatory scientists and are based on analysis of data from monitoring networks, direct observations, and satellite sensors' (as described on the USGS website[1]). The international nature of these aviation products reflects the need for aviation personnel to ascertain the status of volcanic activity across a number of countries and VAL systems (e.g., Guffanti and Miller 2013), which is why a standardised approach is used. While working fine for international air traffic, problems have been encountered for low-level domestic and private aviation with the ACC (Fearnley 2011).

Numerous volcano observatories across the world are now implementing the ACC, and many of these are questioning the adoption of other VAL systems for ground-based hazards and reviewing their volcano early warning systems. This includes reviewing the effectiveness of VAL issuances, as Winson et al. (2014, p. 12) invite "countries to perform their own self-evaluation and weigh the cost of a higher number of alerts against the benefit of a higher accuracy in VAL issuances and to decide how to proceed accordingly with their own local populations".

The benefits of standardisation are principally:

- simplicity through application of common language, frameworks and understanding;
- clarity for emergency responders;

---

[1]https://volcanoes.usgs.gov/vhp/notifications.html accessed 11 May 2017.

- reduced workload for monitoring and emergency management agencies, including education and outreach;
- interoperability of equipment and systems across hazards, and across agencies, countries or internationally.

Warning messages with specific language and approaches can benefit from standardisation. In doing so, it is advisable to publish and make accessible the definitions of technical volcanic terms, and of the words 'risk' and 'hazard' as used by the scientists. This is to ensure local users understand the intention behind the scientist's use of the terms. For example, calculation of 'hazard' can be based on previous events, deterministic and/or probabilistic approaches, and should be described. In addition, the communication of likelihood/probability can also be a point of confusion; ideally numeric values should be mapped against qualitative descriptions and both presented together (e.g., Doyle and Potter 2016). Standardisation can also increase credibility, a known requirement for effective warning, by producing warnings that are clear, and from a trusted source. With a credible source user groups are less likely to look for alternatives or confirmation, leading to faster action. However, end-user decision making and action can be limited in terms of contingent-specific-needs if warning products are standardised too much (e.g., depending on a user's ability to read different types of maps (Haynes et al. 2007); or loss of meaning or relevance in the warning within a local context).

There are various elements of end-to-end warnings that lend themselves to standardisation that can aid the effectiveness of the VAL system:

- Technical: Equipment type, deployment (distribution/location/density), telemetry (radio, wire, internet etc.), visualization (software packages) and analysis all receive some level of standardisation through manufacturing standards, detection limits, and international scientific best practice.

- Analytical tools: Analysis may be further structured through statistical approaches such as expert elicitation, Bayesian event trees or Bayesian belief networks.
- Warning tools: Notification may be standardised through message content (e.g., standard messages, terminology, alert level criteria), packaging (e.g., bulletins, alert levels, maps) and delivery channels (e.g., phone, internet, siren). Some standards lend interoperability, such as a Common Alerting Protocol.
- Response: Decision-making and action by the end-user can be standardised to some extent through communication and education approaches and message content.

Clearly there are cases where standardisation provides many advantages, but the process of standardisation is predominantly triggered and shaped by social, political, and economic factors, rather than in response to scientific needs specific to a region. Standardisation, by definition, tends to exclude the importance of incorporating local factors into a global procedure. Hence, even if standardisation may yield improved strategies for gathering and interpreting warning signals, it will still favour inflexible procedures not designed to accommodate local social and cultural constraints. Challenges are brought about by a range of issues (Fearnley 2011, 2013), especially:

- the realities of varied volcanic systems, each being geophysically unique when examined in detail. The diversity and uncertain nature of numerous hazards that can occur at different temporal and spatial scales require specific EWSs to be developed.
- varied end-users with different needs and perspectives for their decision-making, in terms of the level of volcanic activity, and timing thresholds for response actions.
- multiple local social and cultural contexts and constraints, which presents challenges in relation to the applicability and responsiveness of EWSs to local knowledge and context.

In recent decades, standardisation within VAL systems at a national level has taken place, making provision for consistency of warnings enacted by civil authorities that are required to take action and facilitate national policies for emergency management. VAL systems in a number of countries (including Japan via the Japan Meteorological Agency (JMA), Vanuatu via GeoHazards, the USA via the USGS, and New Zealand via GNS Science) have been standardised in each country for use by all volcanic observatories. Yet, there are variances in the way VAL systems are being standardised. In the USA, for example, two standardised VAL systems are now in place: a textually based version for ground hazards (e.g., watch, warning) and the aviation colour code that uses colours as labels. In New Zealand, the VAL systems adopted in 1994 and reviewed in 1995 also used two standardised VAL systems: one designed for hazards expected at frequently active volcanoes and the other for restless and reawakening volcanoes (see Table 2). Both of the New Zealand VAL systems were numerically based using six levels ranging from 0 to 5 (Scott and Travers 2009). Another review in 2014 resulted in these two systems being combined into one (Potter et al. 2014), whilst also adopting the international aviation colour code (ACC). Notably, both the USA and New Zealand VAL systems are based upon the current activity of a volcano, and neither advocate action nor provide advice to users involved in crisis management and mitigation—this information is provided in other products. In sharp contrast, the Japanese VAL system addresses the measures to be taken for specific areas of danger, indicating extent of evacuation, and outlining the expected volcanic activity.[2]

This chapter focuses on the standardisation of VAL systems using the case study of New Zealand in the revision of the VAL system in 2014, to explore the benefits and challenges in implementing a nationally standardised VAL system. Reflections upon its success will help inform others as to why the devised national alert level in New Zealand is best placed for their nation, and why perhaps an international level of standardisation for VAL systems is still something that is unadvisable, and unfeasible.

1.1 Overview of New Zealand's Volcanic Risk Setting

New Zealand straddles the boundary between the Pacific Plate and the Australian Plate. The resulting subduction zone lies beneath a rifting area of thin crust with magmatic upwelling, called the Taupo Volcanic Zone (TVZ; Fig. 1). The TVZ contains most of New Zealand's active volcanoes, and includes stratovolcanoes and calderas.

New Zealand has a range of volcanic risk settings that creates a challenge in effectively communicating scientific information to stakeholders, including the public. In terms of hazard, the volcanoes have large differences in the potential eruption styles and magnitudes of eruptions, reflected by variations in magma chemistry. The past frequency of eruptions and the date of the most recent eruption also vary considerably between them, which contributes towards a range in the likelihood element of the risk equation. The exposure and vulnerability of communities to unrest and eruptions also differs, with some volcanoes (such as Ruapehu, Ngauruhoe, and Tongariro) situated in a largely unpopulated National Park; and others are islands in the Pacific Ocean with few permanent residents (such as Raoul Island in the Kermadecs). Other volcanoes, such as Auckland Volcanic Field, Taupo Volcanic Centre, Okataina Volcanic Centre and Rotorua Caldera volcano, are in close proximity to cities. A few of the volcanoes receive tens of thousands of visitors each year, and are used commercially by tourist operators (including Whakaari/White Island). Others, such as Taranaki volcano, are surrounded by fertile agricultural land and are near important national infrastructure. Each of these elements of risk (hazard characteristics, likelihood, exposure and vulnerability) influence the type of

---

[2]www.data.jma.go.jp/svd/vois/data/tokyo/STOCK/kaisetsu/English/level.html accessed on 11 May 2017.

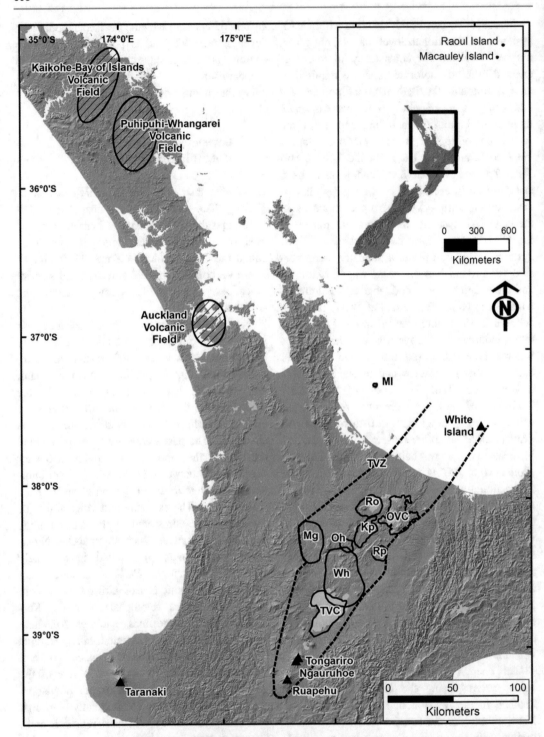

Fig. 1 Map of New Zealand's volcanoes, from Potter (2014), and based on Smith et al. (1993), Nairn (2002), Wilson et al. (2009), and Lindsay et al. (2010). The Taupo Volcanic Zone (TVZ; depicted as a *dashed line*) envelops the majority of the volcanoes. The calderas (*polygons*) are *MI* Mayor Island; *Ro* Rotorua; *OVC* Okataina Volcanic Centre; *Kp* Kapenga; *Rp* Reporoa; *Oh* Ohakuri; *Mg* Mangakino; *Wh* Whakamaru; and *TVC* Taupo Volcanic Centre. The volcanic fields are indicated by *ovals* with *diagonal lines*

Fig. 2 Eruption at Te Maari Crater, Tongariro, on 21 November 2012, captured by the GeoNet Te Maari Crater web camera (GNS Science)

communication and information required, as the local context needs to be considered.

The Tongariro Volcanic Centre is the southernmost volcanic complex of the TVZ, and includes the frequently active andesitic Ruapehu and Ngauruhoe/Tongariro stratovolcanoes. Lahars have frequently occurred, causing a hazard in numerous valleys on the volcanoes (e.g., Leonard et al. 2008). Ruapehu hosts popular ski areas, and last erupted with vigour in 1995–96 (Hurst and McGinty 1999). Small eruptions with short durations also occurred in October 2006 and September 2007. Volcanic unrest is on-going. Ngauruhoe is the most frequently active vent of the Tongariro massif, displaying regular eruptions until 1977 (Scott 1978), but none since. Te Maari Crater and Red Crater on northern Tongariro were active in the late 19th Century, with frequent eruptions in 1896–97 (Scott and Potter 2014). After less than one month of minor unrest, Te Maari Crater was the source of two small, short-lived phreatic eruptions on 6 August and 21 November 2012 (Fig. 2). There were no casualties, however the tourism industry was impacted due to the closure of a popular walking track (the Tongariro Alpine Crossing) by the Department of Conservation (DoC), which manages the National Park.

There are eight areas of known caldera collapse in the central TVZ (Fig. 1), which itself can be considered to be a caldera system similar to Yellowstone in the USA. Although calderas are usually formed in occasional very large eruptions, their magma system can also be the source of many smaller eruptions. The calderas in the TVZ have erupted almost exclusively rhyolitic material in at least 25 caldera-forming eruptions in the last 1.6 million years (Wilson et al. 1984, 2009). Only <0.1% of the volume of deposits in all of the TVZ are from basaltic eruptions (Wilson et al. 1995). The caldera volcanoes have a large range in past eruption magnitudes. For example, the TVZ's most-recent caldera collapse took place at Taupo Volcanic Centre (TVC) in 232 ± 5 AD, erupting 35 km^3 of magma (Wilson 1993; Davy and Caldwell 1998; Self 2006; Hogg et al. 2012). This devastated a significant portion of the central North Island in widespread pyroclastic density currents (Wilson and Walker 1985). However, 26 of the 29 eruptions at TVC in the past 26,000 years (since the last supervolcano eruption) have been much smaller than the most recent eruption (Wilson et al. 2009). Therefore, it is unknown whether future eruptions at TVC will be relatively small, as has been the case most frequently, or devastatingly large, as was the case with the most recent eruption. An additional challenge is managing caldera unrest, which can cause social and economic impacts, without a resulting eruption (Potter et al. 2012, 2015).

The north-eastern extremity of the TVZ contains White Island (Whakaari), a privately owned andesitic stratovolcano located 50 km from the Bay of Plenty coastline. It is currently New Zealand's most frequently active volcano, and a popular tourist destination with approximately 25,000 tourists and tourist operators visiting the island per year. Frequent eruptive sequences have been documented since written records began in 1826 (Nairn et al. 1991), with the most recent eruptions occurring in 2016. New Zealand is also responsible for a number of other island volcanoes, including Mayor Island in the Bay of Plenty, and Raoul and Macauley Calderas in the Kermadec Island chain, 750–1000 km northeast of New Zealand. These islands have very few visitors or residents. About 30 submarine volcanoes are also known in the Kermadec area and some exhibit eruptive activity. Due to the lack of monitoring data, a VAL is not allocated to them.

Taranaki Volcano is a stratovolcano located in the west of the North Island, outside of the TVZ. It is thought to have last erupted in 1755 AD (Druce 1966), but it may have subsequently extruded lava, forming a dome after this date (Platz 2007). It is capable of fairly large eruptions, and has a history of sector collapse (e.g., Neall 2003). Taranaki Volcano is surrounded by productive agricultural land and is in a major hydrocarbon (gas and oil) production region. There is a regional population of just over 100,000 people, most of whom live in the city of New Plymouth.

The intraplate Puhipuhi-Whangarei Volcanic Field (PWVF), the Kaikohe-Bay of Islands Volcanic Field (KBOIVF) and Auckland Volcanic Field (AVF) are in northern New Zealand (Fig. 1). PWVF is thought to have been active as recently as 0.26 Ma, while dating of the KBOIVF indicates an eruption occurred at about 0.05 Ma (Smith et al. 1993), or perhaps as recently as in 200–500 AD (Kear and Thompson

1964). AVF has had many eruptions from at least 53 basaltic vents, most recently about 600 years ago (e.g., Needham et al. 2011). Auckland city hosts a third of New Zealand's population with 1.4 million residents, and is sited directly on top of AVF.

Eruptions from most of New Zealand's volcanoes are likely to impact infrastructure of national importance, including many State Highways and road networks, electricity lines and power stations, train lines, water supplies, and sewage facilities (Wilson et al. 2012). Additionally, industries important to the local, regional, and national economies may be threatened during future eruptions, including the tourism, agricultural, forestry, and hydrocarbon industries.

1.2 Communication of Volcano-Related Information in New Zealand

In New Zealand, GNS Science is the agency appointed by the Government to provide scientific advice to local, regional, and central government organisations for geological hazards, as stated in the Guide to the National Civil Defence Emergency Management (CDEM) Plan (MCDEM 2015a) and a Memorandum of Understanding with the Ministry of Civil Defence and Emergency Management (MCDEM)(GNS Science, The Ministry of Civil Defence and Emergency Management 2015). The MOU outlines the obligations of GNS Science for geohazard warnings, whereas the Guide to the CDEM Plan is to assist New Zealand agencies to achieve the objectives of the National CDEM Plan (MCDEM 2015b). New Zealand's volcanoes are today monitored by GNS Science through the GeoNet project (Scott and Travers 2009), funded primarily by the New Zealand Earthquake Commission (EQC).

Volcano-related information is communicated to stakeholders, including the public, in a variety of formats before, during, and after volcanic crises. The primary tool used is the Volcano Alert Bulletin (VAB), supported by web page information on GeoNet News,[3] blogs, news items and social media tools (Facebook, twitter). Volcano hazard and status information is also presented by scientists during meetings, conferences, workshops, and public lectures; websites; in scientific and non-scientific publications; and via the media. Smartphone app push alerts, emails, faxes, pager alerts, and text messages provide one-way information to registered end-users during crises or changes in volcanic activity. Volcanic ash impact posters (a product of the Volcanic Impact Study Group, commissioned by the Auckland Lifelines Group) provide accessible information for critical infrastructure stakeholders (Wilson et al. 2014). Social media and 'ask an expert' interactive online sessions allow questions to be asked by the public and answered by scientists in real-time. Informal conversations during meetings, workshops, or by telephone provide end-users with more specific information from volcanologists, with the opportunity for two-way communication. An example of this is during Volcanic Advisory Group meetings, which are attended by key stakeholders and volcanologists to discuss volcano- and response-related information (Doyle et al. 2011). Long-term hazard maps have been created for some of the more active volcanoes, based on geological evidence of past eruptions (Neall and Alloway 1996; Scott and Nairn 1998). Event-specific hazard maps are created during unrest depending on the situation, likely vent location, and the style and magnitude of the potential eruption, etc. Event-specific hazard maps were created prior to and after the Tongariro eruption in 2012 (Leonard et al. 2014).

New Zealand's VAAC, based at the Meteorological Service of New Zealand Ltd. (MetService) office in Wellington, is designated by the International Airways Volcano Watch system to communicate ash information for a large section of the southwest Pacific, including New Zealand's active volcanoes (Lechner 2012). MetService issues Volcanic Ash Advisories in a text and graphic form, and disseminates a Significant Meteorological Information (SIGMET) message, while Airways Corporation issue Notice to Airmen (NOTAM), which draws attention to volcanic ash hazards within the NZ VAAC area. These globally standardised messages are also issued when New Zealand's VAL changes, prompting restrictions to local air space. After consultation with GNS Science for NZ volcanoes, Volcanic Ash Advisories are communicated by the VAAC to MCDEM, in addition to being provided to international aviation agencies and meteorological communities (MCDEM 2015a). Volcanic Ash Advisories from MetService forecast the distribution of volcanic ash in the atmosphere for the purpose of aviation safety, whereas GNS Science issues ashfall prediction maps as a VAB, relating to the distribution and thickness of tephra deposits at ground level. In addition to these systems, GNS Science issues Volcano Observatory Notices for Aviation (VONA) to the VAAC to report on ground-based volcanic activity whenever they change the Aviation Colour Code (ACC; Table 1). The ACC is used by the Civil Aviation Authority of New Zealand to alert the aviation industry to changes in the status of volcanoes within the designated coverage area (Lechner 2012).

1.3 New Zealand's Past VAL Systems

In New Zealand, scientists at GNS Science determine the VAL as mandated in the Guide to the National CDEM Plan (MCDEM 2015a), with consideration of monitoring data and using their experience and knowledge. When decisions need to be made rapidly (e.g., if an eruption has taken place), the Volcano Duty Officer can make the VAL decision alone. GNS Science, through GeoNet, communicates this information to MCDEM using Volcanic Alert Bulletins. MCDEM forwards this information on to local authorities and CDEM Groups through the National Warning System. GNS Science also disseminates this information to other agencies, the public, and the media (Scott and Travers 2009).

New Zealand's first VAL system (known as the 'Scientific Alert Level' or SAL table;

---

[3]GeoNet News web page can be found at: http://info. geonet.org.nz/.

Table 2) was introduced in 1994. It was designed by local volcanologists, and included descriptions for different levels of activity for several types of volcanoes, sometimes within a single level of the table. For example, some levels included descriptions for both unrest and eruptions. Later this was to cause confusion as people were not sure which description the assigned VAL referred to.

Several teething issues (including media scrutiny) arose during volcanic unrest at Mt. Ruapehu in late 1994, in part due to the conflicting definitions causing confusion. The original SAL system was therefore reviewed, and a significantly revised and amended version was adopted in September 1995, just one week before the 1995–96 Mt. Ruapehu eruption episode started. The resulting system (Table 3), renamed as the VAL system in 2008, was divided into two separate sections, one for frequently active volcanoes, and the other for reawakening volcanoes. In addition to a level for 'background activity' (VAL 0), the frequently active volcanoes system included one level for unrest (VAL 1) and four levels of increasing magnitudes of eruption (VAL 2–5), whereas the reawakening volcanoes system included two levels for unrest (VAL 1 and 2) and three eruption levels (VAL 3–5). The VAL system was used until it was again revised in 2014.

The division of the VAL system based on eruptive activity was deemed to be beneficial during the creation of this system because the outcome of unrest was perceived to be more uncertain for reawakening volcanoes than frequently active volcanoes due to no eruptions being witnessed (except for the 1886 eruption at Okataina). Similarly, calderas (which were predominantly in the reawakening volcanoes group) were seen as more likely to exhibit unrest without resulting in an eruption than stratovolcanoes. Many scientists have the perception that end-users, who are more familiar with stratovolcano eruptions, think unrest will predominantly result in an eruption (Potter 2014). Thus, by separating reawakening volcanoes from frequently active volcanoes, it is implied that the volcanoes behave differently, and that unrest at reawakening volcanoes may not result in an

eruption. An additional level of heightened unrest was inserted into the reawakening system to help reinforce this meaning.

2 Reviewing New Zealand's VAL Systems

While the VAL system that was developed in 1995 performed well for fifteen years, it underwent an exploratory review between 2010 and 2014 to ensure it was the best system possible. Potter et al. (2014) used a qualitative ethnographic methodology consisting of interviews, observations and document analysis to investigate the VAL system, with the involvement of both scientist and end-user groups. For further details on the methodology and full results of this research, refer to Potter et al. (2014) and Potter (2014). Based on the results of this research, the VAL system was revised (Fig. 3) and implemented in collaboration with MCDEM on 1 July 2014.

2.1 Standardising Multiple Systems into One for All Volcanoes

Many of the research participants identified the division between frequently active volcanoes and reawakening volcanoes in the 1995–2014 VAL system as a concern (Potter 2014). It was recognised that the use of two systems:

- Complicated a system that was intended to be a simple communication tool.
- May cause confusion in the future if two volcanoes exhibiting different levels of surface activity were allocated the same VAL system.
- May cause confusion because as reawakening volcanoes become more active they may switch sides to become frequently active (and vice versa). This was the case in 2006 when an eruption occurred at Raoul Island.

It was undefined whether the volcanoes were grouped according to the time since the last eruption and/or the recurrence rate of eruptions.

Table 1 The International Civil Aviation Organization (ICAO) Aviation Colour Code for volcanic activity (ICAO 2004)

ICAO Colour Code	Status of activity of volcano
Green	Volcano is in normal, non-eruptive state
	Or, after a change from a higher alert level:
	Volcanic activity considered to have ceased, and volcano reverted to its normal, non-eruptive state
Yellow	Volcano is experiencing signs of elevated unrest above known background levels
	Or, after a change from a higher alert level:
	Volcanic activity has decreased significantly but continues to be closely monitored for possible renewed increase
Orange	Volcano is exhibiting heightened unrest with increased likelihood of eruption. *Or,* volcanic eruption is underway with no or minor ash emission *[specify ash-plume height if possible]*
Red	Eruption is forecasted to be imminent with significant emission of ash into the atmosphere likely
	Or, eruption is underway with significant emission of ash into the atmosphere *[specify ash-plume height if possible]*

Table 2 Scientific Alert Level table introduced in 1994 (sourced from Annexe C from the CDEM Plan)

Scientific Level	Phenomena Observed	Scientific Interpretation
1	Abnormal seismic, hydrothermal or other signatures	Initial sign of volcano reawakening. No eruption imminent. Possible minor activity
2	Increase in seismic, hydrothermal and other unrest indicators. Increase from usual background weak eruptions	Indicators of intrusion process or significant change in on-going eruptive activity
3	Relatively high and increasing unrest shown by all indicators. Commencement of minor eruptive activity at reawakening vent(s) or increased vigour of on-going activity	If increasing trends continue there is a real possibility of hazardous eruptive activity
4	Rapid acceleration of unrest indicators. Established magmatic activity at reawakening vents or significant change to on-going activity	Hazardous volcanic eruption is now imminent
5	Hazardous volcanic eruption in progress	Destruction within the Permanent Danger Zone (red zone) and significant risk over wider areas

Other criteria for grouping volcanoes were also considered. Options identified included grouping volcanoes by their:

- Type (such as volcanic fields vs. calderas vs. stratovolcanoes)
- Potential size of eruption (however even the most explosive volcanoes predominantly have small eruption sizes)
- Tectonic setting (intraplate vs. subduction zone)
- Typical risk from an eruption (e.g., Auckland Volcanic Field vs. Raoul Island)

Table 3 New Zealand's VALS used between 1995 and 2014. Reproduced from the MCDEM (2006) Guide to the National CDEM Plan, prior to its 2014 revision

Frequently active cone volcanoes White Island, Tongariro/Ngauruhoe, Ruapehu, Kermadecs		VOLCANIC ALERT LEVEL	Reawakening volcanoes Northland, Auckland, Mayor Island, Rotorua, Okataina, Taupo, Egmont/Taranaki	
Volcano status	Indicative phenomena		Indicative phenomena	Volcano status
Usual dormant, or quiescent state	Typical background surface activity, seismicity, deformation and heat flow at low levels.	0	Typical background surface activity; deformation, seismicity, and heat flow at low levels.	Usual dormant, or quiescent state.
Signs of volcano unrest	Departure from typical background surface activity.	1	Apparent seismic, geodetic, thermal or other unrest indicators.	Initial signs of possible volcano unrest. No eruption threat.
Minor eruptive activity	Onset of eruptive activity, accompanied by changes to monitored indicators.	2	Increase in number or intensity of unrest indicators (seismicity, deformation, heat flow and so on).	Confirmation of volcano unrest. Eruption threat.
Significant local eruption in progress.	Increased vigour of ongoing activity and monitored indicators. Significant effects on volcano, possible effects beyond.	3	Minor steam eruptions. High increasing trends of unrest indicators, significant effects on volcano, possible beyond.	Minor eruptions commenced. Real possibility of hazardous eruptions
Hazardous local eruption in progress.	Significant change to ongoing activity and monitoring indicators. Effects beyond volcano.	4	Eruption of new magma. Sustained high levels of unrest indicators, significant effects beyond volcano.	Hazardous local eruption in progress. Large-scale eruption now possible.
Large hazardous eruption in progress.	Destruction with major damage beyond volcano. Significant risk over wider areas.	5	Destruction with major damage beyond active volcano. Significant risk over wider areas.	Large hazardous volcanic eruption in progress.

- Geographical region (or existing Volcanic Advisory Group).

One VAL system for each volcano was also considered, with the perceived benefit of being locally appropriate. However, this would result in at least 15 systems in New Zealand, most of which require a response by the same group of stakeholders and scientists due to the relatively small population size and land area, and having only one volcano observatory (at GNS Science near the township of Taupo). It is more likely

that having multiple systems in this situation will lead to confusion and mismanagement than in a larger country where separate groups of people are responding to the same, familiar volcano over time. Many participants specifically stated that they would not want the over-complication of having too many VAL systems.

The division of volcanoes into separate VAL systems should be considered very carefully. The need for the VAL system to be used as a simple communication tool very likely outweighs any benefits of multiple tailored and more detailed

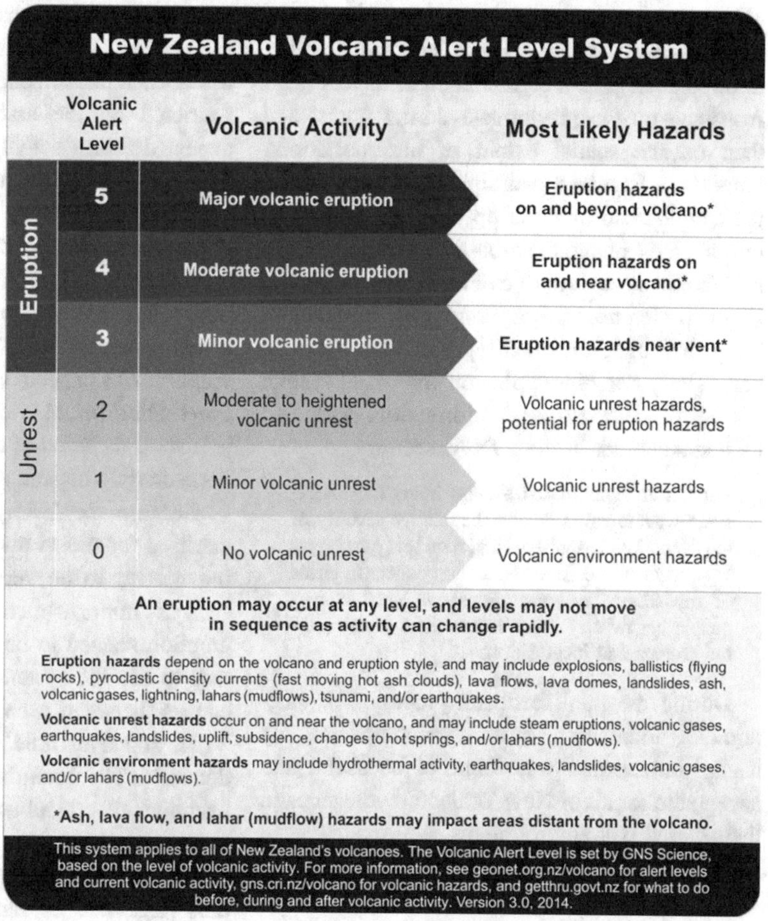

New Zealand Volcanic Alert Level System

Volcanic Alert Level	Volcanic Activity	Most Likely Hazards
5	Major volcanic eruption	Eruption hazards on and beyond volcano*
4	Moderate volcanic eruption	Eruption hazards on and near volcano*
3	Minor volcanic eruption	Eruption hazards near vent*
2	Moderate to heightened volcanic unrest	Volcanic unrest hazards, potential for eruption hazards
1	Minor volcanic unrest	Volcanic unrest hazards
0	No volcanic unrest	Volcanic environment hazards

Levels 3–5 labelled "Eruption"; Levels 0–2 labelled "Unrest".

An eruption may occur at any level, and levels may not move in sequence as activity can change rapidly.

Eruption hazards depend on the volcano and eruption style, and may include explosions, ballistics (flying rocks), pyroclastic density currents (fast moving hot ash clouds), lava flows, lava domes, landslides, ash, volcanic gases, lightning, lahars (mudflows), tsunami, and/or earthquakes.

Volcanic unrest hazards occur on and near the volcano, and may include steam eruptions, volcanic gases, earthquakes, landslides, uplift, subsidence, changes to hot springs, and/or lahars (mudflows).

Volcanic environment hazards may include hydrothermal activity, earthquakes, landslides, volcanic gases, and/or lahars (mudflows).

*****Ash, lava flow, and lahar (mudflow) hazards may impact areas distant from the volcano.**

This system applies to all of New Zealand's volcanoes. The Volcanic Alert Level is set by GNS Science, based on the level of volcanic activity. For more information, see geonet.org.nz/volcano for alert levels and current volcanic activity, gns.cri.nz/volcano for volcanic hazards, and getthru.govt.nz for what to do before, during and after volcanic activity. Version 3.0, 2014.

VAL system. For these reasons, the revised VAL system was designed to be used for all of New Zealand's volcanoes, regardless of factors such as the type, setting, frequency of eruptions, or typical eruption style.

The foundation of the VAL system was also explored in order to determine how the level of volcanic activity could best be communicated (Potter et al. 2014). The 1995–2014 VAL system was based on the severity of the volcano phenomena (e.g., magnitude of eruption); it ranged from 'background activity' to 'large hazardous eruption'. The perceived benefits of this foundation included:

(a) Scientists were most knowledgeable and "comfortable" in determining the severity of phenomena, as opposed to considering other elements of risk. This would lead to less uncertainty and shorter warning times.

(b) The severity of phenomena was seen as the first step in communication, and as being more relevant for a wider range of stakeholders. Interpretation and forecasting information can subsequently be tailored to various audiences, environments, and situations in other communication products.

Various other foundations were considered, including: the level of hazard (taking into account the geological and recent eruptive history of a volcano and spatial extent of hazards, but not the exposure and vulnerability of populations); volcano processes and state of the underlying magma system (ranging from 'no magma' to 'large extrusion of magma'); the level of risk (taking into account the severity of the hazard as well as the exposure and vulnerability of populations); or a combination of factors (e.g., focussing on the phenomena during unrest and then on the spatial extent of hazards during eruptions). Research participants were asked for their preference of these options; scientists preferred the phenomena-based system, while stakeholders were more evenly spread but had a slight preference for a combined foundation. They also suggested other types of VAL systems, particularly retaining a phenomena-based system that also included hazard information. As stated by a stakeholder in the CDEM sector:

> The phenomenon-based system helps me understand what is going on and the relative severity of the event. The hazard-based system sets out clearly what needs to be done as a consequence. In terms of my CDEM responsibilities, we need both—people get twitchy about instructions given without context and justification.

During the final feedback process, this phenomena foundation system accompanied by hazard information was found to be useful and acceptable for all of New Zealand's volcanoes in their varied risk environments.

2.1.1 Other Considerations When Standardising the VAL System

Very careful consideration was given to all words in the VAL system (Fig. 3) by its developers (Potter 2014; Potter et al. 2014). Not only did it need to be effective during escalation, de-escalation and static levels of volcanic activity, it also needed to be appropriate for the wide range of volcano types and settings in New Zealand. For example, the term 'vent' was used in VAL 3 in the hazard column instead of 'crater' because some volcanoes have very large craters (e.g., Taupo), or an eruption may occur from a vent on a volcano flank. The use of 'vent', 'near volcano' and 'beyond volcano' is part of the introduction of a dimensionless nomenclature to the VAL system.

No eruption forecasting language was included in the VAL system (beyond 'potential for eruption hazards' in VAL 2), because the capabilities and experience of the volcanologists in forecasting at each of New Zealand's volcanoes is unequal. The use of each phenomena-based alert level would be restricted by the associated description of the expected future activity, which at some of New Zealand's volcanoes, will be very uncertain. For example, if statements such as 'eruption expected within the next two weeks' were included in VAL 2, volcanologists would not be able to communicate that a volcano was showing heightened levels of unrest unless they also thought that an eruption was expected within the next two weeks. Forecasting information specific to each volcano is instead included in supplementary information, particularly Volcanic Alert Bulletins (VABs).

Because the VAL system needs to be standardised for use at multiple volcanoes, the wording needed to be very simple. Therefore, terms such as 'minor', 'moderate' and 'major' volcanic eruption needed to be defined in order for scientists to use the system consistently between volcanoes, between each other when voting on the VAL, and over time. A GNS Science guideline document has been drafted for this purpose with examples of typical activity shown at each VAL by various volcanoes. This approach was taken rather than describing monitoring criteria thresholds (e.g., rate of earthquakes or rate of deformation), to ensure the system can be used for every volcano regardless of its setting, and to give the scientists more flexibility. A GNS Science YouTube video[4] was developed to help communicate the typical levels of activity for each of the VAL systems to the public and stakeholders.

3 Lessons Learnt from the NZ Case Study in Relation to the Standardisation of VAL System

When considering whether to utilise a standardised warning approach, Potter et al. (2014) explored the purpose of the VAL system, the information needs of New Zealand's

---

[4]www.youtube.com/watch?v=WeZxW2xyam0&list=UUTL_U_K1eP4T885-JL3rVgw, accessed on 11 May 2017.

stakeholders, and the capabilities of the volcano monitoring system. They paid particular attention to the benefits and challenges of combining warning systems for all of the volcanoes into one. This included determining the foundation of the VAL system, the words used in the table, whether forecasting language should be included, and how the system was going to be used consistently over time and at multiple volcanoes. As identified by the IDNDR Early Warning Programme Convenors (1997), locally appropriate communication methods should be established for the distribution of warnings. The social science research used for this investigation was a robust process that enabled the revision of the VAL system to be based on evidence, as advocated by Leonard and Potter (2015). The resulting VAL system has been used for all of New Zealand's volcanoes since June 2014. It has worked satisfactorily to date. For example, volcanic unrest or eruptions with which the revised system has been successfully used have included: unrest at Ngauruhoe (VAL raised to 1 and later lowered); decrease in unrest at Te Maari (VAL lowered to 0); and small eruptions at White Island (VAL raised to 3 and lowered later). Volcanologists at GNS Science have found the simple descriptions in the revised VAL table beneficial for allowing flexible decision-making when determining the level of activity. Having just one system for all NZ volcanoes has also improved clarity. Regular evaluations of this warning tool will take place in the future, involving stakeholders, volcanologists, and the public. There is more monitoring data and interpretation relevant to end-users and decisions than the VAL itself and in New Zealand such information is included in the accompanying VAB. Based on the recent experiences with review and implementation of New Zealand's VAL systems, we strongly recommend stakeholders consider exactly what parameters, impacts, uncertainties and lead-times are important to each decision that needs to be made (e.g., evacuation) and not simply tie responses to changes in the VAL.

The revised VAL system has been developed for the New Zealand context, including our volcanic settings and risk environments, the roles and responsibilities of our agencies, and our social and cultural environments including the centralised nature of our volcano monitoring and warning system. As such, it is unlikely that it is able to be directly copied for other countries. However, the process that we followed can be used, as summarised in the next section.

4 Recommendations for Reviewing or Developing a VAL System

This chapter has focussed on aspects relating to standardisation, when reviewing a VAL system. There are other considerations to take into account as well. We describe below our recommended processes and considerations when reviewing a VAL system (or developing a new one), based on our research and experience.

(1) Understand the context

It is vital to understand the physical, cultural, social, organisational and historical context of the VAL and related systems. Potter et al. (2014) found that using the qualitative methodology of ethnography allowed a deep understanding of the culture of the volcanologists to be built to address many of the following factors. However, we recognise that this is a time-consuming process that is not an option for many observatories looking to revise their VAL systems. If this is the case, then drawing on published material, attending volcano monitoring meetings, and holding discussions with those familiar with the various environments should be sufficient. We recommend understanding as much as possible about the:

- Range of potential volcanic activity at every volcano that the VAL system may be used for, including frequency of eruptions, level of ongoing unrest or eruptions, potential magnitude of eruptions (and unrest phenomena), and severity of all possible hazards.
- Volcano monitoring system to understand capabilities and factors such as timing, uncertainties and content of incoming data.

- Level of exposure and vulnerability of elements at risk to volcanic hazards, including the built, social, economic and natural environments.
- Roles and responsibilities, including legislative requirements, of scientific advisors from all institutions, and organisations with the role of planning, education, response and recovery from volcanic events (including governmental and civil defence agencies, infrastructure/lifelines, emergency services, health, agricultural/horticultural and business sectors). For example, understand which agencies have the responsibility to communicate directly to the public.
- Influences on the VAL decision-making process. This includes understanding the cultures of people/groups determining the VAL system, and the receivers of the information including decision-makers and stakeholders, and the public. For example, influences on the VAL decision-making process may include experience, external pressure, peer pressure and other social psychology biases, internal voting guidelines, how individuals interpret the content and structure of the VAL system, and the desire to maintain credibility or conduct fieldwork (Potter 2014). Factors such as these contributed towards the design of the revised NZ VAL system. Additionally, understanding the way stakeholders read volcano-related information and use it in their decision-making contributed towards determining what information was included in the NZ VAL system, how it is communicated to them.
- Previous VAL systems, and any other existing alerting systems for volcanic or other hazards used in the country or for the volcano in question. What were people's experiences with those systems? What worked well or didn't work well? Are there any other warning systems being used on or near the volcano, including the international ACC? Understanding VAL systems that have been implemented in other parts of the world is also useful. Being familiar with the challenges and benefits of standardisation as

outlined in this chapter is relevant for this point.
- What other communication avenues exist for related information from all agencies? For example, Volcanic Alert Bulletins, phone calls, meetings, websites, emails, social media. These provide the context of whether the VAL system will need to include all important information as a standalone system, or if it can be supported by other channels.

(2) Understand the challenges and benefits of the existing VAL system

It is important to know who the audience is for the VAL system, and what they use it for. If the system is targeted at stakeholders and decision-makers, perhaps more technical and specific information could be included than if the audience includes the public. This information might also inform the position of the divisions between alert levels by matching it to their decision-making needs. However, due to the wide range of stakeholder needs and the differing points at which they need to take action, coupled with a system that communicates to multiple audiences, it is likely that discussions will need to be held to encourage stakeholders to determine their own decision points, rather than them relying on changes in alert levels. Understand their perception of the purpose of the VAL system, and their experiences with any existing VAL systems, through methods such as interviews, open-ended questions in surveys, or workshops/focus groups. Ask the volcanologists (or whomever determines the VAL) what they find useful or challenging, and see if their perceived purpose of the system matches that of the stakeholders/public. Analyse the existing system to identify jargon, unclear meanings, and the foundation of the system. Understand what channels are used to communicate it (such as websites, social media, Bulletins), as this may pose opportunities or limitations in designing a revised system. Ask all parties what they would like to see in a revised system. Determine factors such as whether it should include eruption fore-casting messages, what its foundation should be based on, and whether it is a standardised system for multiple volcanoes, or specific to a single volcano.

(3) Produce a draft version of the revised VAL system and seek feedback

By considering the context, and collating the information outlined in step 2, recommendations for a revised system can be developed. Based on those recommendations, a draft version (or multiple options) can be developed. Both the summarised findings and the draft(s) can then be circulated back to participants to ensure their perceptions and needs have been accurately captured. For the NZ revision, Potter (2014) asked participants to rank five draft versions, which helped to determine the most appropriate foundation and structure of the revised VAL system. Multiple iterations then occurred to produce the final version. Do not underestimate the amount of time needed for this process! A final version of the VAL system can now be developed (consider utilising graphics specialists).

(4) Release revised system in collaboration with stakeholders

In conjunction with key stakeholders, deter-mine a date on which the new system will be used, taking into account the length of time all parties need to update documentation, websites, etc. Release communications to circulate the change in VAL system through the media, stakeholder newsletters, meetings, etc. In NZ, GNS Science developed a media release six weeks prior to the start date of the system, and on the day of the changeover, with support from MCDEM. Make a plan for if a new eruptive episode should start close to the changeover date (thankfully no eruptions occurred for the chan-geover date in NZ!). Write any supporting doc-umentation or procedures, such as a guideline for consistent use by the volcanologists, or whether the VAL system will be used exactly as written or more flexibly.

(5) Evaluate the revised VAL System

Conduct regular evaluations to ensure the revised (or new) VAL system is effective, and meeting the needs of stakeholders, the public, and volcanologists. Real events and exercises can be used.

5 Volcanic Crisis Communication

Warnings about natural hazard events are com-municated in order to minimise losses (Newhall 2000). However, trust and communication net-works must already be in place prior to a crisis for effective planning and response. This can be achieved by developing networks, ascertaining information needs and establishing methods of effective communication (Paton et al. 1998). VAL systems are just one aspect of an EWS, and by design are the simplest tool to communicate the status of the volcano (or level of response required, etc.). Due to this overarching purpose, standardising VAL systems can help ensure a consistent, simple, and understandable design. As we have outlined in this chapter however, there are issues with using a one-size-fits-all communication product. This was also identified by Thompson et al. (2015) in using probabilistic volcanic hazard maps. We found that taking into account the local context is vital, which supports the findings of numerous recent research in vol-canic crisis communication (Haynes et al. 2007; Fearnley et al. 2012; Potter et al. 2014). Pro-viding supporting information using other means helps to alleviate these issues.

Over the past two decades, social science has increasingly played a valuable role in mitigating volcanic risks by providing evidence-based links between communities, stakeholders and scien-tists (Barclay et al. 2008; Leonard and Potter 2015). We recommend that in the future these robust methodologies are embraced by volcano observatories, such as when revising communication strategies and products, to help effectively share information and reduce the risk to society.

Acknowledgements The authors would like to thank all research participants, supervisors and colleagues for their contribution to this research. The research was primarily funded by the Government of New Zealand through the Natural Hazards Research Platform, and through the Earthquake Commission.

References

Barclay J, Haynes K, Mitchell T, Solana C, Teeuw R, Darnell A, Crosweller HS, Cole P, Pyle D, Lowe C, Fearnley C (2008) Framing volcanic risk communication within disaster risk reduction: finding ways for the social and physical sciences to work together. Geol Society Lond Spec Publ 305(1):163–177

Carsell KM, Pingel ND, Ford DT (2004) Quantifying the benefit of a flood warning system. Nat Hazards Rev 5(3):131–140

Davy BW, Caldwell TG (1998) Gravity, magnetic and seismic surveys of the caldera complex, Lake Taupo, North Island, New Zealand. J Volcanol Geoth Res 81:69–89

Doyle EEH, Potter SH (2016) Methodology for the development of a probability translation table for GeoNet. GNS Science report 2015/67. GNS Science, Lower Hutt, New Zealand, p 18

Doyle EE, Johnston DM, McClure J, Paton D (2011) The communication of uncertain scientific advice during natural hazard events. NZ J Psychol 40(4):39–50

Druce AP (1966) Tree-ring dating of recent volcanic ash and lapilli, Mt Egmont. NZ J Bot 4(1):3–41

Fearnley CJ (2011) Standardising the USGS volcano alert level system: acting in the context of risk, uncertainty and complexity. Ph.D thesis, University College London, London, UK

Fearnley CJ (2013) Assigning a volcano alert level: negotiating uncertainty, risk, and complexity in decision-making processes. Enviro Plan A 45(8):1891–1911

Fearnley CJ, McGuire WJ, Davies G, Twigg J (2012) Standardisation of the USGS Volcano Alert Level System (VALS): analysis and ramifications. Bull Volc 74(9):2023–2036

Garcia C, Fearnley CJ (2012) Evaluating critical links in early warning systems for natural hazards. Environ Hazards 11(2):123–137

Gardner CA, Guffanti MC (2006) U.S. geological survey's alert notification system for volcanic activity. U.S. Geological Survey Fact Sheet 2006–3139. U S Geol Surv

GNS Science, The Ministry of Civil Defence and Emergency Management (2015) Memorandum of understanding between the Ministry of Civil Defence

& Emergency Management (MCDEM) and Institute of Geological and Nuclear Sciences Limited (GNS Science) for the engagement of geoscience and Civil Defence Emergency Management. pp 18

Guffanti MC, Miller TP (2013) A volcanic activity alert-level system for aviation: review of its development and application in Alaska. Nat Hazards 69:1519–1533. doi:10.1007/s11069-013-0761-4

Haynes K, Barclay J, Pidgeon N (2007) Volcanic hazard communication using maps: an evaluation of their effectiveness. Bull Volc 70(2):123–138. doi:10.1007/s00445-007-0124-7

Hogg A, Lowe DJ, Palmer J, Boswijk G, Ramsey CB (2012) Revised calendar date for the Taupo eruption derived by ^{14}C wiggle-matching using a New Zealand kauri ^{14}C calibration data set. Holocene 22(4):439–449

Hurst AW, McGinty PJ (1999) Earthquake swarms to the west of Mt Ruapehu preceding its 1995 eruption. J Volcanol Geoth Res 90(1–2):19–28

ICAO (2004) Handbook on the International Airways Volcano Watch (IAVW). Doc 9766-AN/968: International Civil Aviation Organization (ICAO)

IDNDR early warning programme convenors (1997) Guiding principles for effective early warning. Convenors of the international expert groups on early warning. United nations international decade for natural disaster reduction, Geneva, Switzerland

Kear D, Thompson BN (1964) Volcanic risk in Northland. NZ J Geol Geophys 7(1):87–93

Lechner P (2012) Living with volcanic ash episodes in civil aviation: the New Zealand Volcanic Ash Advisory System (VAAS) and the International Airways Volcano Watch (IAVW) Civil Aviation Authority of New Zealand

Leonard G, Potter S (2015) Developing effective communication tools for volcanic hazards in New Zealand, using social science. In: Loughlin SC, Sparks S, Brown SK, Jenkins SF, Vye-Brown C (eds) Global volcanic hazards and risk. Cambridge University Press, Cambridge, UK, pp 305–310

Leonard GS, Johnston DM, Paton D, Christianson A, Becker J, Keys H (2008) Developing effective warning systems: ongoing research at Ruapehu volcano, New Zealand. J Volcanol Geoth Res 172(3–4):199–215. doi:10.1016/j.jvolgeores.2007.12.008

Leonard GS, Stewart C, Wilson TM, Procter JN, Scott BJ, Keys HJ, Jolly GE, Wardman JB, Cronin SJ, McBride SK (2014) Integrating multidisciplinary science, modelling and impact data into evolving, syn-event volcanic hazard mapping and communication: a case study from the 2012 Tongariro eruption crisis, New Zealand. J Volcanol Geoth Res 286:208–232. doi:10.1016/j.jvolgeores.2014.08.018

Lindsay J, Marzocchi W, Jolly G, Constantinescu R, Selva J, Sandri L (2010) Towards real-time eruption forecasting in the Auckland Volcanic Field: application of BET_EF during the New Zealand national disaster exercise 'Ruaumoko'. Bull Volc 72(2):185–204. doi:10.1007/s00445-009-0311-9

Maskrey MA (1997) IDNDR Secretariat, Geneva, Oct 1997

MCDEM (2015a) The guide to the National Civil Defence Emergency Management Plan 2015. Ministry of Civil Defence & Emergency Management, Wellington. Retrieved from http://www.civildefence.govt.nz/assets/guide-to-the-national-cdem-plan/Guide-to-the-National-CDEM-Plan-2015.pdf

MCDEM (2015b) National Civil Defence Emergency Management plan order 2015. (2015/140) Ministry of Civil Defence and Emergency Management,New Zealand. Retrieved from http://www.legislation.govt.nz/regulation/public/2015/0140/latest/DLM6486453.html?src=qs%20

Mileti DS, Sorensen JH (1990) Communication of emergency public warnings—a social science perspective and state-of-the-art assessment. Laboratory, Oak Ridge National, p 166

Nairn IA (2002) Geology of the Okataina Volcanic Centre. Geological map 25, scale 1:50 000, 1 sheet + 156p. Institute of Geological & Nuclear Sciences Limited, Lower Hutt

Nairn IA, Houghton BF, Cole JW (1991) Volcanic hazards at White Island. Volcanic hazards information series, vol 3. Ministry of Civil Defence

Neall VE (2003) The volcanic history of Taranaki. Massey University, Palmerston North, New Zealand, Institute of Natural Resources

Neall VE, Alloway BV (1996) Volcanic hazards at Egmont Volcano; volcanic hazard map of western Taranaki. Volcanic hazards information series, vol 12. Occasional Report. Massey University, New Zealand

Needham AJ, Lindsay JM, Smith IEM, Augustinus P, Shane PA (2011) Sequential eruption of alkaline and sub-alkaline magmas from a small monogenetic volcano in the Auckland Volcanic Field, New Zealand. J Volcanol Geoth Res 201(1–4):126–142. doi:10.1016/j.jvolgeores.2010.07.017

Newhall CG (2000) Volcano warnings. In: Sigurdsson H, Houghton BF, McNutt SR, Rymer H, Stix J (eds) Encyclopedia of Volcanoes. Academic Press, San Diego, CA, pp 1185–1197

Paton D, Johnston DM, Houghton B (1998) Organisational response to a volcanic eruption. Disaster Prev Manage 7(1):5–13

Platz T (2007) Aspects of dome-forming eruptions from andesitic volcanoes through the Maero Eruptive Period (1000 yrs B.P. to present): activity at Mt. Taranaki, New Zealand. Ph.D thesis, Massey University, Palmerston North, New Zealand

Potter SH (2014) Communicating the status of volcanic activity in New Zealand, with specific application to caldera unrest. Ph.D thesis, Massey University, Wellington, New Zealand. Retrieved from http://mro.massey.ac.nz/handle/10179/5654

Potter SH, Scott BJ, Jolly GE (2012) Caldera unrest management sourcebook. GNS Science Report 2012/12. GNS Science, Lower Hutt, New Zealand, p 73

Potter SH, Jolly GE, Neall VE, Johnston DM, Scott BJ (2014) Communicating the status of volcanic activity: revising New Zealand's volcanic alert level system. J Appl Volcanol 3(13)

Potter SH, Scott BJ, Jolly GE, Johnston DM, Neall VE (2015) A catalogue of caldera unrest at Taupo Volcanic Centre, New Zealand, using the Volcanic Unrest Index (VUI). Bull Volc 77:78. doi:10.1007/s00445-015-0956-5

Scott BJ (1978) Volcano and geothermal observations 1977: observed activity at Ngauruhoe. New Zealand Volcanological Record (pp 44–45): NZ Geol Surv, DSIR

Scott BJ, Nairn IA (1998) Volcanic hazards: Okataina volcanic centre. Scale 1:100,000: Bay of Plenty Regional Council. Resource planning publication/Bay of Plenty Regional Council, New Zealand 97/4

Scott BJ, Potter SH (2014) Aspects of historical eruptive activity and volcanic unrest at Mt. Tongariro, New Zealand: 1846–2013. J Volcanol Geoth Res 286 (Special issue: Tongariro Eruption 2012):263–276

Scott BJ, Travers J (2009) Volcano monitoring in NZ and links to SW Pacific via the Wellington VAAC. Nat Hazards 51(2):263–273. doi:10.1007/s11069-009-9354-7

Self S (2006) The effects and consequences of very large explosive volcanic eruptions. Philos Trans R Soc A Math Phys Eng Sci 364(1845):2073–2097. doi:10.1098/rsta.2006.1814

Smith IEM, Okada T, Itaya T, Black PM (1993) Age relationships and tectonic implications of late Cenozoic basaltic volcanism in Northland, New Zealand. NZ J Geol Geophys 36(3):385–393

Thompson MA, Lindsay JM, Gaillard JC (2015) The influence of probabilistic volcanic hazard map properties on hazard communication. J Appl Volcanol 4(1):1–24

UN ISDR (2003) Terminology: basic terms of disaster risk reduction retrieved 25/08/2009, from http://www.unisdr.org/eng/library/lib-terminology-eng%20home.htm

United Nations (2006) Global survey of early warning systems. Final Version. 46 p. http://www.unisdr.org/2006/ppew/info-resources/ewc3/Global-Survey-of-Early-Warning-Systems.pdf (accessed 20 March 2010)

Wilson CJN (1993) Stratigraphy, chronology, styles and dynamics of late quaternary eruptions from Taupo volcano, New Zealand. Philos Trans Phys Sci Eng 343 (1668):205–306

Wilson CJN, Walker GPL (1985) The Taupo eruption, New Zealand I. General aspects. Philosophical Transactions of the Royal Society of London. Ser A Math Phys Sci 314(1529):199–228

Wilson CJN, Rogan AM, Smith IEM, Northey DJ, Nairn IA, Houghton BF (1984) Caldera volcanoes of the Taupo Volcanic Zone. J Geophys Res 89(B10): 8463–8484

Wilson CJN, Houghton BF, McWilliams MO, Lanphere MA, Weaver SD, Briggs RM (1995) Volcanic and structural evolution of Taupo Volcanic Zone, New Zealand—a review. J Volcanol Geoth Res 68(1–3): 1–28

Wilson CJN, Gravley DM, Leonard GS, Rowland JV (2009) Volcanism in the central Taupo Volcanic Zone, New Zealand: tempo, styles and controls. Studies in Volcanology: the legacy of george walker. Spec publ IAVCEI 2:225–247

Wilson T, Stewart C, Sword-Daniels V, Leonard GS, Johnston DM, Cole JW, Wardman J, Wilson GJ, Barnard ST (2012) Volcanic ash impacts on critical infrastructure. Phys Chem Earth Parts A/B/C 45–46:5–23

Wilson T, Stewart C, Wardman JB, Wilson G, Johnston DM, Hill D, Hampton SJ, Villemure M, McBride S, Leonard G (2014) Volcanic ashfall preparedness poster series: a collaborative process for reducing the vulnerability of critical infrastructure. J Appl Volcanol 3(1):1–25

Winson AE, Costa F, Newhall CG, Woo G (2014) An analysis of the issuance of volcanic alert levels during volcanic crises. J Appl Volcanol 3(1):14

More Than Meets the Eye: Volcanic Hazard Map Design and Visual Communication

Mary Anne Thompson⬛, Jan M. Lindsay
and Graham S. Leonard

Abstract

Volcanic hazard maps depict areas that may be affected by dangerous volcanic processes, such as pyroclastic density currents, lava flows, lahars, and tephra fall. These visualisations of volcanic hazard information are used to communicate with a wide variety of audiences both during times of dormancy and volcanic crisis. Although most volcanic hazard maps show similar types of content, such as hazard footprints or zones, they vary greatly in communication style, appearance, and visual design. For example, maps for different volcanoes will use different combinations of graphics, symbols, colours, base maps, legends, and text. While this variety is a natural reflection of the diverse social, cultural, political, and volcanic settings in which the maps are created, crises and past work suggest that such visual design choices can potentially play an important role in volcanic crisis communication by influencing how people understand the hazard map and use it to make decisions. Map reading is a complex process, in which people construct meaning by interpreting the various visual representations within the context of their information needs, goals, knowledge, and experience. Visual design of the map and the characteristics of the hazard map audience can therefore influence how hazard maps are understood and applied. Here, we review case studies of volcanic crises and interdisciplinary research that addresses the relationship between hazard maps, visual design, and communication. Overall,

M.A. Thompson (✉) · J.M. Lindsay
School of Environment, University of Auckland,
Private Bag 92019, Auckland, New Zealand
e-mail: m.thompson@auckland.ac.nz

J.M. Lindsay
e-mail: j.lindsay@auckland.ac.nz

G.S. Leonard
GNS Science, P.O. Box 30-968, Lower Hutt 5040,
New Zealand
e-mail: g.leonard@gns.cri.nz

Advs in Volcanology (2018) 621–640
https://doi.org/10.1007/11157_2016_47
© The Author(s) 2017
Published Online: 15 March 2017

this growing body of work suggests that volcanic hazard maps can be very useful visual tools for crisis communication if they are designed in a way that provides clear and useful information for the audience. Further, while it is important that each map is designed for its unique situation and setting, engaging with hazard map audiences to better understand their information needs and considering lessons learnt from interdisciplinary work on visual communication can help inform and guide knowledge exchange using maps.

1 Introduction

As a volcanic crisis begins to unfold, demand for information about when and where dangerous volcanic hazards might impact increases. A key medium for communicating this information is a volcanic hazard map—a visual, spatial depiction of where volcanic phenomena might occur within a certain time frame. While hazard maps play a role in managing many elements of a volcanic crisis, such as understanding relationships between hazards, identifying areas of potential danger, informing risk assessments, and planning evacuation routes, they serve as an important tool in crisis communication.

We live in an increasingly visual society, where most of us see and process images more than we read words (Lester 2014). In many cases, images can attract visual attention (Carrasco 2011), trigger information-processing (Domke et al. 2002), stimulate emotional response (Mould et al. 2012; Lester 2014), and influence decision-making (Tufte 1997; Daron et al. 2015) more than other types of media. Images can be concisely delivered in many different formats, through many different channels, and can communicate across lexical and linguistic boundaries. Hazard maps are common images used by scientists to communicate information about volcanic hazards with a wide range of audiences. These maps, and the inferences and responses that they elicit, become particularly important during crisis situations when they may become read and circulated widely. During such high-stakes, high-pressure situations, people tend to rely more

on their initial impressions and intuitive feelings about hazard and risk than on exhaustive analytical evaluation of hazard and risk information (Finucane et al. 2000). Accordingly, the way that a hazard map captures visual attention and conveys affective meaning could have a significant impact on decisions made during a volcanic crisis. It is therefore important to understand how people interact with hazard map images, and how visual communication processes influence the messages that audiences take away.

Volcanic hazard maps are created by scientists across the world, using a number of different types of datasets, methodologies, and approaches (Calder et al. 2015). For example, a map could show only one volcanic hazard (e.g., ash fall), or multiple volcanic hazards (e.g., ash fall, lava flow, and ballistic ejecta). These hazards may be depicted as intensities (e.g., centimetres of ash that are likely to accumulate) or as a set of nested or cumulative zones (e.g., high, medium, and low hazard zones). The hazard map may be based on observation of past volcanic hazard deposits, probabilistic hazard modelling, simulation of a particular hazard scenario, or information drawn from an analogue volcano. The high degrees of freedom mean that volcanic hazard maps can represent many different types of information. In reviewing 120 volcanic hazard maps from around the world, Calder et al. (2015) identify five different hazard map "types" which describe these various combinations: geology-based maps, integrated qualitative maps, administrative maps, modelling-based maps, and probabilistic maps (Fig. 1). The classification provides a way to categorise and consider the types of inputs used in developing volcanic hazard maps around the world. The way that these inputs are visualised into a final map design output are similarly diverse.

Volcanic hazard maps are traditionally created by the scientists who carry out volcanic hazard assessments. Visual design of a hazard map is therefore typically governed by factors such as the specific methodology used, common scientific and cartographic practice at the time, status of volcanic activity, social and cultural setting, and local agency standards or policy requirements in place. Variation in these factors over time and place has resulted in the vastly different layouts, formats, colour schemes, data representations, symbology, and hazard map styles being

used around the world today. While this visual diversity reflects important and unique differences in map purpose and social and volcanic setting, past crises and research over the last few decades have highlighted that such visual design choices can also carry great importance for communication, as they may influence how different audiences interpret the map and use it to make decisions regarding hazard and risk.

Maps communicate more than meets the eye. Each reader individually constructs meaning from the map through visual cognition and interpretation of the various symbols, colours, shapes, and text within the context of his or her prior knowledge and experience (MacEachren 1995; Perkins et al. 2011). Map reading is thus a

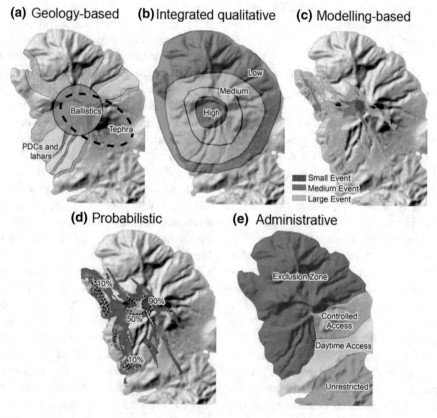

Fig. 1 Hypothetical examples of the five types of volcanic hazard map identified by Calder et al. (2015). Each map type represents a different type of input information: **a** geology-based maps, the most common type of volcanic hazard map, are based on hazard footprints of past events; **b** integrated qualitative maps are based on amalgamation of many different types of hazard information; **c** modelling-based maps are based on simulation of certain hazard scenarios; **d** probabilistic maps are based on probabilistic assessment of hazards; and **e** administrative maps are based on both hazard information but also on emergency management and administrative information. Modified from Calder et al. (2015)

complex information-processing exercise, and visual design and audience background can have a strong effect on the messages that people take away (Robinson and Petchenik 1976; Bertin 1983; MacEachren 1995; Monmonier 1996; Lloyd 2011; Perkins et al. 2011). During a crisis, hazard maps can become widely distributed and used for communicating with many different audiences. In these rapid mass communication contexts, audiences may not always consult supporting resources beyond the map image itself (e.g., Leonard et al. 2014). In such contexts, it is important to consider how visual design and communication factors influence hazard map reading, knowledge exchange, and decision-making. Here, we draw upon case studies and past work to review how volcanic hazard maps are used to visually communicate with difference audiences, and how visual design plays a role in this communication.

2 Visual Communication

Volcanic hazard maps synthesize a wealth of information about individual processes and interdependent phenomena over a range of spatial and temporal scales. As with all cartographic representations, a number of generalisations therefore have to be carried out in order to visually communicate this complex data in a clear and concise way in two dimensions. This often requires simplifying complicated physical and numerical concepts, such as particle and flow dynamics and probabilistic uncertainty. Deciding upon the most salient and useful content, and the clearest and simplest way to display that content, is a challenging, but important task. Highly complex maps are often difficult for most audiences to understand (MacEachren 1982). However, past crises and work have shown that engaging with audiences to understand the way that they perceive hazards can help guide generalisation of complex content and inform communication approaches.

2.1 Communicating Complex Content

On 13 November 1985, after a year of awakening, but with little short-term warning, the ice-capped volcano Nevado del Ruiz erupted. The eruption sent devastating lahars—turbulent mixes of snow, ice, meltwater and pyroclastic debris—down valleys and channels to the Colombian town of Armero, causing one of the worst volcanic disasters in history (Pierson et al. 1990). The Nevado del Ruiz tragedy was the result of a complex interplay between a number of technological, political, and social circumstances (Voight 1990). However, retrospective accounts recall the "state of frustration and confusion" (Voight 1990, p. 180) that arose from a "poorly understood" (Parra and Cepeda 1990, p. 117) hazard map (Fig. 2a). Although a revised hazard map was being prepared, the lahars struck two days before the planned release of the new map. Although the available map showed overall accurate content, it was displayed using scientific and probabilistic concepts that were unfamiliar to many map audiences, leading to miscommunication among authorities, the media, and the public (Parra and Cepeda 1990; Voight 1990).

In 1990, the post-event hazard map was simplified by replacing individual probabilistic hazard paths with generalized hazard zones (high, moderate, and low) (Parra and Cepeda 1990). The revision aimed to develop a map that was "easily comprehensible to non-specialists and therefore less susceptible to misinterpretation" (Parra and Cepeda 1990, p. 117). Today, the most recent Nevado del Ruiz hazard map (SGC 2015; Fig. 2b) continues this generalisation approach. Efforts to design an "intuitive" (Parra and Cepeda 1990, p. 117) map for non-scientific audiences acknowledged the important crisis communication role of volcanic hazard maps and brought attention to the importance of considering audience perspectives in designing maps. The experience led to reflection about hazard map design in other volcanically active parts of

(a) Nevado del Ruiz hazard map *October 1985*

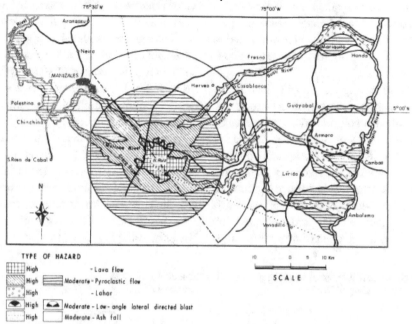

(b) Nevado del Ruiz hazard map *2015*

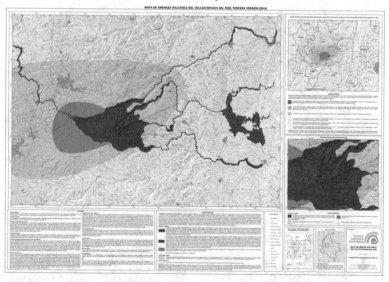

Fig. 2 Evolution of the volcanic hazard map for Nevado del Ruiz, showing **a** a simplified black-and-white version of the hazard map that was available during the time leading up to the November 1985 crisis (modified from Parra and Cepeda 1990), and **b** the current, revised hazard map produced by the Colombian Geological Survey in 2015 (SGC 2015)

the world. For example, Nakamura et al. (2008) note that the crisis sparked an evaluation of Japanese volcanic hazard maps, resulting in a design change "from being specialist-oriented to being designed to be more easily understood" (p. 297).

The value of having simple and clear hazard maps for use in crisis communication has emerged in a number of other volcanic crises, including the eruption crisis on the Caribbean island of Montserrat. On 18 July 1995, a small phreatic explosion on Soufrière Hills volcano marked the start of an eruption that would go on to continue for nearly two decades. Episodes of andesitic dome-building and collapse produced rapid, hot pyroclastic flows that devastated nearly two-thirds

of the island (Aspinall et al. 2002). The people of Montserrat were badly affected by the disaster. More than 90% of the population was displaced, and communities suffered ongoing distress and uncertainty (Kokelaar 2002; Sword-Daniels et al. 2014). Over the course of the eruption, hazard maps and risk management maps were widely used in communication with authorities and local communities (Aspinall et al. 2002).

In an effort to minimise disruption and keep as much land open to utilisation as possible, early maps used a microzonation approach, where the island was divided into seven different zones reflecting gradual levels of risk, from A (more risk) to G (less risk) (Aspinall et al. 2002; Kokelaar 2002; Fig. 3a). Microzones were tied to

(a) Montserrat microzonation map
November 1996

(b) Revised Montserrat map
September 1997

(c) Examples of aerial and perspective photos of Montserrat tested in *Haynes et al. (2007)*

Fig. 3 Black-and-white versions of maps used to communicate with the public during the Soufrière Hills eruption crisis on Montserrat in **a** November 1996 and **b** September 1997 (modified from Kokelaar 2002); and **c** examples of the aerial and perspective photographs of Montserrat, which were easier for participants to read and use than plan view maps (modified from Haynes et al. 2007)

access restrictions, which varied based on changes in an associated volcanic alert level system. However, the complex maps, together with their dynamic relationship to alert levels, were sometimes found to be "difficult to communicate to the public" (Kokelaar 2002, p. 12). Alert levels alone can be complex concepts to communicate (Fearnley et al. 2012; Potter et al. 2014). Recognising a need to simplify the maps for visual communication purposes, later versions of the map (September 1997 onwards) generalised the microzones into two to three larger zones representing different levels of access, including an exclusion zone around the volcanic edifice (Fig. 3b). The responsive change illustrated an audience-driven shift in map design, but also highlighted the challenges associated with communicating complex and interdependent content about hazard and risk.

While the Montserrat experience highlighted the importance of considering how key volcanic hazard and risk information is generalised and displayed on a map, Haynes et al. (2007) found that other fundamental elements of hazard map design can also play a role in crisis communication. Haynes et al. (2007) developed several different versions of the Montserrat hazard and risk maps that utilised a variety of different visual formats. They found that visual design elements, such as the choice of base map, influenced how local audiences used and understood the information. For example, participants were able to better identify spatial features and orient themselves with the information when it was portrayed on aerial or perspective photographs (Fig. 3c). While plan view or topographic contour maps may be an intuitive choice for an earth scientist, it may not be the most suitable choice for communicating spatial hazard information with other audiences (Haynes et al. 2007). Nave et al. (2010) found similar results in a study of Stromboli volcano hazard map styles, recommending plan view contour hazard maps for government officials, but perspective displays for non-specialist audiences. Collectively, these, and many other past volcanic crises have contributed valuable knowledge about the ways that different

audiences respond to certain hazard visualisation approaches and how this may influence crisis communication efforts.

2.2 Considering Audience Perspectives

In order to share valuable and useful knowledge about a hazard or risk with an audience, it is first important to understand the audience's existing knowledge and perspectives regarding the hazard or risk, and what information is valued and needed (Bostrom and Löfstedt 2003; Perry et al. 2016). The way that different audiences perceive volcanic hazard and risk can have an important influence on how they respond to hazard and risk communication efforts (Johnston et al. 1999; Paton et al. 2008; Gaillard and Dibben 2008; Doyle et al. 2014). Engaging with hazard map audiences to better understand their existing knowledge and perceptions of volcanic hazard and risk can therefore help guide and inform approaches to hazard and risk communication, including hazard map design. Integrative engagement with audiences can also facilitate constructive dialogue about volcanic hazards and help engender trust in the resulting maps and communication products (Cronin et al. 2004; Haynes et al. 2008; Leone and Lesales 2009; Pierson et al. 2014).

Audience perception of volcanic hazards played a key role in the redesign of the volcanic hazard maps for Mt. Ambae, the largest active volcano of the Pacific island nation of Vanuatu. Cronin et al. (2004) found that the existing scientific volcanic hazard map (Fig. 4a) was poorly understood by most people living near the volcano because of differences in the ways that the scientists and local communities perceived hazardous volcanic phenomena. In order to create a hazard map design which better aligned with audience perspectives, the scientists engaged with the local communities to better understand how locals viewed and conceptualised volcanic hazard. The hazard map was then revised to assimilate both local and scientific worldviews. For example, while scientists and locals believed the

(a) Mt. Ambae volcanic hazard map *1995*

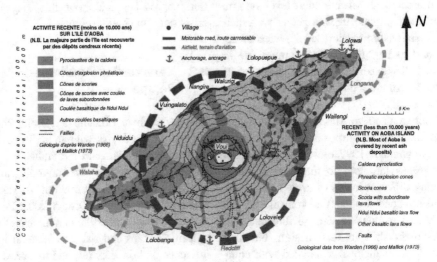

(b) Revised Mt. Ambae hazard map *Cronin et al. 2004*

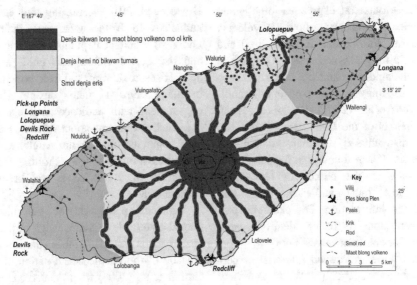

Fig. 4 Volcanic hazard maps for Mt Ambae. **a** The former scientific-style hazard map (modified from Monzier and Robin (1995)), and **b** The revised hazard map, which was developed through engagement with communities, and represents an integration of both local, traditional perspectives and outside, scientific perspectives (modified from Cronin et al. 2004)

summit area of the volcano was dangerous for different reasons, both groups acknowledged that the summit crater was a highly dangerous place. Similarly, although the scientists and locals believed in different causes of lahars, valleys were seen as particularly dangerous areas by both groups (Cronin et al. 2004). The resulting map (Fig. 4b) represents a visual integration of both traditional and outside scientific worldviews about hazardous volcanic areas, and is an example of the how engagement can help achieve common ground for visual communication.

While technical scientific hazard maps are still an essential tool for certain specialist tasks and stakeholders, different types of hazard map content may be prioritised for communication with other audiences who visit, work, and live in volcanic areas. For example, engagement with audiences in outdoor recreation areas near volcanoes in New Zealand and the United States has led to an emphasis on including life safety advice on volcanic hazard maps to share knowledge about what to do in the event of volcanic activity. Ruapehu is an active volcano in New Zealand with ski fields located on its summit and flanks. Annual engagement with audiences on the ski slopes has been used to guide visual design and content of the volcanic hazard map posters displayed in ski areas on Ruapehu (Leonard et al. 2008). The hazard maps are tailored specifically for winter sport audiences, illustrating valley areas exposed to lahar hazard and providing advice about how to evacuate valleys in the event of an eruption. Engagement with local audiences also led to integration of preparedness and evacuation advice into large interpretive outdoor signs about volcanic hazards for volcanoes of the Cascade Range in the United States, such as Mount Baker, Glacier Peak (Eske et al. 2015), and Mount Rainier (Schelling et al. 2014) (Cadig et al. *this volume*, Driedger et al. *in prep*). Combining volcanic hazard maps with supporting information about hazard phenomena and advice for increasing personal response capacity may encourage engagement and elaboration with hazard map information among some audiences (Paton 2003; Rakow et al. 2015). Although engaging with audiences can be time and resource intensive, carrying out work to understand audience perspectives in times of dormancy may prove useful in times of crisis communication.

In 2012, consideration of audience communication needs became a key consideration during response to the Te Maari eruption crisis at the Tongariro Volcanic Centre (TgVC) in New Zealand. While the Te Maari eruption was altogether small in scale, consisting of two phreatic explosions several months apart (Jolly et al. 2014a), it generated a high level of stress and uncertainty surrounding a potential increase in volcanic risk. The eruption vents were located within 2 km of New Zealand's most popular day track, the Tongariro Alpine Crossing, which averages up to 1500 visitors a day during the peak summer season. The TgVC is also a complex stratovolcano system capable of much larger, sub-Plinian eruptive activity (Moebis et al. 2011; Jolly et al. 2014b). While there was an existing series of background hazard maps that were designed for communicating with non-specialist audiences (Fig. 5a) (Leonard et al. 2008, 2014), the eruption meant that a new, event-focussed, crisis hazard map needed to be developed rapidly in order to provide information and life-safety advice directly related to the activity unfolding at the Te Maari vents (Fig. 5b).

Developing an audience-focussed hazard map under the stress and time pressures of a crisis situation was complex and taxing. Leonard et al. (2014) note that between the first and second versions of the map "at least 147 emails were sent by 23 different people across 9 different agencies and groups over a 24 day period" (p. 219). These numbers reflect the high level of engagement and interaction between various groups involved in management of the crisis, but also the complicated nature of rapidly compiling, synthesizing, and deciding on the content, messaging, and design of a hazard map during a crisis. The rapid, high-stakes nature of volcanic crises means that there are often limited resources to dedicate towards revising hazard map design during an actual event. During the Te Maari crisis, the relationships formed through past engagement were valuable in facilitating map development and design (Jolly et al. 2014b; Leonard et al. 2014). For example, one of the main populations affected by the event was a local indigenous group who provided valuable feedback into the final map design style (Leonard et al. 2014). In addition, the team relied heavily on resources that had been pre-prepared, emphasising the value in planning and considering approaches to map design during times of dormancy.

(a) Background hazard map TgVC *2005*

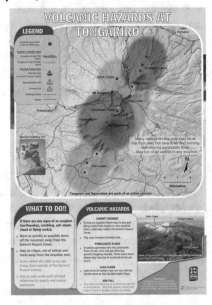

(b) Crisis hazard map Te Maari *2012*

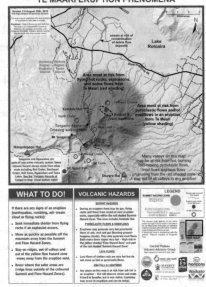

Fig. 5 Tongariro Volcanic Centre volcanic hazard maps for **a** typical background levels of activity (GNS Science 2005) and **b** during the Te Maari eruption crisis in 2012 (GNS Science 2012). Background hazard maps are long-term maps that are used to communicate potential hazards during times of volcanic dormancy, while crisis hazard maps are temporary event-specific maps that are developed in response to imminent hazards (modified from Leonard et al. 2014)

A number of lessons regarding crisis hazard maps emerged from the Te Maari experience, including the value of using version numbers, disclaimers, and providing metadata, and these are put forth as a set of recommendations by Leonard et al. (2014) (p. 225). In this list, they note the importance of considering the visual design of the map image itself. While the map images were presented with legends, descriptive text, and explanatory information, a number of media outlets clipped away this accompanying documentation and context when circulating and disseminating the map. Accordingly, in some cases, interpretation of the hazard map information relied almost wholly on the map image alone (Leonard et al. 2014).

3 Visual Design

Communicating with map images relies on visual perception and cognition. The map reader's eyes must sense and interpret visual variables such as shape, size, colour, texture, and orientation, and then cognitively process this information to create meaning (Bertin 1983; MacEachren 1995; Perkins et al. 2011). Well-designed visualisations can augment and enhance this cognitive processing by reducing cognitive load and facilitating inductive reasoning (Hegarty 2011; Patterson et al. 2014). Accordingly, visual map products have been found to improve comprehension of hazard information when compared to non-visual communication formats such as text and tables (Severtson and Vatovec 2012; Cheong et al. 2016; Cao et al. 2016). However, there are many variables to consider when visually designing a map, and it can often be difficult to determine which combination of variables will support cognition and reasoning. Further, map designs which are aesthetically appealing or intuitively preferred by map makers and users are not necessarily the most effective for decision-making tasks (Hegarty et al. 2009; Mendonça and Delazari 2014). Engaging with map audiences and carrying out empirical research into how people read and process map information can help confront these challenges by giving insight into the ways that different variables in visual design influence communication and decision-making. In addition, experimenting with innovative map visualisation formats can help also help create new ways of capturing audience attention and facilitating engagement with hazard information.

3.1 Exploring and Testing Different Designs

Research surrounding information visualisation is carried out in many different disciplines, including human computer interaction, human factors, cognitive psychology, semiotics, visual analytics, graphic design, cartography, and geo-visualisation. Across these fields, a simple method used for evaluating the effectiveness of visual designs is to test how audiences perform in task-based exercises using different visualisations. However, in order test effectiveness, a defined, measurable communication goal needs to be identified (MacEachren 1982), and in the case of volcanic hazard maps, this can often be multidimensional and nuanced. Accordingly, Haynes et al. (2007) propose a mixed methods approach for evaluating volcanic hazard map design that combines quantitative performance evaluations with qualitative investigations. Using this approach, Haynes et al. (2007) were able to capture the complexity of how local audiences engaged with volcanic hazard and risk maps on Montserrat.

Thompson et al. (2015) adopted a similar mixed methods approach to explore the influence of visual design on volcanic hazard map communication in New Zealand. Thompson et al. (2015) took one dataset, which showed the probability of accumulating volcanic ash in the event of a hypothetical eruption, and displayed it using several different visual design variables. More than 100 scientists and organisational stakeholders (e.g., emergency managers, government officials) in New Zealand responded to

quantitative and qualitative survey questions about the volcanic ash hazard using the different maps. The results showed that changing visual design elements, such as the data classification style or colour scheme, can have a significant effect on the way people understand the hazard. For example, participants were more accurate in quantitatively estimating the average probability of accumulating 1 mm of ash when they used a map that classified hazard data into discrete zones of probability (e.g., 5–14, 15–24 … 65–75%) compared to a map that classified the data using gradational shading (Fig. 6a, b). Participants were most precise when these two approaches were combined, with discrete probability isarithms (e.g., 15, 25 … 65%) overlain onto a gradational shading classification (Fig. 6 c). Participants also had strong feelings about the user-friendliness of the different maps styles. Map which were easier to read were associated with increased confidence in ability to use and apply the hazard information. The findings suggest that simple choices in data classification could have a significant influence on the way people understand, interpret, and apply probabilistic hazard information (Thompson et al. 2015).

Thompson et al. (2015) also conclude that it is important to consider colour scheme choices when representing volcanic hazard information on a map. Colour is an important visual design variable, which can guide attention, emotional response, and interpretation of map features (Robinson 1967; Bertin 1983; Wolfe and Horowitz 2004). However, the strong connotations and meaning that colours often carry for map readers introduces potential for miscommunication (Monmonier 1996; Brewer 1994). For example, Thompson et al. (2015) found that using a red-to-blue diverging-hue colour scheme (Fig. 7a) communicated a qualitatively different type of message than a red-to-yellow sequential-hue colour scheme (Fig. 7b). Participants tended to make interpretations about hazard state (presence/absence) when reading the diverging colour scheme map, but tended to

(a) Gradational shading

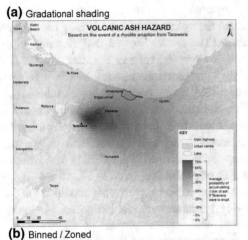

(b) Binned / Zoned

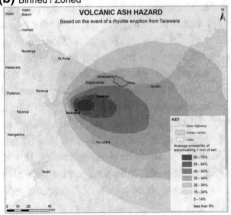

(c) Gradational shading with isolines

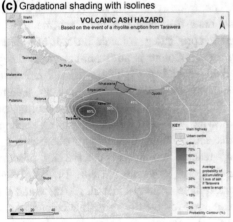

◀ **Fig. 6** Three types of probabilistic volcanic hazard map data visualisations tested by Thompson et al. (2015). Participants struggled to read accurate probability values for the area outlined in *blue* when using **a** a gradational shaded data classification. Participants performed better using **b** a binned (zoned) data classification, and they performed best, with the most accurate and precise estimates of hazard values using **c** gradational shading with isolines. The results suggest that visualisation of hazard data on a map can influence the information that people take away. Modified from Thompson et al. (2015)

make interpretations about hazard degree (less/more) when reading the sequential colour scheme map. In addition, more than two-thirds of survey participants self-reported that the colour scheme influenced the level of hazard they perceived from the map (Thompson et al. 2015).

Similarly, while red-yellow-green "stoplight" colour schemes are applied in a number of volcanic hazard maps around the world, Olson and Brewer (1997) and Jenny and Kelso (2007) warn that red-and-green colour schemes may introduce problems for colour vision deficient map readers. Up to 8% of males have some form of colour-vision deficiency, with difficulty distinguishing between red and green colours being the most common type (Delepero et al. 2005). To this population, the highest (red) and lowest (green) hazard areas may appear the same colour and cause confusion. To assist map makers in choosing appropriate colour schemes for maps, Harrower and Brewer (2003) developed Color-Brewer (www.ColorBrewer.org), a research-backed tool for selecting map colour schemes with appropriate hue, saturation, and contrast to enhance visualisation of map information and prevent potential issues for colour vision deficient users.

Many of the challenges associated with visually designing volcanic hazard maps are faced in other fields of hazard and risk research, such as

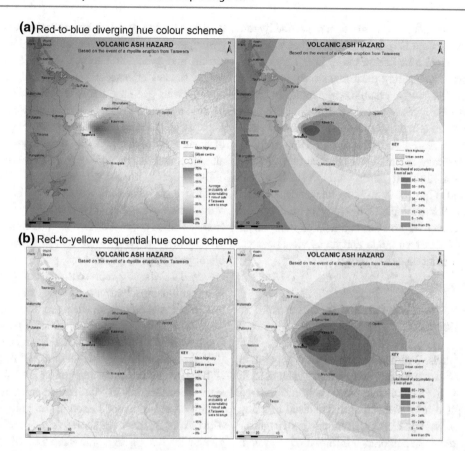

Fig. 7 Two hazard map colour schemes tested by Thompson et al. (2015). Participants were more likely to discuss a hazard state (e.g., present or absent) when viewing **a** the *red-to-blue* diverging hue colour scheme, and were more likely to discuss hazard degree when viewing **b** the *red-to-yellow* sequential hue colour scheme map. Modified from Thompson et al. (2015)

wildfire (e.g., Cheong et al. 2016), hurricane (e.g., Broad et al. 2007; Sherman-Morris et al. 2015), flooding (e.g., Strathie et al. 2015), sea level rise (e.g., Retchless 2014), and health (e.g., Severtson and Myers 2013). For example, researchers have found that visual design of wildfire hazard maps can influence people's interest and engagement with the hazard information and also how they use it to make decisions about evacuation (Cao et al. 2016; Cheong et al. 2016). Similarly, visual design has been found to influence emotional and behavioural responses to tornado warning maps (Ash et al. 2014). Visualising and communicating uncertainty in geospatial data also remains an ongoing challenge across many different fields (Aerts et al. 2003; Spiegelhalter et al. 2011; Kinkeldey

et al. 2014). As work in volcanic hazard map design continues to evolve, it is important to consider lessons learned from research and experience in these diverse fields, and drawn upon them to help inform and guide investigations of volcanic hazard communication.

In addition, contributions from cognitive science research can add new dimensions to understanding how visual design and visual perception of volcanic hazard maps can influence hazard interpretation and decision-making. For example, Hegarty (2011) summarises sixteen "principles of effective graphics" based on decades of cognitive science research into visual-spatial displays. Such principles, such as the relevance principle (Kosslyn 2006), which proposes that visual displays should present no more or no less information than is needed by the audience, could help inform approaches to visual design of volcanic hazard maps. Investigations of weather map reading performance suggests that considering such graphic principles in map design can affect visual processing of map information (Hegarty et al. 2010; Fabrikant et al. 2010). Similarly, Patterson et al. (2014) propose six "leverage points" for augmenting human cognition through information visualisations, which also are likely to have relevancy for map design. For example, they suggest that certain visual design approaches can help capture visual attention and also guide and focus visual search for information.

3.2 Visualising Hazard in Different Formats

Although the 2-dimensional plan view paper map remains a common and useful visualisation method in hazard mapping, volcanic hazard concepts can be mapped and visualised in many other ways. For example, modern geovisualisation and geographic information system (GIS) techniques, such as interactive interfaces (Çöltekin et al. 2009; Roth 2013) will play a significant role in shaping the future of hazard map communication. Interactive and 3D visualizations can add new dimensions to natural hazard and risk maps, through providing location-aware, user-centred data, although further research about these emerging technologies is needed to better understand the way they affect hazard and risk communication (Lonergan and Hedley 2015).

As improved workflows and accessibility of such methods continue to be developed, new opportunities will arise for communicating volcanic hazard in innovative and engaging ways. In a study at Mount Hood volcano in Oregon, USA, Preppernau and Jenny (2015) tested new methods of visualising lahar hazard using 3-dimensional (3D) oblique perspective base maps with isochrones that represented lahar travel time. They found that participants preferred 3D isochrone lahar hazard maps to

traditional plan view contour maps, and that participants' performed better in interpreting terrain and evacuation routes with the 3D displays. Recent advancements in visual technology, such as eye-gaze trackers (devices that can be used to record a readers' eye movement across a visual or graphic), can also enable new forms of insight into map reading behaviour, visual attention, and understanding (e.g., Çöltekin et al. 2009; Meyer et al. 2012; Hegarty et al. 2010).

Innovative visualisations can also be developed with traditional, low-technology approaches. Hands-on, bottom-up, community-led participatory mapping exercises, in which tangible objects such as paper, pens, paint, and stones are used to visually represent and contextualise interrelationships between hazards and society, can help foster important dialogue about natural hazards and risk (Chambers 2008; Cadag and Gaillard 2012). For example, Cadag and Gaillard (2012) outline how participatory 3D mapping (P3DM) was used as an integrative tool for disaster risk reduction in Masantol, a small municipality on the island of Luzon in the Philippines. They found that collaboratively building a physical, 3D geographic model of place empowered the community to engage in constructive dialogue about hazard and risk. Such approaches also offer a way to integrate scientific and local knowledge in a way that is tangible and meaningful for many different people in the community, from government officials to school students (Cadag and Gaillard 2012).

4 Volcanic Hazard Maps into the Future

Volcanic hazard maps have transformed over the past several decades due to advances in hazard analysis methods, lessons learnt through past crises, and ongoing interdisciplinary research into how audiences engage and interact with hazard maps. Modern volcanic hazard maps will continue to evolve into the future as digital technologies, GIS, social media, citizen science, and globalisation have a growing impact on science communication and disaster management

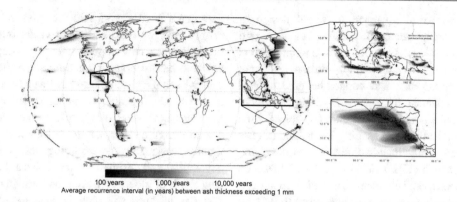

100 years   1,000 years   10,000 years
Average recurrence interval (in years) between ash thickness exceeding 1 mm

Fig. 8 Probabilistic volcanic hazard map showing global volcanic ash fall hazard. The map, which was produced for the UNISDR 2015 Global Assessment Report, is based on large-scale quantitative modelling, and shows average recurrence interval (in years) between accumulating ash thicknesses exceeding 1 mm. Modified from Jenkins et al. (2015)

(e.g., Webley and Watson *this volume*, Kuhn et al. *this volume*). In 2015, one of the first global volcanic hazard maps was created as part of the UNISDR (United Nations Office for Disaster Risk Reduction) Global Assessment Report (Jenkins et al. 2015; Fig. 8). The map represents the growing global collaboration effort in volcanic hazard analysis, as well as the expanding capabilities of hazard modelling and computation. Collaborative international online volcanology networks, such as Vhub (www.Vhub. org; Palma et al. 2014), local online hubs, such as wikis (Leonard et al. 2014), and interactive online tools for volcanic hazard assessment, such as G-EVER (Tsukuda et al. 2012), are helping to facilitate data-sharing and improve access to hazard modelling, enabling new levels of engagement and access to tools and information for map-making and design.

While visual design of hazard maps will continue to evolve with innovation of new technologies and hazard mapping approaches, hazard map audiences will also evolve. As globalisation and population growth continues, hazard map audiences will dynamically shift and become more diverse. For example, growth in volcano tourism could lead to higher numbers of tourists and non-native speakers in hazardous areas, and these populations are likely to have different perceptions of hazard and risk than local audiences (Bird et al. 2010). Future work exploring differences in the information needs and cultural communication styles within and among these diverse audiences will be important for understanding how to adapt and grow approaches to volcanic hazard knowledge exchange.

Interdisciplinary research is becoming increasingly embraced within the field of volcanic hazard and risk (Barclay et al. 2008), and future volcanic hazard maps should continue to work towards integrating new interdisciplinary concepts from research fields such as sociology, communications, human factors, geography, design, and psychology to develop intuitive designs which maximise visual cues, minimise cognitive load, and increase the effectiveness of visual communication. Integrating tacit knowledge from relevant areas of practice (e.g., emergency management, national parks, conservation) in addition to theories and concepts from different areas of research can help ensure that volcanic hazard maps of the future are optimally designed in a way that makes them useful, usable, and used.

It is important to acknowledge that the case studies and research covered in this chapter rely principally on work and experiences published in the academic literature, and are not comprehensive. There is a wealth of tacit knowledge on audience-map engagement and hazard communication gained from practice that is not captured in this summary. However, a key theme which emerges from the case studies and work reviewed in this chapter is that an audience-based, evidence-backed approach to visual design of hazard maps can help facilitate clear hazard communication. As with most communication approaches, hazard map design is not "one-size-fits all", and cannot be guided by a single universal framework or design solution. Nevertheless, hazard maps that: (A) consider the audience and their messaging needs, and (B) use evidence from interdisciplinary research and experience to inform visual map design based on these needs, can help communicate information in a way that is accessible and useful to those who need it. While available resources, target audiences, and volcanic setting will uniquely guide and shape this process for each map, adopting such an approach can help create end results that are grounded in meaningful communication goals.

5 Summary

Volcanic hazard maps that are designed based on both the needs of the audience and evidence from practice and research can help support clear and effective messaging of critical hazard information. Engaging with audiences to explore how they understand and create meaning from hazard maps can foster constructive multi-way dialogue about volcanic hazards, and also help ensure that important messages are visually communicated in a way that is transparent and trusted by those potentially affected by a volcanic crisis. Although hazard maps represent just one component of a hazard assessment, their ability to comprise many types of information into a concise, visually salient graphic that can be shared across many types of media means that they are often used widely in crisis communication. Past volcanic crises across the world have underscored the important communication role of hazard maps, but have also highlighted the significant impact that visual design has on this exchange.

Visual representation of hazard information on a map can influence the way that people engage with the information, as well as the messages that people take away, and decisions

they make. Future work into the ways in which people read, process, and share visual information will open new opportunities for optimising volcanic hazard content for different audiences. This will continue to be important as advances in hazard modelling and visualisation technology introduce new ways of visually communicating hazard during a crisis. As the volcanology community works towards exploring new ways of developing and designing volcanic hazard maps, new levels of global collaboration through online data-sharing hubs will provide ways to connect, share, and integrate these emerging approaches. By considering audience needs and perspectives—how the information might be used, read, understood, and applied—hazard maps can be designed in a way that makes them accessible, relevant, and clear for the people who need them.

Acknowledgements The authors would like to acknowledge the many generations of volcanic hazard maps which have contributed valuable insight and knowledge regarding both the advantages and challenges of visually communicating volcanic hazard. The authors would also like to thank M Monsalve, H Murcia, and C Driedger for reviewing parts of this manuscript, and two anonymous reviewers for constructive and valuable comments on an earlier version of this chapter. MAT and JML gratefully acknowledge support from the New Zealand Earthquake Commission.

References

Aerts J, Clarke K, Keuper A (2003) Testing popular visualization techniques for representing model uncertainty. Cartogr Geogr Inf Sci 30:249–261. doi:10.1559/152304003100011180

Ash KD, Schumann RL III, Bowser GC (2014) Tornado warning trade-offs: evaluating choices for visually communicating risk. Weather Clim Soc 6:104–118. doi:10.1175/WCAS-D-13-00021.1

Aspinall WP, Loughlin SC, Michael FV, Miller AD, Norton GE, Rowley KC, Sparks RSJ, Young SR (2002) The Montserrat Volcano observatory: its evolution, organization, role and activities. In: Druitt TH, Kokelaar BP (eds) The eruption of Soufrière Hills Volcano, Montserrat, from 1995 to 1999. Geological Society of London Memoirs, vol 21, pp 71–91

Barclay J, Haynes K, Mitchell T, Solana C, Teeuw R, Darnell A, Crosweller S, Cole P, Pyle D, Lowe C, Fearnley C, Kelman I (2008) Framing volcanic risk communication within disaster risk reduction: finding ways for the social and physical sciences to work together. Geol Soc Lond Spec Publ 305:163–177. doi:10.1144/SP305.14

Bertin J (1983) Semiology of graphics (trans: Berg WJ). University of Wisconsin Press, Madison, WI

Bird D, Gisladottir G, Dominey-Howes D (2010) Volcanic risk and tourism in southern Iceland: implications for hazard, risk and emergency response education and training. J Volcanol Geoth Res 189:33–48. doi:10.1016/j.jvolgeores.2009.09.020

Bostrom A, Löfstedt RE (2003) Communicating risk: wireless and hardwired. Risk Anal 23:241–248. doi:10.1111/1539-6924.00304

Brewer CA (1994) Color use guidelines for mapping and visualization. In: MacEachren AM, Taylor DRF (eds) Visualization in modern cartography. Elsevier, Tarrytown, New York, pp 123–147

Broad K, Leiserowitz A, Weinkle J, Steketee M (2007) Misinterpretations of the "Cone of Uncertainty" in Florida during the 2004 hurricane season. Bull Am Meteorol Soc 88(5):651–667

Cadag J, Gaillard J (2012) Integrating knowledge and actions in disaster risk reduction: the contribution of participatory mapping. Area 44:100–109. doi:10.1111/j.1475-4762.2011.01065.x

Calder ES, Wagner K, Ogburn SE (2015) Volcanic hazard maps. In: Loughlin SC, Sparks S, Brown SK, Jenkins SF, Vye-Brown C (eds) Global volcanic hazards and risk. Cambridge University Press, Cambridge UK

Cao Y, Boruff B, McNeill I (2016) Is a picture worth a thousand words? Evaluating the effectiveness of maps for delivering wildfire warning information. Int J Disaster Risk Reduct 19:179–196. doi:10.1016/j.ijdrr.2016.08.012

Carrasco M (2011) Visual attention: the past 25 years. Vis Res 51:1484–1525. doi:10.1016/j.visres.2011.04.012

Chambers R (2008) Revolutions in development inquiry. Earthscan, London

Cheong L, Bleisch S, Kealy A, Tolhurst K, Wilkening T, Duckham M (2016) Evaluating the impact of visualization of wildfire hazard upon decision-making under uncertainty. Int J Geogr Inf Sci 1–28. doi:10.1080/13658816.2015.1131829

Çöltekin A, Heil B, Garlandini S, Fabrikant S (2009) Evaluating the effectiveness of interactive map interface designs: a case study integrating usability metrics with eye-movement analysis. Cartogr Geogr Inf Sci 36:5–17. doi:10.1559/152304009787340197

Cronin S, Gaylord D, Charley D, Alloway BV, Wallez S, Esau JB (2004) Participatory methods of incorporating scientific with traditional knowledge for volcanic hazard management on Ambae Island, Vanuatu. Bull Volc 66:652–668. doi:10.1007/s00445-004-0347-9

Daron J, Lorenz S, Wolski P, Blamey R, Jack C (2015) Interpreting climate data visualisations to inform adaptation decisions. Climate Risk Manage 10:17–26. doi:10.1016/j.crm.2015.06.007

Delepero WT, O'Neill H, Casson E, Hovis J (2005) Aviation-relevant epidemiology of color vision deficiency. Aviat Space Eviron Med 76:127–133

Domke D, Perimutter D, Spratt M (2002) The primes of our times? An examination of the "power" of visual images. Journalism 3:131–159. doi:10.1177/146488490200300211

Doyle EE, McClure J, Johnston DM, Paton D (2014) Communicating likelihoods and probabilities in forecasts of volcanic eruptions. J Volcanol Geoth Res 272:1–15. doi:10.1016/j.jvolgeores.2013.12.006

Driedger C, Ramsey D, Faust L (in preparation) Following the tug of the audience—from complex to simplified hazard maps at Cascade Range volcanoes. Front Volcanol Spec Issue Volc Hazard Assess

Ekse W, Burkhardt F, Kloes D, Driedger CL, Faust L, Nelson D (2015) Are you ready for an eruption? Mount Baker and Glacier Peak. Accessed 24 Sept 2016 at: http://www.dnr.wa.gov/programs-and-services/geology/geologic-hazards/volcanoes-and-lahars#volcano-preparedness-posters

Fabrikant SI, Hespanha SR, Hegarty M (2010) Cognitively inspired and perceptually salient graphic displays for efficient spatial inference making. Ann Assoc Am Geogr 100:13–29. doi:10.1080/00045600903362378

Fearnley CJ, McGuire WJ, Davies G, Twigg J (2012) Standardisation of the USGS volcano alert level system (VALS): analysis and ramifications. Bull Volc 74(9):2023–2036. doi:10.1007/s00445-012-0645-6

Finucane ML, Alhakami A, Slovic P, Johnson SM (2000) The affect heuristic in judgments of risks and benefits. J Behav Decis Mak 13(1):1–17

Gaillard J-C, Dibben C (2008) Volcanic risk perception and beyond. J Volcanol Geoth Res 172:163–169. doi:10.1016/j.jvolgeores.2007.12.015

GNS Science (2005) Volcanic hazards at Tongariro (map). Wellington, NZ

GNS Science (2012) Te Maari Eruption Phenomena (map). Wellington, NZ

Harrower M, Brewer CA (2003) ColorBrewer.org: an online tool for selecting colour schemes for maps. Cartogr J 40:27–37. doi:10.1179/000870403235002042

Haynes K, Barclay J, Pidgeon N (2007) Volcanic hazard communication using maps: an evaluation of their effectiveness. Bull Volc 70:123–138. doi:10.1007/s00445-007-0124-7

Haynes K, Barclay J, Pidgeon N (2008) The issue of trust and its influence on risk communication during a volcanic crisis. Bull Volc 70:605–621. doi:10.1007/s00445-007-0156-z

Hegarty M (2011) The cognitive science of visual-spatial displays: implications for design. Top Cogn Sci 3:446–474. doi:10.1111/j.1756-8765.2011.01150.x

Hegarty M, Smallman H, Stull A, Canham M (2009) Naïve cartography: how intuitions about display configuration can hurt performance. Cartographica 44(3):171–186. doi:10.3138/carto.44.3.171

Hegarty M, Canham M, Fabrikant S (2010) Thinking about the weather: how display salience and knowledge affect performance in a graphic inference task. J Exp Psychol Learn Mem Cogn 36:37–53. doi:10.1037/a0017683

Jenkins SF, Wilson TM, Magill CR, Miller V, Stewart C, Marzocchi W, Boulton M (2015) Volcanic ash fall hazard and risk: technical background paper for the UNISDR 2015 global assessment report on disaster risk reduction. Global Volcano Model and IAVCEI

Jenny B, Kelso N (2007) Color design for the color vision impaired. Cartogr Perspect 61–67. doi:10.14714/CP58.270

Johnston DM, Lai MS, Houghton BF, Paton D (1999) Volcanic hazard perceptions: comparative shifts in knowledge and risk. Disaster Prev Manag 8:118–126. doi:10.1108/09653569910266166

Jolly AD, Jousset P, Lyons JJ, Carniel R (2014a) Seismo-acoustic evidence for an avalanche driven phreatic eruption through a beheaded hydrothermal system: an example from the 2012 Tongariro eruption. J Volcanol Geoth Res 286:331–347. doi:10.1016/j.jvolgeores.2014.04.007

Jolly GE, Keys H, Procter JN, Deligne NI (2014b) Overview of the co-ordinated risk-based approach to science and management response and recovery for the 2012 eruptions of Tongariro volcano, New Zealand. J Volcanol Geoth Res 286:184–207. doi:10.1016/j.jvolgeores.2014.08.028

Kinkeldey C, MacEachren AM, Schiewe J (2014) How to assess visual communication of uncertainty? A systematic review of geospatial uncertainty visualisation user studies. Cartogr J 51:372–386. doi:10.1179/1743277414Y.0000000099

Kokelaar BP (2002) Setting, chronology and consequences of the eruption of Soufrière Hills Volcano, Montserrat (1995–1999). In: Druitt TH, Kokelaar BP (eds) The eruption of Soufrière Hills Volcano, Montserrat, from 1995 to 1999. Geological Society of London Memoirs, vol 21, pp 1–44

Kosslyn SM (2006) Graph design for the eye and mind. Oxford University Press, New York

Leonard G, Johnston D, Paton D et al (2008) Developing effective warning systems: ongoing research at Ruapehu volcano, New Zealand. J Volcanol Geoth Res 172:199–215. doi:10.1016/j.jvolgeores.2007.12.008

Leonard GS, Stewart C, Wilson TM, Procter JN, Scott BJ, Keys HJ, Jolly GE, Wardman JB, Cronin SJ, McBride SK (2014) Integrating multidisciplinary science, modelling and impact data into evolving, syn-event volcanic hazard mapping and communication: a case study from the 2012 Tongariro eruption crisis, New Zealand. J Volcanol Geoth Res 286:208–232. doi:10.1016/j.jvolgeores.2014.08.018

Leone F, Lesales T (2009) The interest of cartography for a better perception and management of volcanic risk: from scientific to social representations: the case of Mt. Pelée volcano, Martinique (Lesser Antilles). J Volcanol Geoth Res 186:186–194. doi:10.1016/j.jvolgeores.2008.12.020

Lester PM (2014) Visual communication: images with messages, 6th edn. Wadsworth, Cengage Learning, Boston, MA

Lloyd R (2011) Understanding and learning maps. In: Dodge M, Kitchin R, Perkins C (eds) The map reader: theories of mapping practice and cartographic representation. Wiley, Hoboken, NJ

Lonergan C, Hedley N (2015) Navigating the future of tsunami risk communication: using dimensionality, interactivity and situatedness to interface with society. Nat Hazards 78:179–201. doi:10.1007/s11069-015-1709-7

MacEachren A (1982) The role of complexity and symbolization method in thematic map effectiveness. Ann Assoc Am Geogr 72:495–513. doi:10.1111/j.1467-8306.1982.tb01841.x

MacEachren A (1995) How maps work: representation, visualization, and design. Guilford Press, New York

Mendonça A, Delazari L (2014) Testing subjective preference and map use performance: use of web maps for decision making in the public health sector. Cartographica 49:114–126. doi:10.3138/carto.49.2.1455

Meyer V, Kuhlicke V, Luther J, Fuchs S, Priest S, Dorner W, Serrhini K, Pardoe J, McCarthy S, Seidel J, Palka G, Unnerstall H, Viavattene C, Scheuer S (2012) Recommendations for the user-specific enhancement of flood maps. Nat Hazards Earth Syst Sci 12:1701–1716. doi:10.5194/nhess-12-1701-2012

Moebis A, Cronin SJ, Neall VE, Smith IE (2011) Unravelling a complex volcanic history from fine-grained, intricate holocene ash sequences at the Tongariro Volcanic Centre, New Zealand. Quatern Int 246:352–363. doi:10.1016/j.quaint.2011.05.035

Monmonier M (1996) How to lie with maps. University of Chicago Press, Chicago

Monzier M, Robin C (1995) Volcanic hazard map for Aoba Island. ORSTOM (Institut de recherche pour le developpement)

Mould D, Mandryk RL, Li H (2012) Emotional response and visual attention to non-photorealistic images. Comput Graph 36:658–672. doi:10.1016/j.cag.2012.03.039

Nakamura Y, Fukushima K, Jin X, Ukawa M, Sato T, Hotta Q (2008) Mitigation systems by hazard maps, mitigation plans, and risk analyses regarding volcanic disasters in Japan. J Disaster Res 3(4):297–304

Nave R, Isaia R, Vilardo G, Barclay J (2010) Re-assessing volcanic hazard maps for improving volcanic risk communication: application to Stromboli Island, Italy. J Maps 6:260–269. doi:10.4113/jom.2010.1061

Olson JM, Brewer CA (1997) An evaluation of color selections to accommodate map users with color-vision impairments. Ann Assoc Am Geogr 87:103–134

Palma J, Courtland L, Charbonnier S, Tortini R, Valentine GA (2014) Vhub: a knowledge management system to facilitate online collaborative volcano modeling and research. J Appl Volcanol 3:2. doi:10.1186/2191-5040-3-2

Parra E, Cepeda H (1990) Volcanic hazard maps of the Nevado del Ruiz volcano, Colombia. J Volcanol Geoth Res 42:117–127. doi:10.1016/0377-0273(90)90073-O

Paton D (2003) Disaster preparedness: a social-cognitive perspective. Disaster Prev Manag 12(3):210–216. doi:10.1108/09653560310480686

Paton D, Smith L, Daly M, Johnston D (2008) Risk perception and volcanic hazard mitigation: individual and social perspectives. J Volcanol Geoth Res 172:179–188. doi:10.1016/j.jvolgeores.2007.12.026

Patterson RE, Blaha LM, Grinstein GG et al (2014) A human cognition framework for information visualisation. Comput Graph 42:42–58

Perkins C, Kitchin R, Dodge M (2011) Introductory essay: cognition and cultures of mapping. In: Dodge M, Kitchin R, Perkins C (eds) The map reader: theories of mapping practice and cartographic representation. Wiley, West Sussex, UK, pp 298–303

Perry SC, Blanpied ML, Burkett ER, Campbell NM, Carlson A, Cox DA, Driedger CL, Eisenman DP, Fox-Glassman KT, Hoffman S, Hoffman SM, Jaiswal KS, Jones LM, Luco N, Marx SM, McGowan SM, Mileti DS, Moschetti MP, Ozman D, Pastor E, Petersen MD, Porter KA, Ramsey DW, Ritchie LA, Fitzpatrick JK, Rukstales KS, Sellnow TL, Vaughon WL, Wald DJ, Wald LA, Wein A, Zarcadoolas C (2016) Get your science used—six guidelines to improve your products. USGS Circular 1419:37. doi:10.3133/cir1419

Pierson TC, Janda RJ, Thouret JC, Borrero CA (1990) Perturbation and melting of snow and ice by the November 1985 eruption of Nevado del Ruiz, Colombia, and consequent mobilization, flow and deposition of lahars. J Volcanol Geoth Res 41:17–66. doi:10.1016/0377-0273(90)90082-Q

Pierson TC, Wood NJ, Driedger CL (2014) Reducing risk from lahar hazards: concepts, case studies, and roles for scientists. J Appl Volcanol 3:16. doi:10.1186/s13617-014-0016-4

Potter SH, Jolly GE, Neall VE, Johnston DM, Scott BJ (2014) Communicating the status of volcanic activity: revising New Zealand's volcanic alert level system. J Appl Volcanol 3:12. doi:10.1186/s13617-014-0013-7

Preppernau C, Jenny B (2015) Three-dimensional versus conventional volcanic hazard maps. Nat Hazards 78:1329–1347. doi:10.1007/s11069-015-1773-z

Rakow T, Heard CL, Newell BR (2015) Meeting three challenges in risk communication: phenomena, numbers, and emotions. Policy Insights Behav Brain Sci 2:147–156. doi:10.1177/2372732215601442

Retchless DP (2014) Sea level rise maps: how individual differences complicate the cartographic communication of an uncertain climate change hazard. Cartogr Perspect 77:17–32. doi:10.14714/CP77.1235

Robinson AH (1967) Psychological aspects of color in cartography. Int Yearb Cartogr 7:50–59

Robinson AH, Petchenik (1976) The nature of maps: essays toward understanding maps and mapping. University of Chicago Press, Chicago

Roth RE (2013) Interactive maps: what we know and what we need to know. J Spat Inf Sci 6:59–115. doi:10.5311/JOSIS.2013.6.105

Schelling J, Prado L, Norman D, Walsh T, Driedger C, Faust L, Westby L, Schroedel R, Lovellford P (2014) Are you ready for an eruption? Washington Department of Natural Resources. Accessed 24 Sept 2016 at: http://www.dnr.wa.gov/programs-and-services/geology/geologic-hazards/volcanoes-and-lahars#volcano-preparedness-posters

Servicio Geológico Colombiano (2015) Mapa de amenaza volcánica del Nevado del Ruiz. Escala 1:120,000 (http://www2.sgc.gov.co/Manizales/Imagenes/Mapas-de-Amenaza/VNR/v3_img/Mapa_de_Amenaza_v3-2015-50.aspx)

Severtson D, Myers J (2013) The influence of uncertain map features on risk beliefs and perceived ambiguity for maps of modeled cancer risk from air pollution. Risk Anal 33:818–837. doi:10.1111/j.1539-6924.2012.01893.x

Severtson D, Vatovec C (2012) The theory-based influence of map features on risk beliefs: self-reports of what is seen and understood for maps depicting an environmental health hazard. J Health Commun 17:836–856. doi:10.1080/10810730.2011.650933

Sherman-Morris K, Antonelli KB, Williams CC (2015) Measuring the effectiveness of the graphical communication of hurricane storm surge threat. Weather Clim Soc 7:69–82. doi:10.1175/WCAS-D-13-00073.1

Spiegelhalter D, Pearson M, Short I (2011) Visualizing uncertainty about the future. Science 333:1393–1400. doi:10.1126/science.1191181

Strathie, Netto, Walker GH, Pender (2015) How presentation format affects the interpretation of probabilistic flood risk information. J Flood Risk Manage. doi:10.1111/jfr3.12152

Sword-Daniels VL, Wilson TM, Sargeant S, Rossetto T, Twigg J, Johnston DM, Loughlin SC, Cole PD (2014) Consequences of long-term volcanic activity for essential services in Montserrat: challenges, adaptations and resilience. Geol Soc Spec Publ 39:471–488. doi:10.1144/M39.26

Thompson MA, Lindsay JM, Gaillard J (2015) The influence of probabilistic volcanic hazard map properties on hazard communication. J Appl Volcanol. doi:10.1186/s13617-015-0023-0

Tsukuda E, Eichelberger J et al (2012) The G-EVER1 accord. G-EVER consortium. Retrieved from http://g-ever.org/en/accord/index.html

Tufte ER (1997) Visual explanations: images and quantities, evidence and narrative. Graphics Press, Chesire, CT, p 156

Voight B (1990) The 1985 Nevado del Ruiz volcano catastrophe: anatomy and retrospection. J Volcanol Geoth Res 44:349–386. doi:10.1016/0377-0273(90)90027-D

Wolfe JM, Horowitz TS (2004) What attributes guide the deployment of visual attention and how do they do it? Nat Rev Neurosci 5:495–501

The Role of Geospatial Technologies in Communicating a More Effective Hazard Assessment: Application of Remote Sensing Data

P. W. Webley and I. M. Watson

Abstract

Remote sensing data and the application of geo-spatial technologies have progressively been built into real-time volcanic hazard assessment. Remote sensing of volcanic processes provides a unique synoptic view of the developing hazard, and provides insights into the ongoing activity without the need for direct, on-the-ground observations. Analysis and visualization of these data through the geospatial tools, like Geographical Information Systems (GIS) and new virtual globes, brings new perspectives into the decision support system. In this chapter, we provide examples of (i) how remote sensing has assisted in real-time analysis of active volcanoes; (ii) how by combining multiple sensors at different spatial, spectral and temporal resolutions one is able to better understand a given hazard, leading to better communication and decision making; and (iii) how visualizing this in a common platform, like a GIS tool or virtual globe, augments effective hazard assessment system. We will illustrate how useful remote sensing data can be for volcanic hazard assessment, including the benefits and challenges in real-time decision support, and how the geo-spatial tools can be useful to communicate the potential hazard through a common operation protocol.

List of Acronyms

AIRS	Airborne Infrared Remote Sounder
ASTER	Advanced Spaceborne Thermal Emission and Reflection Radiometer

P. W. Webley (✉)
Geophysical Institute, University of Alaska Fairbanks, Fairbanks, AK 99775, USA
e-mail: pwwebley@alaska.edu

I. M. Watson
School of Earth Sciences, University of Bristol, Bristol, UK

Advs in Volcanology (2018) 641–663
https://doi.org/10.1007/11157_2017_7
Published Online: 22 June 2017

AVHRR	Advanced Very High Resolution Radiometer
CIMSS	Cooperative Institute for Meteorological Satellite Studies
CNES	*Centre National d'Etudes Spatiale*
COMET	Centre for Observation and Modelling of Earthquakes, Volcanoes, and Tectonics
CONAE	Comision Nacional De Actividades Espaciales
EDS	Expedited Data Set
EM	Electromagnetic
ENVI	Environment for Visualizing Images
ERDAS	Earth Resources Data Analysis System
ESA	European Space Agency
EUMETSAT	European Organisation for the Exploitation of Meteorological Satellites
GEOTIFF	Geostationary Earth Orbit Tagged Image File Format
GIS	Geographical Information Systems
GOES-R	Geostationary Operational Environmental Satellite-R
GOME-2	Global Ozone Monitoring Experiment-2
GPS	Global Positioning System
HDF	Hierarchical Data Format
IASI	Infrared Atmospheric Sounding Interferometer
ILWIS	Integrated Land and Water Information System
IMO	Icelandic Meteorological Office
KML	Keyhole Markup Language
KMZ	Keyhole Markup Zipped
KVERT	Kamchatka Volcanic Eruption Response Team
LANDSAT	Land Remote-Sensing Satellite (System)
MIR	Mid-Infrared
MODIS	Moderate Resolution Imaging Spectroradiometer
MTSAT	Multifunctional Transport Satellites
NASA	National Aeronautics and Space Administration
NERC	Natural Environment Research Council
NOAA	National Oceanic and Atmospheric Administration
NRT	Near Real-Time
OMI	Ozone Monitoring Instrument
OMPS	Ozone Mapping Profiler Suite
ORFEO	Optical and Radar Federated Earth Observation
RADAR	Radio Detection and Ranging
RGB	Red, Green, Blue
ROI	Region of Interest
SACS	Support to Aviation Control Service
SEVIRI	Spinning Enhanced Visible and Infrared Imager
SHP	Shape
SO2	Sulfur dioxide
SOPI	Software de Procesamiento de Imágenes
SPIRITS	Software for the Processing and Interpretation of Remotely sensed Image Time Series
SWIR	Short Wave Infrared
TIR	Thermal Infrared

URP	Urgent Request Protocol
USGS	United States Geological Survey
UV	Ultraviolet
VEI	Volcano Explosivity Index
VIIRS	Visible Infrared Imaging Radiometer Suite
VLP	Very long Period
VNIR	Visible Near Infrared

1 Introduction

Timely analysis of volcanic activity makes for enhanced hazard assessment, and provides the best available information to both mitigate the hazard and reduce the risk it could pose. For example, seismic data can be collected and processed with sub-second acquisition frequency (McNutt 1996) to assess earthquake location and signal changes relating to volcanic tremor (Aki et al. 1977; McNutt 1986; Zuccarello et al. 2013), and very long period (VLP) earthquakes (Jousset et al. 2013). Global positioning system (GPS) data can produce time series of vertical and horizontal motion at similar frequencies (e.g. Owen et al. 2000). These two datasets provide point location information of volcanic activity and often require modeling of the volcanic edifice (e.g. Poland et al. 2006) to correlate the data to large changes in the volcanic system. While not having the sub-second periodicity of these geophysical data, satellite-based remote sensing data can be used to examine the large scale spatial changes at a volcano, before, during and after a volcanic event (Rothery et al. 1988; Prata et al. 1989; Guo et al. 2004; Bailey et al. 2010; Thomas et al. 2011; Webley et al. 2013).

Real-time satellite data and new data visualization techniques and software allow the decision maker to evaluate the local, regional, country-wide, continental, and global scales of the volcanic events and impending volcanic crises. Enhanced visualizations of the remote sensing data can be critical to communicate the scale and impact of the volcanic events and connect with those most at risk. Satellite remote sensing data have been readily available for several decades, and while no sensor has been specifically designed for volcanoes only, there are a large number of satellite-based sensors used to analyze volcanoes and detect and map their associated hazards. While listing all past, current and future sensors is beyond this chapter, we discuss the different regions of the electromagnetic spectrum that can be useful for monitoring volcanoes and the tradeoffs between spectral, temporal and spatial resolutions when assessing the most appropriate sensor for the type of hazard assessment. Some satellites and their sensors are useful for mapping ground deposits and studying the volcano as it inflates and deflates under changing volcanic activity, while others are most applicable to detecting atmospheric emissions, such as volcanic ash and sulfur dioxide (SO_2).

2 Remote Sensing Basics

Lillesand et al. (2015) define remote sensing as "the science and art of obtaining information about an object, area or phenomena through the analysis of data acquired by a device that is not in contact with the object, area or phenomena under investigation". Satellite remote sensors are either an active system, like RADAR (Radio Detection and Ranging), that sends a pulse and

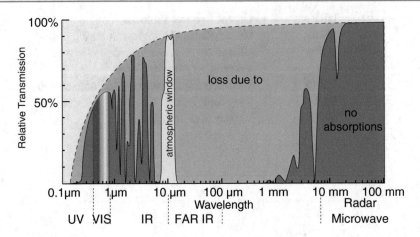

Fig. 1 Electromagnetic spectrum highlighting the regions, UV, IR and microwave, where remote sensing analysis from space is possible; last viewed February 17, 2015 (http://upload.wikimedia.org/wikipedia/commons/ thumb/d/d9/Atmospheric_window_EN.svg/2000px-Atmospheric_window_EN.svg.png). *Dashed line* represents the scattering losses due to absorption across the full spectrum defined in the figure

waits for a return, or a passive system, that measures the response at a specific wavelength as radiation passes through the Earth's atmosphere. Wavelengths of emitted and reflected electromagnetic (EM) radiation, as measured by current satellite sensors, are divided into regions from ultraviolet (UV) to microwave (Fig. 1). Most, though not all observations, are made through 'atmospheric windows' where the absorption by the atmosphere is minimal. For some observations high opacity is useful such as infrared wavelengths around 7.3 µm where the absorption features of SO_2 can be used to measure the amount of SO_2 released by a volcano (Thomas et al. 2009, 2011). In order to undertake a volcanic hazard assessment, the type of satellite sensor is therefore determined by the process being examined and the availability of atmospheric windows (Thomas and Watson 2010).

When describing remote sensing satellites and their sensors, the terms 'spectral', 'spatial' and 'temporal' resolutions are often used to assess if the available data can provide the required coverage for the hazardous event. Sensors can have low spectral resolution, such as the National Oceanic and Atmospheric Administration (NOAA) Advanced Very High Resolution Radiometer (AVHRR) with three broad range channels in the mid and thermal infrared wavelengths of the EM spectrum (Fig. 2a). This is compared to the low spectral resolution sensors, like NASA's Moderate Resolution Imaging Spectroradiometer (MODIS) with 11 broad and narrow range channels across this portion of the EM spectrum. These are compared to multi- or hyperspectral sensors, like the NASA Airborne Infrared Remote Sounder (AIRS), which has over 2000 channels from 3.7 to 15.4 µm.

Trade-offs in the best sensor to use will occur between pixel size and detail level required (Fig. 2b). Finer spatial resolutions are often a function of a narrower swath width, i.e. coverage across the satellite track. Therefore, there is a trade-off between the spatial resolution and the repeat rate of measurements from the same sensor, i.e. its temporal resolution. Differences in temporal resolution generally relate to the chosen orbits of the different satellites (Fig. 2c). Sensors on satellites with a polar orbit, like AVHRR and MODIS, have a longer repeat period than those from a geostationary orbit, such as the Geostationary Operational Environmental Satellite (GOES-R). However, these geostationary satellites, which remain stationary at one point above the equator, have as a consequence a coarser spatial resolution than from a polar orbit. The

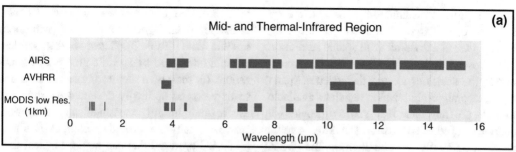

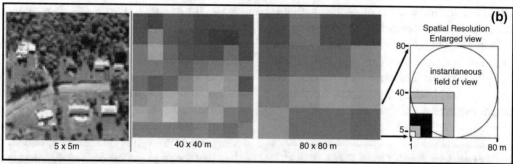

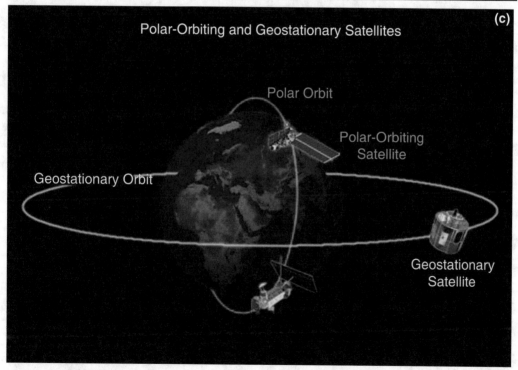

Fig. 2 **a** Varying spectral ranges of satellite channels for AIRS, AVHRR and MODIS (NASA 2003)—note that AIRS data represents multispectral data with narrow bands while AVHRR and MODIS represent individual broad band channels; **b** Changes in image resolution 'smooths' out useful data (the same image is shown at 5 × 5, 40 × 40 and 80 × 80 m pixels to illustrate the data smoothing at coarser resolutions); and **c** differences in polar and geostationary orbits for repeat coverage (adapted from http://www.meted.ucar.edu/satmet/npp/media/graphics/geo_leo_satellites.jpg)

spectral resolution is limited by available energy, and hence, as larger footprints provide more energy there is a trade-off with spatial resolution. The chosen applications of the satellite sensor inform the decisions around the spectral, spatial and temporal resolutions of the data available. NOAA (2015a) and JPL (2015) provide an overview of the NOAA and NASA satellite missions currently in operation, while WMO (2013) provide information on different satellites and their sensors specifically for detecting volcanic ash and SO₂ clouds. In the following section, we highlight some of the online tools currently available that routinely process satellite data and provide derived products for volcanic hazard assessment and decision support.

3 Accessing Real-Time Data

Gaining timely access to remote sensing for hazard assessment is essential for rapid analysis. Users can download available data from online sites provided by the satellite data providers (Table 1). To fully analyze the available data often requires access to expensive software, such as ENVI® (Environment for Visualizing Images), as well as proficiency in remote sensing image analysis (Table 2). Other projects and agencies

have provided online access to products from available polar, geostationary/geosynchronous satellite data (Table 3). These online services are often designed by specific groups or organizations for their own monitoring capabilities and so may not be optimally designed for others to use in their hazard assessment and mitigation. However, they do provide excellent routinely processed products that can be used in real-time to assess the hazardous events and mitigate their impact.

Other groups have set up their own receiving stations (Kaneko et al. 2002; Webley et al. 2008) for local access to data to reduce the time delay between acquisition and data processing. Again, as with accessing data from the services listed in Table 1, there is need for either a local remote sensing expert or a fully automated system and online tool for all to use the derived data during a volcanic event.

4 Applications During Volcanic Eruptions

To illustrate how remote sensing data have been successfully used during volcanic eruptions, the following section focuses on three different eruptive events (Fig. 3) with examples from

Table 1 Some of the available online tools available for accessing the multispectral and high spatial resolution remote sensing data

Primary Org.	Website/Tool name	Capabilities for the user
NASA	https://earthdata.nasa.gov/data/ data-tools/search-and-order-tools	Lists all possible online tools for search and ordering of data from NASA sensors
USGS	EarthExplorer http://earthexplorer.usgs.gov/	Search for data from archival and NRT data, view premade images, download data in HDF form
NASA	Reverb http://reverb.echo.nasa.gov/	Can search any data from NASA or affiliate agencies; download data in HDF form
USGS/NASA	ASTER EDS data https://astereds.cr.usgs.gov/	View all newly acquired ASTER data from urgent request, premade browse images, download HDF data
USGS	GloVis http://glovis.usgs.gov/	Java based interface for viewing archival data, premade browse images, request data in HDF format.
EUMETSAT	Earth Observation Portal https://eoportal.eumetsat.int/	Requires account, can view all EUMETSAT data, need specific agreement to download data

ASTER Advanced Spaceborne Thermal Emission and Reflection Radiometer, *EDS* Expedited Data Sets, *NRT* Near Real-Time and *EUMETSAT* European Organisation for the Exploitation of Meteorological Satellites

Table 2 Example of remote sensing software that can be used to analyze spaceborne data and produce two-dimensional datasets and imagery per channel and satellite overpass

Software	Manufacturer	Platform	Capabilities for the user
BEAM/SNAP	ESA	Unix	Display, basic image processing, band arithmetic, rectification, ROI statistics
ENVI	Harris	Unix, Windows, MacOSX	Display, band arithmetic, rectification, ROI statistics, advanced algorithms for image processing
ILWIS	52 North	Windows	Digitizing, editing, display, GIS, basic analysis
Imagine	ERDAS	Unix, Windows, MacOSX	Display, band arithmetic, rectification, ROI statistics, advanced algorithms for image processing
ORFEO	CNES	Unix, Windows, MacOSX	Display, band arithmetic, rectification, ROI statistics
Opticks	BATC	Unix, Windows	Display, image manipulation, statistics
Geomatica	PCI	Unix, Windows, MacOSX	Display, Mosaics, DEM extraction, rectification
REMOTE-VIEW	Textron	Windows	Display, multispectral analysis, calibration
SOPI	CONAE	Unix, Windows	Display, basic image processing, rectification
SPIRITS	JRC	Windows	Display, basic image processing, band arithmetic, rectification

Table 3 Sample list of online resources, specifically designed for volcanoes, for access to derived products from satellite data

Name	Website/Tool name	Sensor/Satellite analyze	Focus/Details
Volcview	http://volcview.wr.usgs.gov/	AVHRR/MODIS/MTSAT/GOES	Access to a number of satellites in near real time for the North Pacific region
Volcanic Cloud Monitoring	http://volcano.ssec.wisc.edu/	AVHRR/MODIS/MTSAT GOES/SEVIRI/VIIRS	NOAA-CIMSS satellite imagery pages. Includes a range of single, two and multi-channel products
ASTER Volcano Archive	http://ava.jpl.nasa.gov/	ASTER	Provides users access to search for all ASTER data for volcanoes
SACS	http://sacs.aeronomie.be/nrt/	GOME-2, IASI, OMI, OMP, AIRS	Near real time SO_2 and ash alert notification and imagery
NASA SO_2	http://so2.gsfc.nasa.gov/	OMI, OMPS	NASA Global SO_2 monitoring page

SEVIRI Spinning Enhanced Visible and Infrared Imager, *IASI* Infrared Atmospheric Sounding Interferometer, *MTSAT* Multifunctional Transport Satellites, *VIIRS* Visible Infrared Imaging Radiometer Suite and *OMI* Ozone Monitoring Instrument

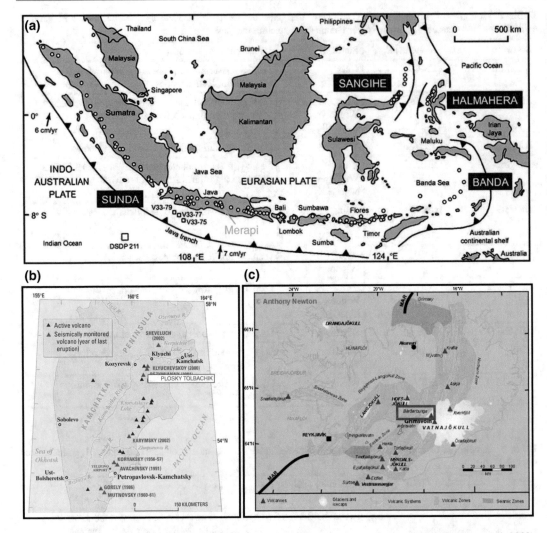

Fig. 3 **a** Indonesia and its volcanoes, highlighting Merapi volcano (modified from Gertisser and Keller 2003); **b** The volcanoes of Kamchatka emphasizing the location of Plosky Tolbachik (adapted from Kirianov et al. 2002); and **c** Icelandic volcanoes and associated geology showing the site of Bárðarbunga volcano (GVP 2011)

across the EM spectrum. Through these case studies, we show examples of available online data as well as post-processed data and imagery that encapsulate multiple sources (Table 4) and different types of analyses.

4.1 Merapi Volcano and Its 2010 Eruption

Merapi Volcano in Indonesia (Fig. 3a) has, since 1900, typical recurrence intervals of 4–6 years with lava dome generation that collapse to generate pyroclastic flows and lahars (Surono et al. 2012). In 2010, the volcano erupted explosively with the main explosive events on October 26, 29, 31 and November 1, 3, and 4. Surono et al. (2012) along with Jousset et al. (2013) provide an excellent overview of the eruption. Here, however, we focus on the use of radar data, as this provided a spectacular view of the growing dome and was essential in aiding forecasting of the eruptive crisis in terms of predicting the impact that could be generated if (or when) the dome collapsed. After the first phreatomagmatic explosive phase on October 26,

Table 4 All sensors described in chapter, for full names of sensors and agencies see descriptions within the chapter

Sensor (Agency)	Spatial[a]	Spectral	Temporal[b]	References
SEVIRI (EUMETSAT)	1–3 km	0.635–12.4 µm	15 min	EUMETSAT (2015a)
MTSAT (JMA)	1–4 km	0.55–12.5 µm	30 min	BOM (2015)
IASI (EUMETSAT)	25 km	3.6–15.5 µm	Twice a day	EUMETSAT (2015b)
OMI (NASA)	13 × 24 km (nadir)	0.27–0.5 µm	Daily	Krotkov et al. (2006)
AIRS (NASA)	13.5 km	3.7–15.4 µm	16 days	NASA (2003)
GOME-2	40 km (nadir)	0.25–0.79 µm	29 days[c]	EUMETSAT (2014)
OMPS (NOAA)	50 km (nadir)	0.3–0.38 µm	16 days	NASA (2015a)
ALI (NASA)	10–30 m	0.48–2.35 µm	16 days	NASA (2011a)
HYPERION (NASA)	30 m	0.4–2.5 µm	16 days	NASA (2011b)
ASTER (NASA)	15–90 m	0.52–11.65 µm	16 days[d]	USGS (2014a)
AVHRR (NOAA)	1 km	0.58–12.5 µm	Multiple a day	NOAA (2013)
MODIS (NASA)	250 m–1 km	0.459–14.3 µm	Twice a day	NASA (2015b)
GOES-R (NOAA)	500 m–2 km	0.47–13.3 µm	15 min	NOAA (2014)
TerraSAR (Astrium)	1–18 m	3.1 cm	2.5 days	Astrium (2015)
RadarSAT-2 (CSA)	1–8 m	5.6 cm	24 days	CSA (2015)
VIIRS (NOAA/NASA)	375 m (nadir)	0.412–12.013 µm	Twice a day	NOAA (2015b)
LANDSAT-8 (USGS/NASA)	15–100 m	0.43–12.5 µm	16 days	USGS (2014b)

[a]Spatial resolution is defined as the range across swath unless specified
[b]THIS is the orbital repeat time, rather than the time it takes to view the same point on the earth. Also, note that for several sensors, such as AVHRR, MODIS and GOME-2, the time is specified per sensor and there are multiple sensors in orbit on different satellites
[c]GOME-2 images the same location more than once every 29 days but at different locations within the swath
[d]ASTER sensor can be tasked and data can be requested to view specific locations more frequently than the 16 day orbit repeat time

remote sensing data became a critical tool in monitoring and guiding those in crisis management (Pallister et al. 2013). The availability of commercial satellite radar data from RADARSAT-2 and TerraSAR-X, at 0.5–3 m pixel resolution, provided a unique view of the summit activity and the data to analyze the dome growth occurring at the summit between each explosive event (Fig. 4).

Pallister et al. (2013) stated that the speed of availability for these data was critical in providing the best available analysis of the growing dome for life-saving crisis response. For example, estimates based on radar data suggest that the dome grew 5 million m^3 between October 26 (Fig. 4a) and November 4 and by the early hours of November 6, the dome was gone (Fig. 4b). However, a new dome of approx. 1.6 million m^3 grew in the space of 11 h that same day (Fig. 4c). The fact that local volcanologists and emergency managers were able to examine the activity at the summit of Merapi at sub 3 m resolution enabled them to make short term assessments on the dome's stability. By

Fig. 4 Merapi RADARSAT-2 and TerraSAR data, adapted from Pallister et al. (2013) showing data from **a** October 11 and 26, 2010; **b** November 4 and 6, 2010; and **c** November 6, 2010 at 11:08 and 22:07 UTC where *PF* pyroclastic flow, *K* Kali Kuning, *Kj* Kinorejo, *C* empty crater, *D* lava dome, and *S* surge deposits. On October 11 and 26, 2010, the *red arrow* highlights the 2006 dome while 'G' indicates the location of Kali Gendoi, where a pyroclastic density current flowed down. The 1 and 2 km distances are equivalent to the length of the *arrow line* segment

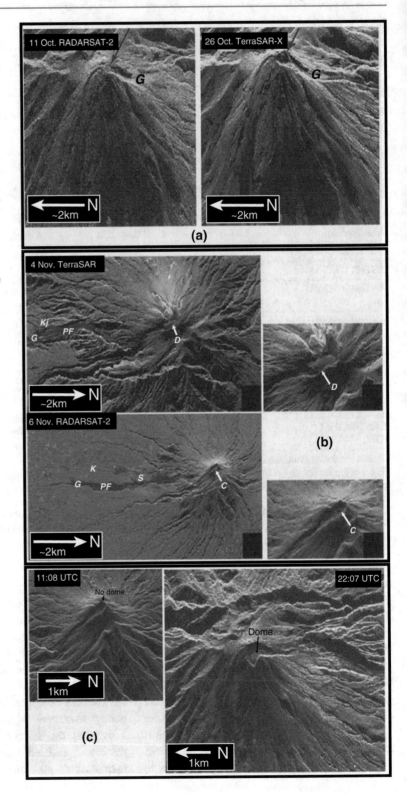

measuring the extrusion rate of the dome as it grew between explosive events volcanologists were able to assess the likelihood of dome destabilization.

Radar data were just one form of remote sensing data that were available during the eruption. High temporal frequency visible and infrared data (see Sect. 5) were available to detect and analyze the location of the dispersing ash clouds for aviation safety. Time series analysis of the dome growth and its potential to destabilize provided the inputs to build probabilistic scenarios of the likelihood for an eruption (Pallister et al. 2013) and for a significant dome collapse that could impact the surrounding population centers.

4.2 Tolbachik Volcano and Its 2012–2013 Eruption

Tolbachik or Plosky Tolbachik volcano, situated on the Kamchatka Peninsula, is a massive basaltic volcano sitting at the southern end of the Klyuchevskaya volcano group. GVP (2017) state that the majority of its activity is relatively low level with volcano explosivity indices (VEI) of 1 and 2 (see Newhall and Self (1982) for explanation of VEI). However, it had a major fissure eruption in 1975–1976 (Fedotov et al. 1980) that was classified as a VEI 4. Tolbachik was quiet until late 2012, when the local volcano monitoring group, Kamchatka Volcanic Eruption Response Team (KVERT), reported that an eruption had begun on November 27.

We examine the mid and thermal infrared wavelength data from both high temporal (multi views a day), low spatial (one [1] km in IR) data and low temporal (one overpass every few days), high spatial (<100 m in IR). Figure 5a shows images from the NASA Advanced Land Imager (ALI) sensor, where differences in the extent of the fissure can be seen from choice of band combinations. Bands 3, 2, 1 represent a true color image. Switch the combination to bands 5′, 4′, and 4, and the red color in the RGB composite is enhanced. Finally, with the composite of Bands 7, 5′, and 5 the thermal signals from the fissures

at Band 7, 2.35 µm, are close to saturation and therefore a 'red glow' effect occurs. These ALI data (Figs. 5a–d) illustrate that by using different band combinations and comparing signals at different wavelengths it is possible to determine the hottest part of the eruptive feature.

For example, Fig. 5b shows ALI data from December 1, five days after the first reported activity. The visible data show the steaming flows as they interact with the surrounding snow and ice, while the SWIR, right-hand side of Fig. 5b, shows the thermal signals being emitted from the flows. Combining visible and SWIR imagery, the ALI data is able to capture the location and intensity of- thermal signals from the fissure eruption. Figure 5c shows visible and thermal infrared (TIR) data from the NASA Advanced Spaceborne Thermal Emission and Reflection Radiometer (ASTER) sensor on February 13, 2013 'tasked' using the ASTER Urgent Request Protocol (URP), see Ramsey and Dehn (2004). The TIR data shows the hottest part of the flow on the eastern side, noting the earlier western erupted flows are still warmer than the surrounding region. Compare to this to SWIR from the ALI data on December 1, 2012, Fig. 5b. This cooler material on February 13 corresponds to the hot active flows six weeks earlier. With its 10.6 µm data, the ASTER URP data could be used to determine the effusion rates from the fissure eruption and assess if the activity was increasing or waning.

With the chosen channels for the RGB composite (Fig. 5d), it is possible to differentiate the higher radiances at 2.193 µm, indicating warmer temperatures. This composite analysis for HYPERION shows how with hyperspectral data it is possible to assess subtle variations in the thermal signals of active flows. Additional analyses can include examining the full suite of 220 channels to determine the surface characteristics, such as rock type and mineral composition (see Hubbard et al. 2003).

The high spatial resolution imagery from ASTER, ALI and HYPERION were not the only thermal infrared data to capture the activity at Tolbachik volcano. The high temporal, lower spatial resolution data from the NOAA AVHRR

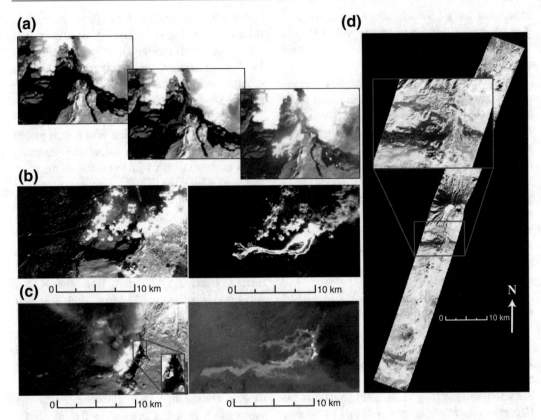

Fig. 5 Different sensors over time for Tolbachik volcano in 2012–2013 where differences in the extent of the fissure can be seen based on the choice of band combinations. **a** NASA ALI on February 14, 2013 of Bands 3, 2, 1 (0.63–0.69 μm; 0.525–0.605 μm; 0.45–0.515 μm) showing a true color image; Bands 5′, 4′ and 4 (1.1–1.3 μm; 0.845–0.89 μm; 0.775–0.805 μm) and Bands 7, 5, and 5′ (2.08–2.35 μm; 1.55–1.75 μm; 1.2–1.3 μm); **b** NASA visible and SWIR ALI from December 1, 2012; **c** NASA ASTER Urgent Request Protocol (URP) data (visible and TIR) from February 13, 2013 with subset in the *red box* showing a thermal signal from the most active part of the fissure as the *red color* is elevated; and **d** NASA HYPERION RGB composite with Channels 204, 150 and 93 at 2.193, 1.649 and 1.074 μm from February 14, 2013 data showing the most active portion of the flows (*red* and *yellow regions* in the *inset*)

and NASA MODIS sensors provided a vital tool in monitoring the fissure eruption. March 2, 2013 provided an ideal opportunity to compare the thermal signals available from the three different sensors (Fig. 6) The ASTER data resolution allows finer details in the flow pattern to be captured and therefore is closer to the actual spatial extent of the flow. The MODIS and AVHRR data have much larger pixels, 1.1 km^2 at nadir that grows in size towards the swath edge impacting the ability to detect small scale (spatial and thermal output) thermal events.

The timely data from the MODIS and AVHRR sensor meant that they could be used to assess relative changes from one overpass to the next, given the sub-daily temporal frequency of the available data. The ASTER data, while more accurate given its pixel size, is one point in time. Therefore, for enhanced hazard assessment the MODIS and AVHRR data, coupled to the finer spatial resolution ASTER for comparison, became a critical tool in the daily monitoring of the ongoing hazard. The MODIS and AVHRR data (Fig. 6) were routinely collected by KVERT for analysis of the ongoing hazard, while the ASTER data were acquired by NASA URP and the ALI/HYPERION data were available via the online tools from satellite providers. This satellite

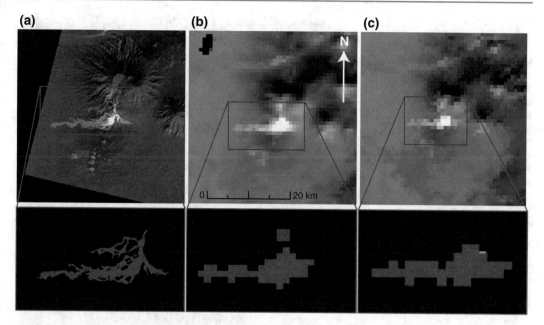

Fig. 6 Intercomparison of sensors on the same date: **a** ASTER Band 13 at 00:32 UTC; **b** AVHRR Band 4 at 01:18 UTC; and **c** MODIS Band 31 at 00:30 UTC on March 2, 2013. Here ASTER Band 13 data, 9.1 μm, is compared to AVHRR Band 4, 10.6 μm, and MODIS Band 31, 11.02 μm. ASTER with 90 m spatial resolution captures more of the flow's details, and shows an area of 30 km². Compare this with the MODIS data with an area of 54.8 km² and AVHRR with 47.3 km²

remote sensing data was useful to analyze the relative change in the fissure eruption and showed how, by coupling multi resolution data, one could assess if the eruption was over or if it was increasing towards a new level of heightened activity.

4.3 Bárðarbunga Volcano and Its 2014–2015 Eruption

Bárðarbunga volcano is situated beneath the northwestern part of Vatnajökull ice cap in south-eastern Iceland. In August 2014, the Icelandic Meteorological Office (IMO) reported increased seismicity at the volcano. By August 23, this activity had developed into a 150–400 m long dyke beneath the Dyngjujökull glacier, prompting a change in the aviation color code to red (GVP 2014). By August 29, IMO reported that a small fissure eruption had started in Holuhraun and lava started to erupt on August 31 along a 1.5 km long fissure (GVP 2014).

Following the eruptions from Eyjafjallajökull 2010 (Gudmundsson et al. 2010, 2012) and Grimsvötn 2011 (GVP 2011), there had been a lot of focus on the potential impact of another Icelandic eruption on Europe and its transportation infrastructure. Routine satellite remote sensing data were collected and processed to build derived products from ultraviolet data with Ozone Mapping Profiler Suite (OMPS) and Global Ozone Monitoring Experiment-2 (GOME-2) sensors, in addition to infrared data from VIIRS, AIRS and other geosynchronous/polar-orbiting sensors.

IMO provided daily online updates on the ongoing eruption from Bárðarbunga volcano (Holuhraun fissure) (IMO 2015), including any field observations and notes from meetings of the Scientific Advisory board. Satellite remote sensing data was an integral part of this monitoring and daily reporting of volcanic activity. For example, IMO used LANDSAT-8 (Land Remote-Sensing Satellite (System)-8) data to analyze the spatial footprint of lava flows,

Fig. 7 Data from the Icelandic Meteorological Office (IMO) website **a** LANDSAT data as RGB composite of the flows on January 3, 2015; **b** as GIS layer see http://en.vedur.is/media/jar/myndsafn/full/20150115_afstodukort_ornefni_300dpi.png; and **c** ALI thermal from January 16, 2015 using the same geographical projection to allow comparison to the LANDSAT-8 data from January 3

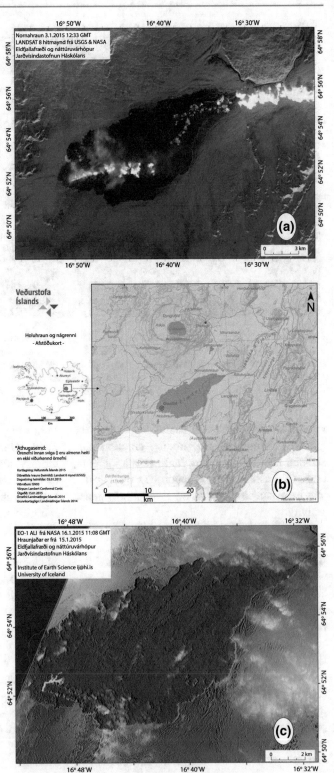

determine the most active portion (Fig. 7a) and develop a map of the ongoing activity for decision making (Fig. 7b). The combined images (Fig. 7a–c) show how IMO and other local authorities were able to use the high spatial resolution data to map out the full flow extent and also generate GIS data layers for use in comparison to other data from field observations and or seismic activity.

The activity from Bárðarbunga volcano along the Holuhraun fissure generated more satellite derived SO_2 imagery than for volcanic ash (Fig. 8). Figure 8a for AIRS on September 1 shows less detectable SO_2 than from GOME-2, Fig. 8b, and OMPS, Fig. 8c. Figure 8d–f, illustrate the influence of the sunlight on deriving UV satellite based measurements. For the OMPS sensor, Fig. 8f, the derived SO_2 is cut off as a result of the lack of sunlight. This impact can be significant for any northerly drift of the SO_2 cloud from the Holuhraun fissure.

Figure 8 shows the usefulness of using common spatial domains and temporal averaging of

satellite data to analyze for relative change. However, the background information on local infrastructure, populated locations, and transportation links as well as the ability to overlay mutli satellite data together is limited. Often Geographical Information Systems (GIS) are used to display different datasets together or to generate maps for hazard assessment. The need to compare different data sources and include supplementary information has led to the use of Virtual Globes.

5 Geo-Spatial Tools and Virtual Globe Applications

Satellite remote sensing has become an integral part of daily monitoring of volcanic activity (Webley et al. 2008) and is part of a suite of geophysical data utilized by volcano observatories (Webley et al. 2013). To combine all of these datasets together requires a Geospatial tool, remote sensing processing software or a virtual

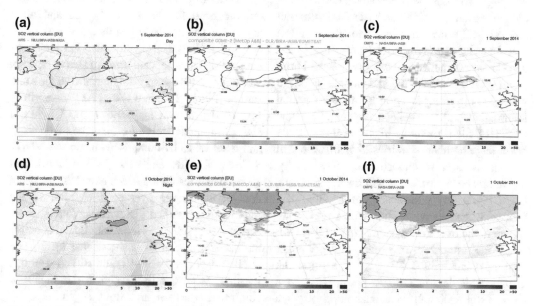

Fig. 8 Real-time daily Support to Aviation Control Service (SACS, http://sacs.aeronomie.be/) products, AIRS, GOME-2 and OMPS on September 1 (**a, b, c**) and October 1, 2014 (**d, e, f**) from Bárðarbunga/Holuruhan volcanic activity. Note these are daily composites given the polar orbiting nature of the satellites and that

SACS (2015) have set both common geographical domains and measurement scales to ensure quick comparisons. These common geographical regions and temporal periods allow those in hazard assessment to compare the relative change in the SO_2 output without the need to process and/or re-project the data globe. These Geospatial or Virtual Globe tools allow the user to overlay many different datasets and add in other geophysical data such as earthquake locations or GPS point data. To display remote sensing data within a Geospatial tool, the satellite imagery (often in Hierarchical Data Format [HDF], or Geostationary Earth Orbit Tagged Image File Format [GEOTIFF]) needs to be converted into shape (SHP) files with associated metadata. These files can be displayed along with other GIS data layers to produce hazard maps which can be used to assess the risk and mitigate impacts.

It is possible to use Virtual Globes to display the imagery and make rapid assessments of the hazard and its potential impact. Recently, Google Earth has been used for visualizing geospatial data (Webley et al. 2009; Prata and Prata 2012) but it requires the user to generate a Keyhole Markup Language (KML) or Keyhole Markup Zipped (KMZ) file of their own satellite imagery (Bailey et al. 2010). In early 2015, the ability for all to use Google Earth Pro was made possible (Google 2015) as it was publically available free of charge. This has the added advantage of being able to read in GEOTIFF and GIS SHP files, but there are still some limitations. Google Earth Pro will open automatically to its pre-defined start up. The user is then able to simply drag and drop their GEOTIFF or SHP file and Google Earth Pro will fly to the center location from the geospatial information in the associated file (Fig. 9). If the GEOTIFF file is too large in size or the SHP file has a large number of entries, the software will request the user to choose the desired image resolution. Once chosen, it will display the data, but as Fig. 9 shows, the scaling of the GEOTIFF data may lead to no useful information being shown in the overlay. By generating a GEOTIFF *image* file from ENVI or similar software rather than the default GEOTIFF *data* file, the user can display their data in Google Earth Pro with a more useful scaling for visualization. ENVI® is a geospatial product (see Table 2 for list) provided by Harris Geospatial (Harris Geospatial 2016) that combines advanced image processing and geospatial technology to allow the user to extract useful information from remote sensing data and make better informed decisions.

Once the data are available in either a Geospatial tool or Virtual Globe, users can map the hazard and measure its spatial extent and potential for further impacts. Virtual Globes like Google Earth Pro bring a new mapping capability not available in Google Earth. Figure 10 shows how a hazard manager would load the data into ENVI or a similar image processing tool, and then map out a region of interest (ROI) using either points (red) or as a polygon (green). This can also be done in Google Earth Pro where the imagery is loaded and displayed as an overlay. Using the polygon tool, the hazard manager can map out the edge of the flow. Comparing the area from ENVI with its ROI tool and a pixel by pixel analysis to the results from Google Earth Pro with its polygon tool, the total mapped area for this flow was calculated as 76.13 and 76 km^2 respectively. The relative different was 0.13 km^2 or 0.02% difference, thus highlighting that Google Earth Pro allows those in hazard assessment and mitigation to quickly and accurately map out volcanic features, without the need for access to specific and often expensive remote sensing software.

Virtual Globes also allow for the intercomparison of satellite data from different sensors and across different spectral ranges with the same sensor. Figure 11 shows the ALI data, from Fig. 10, by displaying Band 1 (0.48–0.69 μm and 10 m spatial resolution panchromatic data) and a set of RGB composites using the VNIR, SWIR and MIR channels (0.43–2.35 μm and 30 m spatial resolution data). Note that the black area in the image overlay is a result of the satellite overpass and therefore no data is recorded in this region. It is important information for the decision maker as it shows that there is no satellite data but the flow could still exist in this location. More data is needed in this region to confirm the lack of lava flow. As the data are geolocated, the relative differences between channels can be easily compared and used to assess the location of the hottest part of the flow. Small, hot targets will produce the strongest

Fig. 9 How to display GEOTIFF data in Google Earth Pro, using example from ALI Band 1 on January 16, 2015. Google Earth Pro allows the use of 'drag and drop', where the user can simply drop the GEOTIFF file into software. © 2016 Google Inc. All rights reserved. Google and the Google Logo are registered trademarks of Google Inc.

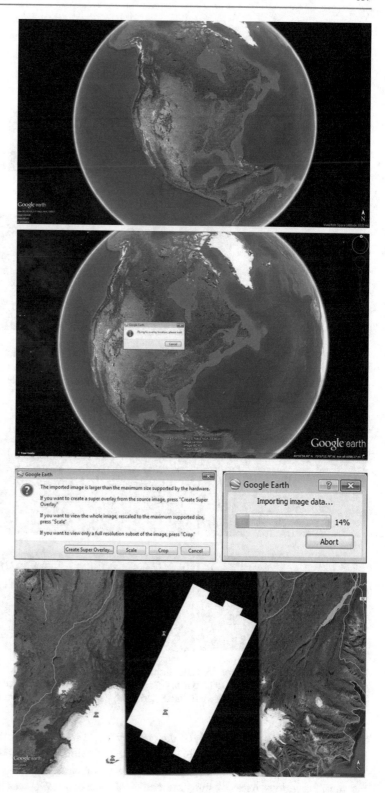

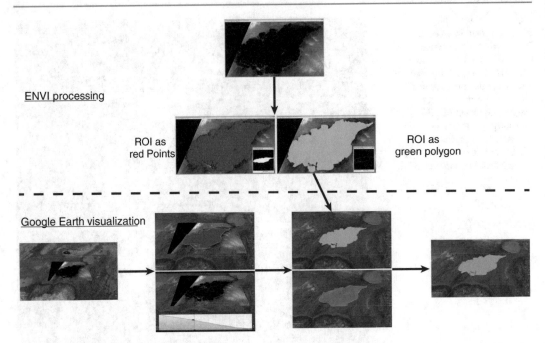

Fig. 10 ENVI® and Google Earth Pro workflow to analyze the spatial footprint of the flow from Band 1 ALI data on January 16, 2015. ENVI® analyzed area was 761,359 points or area of 76.13 km², Google Earth Polygon was 76 km². © 2016 Google Inc. All rights reserved. Google and the Google Logo are registered trademarks of Google Inc.

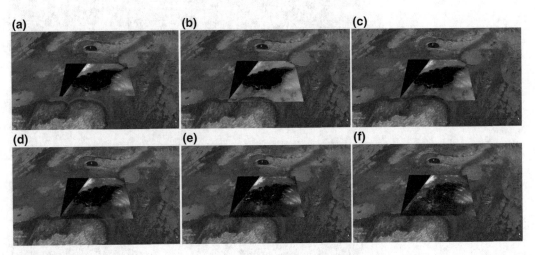

Fig. 11 ALI data from January 16, 2015 displayed in Google Earth Pro®. **a** Band 1; **b** RGB Bands 432; **c** RGB Bands 543; **d** RGB Bands 654; **e** RGB Bands 765; and **f** RGB Bands 876. © 2016 Google Inc. All rights reserved. Google and the Google Logo are registered trademarks of Google Inc.

radiant signal in the SWIR and MIR data and by comparing the RGB data from the VNIR to the SWIR, operational hazard managers can assess the most hazardous parts of the flow and thus where to prioritize resources.

For some of the shorter wavelengths, the resultant satellite data may be saturated so longer wavelength data, greater than 2–3 μm, may be needed to assess subtle changes (see use of thermal infrared data (10 μm) over mid-infrared data (3.7 µm) in Webley et al. (2013)). By combining the data together (both across different satellite sensors and across different spectral ranges), hazard managers can make a quick assessment that can assist those in the field and reduce the potential impact to local population or infrastructure.

Virtual Globes, like Google Earth Pro highlighted here, and GIS tools (see Fig. 7b) can be used as the interfaces for those in hazard assessment to visualize and analyze the available satellite data. From the analysis in Figs. 10 and 11, KMZ and KML data files can be generated for comparison to other geophysical data along with georeferenced maps, with associated data such as latitude/longitudinal grids, distance scale bars, population centers and transportation links (Fig. 12). As with Fig. 11, the black area highlights the region of no data due to the swath edges from the satellite data.

The insert in Fig. 12 shows how a user is able to define site visit locations using the place marker option in Google Earth Pro. The latitude and longitude location can be extracted and used for fieldwork planning and logistics. With more real-time geophysical data being developed for the Virtual Globe market, as well as the development of online geospatial tools, such as the Google Earth Engine and Google Maps Engine,

the use of the tool will continue to increase among the real-time hazard assessment community.

6 Summary

There has been an integration of remote sensing data and geospatial tools into day-to-day operations for volcano monitoring. They are used to perform a more effective hazard assessment of ongoing volcanic activity. Remote sensing data are now available from ultraviolet to microwave wavelengths (spectral); from m's to km resolutions (spatial) and from daily to weekly repeat times (temporal). There is a trade-off between the spectral, spatial and temporal and it has been the combination of all available data that has increased the uptake of the remote sensing data into routine volcano monitoring. Satellite data provides an overview of the local, regional, country-wide, continental, and global scale of the volcanic eruptive events. Effective visualization of these data can then significantly help decision makers to communicate the impact to those at risk in an effective and efficient manner.

This chapter highlights examples where satellite remote sensing was used during volcanic eruptions to: (1) improve the assessment of

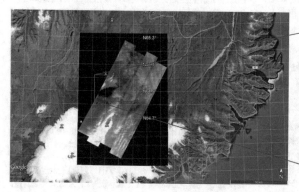

Fig. 12 ALI data from January 16, 2015 map as generated from Google Earth Pro. Map shows latitude and longitude grid, scale bar to show distances in km, North arrow to align image N–S, the locations of all the volcanoes, airports, roads (*yellow lines*) and also any population centers. *Inset* shows area zoomed in on the flow with user defined locations for site visits of Flow edge-North and Flow edge-South. © 2016 Google Inc. All rights reserved. Google and the Google Logo are registered trademarks of Google Inc.

potential for dome collapse using meter resolution radar data (Merapi 2010); (2) examine the changing shape and thermal properties of a large fissure eruption using multi wavelength and resolution data (Tolbachik 2012–2013); and (3) map the spatial extent of thermal activity from volcanic flows using high resolution data and routinely available volcanic gas measurements (Bardarbunga/Holuhraun 2014–2015). Additionally, the chapter shows how Virtual Globes have become an integral tool for hazard assessment and new developments will lead to a greater use of these in real-time assessments.

Remote sensing data and geospatial technologies are now critical parts of the monitoring toolbox. To be used in a real-time event, the data needs to be collected, processed and displayed routinely, i.e. for every possible satellite overpass and using the same geographical domain. Satellite data providers have created several online tools to undertake the initial hazard assessment. To combine all the data together, they need to be on common projections and have associated metadata on the spectral, spatial and temporal resolutions. Then the choice is on the best geospatial technology for the application and/or assessment required. This can be a GIS tool where ancillary data is already in a GIS format, or it could be a Virtual Globe where other data is in a KML/KMZ format. Using the new geospatial technologies to display, visualize and analyze the routinely (automatically) acquired and processed remote sensing data will provide the best tools for real-time hazard assessment, in essence building a system ready if an eruption occurs, rather than building the system as it is developing.

Acknowledgements We would like to thank the remote sensing students of the Geophysical Institute, University of Alaska Fairbanks for assisting us in acquiring and analyzing the Tolbachik AVHRR, ASTER and MODIS data. This work was undertaken partially as part of a graduate level class and allowed us to further analyze the most appropriate data for this manuscript. We thank the Natural Environment Research Council (NERC) Centre for Observation and Modelling of Earthquakes, Volcanoes and Tectonics (COMET) for support for Dr. Watson.

References

Aki K, Fehler M, Das S (1977) Source mechanism of volcanic tremor: fluid-driven crack models and their application to the 1963 Kilauea eruption. J Volcanol Geotherm Res 2(3):259–287

Astrium (2015) TerraSAR-X specifications, http://www.satpalda.com/product/terrasar-x/, last viewed Apr 13, 2015

Bailey JE, Dean KG, Dehn J, Webley PW (2010) Integrated satellite observations of the 2006 eruption of Augustine Volcano, Chap. 20. In: Power JA, Coombs ML, Freymueller JT (eds) The 2006 eruption of Augustine Volcano, Alaska: U.S. Geological Survey Professional Paper, vol 1769, pp 481–506

Bureau of Meteorology (BOM) (2015) Multi-functional transport satellite (MTSAT). http://www.bom.gov.au/australia/satellite/mtsat.shtml, last viewed Apr 13, 2015

Canadian Space Agency (CSA) (2015) RADARSAT satellite characteristics. http://www.asc-csa.gc.ca/eng/satellites/radarsat/radarsat-tableau.asp, last viewed Apr 13, 2015

European Organization for the Exploitation of Meteorological Satellites (EUMETSAT) (2014) Global ozone monitoring experiment-2 (GOME-2) fact sheet. http://www.eumetsat.int/website/wcm/idc/idcplg?IdcService=GET_FILE&dDocName=PDF_GOME_FACTSHEET&RevisionSelectionMethod=LatestReleased&Rendition=Web, last viewed Feb 17, 2015

European Organization for the Exploitation of Meteorological Satellites (EUMETSAT) (2015a) IASI mission. http://www.eumetsat.int/website/home/Satellites/CurrentSatellites/Metop/MetopDesign/IASI/index.html, last viewed Apr 13, 2015

European Organization for the Exploitation of Meteorological Satellites (EUMETSAT) (2015b) Meteosat 2nd generation channels for the SEVIRI sensors. http://oiswww.eumetsat.org/WEBOPS/msg_interpretation/PowerPoints/Channels/MSGChannels.htm, last viewed April 13, 2015

Fedotov SA, Chirkov AM, Gusev NA, Kovalev GN, Slezin YB (1980) The large fissure eruption in the region of Plosky Tolbachik volcano in Kamchatka, 1975–1976. Bull Volcanol 43(1):47–60

Gertisser R, Keller J (2003) Trace element and Sr, Nd, Pb and O isotope variations in medium-K and high-K volcanic rocks from Merapi Volcano, Central Java, Indonesia: evidence for the involvement of subducted sediments in Sunda Arc magma genesis. J Petrol 44(3):457–489

Global Volcanism Program (GVP) (2011) June 2011 report of the Bulletin of Global Volcanism Network (BVGN). http://www.volcano.si.edu/volcano.cfm?vn=373010#bgvn_201106, viewed Feb 17, 2015

Global Volcanism Program (GVP) (2014) August 2014 report of the Bulletin of Global Volcanism Network

(BVGN). http://www.volcano.si.edu/volcano.cfm?vn=373030#August2014

Global Volcanism Program (GVP) (2017) Eruptive History of Tolbachik Volcano provided by the Bulletin of Global Volcanism Network (BVGN). https://volcano.si.edu/volcano.cfm?vn=300240&vtab=Eruptions, viewed May 26, 2017

Google (2015) Google earth for desktop. http://www.google.com/earth/explore/products/desktop.html, last viewed Feb 17, 2015

Gudmundsson MT, Pedersen R, Vogfjörd K, Thorbjarnardóttir B, Jakobsdóttir S, Roberts MJ (2010) Eruptions of Eyjafjallajökull Volcano, Iceland. EOS Trans AGU 91(21):190–191

Gudmundsson MT, Thordarson T, Höskuldsson Á, Larsen G, Björnsson H, Prata FJ, Oddsson B, Magnusson E, Hognadottir T, Petersen GN, Hayward CL, Stevenson JA, Jónsdóttir I (2012) Ash generation and distribution from the April-May 2010 eruption of Eyjafjallajökull, Iceland. Scientific reports, 2

Guo S, Bluth GJ, Rose WI, Watson IM, Prata AJ (2004) Re-evaluation of SO_2 release of the 15 June 1991 Pinatubo eruption using ultraviolet and infrared satellite sensors. Geochem Geophys Geosyst 5(4)

Harris Geospatial (2016) ENVI geospatial products. http://www.harrisgeospatial.com/ProductsandSolutions/GeospatialProducts/ENVI.aspx, viewed Sept 23, 2016

Hubbard BE, Crowley JK, Zimbelman DR (2003) Comparative alteration mineral mapping using visible to shortwave infrared (0.4–2.4 μm) Hyperion, ALI, and ASTER imagery. IEEE Trans Geosci Remote Sens 41(6):1401–1410

Icelandic Meteorological Office (IMO) (2015) Bárðarbunga—update. http://en.vedur.is/earthquakes-and-volcanism/articles/nr/2947, last viewed Feb 17, 2015

Jet Propulsion Laboratory (JPL) (2015) All current satellite missions. http://www.jpl.nasa.gov/missions/, last viewed Feb 18, 2015

Jousset P, Pallister J, Surono (2013) The 2010 eruption of Merapi volcano. J Volcanol Geotherm Res 261:1–6

Kaneko T, Yasuda A, Ishimaru T, Takagi M, Wooster MJ, Kagiyama T (2002) Satellite hot spot monitoring of Japanese volcanoes: a prototype AVHRR-based system. Adv Environ Monit Model 1 (1):125–133

Kirianov VY, Neal CA, Gordeev EI, Miller TP (2002) The Kamchatkan Volcanic Eruption Response Team (KVERT). US Geological Survey Fact Sheet, 64-02

Krotkov NA, Carn SA, Krueger AJ, Bhartia PK, Yang K (2006) Band residual difference algorithm for retrieval of SO_2 from the Aura Ozone Monitoring Instrument (OMI). IEEE Trans Geosci Remote Sens (AURA Special Issue) 44:1259–1266. doi:10.1109/TGRS.2005.861932

Lillesand T, Kiefer RW, Chipman J (2015) Remote sensing and image interpretation, 7th edn. Wiley, London

McNutt SR (1986) Observations and analysis of B-type earthquakes, explosions, and volcanic tremor at Pavlof Volcano, Alaska. Bull Seismol Soc Am 76(1):153–175

McNutt SR (1996) Seismic monitoring and eruption forecasting of volcanoes: a review of the state-of-the-art and case histories. In: Monitoring and mitigation of volcano hazards. Springer, Berlin, pp 99–146

National Aeronautical and Space Administration (NASA) (2003) Atmospheric Infrared Sounder (AIRS) Instrument Guide. http://disc.sci.gsfc.nasa.gov/AIRS/documentation/airs_instrument_guide.shtml, last viewed Feb 17, 2015

National Aeronautics and Space Administration (NASA) (2011a) Sensors—Advanced Land Imager (ALI). http://eo1.usgs.gov/sensors/ali, last viewed March 23, 2015

National Aeronautics and Space Administration (NASA) (2011b) Sensors hyperion. http://eo1.usgs.gov/sensors/hyperion, last viewed March 23, 2015

National Aeronautical and Space Administration (NASA) (2015a) Ozone mapping profiler suite. http://npp.gsfc.nasa.gov/omps.html, last viewed Feb 17, 2015

National Aeronautics and Space Administration (NASA) (2015b) MODIS specifications. http://modis.gsfc.nasa.gov/about/specifications.php, last viewed Apr 13, 2015

National Oceanic and Atmospheric Administration (NOAA) (2013) Advanced Very High Resolution Radiometer—AVHRR. http://noaasis.noaa.gov/NOAASIS/ml/avhrr.html, last viewed Apr 13, 2015

National Oceanic and Atmospheric Administration (NOAA) (2014) GOES-R series concept of operations. http://www.goes-r.gov/syseng/docs/CONOPS.pdf, last viewed Apr 13, 2015

National Oceanic and Atmospheric Administration (NOAA) (2015a) NESDIS NOAA: satellite missions. http://www.nesdis.noaa.gov/about_satellites.html; last viewed Feb 18, 2015

National Oceanic and Atmospheric Administration (NOAA) (2015b) Suomi NPP VIIRS land. http://viirsland.gsfc.nasa.gov/index.html, last viewed Apr 13, 2015

Newhall CG, Self S (1982) The volcanic explosivity index (VEI) an estimate of explosive magnitude for historical volcanism. J Geophys Res Oceans (1978–2012) 87(C2):1231–1238

Owen S, Segall P, Lisowski M, Miklius A, Murray M, Bevis M, Foster J (2000) January 30, 1997 eruptive event on Kilauea Volcano, Hawaii, as monitored by continuous GPS. Geophys Res Lett 27(17):2757–2760

Pallister JS, Schneider DJ, Griswold JP, Keeler RH, Burton WC, Noyles C, Newhall CG, Ratdomopurbo A (2013) Merapi 2010 eruption—chronology and extrusion rates monitored with satellite radar and used in eruption forecasting. J Volcanol Geotherm Res 261:144–152

Poland M, Bürgmann R, Dzurisin D, Lisowski M, Masterlark T, Owen S, Fink J (2006) Constraints on the mechanism of long-term, steady subsidence at Medicine Lake volcano, northern California, from

GPS, leveling, and InSAR. J Volcanol Geotherm Res 150(1):55–78

Prata AJ (1989) Observations of volcanic ash clouds in the 10-12 μm window using AVHRR/2 data. Int J Remote Sens 10(4–5):751–761

Prata AJ, Prata AT (2012) Eyjafjallajökull volcanic ash concentrations determined using Spin Enhanced Visible and Infrared Imager measurements. J Geophys Res Atmos (1984–2012) 117(D20)

Ramsey M, Dehn J (2004) Spaceborne observations of the 2000 Bezymianny, Kamchatka eruption: the integration of high-resolution ASTER data into near real-time monitoring using AVHRR. J Volcanol Geotherm Res 135(1):127–146

Rothery DA, Francis PW, Wood CA (1988) Volcano monitoring using short wavelength infrared data from satellites. J Geophys Res Solid Earth (1978–2012) 93:7993–8008

Support to Aviation Control Service (SACS) (2015) SACS SO₂ and ash notification service. http://sacs.aeronomie.be/, last viewed Feb 17, 2015

Surono JP, Pallister J, Boichu M, Buongiorno MF, Budisantoso A, Costa F, Andreastuti S, Prata F, Schneider D, Clarisse L, Humaida H, Sumarti S, Bignami C, Griswold J, Carn S, Oppenheimer C, Lavigne F (2012) The 2010 explosive eruption of Java's Merapi volcano—a '100-year' event. J Volcanol Geotherm Res 241:121–135

Thomas HE, Watson IM (2010) Observations of volcanic emissions from space: current and future perspectives. Nat Hazards 54(2):323–354

Thomas HE, Watson IM, Kearney C, Carn SA, Murray SJ (2009) A multi-sensor comparison of sulphur dioxide emissions from the 2005 eruption of Sierra Negra volcano, Galápagos Islands. Remote Sens Environ 113(6):1331–1342

Thomas HE, Watson IM, Carn SA, Prata AJ, Realmuto VJ (2011) A comparison of AIRS, MODIS and OMI

sulphur dioxide retrievals in volcanic clouds. Geomat Nat Hazards Risk 2(3):217–232

United States Geological Survey (USGS) (2014a) ASTER sensor overview, USGS Land Processes Distributed Active Archive Center. https://lpdaac.usgs.gov/products/aster_products_table/aster_overview, last viewed March 23, 2015

United States Geological Survey (USGS) (2014b) Frequently asked questions about the landsat missions. http://landsat.usgs.gov/band_designations_landsat_satellites.php, last viewed March 23, 2015

Webley PW, Wooster MJ, Strauch W, Saballos JA, Dill K, Stephenson P, Stephenson J, Escobar Wolf R, Matias O (2008) Experiences from near-real-time satellite-based volcano monitoring in Central America: case studies at Fuego, Guatemala. Int J Remote Sens 29(22):6621–6646

Webley PW, Dean K, Bailey JE, Dehn J, Peterson R (2009) Automated forecasting of volcanic ash dispersion utilizing virtual globes. Nat Hazards 51(2):345–361

Webley PW, Lopez TM, Ekstrand AL, Dean KG, Rinkleff P, Dehn J, Cahill CF, Wessels R, Bailey JE, Izbekov P, Worden A (2013) Remote observations of eruptive clouds and surface thermal activity during the 2009 eruption of Redoubt volcano. J Volcanol Geotherm Res 259:185–200. doi:10.1016/j.jvolgeores.2012.06.023

World Meteorological Organization (WMO) (2013) Consensual documents of the 2nd IUGG-WMO workshop on ash dispersal and civil aviation, Geneva, Switzerland, 35 p. http://www.unige.ch/sciences/terre/mineral/CERG/Workshop2/results-2/2nd-IUGG-WMO-WORKSHOP-CONS-DOC2.pdf, last viewed Feb 18, 2015

Zuccarello L, Burton MR, Saccorotti G, Bean CJ, Patanè D (2013) The coupling between very long period seismic events, volcanic tremor, and degassing rates at Mount Etna volcano. J Geophys Res Solid Earth 118 (9):4910–4921

Re-enchanting Volcanoes: The Rise, Fall, and Rise Again of Art and Aesthetics in the Making of Volcanic Knowledges

Deborah P. Dixon and Daniel J. Beech

Abstract

Current day volcanology largely tends to an instrumentalist view of art as, in its mimetic form, capable of providing proxy data on the timing and unfolding of particular volcanic events and, in its impressionistic form, of conveying the sublime grandeur of volcanic events and scenes. In this chapter, we note that such a reductionist view of what science *is* unhelpfully glosses over a much more complex disciplinary lineage, wherein both art and aesthetics played a key role in knowledge production concerning volcanoes. Using the work of Sir William Hamilton and Mary Somerville as case studies, we emphasise that art and aesthetics were part and parcel of both an 18th and 19th century approach to the study of volcanoes, and the making of particular scientific audiences. What is more, it is this lineage that provides a creative reservoir for more recent efforts that cut across scientific and arts divides, such that the 'communication' of the nature of volcanoes becomes a multi-media, multi-affective endeavour that speaks to a diverse range of publics.

Keywords

Art · Aesthetics · Hamilton · Somerville · Art-science collaborations

D.P. Dixon
School of Geographical and Earth Sciences,
University of Glasgow, Glasgow, UK
e-mail: Deborah.Dixon@Glasgow.ac.uk

D.J. Beech (✉)
Department of Geography and Earth Science,
Aberystwyth University, Aberystwyth, UK
e-mail: dib8@aber.ac.uk

1 Introduction

As many of the authors in this edited collection attest, communication is an embodied practice that serves to impart information, allowed for by sensory and cognitive modes of knowledge making. These modes are biologically as well as socially embedded, framing and valuing particular ways of knowing (or epistemologies) that

Advs in Volcanology (2018) 665–675
https://doi.org/10.1007/11157_2016_41
© The Author(s) 2017
Published Online: 15 March 2017

tell us how to 'see' the world; they operate to sustain or frustrate power relations between and amongst people and things, mobilising a range of affective relations and emotions. In this chapter, we bear this expanded notion of communication in mind as we outline something of the history of how a knowledge of volcanoes was produced and disseminated during the 18th and 19th centuries—a crucial period during which volcanology was to became a modern-day scientific enterprise. As we go on to outline below, this making modern of a discipline was predicated on a shift in the way in which knowledge was collected and disseminated. A combination of institutional entities, theoretical frameworks and methodological devices (Good 2000) led to a reorganising of scientific communities and audiences, but also the process through which scientific practice was undertaken.

In the 18th century we could still observe a Renaissance 'truth-to-nature' approach to knowledge gathering that made little separation between science and art. Where the artist provided a means of accurately observing and capturing the complexity of a physical scene, the scientist's finer 'touch' teased out details that escaped such vistas. Complementary to each other, these efforts not only allowed for the production of knowledge concerning the realities of Nature, but its particular mode of dissemination, such as the exchange of letters and sketches, as well as interesting objects, via scientific societies. Many scientific artefacts, representations and antiquities emerged from the expeditions embarked upon across Europe, and particularly the Mediterranean, by scientists such as Sir William Hamilton (1730–1803). Field-walking in the volcanic landscapes of central and southern Italy provided him the grounded knowledge required to recreate volcanic environments in the form of textual and visual illustrations, providing a mechanism though which to report the knowledge upon which volcanology is now largely predicated (Schnapp 2000; Vaccari 2008).

In the 19th century, however, we can discern more of an emphasis upon a systematised description and explanation of natural phenomena that glossed over the individualised and embodied presence of the scientist. Mary Somerville's (1780–1872) letters, for example, remained testament to the visceral impact of volcanic landscapes, but her textbooks—*Physical Geography* (1854) and *On the Connexion of the Physical Sciences* (1858)—conveyed more of a Gods-eye view of the cause-effect processes at work. The cultivation of a mechanical mode of objectivity reworked the communication of scientific knowledge more broadly, as indicators of the imaginative work of the scientist were erased from the writing up process (as observed in the emergence of the passive tense, for example, and the valuing of a prescriptive report template), but also in image-making. As the now iconic article by Daston and Galison (1992) makes clear, there was to emerge a polarisation of the personae of the artist and the scientist; ideals of actuality, accuracy and credibility had begun to be set apart from the imaginative, descriptive realms of the arts and humanities (Sigurdsson 1999). As the 21st century unfolds, however, we are witness to more inter-disciplinary collaborations that do not so much 'bridge' the arts and sciences divide, as worry at their differences, and, often times, look back into the history of both for inspiration.

Such a history is of interest in and of itself insofar as it reveals the often overlooked yet crucial role of not only sensory (or aesthetic) experience in the making of a knowledge of volcanoes, but of artistic practices also. The latter not only provided for the dissemination of a visual literacy concerning volcanic forms and attributes, but also facilitated a scientific curiosity and wonder. As Atkinson (1998) remarks, the 'scientific approach' that is so often taken for granted today as a common sense, seamless mode of knowledge production belies a rich, often tension-ridden heritage of 'science in the making,' as various tropes and techniques emerged and became standardised at the expense of others. The adventures of Hamilton and Somerville were not the exception, but instead typify modes of scientific exploration and knowledge discovery of the 18th and 19th centuries within the field of volcanology, each helping to shape 'a new science of mankind' (Schnapp 2000: 123). The explorations of

Hamilton and Somerville thus offer insights into the role of aesthetics and art in volcanic communication, and the nature of volcanology as a science (Sigurdsson 1999; Von Der Thüsen 2003). Throughout the 20th and 21st centuries, these mechanisms of representing volcanic landscapes have been further transformed, notably through the rise and expansion of cyber and digital technologies. Epistemically, the making of knowledge has used the middle-ground between art and science to garner and further the immersive, virtual and multi-perspectival exploration of the landscapes that Hamilton and Somerville previously conveyed.

In what follows, we illustrate, through the expeditions of Hamilton and Somerville, how aesthetics (in the form of an embodied, sensuous encounter with Nature) and art (in the form of imaginative visualisations of Nature), were a crucial part of knowledge-making within volcanology. Firstly, we outline the work of Hamilton, and his emphasis upon diligent observation in the field, made possible via the sensibilities of an educated elite. Next, we turn to Somerville's output, which encompasses her representations of volcanic environments in the form of letters, but which also notes the significance of the new literary space created by the scientific textbook. With this history in mind, we go on to outline something of the contemporary movement to collaborate across art and science, on the subject of volcanoes, drawing out how their aesthetics allow pause for thought on the hesitancies and complexities of volcanic science itself, as well as the manner in which key concepts such as 'deep time' become sedimented in broader understandings of human-environment relations (Wilkinson 2005).

2 A Gentleman's Report

Over the course of the 18th century a plethora of what Edmund Burke (see Phillips 2008) was to refer to, in his classic texts of 1770 and 1796, as 'sublime' landscapes—that is, capable of producing awe and even terror in the observer—were to be configured not simply as visceral experiences, but also as complex scenes that could be described and even explained by virtue of a close attentiveness to both their sweep and their detail. These landscapes were encountered in the rural margins of Europe, as well as the newly colonised lands of America and Austral-Asia, by discerning, wealthy visitors who, cognizant of philosophical discussions on the role of the intellect in making sense of sensual phenomena, as well as rapidly developing techniques of recording and mapping physical phenomena, viewed themselves as having the subtle imagination necessary to appreciate what lay before them. Most of the British visitors to the Continent were preoccupied with the 'Grand Tour' and the cultivation of a sense of cultural superiority through travel, but some, such as Sir William Hamilton (the British Envoy to the Spanish Court at Naples, 1764–1800), were on official, diplomatic business, and took the opportunity to enhance their residence abroad by reporting on the scientific value of nearby landscapes, such as the volcanic landscapes of Vesuvius, to an organised scientific community (Ramage 1990; Vaccari 2008).

One of the key nodes of scientific knowledge making at this time was the Royal Society (or, 'The President, Council, and Fellows of the Royal Society of London for Improving Natural Knowledge'), established in 1660 following a charter by King Charles II, and granted funds from Parliament in return for providing the government scientific advice on a wide variety of subjects. Inspired by the empiricism of Baconian science, the Royal Society was constituted by wealthy gentlemen interested in experimentation and detailed observation as a means of grasping the natural laws that underlay the human condition, as well as the universe at large (Atkinson 1998); research that was published in regular editions of the *Philosophical Transactions of the Royal Society* (1675 onwards), distributed to fee-paying members. These articles were in turn constituted in large part from letters sent to the society's headquarters, and read aloud to an audience that could also include interested non-members. These missives imparted content, to be sure, but in a format that made clear that this was an exchange of ideas between colleagues; a communicative procedure that expressed the immediacy of a face-to-face conversation, and even the rhythm and tone of the spoken word, even as it made clear the often immense physical separation between author and audience (Redford 1986). Often, they were accompanied by objects of interest, such as mineralogical displays or fossils, that provided a tangible focus for the play of intellectual curiosity (Da Costa 2002).

Such letters were by far the most important medium for the production of scientific knowledge in the 18th century. In the context of the Royal Society, these letters made clear an intellectual curiosity on behalf of the scientist, but also a rendering of scientific practice as a moral as well as instrumental enterprise, characterised by diligence, a discerning judgement, and skilled workmanship (Sorrenson 1996). It is these qualities that ensured the production of accurate, 'true-to-nature' observations. Crucially, these observations were achieved not by erasing either the individuality or situatedness of the scientist, nor by imagining the body of the scientist as somehow separate from Nature, but via the careful cultivation of one's immersion in Nature. The experiments that scientists conducted were as much a matter of a mannered sensibility, then, as the correspondence via which they were reported.

This careful cultivation of a scientific aesthetic—by which is meant a sensuous as well as cognitive engagement with Nature—can be clearly discerned in the work of Hamilton, one of the most famous early correspondents on volcanic landscapes, and whose communications so neatly befitted the mannerism of social attitudes towards science at the time. Hamilton was elected a Fellow of the Royal Society in 1766; a long-distance membership that complemented his antiquarian interests, as well as his curiosity concerning the forces that shaped the Earth's surface over substantial periods of geologic time. Contemporaneous explorations of Pompeii and Herculaneum had, of course, stimulated interest in the Classics, but had also raised the issue of the formative role of volcanoes, and in particular the vexing question of where the vast heat sources needed for eruption were located. Were these deep underground, or much closer to the summit? Such a question was in turn situated within a rapidly unfolding earth science that, whilst it left the origins of the Earth to religion, nevertheless sought to describe and analyse mountains in terms of an internal examination of the lithology and the fossil content of rocks, the geomorphology of an area (as read though its dips and inclines, for example), and of geologic strata as providing a chronology of the Earth's formation.

With a house in the foothills of Vesuvius, as well as a residence in Naples, Hamilton was to explore this volcano's craters and associated features time and again, all the while collecting antiques and landscape paintings as well as local gemstones and geological samples (Sleep 1969; Ramage 1990). Letters recounting his experiences were read at weekly meetings in London, where the audience could peruse the paintings, sketches, lava and soil samples that Hamilton had also sent over. Some of these images were adapted and published alongside the letters in *Philosophical Transactions*. For Hamilton, such images of the landscape, as well as tangible objects collected from them, were very much a key component of knowledge making. That is, these did not simply capture the interest of others, but were able to sensually convey, in a visual, material form, something of the eye witness, aesthetic experience he himself had cultivated.

Hamilton's production in 1776 of the richly illustrated folio collection of letters, *Campi Phlegraei: observations on the volcanos of the two Sicilies*, is testament to this desire to more fully immerse interested observers in volcanic landscapes, as well as to inform them of his own commentaries. A sensuous engagement with landscape, via the cultivated senses, was a far superior route to knowledge, he wrote, than "systems, which other ingenious and learned men, have perhaps formed in their closets, with as little foundation of felt experience." These were likely to "heap error upon error" (p. 5). Though Hamilton was well aware of the visceral effect of such scenes, as well as the emotions raised by a consideration of how past towns and villages had suffered, he did not consider these to be sublime in the sense that they indicated an awesome, indifferent Nature, as Kant would have it. Neither did Hamilton emphasise the overpowering forces of Nature in the manner of Burke's narrative on the sublime (1770, 1796), which unnervingly depicted the ease with which Nature could strike. Instead, Hamilton harkened to a more traditional view of Nature, whereby such tumultuous episodes were yet part of a fundamentally harmonious relationship between environment and society. "There is no doubt, but that the neighbourhood of an active Volcano, must suffer from time to time the most dire calamities, the natural attendants of earthquakes, and eruptions...," he wrote, "But to consider such misfortunes, on the great scale of nature, it was no more than the chance or ill fate of those cities to have stood in the line of its operations; intended perhaps for some wise purpose, and the benefit of future generations" (p. 3–4) including, he noted, rich soils for agriculture.

The cost and rarity of this text, and its appendix of images, also, it must be noted, enhanced Hamilton's own dilettante status. Overseeing the entire publication process, Hamilton employed the Naples-based painter Peter Fabris to sketch the locales described in his letters, "being still sensible of the great difficulty of conveying a true idea of the curious country I have described, by words alone" (p. 5); these sketches were then engraved, printed, and hand-painted in gouache by a cohort of local artists (for example, Fig. 1). In most of these plates, Hamilton himself is drawn, pointing out key features, or resting on his walking stick; a technique that reiterated once more the importance of a truth-to-nature, eye-witness account by a gentleman observer capable of discerning the subtleties of Nature. In *Campi Phlegraei*, Karen Wood writes,

> Readers could view the separate volumes of plates and letters in parallel. Thus integrated, Campi Phlegraei's literary and visual technologies contributed simultaneously to achieving a more complete witnessing experience... Whereas the pictures in Philosophical Transactions had depicted only distant eruptions, Campi Phlegraei's plates had more varied subject matter. Many placed foreground human figures in specific, visible relationships with the landscape, dramatizing the practices embodied by the book. People positioned on the mountainside stressed direct interaction with nature, while those absorbed in concentrated thought demonstrated focused attention... Using colour, lighting and pose, these plates also distinguished visually the rational behaviour of curious aristocrats from the captivated fear and wonder of awe-struck peasants... His final plates depicted examples of volcanic matter in remarkable detail, using trompe l'oeil techniques that

Fig. 1 Hamilton at the crater of Forum Vulcani, examining the sulphur and arsenic that emerged from craters near the source of hot springs. *Source* Campi Phlegraei (1776)

invited beholders to inspect the images just as they would specimens in a collector's cabinet (2006, p. 90–91).

3 A Scientist's Synthesis

Though the importance of witnessing phenomena first-hand, and of accurately describing them, was to remain a pivotal objective of scientific practice into the 19th century, we can also track a gradual erasure of the sensuous, embodied presence of the scientist from the communicative process, and a simultaneous celebration of the intellect as a means of synthesising information and organising it into a comprehensible whole for others. In the work of Mary Somerville, we can discern the first major English-language effort to create such a 'modern' synthesis for a rapidly evolving geographic discipline, one that strove to distinguish its practitioners as experts in human-environment relations, but was yet populated by academics whose research, like Somerville's, extended across what we now think of as the sciences and arts.

Despite a highly gendered academic and social environment, Somerville was able to make use of personal networks in both an 'Enlightened' Edinburgh and Paris, and the Royal Society (of which her second husband was a member) to further her research into mathematics, astronomy and geology, but also to practice painting and Greek (see Patterson 1974). She was elected honorary membership of Société de Physique et d'Histoire Naturelle de Genève, and the Royal Irish Academy in 1834, was a member of the Royal Astronomical Society in 1835. She went on to be elected to the American Geographical and Statistical Society in 1857, and the Italian Geographical Society in 1870, receiving the Victoria Gold Medal of the Royal Geographical Society that same year. Like many of her British contemporaries, Somerville thus straddled two worlds, and thrived upon contestation; on the one hand, she was the product of what was becoming known as a scientific 'amateurism' characterised by self-funded and self-governed scientific assemblies; on the other hand, her expertise was

formally recognised by the 'professional,' continental scientific circles residing in university departments and granted stature through a promotion and awards system.

Mary Somerville moved to Rome for her husband's health in 1838, where she began work on the two-volume *Physical Geography* (1858). She lived in Italy until her death, travelling throughout the country and maintaining a travel diary as well as correspondence with scientists across Europe, and family; some of these are collected in Somerville's *Personal Recollections* (1874). These documents recorded her descent into the crater of Vesuvius in 1818, and her witnessing of the April 1872 eruption. And, in them can be discerned her interest in the Earth as a dynamic planetary body, but also of the aesthetic, visceral impact of volcanic processes. That is, Somerville is at pains to convey the first-hand experience of the field, noting scenery, sounds, smells and even the touch of the landscape. Somerville uses the power of the textual narrative to initiate the reader's senses and, in due course, educate the reader on the characteristics and processes of the natural, geomorphological environment. Many times within the text her tone is scenic, reflecting her painterly eye; she notes, for example, "We have bright sun-shine with bitterly cold wind and frost, Vesuvius has been powdered with snow but still sends out vapour" (January 1869).

During the 1872 eruption of Vesuvius, however, Somerville's description becomes more visceral and sublime. Somerville's account is geared more towards the awe-inspiring nature of volcanoes, rendered in terms of their cataclysmic impacts, and their colouring of landscapes as gloomy and risk-ridden. "On Sunday, 28th," she wrote, "I was surprised at the extreme darkness … the fall was a little less dense during the day, but at night it was worse than ever … certainly the constant loud roaring of Vesuvius was appalling enough amidst the darkness and gloom of the falling ashes" (1874, p. 369). "In sunshine the contrast was beautiful," she continued, "between the jet-black smoke and the silvery-white clouds of vapour … At length, the mountain returned to apparent tranquillity, though the violent detonations occasionally heard gave warning that the calm might not last long" (1874, p. 125).

Somerville's observations may well be read as resonating with a Kantian notion of how particular, complex landscapes produced particular responses in observers, as manifest in his (1764) text *Observations on the feeling of the Beautiful and Sublime*; certainly, Somerville's incorporation of the knowledge gained through her field experiences in her text books implies a similar celebration of the human intellect in making sense of such states. As Fara (2008: 83) observes, "Somerville conveyed to wide Victorian audiences not only the impact of the latest scientific discoveries, but also the sublime thrill of engaging with cosmological mysteries". Somerville's text *Physical Geography*, however, is perhaps more usefully regarded as a 'Humboldtian' mode of knowledge production—after the geographer Alexander Von Humboldt, with whom she corresponded—whose work strove to present the fundamentally harmonious nature of the *Kosmos*, and the need to combine mathematical precision in recording processes with an aesthetically-sensitive recognition of their role in creating distinct landscapes (Dixon et al. 2012). Somerville's turn towards the aggregation of facts, scientific truths, relationality and the processual structure of systems emphasised knowledge as a matter of bringing together seemingly disparate elements, and of explaining their import in terms that were accessible to a large audience. Importantly also, and *contra* the work of Hamilton and the Royal Society, such knowledge, she argued, must needs be "widely diffused amongst all ranks of society" (1854: 395).

At times, this mode of communication is achieved through the use of analogy, as when, "An internal expansive force acts upwards upon a single point in the earth's crust, the splits or cracks must all diverge from that point like radii in a circle, which is exactly the case in many volcanic districts" (*Physical Geography*, 1854, p. 46). Here, the embodied perspective of the scientist has been replaced by an 'objective' Gods-eye view—facilitated by reference to the 'pure' language of geometry—that relates the universal 'fact' of what a process consists of and does. Elsewhere, Somerville relies on recreating, with emotion-laden rhetoric, a visual spectacle for the edification of the reader, such that, "The desolation of this dreary waste, boundless to the eye as the ocean, is terrific and sublime ..." (1854, p. 91), while, "The chasms yawning into dark unknown depths, strike the imagination ..." (1854, p. 103). Here, the reader is one with the scientist, both immersed in the experience of a volcanic eruption. Though she did not credit herself (as an intellectual *woman*) with the capacity for originality (see Creese 1998: 204), Somerville's work firmly argued for volcanology as an inter-disciplinary endeavour, belonging to the Natural and Earth Sciences. What is more, these sciences were understood as having an explicit aesthetic dimension, whereby a sensuous engagement with Nature lent itself to, and certainly did not preclude, an understanding of the cause-effect relations behind particular events.

There is no doubting that the emergence of scientific textbooks such as *Physical Geography* has become a key means of democratising knowledge (Grinstein 1987), somewhat distancing Somerville's methods from those of Hamilton. Indeed, for Richard Holmes (2014), Somerville's *On the connexion of the physical sciences* plays a crucial role in ushering in the textbook as a means of popular education about the world. They have also, it must be added, facilitated in the process a translation of information into what Neeley (2001) terms a 'general enlightenment,' whereby a clear, logical rhetoric—which, of course, developed out of a particular place and time—is used to synthesise knowledge, narrowing it down to a simple, coherent 'message' that prevents contradictions. The structure, linguistics and content of such textbooks articulated a more holistic view of science, transforming, rather than connecting with, the attitudes and expectations of a society that had previously revelled in a plethora of sensuous, sublime and wondrous narratives. The writing of such textbooks both presumes and reproduces a replicable literary space wherein this message remains, hopefully, consistent across context, regardless of who the reader is, or of what the format of the book itself looks like.

4   The Fall and Rise Again of Aesthetics

For much of the Earth Sciences, today, the purity of the 'message' promised by the mass media textbook is matched by the 'objectivity' of data produced through field-based research. To be sure, observing in the field remains a crucial part of Earth Science training; but, this is now rigidly systematized as the technologically enhanced, error-free measurement and recording of variables. Such aesthetic encounters, in effect, allow for the production of data. On occasion, an interest in art can still be discerned. But, such works are valued because they presumably 'capture' something of the 'real' world nature of events.

A key example of this instrumental approach is Zerefos et al.'s (2007) reconstruction of past aerosol optical depth, before, during and after major volcanic eruptions, using as data proxies the coloration of the atmosphere in 500 paintings that depicted sunsets between 1500 and 1900. They note that the artists under study "appear to have simulated the colours of nature with a remarkably precise coloration", and conclude that their study provides a "basis for more research that can be done on environmental information content in art paintings" (p. 4033). The fundamental separation of art and science posited here can also be found, it must be noted, in numerous art statements and curated exhibitions. The recent major retrospective Volcano: Turner to Warhol, held in 2010 at the Compton Verney Gallery, for example, juxtaposed iconic and new art works that emphasise the 'mystery' of these phenomena with the comments of volcanologists, the latter pointing out the 'realities' of the processes at work.

Yet, one can also discern a burgeoning 'art-science' movement that eschews the modern-day, institutional compartmentalization that distances the arts, as a subcomponent of the humanities, from the natural sciences, and looks instead to how these revolve within a shared history characterized as much by negotiation, mutual learning, and symbiosis as by the search for fundamental difference. A prominent example of this kind of work is The Other Volcano, produced under the auspices of volcanologist Carina Fearnley and designer/performer Nelly Ben Hayoun. The project takes the form of what the artist describes as 'semi-domesticated volcanoes', each capable of 'randomly' erupting dust and gloop (from a combination of gunpowders, potassium nitrate and sugar), and housed in galleries—such as the Welcome Trust, London (October–December 2010) and the Central Booking Gallery, New York (April–June 2011)—and the living spaces of volunteers. As Dixon et al. describe it,

> It is while 'waiting' for the eruption, which sometimes does not happen, that observers are presented with the complexity of natural disasters, as well as the challenges faced by those who predict natural hazards (2012, p. 12).

A key concern here is not to present the work as an 'alternative' form of science communication concerning environmental risk—one that has public appeal because of its spectacular imagery. Rather, the intent is to convey a state of anticipation and anxiety that cuts across scientific and artistic cohorts, and extends to the public at large; while these motions may well be experienced in singular fashion, there is yet a shared awareness of the affective capacity of such earth shuddering events.

One of the 'radical' aspects of Mary Somerville's work was her acceptance of what Scottish geologist James Hutton referred to as a time with "no vestige of a beginning, no prospect of an end" (1785, 30). And, artists working on volcanic landscapes have for the most part attempted to convey something of this abyssal sense of time past, and time future, punctuated by blasts of movement and energy during eruptions that fracture and warp physical space, by drawing attention to the differing materialities of sedimentary layers and lava in motion. One can also discern, however, an effort to blur, entangle and generally probematise the notion of a human versus an Earthly time. Ilana Halperin's *Meeting on the Mid-Atlantic Ridge* (1999) and *Integrating Catastrophe* (2000), for example, both explore the fleshy and elemental temporalities at work in encountering volcanoes. The former, a photograph, features two pairs of feet standing on either side of a fissure. The latter, a series of sketches, builds on this, imagining that,

> In the North of Iceland along the Mid-Atlantic Ridge, two newlyweds move into their first house. They are very excited - new house/new life. No one tells them when they move into the house that it sits on a fault line. There is a massive volcanic eruption followed by an earthquake. Their house splits in two. Their living room has a huge gash straight through it. They are horrified—devastated. What does this mean? Their house is destroyed. Their marriage had only just begun, and the chasm running through their marital bed does not bode well for their future. They realise that actually, the house has a clean break down the middle, and instead of devastation it could be a sign for something much better. They build a new room in the space of the gap, transforming a potentially catastrophic situation into an expanded living space. Integrating catastrophe (www.ilanahalperin.com/new/integrating_catastrophe.html, 2014).

Subsequent works, which are undertaken in the field alongside mineralogists, geologists, and archaeologists as well as volcanologists, explore what Halperin terms a 'geologic intimacy' (www.ilanahalperin.com/new/statement, 2014); a project that very much resonates with efforts in human geography to map out a more elementally-aware 'post-humanism' that no longer equates agency with human will. There is no doubt that art has the capacity to heighten an audience's awareness of the emotional connections drawn between humans and their volcanic environments (Sigurdsson 1999). What seems to be stimulating more recent works, however, is a fascination with how scientific notions of a 'deep time' (Wilkinson 2005) provide a contrast with more avowedly social temporalities, from policy time-frames to anniversaries, but also help call into question how these same temporalities are

themselves interwoven with a more elemental Earth History.

As the process of representing volcanic environments has delved into virtual and digital worlds, the rise of the internet has drastically transformed the relationship that society has with such landscapes and their temporalities. The generation and exploitation of social media sites, together with the rise of applications such as Google Earth, allows the audience a new sense of wonder, whereby they can 'journey' through volcanic worlds in real-time. A stark contrast with the texts of Somerville, and the emotive imagery of Halperin, emerging screen-based practices transform the positionality of the traveller, engaging them in a manner that asserts them as an end-user, able to control the pace and prowess of their own movement from the domestic or enclosed space within which they are situated into a mobile, fluid and data-dense virtual field. It is within this multi-faceted spectrum field of vision/action that a new, affective envisioning of volcanic environments can be explored, shaped and interrogated.

5　Concluding Remarks

Collectively, the communicative mediums outlined above provide what Wise (2006) calls a 'materialized epistemology,' by which he means that these are not simply the end-products of a scientific project, but are themselves an essential part of 'doing' the work of knowledge production. And, in drawing out particular examples, we can discern a key shift in how aesthetics, and art, have been considered a part of, and indeed outside of, scientific practice (Sigurdsson 1999). That is, in the 18th century, aesthetics are actually key to the emergence of an observant science, one that makes explicit the role of gentlemanly sensibilities in making sense of Nature. By the turn of the 20th century, however, such an embodied practice has come under suspicion as prone to bias and error; in its place, we see an unfolding God's Eye view of Nature that promises to provide a pure, unsullied message that remains constant across audiences and mediums. With the onset of the 21st century, however, we can also discern something of a retrospective approach to volcanic communication. This art-science movement is by no means restricted to an educated elite—indeed, a variety of audiences are envisioned—but it does seriously re-engage with aesthetics a means of creating and building knowledges concerning volcanoes.

The question emerges, however, as to the value and relevance of such collaborative efforts. On the one hand, there is no doubt that these are receiving greater prominence, and funding, in light of the pervasive argument that environmental problems, including volcanic hazards, are 'wicked,' such that the expertise of several disciplines must needs be brought to bear in their analysis. Critiques of a modern-day 'silo-thinking,' for example, can be found across the UK's physical, engineering, social science and arts-focused Research Councils, as well as an ensuing presumption that solutions lie in 'bridging' such divides. On the other hand, there is also an intellectual impetus at work here. That is, we can also discern what is usefully described as a 'transdisciplinary' effort that seeks out otherwise abandoned and dismissed histories—such as an Earth Science aesthetic—and reanimates these as a means of creating 'new' knowledges. Critical of synthesis as both a method and a goal, such transdisciplinary work helps to recast what we consider 'communication' to be, casting it adrift in a sea of contingency that refuses reduction to a fixed set of practices and effects, yet, we hope, is all the more welcomed for that.

References

Atkinson D (1998) Scientific discourse in sociohistorical context: the philosophical transactions of the Royal Society of London, 1675-1975. Routledge, London

Burke E, Phillips A (eds) (2008) A philosophical enquiry into the sublime and beautiful. Oxford Paperbacks, Oxford

Compton Verney (2010) Volcano: turner to warhol. Available via http://www.comptonverney.org.uk/ events/2/volcano_turner_to_warhol.aspx. Accessed 28 Jan 2015

Creese M (1998) Ladies in the laboratory? American and British women in science, 1800-1900: a survey of their contributions to research. Scarecrow Press, Maryland

Da Costa PF (2002) The culture of curiosity at the Royal Society in the first half of the eighteenth century. Notes Rec Roy Soc London 56(2):147–166. doi:10. 1098/rsnr.2002.0175

Daston L, Galison P (1992) The image of objectivity. Representations 40:81–128. doi:10.1525/rep.1992.40. 1.99p0137h

Dixon DP, Hawkins H, Straughan ER (2012) Wonder-full geomorphology: sublime aesthetics and the place of art. Prog Phys Geogr 37:601–621. doi:10.1177/ 0309133312457108

Fara P (2008) Mary Somerville: a scientist and her ship. Endeavour 32(3):83–85. doi:10.1016/j.endeavour. 2008.05.003

Good GA (2000) The assembly of geophysics: scientific disciplines as frameworks of consensus. Stud Hist Philos Sci Part B Stud Hist Philos Mod Phys 31 (3):259–292. doi:10.1016/S1355-2198(00)00018-6

Grinstein LS (1987) Women of mathematics: a biobibliographic sourcebook. Greenwood Publishing Group, Connecticut

Hamilton W (1776) Observations on the volcanoes of the two Sicilies As they have been commanded to the Royal Society of London by Sir William Hamilton KBFRS His Britannic Majesty's Envoy Extraordinary and Plenipotentiary at the Court of Naples. Naples, Frenel

Holmes R (2014) In retrospect: on the connexion of the physical sciences. Nature 514(7523):432–433. doi:10. 1038/514432a

Kant I (1764) Observations on the feeling of the beautiful and sublime. (trans: Goldthwait JT (1960)). University of California Press, Berkeley

Neeley KA (2001) Mary Somerville: science, illumination, and the female mind. Cambridge University Press, Cambridge

Patterson EC (1974) The case of Mary Somerville: an aspect of nineteenth-century science. Proc Am Philos Soc 118(3):269–275

Ramage NH (1990) Sir William Hamilton as collector, exporter, and dealer: the acquisition and dispersal of his collections. Am J Archaeol 94(3):469–480. doi:10. 2307/505798

Redford B (1986) The converse of the pen: acts of intimacy in the eighteenth-century familiar letter. University of Chicago Press, Chicago

Schnapp A (2000) Antiquarian studies in Naples at the end of the eighteenth century. From comparative archaeology to comparative religion. Naples in the Eighteenth Century, The birth and death of a nation state. Cambridge University Press, Cambridge

Sigurdsson H (1999) Melting the earth: the history of ideas on volcanic eruptions. Oxford University Press, Oxford

Sleep MCW (1969) Sir William Hamilton (1730–1803): his work and influence in geology. Ann Sci 25 (4):319–338. doi:10.1080/00033796900200191

Somerville M (1854) Physical geography. Blanchard and Lea, Philadelphia

Somerville M (1858) On the connexion of the physical sciences. John Murray, London

Somerville M (1869) Letter from Somerville to Murray, 3 Jan 1869. Available via http://www.scottishcorpus.ac.uk/cmsw/document/?documentid=188. Accessed 8 Dec 2014

Somerville M (1874) Personal recollections, from early life to old age, of Mary Somerville: with selections from her correspondence by her daughter Martha Somerville. John Murray, London

Sorrenson R (1996) Towards a history of the royal society in the eighteenth century. Notes Rec Roy Soc London 50(1):29–46. doi:10.1098/rsnr.1996.0003

Vaccari E (2008) Volcanic travels and the development of volcanology in 18th century Europe, In scientific exploration in the mediterranean region: cultures and

institutions of natural history: essays in the history and philosophy of science 37. California Academy of Sciences, California

Von der Thüsen J (2003) Painting the Vesuvius development in eighteenth-century landscape art. ASEAN Female Writers/Dramatists 16:89–100

Wilkinson BH (2005) Humans as geologic agents: a deep-time perspective. Geology 33(3):161–164. doi:10.1130/G21108.1

Wise MN (2006) Making visible. Isis 97(1):75–82. doi:10.1086/512939

Wood K (2006) Making and circulating knowledge through Sir William Hamilton's Campi Phlegraei. Br J Hist Sci 39(01):67–96. doi:10.1017/S0007087405007600

Zerefos CS, Gerogiannis VT, Balis D, Zerefos SC, Kazantzidis A (2007) Atmospheric effects of volcanic eruptions as seen by famous artists and depicted in their paintings. Atmos Chem Phys 7(15):4027–4042. doi:10.5194/acp-7-4027-2007 (European Geosciences Union (EGU))

Living with an Active Volcano: Informal and Community Learning for Preparedness in South of Japan

Kaori Kitagawa

Abstract

In a disaster-prone country like Japan, learning how to live *with* disaster [*kyozon*] has been crucial. Particularly since the Great East Japan Earthquake and Tsunami of 2011, disaster preparedness has been a primary concern of the government. Drawing on Paton's (The phoenix of natural disasters: community resilience. Nova Science Publishers, New York, pp. 13–31, 2008) Community Engagement Theory, which endorses an integrated model that combines risk management with community development, this study discusses the case of Sakurajima Volcano (SV) situated in the south of Japan, with a focus on how the lessons learnt from previous eruption experiences have informed present-day preparedness activities. The study adapts Community Engagement Theory's quantitative framework to a qualitative analysis to consider the preparedness teaching and learning of a population living with the everyday threat of volcanic hazards in the case of SV. The study argues that two particular local lores—'do not rely on authorities' and 'be frightened effectively'—have been the underlying principles in volcanic preparedness in the region. The study also argues that the notion of '*kyojo* [collaborative partnerships]' has been central to the planning and implementation of preparedness programmes, such as the Sakurajima Taisho Eruption Centenary Project, which offered a wide range of informal teaching and learning opportunities. Applying the framework of Community Engagement Theory, the paper suggests that at the individual level, the principles of 'do not rely on authorities' and 'be frightened effectively' form the basis for positive 'outcome expectancy'. At the community level, 'kyojo' is the notion which encompasses both of the community factors—'community participation' and 'collective efficacy'. At the societal level, 'kyojo'

K. Kitagawa (✉)
Cass School of Education and Communities,
University of East London, E15 4LZ London, UK
e-mail: k.kitagawa@uel.ac.uk

Advs in Volcanology (2018) 677–689
https://doi.org/10.1007/11157_2015_12
Published Online: 26 March 2017

contributes to the building of 'empowerment' and 'trust' between citizens and authorities. The paper concludes by proposing that the SV case can be considered as an example of 'the integrated model'.

1 Introduction

Japan is located in the Circum-Pacific Mobile Belt where seismic and volcanic activities occur constantly (Cabinet Office 2011). The teaching and learning of disaster preparedness has been a primary concern of the government, particularly since the Great East Japan Earthquake and Tsunami of 2011 (Kitagawa 2014; The Expert Committee[1] 2012). Where natural disasters are not exceptional (Preston et al. 2014), learning how to live *with* disaster [kyozon] has been crucial. This paper examines preparedness for Sakurajima Volcano, which is an active volcano in Kagoshima City, Kagoshima Prefecture in Japan. In Japanese, 'Sakurajima' refers to both the volcano and the island.[2] In this paper, Sakurajima as a volcano is referred to as 'Sakurajima Volcano (SV)', and Sakurajima as an island is called 'Sakurajima'. Focusing on how lessons have been learned from previous eruption experiences, the study aims to contribute to the development of a preparedness model of 'living with an active volcano'.

In Japan, volcanic preparedness was systematised in a government-led initiative (1974–2008) (Ishihara 2012), which introduced the volcanic warning system and implemented mitigation plans, hazard maps and evacuation plans for major active volcanoes. However, prevention, response and recovery measures for volcanic disaster have not been as advanced as those for earthquakes or floods (Takahashi 2007). This was unfortunately demonstrated by the recent disaster at Ontake Volcano in central Japan, which has become the worst volcanic disaster in Japan in the past 90 years (BBC 2014), since the 1914 eruption of SV.[3] Prior to the eruption, there had been debate about whether to raise the warning level which was Level 1[4] at the time. The incident reconfirmed the complex nature of volcanic preparedness.

This paper discusses the development of the volcanic preparedness in the Kagoshima region (Fig. 1). The purpose is twofold: first, to examine how the lessons learnt from the Taisho Eruption[5] of 1914 have informed present-day preparedness activities; and second, to explore whether the SV case can be considered as 'an integrated model' endorsed by Paton's Community Engagement Theory (CET). The paper first argues that two particular local lores—'do not rely on authorities [riron ni shinraisezu]' and 'be frightened effectively [seito ni kowagaru]'[6]—have taught the local population the underlying principles of volcanic preparedness—the development of

---

[1]The Expert Committee to Discuss Disaster Education and Management Considering the Lessons Learnt From the Great East Japan Earthquake [Higashinihon daishinsai o uketa bosaikyoiku/bosaikanri ni kansuru yushikisha kaigi].

[2]90 % of Sakurajima Island is occupied by the volcano. Sakurajima has been referred to as an island rather than a peninsula although the island was connected to the mainland by a lava flow in 1914.

[3]Ontake Volcano erupted on 27 September 2014, killing 51 people (as of 31 October 2014). The victims who were mostly hikers were said to have covered by dense ash fall and inhaled poisonous fumes or hit by ash deposits thrown out from the crater.

[4]The current official volcanic warnings set by the Japan Meteorological Agency are: Level 1 Normal; Level 2 Do not approach the crater; Level 3 Do not approach the volcano; Level 4 Prepare to evacuate; Level 5 Evacuate.

[5]Japan uses its own name of an era which refers to the Emperor at the time. Taisho Era was 1912–1926, the reign of Taisho Emperor, Yoshihito. Showa Era was 1926–1989, the reign of Showa Emperor, Hirohito. The current era is Heisei Era, the reign of Heisei Emperor, Akihito. The eruption of 1914 occurred during Taisho Era, and therefore it has been referred to as 'Taisho Eruption'.

[6]The former is part of the inscription on the Sakurajima Taisho Eruption Monument, and the latter is the words from an intellectual.

Fig. 1 Map of Kagoshima

Kagoshima Prefecture

Sakurajima

Kagoshima City

agency and the control of emotion. The paper also looks at the implementation of the above two principles through informal[7] teaching and learning programmes. The paper argues that those programmes are based on '*kyojo*'[8]—collaborative partnerships—which has become the key policy term in disaster preparedness in Japan particularly since 2011.

In parallel, the paper discusses the qualitative findings, applying the individual/community/societal framework of CET. The argument of the theory is that 'an integrated model' that combines risk management and community development is an effective and sustainable approach to enhance community resilience to adversity (Paton et al. 2011; Paton 2008; Paton and Johnston 2006). Unlike Paton et al's (2013) and Paton and Jang's (2013) studies, this paper is not measuring the factors that contribute to people's intention to prepare for disaster, rather the variables of the theory—'outcome expectancy', 'community participation', 'collective efficacy', 'empowerment', 'trust'—are employed to consider the preparedness teaching and learning of a

---

[7]Here, 'informal' teaching and learning refers to a form of teaching and learning outside of formal schooling.

[8]'Kyojo' is part of the four forms of aid: kojo, jijo, gojo and kyojo. The common word 'jo' at the end means 'aid'. 'Kojo' is aid provided by governments, both central and local, and 'jijo' is self-help. 'Gojo and 'kyojo' refer to mutual help, however, the difference lies in that the former happens within the people you are familiar with, whereas the latter has a philanthropic nature.

population living with the everyday threat of volcanic hazards in the case of SV.

The paper is structured in the following way. The first section of the paper is about methodological and theoretical considerations of this study. This is followed by a description of the major volcanic activities of SV. The paper then analyses the population response to the Taisho Eruption, lessons learned from it and present-day communication about volcanic conditions in the Kagoshima region. In doing so, the paper connects the findings from the SV case to the key concepts of CET. The challenges that the government and the population face in living with the active volcano are also discussed.

2 Methodological and Theoretical Considerations

This case study particularly focused on two aspects of the disaster cycle: how population response informs disaster prevention and reduction. More concretely, the following six questions were pursued:

1. What lessons were learnt from the Taisho Eruption of 1914?
2. Why did false rumours circulate? What did the phrase 'do not rely on authorities' aim to convey?
3. How is the phrase 'learning to be frightened effectively' understood? What measures have been undertaken to teach the principle?
4. What forms of volcanic preparedness particularly outside of formal schooling are currently in place?
5. SV has been active again since 2006, and the Warning Level has remained at 3 since 2012. The local population are used to regular volcanic hazards, and Level 3 has become a norm to them. Under such circumstances, how do they learn/have they learned 'to be frightened effectively'?
6. What are the examples of kyojo activities for the teaching and learning of 'do not rely on authorities' and 'be frightened effectively'?

Data collection and analysis were arranged to respond to the above questions in the following way. First, the study made use of the rich information available in the documentation of the Taisho Eruption undertaken as part of the Sakurajima Taisho Eruption Centenary Project (Centenary Project).[9] Empirical fieldwork was then organised in order to triangulate the information obtained from the documentary analysis. Sakurajima itself and two museums (the Sakurajima Visitor Center and the Prefectural Museum of Kagoshima) were visited to familiarise myself with the volcanology of SV and to obtain the official archival record of the eruptions, responses and recovery. Semi-structured interviews were conducted with three groups of experts: regional policy makers,[10] academics[11] and a curator/researcher of an archival museum.[12] In addition, online resources such as newspaper articles were used to fill any gaps. Thus, the data were analysed according to the six questions, while being collected at each step of the data collection—documentary analysis, archival research and expert interviews.

CET is based on the perspective that risk management and community development as complementary, and such an integrated model of preparedness permits a resilient community.[13] Paton's (2008, p 3) approach is that 'people's understanding of, and response to, risk is determined not only by scientific information about risk, but also by the manner in which this information interacts with psychological, social,

[9]The Centenary Project was implemented in 2013–14 funded by the prefectural and municipal governments of Kagoshima. The detail of the Centenary Project is introduced in the later section.

[10]From the Risk Management and Disaster Prevention Unit from the Kagoshima City authority. Interviewed on 22 July 2014.

[11]From the Regional Disaster Reduction Education and Research Center at Kagoshima University. Interviewed on 23 July 2014.

[12]From the Prefectural Museum of Kagoshima. Interviewed on 23 July 2014.

[13]In Japan, for example, Shiroshita's (2010) 'collaborative education [kyoiku]' is a similar notion, which emphasises collaborative and democratic partnerships, and ownership by the members of the community.

cultural, institutional and political processes to influence outcome....these factors must be understood and accommodated in risk communication strategies'. The theory suggests that interpretive process at the individual level (outcome expectancy) interacts with communities (community participation, collective efficacy) and societal relationship (empowerment, trust) factors to influence preparedness (Paton 2008). Quantitative methods are usually applied in finding out the interactions between the preparedness variables. As a qualitative study, this paper does not intend to measure those interactions, but to explore how the findings of the case of SV could be interpreted, borrowing those concepts.

Five key concepts of the theory are briefly explained here. 'Outcome expectancy' at the individual level is about a person's belief as to whether her/his actions can effectively mitigate or reduce the problem (Paton et al. 2013, p. 21). At the community level, 'community participation' refers to community members' ability to collaborate to make decisions on effective responses for possible hazards and mitigation strategies. 'Collective efficacy' is community members' ability to assess their capabilities and create plans. The societal-level 'empowerment' and 'trust' variables refer to the relationship between citizens and authorities. When people believe their relationship with authorities is fair, empowering and trustworthy, people are more willing to get involved and take responsibility for their own safety (Paton et al. 2013, p. 22).

The paper now moves on to consider the case of SV with reference to this framework.

3 Sakurajima Volcano's Activities

3.1 Overview of Volcanic Activities

SV emerged about 26,000 years ago at the south end of Aira Caldera. The volcano currently has three peaks—Kitadake (north mountain), Nakadake (middle mountain) and Minamidake (south mountain) which has been most active (Japan Meteorological Agency 2014). In its recorded history, SV has had 17 major eruptions, the ones

in 746, 1472, 1779 and 1914 being the largest four. Except the 19 years after the Taisho Eruption of 1914, SV has continuously been active during the past 100 years. In 1946, the lava flow exploded near the peak of Minamidake, created the Showa Crater[14] and buried two villages. The Minamidake crater has been active since 1955,[15] and between 1972 and the early 1990s, eruptions were frequent[16] (Centenary Committee[17] 2014, p 156; Ishikawa cited in Kagoshima University[18] 2014, p. 39). SV was less active until 2006, eruptions have become more frequent since 2009, and in 2012, the Meteorological Agency announced the volcanic alert Level 3—'do not approach the volcano'—which has remained unchanged since.[19]

Research has confirmed that 90 percent of the magma which erupted in 1914 has already been re-accumulated in the underground of Aira Caldera which is said to be the main magma reservoir of SV. Three concrete scenarios have been proposed: (1) the re-aggravation of the Minamidake crater; (2) the aggravation of the Showa Crater followed by large lava flows; (3) flank eruptions of the scale of the Taisho Eruption from both mountainsides (Centenary Committee 2014, p. 142). Science and technology have advanced understanding of the volcano, and preparedness education for the above scenarios are underway. The circumstances at the time of the Taisho Eruption, however, were very different.

---

[14]As explained above, 'Showa' refers to the era of Showa. Some refer to this eruption as 'the Showa Eruption'.

[15]Its summit erupted causing a 5000 m volcanic fume and a high volume of ash deposits which killed one hiker and injured a few others.

[16]During this period, secondary damage was a concern because rain turned the accumulated ash fall into sediment flows threatening the life of the people of the island.

[17]The Planning Committee of the Sakurajima Taisho Eruption Centenary Project.

[18]The Department of Domestic Science Education, Faculty of Education, Kagoshima University.

[19]In August 2013, there was an explosive eruption, resulting in a 5,000-metre volcanic fume and a large quantity of ash fall which reached the city of Kagoshima. The number of recorded eruptions per year in the past five years has been 896, 996, 885, 835 and for 2014, as of 13 August, 209 had already been recorded.

3.2 Taisho Eruption: The Largest in the 20th Century in Japan

At 10:05 on 12 January in 1914, a flank eruption occurred on the west mountainside, and 15 min later, a larger flank eruption occurred on the east mountainside of SV.[20] The volcanic fume rose more than 8000 m, and the ash fall reached as far as the Kamchatka in Russia. Eight hours after the eruption at 18:30, a Magnitude 7.1 earthquake hit the city of Kagoshima. At 20:14, pyroclastic flow began, which burnt whole villages on the west side of the island. From at around 21:00, the pyroclastic flow turned to lava flows, which reached the coastline by the following evening.[21] The total amount of the ash fall was estimated to be two square kilometres, and that of the lava flows was three billion tons. The accumulated ash fall became sediment flows in the mountains in the region, as well as in Sakurajima (Central Disaster Council 2011, p. 38; Kagoshima University 2014, p. 41).

3.3 Damage Brought About by the Taisho Eruption

In 1914, Sakurajima was made up of 3400 households and 21,300 inhabitants. Given the scale of the eruption, the human damage on the island—30 dead or missing—was considered to be small. Of the 30, only two deaths were directly caused by volcanic ejecta. The majority were drowned trying to escape by sea[22] (Centenary Committee 2014, p. 51; Central Disaster Council 2011, p. 40). The property damage was enormous. On Sakurajima, a number of villages were swallowed completely by lava or covered by ash. 62 percent of the houses were burnt down by lava and pyroclastic flows. Residents in surrounding cities also lost homes which were covered by thick ash fall. The earthquake also destroyed houses in Kagoshima City[23] (Centenary Committee 2014, p. 52; Central Disaster Council 2011, p. 40). Nearly half of the population in total had to migrate from Sakurajima after the Taisho disaster (Central Disaster Council 2011, pp. 44–45).

Infrastructure failure was massive. Roads were closed and bridges were destroyed on Sakurajima. On the mainland, some railway lines were destroyed, and telecommunications were disrupted due to the collapse of stations and electrical failure. Agricultural crops in Sakurajima were significantly damaged as well[24] (Centenary Committee 2014, pp. 40–45, 52; Central Disaster Council 2011, pp. 39–41; Kagoshima University 2014, p. 42). Moreover, secondary damage continued for another 10 years. The mountain, still covered by a large amount of ash fall, was attacked by torrential rains in the following months, which led to frequent sediment flows sweeping away more farms and houses. A number of floods were caused by rising river beds because of the accumulated ash fall (Centenary Committee 2014, pp. 46–50).

4 Population Response of the Taisho Eruption

4.1 Delayed and Divided Responses

Prior to the eruption, many islanders picked up the warning signs of a large-scale eruption. These included the drop in water levels in wells and continuous earth tremors. They began preparing for self-evacuation without waiting for an official instruction. Such lessons had been handed down from previous generations. However, the official

---

[20]The Taisho Eruption was part of the geodetic phenomena in the region surrounding Kagoshima Prefecture since 1913.

[21]The lava flow from the west mountainside continued for two months, whereas the one from the east mountainside continued for nearly a year, which connected Sakurajima and Osumi Peninsula.

[22]The earthquake in Kagoshima City added 29 deaths and 111 injured.

[23]Out of the 13,000 houses, 1.3 % totally collapsed and 70 % partly collapsed.

[24]Agriculture being the primary industry of the island, Kagoshima Prefecture had to face the economic implications of the damage in the following years.

forecast from the local meteorological station stated 'no threat of a SV eruption'. In those days, there was only one old-style seismograph and no expert in volcanology or seismology was based at the station. Following the station's announcement, village authorities advised villagers there would be no need for evacuation. The response of the general public was divided. About half the islanders ignored the official view and followed their own judgement based on the aforementioned warning signs and evacuated the island before the actual eruption. The better-educated, however, followed official advice and waited until it was too late. The 30 fatalities mentioned above were made up people who followed official advice and remained on the island. The official forecast also delayed the response of the prefectural and municipal governments and the police (Centenary Committee 2014, pp. 32–34; Central Disaster Council 2011, pp. 31–42).

4.2 Panic Followed by False Rumours

When the Magnitude 7.1 earthquake hit Kagoshima City, the false rumour spread across the city that 'a massive tsunami and a poisonous gas attack will hit the region soon'. It is said that the rumour was created by college students whose motivation was unknown. However, the rumour contributed to a surge of panic, resulting in extreme behaviours such as people rushing to the main railway station leaving everything behind. Kagoshima City became literally empty at one point. The turmoil continued until 13 days after the eruption when a professor of seismology travelled from Tokyo and convinced the population in Kagoshima, that there was 'no further threat to the city' (Centenary Committee 2014, p. 34; Central Disaster Council 2011, p. 43).

In Kagoshima, two important lessons have been emphasised since the experience of this disaster. The next section explores how those lessons can be understood in the context of CET that connects personal, community and societal factors with preparedness and resilience.

5 Lessons Learnt: 'Do not Rely on Authorities' and 'Be Frightened Effectively'

5.1 'Do not Rely on Authorities'—the Development of Agency

After the Taisho Eruption, 64 stone monuments were built within Kagoshima Prefecture. Some of these were to commemorate migration, but the majority conveyed the lessons learnt from the disaster (Suzuki 2014). One of the best-known monuments is the Sakurajima Taisho Eruption Monument built 10 years after the disaster in a village in east Sakurajima. Called 'The Monument to the Distrust of Science', part of the inscription literally says, 'citizens must not rely on theory [jumin wa riron ni shinraisezu]' (Centenary Committee 2014, p. 57; Central Disaster Council 2011, p. 42). The words refer to the story of the 30 islanders who lost their lives because they did not evacuate, following the authority's instruction and trusting the 'scientific' judgement of the meteorological station. The village mayor who had trusted the forecast regretted advising the villagers to stay on the island and felt it was his responsibility to convey the lesson to future generations.[25]

The meaning of the inscription requires a further explanation. What it intended to convey was not the denying of scientific theories and methods, but the importance of individuals' proactive and responsible attitude towards disaster preparedness (Centenary Committee 2014, p. 57; Central Disaster Council 2011, p. 42). The inscription promoted the development of citizens' 'agency', as an antithesis to the passive attitude of the 30 villagers who completely relied on the authority's instruction. Despite the wording of 'theory' and 'science', the underlying message was that citizens should exercise agency being proactive about and responsible for their own safety. Based on this interpretation, this

---

[25]The village mayor himself could not achieve this goal during his time, but it was succeeded to the next mayor who built the monument.

study has rephrased the translation to 'not to rely on "authorities"'.

5.2 'Be Frightened Effectively'—the Control of Emotion

'It is easy not to be frightened or to be too frightened of disaster, but it is difficult to be frightened effectively' (Terada cited in Centenary Committee 2014). It was Torahiko Terada, a seismologist, biologist and poet at the time of the Taisho Eruption, who left this lore. It has been considered as 'the appropriate framework for disaster preparedness' in Kagoshima Prefecture because the ability to control fear and anxiety is considered crucial in living with an active volcano. The approach taken in the region for the teaching and learning of the control of emotion is the development of two types of knowledge: the balanced knowledge about the volcano and the correct knowledge about the volcano. The former refers to both the benefits that human beings gain from the volcano and the threats that human beings may receive from the volcano. As an official from a municipal government (Interview 2014) indicated, 'people have to understand about the volcano. Without knowledge, they will be indifferent or panic. The emphasis should equally be on the understanding of the awe of nature and of the benefits humans receive from nature'.[26] In terms of the correct knowledge, it refers to the validity and accuracy of the information. A researcher and curator of the prefectural museum argues (Interview 2014) that understanding the volcano should be supported by 'evidence-based information and education, which is the only effective and appropriate means'. Thus, Terada's words have been interpreted in the region that having both the balanced and the correct knowledge would enable citizens to make sound judgements preventing false

rumours and mass panic and to co-habit with an active volcano.[27]

5.3 Agency and Knowledge as a Basis for Positive 'Outcome Expectancy'

It can be considered that the development of agency and the control of emotion, which are the capacity of individuals, would have an impact on outcome expectancy—the individual-level factor in CET. A person with a proactive and responsible attitude is likely to be interested in learning about the volcano and volcanic hazards of the region. If the gained knowledge is balanced and correct, the person will probably build a belief that individual actions can influence her/his own safety. Individuals with such belief tend to show an intention to prepare themselves for disasters, as CET suggests. Conversely, a person who does not have a proactive and responsible attitude is less likely to be interested in learning about local environments and risks. Without the balanced and correct knowledge of the region, the person is apt to develop negative outcome expectancy believing that no individual actions can make difference to personal safety.

5.4 Developing Preparedness Through Collaborative Projects

One of the key means of promoting the importance of agency and the control of emotion in the case of SV is through collaborative projects. The Centenary Project was planned and delivered by a committee comprising the Kagoshima

---

[26]The importance of not starting with the teaching of the dangers of nature has been emphasised by disaster educationalists such as Katada (2012). In tsunami education, he argues that teaching about the sea and its role has to come before talking about a tsunami.

[27]Fostering both the balanced and the correct knowledge is implemented in for example, 'Sakurajima and Us [Sakurajima to watashitachi]', a textbook which was created by a group of researchers at Kagoshima University (2014) for the purpose of disaster education offered at schools in Kagoshima Prefecture. The textbook comprises first, 'Our Sakurajima' in which the beauty of the nature and the benefits that the volcano brings are described, and then 'Knowing Sakurajima and its eruptions' which explains the mechanism and the impact of eruptions.

prefectural and municipal governments, academic experts, museums and libraries, the national and municipal meteorological agencies, the development bureau, the chamber of commerce, the confederation of tourism, neighbourhood associations and the media (Centenary Committee 2014, p. 161). The project included a number of sub-programmes aiming to convey the lessons learned from the volcanic disaster for future generations, and also to raise the awareness of volcanic preparedness amongst citizens.[28] Those programmes can be identified as kyojo projects based on collaboration amongst a wide range of stakeholders. 'Kyojo', which literary means 'helping each other', is a policy term widely used in Japan in the area of disaster management and preparedness. 'Kyojo' stresses the population's commitment and cooperation which has to be mobilised in preventing and mitigating foreseen large-scale disasters (Murosaki 2013).

Kagoshima's kyojo programmes were designed to attract a wide audience outside as well as inside the prefecture in the form of informal learning. The programmes ranged from an exhibition of the sketches and the photographs of the Taisho Eruption, to a 160-page documentation of the knowledge and experience of Sakurajima to date. Two programmes are briefly described here as an illustration of the teaching and learning of 'do not rely on authorities' and 'be frightened effectively'. The first programme is 'Creating a stone-rubbing of the Sakurajima Eruption Monument', which aimed to promote an understanding of the inscription discussed earlier through a stone-rubbing of 'The Monument to the Distrust of Science'. One of the participants commented, 'I hadn't known much about Sakurajima before. Through the stone-rubbing, I have learned about the sad story of the mistake made by the local meteorological agency.... I want to participate in more activities and know more about Sakurajima' (Kagoshima City 2013). 'A stamp rally visiting nine places in Sakurajima' combined tourism with education, including a visit to the 'buried' shrine gate in the Kurokami district.[29] The chairman of the community association [chonaikai] explained, 'the community members are looking after this buried shrine gate for the memory of the Taisho Eruption. We hope to raise awareness of volcanic preparedness' (Kagoshima City 2013).

The Centenary Project ended in March 2014, leaving a legacy of kyojo projects. For example, a not-for-profit organisation and academics collaborated to set up the Sakurajima Museum, based on the idea of 'ecomuseum' (Fukushima and Ishihara 2004), defining the whole island of Sakurajima as a museum. The museum undertakes research and develops the teaching and learning of the history and ecology of the island. Similarly, Geo Park virtually puts Sakurajima and surrounding bays together and promotes 'the links between the volcano, the people and nature' (Geo Park Committee, n.y.). While promoting tourism, lifelong learning and disaster preparedness, both projects aim to contribute the community building and the economic development of the region. It is considered in Kagoshima that making the most of the benefits of the volcano as well as learning to be prepared effectively for volcanic hazards is the way to achieve 'kyozon'—living with the volcano.

5.5 Kyojo Encompassing 'Community Participation' and 'Collective Efficacy'

Linking kyojo with the community-level factors of CET, this paper suggests that 'kyojo' is a notion that encompasses both community participation and collective efficacy. At the community

---

[28]With the slogan, 'Let's rethink the potential of the volcano! Let's benefit from SV!', the project included the following sub-programmes: 'Rethinking SV', 'Benefiting from SV' and 'Conveying the lessons learnt'. The International Association of Volcanology and Chemistry of the Earth's Interior (IAVCEI) Conference 2013—Learning from the world, was also one of the sub-programmes. The IAVCEI was an opportunity for volcanologists from different parts of the world to meet in Kagoshima to share and discuss the findings from cutting-edge research.

---

[29]Two-third of the three-metre shrine gate was covered with ash fall and lava from the Taisho Eruption.

level, participation was fundamental in the Centenary Project. The events were developed by diverse members of the region and participated by wider communities, aiming to raise awareness and build the knowledge-base of the general public. Through community participation, people were involved in the teaching and learning about the Taisho Eruption and volcanic preparedness. Kyojo thus is about community participation. Kyojo is also about collective efficacy. As identified in the comparative study by Paton et al. (2013, p. 29) that examined Kagoshima residents' preparedness intentions, 'the high frequency of experience of volcanic hazards in Kagoshima increases its citizens' familiarity with the hazards and what works to manage their risk' (Paton et al. 2013, p. 29), which has resulted in the community's developing 'building codes, ash removal practices and community attitudes and preparedness to facilitate continuity of societal functions during periodic volcanic episodes'. Based on the assessment of 'what works to manage their risk', this community has planned 'practices' and 'preparedness'. It is suggested that the notion of 'kyojo' includes such assessment and decision-making phase, which is identified as collective efficacy in CET.

5.6 Kyojo as a Contributor to 'Empowerment' and 'Trust'

The relationship between kyojo and the societal-level factors—empowerment and trust—can be considered as the following. The more kyojo programmes—the programmes created on the basis of collaborative partnerships amongst a wide range of stakeholders—are developed and delivered, the more empowering and trustworthy relationships between community members and authorities would be built. Through collaborative working, both parties get to know each other, exchange ideas and information and co-construct preparedness schemes. The significant point in the case of SV is that after the Taisho Eruption in which the islanders who lost their lives following

the village mayor's instruction, trust in the authorities must have decreased. As Paton et al. (2013, p. 22) indicate, 'people's perception of the quality of their (historical) relationship with a source of information influences their interpretation of the value of information in a way that is independent of the information itself'. The two major lessons 'do not rely on authorities' and 'be frightened effectively' have played a pivotal role in the rebuilding of trust in the authority. Today, Kagoshima citizens have become familiar with the hazards and how to cope with them. The high frequency of volcanic hazards has meant the information provided by official agencies to guide their preparing has been critical. This was anecdotally confirmed a number of times while the empirical fieldwork was undertaken in Kagoshima. It would be fair to say that the trust relationship between citizens and the authorities in the Kagoshima region has been rebuilt since the Taisho Eruption. The suggestion here is that the kyojo approach has contributed to the building of the empowering and trustworthy relationship.

The SV case appears to have a relatively high degree of individual, community and societal engagement in preparedness, although this hypothesis requires testing. The challenge in the region is, however, their preparedness also has to take into consideration an infrequent mega-scale eruption, which requires evacuation from the island before the eruption starts. This is the topic of the next section.

6 The Challenge of Living Under the Level 3 Warning

As a result of the volcanic warning Level 3 being kept unchanged since April 2012, 'Level 3' has become a norm in Kagoshima Prefecture. Being familiar with small-scale eruptions and ash fall, the local population are capable of facilitating 'continuity of societal functions during periodic volcanic episodes' (Paton et al. 2013, p. 29), not markedly changing their daily routines. Such circumstances could lead to 'the normalcy bias', which was introduced by Omer and Alon (1994, p. 273).[30] Defined as 'underestimating the probability or extent of expected disruption', the term 'the normalcy bias' refers to the perception of the authorities during the crisis that 'it cannot happen to us' or 'life will be unchanged, even after a disaster', resulting in failing to plan for disaster, let alone involving the public in planning for disaster (Valentine and Smith 2002, p. 186). The researchers also identified the so-called 'abnormalcy bias', which refers to the authorities' 'underestimating victims' ability to cope with disaster' (Omer and Alon 1994, p. 273) with a view that 'the people cannot handle a threat of impending disaster' ending up in either panic, shock or looting (Valentine and Smith 2002, p. 186).

As indicated by the curator and researcher of the prefectural museum (Interview 2014), in the case of Kagoshima, 'the normalcy bias' appears to reside more in the public rather than the authorities. Although further investigation is required to confirm this claim, the point to be made here is that 'living *with* an active volcano' seems to promote the normalisation of 'the state of alert', and under such circumstances, there is a risk of people becoming less interested in preparing themselves for a more severe disaster. In other words, Kagoshima residents have high familiarity with preparedness for Level 3 and small-scale volcanic hazards, but that does not necessarily mean they are equally prepared for an eruption of the scale of the Taisho disaster.

7 Conclusion

The teaching and learning of and *for* volcanic hazards, as well as volcanology and seismology have dramatically advanced since SV's Taisho Eruption of 1914. Applying the individual/community/societal framework of CET, this paper has discussed a model for 'living with an active volcano', with a focus on how the lessons

learned from previous experiences have been utilised in present-day preparedness programmes in the Kagoshima region. The overall argument has been that the SV case can be identified as 'the integrated model' endorsed by CET, largely because of their kyojo practices represented in the Centenary Project and the Sakurajima Museum, which incorporated 'disaster management' and 'community development' through 'mobilising community resources to facilitate adaptive capacity rather than having institutional decisions imposed upon a community' (Paton 2008, p. 29). The integrated model entails 'natural coping mechanisms' (Paton 2008, p. 29), or in Yamori's (2012) term, 'everyday preparedness [seikatsu bosai]'. In developing such 'built-in' (Yamori 2010) mechanisms, kyojo becomes critical. This conclusion has been drawn from a qualitative examination of the findings of the SV case against the five factors of CET.

Firstly, at the individual level, the paper has suggested that two principles identified in the SV case form the basis for positive 'outcome expectancy'. One principle is an emphasis on developing agency so that individual citizens develop ownership in making decisions about their behaviour in emergency situations. This has been expressed by a local lore, 'do not rely on authorities'. The other is an emphasis on gaining the balanced and correct knowledge of the volcano and volcanic hazards. The knowledge allows one to 'be frightened effectively' as other local lore advocates, and also prevents one from being distracted by false information. Positive outcome expectancy is likely to develop in a person who has a proactive and responsible attitude to gain the balanced and correct knowledge.

In terms of the community and societal levels, the paper has demonstrated that 'kyojo'—collaborative partnerships—plays a significant role. It is suggested that kyojo encompasses both of the community-level factors—'community participation' and 'collective efficacy'. Participation is a prerequisite for kyojo activities. The efficacy of the community to make collective judgements and reach decisions is likely to be developed through collaboration and exchanging of ideas

---

[30]Their research was in the context of the Gulf War to refer to the response of the authorities to the threat of attacks.

and information. It is also suggested that kyojo is an approach that contributes to the building of societal-level 'empowerment' and 'trust' between community members and authorities.

The above analysis requires strengthening by further researching the relationships between agency, knowledge and kyojo, and outcome expectancy, community participation, collective efficacy, empowerment and trust. Moreover, clarifying the role of agency, knowledge and kyojo in linking 'disaster management' with 'community development' and vice versa is needed in presenting a concrete picture of how 'an integrated model' might look like.

This final part of the paper addresses a specific kind of challenge that the population in the Kagoshima region seems to be facing in living under the continuous Level 3 condition. Even if the population has a high level of preparedness for frequent volcanic hazards, whether the same applies to Taisho-Eruption-scale hazards remains questionable. Having developed individual agency, community participation and trust between citizens and authorities, the population may be immune to 'the abnormalcy bias'; however, a possibility of them growing 'the normalcy bias' appears to be an emerging concern.

Acknowledgments This paper was developed on the basis of the findings from the research project, *Critical Infrastructure Failure and Mass Population Response*, funded by the Economic and Social Research Council in the UK. I would like to thank all of the interviewees for providing me with important information, and Taichi Kaneshiro of MEXT for his co-operation in arranging the interviews.

References

Ishihara K (2012) Understanding of volcanic phenomena and prediction of volcanic eruptions. Annuals Disaster Prevention Res Inst Kyoto Univ 55A:107–115

BBC News (2014) Japan volcano: 16 still missing after Mount Ontake eruption. http://www.bbc.co.uk/news/world-asia-29472384. Accessed 6 Oct 2014

Cabinet Office (2011) Disaster management in Japan. Cabinet Office, Government of Japan, Tokyo

Central Disaster Council, Cabinet Office (2011) Learning from the history of disasters: volcanos [saigaishi ni manabu: kazan hen]. In: Expert Committee on Disaster Lessons. Cabinet Office, Government of Japan, Tokyo

Department of Domestic Science Education, Faculty of Education, Kagoshima University (2014) Sakurajima and us [sakurahjima to watashitachi]. Sakurajima Volcano Education Supplementary Reading Material, Department of Domestic Science Education, Faculty of Education, Kagoshima University, Kagoshima

Fukushima D, Ishihara K (2004) Practical research on educational dissemination for volcanic disaster prevention: a case study based on the Ecomuseum concept. Annuals Disaster Prevention Res Inst Kyoto Univ 47C, n. p

Geo Park Committee (n.y.) The vision of the Geo Park. http://www.kagoshima-yokanavi.jp/shizen/geopark/geo05.html. Accessed 21 Oct 2014

Japan Meteorological Agency (2014). Sakurajima. http://www.data.jma.go.jp/svd/vois/data/fukuoka/eng/506/506-eng.htm. Accessed 21 Oct 2014

Kagoshima City (2013). Sakurajima Taisho Eruption Centenary Project [Sakurajima taisho funka 100 shunen kinen jigyo]. Kagoshima Citizens' Common, 553

Katada T (2012) Disaster information and disaster education [saigai joho to bosai kyoiku]. The record of memorial symposium, 14th conference of the Japan Society Of Disaster Information. 28 Oct 2012

Kitagawa K (2014) Continuity and change in disaster education in Japan. Hist Educ. doi:10.1080/0046760X.2014.979255

Murosaki M (2013) The current situation of disaster projection, volunteering and kyojo, and future expectations [bosai to borantia to kyojo no genjo, soshite kongo eno kitai]', Keynote Lecture, Disaster Projection by All in Kobe—Expanding the Network of Kyojo Meeting, organised by the Cabinet Office, 27 Jan 2013. http://www.bosai-vol.go.jp/kyojo/murosaki.pdf. Accessed 21 Oct 2014

Omer H, Alon N (1994) The continuity principle: a unified approach to disaster and trauma. Am J Community Psychol 22(2):273–286

Paton D (2008) Community resilience: integrating individual, community and societal perspectives. In: Gow K, Paton D (eds) The phoenix of natural disasters: community resilience. Nova Science Publishers, New York, pp 13–31

Paton D, Okada N, Sagala S (2013) Understanding preparedness for natural hazards: a cross cultural comparison. J Integr Disaster Risk Manage 3:18–35

Paton D, Jang L (2013) Increasing community potential to manage earthquake impacts: the role of social and cultural factors. In: Konstantinou KI (ed) Earthquakes: triggers, environmental impact and potential hazards. Nova Publishers, New York, pp 299–320

Paton D, Johnston D (2006) Disaster resilience: an integrated approach. Charles C. Thomas, Springfield

Preston J, Chadderton C, Kitagawa K (2014) The 'state of exception' and disaster education: a multi-level conceptual framework with implications for social justice'. Globalisation Soc Educ 12(4):437–456

Sakurajima Taisho Eruption Centenary Project Planning Committee (2014) The centenary memorial magazine of Sakurajima Taisho Eruption [taisho funka 100 shuunen kinenshi]. Sakurajima Taisho Eruption Centenary Project Planning Committee, Kagoshima

Shiroshita H (2010) Towards the actualisation of collaborative learning for disaster prevention [bosaikyoiku no jitsugen ni mukete]. In: Safety science for disaster prevention and reduction—a proposal in building a safe and secure society [bosai/gensai no tameno shakaianzengaku—anzen/anshin na shakai no kochiku eno teigen]. Faculty of Safety Science, Kansai University ed., Mineruva Shobo, Kyoto, pp 98–114

Suzuki T (2014) The present situation and assignment of monuments for Mt. Sakurajima Eruption in Taisho 3 (1914) 2. Res Rep Kagoshima Prefectural Mus 33:71–88

Takahashi K (2007) Risk management on volcanic disaster. J Saf Issues, Japan Society of Civil Engineering 2:17–22

The Expert Committee to Discuss Disaster Education and Management Considering the Lessons Learnt From the Great East Japan Earthquake [Higashinihon daishinsai o uketa bosaikyoiku/bosaikanri ni kansuru yushikisha kaigi] (2012) The Expert Committee to discuss disaster education and management considering the lessons learnt from the great East Japan Earthquake—final report [Higashinihon daishinsai o uketa bosaikyoiku/bosaikanri ni kansuru yushikisha kaigi—saishu hokoku]'. http://www.mext.go.jp/b_menu/shingi/chousa/sports/012/toushin/1324017.htm. Accessed 22

Valentine PV, Smith TE (2002) Finding something to do: the disaster continuity care model. Brief Treat Crisis Interv 2(2):183–196

Yamori K (2010) The current state and prospects of disaster education—15 years since the Hanshin/Awaji Earthquake. J Jpn Nat Disaster Sci 29(3):291–302

Yamori K (2012) The great East Japan Earthquake and 'everyday preparedness' [higashinihondaishinsai to "seikatsu bosai"]. Build Maintenance Manage Cent, J Re 175:n. p

Using Role-Play to Improve Students' Confidence and Perceptions of Communication in a Simulated Volcanic Crisis

Jacqueline Dohaney⬤, Erik Brogt, Thomas M. Wilson and Ben Kennedy

Abstract

Traditional teaching of volcanic science typically emphasises scientific principles and tends to omit the key roles, responsibilities, protocols, and communication needs that accompany volcanic crises. This chapter provides a foundation in instructional communication, education, and risk and crisis communication research that identifies the need for authentic challenges in higher education to challenge learners and provide opportunities to practice crisis communication in real-time. We present an authentic, immersive role-play called the Volcanic Hazards Simulation that is an example of a teaching resource designed to match professional competencies. The role-play engages students in volcanic crisis concepts while simultaneously improving their confidence and perceptions of communicating science. During the role-play, students assume authentic roles and responsibilities of professionals and communicate through interdisciplinary team discussions, media releases, and press conferences. We characterised and measured the students' confidence and perceptions of volcanic crisis communication using a mixed methods research design to determine if the role-play was effective at improving these qualities. Results showed that there was a statistically significant improvement in both communication confidence and perceptions of science communication. The exercise was most effective in transforming low-confidence and low-perception students, with some negative changes measured for our higher-learners. Additionally, students reported a comprehensive and diverse set of best practices but focussed primarily on the mechanics of science communication delivery. This curriculum is a successful example of how to improve students' communication confidence and perceptions.

J. Dohaney (✉) · E. Brogt · T.M. Wilson ·
B. Kennedy
Geoscience Education Research Group, University
of Canterbury, Christchurch, New Zealand
e-mail: jdohaney@gmail.com

E. Brogt
Academic Services Group, University of Canterbury,
Christchurch, New Zealand

J. Dohaney · T.M. Wilson · B. Kennedy
Department of Geological Sciences, University of
Canterbury, Christchurch, New Zealand

Advs in Volcanology (2018) 691–714
https://doi.org/10.1007/11157_2016_50
Published Online: 15 March 2017

1 Introduction

Communicating scientific results and recommendations about natural hazards and disasters into language easily understandable by non-experts is a challenging task in the best of circumstances. During an actual natural hazard event, stress levels are high and considerable pressure is put on scientists and emergency managers to communicate a wide variety of information to each other and many different stakeholders (Alexander 2007; Barclay et al. 2008; Haynes et al. 2008; IAVCEI Task Group on Crisis Protocols 2016; Rovins et al. 2015, p. 56).

Many practicing scientists receive no formal training in science communication (including communication with the public and with media) (MORI and The Welcome Trust 2001; The Royal Society 2006) or public engagement (Miller and Fahy 2009). Additionally, embedded training of science communication in undergraduate degree programmes is uncommon, though specific degrees, minors, or postgraduate degrees are offered in a relatively select few institutions and predominantly within Europe (Trench and Miller 2012). Therefore, dedicated science and risk communication training for undergraduates provides a valuable opportunity to instil the next generation of natural hazard scientists and emergency managers with communication strategies and skills which, if informed by established best practices, will aid them to better serve a society that faces increasing risks from natural and manmade hazards.

This chapter describes a case study about an interactive, challenging role-play designed to train students how to forecast volcanic eruptions, manage the impacts from these eruptions, and communicate with the public throughout the simulated crisis. The chapter also introduces the reader to the foundations of instructional communication, education, and risk and crisis communication research and demonstrates how to evaluate communication training pedagogy with an evidence-based approach.

We argue that role-play challenges students and provides them with practical experience that they can utilise in their careers. It also improves learner's confidence in their ability to communicate and improves their overall perceptions of risk and crisis communication best practice. We believe the success of the role-play lies in the explicit practicing of authentic communication tasks in a feedback-rich environment and we hope to should encourage instructors to incorporate more authentic tasks into their curricula. We invite our readers to use and adapt this curriculum in classrooms of all levels of formal and informal education.

1.1 Why Is Volcanic Risk Communication Training Important?

There is a long history and multidisciplinary approach to research of risk and crisis communication. Corporate crisis communication and public relations (e.g., Grunig and Repper 1992; Crane and Livesey 2003), health risk and crisis communication (e.g., Reynolds and Shenhar 2016), and broader risk communication (e.g., Morgan et al. 2002; Glik 2007) communities have all explored the strategies, philosophies and evaluation of these communications and how differing approaches may influence its success.

In general, these communities have advised that we should move away from the old, linear, 'transmission' form of communication (i.e., 'source' to 'receiver' or the Shannon-Weaver model of communication) towards a participatory approach to work with communities to establish a *dialogue* (e.g., Fisher 1991; Fischhoff 1995) that supports diversity in the needs of the audience (McCroskey 2006) preferably in an unofficial and relaxed setting that helps to build trust between scientists and the public (Haynes et al. 2007). The Sendai Framework for Disaster Risk Reduction supports this approach, encouraging the sectors of society (i.e., public, private and academic sectors) to work together in a 'people-centred' approach to DRR (United Nations International Strategy for Disaster Risk Reduction 2015, See point 7, p. 10). This shift is important for the delivery of risk and crisis communications and highlights the importance of knowing, understanding and connecting with your audience.

Volcanologists play a major role in the dialogue that occurs in the long-term and short-term communication of volcanic risk. Pielke (2007) provides an excellent overview of the particular roles that experts may choose to take when science has the potential to impact policy, politics and the public. He proposes that experts (e.g., medical practitioners, engineers or scientists) can act as an 'honest broker' by providing clear options to the person(s) at risk, articulate the specific outcomes, while simultaneously accounting for uncertainties and incorporating the most up to date scientific understanding of the topic at hand.

Essentially, it is our job as scientists to provide clear information to the public on the potential risks that they face from volcanoes. However, as stated above, scientists are rarely trained in communication so the pathways and strategies for achieving this aim is less known. Additionally, there have been very few initiatives that have blended volcanology, risk communication, and education but all of these research areas have much to offer to the teaching of communication in the sciences. This research hopes to bridge this gap and describes a research-informed curriculum that can be used to train future volcanologists in the best practices of volcanic risk and crisis communication.

1.2 Instructional Communication Research

Communication is one of the most commonly mentioned graduate attributes for most undergraduate degrees and is also core to the geology profession (Heath 2000; Jones et al. 2010). A quick sample of several university's graduate attribute profiles will show you that communication, in some defined form, is almost always present. Communication was a main focus (i.e., was among the primary goals and outcomes) in all of the courses (see Sect. 2) that featured the

role-play so as part of our efforts to include authentic communication training we undertook a review of instructional communication (i.e., the teaching of communication skills). Here, we share some of what the research community tells us about teaching communication.

Firstly, there are a wealth of studies that advocate for the benefits of learners undergoing some form of communication education. Morreale and Pearson (2008) state that effective communication skills are needed across many disciplines (e.g., sciences, business, engineering or architecture) and helps them to succeed in a range of careers. Morreale and Pearson (2008) also state that communication training encourages global, socially and culturally-aware citizens, including specific areas of global significance allowing our society to make better decisions in areas like health and medicine, crisis management, and policing.

Secondly, effective communication does not come about by simply practicing a speech in front of a mirror. A recent study by Engleberg et al. (2016) compiled the core competencies of communication to assist in building a standardised introductory course in instructional communication. The seven core competencies (listed here, taken directly from Engleberg et al. 2016) shows the reader the diversity of skills that are needed to be an effective communicator:

1. **Monitoring and Presenting Your Self** (i.e., the ability to monitor and present yourself to others within and across a variety of communication contexts);
2. **Practicing Communication Ethics** (i.e., the ability to identify, evaluate, and demonstrate appropriate ethical behaviour within and across a variety of communication contexts);
3. **Adapting to Others** (i.e., the ability to understand, respect, and adapt messages to a diversity of human characteristics and attitudes in order to accomplish a communication goal within and across a variety of communication contexts);
4. **Practicing Effective Listening** (i.e., the ability to listen effectively and respond appropriately to the meaning of messages

within and across a variety of communication contexts);

5. **Expressing Messages** (i.e., the ability to select, demonstrate, and adapt appropriate forms of verbal, nonverbal, and mediated expression that support and enhance the meaning of messages within and across a variety of communication contexts);

6. **Identifying and Explaining Fundamental Communication Processes** (i.e., the ability to identify and explain how specific communication processes influence the outcome of communication interactions within and across a variety of communication contexts);

7. **Creating and Analysing Message Strategies** (i.e., the ability to create and analyse message strategies that generate meaning within and across a variety of communication contexts).

Thirdly, measuring and assessment of communication competence is different from most learning in the sciences and other disciplines. It is a skill that is highly contextualised (See above) and success is in the mind of the receiver(s)/audience(s), that makes it inherently difficult to judge with objective consistency. Determining whether a learner has shown excellence in communication requires observation of the student's performance across a range of situations and contexts. Though these competencies may seem difficult to assess, communication researchers have developed a series of measures that aim to capture some of the many dimensions of communication competency.[1]

Our research aimed to characterise and measure students' *confidence* and *perceptions* of volcanic crisis communication and to determine if the role-play was effective at improving these qualities. This study occurred at the beginning of a longitudinal programme that is exploring a working model of communication denoted by several dimensions that impact an individual's communication performance: communication *confidence* (discussed here), *perceptions* of science/crisis communication (discussed here), previous *experiences* with communication, and content *knowledge* (i.e., expertise in the topic that is being communicated).

Confidence in one's ability to communicate competently relies on having the knowledge, skills and motivation to communicate (Rubin and Morreale 1996). The knowledge to communicate competently requires learners to select the appropriate information and strategy for the right situation, while the skills come about from having the skills to execute these strategies (Kreps and Query 1990). The motivation to communicate arises from learners choosing to engage after weighing several internal and external factors (e.g., grade incentives; Fortney et al. 2001). Courses in public speaking have been shown to increase student's confidence communicating (Miller 1987; Richmond et al. 1989; Rubin et al. 1997; Ellis 1995). It is worth noting that confidence does not directly translate to effective performance and that overconfidence (e.g., Kruger and Dunning 2009) and compulsive communication (Fortney et al. 2001) can be detrimental to learning and communication. Communication confidence was measured by asking students to self-report their perceived competency to communicate to different receivers and in different contexts (described in Sect. 2).

Another important construct to our study was the perceptions of risk and crisis communication best practice. Perceptions are a selection of attitudes or beliefs that an individual holds and that guides their behaviour. McCroskey (2006) proposes that there are three elements to building communication skills: *desire*, *understanding*, and *experience*. Understanding communication involves knowledge and awareness of the multitude of considerations and strategies that you can employ when crafting and delivering a message. A perceptions survey allows you to check for alignment between the views of the students compared with the views of professionals.

---

[1]Please see http://www.jamescmccroskey.com/measures/ for a list of prominent examples.

Our curriculum was focussed on teaching students' volcanic crisis communication, and so their perceptions were measured by asking students whether they agreed with a series of statements concerning best practice under these circumstances (described in Sect. 2). Though it should be stated that simply because you hold a 'correct' perception does not mean that you will (a) execute the strategy effectively, or (b) decide to use the strategy when the opportunity arises. Holding expert-like perceptions is only one part of the tool kit for becoming an effective communicator.

1.3   Educational Research

Educational research is critical for the development and evaluation of curricula. As our understanding of 'how we learn' becomes more sophisticated, the strategies we use in the classroom allow for more effective learning experiences than traditional, stand-and-deliver teaching. At present, we feel that rigorous education research is an underutilised resource at all levels of volcanology education including formal and informal educational settings.

In practice, curriculum development is often content-driven rather than learning outcome-driven (i.e., focuses on specific aspects of volcanism to cover, rather than on the skills and knowledge that an instructor hopes the students will gain from learning about volcanoes). Additionally, curriculum development is undertaken by academics or secondary school educators who may not be aware of applied volcanology and emergency management practices. Consequently, lessons that are developed may be theory-focussed (not skills-focussed) and lack the authentic challenges that accompany volcanic crises.

Authentic learning focuses on real-world, complex problems and their solutions taught within authentic environments through activity and social interaction (Herrington and Herrington 2006; Lombardi 2007; Herrington et al. 2014). Authentic learning seeks to replicate real-world practices in the classroom including the environment, roles, and responsibilities of

professionals. Role-play is one of the many examples of authentic learning. Other examples include: simulation, role-play, mentoring, debate, case studies, coaching, and reflection (e.g., Brown et al. 1989). Authentic learning offers an opportunity for students to explore communication in its fullest complexity leading to a more befitting assessment of their communication skills.

The effectiveness of role-play and simulation for learning has been reported in a number of studies (e.g., DeNeve and Heppner 1997; van Ments 1999). Simulation is defined as a learning experience that occurs within an imaginary or virtual system or world (van Ments 1999) and 'role-play' as the importance and interactivity of roles in pre-defined scenarios (Errington 1997, 2011). Simulation and role-play require more active participation from students than lecture-based teaching techniques and intend to teach practical and theoretical skills that are transferable to different future situations (Roth and Roychoudhury 1993; Lunce 2006). Research shows that role-play and simulation improve student attitudes towards learning (DeNeve and Heppner 1997; van Ments 1999; Shearer and Davidhizar 2003) and interpersonal interactions (Blake 1987; van Ments 1999; Shearer and Davidhizar 2003), generic transferable skills (problem-solving and decision-making skills (Errington 1997; Barclay et al. 2011), communication skills (Bales 1976; van Ments 1999; Hales and Cashman 2008); and teamwork skills (Maddrell 1994; Harpp and Sweeney 2002), as well as discipline-specific knowledge (DeNeve and Heppner 1997; Livingstone 1999) and volcanic eruption forecasting skills (Harpp and Sweeney 2002; Hales and Cashman 2008).

1.4   Risk and Crisis Communication Best Practices

In order to teach students how to communicate about volcanic risk, we must first understand how experts communicate before, during and after volcanic events. The communication of science (more generally) can take on a multitude

Table 1 Study participants demographics

Cohort	Age (n)	Gender (n)	Nationality (n)	Degree programme (n)
Field-based (23 students) Jan 2012	19–22 (18) ≥ 23 (5)	Female (8) male (15)	United States (13) New Zealand (9) Netherlands (1)	BSc (13) PGDipSci[a] (9) PhD (1)
Lecture-based (20 students) Aug 2012	19–22 (7) ≥ 23 (13)	Female (5) male (15)	United States (1) New Zealand (18) India (1)	BSc (11) PGDipSci (9)
All students (43 students)	19–22 (25) ≥ 23 (18)	Female (13) male (30)	United States (14) New Zealand (27) Netherlands (1) India (1)	BSc (24) PGDipSci (18) PhD (1)

Numbers here represent students who participated in the role-play. Some students did not complete all of the surveys in the study
[a]Students in the PGDipSci programme were in the first year of their postgraduate studies focussed on Geology and/or Hazards and Disaster Management. Some of these students later upgraded to a MSc thesis

of formats, styles, objectives, and outcomes. Burns et al. (2003) defined science communication as the "… use of appropriate skills, media, activities, and dialogue to produce one or more of the following personal responses to science: awareness, enjoyment, interest, opinions, and understanding of science (i.e., its content, processes and social factors)". Volcanic risk and crisis communication may include science communication that can be used to educate and promote risk-reducing behaviours to the public (Barclay et al. 2008).

We differentiate between risk and crisis communication using criteria laid out by Reynolds and Seeger (2005; Table 1): *Risk communication* uses messages that focus on reducing the consequences of a known threat (i.e., risk is based on projections and long-term forecasts), occurring prior to an event in frequent or routine communication campaigns, relying on technical experts and scientists to deliver the message; while *Crisis communication* uses messages that focus on information regarding a disruptive event, occurring immediately following and in a response to an event,[2] relying on authority figures and technical experts to deliver the message. Reynolds and Seeger (2005) promote an integrated model where the scientific community can view communication as part of

an ever-evolving cycle around risk factors that must adapt and match to the situation and context. This allows communicators to approach both risk and crisis communication with a set of tools (i.e., best practices) that must be carefully selected and suit the context and needs of the audience. We welcome this way of thinking, and seek to undertake communication training of students and practitioners within this framework.

For the purposes of teaching, we wanted to have a concise set of best practices that incorporated scholarly work but was comprehensible to our students allowing them to pick them up in the short time frame allocated by our curriculum. A colleague at the University of Otago developed a distinct set of rules for risk and science communication, which was derived from research on media from the Canterbury Earthquake sequence, that she called the 7Cs (Taken from Bryner 2012; Ideas influenced from the 10Cs Weingart et al. 2000; Miller 2008). These best practices were explicitly given to students prior to participating in the role-play, and were a part of the theoretical foundation for the perceptions survey used in this study and is described further in Sect. 2. The 7C's say that risk and science communication should be:

> **comprehensible** (i.e., simple, jargon-free, clear and concise),

> **contextualised** (i.e., acknowledges and reflects diversity of your audience),

---

[2]We should acknowledge that volcanic events can become a 'crisis' even before any eruptive activity occurs.

captivating (i.e., entertaining, engaging, salient, and relevant to everyday life),

credible (i.e., open, does not overpromise, acknowledges uncertainty),

consistent (i.e., backed by evidence, confirmable, coordinated and collaborated sources of information),

courteous (i.e., compassionate, empathetic and respectful), and

addresses **concerns** (i.e., empowers action and response, forms a dialogue).

We hope that the literature provided in the above sections has helped to prove to the reader that communication and education research communities have much to offer to the teaching of communication skills in volcanology and hazard and disaster management students. These fields provide the underlying framework and foundation (i.e., the stage and theatre) in which the volcanologists and emergency managers (i.e., the characters) will work through a crisis (i.e., the narrative) and avoid a potential disaster (i.e., the climax) in a role-play. To exemplify these theories in practice, we share with you a pilot study of an authentic role-play, training exercise that specifically aimed to improve university-level students' communication skills during a mock volcanic crisis (described in detail, below).[3]

1.5 The Volcanic Hazard Simulation

1.5.1 Design and Development of the Volcanic Hazard Simulation Role-Play

For some time, training exercises have been used in the emergency management community to simulate real world crises in order to upskill practitioners (Borodzicz and van Haperen 2002). We partnered with experts in the field (e.g., volcanologists, emergency managers and decision-makers) through action research and interviews to develop an authentic role-play and to deduce best practices in volcanic crisis communication. Additionally, we worked closely with instructors to assess the classroom setting, cultures and logistics to be sure that the role-play suited their needs and fitted into their curricula. Such a process allows for effective curriculum development geared towards learners', instructors' and industry needs and builds relationships within different sectors that supports long-term, sustainable teaching practices, and ensures that the curriculum will continue to be used after the educational specialist is out of the picture.

The Volcanic Hazard Simulation role-play was designed and developed by a team of researchers from the geosciences, hazards and disaster management and education disciplines at the University of Canterbury in Christchurch, New Zealand. Emphasis in the early phases of the project was placed on developing authenticity of the roles and teams and ensuring that the simulation was successful at achieving the desired learning goals. Evaluation of the simulation indicated that students found the simulation to be a highly challenging and engaging learning experience and self-reported improved skills (Dohaney et al. 2015). Classroom observations and interviews indicated that the students valued the authenticity and challenging nature of the role-play although personal experiences and team dynamics (within, and between the teams) varied depending on the students' background, preparedness, and personality (Dohaney et al. 2015). For a more detailed discussion on the design and development of the Volcanic Hazards Simulation role-play we refer the reader to Dohaney (2013) and Dohaney et al. (2015) and for instructors who are interested in running the role-play in their course, an instructor manual is freely available for educational use online.[4]

---

[3]The role-play discussed here does not include the risk communication practices that occur over longer time frames or in ongoing volcanic events. The learning goals for our activity were limited to volcanic forecasting, decision-making, and managing community concerns throughout a crisis. For a further explanation of our learning goals and motivations for building this scenario, please see Dohaney et al. (2015).

[4]You can find the user manual in two places on VHUB (https://vhub.org/resources/3395; Dohaney et al. 2014) or on SERC (http://serc.carleton.edu/introgeo/roleplaying/examples/125523.html).

Two eruption scenarios have been built and tested. The first is a large explosive scenario based on a VEI6 eruption from Tongariro Volcanic complex eruption (Cole 1978; Hobden et al. 1999) that is modelled on the 1991 Mt. Pinatubo eruptions (e.g., Wolfe and Hoblitt 1996). The second scenario is an explosive and effusive eruption of the Auckland Volcanic Field that focuses on the science and impacts from monogenetic volcanism in an urban environment. In both cases, the scenarios were chosen as there were existing volcanic monitoring data available to build our models on, and because they had all the pedagogically-relevant stages; from forecasting (that can be denoted by precursors that students could identify), to minor eruption events and results in an exciting, 'blockbuster' climax (major eruption). In the scenario presented here (i.e., the Tongariro scenario), students are presented with real-time, streamed datasets that take the volcano from a quiescent stage, small eruptions (i.e., 'unrest'), and concluding with a very large eruption. The initial design and timeline for the role-play was taken from Harpp and Sweeney (2002) and was subsequently improved through multiple design phases to optimise the exercise and meet the learning goals.

1.5.2 What Happens During the Volcanic Hazards Simulation?

The Volcanic Hazard Simulation is designed for 300–400 level (i.e., upper-year) undergraduate science students from geology, natural hazards, disaster risk reduction, and emergency management. The simulation takes 4–6-h and can accommodate between 15 and 40 students. Students are divided into two teams: the Geoscience team and the Emergency Management team. All students have an authentic role that they are required to research prior to participation in the simulation, such as the field geologist, geodesist, public information manager, or the welfare manager, etc.

The students within the Geoscience team interpret the streamed datasets (e.g., ground deformation, gas, seismicity; see Dohaney et al. (2015) for more details) and communicate

science advice to the emergency management team and to the 'public'. The Emergency Management team is responsible for managing the impacts that the volcanic eruption poses to communities and infrastructure. This set-up is adapted from the organisational structure of operational emergency management in New Zealand dictated by the most recent version of the national guidelines (Ministry of Civil Defence and Emergency Management 2009) and this structure is comparable to other emergency management structures used, globally [e.g., the National Incident Management System (Department of Homeland Security 2008)]. It is important to note that the learning goal for the exercise is not to *replicate* protocols, but to introduce students to the roles and responsibilities in these important events and to improve their skills sets. We emphasise this distinction to the students, and this allows students to free up their cognitive resources to focus on teamwork, decision-making and the communication tasks, rather than perfecting organisational procedures. The simulation is a reasonably fast-paced environment, with events happening in quick succession to mimic the stresses of a real natural hazard crisis.

Students respond to emergency management and the public's information needs via a 'Newsfeed' data stream (i.e., a stream of prompts that replicate common views and needs during a crisis) and communicate to policy-makers and to members of the public (played by facilitators). Students need to be able to adapt both the content and style of the communication appropriately to serve the intended target audience. During the role-play, we included structured communication tasks that incorporate different communication goals, formats, contexts, and receivers (i.e., different audiences):

Students do the following structured communication events or tasks:

1. Media releases (written)
2. Volcanic impact reports (written)
3. Team discussions: Both within the team (intra-team) and between the groups (inter-team) (oral, group)

4. On-the-spot 'dynamic' information requests (written and oral, individual and group)
5. Media TV interviews (oral, public)
6. Press conferences (oral, public)

It should be noted that not all students will directly participate in each task, as these are team tasks in which some students will choose to communicate to the class, and others will not. We aimed to model authentic and effective team behaviour that requires the group to manage the incoming and outgoing communications, as well as adhering to the appropriate responsibilities of individual roles (i.e., team leaders typically volunteered to take on more frequent public speaking tasks).

Students prepare for the role-play through several preparatory activities including: a volcanic hazards mapping activity, pre-readings (with content specific to their role), an exercise instruction document (with learning goals, the rules, and flow of communication maps), and a science communication lecture and homework assignment including reviewing the 7C's (described above) that we used as crisis communication 'best practice'. We expect students to be comfortable with the basics of volcanic monitoring and emergency management, but additional introductory lectures are available for revision.

2 Methods

The current study explores the evaluation of students' communication confidence and perceptions of crisis communication best practices. Below we discuss the study participants, data collection and data analysis procedures.

2.1 Study Participants

Participants (n = 43; Table 1) were recruited from 300- and 400-level physical volcanology and hazards management courses that hosted the Volcanic Hazards Simulation as part of their curricula. The role-play was assessed using a self- and peer-evaluation rubric that accounted

for a small percentage of their grade ($\sim$1% of their total grade). Students were mixed cohorts of American study-abroad students and New Zealand students who attended the University of Canterbury. They ranged in gender [female (13) and male (30)], nationality [New Zealand (27), United States of America (14), Netherlands (1) and India (1)], and age [aged 19–22 (25) and >23 years old (18)].

2.2 Data Collection

Two iterations of the role-play were tested for communication perceptions and confidence; One role-play was embedded at the end of a 7-day field course (January 2012; n = 23) the other was embedded within a lecture-based course (August 2012; n = 20). The nature of the intervention was slightly different in terms of what was covered prior to the exercise. The Field-based cohort carried out a hazards mapping exercise (studying the volcanology and hazards of Tongariro) and reviewed the best practices of science communication in a short lecture, followed by a media release critique [both of which were assessed for a small amount ($\sim$1% of their total grade)] to encourage students to prepare for the role-play. While the Lecture-based cohort received the same science communication lecture but no other activities. These differences in treatment were controlled by course design and allowed the researchers to explore if different treatments of the student groups elicited different communication results.

We used a mixed methods approach in our investigation of the effectiveness of the role-play on science communication using pre- and post-questionnaires that included multiple choice and open-ended questions. The Field cohort was surveyed using hardcopy questionnaires two days before the role-play (Jan 28) while the Lecture cohort was surveyed up to a week prior (Aug 7–13) using email and hardcopies. Both cohorts were surveyed with hardcopy post-questionnaires immediately after the exercise to ensure a high response rate as the study relies on paired data (pre- and post- results).

The questionnaires included several components: the self-reported communication competence instrument (SPCC), a perceptions of crisis communication instrument (PCC), demographics, and open-ended questions.

SPCC is a validated instrument (with a high internal consistency, Cronbach's alpha of 0.92) that measures communication confidence and is guided by the earlier works of McCroskey (e.g., McCroskey et al. 1977; McCroskey 1982). McCroskey and McCroskey (1988) investigated communication competence through self-reported evaluation of one's ability to communicate (i.e., communication confidence). The SPCC instrument considers several dimensions of communication: communication contexts [public, meeting, group, and dyad (or pair; one-on-one)] and receivers of the communication (strangers, acquaintances, and friends). While this measure (and others like it) is not a true characterisation of actual communication competency, it has been used in the discipline to measure gains (i.e., testing of communication competency before and after an intervention) (Fortney et al. 2001) and researchers indicate it is a good predictor of actual communication competence (McCroskey and McCroskey 1988).

The PCC survey (Table 2) was built and piloted for this study. We composed the statements with support from risk communication literature (see Sect. 1), expert views on volcanic crisis communication, and our practices with teaching science communication. The attitudes and beliefs covered by the survey are not exhaustive, but we feel that it covers the common best practices and appropriate behaviours when communicating science during crisis. Further research on the instrument will allow us to refine the statements and to incorporate all the important aspects of science communication. This survey was checked for content validity, but not examined with interview techniques (e.g., Adams and Wieman 2010). The questionnaire also included demographic information and open-ended questions that were designed to gather feedback about the student experience and science communication.

2.3 Data Analysis

The SPCC consists of 12 statements (McCroskey and McCroskey 1988) asking the participant to rate their perceived ability to communicate in different situations and contexts (on a 0–100 scale). The higher the total score, the higher the participant's confidence. We changed the phrasing from "competent" to "ability" and used a 5-point scale in our version (very strong ability, strong ability, average ability, poor ability, very poor ability). We felt this phrase change would be more comprehensible to our students. For further information on the design and scoring of the instrument please see the publication noted above.

The PCC instrument is composed of 17 5-point Likert statements (Table 2). Experts were surveyed in a small, convenience sample (n = 7) made of volcanology, emergency management and geology faculty at the authors' institution to assess expert opinion or 'the right answer'. The responses to the statements can be collapsed to agree, neutral and disagree, to reduce effects of participants preferring less or more conservative use of agreement/disagreement. The student responses can then be assessed as being in agreement or disagreement with the experts (Adams et al. 2006). Neutral responses are not weighted in the calculation.

SPCC and PCC survey results were analysed using the open source PAST statistics programme (Hammer 2015) to determine potential differences or associations with variables within the dataset. SPCC data are treated as interval and groups (i.e., subpopulations) within the dataset were compared using t-tests and one-way ANOVAs. The individual students' % agreement scores are interval data and so typical parametric tests were carried out, however the individual statement data (i.e., all students' responses for one statement) are ordinal data [agree (1), neutral (0) and disagree (−1)] and so were treated with non-parametric tests.

Reponses to an open-ended question (Table 3) in the questionnaire were transcribed and coded using qualitative software (ATLAS.ti, Friese and Ringmayr 2011) by the first author. We used

Table 2 PCC survey results for all students

	Statements	Pre-scores		Post-scores	
		Frequencies	%A	Frequencies	%A
Skills	1. To be a successful scientist, I need to be an effective communicator. (Expert answer: Agree)	A(33), N(2), D(1)	92	A(38), N(0), D(1)	97
	16. To be an effective scientist, I need to practice my communication skills. (Agree)	A(37), N(1), D(1)	95	A(38), N(1), D(0)	97
Scientists	2. Using scientific jargon [discipline-specific words/phrases] makes me sound more professional when communicating with other geologists. (Agree)	A(27), N(5), D(7)	69	A(32), N(3), D(4)	82
	4. I think that using scientific jargon is better for explaining science to geologists. (Agree)	A(29), N(4), D(6)	74	A(32), N(4), D(3)	82
	11. Using numbers, drawings and probabilities is a good method of communicating scientific principles to other scientists. (Agree)	A(35), N(2), D(1)	92	A(37), N(2), D(0)	95
	13. When talking to my science colleagues, it is best to assume that they know nothing about my topic. (Neutral)	NS	NS	NS	NS
	14. When communicating science to other scientists, I think it is best to behave objectively, without emotion or feelings. (Agree)	A(9), N(13), D(17)	23	A(17), N(10), D(12)	44*
Public	3. Using scientific jargon makes me sound more professional when communicating with non-geologists. (Disagree)	A(19), N(10), D(10)	26	A(16), N(8), D(15)	38
	5. I think that using scientific jargon is better for explaining science to non-geologists. (Disagree)	A(0), N(3), D(36)	92	A(0), N(6), D(33)	85
	6. I feel that the public is better left in the dark about the scientific details of a natural hazard event. (Disagree)	A(0), N(3), D(36)	92	A(1), N(5), D(33)	85
	7. I feel that the public is better left in the dark about the level of uncertainty that scientists have about their data, during a natural hazards event. (Disagree)	A(2), N(8), D(29)	74	A(6), N(10), D(23)	59*
	8. I think that the public does not need to understand why volcanoes erupt. (Disagree)	A(0), N(3), D(36)	92	A(0), N(2), D(37)	95
	9. I think that social media (e.g., facebook) is an effective method of communication during a natural hazard event. (Agree)	A(26), N(6), D(6)	68	A(27), N(9), D(3)	69
	10. When a non-scientist expresses an incorrect statement to the media, I believe that scientists have a responsibility to correct this statement. (Agree)	A(36), N(1), D(0)	97	A(34), N(2), D(3)	87
	12. Using numbers, drawings and probabilities is a good method of communicating scientific principles to non-scientists. (Agree)	A(23), N(8), D(8)	59	A(26), N(9), D(4)	67
	15. When communicating science to the public, I think it is best to behave objectively, without emotion or feelings. (Disagree)	A(8), N(15), D(15)	39	A(8), N(9), D(21)	55
	17. When trying to explain a complicated topic, I think an analogy (i.e., a relatable example) can be an effective way to communicate. (Agree)	A(39), N(0), D(0)	100	A(37), N(1), D(1)	95

*Wilcoxon signed rank test for different medians in paired pre- and post-survey results where p≤0.05.

(N=39). No significant differences were found between the Field and Lecture cohorts. The number of agree (A), neutral (N) and disagree (D) responses are shown as well as the overall % agreement with experts (%A) responses for each statement. Overall, most statements show positive changes, few show negative changes (shaded green rows). Two statements were shown to be statistically different from pre- to post-survey (* symbol). These differences were calculated using the Wilcoxon signed rank test for different medians in paired pre- and post-survey results where p (equals less than symbol) 0.05

content analysis that is defined as the process of using systematic and verifiable means of summarising qualitative data (Cohen et al. 2007). In the first pass of the responses, the researcher identified different units for analysis (individual and separate items). Codes were initially taken as verbatim quotes, to denote, as much as possible, the student's meaning. In a second pass, the results were viewed in a network (i.e., a map that shows all the responses and allows the user to group similar phrases). The items were grouped and categorised together (i.e., units of data into meaningful clusters; Lincoln and Guba 1985), where like statements could be assigned to code families. The code families were constructed around the act of communication: the *knowledge*, *skills*, and *attitudes*, needed for *actions* (i.e., strategies) to create an *appearance* to lead to successful *outcomes* when communicating. The data were reviewed in a third pass to refine and check for redundancy within and between the code families. 42 student surveys were evaluated, but the question allowed students to respond to as many items as they wanted. Therefore, frequencies of mentions do not represent individual student responses.

Table 3 Results from a post-survey (n = 42): Students' perceptions of science communication best practices

Question (open-ended): List the most important 'best practices' (or good methods) of communication that scientists should use when talking with the public.
(Categories are capitalised and bolded; "representative student quote" (n of items); added or altered words are in { })

Knowledge and skills (7)
"Communicate frequently" (2)
"Have a format"[a] (1)
"Know the {correct} information and facts" (1)
"{Have} general public speaking abilities" (1)
"{Understand} the topic" (1)
"Refer to the experts" (1)

Attitudes and framing (11)
Be sensitive to the public's concerns: "Be sensitive when correcting false statements" (4)
"Be respectful" (3)
"Be polite" (1)
Be honest about the situation: "Be straight up and honest" (2)
"Put a positive spin on things" (1)

Behaviour (21)
Show emotions, as appropriate: "Show some emotion" (6)
Don't show emotions: "Not getting emotional" (1)
Engage with the audience: "Put the audience in the scene" (6)
Use appropriate body language: "Use good posture" (6)
Dress and behave professionally: "Be professional"(2)

Appearance (20)
Don't appear condescending or patronising: "Not being patronising" (7)
Appear confident: "Sound like you know what you're talking about" (5)
Appear approachable and relatable: "{Be} down to earth" (5)
Appear calm: "Be calm" (2)
Appear authoritative: "{speak} with authority" (1)

Outcomes (5)
Don't increase panic or the public's concerns: "Share concerns without increasing panic or public concern" (4)
"Make the public feel safe" (1)

Strategies (134)
Speech quality (13)
"Speak slowly" (3)
"Speak clearly" or "Be clear"[a] (9)
Repeat the information: "repetition" (1)

Jargon (35)
Explain or define jargon, which is used: "Use some jargon, but explain it" (13)
Use jargon appropriately: "Using appropriate jargon" (9)
Don't use jargon: "Not use jargon" (7)
"Avoid {using} jargon" (6)
Minimise use of jargon: "Minimise technical jargon" (5)

Language and figures (35)
"Use analogies" (13)
Use simple terminology: "Keep things simple" (11)
"{Use} simple explanations" (4)
"{Use} numbers" and "statistics" (4)
"Use examples of everyday things" (3)

Information quantity and specificity (14)
Be concise: "Speaking concisely" (8)
Be specific and precise: "{Keep things} precise" (3)
"Not going into too much detail…" (2)
"Give as much information as possible" (1)

Transparency and uncertainty (6)
Explain "what is known and what is not" (3)
"Explain what science can tell us and it's limitations" (2)
"Don't make statements that are not certain" (2)
"Back up your observations with data" (1)

Content (9)
Careful wording to avoid panic and fear: "Be careful when using words that might 'incite' fear" (4)
Explain what is happening: "Explain what we know" (2)
Explain why things are happening: "To convey the "why" of the situation" (2)
"Consider facts, not opinions" (1)

Use of visual aids (22)
Diagrams (7), maps (5), figures (3), graphs (2), pie charts (1), media (1), charts (1), graphics (1), and drawings (1)

[a]"Speak clearly' and 'be clear' could be two different aspects, but are presented here together

3   Results

3.1   Improvement of Students' Communication Confidence

Figures 1 and 2 show changes in students' self-reported competence (i.e., confidence; SPCC) with communication. In both pre- and post-surveys, most students fell within the 'average' confidence zone, with several students reporting low or high confidence. Altogether, the students showed a positive mean change in confidence (Fig. 1b; Paired t-Test of pre and post-scores, $t = -2.07$, $p = 0.046$). An equal number of individuals showed positive and negative shifts in confidence after participating in the exercise, but the largest observable changes were positive (i.e., changes

Fig. 1 Students' self-reported communication competence before and after the Volcanic Hazards Simulation. **a** A plot showing pre-test versus post-test SPCC scores for individual students and the cohorts of which the means are not statistically different. **b** A table showing SPCC basic statistics. Overall, students showed positive and negative changes, but the positive changes were greater, on average

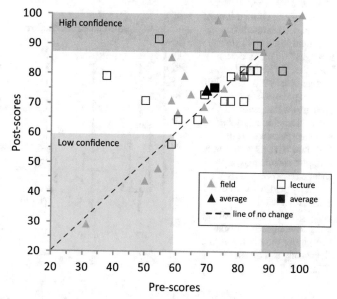

(a) SPCC Total Scores

(b) Basic Statistics

	Field ▲ (n = 19)	Lecture ☐ (n = 18)	All students (n = 37)
Pre-test*	69 ± 16	72 ± 15	71 ± 15
Post-test	74 ± 19	75 ± 9	74 ± 15
Change (Post-test – Pre-test)	5 ± 10	3 ± 15	4 ± 13
n of students showing positive change	8	7	15
average positive change	15 ± 11	16 ± 17	15 ± 13
n of no change	6	1	7
n of negative change	5	10	15
average negative change	-4 ± 2	-5 ± 4	-5 ± 3

* SPCC is scored out of a possible 100. The higher the score, the more competent the speaker feels. The average values are the mean plus or minus the standard deviation.

of >10 points: 8 positive compared to 2 negative). Three 'Low' confidence students showed large positive changes (21, 27, and 42 points). There were no statistically significant differences between the changes achieved by the different cohorts (Unpaired t-Test for same means; $t = 0.37$, $p = 0.71$), but the Field cohort did have lower pre-test scores (average of 69 ± 16). Figure 2a shows the changes for all of the students within each SPCC category (Speaking in public, meetings, groups, or pairs; with strangers, acquaintances, or friends). Overall, the mean changes for the public (5 ± 15) and stranger (7 ± 15) categories were the highest.

We examined the SPCC results for demographic associations with the pre-test scores and changes (Table 1; gender, age, nationality, degree programme, and year of degree programme) as well as curriculum factors [cohort, assigned roles (i.e., data-focussed vs. communication

Fig. 2 a Box and whisker plots of the average change within different dimensions of the SPCC instrument (i.e., communication contexts and receivers) for all students. Note that the highest average change is shown in the public speaking and stranger dimensions that are both emphasised through public speaking tasks within the Volcanic Hazards Simulation. **b** A plot showing the overall change (pre- and post SPCC) sorted by students who did and did not explicitly participate in public speaking tasks. A comparison of the two groups did not result in a statistically significant difference

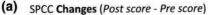

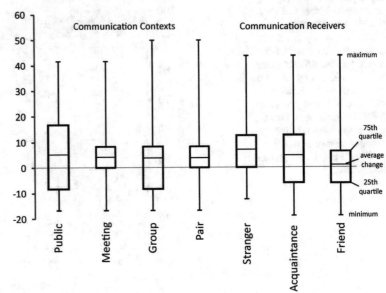

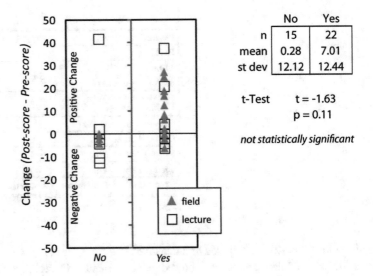

task-focussed) and teams (emergency management or geoscience)]. An interesting relationship surfaced between changes and the pre-test scores and direct participation in the public speaking tasks. Plotting the change scores (post-score minus pre-score) versus pre-test scores showed an inverse relationship (Pearson's product-moment correlation coefficient r = −0.46; p = 0.004); students with lower pre-test scores achieved the

highest changes, and those with the higher pre-test scores achieved the most negative changes. Additionally, we found that students with the greatest individual change in confidence (Fig. 1) participated in the public speaking tasks (i.e., press conferences and media interview) (Fig. 2b; "yes" to participating in public speaking tasks 7.01; "no": 0.28) although the difference was not statistically significant (t = −1.63, p = 0.11).

We would like to explore this affect in the future, with more students and better control over who participates and who does not in the public speaking tasks.

3.2 Improvement of Student Perceptions of Volcanic Crisis Communication

Figure 3 and Table 2 shows the results from the pre- and post-survey (PCC) that measured students' perceptions of communicating during a volcanic crisis. On average, the students' reported statistically significant positive changes (i.e., agreeing with experts) in perceptions (Fig. 3a, b; Paired t-Test, $t = -2.07$; $p = 0.046$) but individual students displayed both increases and decreases in agreement with experts. More students showed positive (17) or no changes (16) than negative shifts in perceptions (7) after participating in the role-play with the largest observable changes being positive (changes of >10 points; 7 positive, 4 negative).

The analysis of the pre-test scores revealed no significant statistical relationships for curriculum factors and most demographic factors. However, we did find that there was a significant difference

Fig. 3 Students' perceptions of volcanic crisis communication before and after the Volcanic Hazards Simulation. **a** A plot showing pre-test versus post-test PCC scores for individual students and the cohorts. There was no statistical difference between changes within the different cohorts (Paired t-Test to test for same means; $t = 0.07$, $p = 0.95$). **b** A table showing basic statistics of the perceptions survey. Overall, students showed positive and negative changes, but there were more students who exhibited positive changes rather than negative

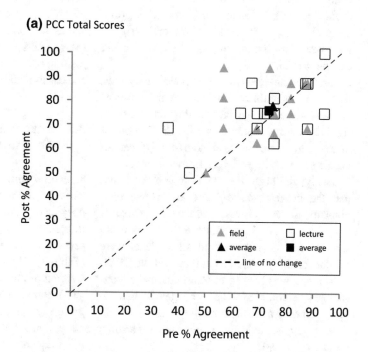

(a) PCC Total Scores

(b) Basic Statistics

	Field ▲ (n = 21)	Lecture □ (n = 18)	All students (n = 39)
Pre-test	75 ± 11	73 ± 15	74 ± 13
Post-test	77 ± 11	76 ± 11	77 ± 11
Change (Post-test – Pre-test)	3 ± 12	3 ± 12	3 ± 12
n of students showing positive change	7	10	17
average positive change	16 ± 13	10 ± 10	12 ± 11
n of no change	7	5	16
n of negative change	4	3	7
average negative change	-10 ± 6	-17 ± 4	-13 ± 6

in pre-test perceptions between students who were in the 300-level, versus the 400-level of their university degree programmes (mean score of 78 and 69%, respectfully; Unpaired t-test for equal means t = 2.18 and p = 0.04).

The changes achieved by students (post-test minus pre-test %) were also examined for curriculum and demographic factors. The cohort, participation in public speaking tasks, year and type of degree programme did not differ. Factors that did differ were: gender (male mean change = 6.7, female = −3.04), age (older students (>23 years of age) mean change = 7.15, younger students = −0.02), nationality (NZ students mean change = 4.30, US students = −1.89), assigned team (Geoscience group mean change = 8.56, EM = −2.3), and assigned role-type (data monitoring-focussed roles mean change = 9.83, communications-focussed = −0.71). However, these results should be considered with caution as none of these change factors showed statistical significance and there is a high likelihood of interacting and mediating factors (e.g., we cannot isolate some of the variables from one another.).

Lastly, similar to the SPCC scores, we found that the pre-test scores show an inverse relationship to the changes achieved (Pearson's r = −0.63; p < 0.001). Additionally, as the mean changes for the cohorts and all students were similar for the perceptions survey and the SPCC instrument, we checked for correlations between changes in confidence and changes in perceptions, but only a weak correlation was found and it was not statistically significant (Pearson's r = 0.30, p = 0.07).

Table 2 illustrates the PCC results broken down by individual statements and grouped together by 'audience'. Changes between the statements within the field and lecture-based cohorts were not shown to be statistically different, and so the combined results are shown. Overall, most statements showed positive changes (i.e., improving the agreement with the experts) from pre to post-survey. In the pre-survey, some statements showed very high agreement with the experts (>90%, statements 1, 16, 5, 6, 8, 10, and 17, bolded). Statement 7 and 14 showed statistically

significant changes from pre to post-survey (Wilcoxon signed rank test, for ordinal data; agree = 1, neutral = 0, disagree = 1 with experts; paired data; p < 0.05). Overall, students had positive changes within the 'skills' and 'communication with other scientists' dimensions, but some negative changes on statements within the 'communication with the public' category. This was surprising, as we were specifically aiming to improve their perceptions of communication with the public. However, a closer look shows that several of the individual statement's negative shifts were from very high values of agreement with experts where the majority of students who agreed with experts shifted into the neutral category (i.e., were questioning their perception). It should be noted here as well, that when 100% of the students agree with experts it can cause a 'ceiling effect', where scores cannot go any higher and can limit the statistical analysis of these results.

3.3 Best Practices of Science Communication

A central aim of the role-play is to enhance students' communication best practices. In a post-survey, students were asked to "list the most important 'best practices' of communication that scientists should use when talking to the public" (Table 3). No significant differences were discovered of the item frequencies between the field and lecture cohorts, and so the results from both groups are presented as a whole. Students views are comprehensive (covering many aspects), but the frequency of items shows a focus on the strategies of communication (134 mentions; e.g., use of jargon, use of analogies, use of visual aids) rather than on how the speaker appears (20), their behaviour (21) and the outcomes of the communication (5). There were a couple of examples of potentially divergent responses within a couple of the categories. For example, in the appearance category, students report that it is important to appear approachable and relatable (5) but another reports that it is important to appear authoritative. Another important example is that students felt it was appropriate to show emotions (6) but another student stated not to. The jargon category was quite popular, and students mentioned a range of recommended approaches including "not using jargon whatsoever", to using it "appropriately".

4 Discussion

4.1 Improvement in Students' Communication Confidence

The overall statistically significant positive changes on the SPCC results (Fig. 1) indicate that the role-play was effective in improving students' communication confidence. Figure 2 showed that the public speaking and stranger (i.e., speaking with strangers rather than someone you know) dimensions were the most positively affected and this result aligns with the learning goals of the role-play (i.e., to improve students' crisis communication skills). Positive changes achieved by students were substantial, however, there were equal numbers of students with small negative changes, and some with no change. This indicates that the role-play may be effective in improving student confidence for some more than others. Changes likely occur when students re-evaluate their abilities based on the performances during the role-play (of themselves and others) and either increase or decrease their confidence in communicating. Research has indicated that self-reported competency (i.e., confidence) is diminished when there are some peers who are compulsive communicators (i.e., dominant and frequent talkers), meaning less frequent speakers may not assess their merit as highly in comparison to their classmates (Fortney et al. 2001). Though we did not survey for compulsive speakers, all cohorts included some frequent and dominant speakers and that aspect could have potentially negatively influenced some student's appraisals of their own abilities. To reduce peer comparison effects, some scholars suggest encouraging students to focus on one's own progress (i.e., self-comparison), rather than comparing their performance with others, thereby reducing social comparison effects (e.g., Luk et al. 2000).

The change in scores can also potentially be attributed to (positive or negative) feedback provided by instructors and peers during the role-play. Feedback (i.e., self-, peer- and instructor feedback) is vital for communication improvement (e.g. Maguire et al. 1996; Maguire and Pitceathly 2002) and it is likely that some of the participants received more meaningful feedback (i.e., explicit guidance on how to improve and what to consider) during the simulation than others. Additionally, some students may shy away from perceived criticism which could result in negative self-appraisals.

It is worth noting that the SPCC scale and other communication instruments (e.g., PRCA-24; McCroskey et al. 1985) were designed and typically used to record longer interventions (over semesters rather than after one, multi-hour event). Some students in this study reported changes of $\sim$2–5% shift, while changes in competency from an entire semester of communication class (e.g., Rubin et al. 1997) resulted in similar magnitude of change. We propose that even small changes may be influential in a student's communication confidence over time and that the role-play has been shown here to have similar affects when compared to longer treatments.

Based on the divergent change results, we checked to see what factors may be influencing the individual student's experiences in different ways. A plot of the change scores versus pre-test scores revealed an inverse relationship (Pearson's product-moment correlation coefficient $r = -0.46$; $p = 0.004$) where students with lower pre-test scores achieved the highest changes, and the higher pre-test scores achieved the most negative changes. This indicates that this exercise is particularly effective at improving student confidence for those with mild communication apprehension. This relationship also indicates that our higher confidence students are becoming less confident. This may be due to a lack of accurate 'benchmarks' for effective competence, in that students with less academic maturity/experience may be overestimating their ability to communicate, and when confronted with a challenging exercise, may have a more realistic assessment of their abilities when compared to other students.

There were no notable differences in demographics (age, year of study, gender, nationality, etc.) in contrast to prior communication research that reports that males tend to have higher confidence in transferable skills and communication than females (Lundeberg et al. 1994; Whittle and Eaton 2001; Donovan and MacIntyre 2004) and that people from different cultures and nationalities are more confident with public speaking than others (Lundeberg et al. 2000). We did not observe these attributes in our study population, however the total sample size was small (n = 37) and these factors may only become apparent with larger groups.

There was no statistical difference in changes between the two cohorts, and for the different roles and teams. This indicates that regardless of the learning environment, the extent of the intervention, or the assigned roles and team (i.e., the specific tasks) the affect was equal on students' confidence. However, as noted in Fig. 2, students who directly participated in the public speaking tasks (i.e., press conferences and media TV interviews) showed more positive changes but this may be due to self-selection (i.e., students who volunteered to speak for the team may be less public-speaking averse than those that passed on the opportunity).

In the future, we may use a more equitable and structured approach to participation in the public speaking tasks (i.e., where all roles are noted and 'called on' by the facilitators or team leaders to speak), but presently we did not want to force students to participate. This approach may encourage students to overcome their perceived aversion to public speaking and improve their confidence. It should be noted that the treatment was not set up to specifically control for students participating in the public speaking tasks and future research will explore this variable further.

4.2 Student Perceptions of Best Practice in Volcanic Crisis Communication

Two datasets were considered to explore students' perceptions of crisis communication best practice: the PCC instrument (Table 2 and Fig. 3) and an open-ended question (Table 3). Overall, the PCC results students showed positive perception changes (i.e., increases in percent agreement with experts; Fig. 3) and more individual positive changes than negative changes, with some students achieving large shifts of >10 points. This indicates that the role-play was effective in enhancing students' perceptions (becoming more expert-like).

The data shows that the 300-level students had higher pre-test scores than 400-level students. This is separate from nationality, age, and cohort (which showed no differences) indicating that there is an element of academic maturity/ experience that is having an effect on their initial perceptions. It is not possible at this stage to differentiate specific reasons why these levels of students had different pre-test scores and will explore it further in our future work.

Overall, several factors (curriculum and demographic) may be impacting the amount of changes in student perceptions: gender, age, nationality, assigned team and role-type though these differences are not statistically significant and we did not observe (i.e., noted during observations of the role-play) distinguishing affects during the role-play. However, given the likelihood that these factors may be interacting, and that mediating variables (such as group socio-dynamics) might be present, causal inferences are difficult to make. A larger sample and more controlled design could plan for these factors.

However, the changes in perceptions associated with assigned role and team could potentially be due to group dynamics. The exercise is challenging with complex social dynamics within the teams and between. The Geoscience team (predominantly data-focussed students) had higher changes than the EM team. This is surprising, as these students are more concerned with data analysis and interpretation than the other team, because this group focuses on receiving science advice and prioritising and communicating impacts of the volcanic crisis. However, the perceptions survey is focussed on the *communication of science*, and not specifically on advice and actions for the public. It is likely that the Geoscience teams discussed the nuances of science communication at a deeper level than the EM team. This is important consideration when considering your evaluation of these exercises (Does your measure/instrument suit one context over another?).

Results from Table 3 show that students illustrated a comprehensive view of the strategies that you should employ when communicating science, but focussed more on the mechanics of communicating (i.e., the How To's). This indicates that our participants understand that there are many things to consider when communicating, appreciating the complexity of the task. The responses are all consistent with up to date approaches in rhetorical communication in instructional communication texts (e.g., McCroskey 2006). The frequency of mentions that focuses on the mechanics of science communication is not surprising, given their level of academic maturity and previous experiences (i.e., learning the initial skills, before moving on towards more sophisticated elements of the trade). The lesser but somewhat divergent responses (i.e., 'appear authoritative' vs. 'appear relatable') is additional evidence for students valuing different approaches to best practice. The undergraduate teaching community should be assured that students need to walk before they can run, and acknowledging where they are in their communication training can help them to understand where they should aspire to be (i.e., considering more situational aspects of communication). The volcanology community can benefit from this finding in that it may be important to acknowledge that practitioners may also hold divergent views on what is best practice, and that organisations would benefit from discussing the merits of specific approaches in specific circumstances. The risk and crisis communication community has much research for almost each individual statements in Table 2 specific areas [e.g., topics like uncertainty (e.g., Hudson-Doyle et al. 2011) and the importance of building and establishing trust through communication (e.g., Haynes et al. 2007)] and applying a one-dimensional approach to crisis communication is not advised.

It should be noted, that this perceptions survey is a pilot version and it has not yet been rigorously validated. Current research on a new version indicates that some of the statements may be asking about more than one concept (e.g., "Using *numbers, drawings* and *probabilities* is a good method of communicating scientific principles to other scientists"). New results from experts indicate that they may confuse some statements in terms of what is intended by the approach, versus its' effectiveness. Meaning that some strategies or perceptions may be valid in theory, but may not be helpful in practice (e.g., disclosing all of your results to show transparency, versus disclosing only the most important results to create a coherent message to the public). These ideas are somewhat opposed and in conflict with one another, causing a tension for the communicator to overcome. Additionally, our list of perception statements is not exhaustive. There is such a diversity and complexity to communicating during crisis and that is evident in the student responses in Table 3. However, it becomes difficult to capture this complexity in a series of closed statements. Further research into student and expert perceptions through interviewing techniques will allow us to characterise risk and crisis communication best practice.

Further work will validate our measure of communication perceptions (i.e., further refine the instrument and comprehensively define crisis communication best practice with the help of experts and practitioners), and focus on assessment of all of the above dimensions to ascertain the relationship *between* factors that lead to successful communication performance. If we know pedagogical factors influences a student's ability to learn about crisis communication, then we can provide practical suggestions to improve the teaching of communication in the classroom. We would also like to investigate risk and crisis communication in alternate natural hazards scenarios (e.g., earthquakes, Dohaney et al. 2016 and hydroelectric dam failure) to help students diversify their approaches to risk and crisis communication. Additionally, we would like to develop volcanic scenarios over longer mock time frames (e.g., following a community engagement initiative as it progresses through stages of learning about volcanic risk) to help students understand that risk communication occurs through all stages of the 4 R's and cultivating relationships with communities provides the foundation for making crisis communication possible.

4.3 Implications for the Teaching of Volcanic Crisis Communication and Future Work

In this final section, we would like to share with the community some lessons learned from our use of training exercises and teaching about communication, as well as outline our future research into the measurement of communication performance.

The use of training exercises is not uncommon in the emergency management sector, however, it is less used in formal education settings because of the significant time investment that goes into building an authentic scenario, organising a robust curriculum plan, and evaluating and testing whether it is effective. We believe an evidence-based approach to the building and testing of such curricula should include specialists in education and communication research. A partnership among these professionals allows content experts (i.e., volcanologists and emergency managers) to learn about pedagogy of training exercises and the art of evaluating such complex learning activities. Input from communication researchers can further enhance the inclusion of specific communication contexts and tasks, as well as help to guide instructors and students in delving deeper into how messages are constructed and received by diverse audiences. In our case, previous research into the design of this exercise (Dohaney et al. 2015) meant that we could move away from the intricate task of 'tweaking' our exercise and look at the impact that it has had on our students' abilities to communicate. Such alliances create powerful and engaging learning experiences that create memorable and lasting influence on student's ongoing career development.

The results discussed above illustrate that the Volcanic Hazards Simulation has influenced our student's perceptions and confidence with communicating during a mock volcanic crisis. But, does this translate to transferable communication skills moving forward? What we do know is that often knowledge and awareness of best practice (i.e., 'expert-like' perceptions) is the first step towards utilising these communication behaviours and strategies (e.g., McCroskey 2006). And what about communication confidence? Do our high confidence students actually communicate more effectively? Recent research by Kruger and Dunning (2009) suggests that overconfidence and ignorance are not a good thing, however, students with high confidence paired with expert-like perceptions of crisis communication best practice have the tools at their disposal, we hope that as they move forward in their careers they can continue to practice and improve, ultimately leading to better crisis communication practitioners (should they choose to follow that career path).

5 Conclusion

Our study set out to examine whether an authentic volcanic crisis role-play could improve students' communication confidence and their perceptions of science communication. In the role-play, students challenged themselves and moved outside of their 'academic comfort zone' when required to rapidly synthesize new information and communicate the information to differing stakeholders and in different formats. On average, our results indicate that the role-play does improve both confidence and perceptions for our students. In particular, this exercise is most effective for students who have low confidence and low perceptions of communicating science. Students with improved and high confidence in their abilities are more likely to engage in communication experiences (McCroskey et al. 1977), which leads to further improvement, so even a small number of positive shifts in confidence are a success.

However, some students showed both positive and negative changes in confidence and perceptions. Negative appraisals of confidence may be due to peer comparison effects and negative perceptions shifts may be due to shifting from agreeing with experts to neutral responses (i.e., questioning their current perceptions). In future work, we will try and minimise negative experiences and increase the positive experiences for all students. There were no significant differences with regard to students' confidence and perceptions between the cohorts indicating that despite slightly different intervention (one more extended than the other) students achieved positive changes. This indicates that role-play as a standalone part of an instructor's curriculum is flexible enough to accommodate different schedules while still reaching its outcomes.

Results from the open-ended question show that our students illustrated a comprehensive range of views on the best practices of science communication, but focussed primarily on the mechanics of delivery, which is unsurprising as most students are still relatively inexperienced and are continually developing these skills. New scenarios for earthquakes will be tested to improve on our findings. This approach to learning skills through authentic challenges builds confidence and resilience in undergraduate students who are likely to become a part of the geologic and emergency management community.

References

Adams WK, Wieman CE (2010) Development and validation of instruments to measure learning of expert-like thinking. Int J Sci Educ 33(9):1–19. doi:10.1080/09500693.2010.512369

Adams WK, Perkins K, Podolefsky N, Dubson M, Finkelstein N, Wieman CE (2006) New instrument for measuring student beliefs about physics and learning physics: The Colorado learning attitudes about science survey. Phys Rev Spec Top Phys Educ Res 2(1):1–14. doi:10.1103/PhysRevSTPER.2.010101

Alexander D (2007) Making research on geological hazards relevant to stakeholders' needs. Quat Int 171:186–192

Bales RF (1976) Interaction process analysis: a method for the study of small groups. University of Chicago Press, Cambridge, Chicago

Barclay EJ, Haynes K, Mitchell T, Solana C, Teeuw R, Darnell A, Crosweller HS, Cole P, Pyle D, Lowe C, Fearnley C, Kelman I (2008) Framing volcanic risk communication within disaster risk reduction: finding ways for the social and physical sciences to work together. Geol Soc Lond Spec Pub 305(1):163–177. doi:10.1144/SP305.14

Barclay EJ, Renshaw CE, Taylor HA, Bilge AR (2011) Improving decision making skill using an online volcanic crisis simulation: impact of data presentation format. J Geosci Educ 59(2):85. doi:10.5408/1.3543933

Blake M (1987) Role play and inset. J Furth High Educ 11(3):109–119

Borodzicz E, van Haperen K (2002) Individual and group learning in crisis simulations. J Conting Crisis Manag 10(3):139–147. doi:10.1111/1468-5973.00190

Brown JS, Collins A, Duguid P (1989) Situated cognition and the culture of learning. Educ Res 18(1):32–42. doi:10.3102/0013189X018001032

Bryner V (2012) Science communication vodcast assignment: 7 cs of science communication Youtube user: science with Tom. http://www.youtube.com/watch?v=grhrLT8tfjg. Accessed 15 July 2016

Burns T, O'Connor D, Stocklmayer S (2003) Science communication: a contemporary definition. Publ Underst Sci 12(2):183–202

Cohen L, Manion L, Morrison K (2007) Research methods in education, 6th edn. Routledge, Taylor and Francis Group, New York. doi:10.1111/j.1467-8527.2007.00388_4.x

Cole JW (1978) Andesites of the Tongariro Volcanic Centre, North Island, New Zealand. J Volcanol Geotherm Res 3:121–153

Crane A, Livesey S (2003) Are you talking to me? Stakeholder communication and the risks and rewards of dialogue. In: Andriof J, Waddock S, Rahman S, Husted B (eds) Unfolding stakeholder thinking 2: relationships, communication, reporting and performance. Greenleaf, Sheffield, pp 39–52

DeNeve KM, Heppner MJ (1997) Role play simulations: the assessment of an active learning technique and comparisons with traditional lectures. Innov High Educ 21(3):231–246. doi:10.1007/BF01243718

Department of Homeland Security (2008) National incident management system, p 156. https://www.fema.gov/pdf/emergency/nims/NIMS_core.pdf. Accessed 15 July 2015

Dohaney J (2013) Educational theory and practice for skill development in the geosciences. Dissertation, University of Canterbury

Dohaney J, Brogt E, Kennedy B, Wilson TM, Fitzgerald R (2014) The volcanic hazards simulation. VHUB: collaborative volcano research and risk mitigation. https://vhub.org/resources/3395. Accessed 15 July 2016

Dohaney J, Brogt E, Kennedy B, Wilson TM, Lindsay JM (2015) Training in crisis communication and volcanic eruption forecasting: design and evaluation of an authentic role-play simulation. J Appl Volcanol 4(1):1–26. doi:10.1186/s13617-015-0030-1

Dohaney J, Brogt E, Wilson TM, Hudson-Doyle E, Kennedy B, Lindsay J, Bradley B, Johnston DM, Gravley D (2016) Improving science communication through scenario-based role-plays. National project fund research report, Ako Aotearoa, National Centre for Tertiary Teaching Excellence, Wellington, New Zealand

Donovan LA, MacIntyre PD (2004) Age and sex differences in willingness to communicate: communication apprehension and self-perceived competence. Commun Res Rep 21(4):420–427

Ellis K (1995) Apprehension, self-perceived competency, and teacher immediacy in the laboratory-supported public speaking course: trends and relationships. Commun Educ 44:64–77

Engleberg IN, Ward SM, Disbrow LM, Katt JA, Myers SA, O'Keefe P (2016) The development of a set of core communication competencies for introductory communication courses. Commun Educ 4523 (July):1–18. doi:10.1080/03634523.2016.1159316

Errington EP (1997) Role-play. Higher Education Research and Development Society of Australasia Incorporated, Australia

Errington EP (2011) Mission possible: using near-world scenarios to prepare graduates for the professions. Int J Teach Learn High Educ 23(1):84–91

Fischhoff B (1995) Risk perception and communication unplugged: twenty years of process. Risk Anal 15 (2):137–145. doi:10.1111/j.1539-6924.1995.tb00308.x

Fisher A (1991) Risk communication challenges. Risk Anal 11(2):173–179

Fortney SD, Johnson DI, Long KM (2001) The impact of compulsive communicators on the self-perceived competence of classroom peers: an investigation and test of instructional strategies. Commun Educ 50 (4):357–373. doi:10.1080/03634520109379261

Friese S, Ringmayr TG (2011) ATLAS.ti6 User Guide. http://atlasti.com/wp-content/uploads/2014/05/atlasti_v6_manual.pdf. Accessed 15 July 2016

Glik DC (2007) Risk communication for public health emergencies. Annu Rev Publ Health 28(1):33–54. doi:10.1146/annurev.publhealth.28.021406.144123

Grunig JE, Repper FC (1992) Strategic management, publics and issues. In: Grunig JE (ed) Excellence in public relations and communication management. Lawrence Erlbaum Associates, Hillsdale, NJ, pp 117–157

Hales TC, Cashman KV (2008) Simulating social and political influences on hazard analysis through a classroom role playing exercise. J Geosci Educ 56 (1):54–60

Hammer O (2015) PAST: Paleontological Statistics instruction manual, Version 3. Natural History Museum, University of Oslo, Norway, p 243. http://folk.uio.no/ohammer/past/

Harpp KS, Sweeney WJ (2002) Simulating a volcanic crisis in the classroom. J Geosci Educ 50(4):410–418

Haynes K, Barclay J, Pidgeon N (2007) The issue of trust and its influence on risk communication during a volcanic crisis. Bull Volcanol 70(5):605–621

Haynes K, Barclay J, Pidgeon N (2008) Whose reality counts? Factors affecting the perception of volcanic risk. J Volcanol Geotherm Res 172(3):259–272

Heath C (2000) The Technical and non-technical skills needed by Canadian-based mining companies. J Geosci Educ 48(1):5–18

Herrington A, Herrington J (2006) Authentic learning environments in higher education. IGI Global, Hershey, PA, USA. doi:10.4018/978-1-59140-594-8

Herrington J, Reeves TC, Oliver R (2014) Authentic learning environments. In: Spector JM, Merrill MD, Elen J, Bishop MJ (ed) Handbook of research on educational communications and technology. Springer, New York, NY, pp 401–412. doi:10.1007/978-1-4614-3185-5

Hobden BJ, Houghton BF, Davidson JP, Weaver SD (1999) Small and short-lived magma batches at composite volcanoes: time windows at Tongariro volcano. N Z J Geol Soc 156(5):865–868. doi:10.1144/gsjgs.156.5.0865

Hudson-Doyle EE, Johnston DM, McClure J, Paton D (2011) The communication of uncertain scientific advice during natural hazard events. N Z J Psychol 40 (4):39–50

IAVCEI Task Group on Crisis Protocols (2016) Toward IAVCEI guidelines on the roles and responsibilities of scientists involved in volcanic hazard evaluation, risk mitigation, and crisis response. Bull Volcanol 78:31

Jones F, Ko K, Caulkins J, Tompkins D, Harris S (2010) Survey of hiring practices in geoscience industries. CWSEI Report, University of British Columbia, p 18

Kreps GL, Query JL (1990) Health communication and interpersonal competence. Speech communication essays to commemorate the 75th anniversary of the Speech Communication Association, pp 293–323

Kruger J, Dunning D (2009) Unskilled and unaware of it: how difficulties in recognizing one's own incompetence lead to inflated self-assessments. Psychology 1:30–46. Retrieved from http://www.ncbi.nlm.nih.gov/pubmed/10626367

Lincoln YS, Guba EG (1985) Naturalistic inquiry. Sage Publications

Livingstone I (1999) Role-playing planning public inquiries. J Geogr High Educ 23(1):63–76. doi:10.1080/03098269985605

Lombardi MM (2007) Authentic learning for the 21st century: an overview. Educause Learning Initiative

Luk CL, Wan WWN, Lai JCL (2000) Consistency in the choice of social referent. Psychol Rep 86:925–934

Lunce LM (2006) Simulations: bringing the benefits of situated learning to the traditional classroom. J Appl Educ Technol 3(1):37–45

Lundeberg MA, Fox PW, Puncochar J (1994) Highly confident but wrong: gender differences and similarities in confidence judgments. J Educ Psychol 86 (1):114–121

Lundeberg MA, Fox PW, Brown AC, Elbedour S (2000) Cultural influences on confidence: country and gender.

J Educ Psychol 92(1):152–159. doi:10.1037//0022-0663.92.1.152

Maddrell AMC (1994) A scheme for the effective use of role plays for an emancipatory geography. J Geogr High Educ 18(2):155–163

Maguire P, Pitceathly C (2002) Key communication skills and how to acquire them. Br Med J 325(September):697–700

Maguire P, Booth K, Elliott C, Jones B (1996) Helping health professionals involved in cancer care acquire key interviewing skills: the impact of workshops. Eur J Cancer 32A(9):1486–1489

McCroskey JC (1982) Communication competence and performance: a research and pedagogical perspective. Commun Educ 31(January):102–109

McCroskey J (2006) An Introduction to rhetorical communication: a western rhetorical perspective, 9th edn. In: Bowers K, Wheel B (eds). Englewood Cliffs, New Jersey, p 333

McCroskey JC, McCroskey LL (1988) Self-report as an approach to measuring communication competence. Commun Res Rep 5(2):108–113

McCroskey JC, Daly JA, Richmond VP, Falcione RL (1977) Studies of the relationship between communication apprehension and self-esteem. Hum Commun Res 3(3):269–277

McCroskey JC, Beatty MJ, Kearney P, Plax TG (1985) The content validity of the PRCA-24 as a measure of communication apprehension across communication contexts. Commun Q 33(3):165–173. doi:10.1080/01463378509369595

Miller MD (1987) The relationship of communication reticence and negative expectations. Commun Educ 36:228–235

Miller S (2008) So where's the theory? On the relationship between science communication practice and research. In: Cheng D, Claessens M, Gascoigne T, Metcalfe J, Schiele B, Shi S (eds) Communicating science in social contexts. Springer Science and Business Media, Netherlands, pp 275–287

Miller S, Fahy D (2009) Can science communication workshops train scientists for reflexive public engagement? The ESConet experience. Sci Commun 31(1):116–126. doi:10.1177/1075547009339048

Ministry of Civil Defence and Emergency Management (2009) The guide to the national civil defence emergency management plan, 3rd edn, 266 pp

Morgan MG, Fischhoff B, Bostrom A, Atman CJ (2002) Risk communication: a mental models approach, 1st edn. Cambridge University Press, New York, NY

Morreale S, Pearson J (2008) Why communication education is important: the centrality of the discipline in the 21st Century. Commun Educ 57(2):224–240

Pielke RA (2007) The honest broker: making sense of science in policy and politics, 1st edn. Cambridge University Press, Cambridge, UK. http://doi.org/10.1017/CBO9780511818110

Reynolds B, Seeger M (2005) Crisis and emergency risk communication as an integrative model. J Health Commun 10(1):43–55

Reynolds BJ, Shenhar G (2016) Crisis and emergency risk. In: Koenig and Schultz's Disaster medicine: comprehensive principles and practices, p 390

Richmond VP, McCroskey JC, McCroskey LL (1989) An investigation of self-perceived communication competence and personality orientations. Commun Res Rep 6(1):28–36

Roth WM, Roychoudhury A (1993) The development of science process skills in authentic contexts. J Res Sci Teach 30(2):127–152. doi:10.1002/tea.3660300203

Rovins JE, Wilson TM, Hayes J, Jensen SJ, Dohaney J, Mitchell J, Johnston DM, Davies A (2015) Risk assessment handbook. GNS science miscellaneous series. 84:67 pp

Rubin RB, Morreale SP (1996) Setting expectations for speech communication and listening. New Dir High Educ 96:19–29

Rubin RB, Rubin AM, Jordan FF (1997) Effects of instruction on communication apprehension and communication competence. Commun Educ 46(2):104–114. doi:10.1080/03634529709379080

Shearer R, Davidhizar R (2003) Using role play to develop cultural competence. J Nurs Educ 42(6):273–276. http://www.ncbi.nlm.nih.gov/pubmed/12814218

Trench B, Miller S (2012) Policies and practices in supporting scientists' public communication through training. Sci Publ Policy 39(6):722–731

United Nations International Strategy for Disaster Risk Reduction (2015) Sendai framework for disaster risk reduction 2015–2030. Geneva, Switzerland

Van Ments M (1999) The effective use of role-play: practical techniques for improving learning, Kogan Page Publishers

Weingart P, Engels A, Pansegrau P (2000) Risks of communication: discourses on climate change in science, politics, and the mass media. Publ Underst Sci 9(3):304. doi:10.1088/0963-6625/9/3/304

Whittle SR, Eaton DGM (2001) Attitudes towards transferable skills in medical undergraduates. Med Educ 35(2):148–153. doi:10.1046/j.1365-2923.2001.00773.x

Wolfe EW, Hoblitt RP (1996) Overview of eruptions. In: Fire and mud: eruptions and lahars of Mount Pinatubo, Philippines, vol 18, pp 3–20. doi:10.2307/3673980

Learning to Be Practical: A Guided Learning Approach to Transform Student Community Resilience When Faced with Natural Hazard Threats

Justin Sharpe

Abstract

This chapter seeks to explore how creative use of educational resources can challenge students to take responsibility for their own preparedness and safety in response to natural hazard risks. A brief context for the need for learning-focused rather than education-focused curriculum is explored before the England and Wales context is brought into focus. Two methods for transforming learning around the theme of natural hazard risk and response are offered: A film project in which students produce films by and for children and youth and a 'Go-Bag' project in which students take on a practical task of making up a real emergency bag. By guiding student learning, but allowing it to develop inside a reasonable framework, student learning was not only deeper on a cognitive level, but also allowed students to understand their own roles and responsibilities in responding to natural hazard threats. The combination of both is explored through the use of an online questionnaire (n = 176) in which the impact of the learning on students and their families are explored. The classroom and individual learning activities' impact on student efficacy are discussed alongside the results from the questionnaire. Findings included support for prior assumptions about the impact of school-based learning on the family with regard to disaster preparedness as well as deeper cognition regarding the risks and increased self-efficacy in students. The implications for these findings and their role in transforming learning to enhance community resilience that starts with the family are discussed with the door to future research nudged open.

J. Sharpe (✉)
Department of Geography, King's College London, London, UK
e-mail: justin.sharpe@kcl.ac.uk

1 Introduction

Education for disaster risk reduction and resilience was a central tenet of the Hygo Framework for Action which ran from 2005 to 2015 (HFA 2005),

Advs in Volcanology (2018) 715–731
https://doi.org/10.1007/11157_2017_1

and during which time, the classroom activities outlined in this chapter took place. The HFA was itself, partially drawn up in response to the number of losses and damages wreaked by disasters on human populations and the environment. A key tenet of the HFA involved the use of: 'knowledge, innovation and education to build a culture of safety'. It is argued here that this was based on an assumption that if individuals have the 'appropriate' knowledge and education, a culture of safety will follow. However in practice, information alone is insufficient to lead to action (e.g. Kolmuss and Agyeman 2002, Demos/Green Alliance 2003, Talbot et al. 2007) leading to a Value Action Gap (Blake 1999).

Consequently, there is a strong argument that learning is an essential starting point for allowing resilient individuals and communities to develop. The term 'learning' is purposefully used here. The definition of learning taken, is one in which the outcome leads to: 'a change in knowledge, beliefs, behaviours or attitudes that is the result of experience' (e.g. Ambrose 2010). Moreover, learning is also understood to be both experiential and socially constructed with the power to become transformative when learners are challenged and given the expertise, knowledge and time for reflection (Sharpe 2016).

Further research by Sharpe and Kelman (2011), notes that in relatively affluent places such as England, doubt about the effectiveness of particular measures, lack of belief in one's personal ability to carry them out and apathy in considering a situation deemed to be unlikely to occur, provide a number of obstacles to successful education and learning regarding DRR, something supported by other researchers (e.g. Lindell and Perry 2000; Mulilis and Duval 1995; Ronan et al. 2001).

Overcoming such barriers can be helped by drawing on a body of pedagogical research that indicates that experiential learning has the potential for motivating people to action (Dewey 1938; Kolb 1984). Much of the impetus for experiential learning has come as a reaction to overly didactic, teacher-controlled learning. Assumptions about learning from experience are that experience provides the foundation of, and

the stimulus for learning; that learning is a socially and culturally constructed process influenced by the socio-emotional context in which it occurs; and that learners actively construct their own learning experience (Boud et al. 1993). Those assumptions have been challenged through critical analysis of experiential learning theory (e.g. Fenwick 2001) but the theory has then been extended and reworked to try to overcome those challenges, such as for management learning (Kayes 2002).

Consequently, activities that allow for the experiential and explorative learning, might be more meaningful at transforming the views, attitudes and behaviours of students. Furthermore, other researchers examining the impact of hazards education on children in formal education have assumed that there is some form of transmission to family and the wider community (Gordon et al. 1999; Peek 2008). However, this has not been widely tested or reported while, moreover, parents might not be engaged in the learning in the same way as their child. This has the potential to undermine school learning and is why it has been argued that it is important that students of school age have access to: "information that helps a child understand what he or she can do relatively independently to be prepared physically and emotionally" (Ronan and Johnston 2003, p. 1011). As a result, the curriculum by the author of this chapter was developed with the intension of addressing these issues, while making the learning fun, engaging and useful for students, who are shrewd observers and assessors of what is relevant to them.

This chapter therefore outlines how the curriculum was developed and employed to improve communication of hazards in general, including those of a volcanic nature to both children and by extension their family and the wider community. It does this by detailing how the curriculum was designed and delivered in a manner that allowed for guided learning that permitted students to explore key issues of risk, response and preparedness for themselves and their family. Furthermore, it provides evidence of how students developed self-efficacy (belief in their ability to carry out a task) through a curriculum that married guided learning with practical tasks. It is argued that this engaged students in disaster learning, while enabling them to become agents of change in their community, starting with their family as suggested by other researchers (e.g., Gordon et al. 1999; Peek 2008), but not widely tested. The curriculum development and subsequent research methodologies outlined below have strong roots in prior pedagogies for tackling disaster risk reduction education. These have been expanded upon through the development and use of novel and engaging teaching and learning. This includes practical approaches to embedding critical reflection as a way of strengthening understanding of hazard risks (including volcanic ones) and appropriate responses to them. It is contended that this creative and guided learning, opens up avenues of understanding that can make crisis communication more readily accepted and acted upon by the wider community, building outwards from school based education to engage students and their families.

2 Disasters and Geography—The UK Context

The study of school Geography in the United Kingdom is governed by the National Curriculum, which sets out what must be studied, but not the format or style of delivery. At key stage three (11–14 years of age) students are regularly assessed on their progress via a means of testing, essay writing, or project based homework in order to ascertain a National Curriculum Level mark. However, although this tests knowledge, understanding and retention of facts/figures, it does not allow for all learning styles to be accommodated, nor does it test for learning or a change in attitude or behaviour, which is often the goal of disaster risk reduction and resilience educators. (e.g. HFA 2015).

Disasters themselves can be included in curriculum, but often take a top-down approach of response or governmental planning approach to the mitigation of tectonic hazards in particular while not allowing for the roles and responsibilities of the individual to come to the fore (e.g.

Sharpe and Kelman 2011). The following section outlines how a curriculum was planned and executed in response to this gap.

3 The School Context

The activities outlined here, were developed for students in early high school (aged 11–14; known in the UK as key stage 3) for inclusion in mainstream natural hazards focused geography lessons. The teaching activities, learning outcomes and survey were all carried out at Beal High school, Ilford, Essex, while the author was working as a teacher of geography.

The school's ethnic mix at the time was approximately 65% Southern Asian origin (Indian, Pakistani and Bengali), with the children often the first generation born in the UK; 25% White (including those from across Europe) and 8% Afro-Caribbean and approximately 2% of Chinese origin. Many of the children of South Asian origin visited areas of the world where tectonic and hydro-meteorological hazards were both more frequent and of a higher magnitude and yet were unaware that these dangers might present themselves when visiting grandparents and other family in their parents' country of origin. This was a primary concern when setting out the curriculum.

The lessons were planned in order to fit within the National Curriculum while encouraging exploration through provided mediums such as web sites, video and mapping exercises, experimentation through the creation of student films and learning by doing, which allowed for real emergency 'Go-Bags' to be produced as part of a homework task. The students followed a unit of the geography curriculum (approximately eight to nine weeks) called 'Dangerous Geography' in which the geography of hazards, response and preparedness were to be explored.[1] This was

---

[1]A complete description of the Unit 1 'Hazards and How to Prepare for Them' module, lesson plans and resources can be found on the authors website, created to help teachers and educators find creative and effective ways of enaging children of all ages in hazards education. This can be viewed at: http://www.edu4drr.org/page/curriculum-1.

primarily carried out in year seven (11–12 year olds) while future learning in year eight (12–13 year olds) addressed the scientific aspects of plate tectonics, earthquakes and volcanoes. This allows for learning to be built slowly rather than attempting to cram everything together in one unit that students become tired of.

4 The Approach

Being a classroom practitioner allowed me to take an action research approach while forming and developing curriculum and pedagogies as part of my ongoing professional practice. This afforded a unique perspective, in so much as allowing the author's pedagogical approach to being a reflective practitioner with regards to curriculum development and delivery, to sit alongside that of a researcher whose role is more analytical in observing and evaluating the various nuances of learning and their role in changing cognition and behaviour towards disaster risk.

The research decision to collate data was not made when the curriculum was initially in development, but when the curriculum had been running for two years. This allowed for a period of 'bedding in' of the curriculum with small tweaks made in collaboration with departmental colleagues in order to have a greater impact on knowledge and understanding that might unlock deeper learning. The decision to initiate a questionnaire was taken to inform our teaching practice by capturing the impact of the overall curriculum in terms of preparedness and efficacy of response of children and their families. The study was a pilot study, in that although the curriculum had been taught through several times, there was only one set of data collected (n = 175) at the time as the author moved on the following year. The use of data collection via an online survey was agreed with departmental staff as well as senior management in the school. It was agreed that an online format questionnaire would allow students to retain anonymity, permitting them to be honest with their feedback. The questionnaire therefore served a dual role: on the one hand it allowed teaching professionals to evaluate the impact of the curriculum in order to continue to develop, it, while on the other hand it allowed for research into wider issues of impact on family preparedness that have been assumed, but not more widely investigated.

The author was the primary researcher. There were no other researchers in a formal sense, but discoveries, information, lessons learned were shared informally by the other geography teachers (there were three others) within department meetings, but also on a day-to-day basis during break or lunch times. This is an important and neglected area of learning—the contribution of informal feedback with colleagues that allow for curricular to be honed, improved and extended to allow for deeper level learning. This shows the importance of informal socially constructed learning that takes place among reflective practitioners in the teaching of geography.

In particular, this might be viewed as a type of as social learning, of which there are many different descriptions, but the one used here is what McCarthy et al. (2011) summarised as an on-going, adaptive process of knowledge creation that is scaled-up from individuals though social interactions fostered by critical reflection and the synthesis of a variety of knowledge types that result in changes to social structures (e.g. organizational mandates, policies, social norms). In this case, it would mean the tweaking of the curriculum to help students engage with it more readily or to adapt certain parts for students with different competencies of literacy and numeracy or other cognitive needs.

The curricula was created and shared by myself with the other teachers in the department and advice was offered and support given when teachers were challenged by any of the activities. Each teacher followed the curricular as lesson plans and resources were provided and talked about prior to delivery. The questionnaire was deemed to be a reasonable way of allowing the department to ascertain the impact on cognition over a longer time, rather than just after a test or other formalised evaluation. The questions were shared with the other teachers and they were given access to the questionnaire prior to it going live and students asked to complete it three weeks following the last lesson in the 'Dangerous Geography' unit of work.

5 Methods

There were 246 students that this particular study assessed as part of an evaluation of the effectiveness of curriculum based methods for learning how to acknowledge, respond to and prepare for a range of hazards, including those of a volcanic nature. During the filming stage all students participated in the storyboarding, filming and editing of films. Students were split into groups of five or six to achieve this task and while some students were more assertive on the filming tasks, others were more so on the planning or editing phases. This allowed for students with stronger skillsets, talents and learning styles (Gardner 1993, 1999) to develop roles that they felt comfortable in while also challenging others to learn new skills and develop communication and team-working skills as part of the learning process.

Approximately 40 videos were produced in total, but only a small number (12) were added to the YouTube channel (www.youtube.com/edu4hazards) as these were deemed to be accurate in their advice and representation of the correct behaviours and actions.

All students (246 = total) were asked to complete a survey three weeks following the end of the unit of work, in order for them to have time for reflection as well as attempting to ascertain how much had been retained (not just mentally but physically in terms of emergency Go-Bags and their role in family life). 175 students out of 246 (71%) completed the online survey, which was completed at home or in the school library if internet access was not available at home. This was chosen for ease of answering the questions, as well as to limit bias by having a teacher nearby or looking over student shoulders when they carried out the survey. The questions were also asked at random, rather than in the same order each time so that if students worked together on laptops when at a friend's house they would find it harder to crib from their friend's answers.

6 The Lessons

A complete description of the Unit 1 'Hazards and How to Prepare for Them' module, lesson plans and resources can be found at: http://www.edu4drr.org/page/curriculum-1. However a brief outline is given below in order to provide context for this approach as a method for engaging and developing the minds of adolescents for learning to cope with the dangers of a variety of hazards (Fig. 1).

In lesson one, students identified a number of hazards from a word puzzle while the differences between hazards and disasters were explored, before examining a video about Hurricane Katrina in which students learned what was known before and what occurred afterwards in order to explore the impact of humans in the causation of disasters. Following a mapping exercise of disasters in the USA in lesson two, lesson three introduces the concept of how geography can save lives. Students examined the example of Tilly Smith, an English school girl who was the same age as them when she recognised the signs of a tsunami in Thailand and warned her parents who alerted hotel staff and evacuated the beach in 2004, saving many lives. Subsequently, time was given over to what an emergency Go-Bag might be, what it might contain and what could and should be included in their own one. This lesson and subsequent individual 'thinking and doing' tasks are laid out as a separate case study in this chapter.

The lessons following this allowed for student creativity and exploration to become part of the learning process as students storyboard, planned, filmed, edited and presented their films to the class. This took place over three one-hour lessons and the principal guidance was that students must take information from the edu4hazards.org website, created by Sharpe (2007), (see Fig. 2) which was researched and cross-checked with sources such as the International Federation of Red Cross websites and publication (IFRC 2013)

Fig. 1 Film still from student film about what to do in case of a volcanic eruption

and other factually reliable websites in order for the most current information regarding appropriate responses to a range of hazards to be included.

The learning material supporting the edu4-hazards website allows for a variety of learning strategies to be utilised to allow students to engage with the material in different ways (Sharpe and Kelman 2011). Furthermore the production of the films was to ascertain the level of knowledge and understanding about the hazards, risk and response, with students informed that films with misleading or inaccurate information could not be shared on the YouTube channel. It was explained that films would be evaluated by their peers for accuracy, creativity and usefulness to others.

Following these lessons a further set of six to eight one hour lessons examined flood causes, risk and response, and preparation, as this was considered to be most relevant to a local context, with the nearby River Roding having flooded in

the past with relative frequency (10–20 years). At the end of the entire *dangerous geography* unit the students were tested using an end of unit examination with overall grades moderated and decisions made about grade boundaries and how these might equate to National Curriculum levels.

7 Case Study 1: The Volcanic Hazard Films

Students were allowed to work with their friends and decided for themselves what hazards interested them most. Some demonstrated an interest in volcanic hazards as this appealed to them as individuals and volcanoes and their associated hazards are quite an abstract concept for students in the UK.

The first task was the production of a storyboard. This required inter-personal and intra-personal learning skills as well as a degree of creativity. This stage is also important for a

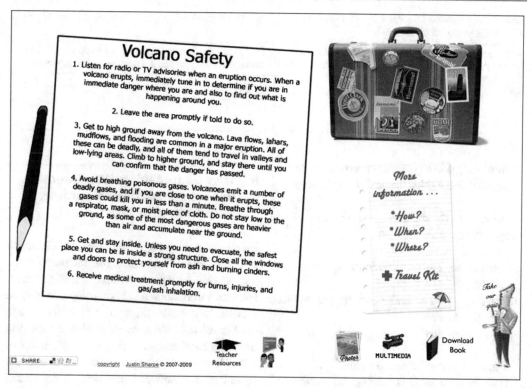

Fig. 2 Screengrab from edu4hazard.org website regarding volcano saftety

teacher to help students negotiate potential pitfalls as well as guiding students to stay focussed on the task at hand. Each location (around the school grounds) was identified by students as well as alternatives if the weather was inclement on the day of filming. Storyboards were viewed by the class teacher and students were allowed to complete outside class if they wished, but a photograph of each group's storyboard was also taken so that if a child was absent with illness for instance, a copy could be printed off for use by the group. Students were then allowed to go and film what they needed, following and using their storyboards which provided structure and allowed for an easier editing process as a result.

Following editing, student films were imported to the class computer and shared with the class. The group that had made the film answered questions about their film, as part of the assessment and evaluation phase of the lesson.

Students films were assessed (by the students, and the teacher) on:

- the basis of content knowledge (e.g., what they remembered about volcanic hazards). This was carried out by comparing the information about the volcanic hazard as an image file on the board as a means of cross referencing that the information given was accurate and well presented.

- the 'correct response' to natural hazards events. This was assessed in terms of how seriously protective action was carried out and whether students had made assertions that were incorrect, misleading or dangerous. Again students and occasionally the teacher highlighted these points as part of the 'two stars and a wish' (see Table 1) process of evaluation. This requires students to make two positive comments about what they had seen and one suggestion to improve it. This allows for reflection and framing of their thoughts in a constructive manner.

- how they communicated the potential dangers. For instance, the actions shown in the

Table 1 A pre-defined feedback matrix given to students to allow them to critically evaluate the success of disaster safety films produced by their peers

What examples from the film helped you understand how to respond to and prepare for the hazard?	What might the film-makers do to be more successful in communicating appropriate responses to the hazard?

film alongside any narration needed to match up well and exhibit creativity and realism.

In the films the students acted out many of the mitigation practices such as:

(a) getting to higher ground away from the volcano;
(b) covering mouths and noses to prevent ash inhalation, and
(c) evacuation tips.

These films can be found online.[2]

Following the making, editing and showing of the films, students provided feedback to each other using peer-evaluations from a pre-defined matrix (Table 1). The goals were to:

• critique each others as well as their own films;
• identify mistakes and to take away lessons learned.

Students were encouraged to write their own responses first, before agreeing and providing feedback in the 'two stars and a wish' format.

Students thought about how such films might be used in a real crisis by scientists and disaster managers to inform the public. This was brought out via questioning in the classroom by teachers and was used as an extension to the main learning in an effort to stretch the critical thinking skills of the more able students and is evidence of planning for students with various levels of academic ability.

8 Why Film-Making Was Deemed Important for Understanding Responding to Risk

Using video for pedagogical purposes, Goodyear and Steeples (1998), suggest that the use of video can provide a platform through which to articulate tacit information and knowledge that may be difficult to describe through text, while Hempe (1999), refers to the strength of video as a visual demonstration, dramatization or presentation of visual evidence, while making an emotional appeal to the viewer. Both were demonstrated through watching student-made films. This outcome was not discussed or shared with the students at the time, due to time constraints of the curriculum, but could certainly be developed further for cross-curricular days in schools, when more time is allowed for critical thinking and reflection.

A further reason for developing films that advised how to survive tectonic or hydro-meterological hazards was in response to the way that students used YouTube and Tumblr on their tablets and mobile phones for both fun, interaction with their peers and as sources of information. Conversations with students in class (by the researcher and classroom teacher) revealed that they saw these sites as trustworthy sources of information, much like previous generations may have gleaned information from documentaries on TV and prior to that, in the cinema. Furthermore, Wahlberg and Sjoberg (2000); Weingart and Pansegrau (2003), note that in post-industrial society the media have become highly influential and powerful in the communication of complex scientific, technological and political changes. However, media organisations

---

[2]YouTube (See: French version: https://www.youtube.com/watch?v=2jSLqlE0TWw; English version: https://www.youtube.com/watch?v=xljmjuVwJy8&list=UUarOrdsyYX7rTWV-9cKH3mw).

may also have political and financial motives for distorting or 'spinning' the scientific, technological or political changes or discoveries: e.g. climate change denial.

Consequently, films by children and youth, for children and youth allow students to take control of the process for themselves, with no other motive other than helping others. This is a yet unexplored realm of research within disasters and yet has been explored in other fields. However, Sharpe (2016) points to the work of Taylor (2002) who cites studies that go beyond an ego-centred motivations, such as: inclusive of spirituality, a transpersonal realm of development (Cochrane 1981; Hunter 1980; Scott 1991; Sveinunggaard 1993; Van Nostrand 1992), compassion for others (Courtenay et al. 1998; First and Way 1995; Gehrels 1984), and a new connectedness with others (Gehrels 1984; Laswell 1994; Weisberger 1995).

Furthermore, the role of the teacher in the transmission of curricula, can be a positive influence in development of critical thinking. A good teacher should allow for the provision of interesting counterpoints to certain arguments, while the inclusion of opposing views are given space with students given the time to explore, debate and decide for themselves. Having said this, the role of the curricula can be used to address any factual misconceptions (about natural hazards, the risk they pose and appropriate responses) in order to contribute to students '*schema*' of understanding (existing frameworks of ideas and concepts; e.g., Rumelhart and Norman 1978) of hazards and human response to disaster. If students have realistic expectations about the nature of natural hazards and the steps that can be taken towards preparing for them, it is more likely that they will do so in the future when they are adults, while also developing their sense of self-efficacy as adolescents.

These issues will be explored further in the discussion section, which follows the second case study in which emergency go-bags were created and the impact of the overall curricula and practical implications and learning was tested using an online survey.

9 Case Study Two: 'Go-Bags' and Post Curricula Survey

During the 'Go-Bag' lessons students were encouraged to talk about what might be useful to include in an emergency 'Go-Bag'. This took the form of a 'Think, Pair, Share' exercise in which time was given over to individual thinking before student ideas are shared with a neighbour and then with the class. The contents of two emergency go-bags held in the geography department were examined before students voted on which was better and why. This allowed for the development of critical thinking as well as reflection and re-purposing of thoughts in the light of new learning. A homework task was set out introducing the edu4hazards.org website and focusing on the emergency kit or 'Go-Bag' page. Students were informed that these items are 'ideal' for a Go-Bag, but they might think of others too.

A further caveat was introduced in order to be as inclusive as possible: students and their families were encouraged to keep costs low and not spend more than £5.00–£8.00 on their kits or 'Go-Bags'. Suggestions were shared with students about the likely places to find such items, including online sources. A week or two was allowed for the completion of homework tasks as it was understood that if resources were sourced on the internet using free delivery it might take longer. It is therefore important that practical considerations are thought out before setting such tasks and discussing these with colleagues prior to the lessons taking place allows for a consistent expectations across the year group.

Lesson time was given over to students bringing in and talking through what they had in their kits, with students asking questions and making suggestions for improvement. Again the method used was 'two stars and a wish' (see Table 1) so that the process is not about overtly criticising, but learning from and developing ideas further. In this way, students self-evaluate with guided learning facilitated by the teacher. This allows for a more equal student-teacher and student-student interaction facilitated by reducing any power relationships and hierarchy, in order to promote more of an exchange, with each party supporting and gaining from the other. Again critical thinking is fundamental, with learning process facilitated by the teacher, similar to the 'guided discovery' approach used in international development and advocated by Bruner (1961).

The questionnaire approach was used to ascertain the level of understanding students had developed as well as understanding how they did or did not engage with the material, the practical tasks and the impact on extending learning into the family home. While there was a small element of group activity and development of critical thinking in class, the 'Go-Bag' homework exercise was designed to allow students to explore how they might prepare themselves and the extent to which family members became involved in the process.

10   Data Collection and Analysis

10.1   Case Study One: Volcanic Hazard Films

Several types of data was collected in this part of the study, including observations of student learning and an analysis of their knowledge and understanding tested through the making of the volcanic hazard films. The following section describes what was carried out, including an analysis of the significance of the film-making process as a way of framing new learning. As students were responsible for their own films, they had to synthesise what they had learned, thinking though the most appropriate ways to communicate risk both visually and via the narrative. The significance of this and its effect on deeper learning, including the wider impact on community hazard resilience is also encompassed in the interpretation and discussion section that later in the chapter.

Observations/Teacher account of making and editing phases included the authors observations coupled with discussions with the other classroom teachers in the department. Again this was part formal and part informal especially during the chaos of teaching and learning how to edit

videos, which was a new skill for some teachers in the department. Being an observant and reflective practitioner was an effective methodology for analysing pedagogical approaches and learning outcomes. These reflections are also associated with transformational learning practices (Mezirow 1995) that might allow for teachers to make sense of what they are doing in order to learn from, test, reflect, test and learn as a continuation of their professional development.

Classroom observations during the 'premieres' of student-made videos was a purposefully discrete process that allowed students to explore their roles as responsible media, whose role was to inform and engage learners of a similar age by critiquing their own and others films. These self and peer evaluations and discussions of videos were useful for (re)-framing what messages were important, the role played by media in education and learning, as well as the need for accurate, independent and trusted content.

Consequently, discussions about each film became an important part of the evaluative process, as students realised and acknowledged their mistakes in a nurturing environment, in which learning from mistakes became integral to deeper learning outcomes. Discussions of negative implications included:

- 'What would happen if videos were put on YouTube with the wrong or exaggerated information?';
- 'What did you learn or understand about how you can protect yourself in the event of a volcanic hazard?';
- 'What shouldn't you do when treating a burn?'

The videos were evaluated by students using a 'Geography-Factor' (an 'X-Factor' format that provided a framework for them to think differently, allowing them to hold different perspectives and viewpoints as they analysed the videos), in order to hold their attention and to take on various roles for feeding back. This also challenged students to acknowledge their own 'automatic thoughts' (Mezirow 1995) and reactions when seeing something for the first time and allowed for more considered and reflected upon responses. Furthermore they learned to be analytical while also understanding the position of others.

Researcher analysis of videos was carried out in an open manner and shared with students as a way to summarise how an 'expert' might critique the videos and the legitimate reasons for doing so.

Formal assessment of the module was carried out in an end of unit examination and given an England and Wales National Curriculum Level mark. Scores were based on students overall knowledge and understanding, A higher weight being given to understanding borne out by an ability to explain, reason and argue the merits of enacting responsible behaviours and actions regarding hazard preparation, including those for volcanic hazards.

These written tests (including differentiated versions for students of a lower ability) were also used to assess overall knowledge and understanding four weeks after completing this initial part of the unit of work.[3]

10.2 Case Study Two: 'Go-Bags' and Post-curricula Survey

Three weeks after the completion of the scheme of work, which included the unit on flooding following the overall hazards lessons and activities, an online post-hazard survey was produced[4] and students asked to complete. These were completed by 175 students. The aim of the survey was to evaluate their level of engagement with the curriculum, the impact of practical tasks, such as making their own 'Go-Bag' and the extent to which the use of a practical task that required parental input, helped or impacted on discussions of safety and practical preparedness in the home. The most

salient findings from a research perspective are included below:

- 114 students (65%) reported talking to parents about what they had been learning about. This is significant as it shows a high level of engagement, but this was also not an accident, as the homework task actively encouraged students to talk to parents when putting a kit together.
- Consequently, 130 (74.3%) reported putting emergency kits('Go-Bags') together with their parents. Of these, 65 (37.1%) were helped by mothers, 19 (10.9%) by fathers, with 37 (21.1%) helped by both and the remaining 13 (7.4%) helped by a sibling in the family home.
- A further 109 students (65%) discussed what they had been learning in class which included hazards and preparedness generally.
- In terms of the utility of the 'Go-Bags' exercise, only 20 students (10.9%) claimed that they no longer had the bag or it was thrown away, which meant that just over 88% still had their kits and with 47% reporting that the kit was in a place easily accessible to all (e.g. by the front door, in a cupboard under the stairs etc.).

In terms of the curricula leading to better levels of preparedness, students were asked: *Do you think that your household is better prepared for emergencies as a result of what you have learned about and shared with you family?* In response to this, 96 students (54.9%) reported that there household was better prepared, with 27 (15.4%) saying no and 47 (26.9%) not sure. This was in fact very helpful. Because a large proportion of students were not sure, teachers asked why this was the case. Reasons that were fed back, included: being more prepared in some ways but not in others such as not having equipment in a 'Go-Bag' for the whole family; parents not taking the exercise seriously or believing that there might be an occasion to use the kit. This led to conversations about how a go-bag might be useful for holidays in places where there were risks of earthquakes, flooding

---

[3]See: http://www.edu4drr.org/page/curriculum-1; Unit 1; Lesson 7).

[4]See: http://www.esurveyspro.com/SummaryReport.aspx? SurveyId=37283&sel=0.

and volcanic eruptions. This neatly rounded the learning off, as this was also one of the first conversations about hazards and risk that had taken place at the beginning of the unit.

This engagement with the practical task of making an emergency kit/go-bag led to further discussion about what students were learning. This included (according to parents and students at parent-teacher conferences) talking about the video tasks and the reasons for actively preparing for disasters even in a country where disasters are often seen as part of the 'other' because they happen to other people, despite large scale flood events and even events such as heatwaves in 2003, which killed 2091 in England alone (Johnson et al. 2005).

Some students commented that their parents carried emergency kits for the car. A few noted that their parents already had a kit in their car and this led to classroom discussions surrounding the reasons for this (with reasons including that it came with the car or they had decided that it was a prudent precaution when children were born) which also allowed for students to understand how their parents looked after them in previously unseen or unheard ways.

10.3 Post-lesson Parent-Teacher Conferences

There is evidence that the curriculum reached beyond the classroom, as parents from different classes (around twenty across the year group) engaged in animated discussions with all of the teachers from the geography department at parent-teacher conferences about the 'Go Bags' homework exercise. Some reported as liking it for engaging 'their son/daughter in doing something practical', while others reported enjoying being involved with their son/daughter's school work. Some parents reported that *they* also engaged with the practical aspect of preparing kits as well as learning about what happens elsewhere.

Furthermore, some parents also talked about their own experiences of hazards, trauma, and preparedness in their country of origin, but these were not noted down at the time, for ethical reasons.

11 Interpretation/Discussion

Interpretation and discussion of the two case studies are carried out below in order to frame their relative importance to overall impact of the learning on both students and their families.

11.1 Case Study 1: Volcanic Hazard Films

Students responded well to the idea of making videos/films that were made by people their own age, for people their own age. Providing trusted sources of information was discussed as well as *why* these were trusted sources of information, such as use of the latest scientific information to aid student understanding. This allowed students to assimilate information for themselves as well as thinking how they might synthesise it for others. Different learning styles, in particular intrapersonal, interpersonal, artistic, linguistic and kinaesthetic (learning through movement or the physicality of undertaking a task) were accessed and utilised in order to help students learn while also providing them the opportunity to teach others. Playing roles or writing the dialogue required students to think about what they might do in the event of a volcanic crisis.[5] Again, this allows critical thinking and reflection skills to be developed as well as being able to show empathy for others by thinking about the implications of disasters on families in different parts of the world.

Furthermore, by understanding the effects of volcanic eruptions, it also allowed students to ask questions of why and how. These questions are provided as a sub-menu on the www. edu4hazards.org website[6] as well as the answers

---

[5]Student written, directed and edited video in French can be viewed at: https://www.youtube.com/watch?v=2jSLqlE0TWw&list=PL8785FE43DBE2F748&index=4.

[6]This website was created by the author in 2007 as a way of engaging children and youth in learning what simple practical measures could be effectively carried out to prepare for natural hazards. It's development and subsequent early use as an educational tool can be found in the UNISDR publication: Towards a Culture of Prevention: Disaster Risk Reduction Begins at School: Good Practices and Lessons Learnt' UNISDR (2007).

which are provided on web pages when students interact with the menu. This sets up a chain of self-discovery that is not directed by a teacher or educator telling students to look or explore; they do this for themselves! Students would ask questions based on what they had seen and this allowed the author to talk about this with them or to direct them to explore further, suggesting other websites they could look at too.

By involving students on their own terms, allowing them to explore their own interests (through the media and specific hazard-topic) they took charge of the learning and produced their own educational materials. This allowed for a high level of participation, both in the classroom and independently, as students were empowered to make their own choices about what to film, how to film it and the information that it should contain.

- Through active planning and participation, students gained a deeper understanding of why actions (such as preparing 'Go-Bags' specific to volcanic crises) are effective at reducing risk while taking steps to ensure their own safety following a natural hazards event.
- Students shared the 'difficulties' of planning, filming and editing, thus allowing for socially constructed learning to take place and enabling transformative learning through a process of critical reflection. Teachers observed that some groups worked well together immediately, while other's struggled to get the balance right, needing more time to negotiate personality conflicts for instance. This type of group dynamic is important to acknowledge and learn from.
- The film-making making, editing and critiquing allowed students to engage in an evaluative and reflective process. This allowed them to be rigorous with the evaluation of their own and others films, while learning about how to respond to volcanic eruptions, as they thought about and acted out how to successfully demonstrate the correct behaviours.

Central to this pedagogical approach was the empowerment of students as knowledge bearers to others and their reflection on their own learning process (i.e., metacognition; Sterling 2011).

11.2 Case Study Two: Go Bags and Post-lesson Survey

Of the surveys completed, a high proportion (75%) reported that they had collated their emergency kits with their parents. This actively involved parents in asking the reasons for such an exercise, with several parents (six or seven across the department) calling the department to ask for clarification or wanting to know more.

A large proportion of returned surveys (almost 55%) reported that they did feel more prepared *as a family* following the lessons, with their corresponding activities and individual projects (the films) and home learning (the 'Go-Bags'). This was an important finding as it supported earlier suggestions by Gordon et al. (1999) and Peek (2008), that students can be *agents of change* within the family with regard to responding to the hazard risk.

However, there were still almost 27% that were unsure, which initially suggested that perhaps not enough had been done to engage students. However, on further inquiry (asking the students), it was found that the barrier was the result of some parents dismissing the exercise as "not useful", because they didn't believe that a hazard event would precipitate the use of 'go-bags' and evacuations. Students facing such obstacles in their own families might be challenged by these views. However, this should not be seen as a barrier but an opportunity to engage in debate and learning across the family. However, if they can make up their own mind and carry out their own protective behaviours, these thoughts and actions may allow them to gain personal efficacy and slowly transform their own learning in order to be better prepared personally.

11.3 Case Study Two: Parent-Teacher Meetings

Conversations with parents at parent-teacher meetings allowed for further reflection as well as parents to show their interest or ask questions about the learning. The bullet points below summarise the key points from these meetings:

- Sharing lesson(s) with parents and siblings allowed for a wider form of transformative learning to be initiated within the local community as well as within the school and classroom.
- Students acted as important agents of change by discussing the film project and 'Go Bags' with their parents and other relatives.
- This 'bottom-up' approach to learning about volcanic hazards led to a wider discussion about hazards in their families.

12 Conclusions

The filming and 'Go-Bags' lessons incorporated practical skills which used authentic, practical tasks and peer evaluation that could play a role in transformative learning. In particular, having the spaces for relational learning and reflection may allow students to negotiate potential cognitive conflicts that occur when their new knowledge and understanding may be threatened by the views of others who are important influences in their life, such as the views of parents, the content of religious teachings or other socio-cultural influences. Mezirow (2000) refers to this process as *Reflective Dialogue* in which a consensus is negotiated with others regarding prior learning, assumptions and actions. However, Mezirow's transformative learning theory has focused on the development of the adult mind to coping with change brought about through experience of training, but it may well mean that this may occur earlier than previously thought and this study tends to support this premise.

Experiential learning, reflection and the act of playing out the threats from hazards provides a space to move beyond knowledge and towards action. By planning, scripting and then acting out and filming in films about volcanic hazards, students placed themselves in the role of reacting to the threat appropriately, while being in a safe, unthreatening environment. As mentioned earlier this allowed students to draw on a range of learning styles and skills while learning new ones, including developing empathy for others, critical thinking skills, intrapersonal skills (following instructions and thinking about and acting on them in the most appropriate way) as well as opportunity for creativity and more.

In terms of volcanic crises communication, the research presented here illustrated that children can and do act as agents of change for their families, supporting prior research that highlighted the importance of engaging children in learning about disasters (e,g., Gordon et al. 1999; Peek 2008). In particular, students reported a higher belief in the efficacy of the family to face future hazard events as the result of their learning, task completion and practical exercises. However, the use of the questionnaire also allowed for further reflection and for a probing of the reasons for uncertainty that still existed in some cases. This could then be discussed further and resolved to a certain extent. This is significant because it shows that deeper reflection could occur with guidance from the teacher as a way of working through problems and unexpected outcomes together.

Furthermore, reflection and evaluation (peer- and self-) are key building blocks to improving self-efficacy and new ways of thinking which Mezirow (2012), describes as allowing individuals to "...negotiate and act on our own purposes, values, feelings, and meanings rather than those we have uncritically assimilated from others—to gain greater control over our lives as socially responsible, clear thinking decision-makers" (2012, p. 76).

Finally, aside from these media, others such as puppetry, performance, and music are also excellent ways of getting across messages about risk, response and preparedness.[7] To this end, the author has also developed a comic strip to tackle hazard safety and preparedness for very young children which includes volcano safety alongside, earthquake, tsunami, lightning, flood, evacuations and family planning for such events. This has been used as a pilot project in Iran (Sharpe and Izadkhah 2014) before being produced as a book for children in Pakistan by UNESCO.[8] In particular, the research described by Sharpe and Izadkhah (2014) show that even at a young age, complex cognitive process were engaged in order for learners to take their new knowledge, place it within the context of their own experience and re-tell it to others.

This pattern of reflection, reasoning and testing is important for deeper learning, which may hold the key to truly resilient individuals and communities. The work outlined here supports this assertion while providing ideas and resources that might usefully be applied to engaging communities in thinking about their own risks and in response, developing their own efficacies in planning and preparing for them.

It is hoped that education and learning that use creative and engaging lessons, materials and resources will continue to challenge and engage children and youth in learning how to prepare for and respond to natural hazard threats and that this has the ability to translate into home safety when students engage their parents in discussion and more importantly, action through practical exercises such as the creation of emergency 'Go-Bags' for the home.

References

Ambrose SA, Bridges MW, DiPietro M, Lovett MC, Norman MK (2010) How learning works: seven research-based principles for smart teaching. Jossey-Bass: San Francisco

Blake J (1999) Overcoming the 'value-action gap' in environmental policy: tensions between national policy and local experience Local Environ 4(3)

Boud D, Cohen R, Walker D (1993) Understanding learning from experience. In: Boud D, Cohen R, Walker D (eds) Using experience for learning. SRHE & Open University Press, Buckingham, pp 1–17

Bruner J (1961) The act of discovery. Harvard Educational Review 31(1):21–32

Courtenay B, Merriam SB, Reeves PM (1998) The centrality of meaning-making in transformational learning: How HIV-positive adults make sense of their lives. Adult Educ Q 48:65–84

Demos/Green Alliance (2003) Carrots, sticks and sermons: influencing public behaviour for environmental goals. A Demos/Green Alliance report produced for Defra: http://www.demos.co.uk/files/CarrotsSticks-Sermons.pdf

Dewey J (1938) Experience and education. Macmillan, New York

Fenwick TJ (2001) Experiential learning: a theoretical critique from five perspectives. Information Series No. 385

First JA, Way WL (1995) Parent education outcomes: onsights into transformative learning. Fam Relat 44:104–109

Foley G (ed) (1995) Understanding adult education and training, 2nd edn. Allen & Unwin, Sydney, pp 225–239

Gardner H (1993) Multiple intelligences: the theory in practice. Basic Books, New York, NY

Gardner H (1999) Intelligence reframed: multiple intelligences for the 21st Century. Basic Books, New York

Gehrels C (1984) The school principal as adult learner. Unpublished doctoral dissertation, University of Toronto, Toronto

Goodyear P, Steeples C (1998) Creating shareable representations of practice. Adv Learn Technol J (ALT-J) 6(3):16–23

Gordon NS, Farberow NL, Maida CA (1999) Children and disasters. Burnner/Mazel, Philadelphia

Hunter EK (1980) Perspective transformation in health practices: a study in adult learning and fundamental life change. Unpublished doctoral dissertation, University of California, Los Angeles, CA

International Federation of Red Cross and Red Crescent Societies (2013) Public awareness and public education for disaster risk reduction: key messages. Accessed online at http://www.ifrc.org/PageFiles/103320/Key-messages-for-Public-awareness-guide-EN.pdf. Accessed May 2016

Johnson H, Kovaats RS, McGregor GR et al (2005) The impact of the 2003 heatwave on mortality and hospital admissions in England. Health Statistics Q 25:6–11

Kayes DC (2002) Experiential learning and its critics: preserving the role of experience in management learning and education. Acad of Manage Learn Educ 1(2):137–149

Kolb DA (1984) Experiential learning: experience as the source of learning and development. Prentice Hall, Englewood Cliffs

---

[7]The resources and information to use these activities can be found on the edu4drr website (www.edu4drr.org).

[8]See: http://unesco.org.pk/education/documents/2014/publications/Learning_about_disasters.pdf.

Kolmuss A, Agyeman J (2002) Mind the gap: why do people act environmentally and what are the barriers to pro-environmental behavior? Environ Educ Res 8 (3):239–260. doi:10.1080/13504620220145401

Laswell TD (1994) Adult learning in the aftermath of job loss: exploring the transformative potential. In: Hymans M, Armstrong J, Anderson E (eds) 35th annual adult education research conference proceedings. University of Tennessee, Knoxville, pp 229–234 (ERIC Document Reproduction Service No. ED 381 616)

McCarthy DDP, Crandall DD, Whitelaw GS, General Z, Tsuji LJS (2011) A critical systems approach to social learning: building adaptive capacity in social, ecological, epistemological (SEE) systems. Ecol Soc 16(3)

Mezirow J (1995) Transformative theory of adult learning. In: Welton M (ed) In defense of the lifeworld. State University of New York Press, Albany

Mezirow J (2000) Learning to think like an adult: core concepts of transformative theory. In: Mezirow J et al (eds) Learning as transformation. Jossey-Bass, San Francisco, pp 3–34

Mezirow J (2012) Learning to think like an adult: core concepts of transformation theory. In: Taylor EW, Cranton P (eds) The handbook of transformative learning. Jossey-Bass, San Francisco, pp 73–95

Mitchell JT (2009) Hazards education and academic standards in the Southeast United States. Int Res Geogr Environ Educ 18(2):134–148

Moss R (1983) Video, the educational challenge. Croom Helm Ltd, London and Canberra

Peek L (2008) Children and disasters. Child Youth Environ 18. Retrieved Aug 2011, from http://www.colorado.edu/journals/cye/181/index.htm

Ronan KR, Johnston DM (2003) Hazards education for youth: a quasi-experimental investigation. Risk Anal 23:1009–1020

Rumelhart D, Norman D (1978) Accretion, tuning and restructuring: three modes of learning. In: Cotton JW, Klatzky R (eds) Semantic factors in cognition. Erlbaum, Hillsdale

Scott SM (1991) Personal transformation through participation in social action: a case study of the leaders in the Lincoln Alliance. Unpublished doctoral dissertation, University of Nebraska, Lincoln, Nebraska

Sharpe J (2007) Education for hazards website. www.edu4hazards.org Accessed May 2015

Sharpe J (2016) Understanding and unlocking transformative learning as a method for enabling behaviour change for adaptation and resilience to disaster threats. Int J Disaster Risk Reduction 17:213–219

Sharpe J, Izadkhah YO (2014) Use of comic strips in teaching earthquakes to kindergarten children. Disaster Prevent Manage 23(2):138–156

Sharpe J, Kelman I (2011) Improving the disaster-related component of secondary school geography education in England. Int Res Geogr Environ Educ 20(4):327–343

Sterling S (2011) Transformative learning and sustainability: sketching the conceptual ground. Learn Teach High Educ 5(11):17–33

Sveinunggaard K (1993) Transformative learning in adulthood: a socio-contextual perspective, In Flannery D (ed) 35th annual adult education research conference proceedings, Penn State University, University Park, PA, pp 275–280

Van Nostrand JA (1992) The process of perspective transformation: Instrumental development and testing in smokers and ex-smokers. Unpublished doctoral dissertation, Texas Women's University, Denton

Wahlberg AAF, Sjoberg L (2000) Risk perception and the media. J Risk Res 3(1):31–50

Weingart P, Pansegrau P (2003) Perception and representation of science in literature and fiction film. Public Understand Sci 12:227–228

Weisberger RD (1995) Adult male learners in a community college setting: possibilities of transformation. Ed. D. dissertation, University of Massachusetts

Role of Social Media and Networking in Volcanic Crises and Communication

Sally S.K. Sennert, Erik W. Klemetti and Deanne K. Bird

Abstract

The growth of social media as a primary and often preferred news source has contributed to the rapid dissemination of information about volcanic eruptions and potential volcanic crises as an eruption begins. Information about volcanic activity comes from a variety of sources: news organisations, emergency management personnel, individuals (both public and official), and volcano monitoring agencies. Once posted, this information is easily shared, increasing the reach to a much broader population than the original audience. The onset and popularity of social media as a vehicle for eruption information dissemination has presented many benefits as well as challenges, and points towards a need for a more unified system for information. This includes volcano observatories using social media as an official channel to distribute activity statements, forecasts, and predictions on social media, in addition to the archiving of images and other information. This chapter looks at two examples of projects that collect/disseminate information regarding volcanic crises and eruptive activity utilizing social media sources. Based on those examples, recommendations are made to volcano observatories in relation to the use of social media as a two-way communication tool. These recommendations include using social media as a two-way dialogue to communicate and receive information directly from the public and other sources, stating

S.S.K. Sennert (✉)
Department of Mineral Sciences,
U.S. Geological Survey, Smithsonian Institution,
Washington, DC, USA
e-mail: kuhns@si.edu

E.W. Klemetti
Department of Geosciences, Denison University,
Granville, OH, USA

D.K. Bird
Risk Frontiers, Department of Environmental
Sciences, Macquarie University, Sydney, Australia

Advs in Volcanology (2018) 733–743
https://doi.org/10.1007/11157_2015_13
Published Online: 26 March 2017

that the social media account is from an official source, and posting types of information that the public are seeking such as images, videos, and figures.

1 Introduction

Public interest in volcanic eruptions is high, especially on social media. Social media is simply social conversation through web-based platforms, encompassing a variety of examples including social networking platforms (e.g. Twitter, Facebook), media sharing platforms (e.g. YouTube, Instagram), crowdsourcing platforms (e.g. Ushahidi, Crisismappers), and others such as blogs, discussion forums, chat rooms, wikis, and apps. Social media not only allows critical information to be disseminated but also publicly discussed; it provides an opportunity for concerned individuals to communicate related issues, express attitudes and share knowledge and experiences of events through stories, photographs, and video. As such, officials use social media to collect data from people affected by volcanic eruptions (e.g. Carranza Tresold 2013).

However, information shared on social media is often not clearly organized and can cause confusion; it may not always be accurate and the sources can be difficult to identify. Nevertheless, as a two-way communication tool, social media give official agencies and the public an opportunity to dispel rumours circulating via social media and other media sources (Bird et al. 2012; Bruns et al. 2012). Where traditional news sources often sensationalize a volcanic event and provide very little follow-through, social media can be used as a means for community connection and support after an event, or for preparedness in times of quiescence.

Despite the obvious benefits, many official agencies, including most volcano observatories, lack guidelines on how to use these cost-effective communication tools to their full potential (Disaster Management SofS Working Group 2014; Dufty 2015) and have not integrated social media into their core communications strategy. There is a wealth of information disseminated every minute via social media platforms. However, many agencies are unsure about the best methods for capitalizing on this valuable resource. Users want the most up-to-date and accurate information about on-going volcanic eruptions and they want it rapidly and to be easily accessible. Volcano observatories that do use social media, especially to post official information, images, and data from eruptions tend to have a large group of followers (in excess of 10,000), which includes bloggers and traditional media.

We provide examples of two projects that require collecting and then disseminating information regarding volcanic crises and eruptive activity: the Weekly Volcanic Activity Report, a joint product of the Smithsonian Institution and the U.S. Geological Survey (USGS) (http://www.volcano.si.edu/reports_weekly.cfm) written by one the authors (Sennert); and, 'Eruptions', a blog written by another of the authors (Klemetti) (www.wired.com/category/eruptions/). We do not offer a comprehensive look at the use of social media in all aspects of crisis communication; instead, we look at the use of social media by these two projects and offer a set of recommendations for that usage by volcano observatories and other official agencies.

2 The Weekly Volcanic Activity Report

The Smithsonian Institution's Global Volcanism Program (GVP) and the US Geological Survey's Volcano Hazard Program collaborate to produce the Weekly Volcanic Activity Report (WVAR), which summarizes new and on-going volcanic activity globally. It is posted on the GVP website every Wednesday and is widely redistributed online.

WVAR was created for timelier reporting of volcanic activity on a global scale and quickly began serving the needs of humanitarian response agencies, military commands, travellers, businesses, scientists, and the general public. It has become a very popular site because it gives readers a snapshot of worldwide eruptive activity and unrest in one place. On average, WVAR received about 21,900 page views from 7620 unique visitors per week in 2014, with increased usage surrounding notable eruptions. Since its inaugural issue in November 2000 through the end of 2014, the regular, consistent, and thorough reporting has resulted in 10,575 individual summaries included in almost 740 weekly reports on over 270 volcanoes.

WVAR aims to include all volcanic activity that occurred on Earth during the week leading up to its online publication. About 20 sub-aerial volcanoes are erupting at any given time (Siebert et al. 2010). Some of the criteria considered in the report-generating process include: raising or lowering of the hazard status; the release of a volcanic ash advisory by a Volcanic Ash Advisory Centre (VAAC); and/or, a verifiable report of new or changing activity as noted in the media or by observers. It is important to note that volcanic activity meeting one or more of these criteria may occur during the week, but may not be included in the WVAR because details about the event were not available. In addition, more than a dozen volcanoes globally have displayed more-or-less continuous eruptive activity for decades or longer, and such routine activity is typically reported on a monthly basis unless a special report is issued.

The core of the WVAR process consists of rapid information gathering, data evaluation, and summarization. The WVAR editor has relied heavily on combing through electronic information channels such as the official websites of volcano observatories. Some observatories that do not have an on-line presence may distribute activity bulletins through email. To obtain primary source data, around 40 trusted websites are visited almost daily, including observatories, Volcanic Ash Advisory Centres (VAACs), civil protection agencies, and meteorological offices. The majority of the websites (29) are volcano observatories, and are visited first (along with two additional meteorological offices) for accurate and up-to-date eruption information. Therefore the content and reliability of each WVAR depends heavily on the accuracy and timeliness of reports posted to these observatory websites.

In cases where an eruption occurs from a volcano that is poorly or not regularly monitored, or a larger event occurs, the information search is expanded to any and ideally all available sources, including social networking platforms, satellite image analysts, gas emission experts, marine biologists, etc. Eruption information from these more transient sources will likely be included if that is the only source of information and/or the source is deemed credible.

2.1 The Contribution of Social Media to the Weekly Volcanic Activity Report

One of the challenges in assessing the state of world-wide volcanic activity in any given week is managing the amount of information available to a user; for just one volcano the source data may be just a sentence or two to well over 100 pages for large events covered by multiple sources. Therefore the greatest disadvantage posed by social media is the additional glut of unedited information. This includes posts from official sources and amateur volcanologists (e.g. a traveller captures an eruption with their phone camera as they pass by on an airplane and instantaneously uploads it). But how much real and accurate information can we glean from an explosion of details on so many platforms from official and amateur sources? Sifting through thousands of Twitter posts, for example, from official and non-official sources about an eruption is prohibitively time-consuming. Moreover, the information can be: (1) difficult to verify; (2) repetitive; (3) transient/ephemeral; (4) not archived; and, (5) not always searchable. A user needs to know where to go for the correct information; events can be missed if one incorrectly assumes that a lack of information on an observatory page means that no volcanic events are occurring.

The greatest advantage to social media comes when those platforms are the only sources of information about an eruption, or provide additional and critical details of the event. Therefore, the WVAR editor seeks eruption information from various social media platforms such as Facebook, Twitter, Flickr, blogs, and YouTube. In several instances over the past few years, the WVAR has sourced critical information not just about large and/or sensational eruptions [e.g. Eyjafjallajökull (2010), Merapi (2010), and Sinabung (2013)], or more unusual events [e.g. Kverkfjöll (2013)], but also about regularly erupting volcanoes (e.g. Sheveluch and Etna), from sometimes chance encounters with social media posts. In one case, details of an explosion including photographs were garnered from a social media connection with an observatory scientist; the details were not shared on the official observatory website until a later date.

In a second case, the WVAR editor sourced volcanic activity information after being directed to an official civil protection Facebook page from a travel blog. The official website had a very brief summary of the event but more comprehensive information (written details, dates, pictures, and an over-flight video) about the event was gleaned through a Facebook page. However, due to the blog-style of Facebook and lack of sufficient data-searching capabilities and archiving, it was difficult and time consuming to revisit the post a few days later to ensure all relevant information had been collected. Despite this, the Facebook posts yielded a better understanding of the event and allowed a more accurate summary of the eruption. Again important details of the event were discovered by chance and could have easily been missed; the official civil protection website did not offer the same details.

Since timely information about eruptions can be crucial to humanitarian efforts, scientists, report writers, and a variety of other users, it is critical that observatories establish reliable channels for the dissemination of eruption information. The two examples described above expose challenges in using social media as a source for information: when are social media sites complimentary to observatory websites and when are they supplementary? Anecdotal evidence suggests that observatory representatives have varying perceptions of social media: some view the extra effort as a burden and therefore post randomly and sparingly while others prefer to post on social media rather than through the observatory website. This highlights the need for observatories to clarify how their information is disseminated to stakeholders so that users know where to go first for official, accurate, timely, consistent, and archived eruption information.

An October 2014 examination of 33 observatory, meteorological office, and civil protection websites showed that all but three provided another means of information distribution, in addition to the official observatory website, by way of links on the website. These include social networking platforms, email distribution lists, and news feeds. Regarding social networking platforms, almost half the observatories use Facebook (48 %) and Twitter (42 %), followed by YouTube (24 %), Google+ and Google Groups (15 %), photo sharing sites (6 %), and one link to Pinterest (3 %). Email distribution lists accounts for 39 and 18 % use RSS/CAP feeds. For the most part links to these other outlets were visible somewhere on the main page, although not all were easily found, and a few were embedded on sub-pages. Two additional observatories use Twitter (Table 1, data for volcano observatories, monitoring agencies or emergency management groups that utilize Twitter) but do not obviously link to their Twitter accounts from their websites so were not included in this tally. Clearly most observatories are utilising social media, but how and to what degree varies.

3 'Eruptions' Blog

'Eruptions' blog is one of the most popular sources for information on volcanic activity on social media. Information is gathered for

Table 1 Volcano observatories, monitoring agencies, or emergency management on Twitter

Organization	Handle	Country	Followers	Following	Tweets	Frequency	Language	Date joined	Notes
ONEMI*	@onemichile	Chile	602,360	0	13,179	Frequent	Spanish	February 2010	Public relations
USGS*	@usgs	USA	424,310	68	7452	Frequent	English	April 2008	Twitter verified, all geosciences
CONRED*	@conredguatemala	Guatemala	227,139	115	27,522	Frequent	Spanish	February 2010	Some automated tweets (earthquakes)
IG Ecuador	@igecuador	Ecuador	136,909	392	5964	Frequent	Spanish	September 2010	Some automated tweets (earthquakes)
OVSICORI	@ovsicori_una	Costa Rica	94,777	78	4843	Frequent	Spanish	September 2009	Some automated tweets (earthquakes)
PHIVOLCS	@phivolcs_dost	Philippines	74,272	9	1067	Somewhat frequent	English	September 2012	Also reserved @phivolcs, Some automated tweets (earthquakes)
ONEMI*	@onemi	Chile	67,432	0	379	Inactive since 2009	Spanish	December 2007	Official announcements
RSN	@rsncostarica	Costa Rica	54,909	105	3061	Frequent	Spanish	August 2009	
INSIVUMEH	@insivumehgt	Guatemala	45,388	1	7947	Frequent	Spanish	April 2012	
BNPB*	@bnpb_indonesia	Indonesia	32,812	118	3856	Frequent	Indonesian	August 2011	
GNS Science	@geonet	New Zealand	31,243	5	9136	Frequent	English	September 2009	Some automated tweets (earthquakes)
SGC	@sgcol	Colombia	18,883	77	3278	Frequent	Spanish	June 2010	Mainly links to reports

(continued)

Table 1 (continued)

Organization	Handle	Country	Followers	Following	Tweets	Frequency	Language	Date joined	Notes
AVO	@alaska_avo	USA	13,311	3	1050	Sporadic	English	January 2009	More active during eruptions, quiet during other periods
IGN Spain	@ignspain	Spain	6187	349	3198	Frequent	Spanish	September 2010	
GNS Science	@gnscience	New Zealand	2764	22	331	Somewhat frequent	English	April 2010	
CRHED Guatemala	@crhedgt	Guatemala	2283	757	10,456	Frequent	Spanish	April 2011	
CENAPRED	@cenapred	Mexico	2173	0	0	Inactive	Spanish	September 2010	Also @monitoreoV
IGP	@igp_peru	Peru	1973	526	1735	Somewhat frequent	Spanish	July 2011	
PVMBG	@vulkanologi_mbg	Indonesia	1945	14	236	Frequent	Indonesian	August 2013	
MVO	@mvoms	UK (Montserrat)	634	4	717	Sporadic	English	June 2010	
Afar Rift Consortium	@afar_rift	Ethiopia/UK/US	245	54	134	Sporadic	English	June 2010	
INETER*	@ineter	Nicaragua	64	0	0	Inactive	Spanish	June 2010	
SERNAGEOMIN	@sernageomin	Chile	55	0	5	Somewhat frequent	Spanish	November 2014	Biblioteca
POVI	@povi_cl	Chile	43	2	62	Sporadic	Spanish	February 2014	

*Geological surveys or emergency management agencies that tweet often about volcanic activity. # Frequent = multiple times per week; somewhat frequent = multiple times per month; sporadic = multiple times per year, usually associated with elevated activity; inactive = has not tweeted in over a year. All Twitter data from 10 November 2014

'Eruptions' via filtered Google News reports, Twitter, and volcano observatory websites. Readers also leave comments with information of on-going eruptive activity gleaned via traditional news sources, blogs written by amateur enthusiasts, and observations of volcano webcams.

Since its inception in May 2008, the 'Eruptions' blog has received over 5 million visits and from 1 May 2012 to 1 May 2013, the blog had 1.978 million page views. This traffic shows the strong interest in volcanology by the general public and the demand for accurate, scientific information on current activity worldwide along with research in volcanology. A Twitter account (@eruptionsblog) is partially linked to the blog. This account has over 8750 followers (as of 18 November 2014) and is primarily used to tweet information about new and on-going eruptions, both from material produced for the blog and from outside sources.

3.1 Using Twitter to Source Volcanic Crisis Information: From the Public

Twitter is one of the primary sources for information published on 'Eruptions'. Whenever a new eruption occurs, Twitter is one of the first places where images of the activity can be found, typically taken by the general public on their phones or digital cameras. Local media reports often appear on Twitter soon after an eruption has begun, typically much faster than they will appear in a Google News search for the volcano or region (especially if it is not in English).

Crowdsourcing data on volcanic activity or via citizen science activities during volcanic eruptions has been attempted on a number of platforms (Klemetti 2010), such as Twitter, Flickr, and Instagram. For example, Pyle and Oxford University's Earth Sciences Class of 2015 (2014) used data gleaned from a variety of social media platforms along with published photographs of the area to estimate the ash fall for the February 2014 eruption of Kelut in Indonesia. They found exponential decay of ash away from the volcano that was similar but larger

than that from Kelut's 1990 eruption. This suggests that collecting ash fall information from social media might be a quick way to calculate the magnitude of the eruption without sending experts to different locations around the volcano or to get information from officials during a time of crisis.

3.2 Using Twitter to Source Volcanic Crisis Information: From Official Sources

The most effective Twitter accounts from volcano monitoring agencies release the following types of information through their primary account: images of eruptions, links to official releases about the on-going activity, updates from volcanologists (brief and timely), information about precautions and evacuations (especially from disaster agencies), and links to webicorders and webcams. This information is especially important during periods of increased media attention leading up to and during an event. An excellent example is how the Alaska Volcano Observatory (AVO) (@alaska_avo) uses Twitter to quickly send updates on the changes to the alert status of Alaskan volcanoes (Fig. 1a). They also post images of the volcanoes that are linked back to their official sources (with credit) hosted on the AVO website (Fig. 1b). These tweets are clear and succinct, and they direct readers back to the original source of the material for more information. Most importantly, they are timely—typically tweeted within an hour of the change in status of the volcano.

However, very few volcano observatories have Twitter accounts (Table 1). Those that have accounts vary their use from very active (posting multiple times a day or week) to inactive. However, even those that are inactive have significant numbers of followers looking for information. The accounts with the most followers are those not singularly dedicated to volcano monitoring, such Chile's Oficina Nacional de Emergencia del Ministerio del Interior (@onemi—602,360 followers), United States Geological Survey (@usgs—424,310 followers), and Guatemala's

(a)

Alaska AVO @alaska_avo · Nov 13

Pavlof in eruption as viewed from Cold Bay on the evening of November 12, 2014.

avo.alaska.edu/image.php?id=6...

(b)

Alaska AVO @alaska_avo · Oct 28

Shishaldin: Uptick in low-level eruptive activity ORANGE/WATCH See

avo.alaska.edu

Fig. 1 Examples of uses of Twitter by the Alaska Volcano Observatory (@alaska_avo). **a** Image of the November 2014 eruption of Pavlof with link to AVO page; **b** statement on the elevated alert status at Shishaldin on October 28, 2014 with link to additional information

Coordinadora Nacional para la Reducción de Desastres (@conrdguatemala—227,139 followers). However, Ecuador's Instituto Geofisico (@igecuador) and Costa Rica's Observatorio Vulcanológico y Sismológico (@ovsicora_una) have over or nearly 100,000 followers.

Interestingly, Latin American volcano monitoring agencies have embraced the use of Twitter the most, with active Twitter accounts for OVSICORI, IG Ecuador, Instituto Nacional de Sismología, Vulcanología, Meteorología e Hidrología (@insivumehgt), Servicio Geológico Colombia (@sgcol), Instituto Geofisico Peru (@igp_peru), and Proyecto Observación Villarrica Internet (@povi_cl). Adoption of Twitter as an avenue for public outreach and distribution of information in other parts of the world has been much slower, although both the Philippine Institute of Volcanology and Seismology (@phivolcs_dost) and Indonesia's Pusat Vulkanologi dan Mitigasi Bencana Geologi (@vulkanologici_mbg) have recently become more active.

In a study on the media response to the eruption of Iceland's Eyjafjallajökull in 2010, Lee et al. (2012) posits that traditional media

underutilised and under-reported the scientific information from the eruption in their coverage. Instead, they used sources from the travel industry for many of their reports due to a focus on the air travel disruption caused by the eruption. This lack of a scientific voice meant that the public reaction to the decision to close the airspace was heavily influenced by industry voices in the media rather than the scientific data and interpretations (Suw Charman-Anderson, pers. comm.).

This situation could have been mitigated to some degree by use of social media, such as Twitter, for official sources to send timely and accurate information about the eruption directly to the public, rather than relying on more traditional and slower information distribution methods such as press releases and conferences. Additionally, misinformation during the eruption (such as the erroneous reporting of an eruption of Katla) could have been corrected faster if such information was released through official channels on Twitter rather than only adding a statement to the Iceland Meteorological Office (IMO) website (http://en.vedur.is) (Suw Charman-Anderson, pers. comm.).

Another example is the recent eruption near Iceland's Barðarbunga starting in August 2014. Information about the start of the eruption quickly spread on Twitter, but from unofficial sources such as the Icelandic media (e.g., @RUVfrettir) or non-volcano monitoring government agencies such as Iceland's Department of Civil Protection and Emergency Management (@almannavarnir). This led to confusion about when the eruption actually occurred as there were reports of subglacial eruptions under Vatnajökull that ended up being unfounded. Official confirmation of the Barðarbunga eruption was not received until the IMO website was updated. The consequences of a subglacial versus subaerial eruption are very different from a hazard perspective, so clear information dissemination is vital. This confusion may have been avoided if IMO had been using an official Twitter account to disseminate this information directly to the public as soon as it was available.

4 Discussion and Recommendations to Volcano Observatories Regarding the Use of Social Media

Posting to social media platforms can seem like an extraneous and unnecessary activity during a time of volcanic crisis. However, as the above-mentioned case studies highlight, there are real advantages to utilising social media as a tool to convey information quickly and directly. One key aspect to a volcano observatory's use of social media is constant contact. Even when there are no volcanic crises, different audiences are looking for information. These audiences include (but are not limited to): local residents, tourists, students both near and far, researchers, news media, and government officials. Social media should be used to convey authoritative information about volcanic activity, as well as content on how to prevent/mitigate and prepare for, respond to and recover from a volcanic eruption (Fig. 2).

Some best-practice advice and suggestions for volcano observatories or monitoring agencies in relation to social media include:

1. **Use it**. Social media is a two-way dialogue to communicate and receive hazard and risk information. Traditional forms such as press releases or conferences allow for the mainstream media to add their agenda to the information. Social media platforms such as Twitter and Facebook cut out this middleman so that critical information is directly conveyed.
2. **Make it official**. State in the account information that the social media account is official. This can add authority to the information released and stop confusion with people tweeting/posting volcanic information as enthusiasts rather than volcano monitoring officials. Twitter does "verify" accounts, but there is no way to request verification, so each observatory should state it clearly. This also means putting a direct link to the social media feeds on the main website as well, which adds authenticity and authority to the account.
3. **Post images and figures**. The types of information most often sought out by the public are images of the eruption or figures of data (such as webicorder traces). Provide these in the social media feed, but be sure to stamp each figure with a date/time. This prevents older images or images of other volcanoes being distributed across social media platforms.
4. **Update volcano status information**. If volcanoes are moved up or down in alert status, tweet/post a brief statement with this information, even if accompanying information is not ready. However, be sure to tweet/post a link to any additional information when it is ready as a lack of timely information reduces trust that the account is official.
5. **Remain active**. Even if there is no on-going crisis, tweet/post information about volcanoes, images or links to webcams, new research or equipment installation. This allows for the public to know that work is being done between eruptions. Also, if the main website or Facebook page is updated, tweet/post about it and provide a link.
6. **Separate automated tweets**. Some volcano monitoring agencies or observatories post automated tweets for events such as earthquakes. This information is valuable but can easily overwhelm a Twitter feed, especially in seismically active areas. Separate that information into a separate feed that is only used for tweeting seismic (or other) information, especially if it is automatically generated.
7. **Tweet/post links to press releases**. If information is released as press releases, informational statements or if press conferences are webcast, tweet/post those links in a timely fashion.
8. **Sign off on observations and statements**. If possible, direct observations and interpretations of events are very useful for the public, bloggers, and the media. If such information can be tweeted/posted, have the scientist "sign" the tweet/post with their initials [e.g., (EK)] so that verification can be made of the information. This helps add trust to the tweets/posts.

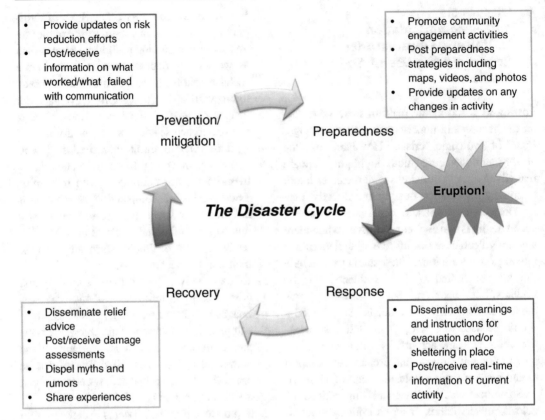

Fig. 2 Examples of how social media can be used throughout the disaster cycle, during periods of heightened activity and quiescence. Please note, some of these activities, such as mythbusting, can be done at any point throughout the cycle

9. **Be aware of myths and, inaccurate views and reporting**. If possible, observatories should monitor social media sites that provide information and/or discussion on an impending or on-going eruption as sometimes well-meaning people share inaccurate views or trolls purposefully offer false information. To counter these, observatories and/or disaster management agencies can post a FAQ (Frequently Asked Questions) Fact Sheet or Mythbusters Fact Sheet in an effort to dispel associated myths.

Volcano observatories must decide how they can most efficiently and effectively disseminate their very important information on eruptions and unrest, let users know where, how, and what kinds of information is being posted, and then follow through with consistency. A statement on the observatories main page describing how they distribute the most up-to-date information (a social media use plan) would be a simple yet effective improvement. The icons for links to social media should be prominent if that is one of the main avenues they use to disseminate information to users. This approach does not lock in an observatory to follow a plan that may not work for them due to staffing issues, degraded internet service, lack of funds, etc. but gives flexibility to change and grow, as long as they communicate those changes in a social media use plan.

5 Conclusion

There is no doubt that a large proportion of the general public has embraced social media and expects to have the ability to communicate information via its various sources. Significantly, social media are two-way communication tools allowing official agencies to rapidly interact with the public and each other in real-time. Social media platforms offer a creative alternative to traditional media (e.g. radio, television) for communicating risk information. They allow volcano observatories to share knowledge and experience of events, and promote through stories, photographs, and videos the utility of volcano monitoring, even during periods of quiescence. As such, social media can be used to generate continued public interest and trust, whether or not an eruption is imminent.

The case studies shown here demonstrate the high demand for accurate and timely volcanic information during a crisis, both pragmatically (for people living nearby) or to satisfy curiosity. Social media such as Twitter can meet those demands as part of an overarching communication strategy. However, Twitter should not be the only social media instrument used and other more traditional forms of media should not be replaced with social media. Overall, social media should be embraced as an additional way to directly communicate with the general public and complement existing strategies. In order to ensure that they remain relevant, however, volcano observatories will have to keep up with the most prominent trends as Twitter and Facebook may be replaced by other platforms in the future.

Disclaimer Any use of trade, firm, or product names is for descriptive purposes only and does not imply endorsement by the U.S. Government.

References

Bird D, Ling M, Haynes K (2012) Flooding facebook—the use of social media during the Queensland and Victorian floods. Aust J Emerg Manage 27 (1):27–33

Bruns A, Burgess J, Crawford K, Shaw F (2012) #qldfloods and @QPSMedia: crisis communication on twitter in the 2011 South East Queensland Floods. ARC Centre of Excellence for Creative Industries and Innovation, Brisbane

Carranza Tresold JA (2013) Social media maps a volcano's aftermath: tracking the consequences of the guatemalan eruption, esri ArcNews

Disaster Management SofS Working Group (2014) Building a system of systems for disaster management workshop: joint issues statement, CSIRO

Dufty N (2015) The use of social media in countrywide disaster risk reduction public awareness strategies. Aust J Emerg Manage 30(1):12–16

Klemetti EW (2010) Using the 2010 Eyjafjallajökull eruption as an example of citizen involvement in scientific research, American Geophysical Union, Fall Meeting 2010, Section: Education and Human Resources, Session: Public Participation in Geoscience Research: Engaging Citizen Scientists III, San Francisco, California

Lee B, Preston F, Green G (2012) Preparing for high-impact, low-probability events. Lessons from Eyjafjallajökull. The Royal Institute of International Affairs, Chatham House, London. ISBN 978 1 86203 253 8

Pyle D and Oxford University's Earth Sciences Class of 2015 (2014) Ash fallout from the 2014 Kelut eruption: a preliminary analysis

Siebert L, Simkin T, Kimberly P (2010) Volcanoes of the world. University of California Press, Berkeley

Part Three Summary: Communicating into the Future

Deanne K. Bird and Katharine Haynes

The chapters that form Part Three explore diverse and wide-ranging communication and engagement processes, where old practices and new technologies are brought together to reduce the impact of volcanic crises in an increasingly globalised world. The chapters highlight challenges of working across disciplines, sectors, institutions, and negotiating differing politics and cultural practices and knowledges. Overall, Part Three demonstrates that trans-disciplinary and multi-organisational partnerships that include social scientists, health professionals, civil defence experts and community members alongside volcanologists, are key to successful volcanic risk reduction. These partnerships must be developed well before a crisis in order that trust, shared meaning and effective working relationships are developed. Part Three also considers progress made and identifies areas that need to be addressed to ensure the success of volcanic crisis communication in the near and long term.

In order to set the scene, Chester et al. in the chapter entitled "Communicating Information on Eruptions and Their Impacts from the Earliest Times Until the Late Twentieth Century" detail a historic overview of the communication and dissemination of volcanic information. Case studies of the Azores, Portugal and Mt Etna, Italy provide contrasting examples of how responses to volcanic crises and their communication have evolved over time.

Pyle in the chapter entitled "What Can We Learn from Records of Past Eruptions to Better Prepare for the Future?" also considers how observations and accounts, before the advent and application of instrumental monitoring, provide unique perspectives of volcanic processes and impacts and how they can aid future decision-making. Through a retrospective analysis of eruptions of the Kameni islands, Santorini, Greece and Soufriere, St Vincent, Pyle looks at the sequence of volcanic processes and the resulting social, economic and political consequences caused by these events. These case studies highlight the valuable detail contained within contextual data on volcanic activity and how that can be applied to enhance our capacity for reducing risk, particularly where there is a lack of instrumental monitoring.

The importance of community participation, particularly the use of local knowledge is well highlighted by Gabrielsen et al. in the chapter entitled "Reflections from an Indigenous Community on Volcanic Event Management, Communications and Resilience". To date, Ngāti Rangi, an indigenous tribe of Aotearoa, New Zealand living on the flanks of Ruapehu, have not always been included in decision-making during volcanic events despite the benefits their local knowledge would bring to the process,

D. K. Bird (✉)
Faculty of Life and Environmental Sciences, University of Iceland, 101 Reykjavik, Askja, Iceland
e-mail: dkb@hi.is

K. Haynes
Department of Geography and Planning, Macquarie University, Sydney 2109, Australia

Advs in Volcanology (2018) 745–749
https://doi.org/10.1007/11157_2017_26
Published Online: 06 December 2017

particularly in relation to communications and warnings. This chapter however, details how the situation is improving through a community-led initiative where the Ngāti Rangi have set up their own monitoring, information collection, and communication systems.

Cadah et al. in the chapter entitled "Fostering Participation of Local Actors in Volcanic Disaster Risk Reduction" discuss, through the lead author's first hand experiences, case studies of community based DRR at Mount Rainier, USA and Bulusan, Philippines. With references to wider case studies and literature, the chapter identifies the key principles and important considerations for fostering community-based participation. However, the chapter concludes by warning against a rigid and standardised procedure for participation as each volcano and at risk community will have their own specific context and needs. Cadah et al. therefore highlight that flexibility and the development of customized approaches is the way forward.

Further highlighting the need to fully appreciate and comprehend the nuances of each particular locale, Wilmhurst in the chapter entitled "There is no Plastic in Our Volcano: A Story About Losing and Finding a Path to Participatory Volcanic Risk Management in Colombia" describes how the best intentions can lead to failure. A stakeholder workshop, bringing together scientists, government officials, emergency service personnel and at-risk communities was conducted in an effort to create dialogue and cooperation and develop a new participatory path for the management of volcanic risk associated with Galeras, Colombia. On the fourth day, however, the workshop took a very different path, one of mistrust and miscommunication. This chapter challenges all stakeholders to critically think about their role in DRR and how their own assumptions, biases, perceptions and beliefs influence their work. This chapter takes the reader on a journey of *getting lost and finding the way again* highlighting the many important lessons learnt along the way, many of which are centred around trust.

Continuing the theme of trust, Newhall in the chapter entitled "Cultural Differences and the Importance of Trust Between Volcanologists and Partners in Volcanic Risk Mitigation" looks at context in relation to cultural differences. As with Wilmhurst, Newhall focuses on the local level, where those in positions of responsibility: scientists, civil defence officials, land managers, engineers, politicians and journalists, as well as at-risk populations, differ greatly with respect to culture. Each of these groups has different concerns, goals, language, approaches, ways of learning and resources. In this chapter Newhall argues that a key to bridging these is trust, which can be earned through understanding and acceptance of, and respect for, the inherent differences. Based on personal experience, Newhall provides a snapshot view of three case studies: Mount St. Helens, USA 1980; Pinatubo, the Philippines 1991; and, Usu, Japan 2000, to highlight the issues of trust and its importance for volcanic risk mitigation. Newhall concludes with some valuable recommendations on how to incorporate professionalism and cross-cultural sensitivity alongside personal touches in every interaction between volcanologists and those who use volcanological information.

In "International Coordination in Managing Airborne Ash Hazards: Lessons from the Northern Pacific", Igarashi et al. examine efforts to bridge cultural divides. Considering air-borne volcanic ash often impacts multiple international jurisdictions, efforts in managing this hazard involve coordination of stakeholders from multiple countries. Through the lens of the 2009 eruption of Sarychev-Peak Volcano in the Kurile Islands and the 2013 eruption of Kliuchevskoi Volcano on the Kamchatka Peninsula, Igarashi et al. identify the challenges of international coordination and strategies to overcome these to ensure the provision of coherent and consistent messaging.

Doyle and Paton further showcase the importance of creating a 'shared meaning' between agencies in "Decision-Making: Preventing Miscommunication and Creating Shared Meaning Between Stakeholders" by examining individual and team decision-making and the critical role of relationships and communication between science advisors, emergency managers and decision makers. Doyle and Paton stress that in order to ensure all stakeholders have a shared understanding of the situation and, each other's functions and responsibilities during a time of crisis, they must have built strong relationships through prior training and development.

The communication of probabilities and uncertainties continues to be the Achilles heel of otherwise successful volcanic risk discourse. In "Using Statistics to Quantify and Communicate Uncertainty During Volcanic Crises" Sobradelo and Marti explore the statistical meaning of probability estimates and their role in the quantification and communication of uncertainty during volcanic crises.

Blong et al. in the chapter entitled "Insurance and a Volcanic Crisis—A Tale of One (Big) Eruption, Two Insurers, and Innumerable Insureds" further our understanding by examining potential issues between the insurance sector and policyholder. Using a realistic disaster scenario of a large eruption, the chapter describes two companies' very different yet possible approaches to managing the situation. This analysis highlights the numerous insurance-related challenges to consider. Blong et al. argue that clear communication between the insurance sector and policyholder is dependent on informed understanding of the nature of volcanic hazards and terminology, insurance wordings and potential losses. The chapter also explores examples of insurance case law from the: 1991 eruption of Pinatubo in the Philippines; 1995 eruption of Ruapehu in New Zealand; 1980 eruption of Mount St. Helens in the USA; and, 2010 Eyjafjallajökull eruption in Iceland, to further highlight issues that arise in interpreting insurance contracts.

Clear communication, through the use of standardised Volcanic Alert Level (VAL) systems, is also central to the next chapter. Used as a communication tool to inform at-risk populations of the status of activity at specific volcanoes, Potter et al. in the chapter entitled "Challenges and Benefits of Standardising Early Warning Systems: A Case Study of New Zealand's Volcanic Alert Level System" explore standardising VAL systems across New Zealand and

internationally. While the benefits are many, the key challenges further highlight the importance of context; that is, the uniqueness of each volcanic system, different needs and perspectives of end-users and, different response actions associated with the spatial and temporal differences in activity.

Volcanic hazard maps are a major tool for the visual communication of hazard and risk to a wide range of audiences. Maps are a coded form of communication, and the visual design and ability of the receiver to decode the message significantly impacts on how hazard maps are understood and applied. Thompson et al. in the chapter entitled "More Than Meets the Eye: Volcanic Hazard Map Design and Visual Communication" review case studies of volcanic crises and interdisciplinary research that addresses the relationship between hazard maps, visual design, and communication. They conclude that future technological innovations will create improved opportunities for people to read, process, and share visual information. However, audience needs and perspectives must continue to be understood in order that hazard maps and visual communications are accessible and meet the needs of users.

To exemplify the importance of the visual representation of hazard and risk, Webley and Watson in "The Role of Geospatial Technologies in Communicating a More Effective Hazard Assessment: Application of Remote Sensing Data" provide a detailed summary of how satellite imagery and geospatial technologies inform real-time analysis, communication and decision-making. Case studies from the 2010 eruption of Merapi, Indonesia; 2012–13 eruption of Tolbachik on the Kamchatka Peninsula; and, 2014–15 eruption of Bárðarbunga, Iceland further illustrate how remote sensing data have been applied during volcanic crises. Overall, Webley and Watson identify important aspects of sensors and technologies relevant to volcanic hazard assessment in a way that is easy to understand with limited knowledge of remote sensing methodologies.

While the science of volcanology rests within a positivist tradition, risk communication does not and is very much socially framed by the values and cultures of both the sender and receiver of the message. Dixon and Beech, in the chapter entitled "Re-enchanting Volcanoes: The Rise, Fall, and Rise Again of Art and Aesthetics in the Making of Volcanic Knowledges", remind us of this and take an expanded notion of communication to outline the history of how knowledge of volcanoes was produced and disseminated during the 18th and 19th centuries, thus bringing Part 3 back to reflect on past experiences and learnings.

The notion of the past informing present-day preparedness activities is explored by Kitagawa in the chapter entitled "Living with an Active Volcano: Informal and Community Learning for Preparedness in South of Japan". Kitagawa provides a local perspective of how residents have learned to live with the everyday threat of volcanic hazards with respect to Sakurajima Volcano in Japan. Emanating from these lessons are two local lores: 'do not rely on authorities' and 'be frightened effectively', which Kitagawa argues, enhance preparedness through the development of agency and being alert but not alarmed. At the community level, collaborative partnerships have had a positive influence on enhancing community participation and collective efficacy, where decisions are made and information and ideas are exchanged. This approach has led to societal-level empowerment and trust between individuals and officials.

Dohaney et al. in the chapter entitled "Using Role-Play to Improve Students' Confidence and Perceptions of Communication in a Simulated Volcanic Crisis" argue that the theory, skills and protocols behind effective volcanic crisis communication are often not taught to young scientists. To address this gap, Dohaney et al. outline and evaluate an innovative role-play exercise where University students are able to practice crisis communication through a realistic real-time scenario. The methodology successfully increases students' confidence to communicate and knowledge of science communication.

Sharpe in the chapter entitled "Learning to Be Practical: A Guided Learning Approach to Transform Student Community Resilience When Faced with Natural Hazard Threats" also investigates the use of a classroom based learning approach to move beyond knowledge creation to promote action. The methodology used experiential learning, reflection and the act of playing out the threats from hazards allowing students aged between 11 and 14 to explore key issues of risk, response and preparedness for themselves and their family. The lessons were planned in order to fit within the National Curriculum and involved guided research and investigation through websites, videos and mapping exercises; the creation of student films; and, a homework task to produce an emergency Go-Bag with parents. A questionnaire completed by the students three weeks after the learning identified that 75% of the sample had collated their emergency kits with their parents and 55% felt more prepared as a family. This aligns with previous research that suggests that with the right support, students can be agents of change within the family.

The final crisis communication tool to be considered is social media. Sennert et al. in "Role of Social Media and Networking in Volcanic Crises and Communication" examine the role of social media as an often preferred source for news. Focusing on two projects, led by the first and second authors of this chapter, a set of recommendations are made in an effort to assist volcano observatories and government officials in using various social media platforms to their advantage. As new technologies and platforms are developed, some of these recommendations may become obsolete. Overall, however, many of them will remain pertinent to volcanic crisis communication, well into the future.

Volcanic Crisis Communication: Where Do We Go from Here?

Carina J. Fearnley[iD], Deanne Katherine Bird,
Katharine Haynes, William J. McGuire and Gill Jolly

This volume brings together a wealth of undocumented knowledge and first hand experience to provide a platform for understanding how volcano crises are managed in practice, with contributions from authors all over the globe ranging from observatory volcanologists and scientists, government and NGO officials and practitioners, the insurance sector, educators, and academics (multiple disciplines), and last but by no means least, vulnerable and indigenous populations. These diverse contributions have provided valuable insights into the various successes and failures of volcanic crises. This final chapter seeks to summarise the key contributions to identify trends and determine the vital future directions for volcanic crisis communications research.

1 Observing the Volcano World

Each part of the volume has explored three key themes: (i) the need to understand the multiple hazards involved in a volcanic crisis; (ii) lessons learned from past crises; and (iii) the tools available for effective communication during a crisis. There are a number of overarching lessons which are discussed below.

1.1 Managing Individual Hazards During a Crisis

Volcanic activity is unique in having a particularly large and diverse portfolio of associated phenomena capable of causing death and injury, societal and economic disruption, and damage to population centres and attendant infrastructure. These hazards vary in scale geographically from those proximal to the volcano, to those that can affect the regional or global climate. Many of these hazards occur during volcanic eruption, but some can occur during times of unrest, or even quiescence. Volcanoes present the ultimate natural hazard challenge with multiple hazards often occurring at the same time requiring bespoke decisions, actions, and warnings for at-risk communities. Given this diversity a 'one size fits all' approach does not provide the most effective means of addressing the communication

C. J. Fearnley (✉)
Department of Science and Technology, University College London, Gower Street, London WC1E 6BT, UK
e-mail: c.fearnley@ucl.ac.uk

D. K. Bird
Faculty of Life and Environmental Sciences, University of Iceland, Askja, 101, Reykjavik, Iceland

K. Haynes
Department of Geography and Planning, Macquarie University, Sydney 2109, Australia

W. J. McGuire
Department of Earth Sciences, UCL Hazard Centre, University College London, London WC1E 6BT, UK
e-mail: w.mcguire@ucl.ac.uk

G. Jolly
GNS Science, 1 Fairway Drive, Avalon, Lower Hutt 5010, New Zealand
e-mail: G.Jolly@gns.cri.nz

Advs in Volcanology (2018) 751–754
https://doi.org/10.1007/11157_2017_27
Published Online: 06 December 2017

of volcanic hazards. Although general principles apply, this generally requires that information is adapted or tailored for the particular hazard in addition to considering the local dynamics of vulnerable communities that need to respond to these warnings. The chapters in Part One of this volume demonstrate how this approach may be utilised successfully to tackle a variety of specific hazards whilst at the same time maintaining trust. The challenge remains that with many people already leading complicated lives, the need to deal with several warnings for just one volcano poses the threat that populations will not engage. Here adopting a range of communication tools appropriate to the particular context can assist in not only communicating the hazard, but also in fostering relationships that provide additional monitoring data and warning information.

1.2 Lessons Identified from Crises Observed

Fundamental lessons have been learnt from a number of volcanic crises. Often these lessons have been learnt the hard way with many relatively minor volcanic events leading to tragedy. It is clear that transparency and solid relationships with stakeholders lead to the success of volcano monitoring and crisis management initiatives. Whilst poor relations may initially prevail, they can be rebuilt over time, through open lines of communication. However, operating in areas of political and social instability presents significant problems in terms of stakeholder collaboration and also partnership with vulnerable communities focused on more pressing concerns to their daily security. Conflict can also emerge amongst those directly responsible for volcanic crisis management; blame and conflict between competing scientific groups in particular, have the potential to be highly disruptive. Frequently this type of conflict is the result of adopting unclear or exclusive protocols. To avoid or ameliorate this situation the use of mediators can be highly successful and should be encouraged where appropriate.

Countries have differing approaches as to whether they integrate the physical and social sciences. Often a barrier is thrown up between these two worlds, with little willingness to breach this divide, or to unite it. In countries where the social and physical perspectives of the volcano are integrated, meaningful communication that adapts over time is developed facilitating a more holistic and robust programme of management that benefits both sides of the divide. Yet, many countries (largely Western) seek to maintain two separate and distinct perspectives. The chapters from Part Two demonstrate that sharing knowledge and experience is important to successfully address any new crisis as long as this is done in a transparent, sensitive manner, and with humility. The best results occur when this is conducted well before a crisis and ensuring that all aspects of risk are addressed; particularly work that reduces underlying vulnerabilities and builds the capacities of the local population, responsible stakeholders and institutions. It is hoped that in time, barriers can be eroded and there can be more scope for discussion, deliberation, and integration of multiple perspectives. It can be done; we only have to look at the successful examples of Mt Pinatubo (1991) and Merapi (2010) to appreciate this. Learning new perspectives, languages, and disciplines requires significant effort and parties on both sides need to have patience and willingness to facilitate this growth.

1.3 Understanding the Role of Communications

A wide range of communication tools cut across the various stakeholders, disciplines, policies, and hazards associated with volcanic crises. Many have been used for centuries; centred around simple forms of communication and collaboration. Indigenous populations today still provide valuable insights that can assist in the management and communication of volcanic crises. There have been some particularly successful examples whereby partnerships between indigenous populations and volcano observatories have enabled different knowledges and values to be brought together in order to find sensible solutions for all, for example as seen at GNS in New Zealand and CVGHM in Indonesia.

Technology is providing an impressive new set of tools to aid in the communication of ideas and practices from using: remote sensing and GIS to help visualise crises and potential scenarios, to developing more engaging maps that enable people to identify areas of danger and safety, to the role of social media. The chapters in Part Three of this volume demonstrate that whilst these tools can give us eyes and ears to information that was previously impossible (e.g. satellite data and social media) there is plenty of evidence to suggest that communication is still largely about trust, building relationships and shared understandings. As revealed in many of the narratives in this book, trust is typically hard to develop and is very easy to lose. Building and maintaining trust and ensuring a shared understanding of processes and outcomes can only occur through effective consultation. Building human relationships is therefore as important for successful volcanic crisis management as the utilisation of new methods and technological advancements.

Social media enables a more democratic multi-directional process of communication empowering the public to be part of a dialogue, as the receivers of the latest data, or the providers of first-hand observations and feedback. However, the abuse of social media can present a threat to an institution's credibility and authority as a knowledge source. It is, therefore, important to be active in the management of social media, even if this requires additional resources.

Clear communication is dependent upon an informed understanding of the nature of volcanic hazards and risks. Increasingly statistics are being used to try and make sense of and communicate the uncertain nature of volcanic hazards to a range of stakeholders. The very process of developing event trees and assessing the likelihood of events can play a vital role in preparing for the unknown, bringing various scientific data sets together to form the basis of coherent decisions. It is important, however, to remember that models are only as good as the data they incorporate and utilise, and that whilst extremely valuable, such tools constitute just one of many. Yet, with healthy levels of scepticism such tools are beneficial and are increasingly being used. Indeed the role of insurance remains significant in providing the security required during such devastating events, and significant modelling and statistical innovation stems from the work in this sector.

Education, via formal or participatory approaches is a vital component of effective communication, involving as it does the receipt or donation of information and learning that is required for a successful outcome to volcanic crises. The use of role-play, simulations, and the adoption of innovative ways of learning for all age ranges and stakeholders enables preparedness actions and facilitates a speedy, more effective, response to a crisis. Giving an individual the power to make informed decisions during a crisis is potentially the most life-saving act possible.

2 Where Are We Now?

This book should be thought of as constituting a first small step in revealing the state of the art, bringing together a collection of revealing and helpful narratives. Many more stories remain to be documented, and future trends and potential directions are constantly emerging. First, there is clearly a need for more data; to gain a better understanding of how the volcano world is observed within different contexts. Continuing to document and publish volcanic crisis communication experiences, both negative and positive, will generate continued data for future analysis on what leads to successes and failures, and what kind of practices may be suitable to adopt in different contexts.

Second, there is a need to build on the narratives presented in this volume, so as to move from the descriptive to the comparative and analytical, in order that lessons identified can become lessons learned. It is hoped that the information and ideas presented here will allow those who wish it, to conduct further scrutiny and analysis. There is always a requirement for continued description of crises, but theoretical analysis is where, ultimately, the most gains will be made.

Third, continuous feedback and engagement are critical requirements in this field of research.

Whilst it is important to recognise their intrinsic academic value, which is frequently overseen, it is also vital to share these stories not just for, but also with various stakeholders, with the goal of moving towards developing more robust volcanic crisis communication and management in the future. To do this, let's make space for these stories; let's value them.

Erratum to: Crisis Coordination and Communication During the 2010 Eyjafjallajökull Eruption

Deanne K. Bird⊙, Guðrún Jóhannesdóttir,
Víðir Reynisson, Sigrún Karlsdóttir,
Magnús T. Gudmundsson and Guðrún Gísladóttir

Erratum to: Chapter "Crisis Coordination and Communication During the 2010 Eyjafjallajökull Eruption" in: D.K. Bird et al., Advs in Volcanology, DOI 10.1007/11157_2017_6

The original version of the chapter was inadvertently published with an incorrect affiliation for the author Deanne K. Bird, which was corrected to "Institute of Life and Environmental Sciences, University of Iceland, Reykjavík, Iceland" in chapter "Crisis Coordination and Communication During the 2010 Eyjafjallajökull Eruption".

The updated online version of this chapter can be found at
http://dx.doi.org/10.1007/11157_2017_6

Advs in Volcanology (2018) E1
https://doi.org/10.1007/11157_2017_22
Published Online: 07 September 2017

Index

Note: Page numbers followed by *f* and *t* refer to figures and tables, respectively

Advs in Volcanology (2018) 755–771
https://doi.org/10.1007/978-3-319-44097-2
© The Authors 2018

Printed in the United States
By Bookmasters